Advances in Intelligent Systems and Computing

Volume 338

Series editor

Janusz Kacprzyk, Polish Academy of Sciences, Warsaw, Poland
e-mail: kacprzyk@ibspan.waw.pl

About this Series

The series "Advances in Intelligent Systems and Computing" contains publications on theory, applications, and design methods of Intelligent Systems and Intelligent Computing. Virtually all disciplines such as engineering, natural sciences, computer and information science, ICT, economics, business, e-commerce, environment, healthcare, life science are covered. The list of topics spans all the areas of modern intelligent systems and computing.

The publications within "Advances in Intelligent Systems and Computing" are primarily textbooks and proceedings of important conferences, symposia and congresses. They cover significant recent developments in the field, both of a foundational and applicable character. An important characteristic feature of the series is the short publication time and world-wide distribution. This permits a rapid and broad dissemination of research results.

More information about this series at http://www.springer.com/series/11156

Suresh Chandra Satapathy · A. Govardhan
K. Srujan Raju · J.K. Mandal
Editors

Emerging ICT for Bridging the Future - Proceedings of the 49th Annual Convention of the Computer Society of India (CSI) Volume 2

Editors
Suresh Chandra Satapathy
Department of Computer Science and Engineering
Anil Neerukonda Institute of Technology and Sciences
Vishakapatnam
India

A. Govardhan
School of Information Technology
Jawaharlal Nehru Technological University Hyderabad
Hyderabad
India

K. Srujan Raju
Department of CSE
CMR Technical Campus
Hyderabad
India

J.K. Mandal
Faculty of Engg., Tech. & Management
Department of Computer Science and Engineering
University of Kalyani
Kalyani
India

ISSN 2194-5357 ISSN 2194-5365 (electronic)
Advances in Intelligent Systems and Computing
ISBN 978-3-319-13730-8 ISBN 978-3-319-13731-5 (eBook)
DOI 10.1007/978-3-319-13731-5

Library of Congress Control Number: 2014956100

Springer Cham Heidelberg New York Dordrecht London

Printed on acid-free paper

Springer International Publishing AG Switzerland is part of Springer Science+Business Media
(www.springer.com)

Preface

This AISC volume-II contains 70 papers presented at the 49th Annual Convention of Computer Society of India: Emerging ICT for Bridging the Future: held during 12–14 December 2014 at Hyderabad hosted by CSI Hyderabad Chapter in association with JNTU, Hyderabad and DRDO. It proved to be a great platform for researchers from across the world to report, deliberate and review the latest progresses in the cutting-edge research pertaining to intelligent computing and its applications to various engineering fields. The response to CSI 2014 has been overwhelming. It received a good number of submissions from the different areas relating to intelligent computing and its applications in main tracks and four special sessions and after a rigorous peer-review process with the help of our program committee members and external reviewers finally we accepted 143 submissions with an acceptance ratio of 0.48. We received submissions from seven overseas countries.

Dr Vipin Tyagi, Jaypee University of Engg and Tech, Guna, MP conducted a Special session on "Cyber Security and Digital Forensics ". Dr. B.N. Biswal, BEC, Bhubaneswar and Prof. Vikrant Bhateja, Sri. Ramswaroop Memorial Group of Professional colleges, Lucknow conducted a special session on "Recent Advancements on Computational intelligence". "Ad-hoc Wireless Sensor Networks" special session was organized by Prof. Pritee Parwekar, ANITS, Vishakapatnam and Dr. S.K. Udgata, Univeisity of Hyderabad. A special session on "Advances and Challenges in Humanitarian Computing" was conducted by Prof. Sireesha Rodda, Dept of CSE, GITAM University, Vishakapatnam.

We take this opportunity to thank all Keynote Speakers and Special Session Chairs for their excellent support to make CSI2014 a grand success.

The quality of a referred volume depends mainly on the expertise and dedication of the reviewers. We are indebted to the program committee members and external reviewers who not only produced excellent reviews but also did in short time frames. We would also like to thank CSI Hyderabad Chapter, JNTUH and DRDO having coming forward to support us to organize this mega convention.

We express our heartfelt thanks to Mr G. Satheesh Reddy, Director RCI and Prof Rameshwar Rao, Vice-Chancellor, JNTUH for their continuous support during the course of the convention.

We would also like to thank the authors and participants of this convention, who have considered the convention above all hardships. Finally, we would like to thank all the volunteers who spent tireless efforts in meeting the deadlines and arranging every detail to make sure that the convention runs smoothly. All the efforts are worth and would please us all, if the readers of this proceedings and participants of this convention found the papers and event inspiring and enjoyable.

We place our sincere thanks to the press, print & electronic media for their excellent coverage of this convention.

December 2014

Volume Editors
Dr. Suresh Chandra Satapathy
Dr. A. Govardhan
Dr. K. Srujan Raju
Dr J.K. Mandal

Team CSI 2014

Chief Patrons

Dr. G. Satheesh Reddy, Director RCI
Prof. Rameswara Rao, VC JNTU-H

Advisory Committee

Sri. S. Chandra Sekhar, NASSCOM
Sri. J. Satyanarayana, A.P. Govt
Sri. Haripreeth Singh TS Govt
Sri. Sardar. G.S. Kohli GNIT
Prof. D.V.R. Vithal, Fellow CSI
Maj. Gen. R.K. Bagga, Fellow CSI
Dr. Ashok Agrwal, Fellow CSI
Dr. D.D. Sharma, GNIT Fellow CSI
Sri. H.R Mohan, President CSI
Sri. Sanjay Mahapatra, Secretary CSI
Sri. Ranga Raj Gopal, Treasurer CSI
Prof. S.V. Raghavan
Prof. P. Trimurty

Conference Committee

Sri. Bipin Mehta
Sri. Raju L. Kanchibotla
Sri. Gautam Mahapatra
Dr. A. Govardhan
Dr. Srujan Raju K.

Conveners

Sri. K. Mohan Raidu Main
Prof. C. Sudhakar
Sri. Chandra Sekhar
Dr. Chandra Sekhar Reddy

Organizing Committee

Sri. J.A. ChowdaryTelent, Sprint
Sri. R. Srinivasa Rao, Wipro
Dr. H.S. Saini, GNIT

Programme Committee

Dr. A. Govardhan
Sri. Bipin Chandra Ikavy
Dr. T. Kishen Kumar Reddy
Prof. N.V. Ramana Rao
Dr. Srujan Raju K.
Prof. I.L. Narasimha Rao
Dr. J. SasiKiran
Prof P.S. Avadhani
Sri. Venkatesh Parasuram
Prof. P. Krishna Reddy IIIT

Finance Committee

Sri. GautamMahapatra
Prof. C. Sudhakar
Sir. Raj Pakala

Publication Committee

Dr. A. Govardhan
Dr. S.C. Satapathy
Dr. Subash C. Mishra
Dr. Anirban Paul
Dr. Srujan Raju K.
Dr. Vishnu
Dr. J.K. Mandal

Exhibition Committee

Sri. P.V. Rao
Sri. Rambabukuraganty
Sri. Balaram Varansi
Sri. VenkataRamana Chary G.
Mr. Hiteshawar Vadlamudi
Dr. D.V. Ramana

Transport Committee

Sri. RambabuKuriganti
Sri. Krishna Kumar
Sri K.V. Pantulu
Smt. P. Ramadevi
Sri. Amit Gupta

Hospitality Committee

Sri. Krishna Kumar Tyagarajan
Sri. P.V. Rao

Sponsorship Committee

Sri. Raj Pakala
Sri. Srinivas konda
Sri. PramodJha

Marketing & PR Committee

Sri. KiranCherukuri
Ms. Sheila P.

Registrations Committee

Sri Ramesh Loganathan
Smt. Rama Bhagi
Sri. Rajeev Rajan Kumar
Sri. Krishna Kumar B.
Sri. Sandeep Rawat
Sri. Ram Pendyala
Sri T.N Sanyasi Rao
Sri D.L. Seshagiri Rao
Sri G. Vishnu Murthy
Prof. Ramakrishna Prasad
Sri. Vijay Sekhar K.S.
Dr. SaumyadiptaPyne

Cultural Committee

Smt. Rama Bhagi
Smt. Rama Devi

International Advisory Committee/Technical Committee

P.K. Patra, India
Sateesh Pradhan, India
J.V.R. Murthy, India
T.R. Dash, Kambodia
Sangram Samal, India
K.K. Mohapatra, India
L. Perkin, USA
Sumanth Yenduri, USA
Carlos A. Coello Coello, Mexico
S.S. Pattanaik, India
S.G. Ponnambalam, Malaysia
Chilukuri K. Mohan, USA
M.K. Tiwari, India
A. Damodaram, India
Sachidananda Dehuri, India
P.S. Avadhani, India
G. Pradhan, India
Anupam Shukla, India
Dilip Pratihari, India
Amit Kumar, India
Srinivas Sethi, India
Lalitha Bhaskari, India
V. Suma, India
Pritee Parwekar, India
Pradipta Kumar Das, India
Deviprasad Das, India
J.R. Nayak, India
A.K. Daniel, India
Walid Barhoumi, Tunisia
Brojo Kishore Mishra, India
Meftah Boudjelal, Algeria
Sudipta Roy, India
Ravi Subban, India
Indrajit Pan, India
Prabhakar C.J, India
Prateek Agrawal, India

Igor Belykh, Russia
Nilanjan Dey, India
Srinivas Kota, Nebraska
Jitendra Virmani, India
Shabana Urooj, India
Chirag Arora, India
Mukul Misra, India
Kamlesh Mishra, India
Muneswaran, India
J. Suresh, India
Am,lan Chakraborthy, India
Arindam Sarkar, India
Arp Sarkar, India
Devadatta Sinha, India
Dipendra Nath, India
Indranil Sengupta, India
Madhumita Sengupta,India
Mihir N. Mohantry, India
B.B. Mishra, India
B.B. Pal, India
Tandra Pal, India
Utpal Nandi, India
S. Rup, India
B.N. Pattnaik, India
A Kar, India
V.K. Gupta, India
Shyam lal, India
Koushik Majumder, India
Abhishek Basu, India
P.K. Dutta, India
Md. Abdur Rahaman Sardar, India
Sarika Sharma, India
V.K. Agarwal, India
Madhavi Pradhan, India
Rajani K. Mudi, India
Sabitha Ramakrishnan, India
Sireesha Rodda, India
Srinivas Sethi, India
Jitendra Agrawal, India
Suresh Limkar, India
Bapi Raju Surampudi, India
S. Mini, India
Vinod Agarwal, India
Prateek Agrwal, India
Faiyaz Ahmad, India
Musheer Ahmad, India
Rashid Ali, India
A.N. Nagamani, India
Chirag Arora, India
Aditya Bagchi, India
Balen Basu, India
Igor Belykh, Russia
Debasish Jana, India
V. Valli Kumari, India
Dac-Nuuong Le
Suneetha Manne, India
S. Rattan Kumar, India
Ch. Seshadri, India
Swathi Sharma, India
Ravi Tomar, India and Many More

Contents

Machine Learning and Computational Intelligence

Data Mining, Data Engineering and Soft Computing

A Non-Local Means Filtering Algorithm for Restoration of Rician Distributed MRI

Vikrant Bhateja, Harshit Tiwari, and Aditya Srivastava

Department of Electronics and Communication Engineering,
Shri Ramswaroop Memorial Group of Profesional Colleges,
Lucknow-226010 (U. P.), India
{bhateja.vikrant,htiwari101092,adityasri1092}@gmail.com

Abstract. Denoising algorithms for medical images are generally constrained owing to the dependence on type of noise model as well as introduction of artifacts (in terms of removal of fine structures) upon restoration. In context of magnitude Magnetic Resonance Images (MRI), where noise is approximated as Rician instead of additive Gaussian; denoising algorithms fail to produce satisfactory results. This paper presents a Non-Local Means (NLM) based filtering algorithm for denoising Rician distributed MRI. The proposed denoising algorithm utilizes the concept of self-similarity which considers the weighted average of all the pixels by identifying the similar and dissimilar windows based on Euclidean distance for MRI restoration. Simulations are carried out on MRI contaminated with different levels of Rician noise and are evaluated on the basis of Peak Signal-to-Noise Ratio (*PSNR*) and Structural Similarity (SSIM) as quality parameters. Performance of the proposed algorithm has shown significant results when compared to the other state of the art MRI denoising algorithms.

Keywords: MRI, Non-Local Means, Rician noise, *PSNR*.

1 Introduction

Digital images are often superimposed by noises during acquisition or transmission. Contamination by noises leads to erroneous intensity fluctuations (pixel values) either due to imperfections of imaging devices or transmission channels. Image denoising is a crucial process which aims to suppress these imperfections coupled with restoration of details and fine structures in the image [1]-[7]. In context of MRI, random variations appears during the process of acquisition due to small acquisition time as well as the interferences from internal components of MR scanner. These variations can be modeled in the form of Gaussian and Rician noises respectively. Gaussian noise is an additive noise and has the probability distribution function equal to the Gaussian distribution. On the other hand, Rician noise is non-additive and tends to make image data Rician distributed. At higher values of Signal-to-Noise Ratios (SNR), the Rician distribution tends to approach Gaussian distribution [8]-[11]. To minimize the effects of noise, it is generally less favorable to exercise control over

S.C. Satapathy et al. (eds.), *Emerging ICT for Bridging the Future – Volume 2*,
Advances in Intelligent Systems and Computing 338, DOI: 10.1007/978-3-319-13731-5_1

acquisition artifacts. However, it is more effective and less time consuming if an appropriate denoising algorithm is applied at the post processing stage. Many denoising algorithms have been proposed in past such as Gaussian filter [12], Anisotropic Diffusion (AD) filter [13]-[15], Directional Filters [16], Morphological Filters [17]-[18], Bilateral Filter [19] and wavelet thresholding [20]-[22], for processing Gaussian noise corrupted images. Spatial filtering via Gaussian, Wiener filters and Bilateral filters are based on weighted averaging within a local window to yield the restored pixel. This tends to remove ample amount of details leading to blurred edges and boundaries. Gerig et al. in their work [23] carried out MRI denoising based on the concept of Partial Differential Equation (PDE) approach, utilizing AD filter of Perona and Malik [13]. This approach by the authors was limited by the incorrect assumption of the noise in MRI; being misinterpreted as purely Gaussian. Sijbers et al. [24] developed the adaptive AD based approach for suppression of Rician noise in MR images. Manjon et al. [25] applied the Non-Local Means (NLM) approach based filter for Gaussian noise in MRI which was based on the advancement of the technique developed by Buades et al. [26]-[27]. The computational complexity of this method was reduced by the fast NLM method [28]-[29]. Later on, Rician noise adaptation of NLM filter was proposed by Wiest-Daessle et al. [30] to suppress noise for DT-MRI images. Liu et al. [31] calculated weights for the Gaussian filtered MRI to minimize noise perturbations and then applied NLM to achieve denoising results. In this paper, a modified NLM approach is used to achieve better results for Rician distributed MRI, where restored values of pixels are calculated on the basis of repeated structures in the whole image. It tends to performs better denoising while preserving fine details for 2D MRI. In remaining part of the paper, Section 2 details the proposed filtering methodology. Section 3 presents the simulation results and their subsequent discussions and Section 4 includes the conclusion part.

2 Proposed Filtering Methodology

In magnitude MRI, noise is governed by Rician distribution as both real and imaginary parts of the complex signal is corrupted with zero mean uncorrelated Gaussian noise [32]. After this, image intensity in presence of noise is given by

$$f(x \mid v,\sigma)=\frac{x}{\sigma^2}\exp(\frac{-(x^2+v^2)}{2\sigma^2})I_0(\frac{xv}{\sigma^2}) \tag{1}$$

where: I_0 is the zeroth order modified Bessel function of the first kind, v is the noise-free signal amplitude, and is the standard deviation of the Gaussian noise in the real and imaginary images. SNR is defined here as v/σ. In the case of squared magnitude images, a signal independent noise bias of $2\sigma^2$ is produced. Thus, for a given MR magnitude image X_n,

$$E(X_n^{\,2}) = X_0^{\,2} + 2\sigma^2 \tag{2}$$

where: X_n is the noisy image and X_0 is the original image. $X_0^{\,2}$ and $X_n^{\,2}$ are the squared image of X_0 and X_n respectively.

In NLM approach, the edges and details can both be well preserved along with reduction of noise by utilizing the similarity of pixels in the MRI. To achieve better results for Rician noises, there is a need to customize the NLM based filtering approach for processing magnitude MR images [33]-[34]. In a given MRI, self-similar structures are present which are well utilized by the NLM algorithm. This algorithm relies on these structures where both symmetry and search windows seek for larger weights depending upon the Euclidean distance calculated between the two similar pixels in the MR image. These weights are then used to calculate the weighted average of all the pixels which yields restored MRI. Mathematically, for a noisy MRI (X), denoised value $X(i)$ for pixel i through calculation of all the pixels within X can be obtained as follows:

$$NLM(X(i)) = \sum_{j \in X} w(i,j)X(j) \tag{3}$$

where, $w(i,j)$ represents the calculated weights which depends upon the similarity between pixel i and pixel j. It is defined as,

$$w(i,j) = \frac{1}{Z(i)} e^{-\frac{\| X(N_i) - X(N_j) \|_{2,\sigma}^2}{h^2}} \tag{4}$$

where, h is a degree of filtering applied to the filter. N_i and N_j are the neighborhood pixels of i and j, respectively, $\| X(N_i) - I(N_j) \|_{2,\sigma}^2$ representing the Euclidean distance and $Z(i)$ is a normalizing constant represented as:

$$Z(i) = \sum_j e^{-\frac{\| X(N_i) - X(N_j) \|_{2,\sigma}^2}{h^2}} \tag{5}$$

In this paper, the magnitude of the complex valued MRI is taken to obtain a magnitude MRI as shown by Eq. (2). On this resultant image, NLM algorithm (Fig. 1) is applied to obtain the filtered image. The quality of the restored MRI is computed objectively using *PSNR & SSIM* as performance parameters [35]-[41].

BEGIN

Step 1 : *Input:* MRI with Rician Noise.
Step 2 : *Process:* MRI in Matrix form of specified dimension.
Step 3 : *Process:* Symmetry Window and Search Window of specified dimension.
Step 4 : *Compute:* Gray Level Vector Intensities.
Step 5 : *Compute:* Euclidean distance using these Vectors
Step 6 : *Compute*: Weight for each pixel.
(a) Large Weights for Small Euclidean distance. (Similar window)
(b) Small Weights for Larger Euclidean distance. (Dissimilar window)
Step 7 : *Process:* Weighted average of all the pixels.
Step 8 : *Output:* Restored MRI.

END

Fig. 1. Non-Local Means (NLM) Algorithm for Rician Distributed MRI Restoration

3 Results and Discussions

In this section, a sample MRI is being used as a reference image. For the simulation process, this image is contaminated with Rician noise of desired variance levels. Once the noisy image is obtained, proposed NLM algorithm is applied on it. For the proposed denoising algorithm, there are certain preliminary parameters which are required to be set such as: search window, symmetry window and degree of filtering. Proper initialization and tuning of these parameters improves the denoising process to reasonable extent. For the simulations in present work, search window is set to 5x5 while the symmetry window is set to 2x2 for better performance. The performance of the denoising algorithm is evaluated and enlisted under Table 1 for differing values of noise variance levels: 5, 9, 13, 17 and 21. The noisy MRI and the corresponding filtered images using proposed algorithm for the purpose of visual assessment are given in figure. 2. It can be interpreted from figure 2 that the restored MRI is devoid of noise contamination and useful in recovering the details and fine structures. This response has been supported with higher values of *PSNR* and *SSIM*.

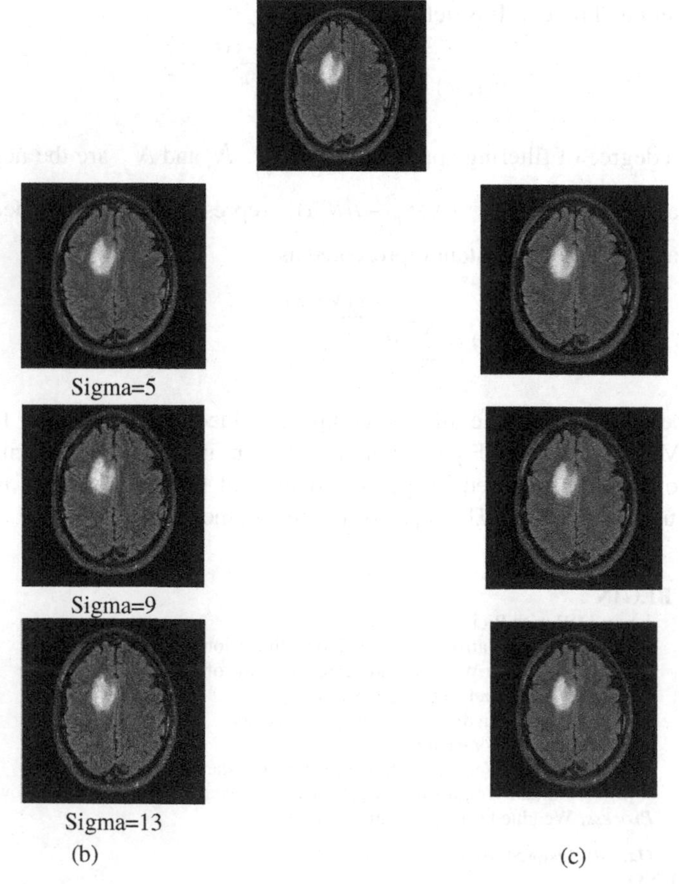

Fig. 2. (a) Original MR Image (b) Noisy MRI with different noise variance levels. (c) Denoised MRI using proposed NLM algorithm.

Table 1. Computation of *PSNR* and *SSIM* for MRI restored using proposed NLM algorithm for different levels of Rician noise

Sigma	*PSNR*(dB)	*SSIM*
5	36.5014	0.9494
9	33.2585	0.9015
13	29.9120	0.8346
17	27.1448	0.7639
21	24.8145	0.6923

The proposed denoising algorithm has been compared with the Rician NLM filter [30]. The obtained results show that the proposed algorithm is visually effective as compared to the preceding approach. It can be visualized from figure 3 that the proposed algorithm results in improved SNR and effective preservation of details and structures. The same has been validated by higher values of PSNR and SSIM, mentioned in figure 3 for MRI restored with proposed NLM filtering algorithm. However, at high values of variance some portion is over-smoothened by the algorithm but loss of details is less as compared to the previous approach. In addition, it is worth noting that the edges of the lesion region in foreground are comparatively sharper with the proposed algorithm. Hence, the proposed restoration algorithm serves as a vital pre-processing tool for computer aided diagnosis [42]-[43].

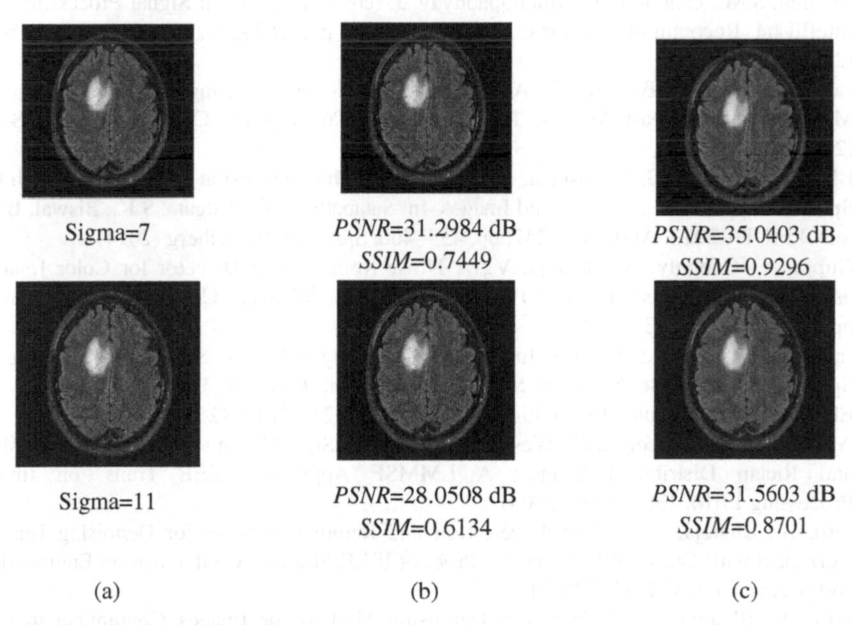

Fig. 3. (a) Rican distributed MRI at noise variance levels of 7 & 11 respectively. (b) Denoised image using existing approach [30]. (c) Denoised image using proposed algorithm.

4 Conclusion

In this paper, a NLM based denoising algorithm suited for Rician noise in MRI has been proposed to improve the noise suppression while preserving the structural details. Simulations were performed on the MRI for different levels of Rician noise and compared with the rician NLM filter. Obtained results showed that the proposed algorithm outperformed the other method in terms of both *PSNR* and *SSIM*. The proposed algorithm managed to recover finer details from the noise corrupted regions. Therefore, the image quality was improved and the structural details were preserved with edges being sharper as compared to the existing denoising algorithms.

References

1. Shukla, A.K., Verma, R.L., Alam, M., Bhateja, V.: An Improved Directional Weighted Median Filter for Restoration of Images Corrupted with High Density Impulse Noise. In: Proc. Int. Conference on Relaibility, Optim., pp. 506–511. IEEE (2014)
2. Bhateja, V., Rastogi, K., Verma, A., Malhotra, C.: A Non-Iterative Adaptive Median Filter for Image Denoising. In: Proc. Int. Conference on Signal Processing and Integrated Networks, pp. 113–118. IEEE (2014)
3. Bhateja, V., Verma, A., Rastogi, K., Malhotra, C., Satapathy, S.C.: Performance Improvement of Decision Median Filter for Suppression of Salt and Pepper Noise. In: Thampi, S.M., Gelbukh, A., Mukhopadhyay, J. (eds.) Advances in Signal Processing and Intelligent Recognition Systems. AISC, vol. 264, pp. 287–298. Springer, Heidelberg (2014)
4. Jain, A., Singh, S., Bhateja, V.: A Robust Approach for Denoising and Enhancement of Mammographic Breast Masses. Int. Journal on Convergence Computing 1(1), 38–49 (2013)
5. Bhateja, V., Singh, G., Srivastava, A.: A Novel Weighted Diffusion Filtering Approach for Speckle Suppression in Ultrasound Images. In: Satapathy, S.C., Udgata, S.K., Biswal, B.N. (eds.) FICTA 2013. AISC, vol. 247, pp. 459–466. Springer, Heidelberg (2014)
6. Gupta, A., Ganguly, A., Bhateja, V.: A Noise Robust Edge Detector for Color Images using Hilbert Transform. In: Proc. of 3rd Int. Advance Computing Conference, pp. 1207–1212. IEEE (2013)
7. Jain, A., Bhateja, V.: A Novel Image Denoising Algorithm for Suppressing Mixture of Speckle and Impulse Noise in Spatial Domain. In: Proc. of 3rd Int. Conference on Electronics & Computer Technology, vol. 3, pp. 207–211. IEEE (2011)
8. Aja-Fernandez, A., Lopez, C., Westin, C.: Noise and Signal Estimation in Magnitude MRI and Rician Distributed Images: A LMMSE Approach. IEEE Trans. on Image Processing 17(8), 1383–1398 (2008)
9. Jain, A., Bhateja, V.: A Novel Detection and Removal Scheme for Denoising Images Corrupted with Gaussian Outliers. In: Proc. of IEEE Students Conference on Engineering and Systems, pp. 434–438 (2012)
10. Jain, A., Bhateja, V.: A Versatile Denoising Method for Images Contaminated with Gaussian Noise. In: Proc. of CUBE International Information Technology Conference & Exhibition, pp. 65–68. ACM ICPS (2012)

11. Gupta, A., Ganguly, A., Bhateja, V.: An Edge Detection Approach for Images Contaminated with Gaussian and Impulse Noises. In: Mohan, S., Kumar, S.S. (eds.) Proc. of 4th International Conference on Signal and Image Processing (ICSIP 2012). LNEE, vol. 222, pp. 523–533. Springer, Heidelberg (2012)
12. Lindenbaum, M., Fischer, M., Bruckstein, A.: On Gabor Contribution to Image Enhancement. Pattern Recog. 27(1), 1–8 (1994)
13. Perona, P., Malik, J.: Scale-Space and Edge Detection using Anisotropic Diffusion. IEEE Trans. on Pattern Anal. and Mach. Intell. 12(7), 629–639 (1990)
14. Gupta, A., Tripathi, A., Bhateja, V.: Despeckling of SAR Images via an Improved Anisotropic Diffusion Algorithm. In: Satapathy, S.C., Udgata, S.K., Biswal, B.N. (eds.) Proceedings of Int. Conf. on Front. of Intell. Comput. AISC, vol. 199, pp. 747–754. Springer, Heidelberg (2013)
15. Bhateja, V., Srivastava, A., Singh, G., Singh, J.: Speckle Reduction in Ultrasound Images using an Improved Conductance Function based on Anisotropic Diffusion. In: Proc. 2014 International Conference on Computing for Sustainable Global Development, pp. 619–624. IEEE (2014)
16. Shukla, A.K., Verma, R.L., Alam Mohd, S., Bhateja, C.: Directional Ordered Statistics Filtering for Suppression of Salt and Pepper Noise. In: Proc. International Conference on Signal Processing and Integrated Networks, pp. 76–81. IEEE (2014)
17. Verma, R., Mehrotra, R., Bhateja, V.: An Integration of Improved Median and Morphological Filtering Techniques for Electrocardiogram Signal Processing. In: Proc. of 3rd International Advance Computing Conference, pp. 1223–1228. IEEE (2013)
18. Verma, R., Mehrotra, R., Bhateja, V.: A New Morphological Filtering Algorithm for Pre-Processing of Electrocardiographic Signals. In: Mohan, S., Kumar, S.S. (eds.) Proc. of 4th International Conference on Signal and Image Processing (ICSIP 2012). LNEE, vol. 221, pp. 193–201. Springer, Heidelberg (2012)
19. Bhateja, V., Misra, M., Urooj, S., Lay-Ekuakille, A.: Bilateral Despeckling Filter in Homogeneity Domain for Breast Ultrasound Images. In: Proc. of 3rd International Conference on Advances in Computing, Communications and Informatics, pp. 1027–1032. IEEE (2014)
20. Donoho, D.L.: Denoising by Soft-Thresholding. IEEE Trans. on Info. Theory 41(3), 613–627 (1995)
21. Srivastava, A., Alankrita, Raj, A., Bhateja, V.: Combination of Wavelet Transform and Morphological Filtering for Enhancement of Magnetic Resonance Images. In: Snasel, V., Platos, J., El-Qawasmeh, E. (eds.) ICDIPC 2011, Part I. CCIS, vol. 188, pp. 460–474. Springer, Heidelberg (2011)
22. Bhateja, V., Urooj, S., Mehrotra, R., Verma, R., Lay-Ekuakille, A., Verma, V.D.: A Composite Wavelets and Morphology Approach for ECG Noise Filtering. In: Maji, P., Ghosh, A., Murty, M.N., Ghosh, K., Pal, S.K. (eds.) PReMI 2013. LNCS, vol. 8251, pp. 361–366. Springer, Heidelberg (2013)
23. Gerig, G., Kubler, O., Kikinis, R., Jolesz, F.: Nonlinear Anisotropic Filtering of MRI Data. IEEE Trans. on Med. Imag. 11(2), 221–232 (1992)
24. Sijbers, J., Den-Dekker, A., Van-Audekerke, J., Verhoye, M., Van-Dyck, D.: Adaptive Anisotropic Noise Filtering for Magnitude MR Data. Mag. Reson. Imag. 17(10), 1533–1539 (1999)
25. Manjon, J., Caballero, J., Lull, J., Gracian, G., Marti-Bonmati, L., Robles, M.: MRI Denoising using Non-Local Means. Med. Image Anal. 12(4), 514–523 (2008)
26. Buades, A., Coll, B., Morel, J.M.: A Review of Image denoising Algorithms, with a New One. Multiscale Model. Simul. 4(2), 490–530 (2005)

27. Buades, A., Coll, B., Morel, J.M.: A Non-Local Algorithm for Image Denoising. IEEE Trans. on Comp. Vision and Patt. Recog. 2, 60–65 (2005)
28. Dauwe, A., Goossens, B., Luong, H.Q., Philips, W.: A Fast Non-Local Image Denoising Algorithm. Image Process. Algo. and Systems 6812, 1–8 (2008)
29. Coupe, P., Yger, P., Prima, S., Hellier, P., Kervrann, C., Barillot, C.: An Optimized Blockwise Nonlocal Means denoising filter for 3-D Magnetic Resonance Images. IEEE Trans. on Med. Imag. 27(4), 425–441 (2008)
30. Wiest-Daesslé, N., Prima, S., Coupé, P., Morrissey, S.P., Barillot, C.: Rician Noise Removal by Non-Local Means Filtering for Low Signal-to-Noise Ratio MRI: Applications to DT-MRI. In: Metaxas, D., Axel, L., Fichtinger, G., Székely, G. (eds.) MICCAI 2008, Part II. LNCS, vol. 5242, pp. 171–179. Springer, Heidelberg (2008)
31. Liu, H., Yang, C., Pan, N., Song, E., Green, R.: Denoising 3D MR Images by the Enhanced Non-Local Means filter for Rician Noise. Magn. Reso. Imag. 28(10), 1485–1496 (2010)
32. Gudbjartsson, H., Patz, S.: The Rician Distribution of Noisy MRI Data. Magn. Reson. Med. 34(6), 910–914 (1995)
33. Manjon, J., Coupe, P., Marti-Bonamti, L., Collins, D., Robles, M.: Adaptive Non-Local Means Denoising of MR Images with Spatially Varying Noise Levels. Mag. Reson. Imag. 31(1), 192–203 (2010)
34. Manjon, J., Coupe, P., Buades, A., Collins, D., Robles, M.: New Methods for MRI Denoising based on Sparseness and Self-Similarity. Med. Image Anal. 16(1), 18–27 (2012)
35. Gupta, P., Srivastava, P., Bharadwaj, S., Bhateja, V.: A HVS based Perceptual Quality Estimation Measure for Color Images. ACEEE International Journal on Signal & Image Processing (IJSIP) 3(1), 63–68 (2012)
36. Jain, A., Bhateja, V.: A Full-Reference Image Quality Metric for Objective Evaluation in Spatial Domain. In: Proc. International Conference on Communication and Industrial Application, vol. 22, pp. 91–95 (2011)
37. Gupta, P., Srivastava, P., Bharadwaj, S., Bhateja, V.: A New Model for Performance Evaluation of Denoising Algorithms based on Image Quality Assessment. In: Proc. of CUBE Int. Info. Tech. Conf. & Exhibition, pp. 5–10 (2012)
38. Bhateja, V., Srivastava, A., Kalsi, A.: Reduced Reference IQA based on Structural Dissimilarity. In: Proc. Int. Conf. on Signal Process. and Integ. Networks, pp. 63–68 (2014)
39. Bhateja, V., Srivastava, A., Kalsi, A.: A Reduced Reference Distortion Estimation Measure for Color Images. In: Proc. Int. Conference on Signal Processing and Integrated Networks, pp. 669–704 (2014)
40. Bhateja, V., Srivastava, A., Singh, G., Singh, J.: A Modified Speckle Suppression Algorithm for Breast Ultrasound Images Using Directional Filters. In: Satapathy, S.C., Avadahani, P.S., Udgata, S.K., Lakshminarayana, S. (eds.) ICT and Critical Infrastructure: Proceedings of the 48th Annual Convention of CSI - Volume II. AISC, vol. 249, pp. 219–226. Springer, Heidelberg (2014)
41. Bhateja, V., Kalsi, A., Shrivastava, A., Lay-Ekuakille, A.: A Reduced Reference Distortion Measure for Performance Improvement of Smart Cameras. IEEE Sensors Journal, Spl. Iss. on Advancing Standards for Smart Transducer Interfaces (2014)
42. Bhateja, V., Devi, S.: An Improved Non-Linear Transformation Function for Enhancement of Mammographic Breast Masses. In: Proc. of 3rd International Conf. on Elec. & Comp. Tech., vol. 5, pp. 341–346. IEEE (2011)
43. Bhateja, V., Misra, M., Urooj, S., Lay-Ekuakille, A.: A Robust Polynomial Filtering Framework for Mammographic Image Enhancement from Biomedical Sensors. IEEE Sensors Journal 13(11), 4147–4156 (2013)

Evaluation of Pheromone Update in Min-Max Ant System Algorithm to Optimizing QoS for Multimedia Services in NGNs

Dac-Nhuong Le

Haiphong University, Haiphong, Vietnam
Nhuongld@hus.edu.vn

Abstract. Next Generation Networks (NGN) is towards providing seamless Quality of Service (QoS) for converged mobile and fix multimedia services anywhere-anytime to users through different access network technologies. Factors affecting the QoS in NGN are speech encoders, delay, jitter, packet loss and echo. The negotiation and dynamic adaptation of QoS is currently considered to be one of the key features of the NGN concept. Optimal QoS for multimedia services in the NGNs for multi-flow resource constrained utility maximization is complex problem. In this paper, we propose a novel Min-Max Ant System (MMAS) algorithm with a pheromone updating scheme to solve it. We compared effective of MMAS to SMMAS (*Smoothed MMAS*), MLAS (*Multi-level Ant System*) and 3-LAS (*Three-level Ant System*). The computational results showed that our approach is currently among the best performing algorithms for this problem.

Keywords: quality of service, multimedia services, next generation network, Min-Max Ant System.

1 Introduction

In NGN, the backbone of the overall network architecture will be IP network, supporting different access network technologies such as WLAN, UTRAN, and WiMax. Moreover, this integrated wireless system, will have to handle diverse types of traffics. NGN will provide advanced services, such as QoS guarantees to users and their applications. Factors affecting the QoS in NGN are speech encoders, delay, jitter, packet loss and echo. In standardized NGN architecture, at the service level specify requirements for entities at the rim of the network to negotiate and select common codecs for each E2E session [1]. The 3GPP project has specified the IMS procedures for negotiating multimedia session characteristics between endpoints [2]. Under these demanding conditions of media-rich services, network service providers must carefully provision and allocate network resource capabilities to providing personalized and optimal service quality implies more advanced negotiation mechanisms matching user preferences/capabilities against service requirements, network resource capabilities, and operator policy.

S.C. Satapathy et al. (eds.), *Emerging ICT for Bridging the Future – Volume 2*,
Advances in Intelligent Systems and Computing 338, DOI: 10.1007/978-3-319-13731-5_2

Related work on QoS adaptation in IMS networks includes solutions for dynamically adapting multimedia content to fit network/system resource availability. Ozcelebi et al have presented a solution to enhancing E2E QoS for multimedia streaming in IMS [3]. Boula et al introduced an enhanced IMS architecture featuring cross-layer monitoring and adaptation mechanisms to maximizing perceived QoS [4]. The DAIDALOS [5] and ENTHRONE [6] projects is further research on multimedia session control and content adaptation. In additional related work, we can be found on profile matching [7], decision-making for the optimization of service parameters [8,9], and the description of various service and transport configurations for service negotiation [10,11,12], limited solutions unite these aspects to support delivery of personalized multimedia services in the NGN. Khan has proposed an heuristic algorithm for finding near optimal solutions for cases with a large number of variables to solve the quality adaptation in a multisession multimedia system problem [13]. A different approach to formulating the optimization problem may be to use tools specified as part of the MPEG DIA standard, Usage Environment Description tools, Universal Constraints Description tool, and Adaptation QoS tools has presented by Mukherjee et al [14]. In [15], the authors have proposed the high-level concept of a QoS Matching and Optimization Function (Q-MOF) to be included along the E2E signaling path in the service control layer of the NGN architecture and heuristic algorithm for finding near optimal solutions.

In this paper, we propose a novel MMAS algorithm to optimal QoS for Multimedia Services (OQMS). The rest of this paper is organized as follows. Section 2 introduces Q-MOF in the NGN architecture and problem formulation. Section 3 presents our new algorithm propose. Section 4 is our simulation and analysis results, and finally, section 5 concludes the paper.

2 Optimal QoS for Multimedia Services Problem

A multimedia service is composed of two or more media components such as audio flows, video flows, graphics data, etc. We assume that a multimedia service exists in one or more versions to meet heterogeneous user and network capabilities. We specify service versions as differing in the included media components. Each media component may be configured by choosing from offered alternative operating parameters. We refer to the overall service configuration as the set of chosen operating parameters for all included media components [13,15]. The problem can be formulated as a multi-choice multi-dimension 0-1 knapsack problem (MMKP), for which we introduce the following notation: $F = (f_1, ..., f_n)$ is set of different media flows, of which flows $f_1, ..., f_h$ are in the *downlink direction* and flows $f_{h+1}, ..., f_n$ are in the *uplink direction*. The flow f_i has p_i operating points. $r_{ij} = (r_{ij1}, ..., r_{ijq}, ..., r_{ijm})$ is set of required resources r for p_j and f_i, of which m is number of resources, $r_{ij1}, ..., r_{ijq}$ are corresponding to bandwidth assigned within different QoS classes. For a single operating point, we assume only one of the values $r_{ij1}, ..., r_{ijq}$ to be greater than zero, while all others are equal to zero. Because only one QoS class is chosen per operating point for a media flow.

$B_{downlink}$ and B_{uplink} are maximum available downlink and uplink bandwidth determined by user terminal capabilities and access network. We represent the resource constraints for $r_{ij1}, ..., r_{ijq}$; $R = (R_{q+1}, ..., R_m)$ is the consumption of a particular resource across all media flows; $R_i = (R_{i1}, ..., R_{im})$ regarding resource consumption per flow f_i (e.g., cost for a media flow must be less than a specified amount); $u_i(r_{ij})$ is the utility value for operating point p_j and media flow f_i; w_i is weight factors assigned to utility values to indicate the relative importance of media flows; $x_{ij} = 1$ if operating point p_j is chosen per media flow f_i, and $x_{ij} = 0$ in otherwise.

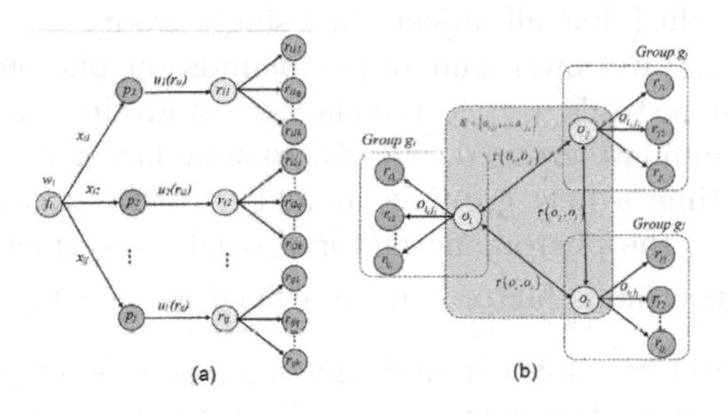

Fig. 1. Flow i^{th} and solution S

Then the formulation of optimal QoS for multimedia services problem for multi-flow resource constrained utility maximization is:

$$\max \sum_{i=1}^{n} \sum_{j=1}^{p_i} w_i x_{ij} u_i(r_{ij}) \tag{1}$$

Subject to:

$$\sum_{i=1}^{h} \sum_{j=1}^{p_i} \sum_{k=1}^{q} x_{ij} r_{ijk} \leq B_{downlink} \tag{2}$$

$$\sum_{i=h+1}^{n} \sum_{j=1}^{p_i} \sum_{k=1}^{q} x_{ij} r_{ijk} \leq B_{uplink} \tag{3}$$

$$\sum_{i=1}^{n} \sum_{j=1}^{p_i} x_{ij} r_{ijk} \leq R_k, \quad \forall k = \overline{q+1, .., m} \tag{4}$$

$$\sum_{j=1}^{p_i} x_{ij} r_{ijk} \leq R_{ik}, \quad \forall k = \overline{1..m}, \forall i = \overline{1..n} \tag{5}$$

$$\sum_{i=1}^{n} \sum_{j=1}^{p_i} x_{ij} = 1 \tag{6}$$

$$x_{ij} \in \{0, 1\}, \forall i = \overline{1..n}, \forall j = \overline{1..p_i} \tag{7}$$

3 Our Approach

The MMAS based on several modifications to Ant System (AS) which aim (i) to exploit more strongly the best solutions found during the search and to direct the ants search towards very high quality solutions and (ii) to avoid premature convergence of the ants' search [16]. There are some AS based solutions for a more restricted variant of Knapsack Problem [17], namely Multi-choice Knapsack Problem (MKP) [18] and Multi-choice Multi-dimension Knapsack Problem (MMKP) [19] when MKP resources have multiple dimensions. However there is no concept of group in MKP. As a result, MKP can be thought of as a restricted version of MMKP, which has all objects in a single group.

Let $\tau_{\min}$ and $\tau_{\max}$ are lower and upper bounds on pheromone values. We initialize all pheromone trails to $\tau_{\max}$ and choose first group g_i has been randomly chosen. For each candidate group o_j, the pheromone factor $\tau_{S_k}(g_j)$ is initialized to $\tau(o_i, o_j)$. Each time a new group g_l is added to the solution S_k, for each candidate group g_j , the pheromone factor $\tau_{S_k}(g_j)$ s incremented by $\tau(o_l, o_j)$. And the highest pheromone factor can be defined as $\tau_{S_k}(g_i) = \sum\limits_{g_j \in S_k} \tau(o_i, o_j)$. When the algorithm starts, ants can perform search in order of group that impose the lowest restriction on the next choices. The key point is to decide which components of the constructed solutions should be rewarded, and how to exploit these rewards when constructing new solutions. A solution is a set of selected objects $S = \{o_{ij} | x_{o_{ij}} = 1\}$. Say $x_{o_{ij}} = 1$ is the mean an object o_{ij} is selected and the corresponding decision variable x_{ij} has been set to 1. We must constructed solutions $S = \{o_{i_1 j_1}, ..., o_{i_n j_n}\}$, pheromone trails are laid on each objects selected in S. So pheromone trail τ_{ij} will be associated with object o_{ij}. Let S_k be the set of the selected objects at the kth iteration. The heuristic factor $s_k(O_{ij})$ is determined by:

$$s_k(O_{ij}) = \frac{w_i u_i(r_{ij})}{\sum\limits_{i=1}^{m} \frac{r_{ijl}}{d_{S_k}(l)}} \tag{8}$$

where, $d_{S_k}(l) = R_l - \sum\limits_{o_{ij} \in S_k} r_{ijl}$; l is index of resource R across all media flows, S_k is the dynamic heuristic information will be changed from step to step.

We select groups in the same way as it selects the objects. We have unique idea of a separate pheromone trail for groups to save the ordering of groups that lead to a good solution. The group pheromone trail also follow a MMAS approach and initialized to the max pheromone value. Once each ant has constructed a solution, pheromone trails laying on the solution objects are updated according to the ACO [23]. First, all amounts are decreased in order to simulate evaporation. This is done by multiplying the quantity of pheromone laying on each object by a pheromone persistence rate $(1 - P)$ such that $0 \leq P \leq 1$. Then, pheromone is increased for all the objects in the best solution of the iteration. When constructing a solution, an ant starts with an empty knapsack. At the first construction step an ant selects a group randomly and at all the latter steps, groups are selected according to their associated pheromone value. After selecting a group, the algorithm removes all the bad *Candidates* that violates resource constraints. The local heuristic information of the remaining candidate objects of the group and selects an object updated following probability equation:

$$P_{S_k(O_{ij})} = \frac{[\tau_{S_k}(O_{ij})]^{\alpha}[\eta_{S_k}(O_{ij})]^{\beta}}{\sum\limits_{O_{ij} \in Candidates} [\tau_{S_k}(O_{ij})]^{\alpha}[\eta_{S_k}(O_{ij})]^{\beta}} \tag{9}$$

in which, *Candidates* are all items from the currently selected group which do not violate any resource constraints; τ_{S_k} is the pheromone factor of the dynamic heuristic information S_k and computed by equation (8); η_{S_k} is the desirability of S_k. The influence of the pheromone concentration to the probability value is presented by the constant α, while constant β do the same for the desirability and control the relative importance of pheromone trail versus local heuristic value. Let $S_{internalbest}$ be the best solution constructed during the current cycle. The quantity of pheromone increased for each object is defined by $G(S_{internalbest}) = \frac{1}{\sum\limits_{k=1}^{n} P_{S_k(O_{ij})}} \times \sum\limits_{o_{ij} \in S_{internalbest}} w_i u_i(r_{ij})$. After each iteration, group pheromone values have much better than the original pheromone trail for ants. The best solution updates the group pheromone trail. All the adjacent groups get the highest amount of pheromone value that gradually diminishes as the distance between groups increases. The *Candidategroups* data structure maintains a list of feasible candidate groups which can be considered next.

Algorithm 1. MMAS algortihm to optimizing QoS for multimedia services

```
Initialize pheromone trails to τ_max for both item and group;
Top_ksolution ⇐ {hold topmost k solutions};
Repeat
  S_globalbest ⇐ ∅;
  For each ant k = 1...N_Max do
    S_internalbest ⇐ ∅;
    Candidategroups ⇐ all the groups;
    While Candidategroups ≠ ∅ do
      C_g ⇐ Select a group from Candidategroups according to group pheromone trail;
      Candidates ⇐ {o_ij ∈ C_g that do not violate resource constraints in (2-5)}
      Update local heuristic values by equation (8);
      Choose an object o_ij ∈ Candidates with probability computed by equation (9);
      S_k ⇐ S_k ∪ o_ij; Remove C_g from Candidategroups;
    Endwhile
    If profit(S_k) > profit(S_internalbest) then S_internalbest ⇐ S_k;
  Endfor
  If profit(S_globalbest) < profit(S_internalbest) then S_globalbest ⇐ S_internalbest;
    Update database and pheromone trails lower and upper bounds by τ_min and τ_max;
Until maximum number of cycles reached (N_max) or optimal solution found;
```

After each ant has constructed a solution, the best solution of that iteration is identified and a random local search procedure and a random item swap procedure is applied to improve it. Then pheromone trail is updated according to the best solution. Also it maintains a database of top k solutions. After each iteration a small amount of pheromone is deposited in the pheromone trails of the objects belonging to the top k solutions. The motivation behind this strategy is to ensure quick convergence on good solutions and to explore better areas more thoroughly. The algorithm stops either when an ant has found an optimal solution, or when a maximum number of cycles has been performed.

4 Experiments and Results

Case Study 1: The parameters are set number of ants is 50, $k = 10$, $\alpha = 1$, $\beta = 5$, $P = 0.01$, $\tau_{min} = 0.01$, $\tau_{max} = 8$, $\tau = 0.5$, $N_{max} = 500$. We implemented the prototype Audio/Video Call (AVC) service allow two end users to engage in an AVC in a laboratory IMS testbed [15]. The AVC service matching parameter and requirement with conversational audio and video, we assume the uplink direction from the call initiating user A to the terminating user B, the user capabilities and operator constraints is shown in Table 1. For illustration, we assume a set of different media flows $F = (f_1, f_2, f_3, f_4)$, f_1, f_2 are audio and video downlink and f_3, f_4 are audio and video uplink flows. $P = (p_1, p_2, p_3, p_4) = (3, 8, 3, 8)$ and $u_i(r_{ij}) \in [0, 1]$, $B_{downlink} = 1200$ kbps and $B_{uplink} = 800$ kbps, a hypothetical price is 10 [monetary unit/bit], maximum cost is 20000 [monetary unit/s], cost as bandwidth class q [bit/s] x price class q [monetary unit/b]. Our operating parameters, resource vectors and corresponding utilities are specified in the service profile and summarized in Table 2. The results of objective value and resource configuration of our algorithm updated step by step show in Table 3. The experimental results with the input data sets in [15] show that resource configurations of our algorithm is similar results solved by GLPK [15].

Case Study 2: We generated random input data to verify the effectiveness of our algorithm follow constraints below. The experiment was conducted on CPU DuoCore 3.0GHz, 2GB of RAM machine. We ran experiment our algorithm, the problems tackled, object value, and time processing shown in Table 4.

Case Study 3: The exist algorithms differ in deciding which component of the problem should be regarded as the pheromone depositing component and in the mechanisms of pheromone updating: (1) pheromone trails on each object to lay pheromone trails on each object belonging to the current solution set the amount of pheromone represents the preference of the object [17]; (2) pheromone trails on each pair, pheromone trails are laid on each pair (o_i, o_j) of successively selected objects of the solution set the idea is to increase the desirability of choosing object o_j when the last selected object is o_i [18]; (3) pheromone trails on all pair to lay pheromone trails on all pairs of different objects of the solution set. Here, the idea is to increase the desirability of choosing simultaneously two objects of S; (4) uses a pheromone diffusion scheme where pheromone trails are laid on objects that tend to occur together in previous solutions [19] the same principle (1). To evaluating pheromone update of MMAS algorithm, we implemented SMMAS [20], MLAS [21] and 3-LAS [22] pheromone update and compared the best solution, the average solution and the worst solution are shown in Table 5.

Table 1. Matching parameter set for AVC services

	Parameter set	Service requirements	User A	User B	Operator constraints
p1	Media Component	audio, video	audio, video, data image, model	audio, video, data image, model	User A and B:audio, video data, image, text
p2	Codecs	audio: mpeg, gsm; video: mpeg, h26	audio: peg, pcm, gsm; video: mpeg mjpeg, h263	audio: peg, pcm, gsm; video: mpeg, mjpeg, h263	Not Allowed: audio, G729, audio G723
p3	Min Bandwidth Downlink	46	1200	1300	1400
p4	Max Delay Downlink	150	150	150	N/A
p5	Max Jitter Downlink	10	N/A	N/A	N/A
p6	Max Loss Downlink	1	N/A	N/A	N/A
p7	Min Bandwidth Uplink	46	800	1000	1400
p8	Max Delay Uplink	150	150	150	N/A
p9	Max Jitter Uplink	10	N/A	N/A	N/A
p10	Max Loss Uplink	1	N/A	N/A	N/A
p11	Resolution Local	176x144	1204x768	N/A	N/A
p12	Resolution Remote	176x144	N/A	1204x768	N/A

Table 2. Operating parameters, resource vectors and utilities for AVC flows

Flow	Parameter	Resource vectors	Bandwidth r_{ij1}	Cost r_{ij1}	Utility value $u_i(r_{ij})$
f_1, f_3	p_1	r_{11}, r_{31}	21	210	0.5
	p_2	r_{12}, r_{32}	34	340	0.8
	p_3	r_{13}, r_{33}	64	640	1
f_2, f_4	p_1	r_{21}, r_{41}	25	900	0.2
	p_2	r_{22}, r_{42}	90	900	0.4
	p_3	r_{23}, r_{43}	370	3700	0.5
	p_4	r_{24}, r_{44}	400	4000	0.6
	p_5	r_{25}, r_{45}	781	7810	0.7
	p_6	r_{26}, r_{46}	1015	10150	0.8
	p_7	r_{27}, r_{47}	1400	14000	0.9
	p_8	r_{28}, r_{48}	2000	20000	1

Table 3. Resource configurations for the MMAS algorithm updated step by step

Number of cycles	Resource Configuration	Objective value
93	$(r_{13}, r_{23}, r_{33}, r_{44})$	2.77
128	$(r_{13}, r_{25}, r_{33}, r_{43})$	2.839
187	$(r_{13}, r_{26}, r_{33}, r_{42})$	2.84
264	$(r_{13}, r_{24}, r_{33}, r_{44})$	2.84
331	$(r_{13}, r_{25}, r_{33}, r_{44})$	2.909
375	$(r_{13}, r_{26}, r_{33}, r_{43})$	2.909
417	$(r_{13}, r_{26}, r_{33}, r_{44})$	**2.98**

Table 4. The problems tackled and results of MMAS algorithm

Problems	No.of flows	Operating parameter Downlink $(p_1, \ldots, p_h)$	Operating parameter Uplink $(p_{h+1}, \ldots, p_n)$	Optimal value	Time processing (second)
#1	2	(2)	(2)	1.752	0.185
#2	2	(3)	(4)	1.965	0.213
#3	4	(4)	(2, 5, 7)	2.786	0.341
#4	4	(3, 8)	(3, 8)	2.980	0.352
#5	4	(3, 7, 4)	(9)	2.863	0.329
#6	6	(6, 8, 9)	(8, 5, 4)	4.792	0.715
#7	6	(10, 15)	(8, 10, 12, 13)	4.937	0.734
#8	8	(10, 15, 20, 25)	(10, 15, 20, 25)	10.573	1.685
#9	8	(15, 20, 30)	(20, 10, 30, 25, 15)	11.218	1.724
#10	8	(20, 25, 30, 35, 45)	(40, 35, 50)	10.983	1.693

Table 5. Comparing results of MMAS, SMMAS, MLAS, 3-LAS algorithms

	MMAS			SMMAS			MLAS			3-LAS		
Problems	best	avg	worst	best	avg	worst	best	avg	worst	best	avg	worst
#1	**1.752**	1.866	1.982	**1.752**	**1.797**	1.891	**1.752**	**1.781**	1.884	**1.752**	1.792	1.863
#2	**1.965**	2.515	3.065	**1.965**	2.325	2.458	**1.965**	2.398	2.578	**1.965**	**2.015**	2.393
#3	**2.786**	3.386	3.836	**2.786**	**2.826**	3.337	**2.786**	3.219	3.339	**2.786**	3.136	3.203
#4	**2.980**	**3.235**	3.905	**2.980**	3.427	3.976	**2.980**	3.281	3.560	**2.980**	3.305	3.861
#5	**2.863**	3.263	3.188	**2.863**	**2.943**	3.423	**2.863**	3.230	3.630	**2.863**	3.013	3.080
#6	**4.792**	5.342	5.492	**4.792**	**4.912**	5.179	**4.792**	5.092	5.112	**4.792**	5.042	5.198
#7	**4.937**	5.537	6.137	**4.937**	5.438	6.048	**4.937**	5.104	5.144	**4.937**	**5.062**	5.573
#8	**10.573**	11.123	11.498	**10.573**	11.093	11.337	**10.573**	11.006	11.466	**10.573**	**10.623**	10.645
#9	**11.218**	11.268	11.343	**11.218**	11.738	12.205	**11.218**	**11.351**	11.671	**11.218**	11.393	11.704
#10	**10.983**	**11.183**	11.418	**10.983**	11.263	11.596	**10.983**	11.183	11.503	**10.983**	11.191	11.552

5 Conclusion and Furture Works

In this paper, we propose a MMAS algorithm to solve the optimal QoS for multimedia services in the NGNs for multi-flow resource constrained utility maximization. We compared effective of MMAS to SMMAS, MLAS and 3-LAS. The computational results showsed that my approach is currently among the best performing algorithms for this problem. Optimizing QoS negotiation models in cases involving multiple providers in service delivery is my next research goal.

References

1. ITU-T Y.2012. Functional requirements and architecture of the NGN R1 (2006)
2. 3GPP TS 23.228: IP Multimedia Subsystem (IMS); Stage 2, Release 8 (2008)
3. Ozcelebi, T., Radovanovic, R., Chaudron, M.: Enhancing End-to-End QoS for Multimedia Streaming in IMS-Based Networks. In: ICSNC, pp. 48–53 (2007)
4. Boula, L., Koumaras, H., Kourtis, A.: An Enhanced IMS Architecture Featuring Cross-Layer Monitoring and Adaptation Mechanisms. In: ICAS, Spain (2009)
5. IST DAIDALOS-EU FP6, http://www.ist-daidalos.org/
6. IST ENTHRONE-EU FP6, http://www.ist-enthrone.org/
7. Houssos, N., et al.: Advanced Adaptability and Profile Management Framework for the Support of Flexible Service Provision. Wireless Communications, 52–61 (2003)
8. Khan, S.: Quality Adaptation in a Multisession Multimedia System: Model, Algorithms and Architecture. PhD Thesis, Univ. of Victoria (1998)
9. Wang, et al.: Utility-Based Video Adaptation for Universal Multimedia Access and Content-Based Utility Function Prediction for Real-Time Video Transcoding. IEEE Trans. on Multimedia 9(2), 213–220 (2007)
10. Guenkova-Luy, T., Kassler, A.J., Mandato, D.: End-to-End Quality-of-Service Coordination for Mobile Multimedia Applications. IEEE J. Selec. Areas Commun. 22(5), 889–903 (2004)
11. Rosenberg, J., et al.: SIP: Session Initiation Protocol. IETF RFC 3261 (2002)
12. Handley, H., et al.: SDP: Session Description Protocol. IETF RFC 2327 (1998)
13. Khan, S.: Quality Adaptation in a Multisession Multimedia System: Model, Algorithms and Architecture. PhD Thesis, Univ. of Victoria (1998)
14. Mukherjee, D., et al.: Optimal Adaptation Decision-Taking for Terminal and Network Quality of Service. IEEE Trans. on Multimedia 7(3), 454–462 (2005)
15. Skorin-Kapov, L., Matijasevic, M.: Modeling of a QoS Matching and Optimization Function for Multimedia Services in the NGN. In: Pfeifer, T., Bellavista, P. (eds.) MMNS 2009. LNCS, vol. 5842, pp. 55–68. Springer, Heidelberg (2009)
16. Stutzle, et al.: Max-min ant system. Future Generation Computer Systems 16, 889–914 (2000)
17. Alaya, I., Solnon, C., Ghédira, K.: Ant algorithm for the multi-dimensional knapsack problem. In: Proceeding of BIOMA 2004, pp. 63–72 (2004)
18. Ji, J., et al.: An ACO Algorithm for Solving the Multidimensional Knap-sack Problems. In: Proceedings of the 2007 IEEE/WIC/ACM International Conference on Intelligent Agent Technology, pp. 10–16. IEEE Computer Society (2007)
19. Shahrear, et al.: A novel ACO technique for Fast and Near Optimal Solutions for the Multi-dimensional Multi-choice Knapsack Problem. In: Proceedings of 13th International Conference on Computer and Information Technology (ICCIT 2010), Dhaka, Bangladesh, pp. 33–38 (2010)

20. Huy, Q.D., et al.: Multi-Level Ant System-A new approach through the new pheromone update for ACO. In: Proc. of the 4th IEEE International Conference in Computer Sciences, Research, Innovation, and Vision for Future, pp. 55–58
21. Dong, D.D., Dinh, H.Q., Xuan, H.H.: On the pheromone update rules of ant colony optimization approaches for the job shop scheduling problem. In: Proc. of the Pacific Rim Int. Workshop on Multi-Agents, pp. 153–160 (2008)
22. Dong, D.D., et al.: Smoothed and Three-Level Ant Systems: Novel ACO Algorithms for the Traveling Salesman Problem. In: Ad. Cont. to the IEEE RIFV 2010, pp. 37–39 (2010)
23. Le, D.-N.: Optimizing QoS for Multimedia Services in NGN Based on ACO Algorithm. Int. Journal of Information Technology and Computer Science 5(10), 30–38 (2013)

Combining Classifiers for Offline Malayalam Character Recognition

Anitha Mary M.O. Chacko and P.M. Dhanya

Department of Computer Science & Engineering
Rajagiri School of Engineering & Technology
Kochi, India
anithamarychacko@gmail.com, dhanya_pm@rajagiritech.ac.in

Abstract. Offline Character Recognition is one of the most challenging areas in the domain of pattern recognition and computer vision. Here we propose a novel method for Offline Malayalam Character Recognition using multiple classifier combination technique. From the preprocessed character images, we have extracted two features: Chain Code Histogram and Fourier Descriptors. These features are fed as input to two feedforward neural networks. Finally, the results of both neural networks are combined using a weighted majority technique. The proposed system is tested using two schemes- Writer independant and Writer dependant schemes. It is observed that the system achieves an accuracy of 92.84% and 96.24% respectively for the writer independant and writer dependant scheme considering top 3 choices.

Keywords: Character Recognition, Fourier Descriptors, Chain Code Histogram, Multiple Classifier Combination, Neural Networks.

1 Introduction

Offline character recognition is the task of recognizing handwritten text from a scanned, digitized or photographed sheet of paper and converting it to a machine editable format. It is an active research field in the domain of pattern recognition and machine vision due to the intrinsic challenges present in them. Further, the large variations in the writing style among different writers and even among the same writers at different times complicate the recognition process. The numerous applications of handwritten character recognition in the domain of postal automation, license plate recognition and preservation of historical and degraded documents makes it an interesting research area.

OCR research is at a well advanced stage in foreign languages like English, Chinese, Latin and Japanese [1]. However, compared to these languages, OCR research in Indic scripts is far behind. The highly complicated and similar writing style among different characters, the large character set and the existence of old and new scripts are some of the factors that attribute to the highly challenging nature of the recognition process in these scripts. Though many works have been reported in the past few years, a complete OCR system for Indian

S.C. Satapathy et al. (eds.), *Emerging ICT for Bridging the Future – Volume 2*,
Advances in Intelligent Systems and Computing 338, DOI: 10.1007/978-3-319-13731-5_3

languages is still lacking. Among the Indian languages, the research on South Indian languages such as Kannada, Tamil, Telugu and Malayalam demands far more attention.

An excellent survey on OCR research in South Indian Scripts is presented in [2],[3]. A detailed survey on handwritten malayalam character recognition is presented in [4]. Rajashekararadhya et al. [5] proposed an offline handwritten numeral recognition system for the four South Indian languages- Malayalam, Tamil, Kannada and Telugu. The proposed system used zone centroid and image centroid based features. Here, Nearest Neighbour and Feedforward neural networks were used as classifiers. The first work in Malayalam OCR was reported by Lajish V.L. [6] using fuzzy zoning and normalized vector distances for the recognition of 44 isolated basic Malayalam characters. A character recognition scheme using Run length count (RLC) and MQDF classifier was proposed in [7]. In [8], a character recognition technique using cross features, fuzzy depth features, distance features and Zernike moments and a Probabilistic Simplified Fuzzy ARTMAP classifier was proposed.

The above HCR systems have proposed many feature extraction techniques and classifiers but a remarkable achievement has not yet been obtained in Malayalam character recognition. All the above works have been based on a single classifier scheme. Even though multiple classifier schemes have been applied to other Indic languages [9], its effect on a complicated language like Malayalam has not been explored till now. This motivated us to investigate the effects of using a multiple classifier system for the recognition of handwritten malayalam characters.

In this paper, we propose a multiple classifier system for the recognition of handwritten Malayalam characters. The extracted features are based on the chain code histogram and fourier descriptors. These features are fed as input to two feedforward neural networks. The final results are obtained by combining the results of individual classifiers using a weighted majority scheme. To the best of our knowledge, the use of a multiple classifier system for the recognition of malayalam characters represents a novelty.

This paper is structured as follows: Section 2 introduces the architecture of the proposed system. Section 3 presents the experimental results and finally, Section 4 concludes the paper.

2 Proposed System

The proposed system consists of mainly 4 stages: Preprocessing, Feature Extraction, Classification and Post Processing. The scanned image is first subjected to preprocessing to remove as much distortions as possible. After preprocessing, chain code histogram features and fourier descriptors are extracted and fed as input to two neural network classifiers. The output of these networks are combined using a weighted majority scheme to obtain the final recognition results. Finally, the characters are mapped to their corresponding Unicode values in the post processing stage. Fig.1 shows the architecture of the proposed system.

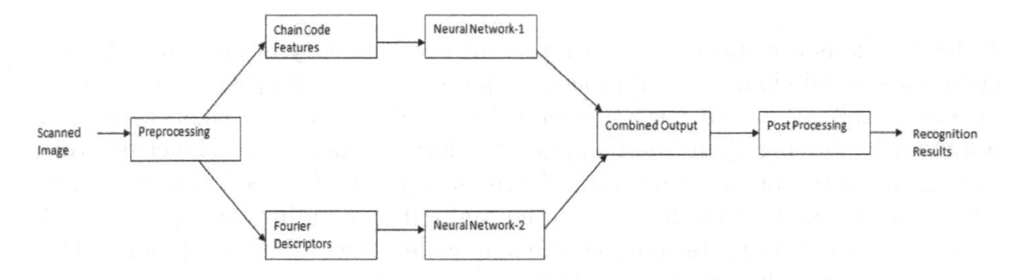

Fig. 1. Architecture of the Proposed System

2.1 Preprocessing

The objective of preprocessing phase is to eliminate as much distortions as possible from the scanned image. These distortions occur due to the poor quality of scanners and degraded documents. A 3x3 median filter is applied to remove salt and pepper noise to a certain extent. The scanned image is converted to binary using Otsu's method of global thresholding. This approach separates the foreground and background pixels. Then the line segmentation and character segmentation of the collected samples are carried out using horizontal and vertical projection profiles. Fig.2 shows the segmentation operation using these methods. Further, the characters are cropped by placing a bounding box around it. These cropped characters are finally normalized to 36x36 using bilinear transformation.

2.2 Feature Extraction

Feature extraction is the process of extracting relevant features from character images which are used as input to the classifier. This is an important phase in character recognition as effective features contribute to the success rate of the classifier. Here, we have used two different feature sets based on chain code histogram and invariant fourier descriptors. These features are obtained after the boundary of the image is extracted.

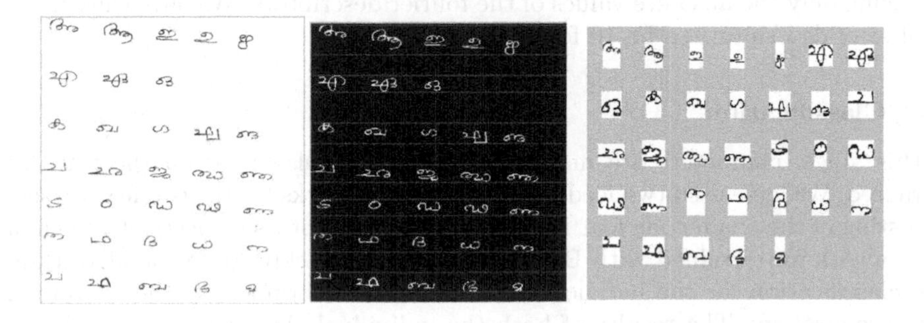

Fig. 2. Line and Character Segmentation

2.2.1 Chain Code Histogram

C hain code is a compact way to represent the boundary of an image. In the eight directional chain code approach proposed by Freeman [10] , the direction of each pixels are coded with integer values of 0-7. Here the chain codes are obtained by moving along the boundary in clockwise direction. The chain code histogram is obtained by computing the frequency of each direction chain codes. These chain code histograms are then normalized by dividing the frequencies of each direction by the total number of chain code values. Thus the input to the first neural network consists of 16 features.

2.2.2 Fourier Descriptors

T he fourier descriptors are powerful boundary based descriptors that represents the shape in the frequency domain. The advantage of fourier descriptors is that the overall shape features can be approximated using only a fixed number of discrete Fourier coefficients. Suppose the set of coordinates(x_k, y_k) describing the contour of a shape consists of N pixels from 0 to N-1. The two dimensional plane can be interpreted as a complex plane where the horizontal axis represents the real part and the vertical axis represents the imaginary part. Thus, the contour can be represented as a sequence of complex numbers of the form:

$$z(k) = x(k) + iy(k) \tag{1}$$

Thus this representation reduces a 2D problem into a 1D problem. The discrete fourier transform of z(k) is:

$$a(u) = \sum_{k=0}^{N-1} z(k) e^{\frac{-j2\pi uk}{N}} \quad u = 0..N-1 \tag{2}$$

The complex coefficients a(u) are called the Fourier descriptors of the boundary. Here, translation invariance is obtained by discarding the zeroth fourier descriptor as it is the only one that contributes to translation. Scale invariance is achieved by dividing all descriptors by the absolute value of the first descriptor. Rotation and shift invariance is achieved by discarding the phase information and using only the absolute values of the fourier descriptors. We have chosen the number of descriptors to be 10 for our experiment.

2.3 Classification

In this stage, characters are mapped to unique labels based on the features extracted. We have used two feedforward neural networks for the two feature sets consisting of the chain code features and fourier descriptors respectively. Each of the network were trained with Bayesian regulation backpropagation algorithm. The classification for the individual classifiers were obtained by the maximum response strategy. The results of both the individual classifiers were combined using a weighted majority scheme to obtain the final recognition results.

2.3.1 Classifier Combination

C ombining individual classifiers overcomes the limitations of single classifiers in solving difficult pattern recognition problems involving large number of output classes. A proper combination of multiple classifiers provide more accurate recognition results than individual classifiers since each classifier offers complementary information about the pattern to be classified. We have used two feedforward neural networks trained with 16 chain code features and 10 fourier descriptors respectively. The weighted majority scheme[9] used to combine results of these neural networks are as follows:The final combined decision d_{com} is:

$$d_{com} = max \sum_{k=1}^{N} w_k * \delta_{ik} \quad 1 \leq i \leq N_c \tag{3}$$

where

$$w_k = \frac{d_k}{\sum_{k=1}^{N} d_k} \tag{4}$$

Here δ_{ik} denotes the kth classifier decision to assign an unknown pattern to class i, N indicates the number of classifiers , N_c indicates the number of output classes, w_k denotes the weight assigned to k^{th} classifier and $d_k$indicates the recognition rate of kth classifier

3 Experimental Results

The experiment was conducted on 33 characters-8 isolated vowels and 25 consonants of Malayalam character set. A standard benchmark database is not available for Malayalam. Fig. 3 shows the sample characters that we have used for our study and the Class-ids assigned to each of them. We have tested the system according to two different schemes: Writer dependant(Scheme 1) and writer independent schemes(Scheme 2). For the writer independent scheme, the database consists of 825 samples of handwritten data collected from 25 people belonging to different age groups and professions and for the writer dependant scheme, the database consists of another 825 samples collected from 5 different people. In the writer dependant scheme, writing samples of people which were used in training were subjected to testing whereas in the writer independent scheme, the writing samples of new users whose samples were not used for training were subjected to testing.

In both schemes, 80% of the samples were used for training and 20% were used for testing. The results are summarized in Fig. 4 and Table 1. The recognition accuracy obtained from the chain code histogram classifier and fourier descriptor classifier are 80.61% and 66.67% respectively for the writer independant scheme.

Class-Id	Character	Class-Id	Chracter	Class-Id	Character	Class-Id	Character
1	അ	2	ആ	3	ഇ	4	ഉ
5	ഋ	6	എ	7	ഏ	8	ഒ
9	ക	10	ഖ	11	ഗ	12	ഘ
13	ങ	14	ച	15	ഛ	16	ജ
17	ഝ	18	ഞ	19	ട	20	ഠ
21	ഡ	22	ഢ	23	ണ	24	ത
25	ഥ	26	ദ	27	ധ	28	ന
29	പ	30	ഫ	31	ബ	32	ഭ
33	മ						

Fig. 3. Database Samples

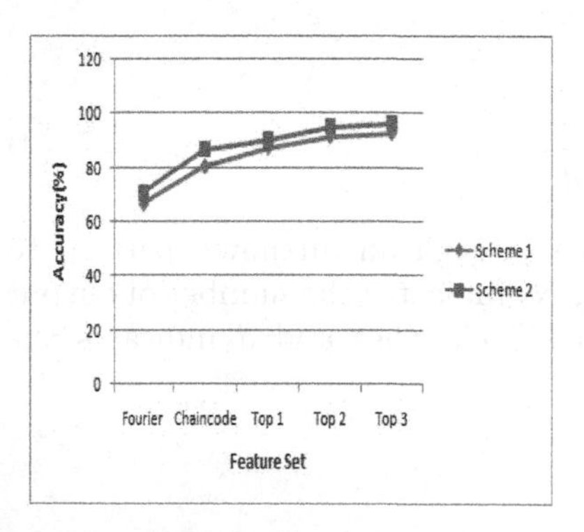

Fig. 4. Performance of Scheme 1 and Scheme 2

Table 1. Experimental Results

Feature Set(%)	Accuracy(%)	
	Scheme 1	Scheme 2
Chain Code	80.61	86.79
Fourier Descriptor	66.67	71.03
Combined(Top 1)	87.63	90.18
Combined(Top 2)	91.63	95.27
Combined(Top 3)	92.84	96.24

The combined classifier gives an overall accuracy of 92.84% as we considered top 3 choices and 87.63% for the top 1 choices for the entire database samples.

Compared to writer independent scheme, a higher recognition accuracy was obtained for the writer dependant scheme. For the writer dependant scheme, a recognition accuracy of 86.79% and 71.03% were obtained for the chain code histogram classifier and fourier descriptor classifier respectively. The combined classifier gives an overall accuracy of 96.24% as we considered top 3 choices and 90.18% for the top 1 choices. Based on the confusion matrix of top 1 choice obtained for both schemes, we have calculated seven useful measures: Precision, Recall(TP Rate/Sensitivity), Specificity(TN Rate), FP Rate and F-score. The results of both schemes are summarized in Table 2.

Table 2. Classification Results: Scheme 1 and Scheme 2

Class-Id	Writer Independant Scheme					Writer Dependant Scheme				
	Precision	Recall	Specificity	FP-Rate	F-score	Precision	Recall	Specificity	FP-Rate	F-score
1	0.6800	0.8947	0.9901	0.0099	0.7727	0.8800	0.9565	0.9963	0.0037	0.9167
2	0.9600	0.9231	0.9987	0.0013	0.9412	0.9600	0.8889	0.9987	0.0013	0.9231
3	0.9600	0.8889	0.9987	0.0013	0.9231	0.8400	0.9130	0.9950	0.0050	0.8750
4	0.8800	0.9565	0.9963	0.0037	0.9167	0.9200	0.9200	0.9975	0.0025	0.9200
5	0.8000	0.8696	0.9938	0.0062	0.8333	0.8800	0.9167	0.9963	0.0037	0.8980
6	0.8000	0.9091	0.9938	0.0062	0.8511	0.9600	0.8571	0.9987	0.0013	0.9057
7	0.8000	0.0062	0.8696	0.0062	0.9938	0.8000	0.8333	0.9938	0.0062	0.8163
8	0.7600	0.7037	0.9925	0.0075	0.7308	0.8400	0.8750	0.9950	0.0050	0.8571
9	1.0000	0.8065	1.0000	0	0.8929	1.0000	0.9615	1.0000	0	0.9804
10	0.8400	0.9545	0.9950	0.0050	0.8936	0.8800	0.8148	0.9962	0.0038	0.8462
11	0.9600	0.9600	0.9988	0.0013	0.9600	0.9600	0.9231	0.9987	0.0013	0.9412
12	0.7600	0.7600	0.9925	0.0075	0.7600	0.9200	0.9583	0.9975	0.0025	0.9388
13	0.9600	0.9600	0.9988	0.0013	0.9600	0.7600	0.9500	0.9925	0.0075	0.8444
14	0.9200	0.8846	0.9975	0.0025	0.9020	0.9200	0.9200	0.9975	0.0025	0.9200
15	0.8400	0.7778	0.9950	0.0050	0.8077	0.9600	0.8889	0.9987	0.0013	0.9231
16	0.8800	0.9565	0.9963	0.0037	0.9167	0.8000	0.8333	0.9938	0.0062	0.8163
17	0.8400	0.9130	0.9950	0.0050	0.8750	0.9200	0.8846	0.9975	0.0025	0.9020
18	0.8800	0.7857	0.9962	0.0038	0.8302	0.9200	0.8846	0.9975	0.0025	0.9020
19	0.9600	0.8276	0.9987	0.0013	0.8889	0.9600	0.9231	0.9987	0.0013	0.9412
20	0.8800	0.8462	0.9962	0.0038	0.8627	0.9200	1.0000	0.9975	0.0025	0.9583
21	0.8400	0.8077	0.9950	0.0050	0.8235	0.9600	0.8571	0.9987	0.0013	0.9057
22	0.9200	0.8214	0.9975	0.0025	0.8679	0.8800	0.8462	0.9962	0.0038	0.8627
23	0.9200	0.9200	0.9975	0.0025	0.9200	0.9200	0.9583	0.9975	0.0025	0.9388
24	0.9200	0.9583	0.9975	0.0025	0.9388	0.9600	0.8571	0.9987	0.0013	0.9057
25	0.8800	1.0000	0.9963	0.0037	0.9362	0.9600	0.9600	0.9988	0.0013	0.9600
26	0.8800	0.8800	0.9962	0.0037	0.8800	0.8400	0.9130	0.9950	0.0050	0.8750
27	0.9600	0.8000	0.9987	0.0013	0.8727	0.8400	0.9545	0.9950	0.0050	0.8936
28	0.8800	0.8462	0.9962	0.0038	0.8627	0.9600	0.9600	0.9988	0.0013	0.9600
29	0.9600	0.8889	0.9987	0.0013	0.9231	0.9200	0.9200	0.9975	0.0025	0.9200
30	0.8400	0.9545	0.9950	0.0050	0.8936	0.9200	0.9200	0.9975	0.0025	0.9200
31	0.8800	1.0000	0.9963	0.0037	0.9362	0.8400	0.8750	0.9950	0.0050	0.8571
32	0.8000	0.9091	0.9938	0.0062	0.8511	0.9200	0.7667	0.9975	0.0025	0.8364
33	0.8800	0.8462	0.9962	0.0038	0.8627	0.8400	0.9545	0.9950	0.0050	0.8936
Avg	0.8764	0.8812	0.9961	0.0039	0.8764	0.9018	0.9044	0.9969	0.0031	0.9016

4 Conclusion

In this paper, we have presented a method for Offline Malayalam Character Recognition using multiple classifier combination scheme. Chain Code Histogram and Fourier Descriptors were extracted from preprocessed character images to form feature vectors. These features were fed as input to two feedforward neural networks. The final outputs were obtained by combining the results of both these neural networks using a weighted majority scheme. The proposed system achieves a recognition accuracy of 92.84% and 96.24% for the writer independant and writer dependant schemes respectively. From the analysis of the confusion matrix obtained for the experiment, we have found that most of the misclassifications are due to confusing pair of characters. So the future work aims at reducing these errors by incorporating additional features in the post processing stage.

References

1. Plamondan, R., Srihari, S.N.: Online and offline handwriting recognition: A comprehensive survey. IEEE Trans. on PAMI 22(1), 63–84 (2000)
2. John, J., Pramod, K.V., Balakrishnan, K.: Handwritten Character Recognition of South Indian Scripts: A Review. In: National Conference on Indian Language Computing, Kochi, February 19-20, pp. 1–6 (2011)
3. Abdul Rahiman, M., Rajasree, M.S.: A Detailed Study and Analysis of OCR Research in South Indian Scripts. In: Proceedings of International Conference on Advances in Recent Technologies in Communication and Computing, pp. 31–38. IEEE (2009)
4. Chacko, A.M.M.O., Dhanya, P.M.: Handwritten Character Recognition in Malayalam Scripts -A Review. Int. Journal of Artificial Intelligence & Applications (IJAIA) 5(1), 79–89 (2014)
5. Rajasekararadhya, S.V., Ranjan, P.V.: Efficient zone based feature extraction algorithm for handwritten numeral recognition of popular south Indian scripts. Journal of Theoretical & Applied Information Technology 7(1), 1171–1180 (2009)
6. Lajish, V.L.: Handwritten character recognition using perpetual fuzzy zoning and class modular neural networks. In: Proc. 4th Int. National Conf. on Innovations in IT, pp. 188–192 (2007)
7. Moni, B.S., Raju, G.: Modified Quadratic Classifier for Handwritten Malayalam Character Recognition using Run length Count. In: Proceedings of ICETECT, pp. 600–604. IEEE (2011)
8. Vidya, V., Indhu, T.R., Bhadran, V.K., Ravindra Kumar, R.: Malayalam Offline Handwritten Recognition Using Probabilistic Simplified Fuzzy ARTMAP. In: Abraham, A., Thampi, S.M. (eds.) Intelligent Informatics. AISC, vol. 182, pp. 273–283. Springer, Heidelberg (2013)
9. Arora, S., Bhattacharjee, D., Nasipuri, M.: Combining Multiple Feature Extraction Techniques for Handwritten Devnagari Character Recognition. In: Proceedings of: IEEE Region 10 Colloquium and the Third ICIIS, Kharagpur, India, December 8-10, pp. 1–6 (2008)
10. Freeman, H.: On the encoding of arbitrary geometric configurations. IRE Trans. on Electr. Comp. or TC(10) (2), 260–268 (1961)

A Probabilistic Based Multi-label Classification Method Using Partial Information

Gangadhara Rao Kommu and Suresh Pabboju

Department of IT, Chaitanya Bharathi Institute of Technology
Hyderabad, Telangana-500075, India
kgr@cbit.ac.in

Abstract. In recent study there exist many approaches to solve multi-label classification problems which are used in various applications such as protein function classification, music categorization, semantic scene classification, etc., It in-turn uses different evaluation metrics like hamming loss and subset loss for solving multi-label classification but which are deterministic in nature. In this paper, we concentrate on probabilistic models and develop a new probabilistic approach to solve multi-label classification. This approach is based on logistic regression.The other approach is based on the idea of grouping related labels. This method trains one classifier for each group and the corresponding label is called as group representative. Predict other labels based on the predicted labels of group representative. The relations between the labels are found using the concept of association rule mining.

Keywords: hamming-loss, multi-label, deterministic, probabilistic.

1 Introduction

There also exist several types in multi-label classification. Some of them are multi-label collective classification, hierarchical multi-label classification etc. In multi-label collective classification, related instances in a dataset are classified together unlike in normal multi-label classification. For example, consider two collaborated researchers. If we know one of them has interests in data mining, then there is a high probability that the other researcher also has interest in data mining. This tells correlation among instances. Similarly correlated labels are also predicted simultaneously. For example, one person may have interest in machine learning as well as artificial intelligence i.e., ML and AI are correlated. To know the correlation among instances and correlation among labels, we need to provide extra input to algorithms which solve multi-label collective classification[4].

The rest of the paper is organised as follows. In section 2, we will look at some of the existing methods of MLC. In Section 3 we will see how we extended LR to multi-class classification. In section 4, we will look at our multi-label classification approaches. In section 5, we will see experimental results of proposed methods. In section 6, we conclude this paper.

S.C. Satapathy et al. (eds.), *Emerging ICT for Bridging the Future – Volume 2*,
Advances in Intelligent Systems and Computing 338, DOI: 10.1007/978-3-319-13731-5_4

2 Literature Review

As we have seen earlier, there exist many multi-label classification methods. We will look at some of the methods. The oldest approach is Binary Relevance (BR) method. This is one of the PT methods. This method trains one binary classifier for each label in the label set L, using modified data.

ECC trains m CC classifiers[3] $C_1, C_2, ..., C_m$. Each C_k is trained with a random chain order and a random subset of dataset D. Each C_k is able to give different multi-label predictions[1]. These predictions are summed by label so that each label receives a number of votes. A threshold is used to select the most popular labels.

The random k-labelset($RAkEL$) method [6]. It constructs an ensemble of LP classifiers in which each LP classifier is trained with a different small random subset of the set of labels. In this way, RAkEL avoids the problem of Label Powerset algorithm and also it manages to take label correlations into account.

A number of methods exist based on the famous k Nearest Neighbours (kNN) lazy learning approach. The first step in all these approaches is same as kNN, i.e. retrieving the k nearest examples. For example in [8], ML-kNN utilizes maximum a posteriori principle to determine the label set for the unseen instance. AdaBoost.MH [5, 7] and AdaBoost.MR [5, 7] are two extensions of AdaBoost for multi-label data. While AdaBoost.MH is designed to minimize hamming loss, AdaBoost.MR is designed to find a hypothesis which places the correct labels at the top of the ranking.

3 LR: Binary Classification

Before seeing the details of logistic regression, we first look at the format of data. The dataset D is of the form (X_i, Y_i) where $X_i = (x_{i1}, x_{i2}, ..., x_{ik})$ is the feature vector of ith example and k is the dimensionality of dataset. Y_i =1 if label is assigned, 0 otherwise.

In logistic regression, the probabilities are defined as follows.

$$P\left(y = 1/\omega, X_i\right) = P_i = \frac{1}{1 + e^{-\omega^T * X_i}} \tag{1}$$

$$P\left(y = 0/\omega, X_i\right) = 1 - P_i = \frac{e^{-\omega^T * X_i}}{1 + e^{-\omega^T * X_i}} \tag{2}$$

From both the above equations, we can generalize the probability formula as:

$$P(y_i/\omega, X_i) = P_i^{y_i} * (1 - P_i)^{1-y_i} \tag{3}$$

Then the likelihood is as follows:

$$P(D/\omega) = \Pi_{i=1}^{n} P(y_i/X_i, \omega) \tag{4}$$

Posterior is proportional to product of likelihood and prior. i.e.,

$$\mathrm{P}(\omega/D) \;\alpha\; P(D/\omega) * P(\omega)$$

We need to maximize the posterior of given dataset D. Assume prior has normal distribution. Instead of maximizing posterior directly, we maximize log of the posterior, i.e.,

$$\max \log[P(D/\omega) * P(\omega)]$$

$$\Rightarrow \max \log(P(D/\omega)) + \log(P(\omega))$$

$$\Rightarrow \max \sum_{i=1}^{n} log\,(P(y_i/X_i, \omega)) - \tfrac{\lambda}{2}\|\omega\|^2 \qquad (from\,(4))$$

$$\Rightarrow \max \sum_{i=1}^{n} log\left(P_i^{y_i} * (1 - P_i)^{1-y_i}\right) - \tfrac{\lambda}{2}\|\omega\|^2$$

$$\Rightarrow \max \sum_{i=1}^{n} [y_i log\,(p_i) + (1 - y_i)\,log\,(1 - p_i)] - \tfrac{\lambda}{2} \parallel w \parallel^2$$

From the fact that maximizing a function f is equivalent to minimizing a function -f, the objective function becomes

Objective function:

$$\mathrm{E} = min\ \tfrac{\lambda}{2} \parallel w \parallel^2 - \sum_{i=1}^{n} [y_i log\,(P_i) + (1 - y_i)\,log\,(1 - P_i)]$$

Differentiation:

$$\nabla \mathrm{E} = \lambda w - \Sigma_{i=1}^{n} (y_i - p_i)\, x_i$$

To solve the above optimization problem E, we used steepest descent method with backtracking line search. The parameters needed are c_1=0.5, δ=0.75 and α=0.05. We have chosen λ value appropriately. For this method, we need differentiation of objective function that we have found already. Our approach is explained in algorithm 1.

Algorithm 1. Logistic Regression for Binary Classification

Input: Train_data,Train_labels, lamda.(*Each example in data set is row − wise*)
Output: Classifier function.

1: E$\longleftarrow$ $Objective function$.
2: F$\longleftarrow$ ∇E.
3: $\lambda \longleftarrow$ lamda.
4: w$\longleftarrow$ Initialize.
5: itr$\longleftarrow$0
6: while($i < 50$)
7: $\quad$ $f_1 \longleftarrow$ compute E at w.
8: $\quad$ g$\longleftarrow$ compute F at w
9: $\quad$ $c_1 \longleftarrow 0.5$
10: $\quad$ $\alpha \longleftarrow 0.005$
11: $\quad$ $\phi_1 \longleftarrow f_1 + c_1 * \alpha * g^t * g$
12: $\quad$ w_{new} =w-$\alpha * g$;
13: $\quad$ $\phi \longleftarrow$ compute E at w_{new}
14: $\quad$ while($\phi_1 < \phi$)
$\quad\quad$ $\alpha \longleftarrow \alpha * 0.75$
$\quad\quad$ w_{new} =w-$\alpha * g$
$\quad\quad$ $\phi \longleftarrow$ compute E at w_{new}
$\quad\quad$ $\phi_1 \longleftarrow f_1 + c_1 * \alpha * g^t * g$
$\quad$ end while
15: $\quad$ w=w-$\alpha * g$
16: $\quad$ itr=itr+1
17: end while

4 Multi Label Classification Method Using Partial Information(MPI)

In multi-label setting, examples are associated with several classes. Formally we define the format of dataset D. It consist of n examples, which are of the form (X_i, Y_i) for i=1 to n where X_i is a feature vector of ith example and $Y_i \in \{0,1\}^q$ and q is number of total classes in a setting.

While solving multi-label classification problem, one may face two main problems. They are, considering label correlation and how many number of labels to be assigned for each test example. We are focusing on the second problem by considering independence among labels.The relations between the labels are found using the concept of association rule mining[9]. To choose how many number of labels to be assigned for test example, we propose a method which can be implemented using multi-class logistic regression and binary logistic regression.

Our approch is based on multi-class logistic regression. The training process of this method is same as the training process of binary relevance method i.e., it learns binary classifier for each class. The training process of our method is shown in algorithm 2.

Algorithm 2. Training process of MPI

Input: Train_data,Train_labels
Output: Classifiers for all classes.
1: For each class i
   $W_i \longleftarrow$ Train a binary classifier using data matrix and i^{th} label vector.
2: end for

There exist three phases in the testing process of this method. The objective of the first phase is to find out the number of labels to be assigned for test examples. The objective of second phase is to apply binary logistic regression on multi-label data to get probabilities of all classes for each test example. The objective of third phase is to assign set of labels to each test example which is based on the results of first two phases.

Algorithm 3. Testing process MPI.

Input: All trained classifiers W, Test_data
Output: Predicted labels of Test_data
1: T $\longleftarrow$ design a vector of length equal to number of train examples and T_i tells how many number of labels assigned with i^{th} train example
2: Train a multi-class classifier on data matrix of train set and vector T using multi-class logistic regression
3: Z $\longleftarrow$ multi-class classifier obtained in above step.
4: V $\longleftarrow$ vector of length equal to number of test examples and V_i tells predicted number of labels assigned with i^{th} test example by invoking Z on i^{th} test example
5: for each test example t_k
   i) Invoke all trained classifiers W and get probabilities for all classes
   ii) Arrange all classes in ascending order of probabilities
   iii) $d \leftarrow V_k$
   iv) Assign top d classes to test example t_k

In first phase, the first step is to create a new vector of length equal to number of training examples in which i^{th} component tells how many number of labels are assigned to i^{th} training example. Train a multi-class classifier on train data and vector created in first step using multi-class logistic regression. It results in classifiers for all classes. Invoke these classifiers on test data to get predicted labels. For example, predicted label of jth test example is 3, which means that j^{th} test example should be tagged with three labels.

In second phase, apply binary logistic regression on actual training data to get probabilities of all class labels for each test example. In third phase, assign labels to each test example using probabilities obtained in second phase. Result of first phase tells that the how many of labels are to be assigned. For example, j^{th} test example is tagged with d in first phase, which means that j^{th} test example should be tagged with d labels. For each test example, arrange all classes in descending order of probabilities obtained in second phase and tag first d classes to it. Testing phase of this method is explained in algorithm 3.

In the above method, we used classification twice. One classification is for getting number of labels to be assigned for test examples. Other classification is used to get the probability values for each test example. Since this method involves classification twice, time taken by these methods is higher than existing approaches like BR method. These methods are well suitable for small datasets. The results of above implementation are shown in section 5.

5 Experimental Results

We used eight different datasets to compare the performance of our proposed approaches with the binary relevance method. The names of those datasets and statistics of those datasets are shown in Table 1. All those datasets are arranged

Table 1. Results of BR and MPI methods

Dataset	Metric	Binary Relevance	MPI
Scene	Hamming Loss	0.1219	**0.0992**
	Subset Loss	0.5602	**0.3311**
	Micro-F1 score	0.6160	**0.7152**
	Macro-F1 score	0.6137	**0.7221**
Emotions	Hamming Loss	0.2987	0.3176
	Subset Loss	0.9158	**0.8713**
	Micro-F1 score	0.3170	**0.6050**
	Macro-F1 score	0.2440	**0.5755**
Yeast	Hamming Loss	0.4284	**0.3065**
	Subset Loss	0.9346	**0.9138**
	Micro-F1 score	0.4695	0.4760
	Macro-F1 score	**0.4286**	0.3972
Corel5k	Hamming Loss	0.0096	0.012
	Subset Loss	0.9980	0.9980
	Micro-F1 score	0.0180	0.0546
	Macro-F1 score	0.0006	0.0007
Delicious	Hamming Loss	0.0216	0.0451
	Subset Loss	0.9987	**0.9541**
	Micro-F1 score	0.2264	0.2154
	Macro-F1 score	0.0503	**0.0725**

Table 2. Number of classifiers learnt in different methods

Dataset	Binary Relevance	MPI
Scene	6	6+3
Emotions	6	6+3
Yeast	14	14+11
tmc2007	22	22+10
MediaMill	101	101+18
Bibtex	159	159+28
Corel5k	374	374+5
delicious	983	983+25

in increasing order of number of labels in it. We also used four different performance metrics to compare the results of our method with BR method.

Now, let us compare the results of our MPI method. The best result for each metric is shown in bold. If we analyse the datasets from their statistics, we can observe that MPI method performs better for small datasets such as scene (2407 *examples with* 294 *dimension*) and emotions (593 *examples with* 72 *dimension*).

Let us define a term called *maximum label cardinality* of dataset. It is defined as the maximum number of labels assigned to any train example in a dataset. In MPI, apart from base classifier we also used LR for multi-class classification. In this muti-class logistic regression setting, the number of classes is equal to *maximum label cardinality* of dataset. The MPI method should train these many number of classes apart from classifier for each label. The total number of classifiers learnt in all three methods are shown in table 2.

6 Conclusion and Future Work

We performed experiments on logistic regression for both binary and multi-class classification. We proposed a new approache to solve multi-label classification which uses pairwise information between the label which is probalistic in nature.This approach gives better results in less time than any deterministic alogirithms.this work can be extended by using group wise information among labels instead of pairwise relation by using more datasets.

References

1. Zhang, Y., Schneider, J.: A Composite Likelihood View for Multi-Label Classification. In: Proceedings of the 15th International Conference on AISTATS 2012, La Palma, Canary Islands (2012)

2. Bian, W., Xie, B., Tao, D.: CorrLog: Correlated Logistic Models for Joint Prediction of Multiple Labels. Appearing in Proceedings of the 15th International Conference on AISTATS 2012, La Palma, Canary Islands (2012)
3. Read, J., Pfahringer, B., Holmes, G., Frank, E.: Classifier Chains for Multi-label Classification
4. Kong, X., Shi, X., Yu, P.S.: Multi-label Collective Classification
5. Tsoumakas, G., Katakis, I., Vlahavas, I.P.: Mining multi-label data. In: Data Mining and Knowledge Discovery Handbook, pp. 667–685 (2010)
6. Tsoumakas, G., Vlahavas, I.: Random *k*-labelsets: An ensemble method for multilabel classification. In: Kok, J.N., Koronacki, J., Lopez de Mantaras, R., Matwin, S., Mladenič, D., Skowron, A. (eds.) ECML 2007. LNCS (LNAI), vol. 4701, pp. 406–417. Springer, Heidelberg (2007)
7. Tsoumakas, G., Katakis, I.: Multi-label Classification: An Overview
8. Zhang, M.L., Zhou, Z.H.: Ml-knn: A lazy learning approach to multi-label learning. Pattern Recognition 40, 2038–2048 (2007)
9. Fujino, A., Isozaki, H.: Multi-label classification using logistic regression models for NTCIR-7 patent mining task. In: Proceedings of NITCIR-7 Workshop Meeting, Tokyo, Japan (2008)

Sensor Based Radio Controlled Multi-utility Solar Vehicle to Facilitate Plough in Agriculture

Vishu, Prateek Agrawal, Sanjay Kumar Singh, Kirandeep Kaur, and Karamjeet Kaur

Department of Computer Science Engineering,
Lovely Professional University, Phagwara, Punjab, India
{Vishumadaan123,prateek061186,sanjayksingh012,
kaur.karam15,kirankambo18}@gmail.com

Abstract. This paper proposes an approach to implement computer driven multi-utility engineering vehicle that can be operated from any location within the radio-frequency to reduce the human efforts. Based on supervise autonomy, tractor is able to plough the field with defined pressure on the soil. A simple desktop computer can operate this radio controlled tractor through arrow keys by viewing the scene from camera headed over front of it. This is a prototype model which can be implemented in real world environment. The experiment with this demonstration shows a significant and positive effect on plowing the field without visiting there with solar charged batteries only. Looking ahead to the futurity, estimates are that farmers will be devoured.

1 Introduction

With today's emphasis on intelligent and sustainable farming methods, some autonomy associated with agricultural machines is very essential. Technology in precision farming is being used to save time and money in the field. Today, farmers are adopting new means of monitoring and controlling their farms. But it's not just about saving time and money for these farmers. Automating processes and machinery allows these young entrepreneurs to have more time for managing the business side of their farms, spending time with their families or doing whatever else they enjoy. In short, it offers them a better quality of life than previous generations farmers have experienced [4].

This is one another advanced step towards more cost-effective ways to help them automate their farms. A characteristic feature of the operation of vehicle–tractor units in plowing the soil is the monotony of the technological processes. In this case, the supervisor can simultaneously perform several operations: controls the direction of travel of the unit and plowing the soil by sitting at any place within radio frequency range of it. As a result, fatigue accumulating from the intensity of functions requiring instantaneous actions to be performed by physical presence of driver can be getting rid of. And as a consequence, same accuracy during manual driving can be achieved through our desktop only. This is purely electronically controlled tractor which will work on batteries being charged by solar plates mounted over its head. Ideologically it

S.C. Satapathy et al. (eds.), *Emerging ICT for Bridging the Future – Volume 2*,
Advances in Intelligent Systems and Computing 338, DOI: 10.1007/978-3-319-13731-5_5

feels nice to have a tractor running completely free of fossil fuels [2]. According to an article the only true scalable and energy-positive system that could supercede a fossil-fuel based lifestyle is SOLAR ELECTRIC [3]. All other "renewable" non-fossil fuel forms are limited, site-specific, compete with food, or are energy negative. Even at its best imaginable scope, a solar electric economy could only constitute a small fraction of our present fossil fuel consumption [1]. To automate this driving, Software "RC Controller" has been used which act as a bridge between PC and tractor.

Objectives of this research work are

- Effective way of farming for helpless and handicapped people.
- Too easy to drive like playing a game.
- Operated by any family member with a little training.
- In dark, lights will automatically be switched on.
- Handling moldboard is very easy, no physical force is required to drag it.
- Ready to be operational in any weather conditions. In the absence of sun-light, batteries can be electrically charged.
- No fossil fuel is required to get it into action.

A huge increase in fuel consumption leads to incline in the graph of diesel and petrol prices. The high and increasing costs associated with fossil fuels used by agricultural machines have negatively affected the farmer's life. This investigation identifies how informed and concerned individuals can consume solar energy to cut down the problem being caused by fuels. Experimentation of Prototype designing of a radio controlled object operated by computer and sensors based on solar energy charged battery are carried out.

Fig. 1. Working Prototype of Multi Utility vehicle

2 Methodology

Working of tractor is purely electronic, based on the embedded circuit getting power from the solar charged batteries. 3 Volt batteries are connected for powering the transmitter. It is built with a modular design and has two solar charged electric front and rear motors that power it up. Table 1 shows rest of its components and details.

2.1 Signal Flow Diagram from PC to Transmission Station

The signal flow from Personal computer to transmitter station is shown in figure 2. To control tractor with computer, software is needed. That's why it is a mandatory condition for desktop computer that RC controller software should be installed on it. The software is written in Visual Basic 6, and is recently updated to support by all windows operating systems [5]. Transistors embedded on PCB are connected with directed link to transmitter circuit. Desktop computer must have a parallel port of 25 pins to connect with interface circuit.

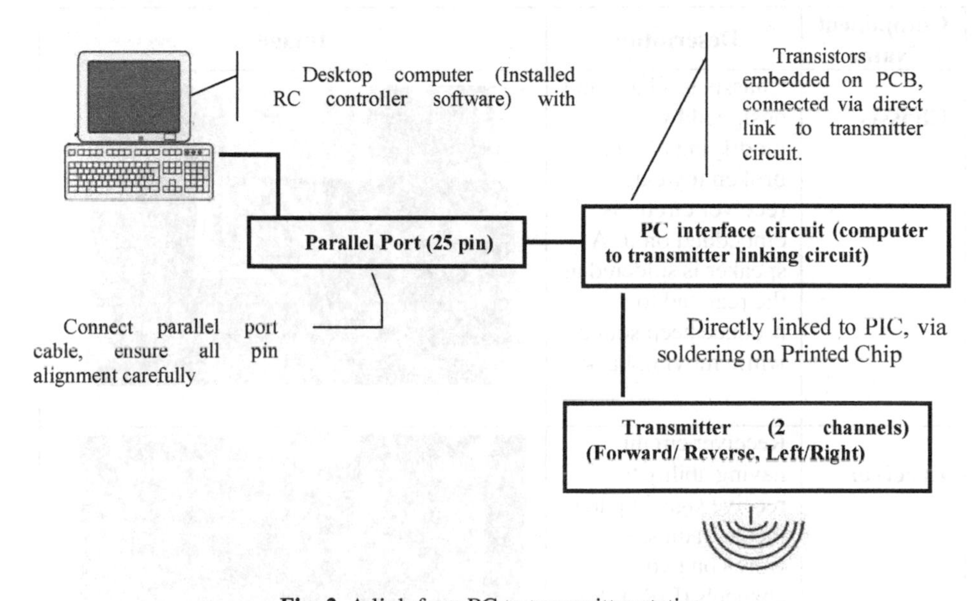

Fig. 2. A link from PC to transmitter station

2.2 Signal Flow Diagram from PC Receiver to Tractor Station

Figure 3 shows the signal transmission from receiver to tractor station. Receiver 1 is associated with two channels for LEFT, RIGHT, FORWARD and BACKWARD movement. Receiver 2 is with single channel for up and down movement of moldboard.

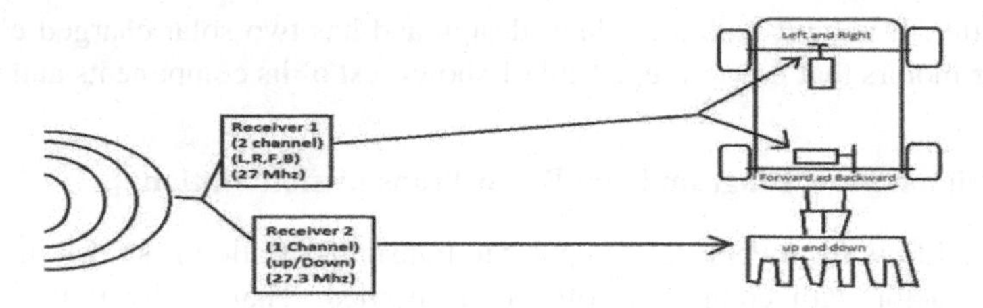

Fig. 3. A link from receiver to tractor station

3 Construction Design Details

Table 1. Components and its details

Component Name	Description	Image
Chassis	Chassis is a base and designed by modifying an old broken toy car, receiver circuit is embedded on it. A speaker is soldered at the rear end to produce beep sound while moving back.	
Receiver	Receiver circuit having ability to receive signal up to 6 meters radius. It works on two channels (L/R, F/R) and having IC-RX28;pins 12,13,14,15 are basic pins which control the action of the tractor. Transistors embedded on it are used to amplify signal waves. coil has almost 27 turns on it of copper wire which acts as antenna.	

Table 1. *(continued)*

Rear Motor	Powerful 6V Motor (DC). Beep speaker (black) for reverse signal. Reverse signal sound generator (beep speaker), is installed with reverse action, to warn someone during reverse movement.	
Front Motor	This is a front motor for left and right movement. 1 volt DC motor is installed for this action.	
Mold Board and its Circuit	Moldboard has been installed at the end of tractor for plowing soil. Movement: up and Down on level press. Moldboard or plowing jack has been installed using a wireless receiver embedded with tractor circuit. Motor is fixed in moldboard to drag it up and down. Motor is attached to a set of gears, which gives RPM of 30:1, Further it is connected to mold levels, which control up/down height.	

Table 1. (*continued*)

Solar cell Plate	Power = 12 volt Max current output = 10-11volt, 1.2 ampere	
Camera	A small pin hole wireless camera is installed to give a realistic view on the desktop screen. Resolution: 628*582 on screen. Range up to: 8 meters. Power required: 9volt	
Camera Receiver kit	USB conversion for pc view, using easy cap software. Easy tune switch and normal lines of view. Powered by 220 volt ac adapter.	
Head lamps and rear lamp	4 LEDs at front and 2 at rear end are installed with an automatic light circuit. LDR is attached to detect light intensity according to that LEDs will be switched on/off when required. It gives good beam at front and back for camera visibility.	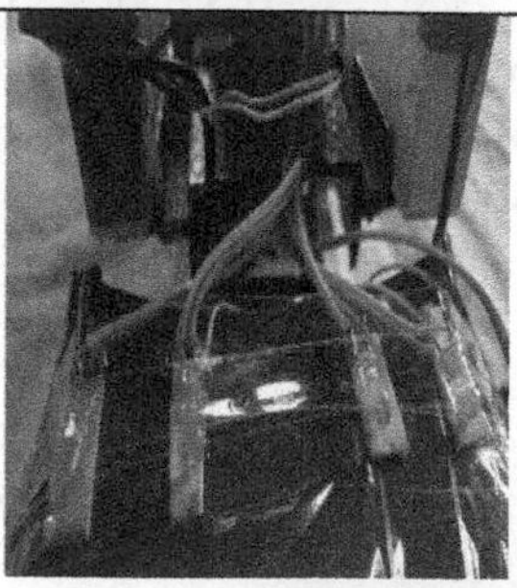
Transmitter PIC (Pc interface circuit):	Transmitters (27 MHz), battery (3V) for powering transmitter, PC interface circuit with Parallel port connector cable. Parallel port cable (25 pin) for interfacing transmitter and PIC circuit. Pin 2,3,4,5 used as movement signals and pin 23 is grounded. 4*2N3904 Switching Transistors for switching between pc and transmitter.	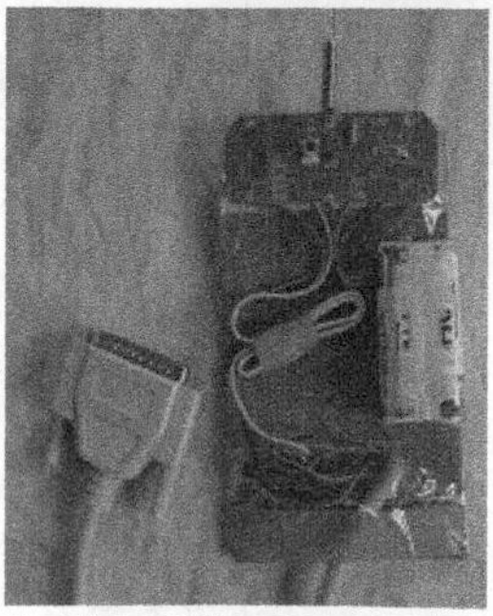

4 Working with PIC Circuit and RC Software

Tractor will be guided by the supervisor who is handling it with keyboard. First of all, Switch off tractor. Insert parallel port into rear slot of CPU. To make the software work, there is a need to know the port address of parallel port. On connecting parallel port cable with desktop computer, port address can be verified from DEVICE MANAGER Option and then check Ports (COM & LTP). After that go to properties select port and its resources to get port address. Start RC controller software, Switch your tractor ON, and use keyboard arrow keys to drive it. As we press button on transmitter to drive tractor, simply PIC circuit works switch for computer keyboard. When the user press "up key " on keyboard, it emits signal on parallel port pin 2, it turns positive, and sends signal to PIC transistor 1, it does switching to transmitter forward switch and turn it ON. Tractor moves forward. Similarly other three movements can also be made. Figure 4 shows the view of RC controller software while it is in active mode.

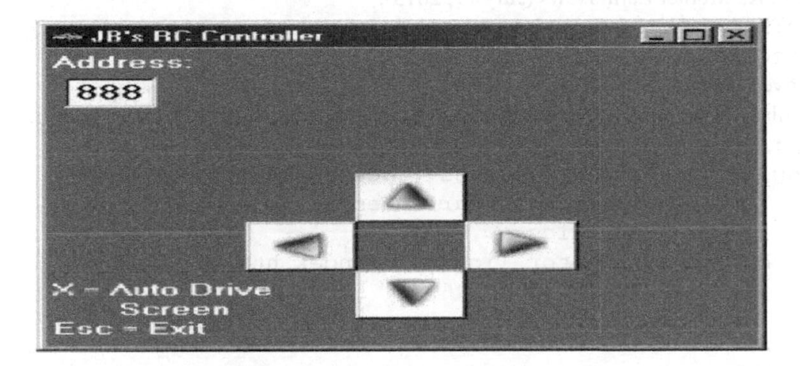

Fig. 4. View of RC controller software [5]

When a key is pressed, the appropriate button indicator button will turn blue and the tractor will move in that direction. Same way moldboard can also be dragged up and down according to our requirement.

5 Inescapable Conclusion and Future Scope

Advancement in already existing tractors being used by today's farmer is really appreciable. This research work can be enhanced using some more advanced technologies like GPS based tractors which are fully autonomous. Like plowing, same way harvesting can also be done. Some sensors can be embedded to detect soil moisture so that required amount of water can be sprinkled. Optical sensors can be set up to measure the nutritional status of the plots. Same concept can be used with sprinkling of fertilizers. By connecting various other components, it can be a multi-purpose tool. We can extend this work by considering number of parameters like environment, climate, terrain, type of land and so on for control. This machine can be converted into an Expert System if it is able to check soil properties and let us know which type of crop will yield more in future. This demonstration in the form of a prototype model can be really converted into a recognized structure in the market. It will be a cost-effective model for poor farmers. It's just a pretty small step to farming from our pickup. Maintenance of tractor will be at the module level, not down to individual parts, so repair time will be minimal. This is not like to get the farmer to take computer science classes, a very little and basic training for 1-2 hours is required.

References

1. Solar car and tractor, http://www.solarcarandtractor.com//Home.html (retrieved June 12, 2014)
2. Converting Your Old Allis-Chalmers "G" Cultivating Tractor to an Electric Vehicle, http://www.flyingbeet.com/electricg/solar.html (retrieved June 15, 2014)
3. Griffin, K.: Mother Earth News (July 17, 2013), http://www.motherearthnews.com/green-transportation/our-solar-powered-tractor-zbcz1307.aspx#axzz34pI6ZTdH (retrieved June 15, 2014)
4. O'Hanlan, T.: Young-farmers-and-driverless-tractors (June 20, 2012), from sealevel: http://www.sealevel.com/community/blog/young-farmers-and-driverless-tractors/ (retrieved June 17, 2014)
5. http://www.jbprojects.net/articles/rc/
6. http://www.electronics-lab.com/projects/motor_light/010/index.html/

Analysis of Performance of MIMO Ad Hoc Network in Terms of Information Efficiency

Swati Chowdhuri[1], Nilanjan Dey[3], Sayan Chakraborty[2], and P.K. Baneerjee[3]

[1] Dept. of ECE, Seacom Engineering College, West Bengal, India
[2] Dept. of CSE, JIS College of Engineering, Kalyani, Nadia, India
[3] Dept. of ETCE, Jadavpur University, Kolkata, West Bengal, India
{swati.chowdhuri,neelanjan.dey,sayan.cb,pkbju65}@gmail.com

Abstract. Real time channel models are employed to account the acceptance of the path loss in terms of frequency and transmission range in Rayleigh fading environment. Different interference levels are encountered across a packet transmission, creating the interference diversity that mitigates the multiple access interference effects. It also improves the spectral efficiency. The information efficiency of the network is evaluated in terms of the product of spectral efficiency and distance. In this paper, different path losses are considered to optimize the transmission range and to enhance information efficiency. The performance of the framework is evaluated for real time channel model.

Keywords: Multi Input Multi Output (MIMO), Ad hoc Network, propagation loss, transmission range, Information Efficiency.

1 Introduction

Information efficiency is an important parameter to analyze a network. If two communicating nodes in the network are not in the range of each other in wireless ad hoc networks, then they must rely on multi-hop transmissions. Under such conditions, packet forwarding, or routing, becomes necessary. The value of the radio transmission's range affects network [1] topology and energy consumption considerably. A larger transmission range increases the distance progress of data packets toward their final destinations. This is unfortunately achieved at the expense of higher energy consumption per transmission. On the other hand, a shorter transmission range uses less energy to forward packets to the next hop, although a larger number of hops are needed for packets to reach their destinations [2,3]. In previous works, the optimum transmission ranges (which mitigate interference) were found by maximizing the expected forward progress, which is also a measurement of distance. Real time channel models are employed to account the acceptance of the path loss in terms of frequency and transmission range in Rayleigh fading environment. Different interference levels are encountered across a packet transmission, creating the interference diversity that mitigates the multiple access

S.C. Satapathy et al. (eds.), *Emerging ICT for Bridging the Future – Volume 2*,
Advances in Intelligent Systems and Computing 338, DOI: 10.1007/978-3-319-13731-5_6

interference effects. The information efficiency of the network is evaluated in terms of the product of spectral efficiency and distance. Different path losses [4,5] can be considered to optimize the transmission range and to enhance information efficiency, in this work. In this paper the transmission range of a network is evaluated in order to improve information efficiency within optimize transmission range.

2 Evaluation of Transmission Range

The transmitted signal of MIMO based Mobile ad hoc network is divided into an in-phase stream and quadrature phase stream. They are modulated with separate spreading sequences. After spreading the in-phase stream, it is modulated by cosine function. Later quadrature phase stream is modulated by the sine function of a carrier wave. The two signals are combined and sent to the receiving node [6] as shown in Figure 1.

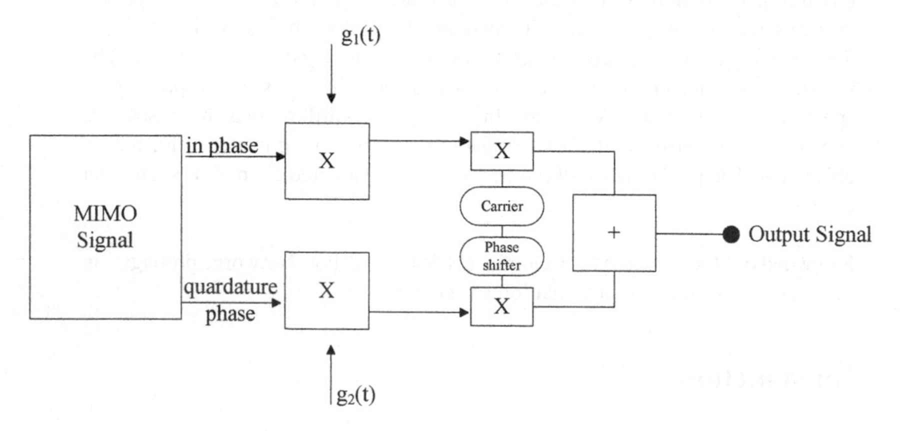

Fig. 1. Transmitted signal from source node

The transmitted signal propagates to the receiver from the transmitting node; its power decaying with distance is given by

$$P_d = P_t K r^{-\mu/2} \tag{1}$$

where K is constant, r is the transmission range of the network and μ is path loss exponent [7,8,9]. Path loss exponent (μ) is the function of frequency of the transmitted signal and distance between the transmitting and receiving antennas. This term exponentially refers to the frequency of the transmitted signal and distance between the transmitting and receiving antennas. As the Rayleigh fading signal amplitude fluctuates, the desired signal at the receiver is corrupted by interference from other active nodes. The active nodes are distributed randomly so data reception depends on the average number of nodes per unit area [10,11,12]. The probability of data reception is given by

$$P(d) = P_r \left[\frac{P_0}{P_i + B} \right] \tag{2}$$

where P_i is the power of the multiple-access interference, P_0 is the received power due to transmitter and B is the power of background noise [13]. In Rayleigh fading environment the Signal to Noise and Interference Ratio (SNIR) due to background noise B can be represented using the following equation,

$$SNIR = \frac{P_t K r^{-\mu/2}}{B} \tag{3}$$

Path loss exponent is a parameter that varies with changes the frequency change of the transmitted signal and the distance between transmitter and receiver [14,15,16]. From equation (1) it is observed that while varying the path loss exponent the transmission range of the network will vary at different Signal to Noise and Interference Ratio [17,18,19]. Thus the transmission range depends directly on path loss exponent and the number of antennas at the transmitting and receiving end. Now MIMO mobile ad hoc network [20] each node consists of more than one antenna where all are used as transmitter as well as receiver in the propagation environment so the transmission range will vary. As the information efficiency is a parameter which depends on transmission range, our main focused endeavor will be to improve the information efficiency within the optimize transmission range.

3 Packet Transmission of Individual Nodes

A network with 40 numbers of nodes was deployed in this work. The parameters set of the network is shown in Table 1. Then the simulation of the network was done in OPNET.

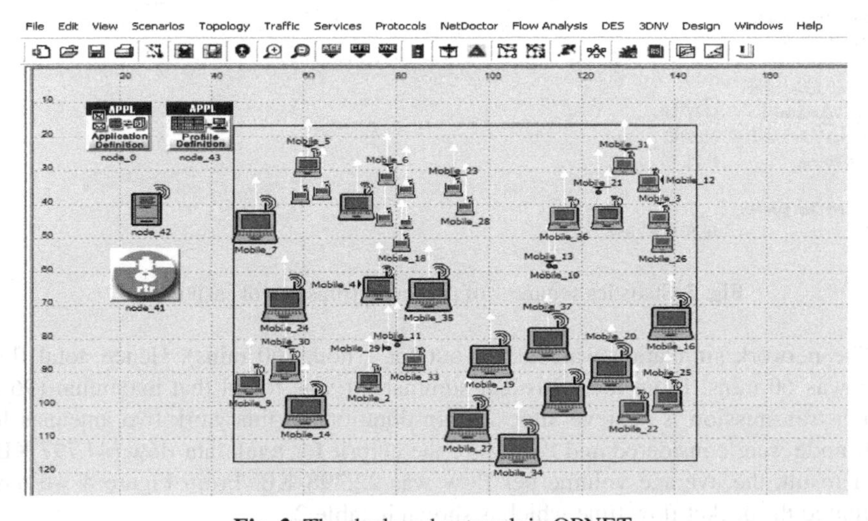

Fig. 2. The deployed network in OPNET

Table 1. Parameter set of deployed network

Node Transmission Power	.005Watt
Operational Mode	802.11b standard
Data rate	11 Mbps
Ad hoc routing protocol	AODV
Technology	WLAN (ad hoc)

Here each node consists of two sets of antenna. All of them can be used as a transmitter as well as receiver. The transmission range depends on the path loss exponent and proportionally varies with the number of antenna in each node. Information efficiency is a term which depends on transmission range and packet transmission rate of individual node. Hence, the packet transmission of individual node is an important parameter that should be evaluated in order to provide information efficiency. The statistical summary of packet transmission and packet transmission rate for a particular node is shown in Fig. 3 and Fig. 4. Let, t_r be the packet transmission rate of the individual node then efficiency should be exponentially dependent on t_r.

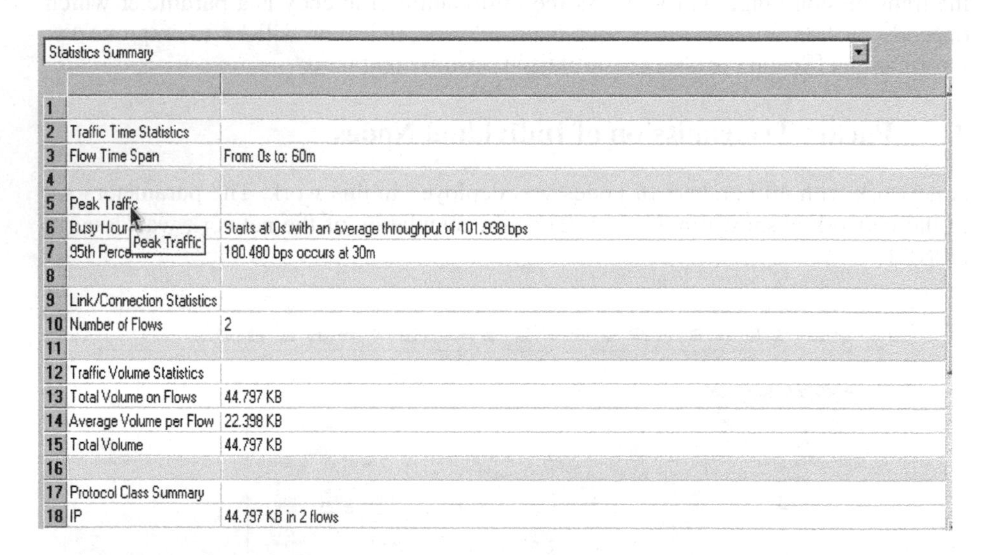

Statistics Summary

1		
2	Traffic Time Statistics	
3	Flow Time Span	From: 0s to: 60m
4		
5	Peak Traffic	
6	Busy Hour	Starts at 0s with an average throughput of 101.938 bps
7	95th Perce[illegible]	180.480 bps occurs at 30m
8		
9	Link/Connection Statistics	
10	Number of Flows	2
11		
12	Traffic Volume Statistics	
13	Total Volume on Flows	44.797 KB
14	Average Volume per Flow	22.398 KB
15	Total Volume	44.797 KB
16		
17	Protocol Class Summary	
18	IP	44.797 KB in 2 flows

Fig. 3. Statistics summary of packet transmission of each node

The network simulation was carried out for 1 hour (60 mins). Hence, total flow time was 60 mins. From the statistical summary it was found that maximum (96%) packet transmission is achieve after 30 min duration. In this work two antennas for each node were considered and they were the culprit for total data flow (44.797 KB). As a result, the average volume per flow was 22.398 KB. From Figure 4 we have evaluated the packet flow time which is shown in table 2.

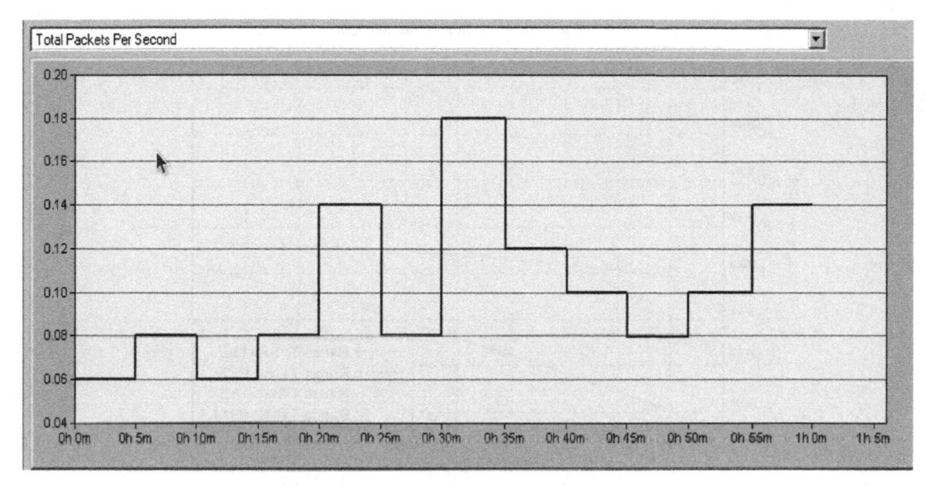

Fig. 4. Transmission rate of a single node

Table 2. Packet transmission rate of single node

Time (mins)	Transmission rate (Kbps)
5	0.08
20	0.14
30	0.18
60	0.14

4 Simulation Result and Discussion

4.1 Optimization of Transmission Range

Each ad hoc nodes of the model network consist of more than one antenna, which are used at the transmitter as well as the receiver end. Path loss depends on different frequency and the distance between transmitting and receiving antennas. Fig. 5 shows the transmission range obtained by varying the path loss with the different transmission frequency and distance between transmitter and receiver.

Fig. 5 shows that different sets of antennas were present in the transmitting and receiving nodes. The figure also shows that the path loss tends to constant for the multiple sets of antennas of mobile ad hoc nodes. The transmission range can be evaluated for different path loss and different sets of antennas considering SINR in Rayleigh fading environment. For the constant path loss the transmission range almost identical over the range of node densities.

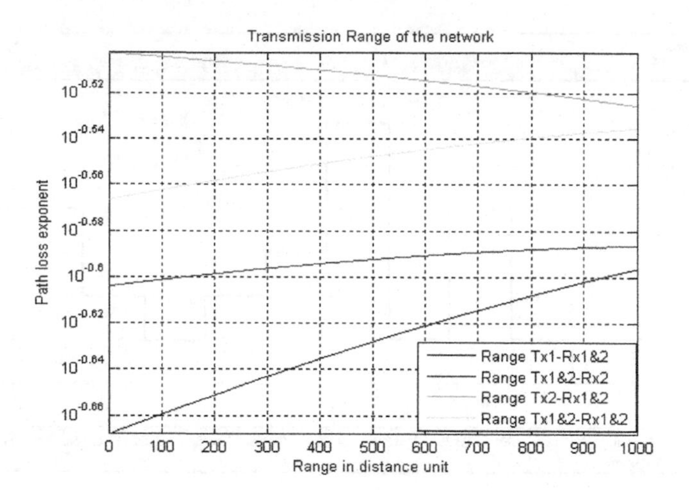

Fig. 5. Transmission Range of the Network

4.2 Information Efficiency Calculation Based on Packet Transmission Rate

Information efficiency is a parameter which depends on transmission range and transmission rate of individual node. The variation of the information efficiency with transmission rate for different transmission range is shown in Fig. 6.

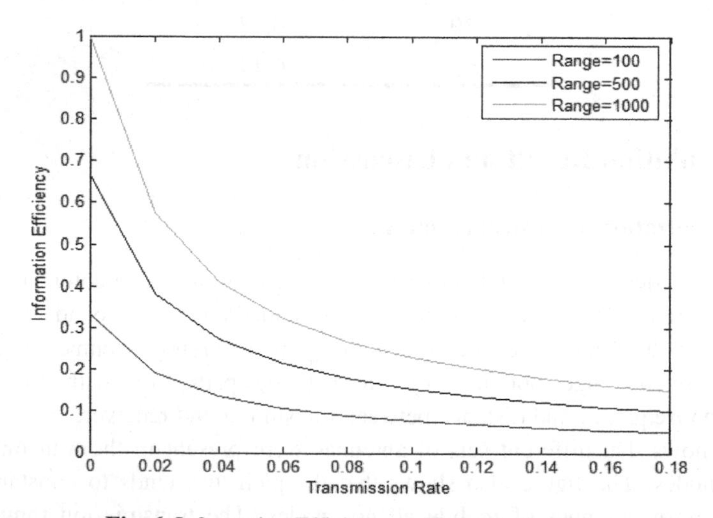

Fig. 6. Information Efficiency vs Transmission Rate

Fig. 6 illustrated that efficiency became maximum with higher transmission range as the interference among multiple traffic flows can occur in the shorter range. Above a certain limit (96% or maximum packet transmission rate of the single node) the information efficiency of the network is more or less constant.

5 Conclusion

The main objective of this work was to evaluate the transmission range and information efficiency. These parameters are the performance matrix of the MIMO based MANET. This paper discussed the mobile ad hoc node with multiple antennas in OPNET TIREM propagation environment and also evaluated the packet transmission rate considering Rayleigh fading and evaluates information efficiency within an optimized transmission range. Different simulation results were also shown to evaluate the transmission range of the network in the specific propagation environment. The propagation environment depends on many parameters such as the path loss exponent, the distance between the transmitter and receiver, the number of antennas on the transmitting and receiving node etc. In this paper transmission range was optimized within a certain limit where transmission range was almost identical. Efficiency is higher for larger transmission range and it degrade for shorter range due to the interference among multiple traffic flows. A short transmission range is required a larger number of hops for a data packet to reach its final destination. Information efficiency of the network is almost constant above 96% or maximum packet transmission rate of the single node. Mobile ad hoc network' nodes consist of limited battery power. A node can effectively increase the transmission range by reducing data rate and by keeping fixed transmission power. Apart from that, power adjustment could be implemented on this work to adapt the transmission range by fixing the data rate, in near future. Future work must focus on improvement of information efficiency to consume minimum power from ad hoc node.

References

1. Chandra, M.W., Hughes, B.L.: Optimizing information efficiency in a direct-sequence mobile packet radio network. IEEE Trans. Commun. 51(1), 22–24 (2003)
2. Subbarao, M.W., Hughes, B.L.: Optimum transmission ranges in multihop packet radio networks in the presence of fading. In: Proc. Conf. Info. Sci. Systems (CISS), pp. 684–689 (1997)
3. Sousa, E.S., Silvester, J.A.: Optimum transmission ranges in a direct-sequence spread-spectrum multihop packet radio network. IEEE J. Select. Areas Commun. 8(5), 762–771 (1990)
4. Zorzi, M., Pupolin, S.: Optimum transmission ranges in multihop packet radio networks in the presence of fading. IEEE Trans. Commun. 43(7), 2201–2205 (1995)
5. Liang, P.C.P., Stark, W.E.: Transmission range control and information efficiency for FH packet radio networks. In: Proc. IEEE Military Commun. Conf. (MILCOM), vol. 2, pp. 861–865 (2000)
6. RFC 802.11n Standard for MIMO communication. www.802.11n standard
7. Stamatiou, K., Proakis, J.G., Zeidler, J.R.: Information efficiency of ad hoc networks with FH-MIMO transceivers. In: Proc. IEEE Int. Conf. Commun. (ICC), pp. 3793–3798 (2007)
8. Sui, H., Zeidler, J.R.: A robust coded MIMO FH-CDMA for mobile ad hoc networks. IEEE J. Select. Areas Commun. 25(7), 1413–1423 (2007)
9. Ephremides, A., Wieselthier, J.E., Baker, D.J.: A design concept for reliable mobile radio networks with frequency hopping signaling. Proc. of IEEE 75(1), 56–73 (1987)

10. Kong, N., Milstein, L.B.: Error probability of multicell CDMA over frequency selective fading channels with power control error. IEEE Trans. Commun. 47(4), 608–617 (1999)
11. Sui, H., Zeidler, J.R.: Information Efficiency and Transmission Range Optimization for Coded MIMO FH-CDMA Ad Hoc Networks in Time-Varying Environments. IEEE Trans. Commun. 57(2), 481–491 (2009)
12. Pursley, M., Taipale, D.: Error probabilities for spread-spectrum packet radio with convolutional codes and Viterbi decoding. IEEE Trans. Commun. 35(1), 1–12 (1987)
13. Torrieri, D.J.: Future army mobile multiple-access communications. In: Proc. IEEE Military Commun. Conf. (MILCOM), vol. 2, pp. 650–654 (1997)
14. Eltahir, I.K.: The Impact of Different Radio Propagation Models for Mobile Ad hoc Networks (MANET) in Urban Area Environment. In: The 2nd International Conference on Wireless Broadband and Ultra Wideband Communications, pp. 1–30 (2008)
15. Gass, J.H., Pursley, M.B.: A comparison of slow-frequency-hop and direct-sequence spread-spectrum communications over frequency selective fading channels. IEEE Trans. Commun. 47(5), 732–741 (1999)
16. Sousa, E.S.: Performance of a spread spectrum packet radio network link in a Poisson field of interferers. IEEE Trans. Inform. Theory 38(6), 1743–1754 (1992)
17. Chandra, M.W., Hughes, B.L.: Optimizing Information Efficiency in a Direct-Sequence Mobile Packet Radio Network. IEEE Transactions on Communications 51(1), 22–24 (2003)
18. Zorzi, M., Pupolin, S.: Optimum Transmission Ranges in Multihop Packet Radio Networks in the Presence of Fading. IEEE Transactions on Communications 43(7), 2201–2205 (1995)
19. Subbarao, M.W., Hughes, B.L.: Optimal Transmission Ranges and Code Rates for Frequency-Hop Packet Radio Networks. IEEE Transactions on Communications 48(4), 670–678 (2000)
20. Chowdhuri, S., Chakraborty, S., Dey, N., Azar, A.T., Chaudhury, S.S., Banerjee, P.: Recent Research on Multi Input Multi Output (MIMO) based Mobile ad hoc Network: A Review. International Journal of Service Science, Management, Engineering, and Technology (IJSSMET) (in press)

Handwritten Kannada Numerals Recognition Using Histogram of Oriented Gradient Descriptors and Support Vector Machines

S. Karthik and K. Srikanta Murthy

P E S Institute of Technology Bangalore South Campus, Bangalore, India
karthiks@pes.edu

Abstract. The role of a good feature extractor is to represent an object using numerical measurements. A good feature extractor should generate the features, which must support the classifier to classify similar objects into one category and the distinct objects into separate category. In this paper, we present a method based on HOG (Histogram of Oriented Gradients) for the recognition of handwritten kannada numerals. HOG descriptors are proved to be invariant to geometric transformation and hence they are one among the best descriptors for character recognition. We have used Multi-class Support Vector Machines (SVM) for the classification. The proposed algorithm is experimented on 4,000 images of isolated handwritten Kannada numerals and an average accuracy of 95% is achieved.

1 Introduction

There are too many applications (i.e. Indian offices such as bank, sales-tax, railway, embassy, etc.) in which both English and regional languages are used. Many forms and applications are filled in regional languages and sometimes those forms have to be scanned directly. If there is no character recognition system, then image is directly captured and there is no option those for editing documents. As large number of such documents has to be processed every day in organizations like banks, post offices, reservation counters, libraries, and publishing houses, automatic-reading systems can save much of the work and time. Even though various advances have been made in recent years, the recognition of handwritten characters still poses an interesting challenge in the field of Pattern Recognition. Many commercial OCR systems are available for Roman, Chinese, Japanese and Arabic languages [1, 2, 3, 4, 5, and 6]. But, limited work can be found about character recognition for Indian languages.

It is observed that, a good amount of work is carried out for the recognition of printed characters of the Indian languages. However, recognition of handwritten characters is still a open challenge. Most of the reported works for the recognition of handwritten characters are found for the Devnagari and Bangla languages [7, 8, 9, 10, 11, and 12]. Few studies are reported for the recognition of other languages like Marathi, Malayalam, Tamil, Telugu, Oriya, Kannada, Punjabi, Gujarathi, etc[13,14,15,16,17,18,19,20,21,22,23,24]. The developments of OCR work for the

S.C. Satapathy et al. (eds.), *Emerging ICT for Bridging the Future – Volume 2*,
Advances in Intelligent Systems and Computing 338, DOI: 10.1007/978-3-319-13731-5_7

Indian scripts are reviewed in detail in [25]. The research in the development of OCR for the south Indian languages was presented in detail in [26].

From the literature it is seen that very few works are reported in the Kannada language. Most of the works are on the recognition of machine printed Kannada characters and handwritten numerals [22, 27, 28, 23, 29,30, 31]. It is clear that in the Indian context, handwritten character recognition is still a open research area, in particular for handwritten Kannada character recognition. The role a good feature extraction method is vital in achieving high recognition accuracy in character recognition systems. In [5] a survey on the feature extraction methods for character recognition is reviewed. Researchers have used many methods of feature extraction for recognition of characters. Fractal code, moments, shadow code, profiles, wavelets, structural (points, primitives), directional feature, templates etc., have been addressed in the literature as feature[6]. It is highlighted from the literature that majority of the methods works on the principle of extraction of structural or statistical features. However, it is noted that these methods are very sensitive to scaling and rotation. Here, we made an attempt to overcome these problems using the HOG features. In this paper we propose a method based on HOG features for the recognition of isolated off-line handwritten Kannada numerals. In the rest of the paper, we have discussed the properties of Kannada characters and the challenges; the feature extraction and classification procedure, the experimental results and discussions followed by the paper's conclusion.

2 Properties of Kannada Language

Kannada is one of the popular south Indian language. The characters of this language exhibits a large number of structural features. Modern Kannada or the Hosagannada has 49 base characters along with 10 numerals. Figure 1 shows the handwritten kannada numerals. One of the tough task in Kannada handwritten character recognition is to differentiation between similar shaped characters. A very small variation between two characters contributes to recognition complexity and also to recognition accuracy. The writing styles and size of the characters vary from person to person. Few samples of the similar characters are shown in figure 2. For the experimentation, we have collected sample data from the authors of [25]. The dataset consists of 400 samples for each of the handwritten numerals.

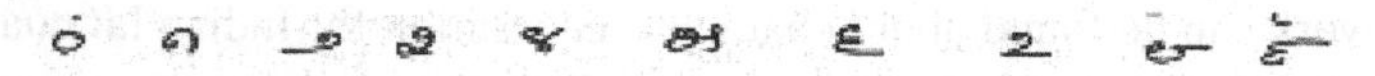

Fig. 1. Handwritten kannada numeral sample

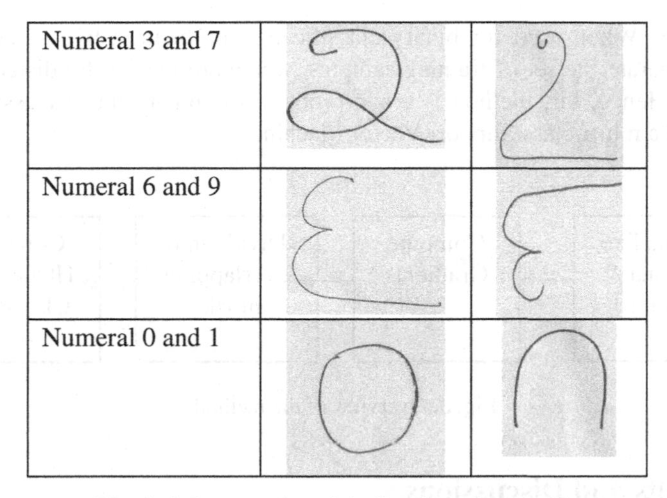

Numeral 3 and 7
Numeral 6 and 9
Numeral 0 and 1

Fig. 2. A few samples of similar shaped numerals

3 Feature Extraction and Classification

In this paper, we propose to use a feature, which is robust against shape variations and geometric transformation. From the literature, it is noted that the Histogram of Oriented Gradient is one such descriptor. Also this is proven to be effective compared to few other descriptors like SIFT, SURF[32]. The distributions of local intensity gradients of ridge directions are very effective in characterizing an object. This is the central idea of histogram of oriented gradients. This is achieved by partitioning an image into smaller blocks called as "cells". For each cell, local 1-D histograms of gradient directions are obtained. The combined histograms of all the cells form the feature vector. To achieve better invariance to illumination, shadowing, etc., it is also useful to contrast-normalize the local responses before using them. This can be done by accumulating a measure of local histogram "energy" over somewhat larger spatial regions ("blocks") and using the results to normalize all of the cells in the block. We will refer to the normalized descriptor blocks as Histogram of Oriented Gradient (HOG) descriptors. The steps involved in this is shown in figure 3.

The HOG features have few advantages over other methods. These features capture the edge information very effectively and also with invariance to geometric and photometric transformations. Further, rotation and scaling of images makes little difference to these features

In machine learning, support vector machines (SVMs) are supervised learning models with associated learning algorithms that analyze data and recognize patterns, used for classification and regression analysis. When used for binary classification problems, the SVM training algorithm builds a model that classifies the new sample into one of the two possible SVM is a learning algorithm originally introduced by Vapnick and co-researchers and later on extended by many other researchers. SVMs are remarkably robust for sparse and noisy data and hence becoming an obvious choice for many applications [33]. Classification problems are one such area. When used for binary classification problem, a hyper-plane will be formed to separate the set of training samples, which are maximally distant from the hyper-plane. Hence, this method is a non-probabilistic binary linear classifier. Here, we have used a multi-class Support Vector Machines.

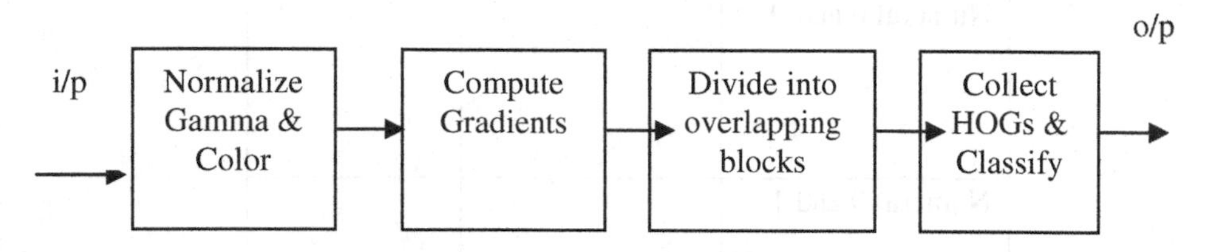

Fig. 3. Overview of the method

4 Results and Discussions

The method is tested on isolated handwritten kannada numerals and the performance of the proposed method is evaluated. A total of 400 different samples of each numeral were used with the total of 4000 samples. We have used 80% of the dataset for the training and the rest for testing. A 5-fold cross validation scheme is adopted to calculate the recognition accuracy. The recognition accuracy of every numeral is shown in the figure 4. We are able to achieve an average accuracy of 95%. The comparison of the proposed method along with some of the existing methods is shown in table 1.

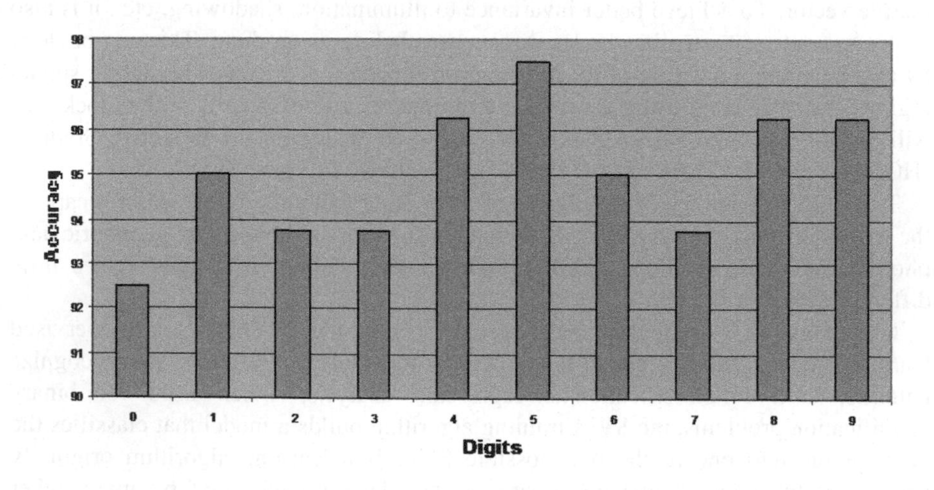

Fig. 4. Recognition accuracy for all the numbers

Table 1. Result comparison

Authors	No. of samples in data set	Feature extraction method	Classifier	Accuracy (%)
Rajashekhara Aaradhya S V et.al[34]	1000	Vertical projection distance with zoning	NN classifier	93
R Sanjeev Kunte et al[35]	2500	Wavelet	Nueral classifier	92.3
G G Rajput et al[23]	1000	Image fusion	NN classifier	91.2
V N Manjunath Aaradhya et al[36]	2000	Radon features	NN classifier	91.2
Dinesh Aacharya U et. al[37]	500	Structural features	K-means	90.5
Proposed	4000	HOG	SVM	95%

5 Conclusion

In this paper, we proposed a new method based on Histogram of Oriented Gradients features and multi class Support Vector Machines for the recognition of handwritten kannada numerals. HOG features are proved to be a scale and rotation invariant. The proposed method attains an average recognition accuracy of 95%. From the comparison study it is noted that, the proposed method has delivered promising results. These features can be used for the recognition of handwritten vowels and consonants.

Acknowledgments. The authors would like to thank Mr. Umapada Pal, Mr. Nabin Sharma, Mr. Fumitaka Kimura and Mr. Tetsushi Wakabayashi for providing the numeral database and also for encouraging us to take up this work.

References

1. Mori, S., Suen, C.Y., Yamamoto, K.: Historical Review of OCR Research and Development. IEEE-Proceedings 80(7), 1029–1057 (1992)
2. Plamandon, R., Lopresti, D., Schomaker, L.R.B., Srihari, R.: Online Handwriting Recognition. In: Webster, J.G. (ed.) Encyclopedia of Electrical and Electronics Eng., vol. 15, pp. 123–146. Wiley, New York (1999)

3. Amin, A.: Off-line Arabic Character Recognition: the State of the Art. Pattern Recognition 31, 517–530 (1998)
4. Nagy: Chinese Character Recognition—A Twenty Five Years Retrospective. In: Proceedings of the Ninth International Conference on Pattern Recognition, pp. 109–114 (1988)
5. Trier, D., Jain, A.K., Taxt, T.: Feature Extraction Methods for Character Recognition—a Survey. Pattern Recognition 29, 641–662 (1996)
6. Plamondon, Srihari, S.N.: On-line and Off-line Handwritten Recognition: a Comprehensive Survey. IEEE Trans. Pattern Anal. Mach. Intel. 22, 62–84 (2000)
7. Majumdar, A., Chaudhuri, B.B.: A MLP Classifier for Both Printed and Handwritten Bangla Numeral Recognition. In: Kalra, P.K., Peleg, S. (eds.) ICVGIP 2006. LNCS, vol. 4338, pp. 796–804. Springer, Heidelberg (2006)
8. Chaudhuri, Majumdar, A.: Curvelet–based Multi SVM Recognizer for Offline Handwritten Bangla: A Major Indian Script. In: Ninth International Conference on Document Analysis and Recognition (ICDAR 2007), pp. 491–495 (2007)
9. Naser, Mahmud, A., Arefin, T.M., Sarowar, G., Naushad Ali, M.M.: Comparative Analysis of Radon and Fan-beam based Feature Extraction Techniques for Bangla Character Recognition. IJCSNS International Journal of Computer Science and Network Security 9(9), 287–289 (2009)
10. Shrivastava, I.K., Gharde, S.S.: Support Vector Machine for Handwritten Devanagari Numeral Recognition. International Journal of Computer Applications 7(11), 9–14 (2010)
11. Arora, S., Bhattacharjee, D., Nasipuri, M., Basu, D.K., Kundu, M.: Combining Multiple Feature Extraction Techniques for Handwritten Devnagari Character Recognition. In: 2008 IEEE Region 10 Colloquium and the Third ICIIS, Kharagpur, INDIA, pp. 342/1–342/5 (2008)
12. Pal, U., Wakabayashi, T., Kimura, F.: Comparative Study of Devnagari Handwritten Character Recognition using Different Feature and Classifiers. In: 10th International Conference on Document Analysis and Recognition, pp. 1111–1115 (2009)
13. Rajput, Mali, S.M.: Fourier Descriptor based Isolated Marathi Handwritten Numeral Recognition. International Journal of Computer Application 10, 5120/724–5120/1017 (2010)
14. Maloo, M., Kale, K.V.: Support Vector Machine Based Gujarati Numeral Recognition. International Journal on Computer Science and Engineering 3(7), 2595–2600 (2011)
15. Lehal, S., Singh, C.: A Post Processor for Gurmukhi OCR. Sadhana 27, 99–111 (2002)
16. Pal, U., Wakabayashi, T., Kimura, F.: A System for Off-line Oriya Handwritten Character Recognition using Curvature Feature. In: 10th International Conference on Information Technology, pp. 227–229 (2007)
17. Srinivas, A., Agarwal, A., Rao, C.R.: Telugu Character Recognition. In: International Conference on Systemics, Cybernetics, and Informatics, Hyderabad, India, pp. 654–659 (2007)
18. Vasantha Lakshmi, C., Jain, R., Patvardhan, C.: OCR of Printed Telugu Text with High Recognition Accuracies. In: Kalra, P.K., Peleg, S. (eds.) ICVGIP 2006. LNCS, vol. 4338, pp. 786–795. Springer, Heidelberg (2006)
19. Suresh, R.M., Arumugam, S.: Fuzzy Technique based Recognition of Handwritten Characters. Image Vision Computing 25, 230–239 (2007)
20. Seethalakshmi, R., Sreeranjani, T.R., Balachandar, T., Singh, A., Singh, M., Ratan, R., Kumar, S.: Optical Character Recognition for printed Tamil text using Unicode. Journal of Zhejiang University SCI 6A(11), 1297–1305 (2005)

21. Abdul Rahiman, M., Rajasree, M.S.: Printed Malayalam Character Recognition Using Backpropagation Neural Networks. In: IEEE International Advance Computing Conference, Patiala, India, pp. 1140–1144 (2009)
22. Kunte, R.S., Samuel, R.D.S.: A Simple and Efficient Optical Character Recognition System for Basic Symbols in Printed Kannada Text. Sadhana 32, 521–533 (2007)
23. Rajput, G.G., Hangarge, M.: Recognition of Isolated Handwritten Kannada Numerals Based on Image Fusion Method. In: Ghosh, A., De, R.K., Pal, S.K. (eds.) PReMI 2007. LNCS, vol. 4815, pp. 153–160. Springer, Heidelberg (2007)
24. Ragha, L., Sasikumar, M.: Adapting Moments for Handwritten Kannada Kagunita Recognition. In: Second International Conference on Machine Learning and Computing, pp. 125–129 (2010)
25. Pal, U., Chaudhuri, B.B.: Indian Script Character Recognition: a Survey. Pattern Recognition 37, 1887–1899 (2004)
26. Abdul Rahiman, M., Rajasree, M.S.: A Detailed Study and Analysis of OCR Research in South Indian Scripts. In: International Conference on Advances in Recent Technologies in Communication and Computing, pp. 31–38 (2009)
27. Sagar, B.M., Shobha, G., Ramakanth Kumar, P.: Character Segmentation Algorithms For Kannada Optical Character Recognition. In: International Conference on Wavelet Analysis and Pattern Recognition, Hong Kong, pp. 339–342 (2008)
28. Sheshadri, K., Ambekar, P.K.T., Prasad, D.P., Kumar, R.P.: An OCR System for Printed Kannada using K-Means Clustering. In: International Conference on Industrial Technology, pp. 183–187 (2010)
29. Niranjan, S.K., Kumar, V., Hemantha Kumar, G., Manjunath Aradhya, V.N.: FLD based Unconstrained Handwritten Kannada Character Recognition. In: International Conference on Future Generation Communication and Networking Symposia, vol. 3, pp. 7–10 (2008)
30. Rajashekararadhya, S.V., Vanaja Ranjan, P.: Support Vector Machine based Handwritten Numeral Recognition of Kannada Script. In: International Advance Computing Conference, Patiala, India (2009)
31. Rajput, G.G., Horakeri, R., Chandrakant, S.: Printed and Handwritten Mixed Kannada Numerals Recognition using SVM. International Journal on Computer Science and Engineering 02(05), 1622–1626 (2010)
32. Dalal, N., Triggs, B.: Histograms of Oriented Gradients for Human Detection. In: IEEE Computer Society Conference on Computer Vision and Pattern Recognition (2005)
33. Bovolo, F., Bruzzone, L., Carlin, L.: A Novel Technique for Sub-pixel Image Classification Based on Support Vector Machine. IEEE Transactions on Image Processing 19(11), 2983–2999 (2010)
34. Rajashekararadhya, S.V., Vanaja Ranjan, P.: Neural Network Based Handwritten Numeral Recognition of Kannada and Telugu scripts. In: TENCON (2008)
35. Sanjeev Kunte, R., Sudhakar Samuel, R.D.: Script Independent Handwritten Numeral recognition. In: International Conference on Visual Information Engineering, pp. 94–98 (2006)
36. Manjunath Aradhya, V.N., Hemanth Kumar, G., Noushath, S.: Robust Unconstrained Handwritten Digit Recognition Using Radon Transform. In: Proc. of IEEE International Conference on Signal Processing, Communication and Networking, pp. 626–629 (2007)
37. Dinesh Acharya, U., Subba Reddy, N.V., Krishnamurthy: Isolated handwritten Kannada numeral recognition using structural feature and K-means cluster. In: IISN, pp. 125–129 (2007)

Design and Utilization of Bounding Box in Human Detection and Activity Identification

P. Deepak* and Smitha Suresh

Research Scholar, STATS, M.G University regional Centre,
Edappally, Cochin, 24, Kerala, India
deepakpeeth@gmail.com
Computer Science Department,
Sree Narayana Gurukulam College of Engineering, Kadayiruppu, Kolenchery, Kerala, India
smithadeepak2006.sngce@gmail.com

Abstract. All video surveillance system uses some important steps to detach moving object from background and thus differentiate all the objects in video image frames. The main endeavor of this paper is to use an object bounding box concept to detect the human blob and human activity detection using a system. The images are captured by using single static camera to make the system cost effective. The bounding box that bounds each object in the image frames can be utilized to track and detect activities of the moving objects in further frames. These bounding boxes can be utilized to detect human and human activities like crowd and the estimation of crowd. This paper gives the implementation results of bounding box for activity detection.

Keywords: Video surveillance, Human blob, Bounding Boxes, Human activity detection, Static camera, Crowd, Crowd estimation, and Tracking.

1 Introduction

In the vast use of video surveillance systems, human recognition and human activity identification like human fall are the well-known areas in which the research is carried out [1]. Nowadays these systems are so much useful for analyzing and recording of all the activities performed by moving objects in a scene and it can also detect some difficult efforts such as human interactions in the presence of crowd. Pulling out of composite information from the video by using a personal computer is an essential part in this cyber age. All surveillance spots consists of a collection of people with diverse pattern of interactions between them and the learning of such interactions in the public venues has great attention in law enforcement and people safety. The people cannot be distinguished using high frequency signals because of the reflection of signal from human body parts is very less. So video surveillance techniques are greatly employed to detect the human activities at streets. In the premature stages, the computer vision system worked with the help of many operators

* Corresponding author.

S.C. Satapathy et al. (eds.), *Emerging ICT for Bridging the Future – Volume 2*,
Advances in Intelligent Systems and Computing 338, DOI: 10.1007/978-3-319-13731-5_8

to detect the suspicious activities of human from the computer screen. This type of semi automated surveillance systems will give false alerts because of the lack of interest of operators in identifying the genuine threats because of sleepiness, tiredness or other factors.

Human activity study is the major area in which a well-groomed computer vision system utilizing for analyzing human video images in real-time. Generally these types of systems are used to observe or trace the activities or any other altering information of human in the video image frames. Today the growth of fully computerized system ensures the human monitoring effectively. The main effort of using such system in public places is to detect the abnormal and suspicious activities.

The necessity of computer vision system to observe the moving targets are good speed and truthfulness. The fast boost of pedestrians in streets creates problems in law enforcement and the mishaps due to crowd are growing in our daily life. So essentially, the human overcrowding must be reduced in narrow streets to ensure traffic safety. An ordered system should watch the moving or non-moving human in the prohibited areas of a street [2]. Suspicious activity can be identified by a system based on the total time in which a person seemed to present in an area of interest or the persons who wander, sit and stop for a long time in any area. Similarly passenger safety measures in railway stations and aerodrome is a crucial issue for all railway and aerodrome agencies. At Present, large advancement in the field of computer vision gives more attention in the control and monitoring of traffic [3].

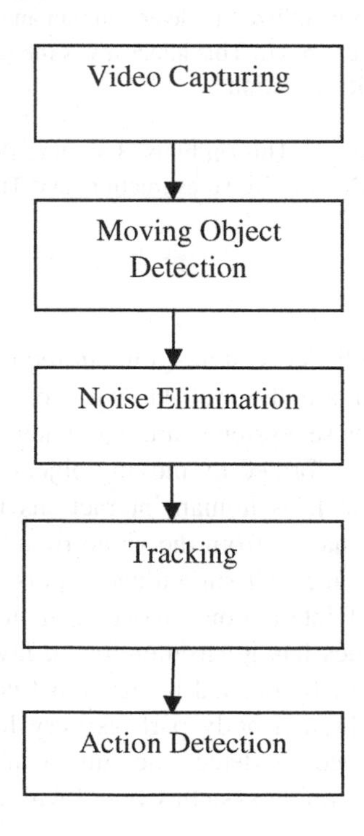

Fig. 1. General Architecture of Computer Vision System used for activity detection

Most of the human detection or monitoring systems struggles to distinguish human action in more complex scenes present at streets. These limitations in the system design are mainly due to the inaccuracy of object detection in the presence of shadow and variation of lighting in the surrounding environment [4]. The individual activity monitoring is much harder in the presence of occlusion and the mass. The section 2 describes the major areas of pedestrian detection that is useful in the proposed concept of this paper and the region in which the human detection are to be considered with great care. Section 3 states the proposed method that is useful for many of the human detection and activity detection applications. Section 4 and 5 shows the results and conclusion respectively.

Figure 1 shows the fundamental steps concerned in all human activity observation system after the video recording using a camera. The input video is taken by video cameras. The main units in the proposed system are object detection, tracking and action identification modules. The recognition unit is mainly used to break up the foreground objects from background in all video image frames. To achieve high-quality result from visual observation system, these units must be vigorous.

2 Related Works

This section gives a few but relevant areas of computer vision that is to be worked out with an improved approach. The main attempt of developing such systems is to identify the scene and estimate the relations of human individuals in a video based data. The factors such as human count, walking rate, flow rate, can be determined in an intellectual system. The researchers functioning in this plane are building a verity of algorithms for numerous functions of human activity detection and tracking. But the attractiveness of the planned system is that, most of the applications of human detection and activity can be solved with the help of a simple rectangular box that bounds object blob in all the frames. The subsequent sentences give some areas of suggestions for using this concept.

These days the majority of the terrorist's attacks were taken place in the existence of intentionally created occlusion or crowd. Occlusion is a chief problem in visual surveillance and it is the hiding of visual parameters of one object by a different object in a video image frame. A terrorist can purposely make occlusion and crowd for the establishment of his or her suspicious activity. The overlapping of bounding box around each object in an image frame can be used to detect the occlusion and crowd [5], [6].

Counting the figure of people in an image frame can be made with the aid of bounding box in such a way that counting the number of bounding boxes present in a frame and will gives the count of the people present in that frame [7]. It is a vital task and getting attractiveness in visual surveillance system if the present situations in surveillance and security sites are considered. Here researchers are capable of labeling every bounding box in an image and it is very easy and useful in walker counting. For detecting and counting the people in a particular area of interest can also be calculated by placing a target contour in any area of the frame and by using this they will be able to calculate how many people reside inside the contour. For creating this contour, bounding box concept can be utilized [8]. If the dimension of a bounding box that contains many objects blobs goes beyond a pre -defined threshold, it can be treated as crowd.

An intelligent home care system which is implemented at homes or surveillance sites can detect human fall and can raise alarm if any abnormality is detected. This can also be completed with the help of bounding box. If a person fell down, most of the cases the bounding boxes shrink in dimension. This can be taken as a symbol of human fall. Human descend may make head injury and it may leads to insensible situations [9] and can be spot effortlessly by identifying the stillness of bounding box. The immobility of a person in surveillance sites or home for an elongated time slot must be detected with care because sometimes that person may plans to do some distrustful activity. If any related movement is detected from any of the bounding box for a period of time, it can be considered as illegal or suspicious action.

Most of the activities of any individual can be recognized by the movement of human body parts and can be detected with more bounding boxes. An enhanced concept with the use of four or more bounding box can be used to supervise human motion of the upper part of the body, middle part and two lower parts for legs [10]. Other daily activities jogging, sitting down, squatting down can also detected using bounding box [11]. All the activities exhibited by a human can be detected by the bounding box concept and further action can be taken if necessary. The proposed system mainly concentrated to create the bounding box.

3 Proposed Method

Detection of moving object, Tracking and analyzing the object in each frame to identify their behavior or activities are the normal steps performed in all video analysis systems. Our aim is to work out the bounding box to address problems specified in the above literature. But the scope of this work is limited to bounding box creation. The main intend of the proposed system is to detect human in each frame using simple object detection and bounding box creation method by utilizing the video from single camera. Simulation of this work is carried out in MATLAB 2013 software.

3.1 Adaptive Background Subtraction

Background subtraction is an effective method to detect the moving object in a video that is recorded using a fixed camera. It offers fine performance in dynamic scenes. But it has poor performance in detecting all appropriate foreground pixels due to various reasons and many noisy pixels will associated in the object detected image. The background subtraction procedures are assessed based on speed, memory requirement and accuracy and it is the most accepted method in object detection because of the less computational complexity.

In the proposed work, an adaptive background subtraction technique was used for object detection. Here, the pixels values in the current image frame are subtracted from the background model for detecting the foreground image and if this pixel value is taken as white or black by using a threshold selection. Threshold selection for a site is to be done in prior to the system installation after enough trial and error threshold hypotheses. The benefit of the proposed system is that the user or the operator of the system can select different threshold to obtain a good object detected image without noise pixels.

To detect the fore ground object, we read each pixels in a frame and checks whether the intensity value of that pixels exceeds the pre defined threshold value. If the intensity value of current pixel falls below the threshold value, that pixel is assumed as black pixel. Similarly, if the intensity value of current pixel exceeds the threshold value, that pixel is assumed as white pixel. The algorithm for adaptive background subtraction is as follows.

Step 1 Read each pixel in a frame
Step 2 If I (n) > TH; Current Pixel= 1(Foreground Pixel)
Step 3 I (n) < TH; Current Pixel= 0(Background Pixel)
Step 4 Repeat the process for all pixels
Step 5 Repeat the process to all image frames

Where I (n) = Intensity value of current pixel in a frame,
TH=Threshold Intensity Value

3.2 Noise Elimination

After object detection, image restoration process is used to fill the missing spaces in moving objects. The noises are normally appeared as white pixels in background image and black pixels in foreground image. Camera noise and irregular object movement noises should be clean out from the image frame with the use of morphological operations. Here also the operator gets a chance to select the no of times the operations are to be performed to avoid the noisy pixels completely. The morphological operations are to search the image with a simple pre- defined shape, illustrating conclusions on how this shape fits or misses the shapes in the image. Morphological operations consists of some structure elements like disk, square and cross shaped elements fixed sizes. The essential operations of binary morphology operations are dilation, erosion, closing, and opening. Dilation operation enlarges the region, while erosion makes the region small. Closing operation is defined as performing erosion after dilation and it can fill the internal holes in the region. In opening operation dilation is performed after erosion and it can clear small portions that are just out from the boundary in to background. The mathematical operators “bwmorph” is used on each image frames perform erosion and dilation. The syntax of morphological operators in commonly used formats is given below. Assume that our current image or frame is stored in a variable (X1)

1. Y1 = bwmorph (X1, ZZ)
2. Y1=bwmorph (X1, ZZ, A)

Where Y2 = Variable for storing O/P image, ZZ= Dilate, Erode etc, A= No of times the process is to be repeated.

3.3 Bounding Box for Object Blob Detection

The bounding box is a single rectangular box that bounds the entire objects in a scene. A bounding box can be represented by giving the coordinates of the upper left and the lower right corners of the box [12]. It is computationally efficient, easy to implement and it can be used to detect the motions associated with human body. The height and width of bounding box can be used detect the human activity. The rectangular bounding boxes are created by taking an approximation that, the height and width of the bounding boxes must embed all objects in that area or scene. If the dimension of the bounding box is very small, then some part of the objects will not be included in the bounding boxes. The whole human body can be bounded using single rectangular bounding box and entire motion has to be detected using this single box. The location of the pedestrian can also be found using bounding boxes. The normally used shapes are sphere, parabolic, square and rectangular. The cylindrical shape also can be used as bounding box. In our previous work, we created the bounding box for detecting occlusion by measuring the distance between the bounding boxes between objects [13]. Now these bounding boxes have found many number of applications addressed in the related area of this paper.

Figure 2 shows the basics steps involved in the proposed system that consists of a camera to capture the image and perform some significant steps to pre process the video using the above algorithms.

Finding of human by means of rectangular bounding boxes is very easy as compared with other shapes. The vertical sides of the bounding boxes remain invariant when a person walks. But in the case of incomplete occlusion or the person may bend, the edges are susceptible to actions. If the dimension of the bounding box vary than the expected extent, then this shows occlusion or segmentation is not proper. Basically the three reasons for using bounding box are it is very easy to create a box that bound human blob, the information extracted from bounding box is easy to process and the pull out of object characteristics using bounding is more stable than the feature extracted directly from the binary image. To create bounding boxes around each object in a frame, we use feature extraction using regionprops. The "regionprops" function gives value of some the parameters like the area, centroid, four corner points of the bounding box and diameter etc of an image. By using the coordinate's value these four points, bounding box can be drawn.

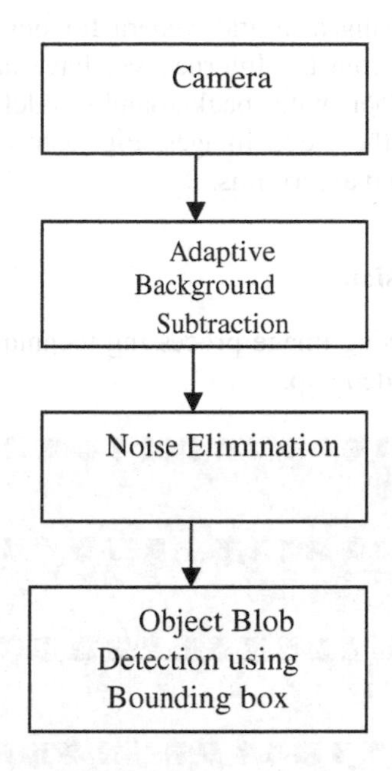

Fig. 2. Important steps involved in the design of the proposed system

4 Result and Discussions

4.1 Video Recording

Camera coordination and camera calibration is a time consuming process, if we are using multiple cameras. So in this work, we use a single stationary camera to capture the video. This system was tested for two video images. The first video was recorded

Fig. 3. Video Input

in an outdoor environment using a digital camera for our previous work [13]. The second video is downloaded from the Internet. We have taken these videos for two different scenarios; 1. Outdoor with background model is already captured, 2. Background is created from the video images. Fig. 3shows one of the input video utilized for testing our proposed algorithms.

4.2 Video Frame Conversion

Video framing can be done using image processing techniques. Fig 4 shows some of image frames from the first video clip.

Fig. 4. Results of Video Framing

4.3 Reference Background Creation

To perform adaptive background subtraction, we need a reference background. Fig.5 shows the implementation result of reference background creation. Here video background is created from the image frames. Here background is created by performing subtraction between all the frames. If we do subtraction operation between all the frames, the foreground object will be eliminated and we get only background pixels.

Fig. 5. Result of Reference Background

4.4 Adaptive Background Subtraction

Object detection is an important step in video surveillance systems. Object detection is nothing but the isolation of objects from back ground. After framing the videos, object detection using adaptive back ground subtraction was performed. In background subtraction each frame is subtracted from the reference background and thus the foreground object can be detected. Fig.6 shows the result of adaptive background subtraction performed on second video. It can be seen from result that, noise pixels are present. Most of the noisy pixels are generated because of the guardrail seemed in the steps of the first video frames. From figure 5 we can see that the histories of guardrail are present in background image ie objects in the background are not fully removed. So we have to use noise elimination techniques.

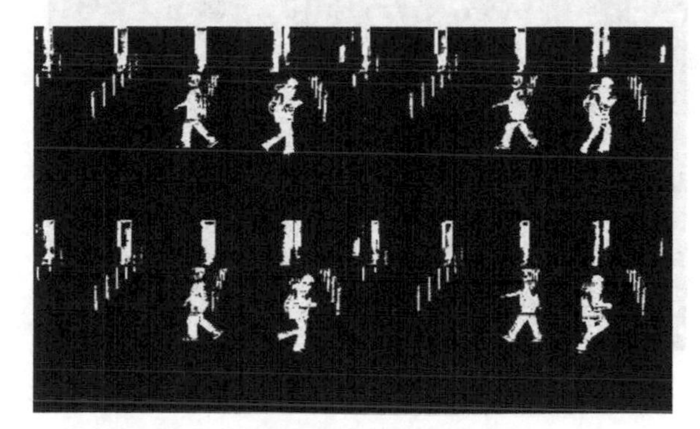

Fig. 6. Result of Background Subtraction

4.5 Noise Elimination

In Fig 6 the white pixels obtained from stationary objects are considered as noise and we can see that, these noise pixels are present in the background. These noise pixels

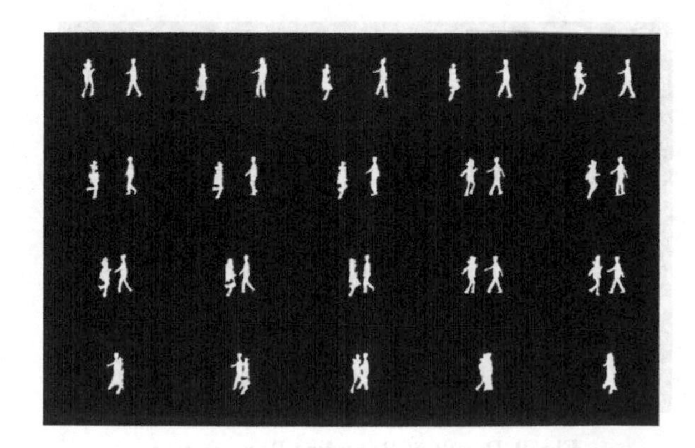

Fig. 7. Result of Noise Elimination

are to be removed from background in each frame. Noises can be removed using morphological operations. After these operations, the noise pixels are removed and the resulting frames are shown in Fig 7.

4.6 Object Separation

Now the object detection results are applied to all the frames. Thus we obtain the result as shown in figure 8. Here the object is separated from the background and so the backgrounds are seemed as black.

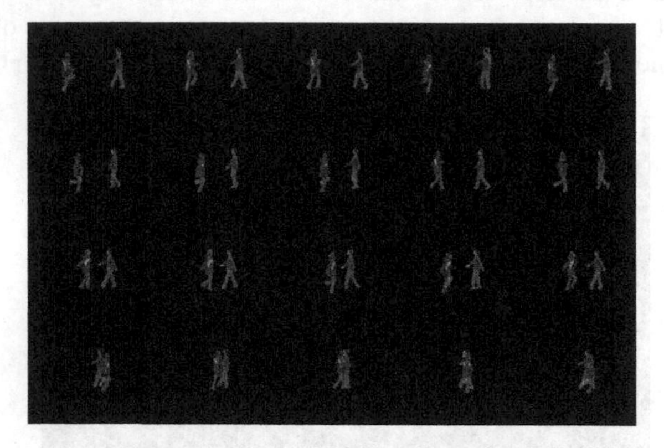

Fig. 8. Result of object separation

4.7 Bounding Box for Blob and Activity Detection

The main purpose of creating bounding box is to track and detect the moving object in the frame and to detect the human activities. If more number of people is present in the frame, there will be number of bounding boxes present in the frame. The width, height, position, distance between the boxes and number of boxes in the frames can

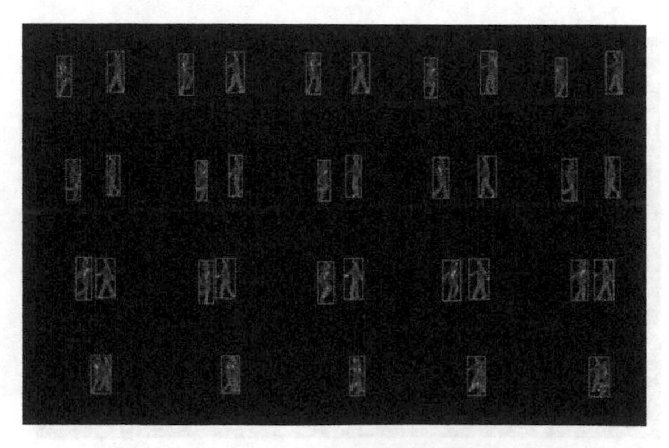

Fig. 9. Result of Bounding Box creation

be used to identify many activities of human. The result of bounding box creation in each frame is shown in figure 9. The shrinking of bounding box is used to detect the object hiding and human walking activity is also detected here and it is shown as result.

5 Conclusion and Future Possibility

Today, computer vision system has been used in many sites. All video surveillance system used in this field can be used for many purposes without troubling hardware but gives minor modification in the software part. Our system can detect human hiding and human walking using a bounding box concept and it is adjustable for detecting many applications such crowd detection, crowd density estimation and much other activity detection. In future, this concept can be extended to use for activity detection in crowd. The attractiveness of this system is that computational complexity of the algorithms is less.

References

1. Charfi, I., Miteran, J., Atri, M.: Definition and Performance Evaluation of a Robust SVM Based Fall Detection Solution. In: International Conference on Signal Image Technology and Internet Based Systems. IEEE (2012)
2. Karpagavalli, P., Ramprasad, A.V.: Estimating the density of people and Counting the number of people in a crowded environment for Human Safety. In: International Conference on Communication and Signal Processing. IEEE (2013)
3. Ashok Kumar, P.M., Vaidehi, V., Chandralekha, E.: Video Traffic Analysis For Abnormal Event Detection Using Frequent Item Set Mining. In: International Conference on Recent Trends in Information Technology. IEEE (2013)
4. Mahendran, S., Vaithiyanathan, D., Seshasayanan, R.: Object Tracking System Based on Invariant Features. In: International Conference on Communication and Signal Processing. IEEE (2013)
5. Sanna, A., Montuschi, P.: On the computation of groups of bounding boxes for fast test of objects intersection. In: IEEE (1995)
6. Madden, C., Piccardi, M.: Detecting Major Segmentation Errors for a Tracked Person Using Colour Feature Analysis. In: 14th International Conference on Image Analysis and Processing, ICIAP 2007 (2007)
7. Xu, L.-Q., Anjulan, A.: Crowd Behaviours Analysi. In: Dynamic Visual Scenes of Complex Environment. In: IEEE (2008)
8. Lien, C.-C., Huang, Y.-L., Han, C.-C.: People Counting Using Multi-Mode Multi-Target Tracking Scheme. In: International Conference on Intelligent Information Hiding and Multimedia Signal Processing. IEEE (2009)
9. Qian, H., Mao, Y., Xiang, W.: Home Environment Fall Detection System Based on a Cascaded Multi-SVM Classifier. In: International Conference on Control, Automation, Robotics and Vision, Vietnam. IEEE (2008)
10. Perrin, N., Stasse, O., Lamiraux, F.: Real-time footstep planning for humanoid robots among 3D obstacles using a hybrid bounding box. In: International Conference on Robotics and Automation. IEEE (2012)

11. Lertniphonphan, K., Aramvith, S., Thanarat, H.: The Region-Based Distance of Oriented Gradient and Motion Direction for Human Action Classification. In: International Sumposium on Communication and Information Technologies. IEEE (2012)
12. Choeychuent, K., Kumhomtand, P., Chamnongthait, K.: An Efficient Implementation of the Nearest Neighbor Based Visual Objects Tracking. In: International Symposium on Intelligent Signal Processing and Communication Systems (ISPACS 2006), Japan (2006)
13. Suresh, S.: Chitra K.: Patch Based frame work for occlusion Handling. In: IEEE (2013)

Performance Analysis of PBlock Algorithm Implemented Using SIMD Model to Attain Parallelism

Disha Handa[1] and Bhanu Kapoor[2]

[1,2] Chitkara University, Himachal Pradesh Campus, Baddi, India
[2] College of Management and Technology, Walden University, Minneapolis, MN, USA
{disha.handa,bhanu.kapoor}@chitkarauniversity.edu.in

Abstract. Software applications are driving the growth of cloud computing era. The security of data over the networks is essential in order to enable cloud computing applications. The security of data is ensured through the use of various types of encryption algorithms. In the current cloud computing era we are witnessing the use of multi core processors which has enabled us to run security applications, simultaneously at both client and the server end. The encryption as well as decryption process of security algorithms is compute intensive and can take significant benefit from parallel implementations that can run on these multi core processors. Moreover these algorithms will consume more energy on uniprocessor systems due to the massive calculations they do, because there is a non-linear relationship between frequency of a core and power supply. This paper introduces a parallel version of Blowfish algorithm using Single Instruction Multiple Data model which is named as PBlock and its implementation on a Symmetric Multi Processor machine along with the results of performance gains that we have obtained on a number of benchmark examples.

Keywords: Parallel algorithms, Encryption, Blowfish, Feistel network, Symmetric, Block cipher.

1 Introduction

To secure information communications over the network, different encryption algorithms have been used from time to time. The encryption algorithms [1] are further categorized into two broad categories: Symmetric and Asymmetric. In symmetric algorithms, the key is common to both encryption and decryption process. Some symmetric block cipher algorithms include DES, 3-DES, AES and Blowfish. RC4 is a symmetric stream cipher algorithm [20]. Asymmetric algorithms use two dissimilar keys for encryption and decryption. The public-key infrastructure-based algorithm such as RSA is an example of an asymmetric encryption algorithm. Apart from the security level, speed of the encryption algorithm is also a very important aspect in the cryptographic world. A slow algorithm can drastically affect the speed of entire application and condenses its effectiveness. Power consumption by electronic devices such as computer systems is another challenging issue

S.C. Satapathy et al. (eds.), *Emerging ICT for Bridging the Future – Volume 2*,
Advances in Intelligent Systems and Computing 338, DOI: 10.1007/978-3-319-13731-5_9

that has evolved as a significant concern at the individual level as well as the community level (the heat produced by these electronic systems raises the temperature of green house gases). In multi core processor models, applications can be executed on n number of cores where n can vary, and these cores can operate at diverse frequencies. The overall performance and power cost of a parallel algorithm will depend on different parameters such as the number of cores used by an algorithm, frequency of cores, and formation of the parallel algorithm.

Sequential security algorithms can be made faster using parallelization. Fortunately, with the advent of parallel processors in computing, we now have easily available means to parallelize the algorithms to make them faster. The symmetric multiprocessors such as those from Intel and AMD can be used in conjunction with parallel programming APIs such as the OpenMP [2][7][10] to make security algorithms parallel and faster. It is feasible to use parallel algorithms for any of the cryptographic techniques currently in use.

There have been many attempts to parallelize security algorithms [3][6-12][21] most of which were done in order to make parallel hardware implementations of security algorithms. Recently, Kama Beady, Victor Pena and Chen Liu [19] presented a pipelined approach for *"High speed execution of Blowfish block cipher Algorithm using Single-Chip Cloud Computer (SCC)"*. In this model, each core is responsible to perform single round of computations and then data is sent to next core for next round computations. According to the paper this approach is 27X faster than the sequential one. However, this approach has the limitation that the algorithm cannot use large data chunks as input due to communication overheads and latency associated with this model. This research has used an experimental processor having 48 cores created by Intel labs for research projects. Each active core is operated at 533 MHz of frequency and 0.9V voltage, which is for a different architecture comparing with ours.

There are many benefits of using parallel programming using Symmetric multiprocessor architecture. The basic idea behind the motivation for this research is to observe if complex security algorithms could broken down their responsibilities as tasks that can be executed in parallel successfully leading to performance gains. In this paper, using OpenMP Application programming interface (API) on multi core system, we present a parallel implementation for Blowfish algorithm. The algorithm divides a message into fix sized large blocks and encrypts these blocks concurrently on multi core machine. We will discuss the implementation of this parallel algorithm on a multi core machine along with the results of performance gains that we have obtained on a number of benchmark examples.

The rest of the paper is structured as follows: In section 2, we briefly introduce existing Blowfish algorithm. Section 3 will present PBlock, which is a parallel implementation of blowfish. In Section 4, the tools and techniques used for parallelization of PBlock will be discussed. Section 5 will present the machine setup that we have used for our experiments and Section 6 discusses the experimental results along with the speedups gained through the use of multi core machines on Linux. We conclude the paper with a discussion on some future work.

2 Blowfish Cryptographic Algorithm

Blowfish is a 64 bit symmetric block cipher security algorithm that uses a single key for encryption and decryption [1] [5] [6][4]. It is also known as Schneier's cipher as it was designed by Bruce Schneier in 1993. This algorithm is based on feistel network, which is a common method of converting any function into a permutation and is utilized in many symmetric key encryption algorithms. The algorithm has two sub parts: Key expansion and Data encryption. Key expansion function will convert the key bits into many sub key arrays that ranges up to 4168 bytes. And data encryption part will encrypt data using P-box: Permutation box, S-box Substitution box and the feistel function. Following algorithm shows the encryption process of the Blowfish algorithm:

The input is 64 bit block
Divide each block into 32-bit halves: Lblock, Rblock
For I=1 to 16
Lblock = Lblock XOR pi
Rblock =F (Lblock) Xor
Rblock Swap Lblock and
Rblock
End For
Swap Lblock and Rblock again to undo the last swap
Then, Rblock=Rblock XOR p17 and Lblock=Lblock XOR p18.
Finally, recombine Lblock and Rblock to get the cipher

Fig. 1. Encryption process in the Blowfish algorithm

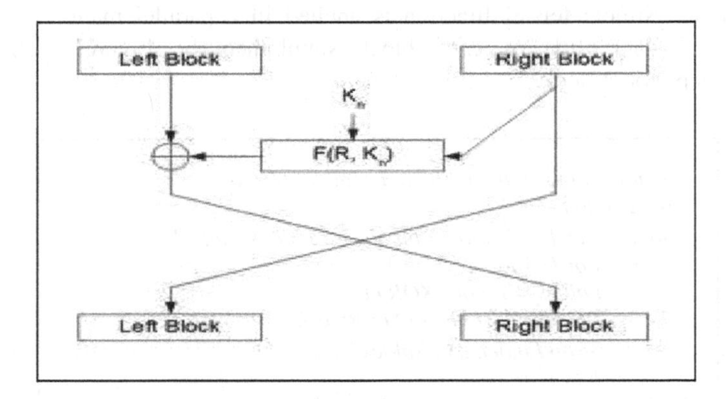

Fig. 2. Functionality of feistel network

Above mentioned encryption process is based on two key methods: *feistel network* and *F function,* where feistel network is the process related to transforming any function into a permutation and F function is a commonly used function in Blowfish. It necessitates a 32 bit input data to be divided into four eight bit blocks. Each block references the S-Box and each entry of the S-box output a 32 bit data. The output of S-box 1 and S-box 2 are added first and then result is xored with S-box 3. Finally, S-box 4 is then added to the output of the Xored operation and it provides a 32 bit data as output.

3 PBlock Cryptographic Algorithm

Depending on the architecture of cryptographic algorithm, there are numerous techniques to attain parallelism. Based on the architectural study of Blowfish security algorithm, we proposed a single instruction multiple data (SIMD) approach in order to get data parallelism on symmetric multiprocessors (SMP) platform. The resulted parallel version is named as PBlock, which is acronym for Parallel Blocks. According to the SIMD model, Single instruction for multiple data sets is executed in parallel on symmetric multiprocessor machine. In this model, single instruction is the complete 16 round function and multiple data is the 64 bit blocks.

3.1 Algorithm Framework

The algorithm is divided into two sub parts: key generation and data encryption. We have applied the parallel mechanism only on data encryption because data in security algorithm is subject to increase. In Key generation process, key is first expanded to 4168 bytes to make it complex [13]. For encryption, the plain text is divided into fixed length blocks. All blocks execute concurrently in order to achieve encryption and decryption. Simple feistel function is applied in a parallel manner to achieve encryption and decryption of each block simultaneously. Figure 3 shows the algorithmic steps for PBlock.

```
Divide plaintext into 64 bit n number of blocks
For I=0 to n-1:
Divide each block into 32-bit halves: Lblock, Rblock
        For I=1 to 16
        Lblock = Lblock XOR pi
        Rblock =F (Lblock) Xor Rblock
        Swap Lblock and Rblock
        End For
    End For
Swap Lblock and Rblock again to undo the last swap
Then, Rblock=Rblock XOR p17 and Lblock=Lblock XOR p18.
Recombine Lblock and Rblock to get the cipher text
Finally recombine the output of all blocks to get cipher text
```

Fig. 3. Algorithm used to encrypt individual block

3.2 Design of Parallel Feistel Network

Parallel feistel network is designed with the basic idea of data parallelism. The data is divided into fixed length blocks. Each block is further divided into Right and Left block. For example: If we have 16,77,7216 bits of data. As we have 64 bit block size, the number of parallel blocks will be 16,77,7216 /64 i.e. 26, 2144.Now each 64 bit block is further divided into 32 bit left half and 32 bit right half. 26, 2144 blocks will run parallel to encrypt 16,77,7216 bits of data. At last step each block is concatenated to make complete cipher text. In this model there is no dependency between the individual blocks. Figure 4 shows the graphical representation of parallel feistel network.

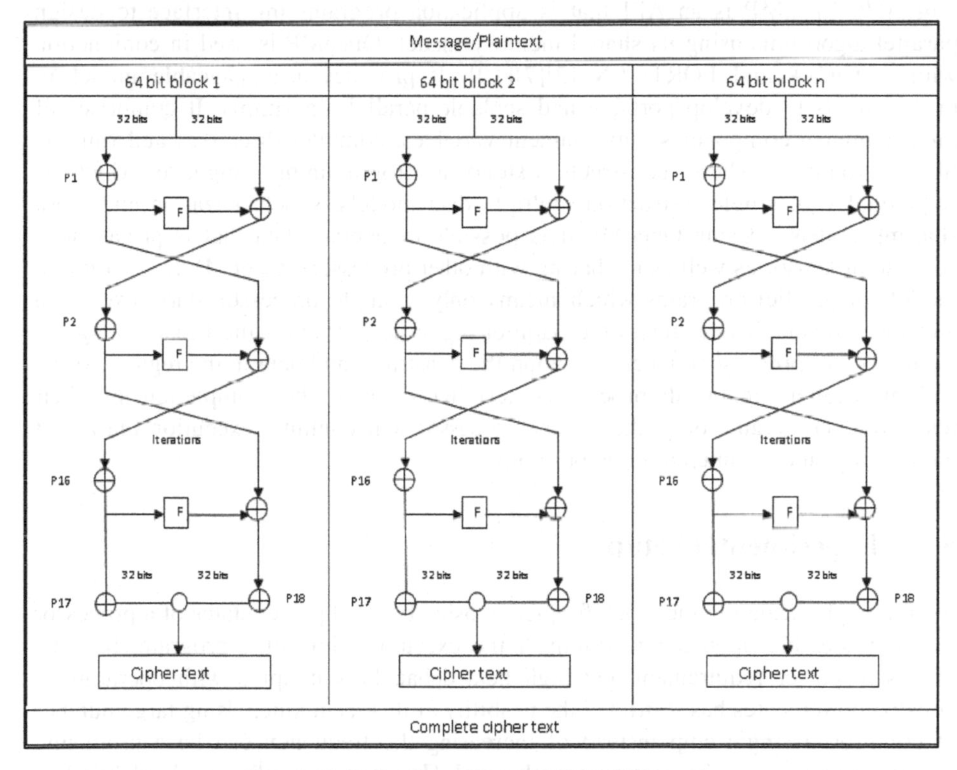

Fig. 4. Functionality of Parallel Feistel Network

4 Techniques Used For Parallelism

There are different steps involved in parallelization of an algorithm. First and foremost condition for parallelism is that the algorithm must be architected in a way so that it can support parallelism. One must consider following points before executing algorithm in parallel:

- Number of time consuming functions in sequential implementation.
- Any data dependency

Above questions can be answered with the help of a profiler tool. The code profiler will tell you how many functions are compute intensive in your algorithm, how many functions are calling other functions and how many times and where are loop dependencies. For our research, we have used GNU's Gprof.

Gprof: This is a profiler program that collects and arranges statistics on programs [14][15]. It generates "gmon.out" data file having all details of program like number of executions of a function, number of times a function calls some another function in the program. It provides many options to get details about the program.

OpenMP: OpenMP is an API that is application programming interface to design parallel algorithms using its shared memory model. OpenMP is used in conjunction with C/C++ and FORTRAN [2][7][10]. It provides a manageable model to programmers to develop portable and scalable parallel algorithms. It comprises of three common components: environment variables, compiler directives and runtime library routines. These constructs extend a programming language which is sequential with single instruction multiple data models, synchronization and work sharing models. Using OpenMP it is possible to operate data that is private to a specific processor as well as in sharing with other processors. OpenMP uses fork join model for parallel programs which means only a single processor starts execution and the moment it encounters the parallel region it distributes the tasks among the team of other processors depending upon the constructs and data in the region. At the end of parallel region, all processors terminated after the completion of their respective tasks and only the master processor will continue execution until next parallel region encountered in the program.

5 Experimental Setup

Even though escalating the clock frequency using boosted power states of a processor is a considerable approach to diminish the execution time of a program, but, the excessive power requirement and high heat debauchery coupled with operating at boosted power states has restricted the usability of this technique. Using large number of cores on a single chip instead of increasing the frequency can be a promising technique to enhance the computational speed. However, according to Amdahl's law [15], potential speedup for a program is restricted by the fraction of the program that cannot be parallelized to execute on Symmetric multiprocessor machine. This means the performance gains are highly dependent on software algorithms and their implementation.

Symmetric multiprocessor is a single computing component with two or more independent actual cores that can run multiple instructions at the same time and increase overall speed for programs. Multi core processors are widely used in market for various purposes. For our research, we have set up the machine with the following configuration:

- Processor - AMD FX(tm) - 8320, eight core processor running @ 3500 MHz
- RAM - 8 GB
- System Type - 64 bit operating system, x64- based processor
- Operating System - Linux/bunt 12.04 version
- OpenMP - 4.0
- Compiler- GNU's GCC
- Programming Language - C with OpenMP

6 Results and Analysis

In our experiment, we first executed the sequential blowfish cryptographic algorithm using single core to evaluate the execution time. The sequential results served as the base parameter for comparison with the results of our improved parallel algorithm PBlock. The following tables demonstrate the performance gains after executing both sequential and parallel algorithm on diverse cores on Linux platform.

Table 1. Speedup gained by eight cores on LINUX platform

File Size [MB]	Execution Time using Single core[ms]	Execution Time using Eight Cores[ms]	Speedup
0.5	0.2308	0.04607	5X
1	0.4612	0.08144	5.6X
1.5	0.6956	0.11934	6.2X
2.5	1.1654	0.18861	6.2X
5	2.5608	0.36410	6.4X

Table 2. Execution time in ms taken by PBlock on multiple cores with different file size

File Size [MB]	Single core	Two Cores	Four Cores	Six Cores	Eight Cores
0.5	0.2308	0.13893	0.07394	0.05004	0.04607
1	0.4612	0.27557	0.14276	0.09527	0.08144
1.5	0.6956	0.41135	0.21143	0.13817	0.11934
2.5	1.1654	0.68537	0.34257	0.22877	0.18863
5	2.5608	1.2804	0.6402	0.42681	0.36410

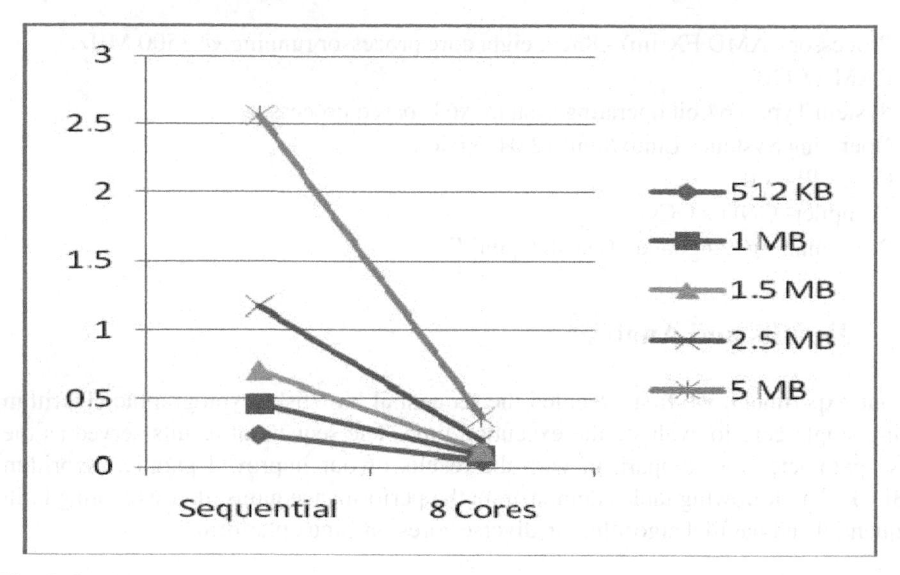

Fig. 5. Graphical representation of speedup gained using 8 core machines on LINUX platform

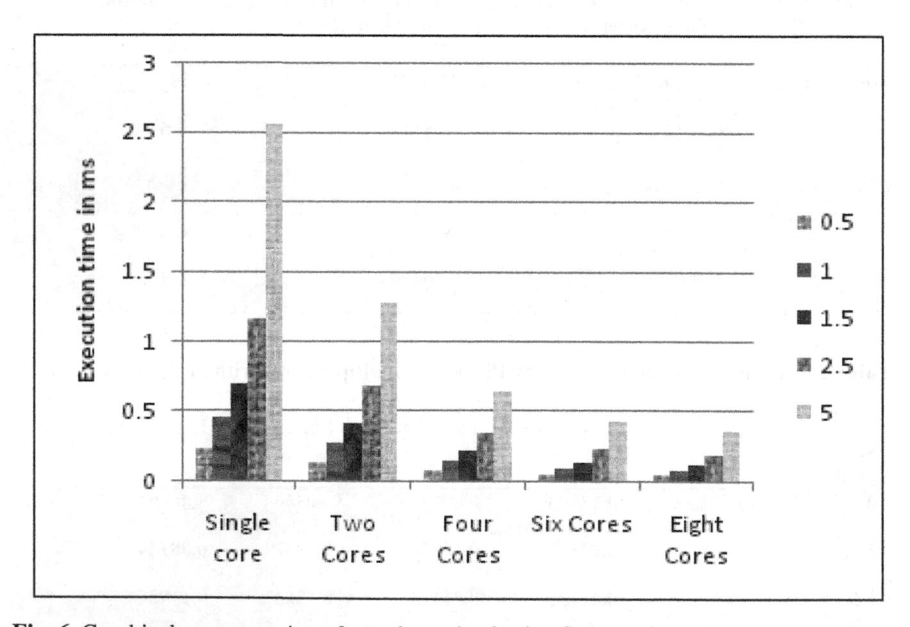

Fig. 6. Graphical representation of speedup gained using Symmetric multiprocessor systems

In table 1 the first column shows the size of the file used for encryption. The second column shows the runtimes when we use a single core and the third column shows the runtimes when we use eight cores. The last column shows the overall speedup gained using multi core processor. It is noticeable from the above table that speedup increases as we increase the file size which was not in the case of "*High speed execution of Blowfish block cipher Algorithm using Single-Chip Cloud Computer (SCC)*"[19].

The data in the second table details the results when we use different number of cores. The performance improvement as we increase the number of cores is evident from this table. We have also plotted the speedups for the eight core situation as well as the speedups obtained for various number cores in Figure 5 and Figure 6.

7 Conclusion and Future Work

In today's complex and cloud computing era, there is an ever increasing requirement to secure data over the channel. To congregate this necessity of data security, security algorithms are getting more complex in terms of mathematical calculations. Conversely, the requirement for a consistent and fast algorithm is an additional necessity of current era. By using multi-core systems, we can achieve noteworthy improvement in performance with respect to execution time. However, potential performance gains are restricted by the fraction of code that can be parallelized, because some portion of code needs to execute sequentially. Moreover the potential speedup is possible only if the calculation overhead of the code is greater than the communication overhead. In this paper, to achieve parallelism on the eight core symmetric multiprocessor we proposed a SIMD model for the implementation and execution of the Blowfish cryptographic algorithm and resulted parallel version is termed as PBlock. As our results proved, this model yields a faster execution time of the program on the multi core systems. This is the initial technique to parallelize blowfish algorithm, which is quite fast, and simple having no communication overhead and latency associated with cores. This parallel algorithm is scalable as there is no need to make changes in code with the increase of cores. Rather it will make it faster as compared to pipelined model. Future work includes the implementation of hybrid parallelism on this algorithm to achieve optimization.

References

1. Stallings, B.: Computer Security: Principles and Practice. Prentice Hall, Upper Saddle River (2008)
2. Chapman, J., Der Pas, R.V.: Using OpenMP: Portable Shared Memory Parallel Programming. MIT Press (2008)
3. Damrudi, M., Ithnin, N.: State of the Art Practical Parallel Cryptographic Approaches. J. Basic and Applied Sciences, 660–677 (2011) ISSN 1991-8178
4. Schneier, B.: Description of a New Variable-Length Key, 64-bit Block Cipher (Blowfish). In: Anderson, R. (ed.) FSE 1993. LNCS, vol. 809, pp. 191–204. Springer, Heidelberg (1994)
5. Elmina, D., Kader, H., Hadhoud, M.: Evaluating the performance of symmetric encryption algorithms. Int. Journal of Network Security 10(3), 213–219 (2010)

6. Krishnamurthy, G.N., Ramaswamy, V., Leela, G.H., Ashalatha, M.E.: Performance enhancement of Blowfish and cast-128 algorithms and security analysis of improved blowfish algorithm using AVALANCHE effect. Int. Journal of Computer Science and Network Security 8(3), 244–250 (2008)
7. https://software.intel.com/en-us/articles/getting-started-with-openmp
8. Dongara, P., Vijaykumar, T.N.: Accelerating private-key cryptography via multithreading on symmetric multiprocessors. In: Proc. IEEE Int'l Symp. Performance Analysis of Systems and Software. IEEE Press (2003)
9. Vanitha, M., Selvakumar, R., Subha, S.: hardware and software implementation for highly secured modified wired equivalent privacy (MdWEP). J. Theoretical and Applied Information Technology 48(2) (2013)
10. http://openmp.org/wp/2013/12/tutorial-introduction-to-openmp/
11. Kholidy, H.A., Alghathbar, K.S.: Adapting and accelerating the stream cipher algorithm "RC4" using "ultra gridsec" and "HIMAN" and use it to secure "HIMAN" data. J. Information Assurance and Security, 274–283 (2009)
12. Tsoi, K.H., Lee, K.H., Leong, P.H.W.: A massively parallel RC4 key search engine (With FPGA). In: Proc. Tenth Annual IEEE Symposium on Field-Programmable Custom Computing Machines. IEEE Press (2002)
13. Sastry, V.U.K., Anup Kumar, K.: A Modified Feistel Cipher Involving Substitution, shifting of rows, mixing of columns, XOR operation with a Key and Shuffling. Int. J Advanced Computer Science and Applications 3(8), 23–29 (2012)
14. http://www.ibm.com/developerworks/library/l-gnuprof.html
15. http://sourceware.org/binutils/docs/gprof/
16. Karthigai Kumara, P., Baskaranb, K.: An ASIC implementation of low power and high throughput blowfish crypto algorithm. Microelectronics Journal 41, 347–355 (2010)
17. Weerasinghe, T.D.B.: Improving throughput of RC4 algorithm using multithreading techniques in multicore processors. Int. J. Computer Applications 51(22), 45–51 (2012) ISSN 0975-8887
18. Karthigaikumar, P., Baskaran, K.: Partially pipelined vlsi implementation of blowfish encryption/decryption algorithm. Int. J. Image and Graphics 10, 327 (2010)
19. Ebadi, K., Pena, V., Liu, C.: High-Performance Implementation and Evaluation of Blowfish Cryptographic Algorithm on Single-Chip Cloud Compute: A Pipelined Approach. In: Proc. International Conference on Applied and Theoretical Information Systems Research (2012)
20. Handa, D., Kapoor, B.: PARC4: High performance implementation of RC4 cryptographic algorithm using parallelism. In: Proc. International Conference on Optimization, Reliabilty and Information, pp. 286–289 (2014)
21. Handa, D., Kapoor, B.: State of the Art Realistic Cryptographic Approaches for RC4 Symmetric Stream Cipher. IJCSA 4(4), 27–37 (2014)

Blind Watermarking Technique for Grey Scale Image Using Block Level Discrete Cosine Transform (DCT)

Ravi Tomar, J.C. Patni, Ankur Dumka, and Abhineet Anand

University of Petroleum & Energy Studies, Dehradun, India
ravitomar7@gmail.com

Abstract. This paper presents a robust watermarking technique to copyright an image. Proposed technique is totally based on DCT and is different from most of the available techniques.We are embedding block-wise watermark against the noise, filtering and cropping attack. Before embedding the watermark for any host image we must calculate the gain factor. According to our approach gain factor will vary for two different host images . The experimental results show that in addition the invisibility and security, the scheme is also robust against signal processing.

Keywords: Blind watermarking, DCT, IDCT, encrypted watermark, digital watermark.

1 Introduction

As with the spread of multimedia over the internet, we are facing lots of issues in copyrighting multimedia contents as a digital duplicate copy can be very easily generated hence resulting in failure to protect ownership of content. In this paper we will be emphasizing & providing a mechanism to protect the ownership of digital images by invisible watermarking. Watermarking is a process of adding some additional information within the image that would be having authenticity of the image and can be retrieved to check its ownership. Watermarking can be visible or invisible, in visible watermarking the copyright text or image is overlapped over visual section of image whereas in invisible watermarking all the data used to copyright image is stored/embedded in the image data itself so as it cant be seen to a normal user but can be retrieved by its owner. This type of watermarking is very useful & is in practice to protect ownership of image. We use different types of algorithms to implement the same but there is no algorithm which could provide robustness to all the attacks, So we would consider to provide robustness for some specific attacks. Using this algorithm we try to achieve invisible digital watermarking that it should be almost unperceable by ordinary human senses also it should provide security from removing it by an unauthorized user. There should be any password mechanism to access or modify this secure information. unobtrusive, readily extractable robust, unambiguous, innumerable Techniques for watermarking as available today can be divided into two

S.C. Satapathy et al. (eds.), *Emerging ICT for Bridging the Future – Volume 2*,
Advances in Intelligent Systems and Computing 338, DOI: 10.1007/978-3-319-13731-5_10

main groups: Firstly those which map image directly onto the spatial domain and other which operate in a transformed domain. e.g. the frequency domain. Out of the two The frequency-domain watermarking approaches[8][10][11] are the most popular for robust image watermarking. here the image is transformed-via some common frequency transform. Watermarking which is achieved by altering the transform coefficients of the image. Commonly used transforms are discrete cosine transform (DCT), or the discrete fourier transform (DFT), and the discrete wavelet transform (DWT). We propose watermarking algorithm using the DCT coefficients for embedding. In order to increase robustness of the process, the high-frequency coefficients are selected for embedding. To give copyright to the image the small amount of information in terms of an image(it may contain signature, text ,logo or even own image) is embedded into the original image. A standard approach is proposed by Feng Liu and Yongtao Qian in[1]. In this approach the original image is bisected into blocks according to watermark's size, each block corresponds to each pixel value of watermark. Secondly, the DCT is applied in each block twice and form new blocks. Then, SVD on the each new block to get matrices U, S and V for each block and the first value of each matrix S is collected together to form a new matrix. Apply SVD on the new matrix again to get the S matrix .The pixel value of watermark is embedded into the new S matrix through some method. And the watermark can be detected with the original image embedding. Shang-Lin Hsieh et al.[3] presented a watermarking scheme for copyright protection of color images. By the scheme the requirement of imperceptibility and robustness for a sensible watermarking scheme. By the experiment it has been demonstrated that the resistance of their scheme in opposition to many attacks for example cropping, scaling, and JPEG compression. Additionally, the capability of the scheme to extract unique features from different images. Raja'S. Al Omari et al. have proposed an approach in[6]. According to this algorithm it uses the Discrete Cosine Transform (DCT) for the watermarking process. It is compared with the well known Spread Spectrum Algorithm. An asymmetrical watermarking method for copyright protection that satisfies the zero knowledge principle with the objective to defeat the weaknesses of contemporary symmetric watermarking methods is offered by Jengnan Tzeng et al.[8]. Owing to the copyright protection in opposition to piracy of digital images an invisible and blind watermarking scheme is offered by M.A. Dorairangaswamy[2]. Most existing watermarking schemes embed their watermarks into the whole image without considering the image content. However, some parts of the image are more important than the rest of the image (i.e. the region of background ROB) in some applications. The more important parts are often called the ROI (region-of interest)[10]. A two-phase watermarking scheme which extracts together the grayscale watermark and the binary one from the protected images was presented by Ming-Chiang Hu et al.[4]. Ming-Shi Wang and Wei-Che Chen[5] proposed a digital image copyright protection scheme based on visual cryptography (VC) and singular value decomposition (SVD) techniques. The SVD is applied at first in their image to a host image to build a master share. In the scheme, the secret image is embedded with no alteration of the host image. It is not essential to employ the original host image and the help of computers to extract the out of sight secret image in addition.

2 Proposed Scheme

Proposed technique is totally based on Discrete Cosine Transform(DCT).we implement block-wise watermark embedding against noise and filtering attacks, however acceptable results are obtained for cropping attacks also but we keep it for future work. We used gain factor calculation for varying between any two host images, this will result in a randomized bits onto which we can append our invisible watermark.In proposed algorithm we used DCT coefficients for embedding the copyright protection watermark. Figure 1 shows the block diagram of entire process which results in a embedded watermark image that carry copyright protection image within.we have also used a special key which will be used to encrypt & decrypt the watermark to from the image.

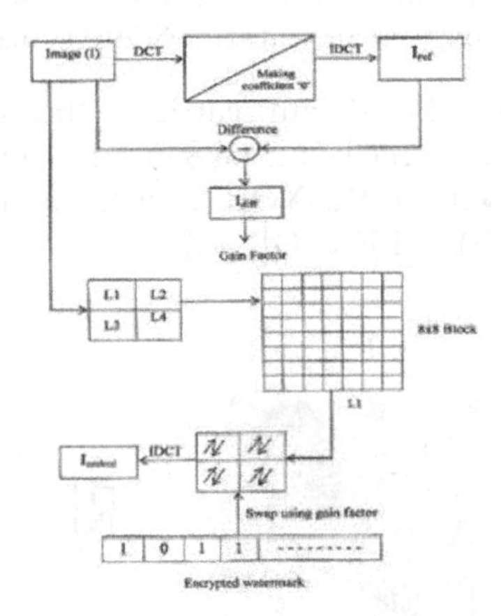

Fig. 1. Block Diagram for Encryption and embedding watermark in image

2.1 Watermark Embedding

A gray-level image with size of 256 X 256 is used to as the original image and logo of size 32 X 32 is used as the watermark. Steps involved in encryption are detailed below:

2.1.1 Gain Factor Calculation

To calculate gain factor we need to follow these steps:

1. As in case of DCT all the low frequencies are scattered at upper left side of matrix, so we take upper left dct values of transformed original image.this is achieved by making a zero filled matrix to be filled by upper left values of dct image.

2. then we will create a idct image of the above produced upper left matrix.this will be called as reference image.
3. Now, we will calculate the difference between the original image and reference image and the difference would be a matrix with some pixel values.This will be called as difference matrix with some pixel values now in order to calculate gain ratio we extract maximum and minimum values from these pixel values.The difference between these wo values will give gain ratio.

- Gain factor = $(max_{difference} - min_{difference})/2$
- Gain factor will be used to find the change in pixel values.

2.2 Watermark Encryption

Watermark and encryption both words differ in form of visibility of embedded information, we are using them combinedly because we are doing hidden watermarking and this hidden watermark is encrypted by a key which later will be used to extract this information(watermark) from the image.

So, Here are the steps involved in watermark encryption:

- We will Generate a pseudo random matrix of the size same as that of watermark size by using a secret key . i.e S = s(i) : $0 \leq i \leq n; s(i)\epsilon[0;1]$
- Then the watermark will be Bit-XORed with the above generated pseudo random matrix S,i.e.S(i) = i EX -OR s(i)

RT

Fig. 2. Original image in which watermark is to be embedded and The watermark which is to be embedded

2.3 Block Division

Now comes the part where we will be finding space within the image to insert our watermarked bits. Here we will take the original image I and will perform the blocking operation i.e. divide the host image into four blocks of size 256 x 256. And within each 256 x 256 block we will divide into 1024 blocks of size 8 8.

2.4 Watermark Embedding

In this section we will discuss how to embed the watermark into the original image whose ownership is to be protected, by now we have a encrypted watermark which is generated by pseudo random matrix ,the gain ratiom and the original image.

Here are the steps involved in embedding watermark:

- As the first most step needed in matlab we convert encrypted watermark image in to vector. Then following operation's will be performed for each 8 8 blocks.
- Now divide these 8x8 blocks into further four equal blocks.
- Skip the 1st block as it is so as to maintain good resolution of the image.
- Take the DCT of remaining three blocks,If watermark bit is 1 (this is the message bit which is obtained after XOR-cd watermark & pseudo random number) then do following Check for r(location value) > c(location value) : r €{15;43; 47}; c €{22; 50; 54}

Table 1. Reference table for changing pixel values

1	2	3	4	5	6	7	8
9	10	11	12	13	14	15	16
17	18	19	20	21	22	23	24
25	26	27	28	29	30	31	32
33	34	35	36	37	38	39	40
41	42	43	44	45	46	47	48
49	50	51	52	53	54	55	56
57	58	59	60	61	62	63	64

  - r(location value) = r(location value) + gain factor/2
  - c(location value) = c(location value) - gain factor/2

  Else swap the r(location value) with n(location value) then
  - r(location value) = r(location value) + gain factor/2
  - c(location value) = c(location value) gain factor/2

  if watermark bit is 0 then do following
  - Check if r(location value) > c(location value) : r €{15;43; 47}; c €{22; 50; 54}
  - r(location value) = r(location value)-gain factor/2
  - c(location value) = c(location value) +gain factor/2
  - Else swap the r(location value) with c(location value)then
  - r(location value) = r(location value)-gain factor/2
  - c(location value) = c(location value)+gain factor/2
- Take IDCT of 8 x 8 block and We will repeat these above step for each 8 x 8blocks, and finally we will get watermarked image.

2.5 Watermark Extraction

Fig. 3. Final watermarked image

The good part of this invisible watermark extraction process is that it does not require the original image. Embedded watermark can be extracted from watermarked image using only the key which was used to encrypt the watermark and decryption algorithm.

Input: A Watermarked Output image: Watermark w- Transform the watermarked image in to DCT coefficient.- Convert DCT coefficient matrix in upper triangular matrix and perform inverse discrete cosine transform (IDCT) and obtain its reference image.- Divide watermarked image in to 4blocks of size 256 x 256.And within each 256 x 256 block we will divide into1024 blocks of size 8 x 8.Take the DCT of 8 8 within each 8 x 8block do following:

Divide 8 x 8 blocks in to 4 blocks. Skip the 1st block as it is. For remaining 3 blocks do We will take two variable c1 and c2 and check Check if m(value) > n(value) then increment c1 by 1 else increment c2 by 1 And check if c1 > c2 then watermark bitmust be 1 else 0.

3 The Experimental Results

In the following experiments, a gray-level Lena image with size of 512 X 512 is used to as the original image and a watermark logo containg RT is used to as the watermark information.After the watermark had been embedded into the original image, the PSNR (Peak Signal to Noise Ratio) of the watermarked image is 5.6617 dB.as we know for identical images PSNR tends to zero. Proposed technique is totally based on Discrete Cosine Transform (DCT).we implement block-wise watermark embedding against noise and filtering attack, As in case of DCT all the low frequencies are scattered at upper left side of matrix, so

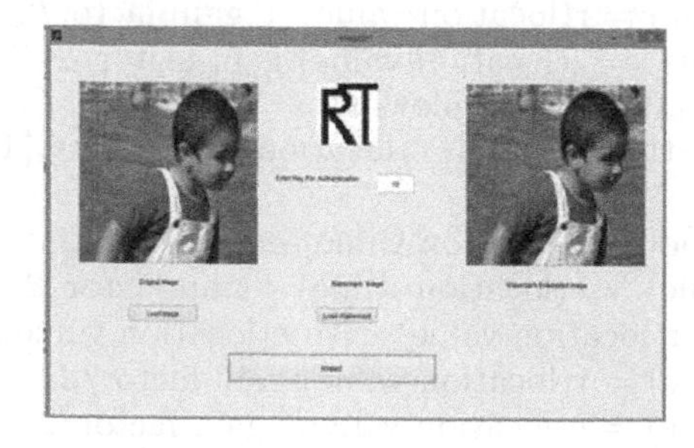

Fig. 4. Interface to embed image with watermark

Tests	Lena Image				Baby Image			
	Gaussian Noise	Smoothing Filter	Sharpening	Salt and pepper Noise	Gaussian Noise	Smoothing Filter	Sharpening	Salt n pepper Noise
PSNR	61.3111	51.4321	54.4946	68.1332	59.6577	51.1573	55.2668	68.1266
Bit Error Rate	1	0.2197	1	0.9795	1	0.1504	1	0.9922
Normalized Correlation	0	0.8138	0	0.0156	0	0.8746	0	0.0078

Fig. 6. Following table depicts the results of applying various filters and adding noise

we take upper left dct values of transformed original image.this is achieved by making a zero filled matrix to be filled by upper left values of dct image. Finally after watermark is embedded we check for PSNR(Peak signal to noise ratio) to describe the quality of watermarked image. In the following experiment, we use several image manipulations, including noise addition and applying filtering, on the watermarked images to evaluate the robustness of the proposed scheme.

Here we load the original image then load watermark and then after clicking we get watermarked image.We use authentication key to encrypt watermark by a pseudo random number XORed with, figure 5 shows what happens if we try decrypting using wrong authentication key which was used during encryption.

Fig. 5. Distorted watermark extracted with wrong input secret key

Fig. 7. Recovered watermark from Gaussian noise added lena image and Recovered watermark from Smoothing Filter applied lena image

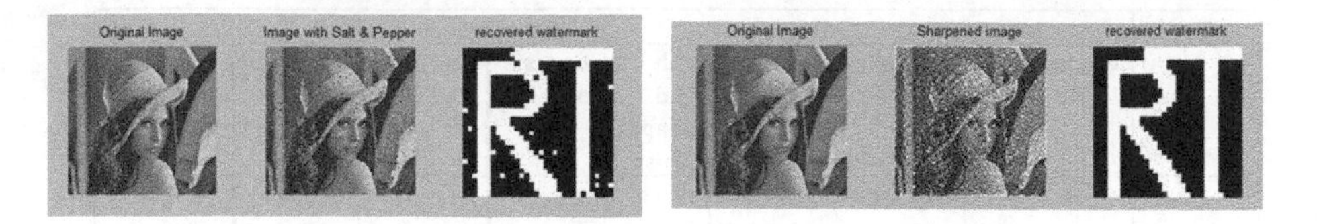

Fig. 8. Recovered watermark from Salt and pepper noise added image and Recovered watermark from adding sharpening filter lena image

Decryption Results: We have taken some measures to evaluate the efficiency of algorithm. They are described below

PSNR(Peak Signal to Noise Ratio): PSNR is most commonly used to measure the quality of reconstruction of lossy compression codecs (e.g., for image compression). The signal in this case is the original data, and the noise is the error introduced by compression.

Normalized Corelation: The quality of the extracted watermark is evaluated using term Normalized cross correlation (NC). The ideal value of the NC is 1 which means the original and the extracted watermarks are exactly the same.

Bit Error Rate: The bit error rate or bit error ratio (BER) is the number of bit errors divided by the total number of transferred bits during a studied time interval.

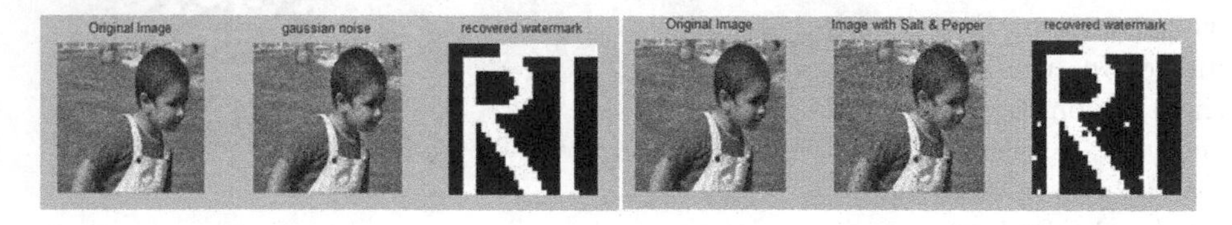

Fig. 9. Recovered watermark from Gaussian noise added baby image and Recovered watermark from Salt and pepper noise added baby image

Figure 7,8,9,10 shows the original image, added noise applied filter & recovered watermark. We have used two images for this purpose and results received are fairly good.

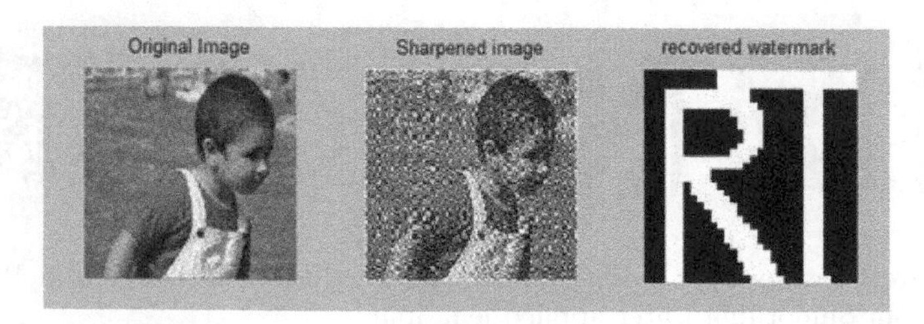

Fig. 10. Recovered watermark from adding sharpening filter baby image

4 Conclusion

A robust invisible watermarking technique is proposed in this paper. We tried to over-come the weak robustness problem that watermark embedded into the spatial domain can be destroyed easily after image processing attacks. The proposed scheme transform the original image in to DCT Coefficient by then, convert that matrix in to upper triangular and obtain its reference image by performing inverse discrete cosine transform. After the watermark is embedded in to the original image, the PSNR (peak signal to noise ratio) is used to evaluate the quality of the watermarked image.

References

[1] Liu, F., Qian, Y.: A Novel Robust Watermarking Algorithm Based On Two Levels DCT and Two Levels SVD. In: Third International Conference on Measuring Technology and Mechatronics Automation, pp. 206–209. IEEE (2011)

[2] Dorairangaswamy, M.A.: Protecting Digital-Image Copyrights: A Robust and Blind Water-marking Scheme. IEEE 9(4), 423–427 (2009)

[3] Hsieh, S.L., Tsai, I.J., Huang, B.-Y., Jian, J.J.: Protecting Copyrights of Color Images using a Watermarking scheme based on secret sharing and wavelet transform. Journal of Multimedia 3(4) (October 2008)

[4] Hu, M.-C., Lou, D.-C., Chang, M.-C.: Dual-wrapped digital water-marking scheme for image copyright protection. Computers & Security 26(4), 319–330 (2007)

[5] Wang, M.-S., Chen, W.-C.: Digital image copyright protection scheme based on visual cryptography and singular value decomposition. Optical Engineering 46(6) (2007)

[6] Al Omari, R., Robust, A.A.: Watermarking Algorithm for Copyright Protection, pp. 1–7. IEEE (2005)

[7] Tzeng, J., Hwang, W.-L., Chern, I.-L.: An Asymmetric Subspace Water-marking Method for Copyright Protection. IEEE Transactions on Signal Processing 53(2) (February 2005)

[8] Lou, D.C., Yin, T.-L.: Adaptive digital watermarking using fuzzy clustering technique. IEICE Transactions on Fundamentals E84-A(8), 2052–2060 (2001)

[9] Su, P.C., Wang, H.-J., Jay Kuo, C.-C.: Digital watermarking in regions of interest. In: IS T Image Processing lmage Quality lmage Capture Systems (PICS), April 25-28 (1999)

[10] Cox, I.J., Kilian, J., Leighton, F.T., Shamoon, T.: Secure spread spectrumwatermarking for multimedia. IEEE Transactions on Image Processing 6(12), 1673–1687 (1997)

[11] Hsieh, M.S., Tseng, D.-C., Huang, Y.-H.: Hiding digital watermarks using multiresolution wavelet transform. IEEE Transactions on Industrial Electronics 48(5)

Recognition and Tracking of Occluded Objects in Dynamic Scenes

Suresh Smitha[1,*], K. Chitra[2], and P. Deepak[3]

[1] Dept. of Computer Science,
Bharathiar University, Coimbatore, Tamilnadu, India
smithadeepak2006.sngce@gmail.com
[2] Department of Computer Science, Govt. Arts Collge, Melur, Tamilnadu, India
manikandan.chitra@gmail.com
[3] Dept. of Electronics and communication Engineering, SNGCE, Ernakulam, Kerala, India
deepakpeeth@gmail.com

Abstract. The object tracking in video sequences is required in high level applications such as video communication and compression, object recognition, robotic technology etc. However; such systems are challenged by occlusions caused by interactions of multiple objects. A visual tracking system must track objects which are partially or even fully occluded. The proposed system uses a robust object tracking algorithm to handle partial and full occlusions. Occlusion handling is based on the concept of patch based framework to identify the parts which are going to occlusion. An adaptive Gaussian Mixture Model is used to extract the visual moving objects and then the noise removal steps are performed. This paper describes a bounding box system for tracking the moving objects. The main objective of the proposed system is to track the human objects and detect occlusion by a patch concept during tracking.

Keywords: Occlusion, Patches, Background Subtraction, Centre of mass, Bounding box.

1 Introduction

Video Surveillance can be mainly used to analyze the video sequences to detect abnormal behavior, activities and other changing information of objects. Object tracking is an essential component of every intelligent video surveillance system. During human tracking, occlusion boundary identification is a tough task in numerous computer vision algorithms. The goal of our work is to detect all the objects using a single stationary camera and track the objects with their boundary patches that are going to formulate occlusion in a crowded environment. The visual surveillance system must record all the visual video data in a compressed manner. But the clarity of the compressed data drops out due to noise. The motion detection algorithm is needed especially in the video image analysis process. Video object detection and

* Corresponding author.

S.C. Satapathy et al. (eds.), *Emerging ICT for Bridging the Future – Volume 2*,
Advances in Intelligent Systems and Computing 338, DOI: 10.1007/978-3-319-13731-5_11

tracking are the most important steps used in the video surveillance processes. Two main approaches used for object detection in video surveillance systems are temporal differencing and background subtraction. Both approaches require one or more static cameras to capture the video. In these methods, moving object extraction is achieved by a pixel based threshold and an assumption made in such that the difference image between background model and current image exceeds the threshold or not.

Background subtraction is the most popular object detection method used in object tracking. Numerous methods are available for object tracking in the presence of occlusion. In work [1], Marques et al. proposed an appearance model to identify the people after the occurrence of occlusions. But this approach provides limited support in complex object interactions. In [2, 3] Chang and Dockstader used a fusion of multiple camera inputs to handle occlusion in multiple object tracking. A.D Jepson et al. [4] propose a target appearance model to characterize the occluded objects by their lost component. This algorithm works well when the statistical properties of occluded objects are happened to agree with their assumptions. Lipton et al. [5] described a method based on template matching and temporal consistency through object classification. But this model deals only partial occlusion. D. Comaniciu et al. [6] projected a blob based tracking techniques called mean shift tracking algorithm. This technique is popular due to its simple implementation. Khalid Housni [7] proposes a graph cut approach for integrated tracking and segmentation of moving objects using a moving camera. By using this graph cut method, the border of predicted object is refined. Sherin M. Youssef et al. [8] introduces a discrete wavelet transform to detect the moving objects by their colour and spatial information. This model can deal with small, partial and full occlusions. B.S.M.Madhavi et.al.[9] proposed a dynamic optimization threshold method to obtain moving objects. Contour projection analysis removes the effect of shadow and occlusion by centroid of each object. Valtteri Takla et al. [10] proposed a method to perform robust tracking that deals some instances of occlusion but cannot handle illumination changes. In work [11], Daw- Tung Lin employs a method called component shape- template particle filter for object tracking and they predicted the occlusion by kalman filter. Some of the above systems achieve very good results.

Our proposed system is overviewed in section 2. In section 3, we described a moving object segmentation using background subtraction method. In section 4 noise cancellation and object extraction methods are explained. In section 5 objects tracking by bounding box is described. The main parts of this work are presented in sections 6 and 7 where we described the concepts of patch creation and centre of mass to handle occlusion. Experimental result and discussions are provided in section 8.Finally, concluding remarks can be found in section 9.

2 System Overview

We propose to detect and track human objects in video surveillance system in the presence of occlusions using a single stationary camera. We believed that this system is applicable in a wide variety of circumstances. The entire system for occlusion detection is shown in fig1.First step is to capture the video and then it is converted in to frames for further processing. The background subtraction algorithm was performed in all image frames to extract the foreground object from background. After that, morphological operations were performed for noise cancellation. Then object separation is performed to extract foreground images. Then bounding box is created around each object for moving object tracking and the centre of mass is calculated for occlusion detection. The boundary patches with fixed size is incorporated for detecting the occluded patches in each object by calculating the distance between the patches of objects in different direction.

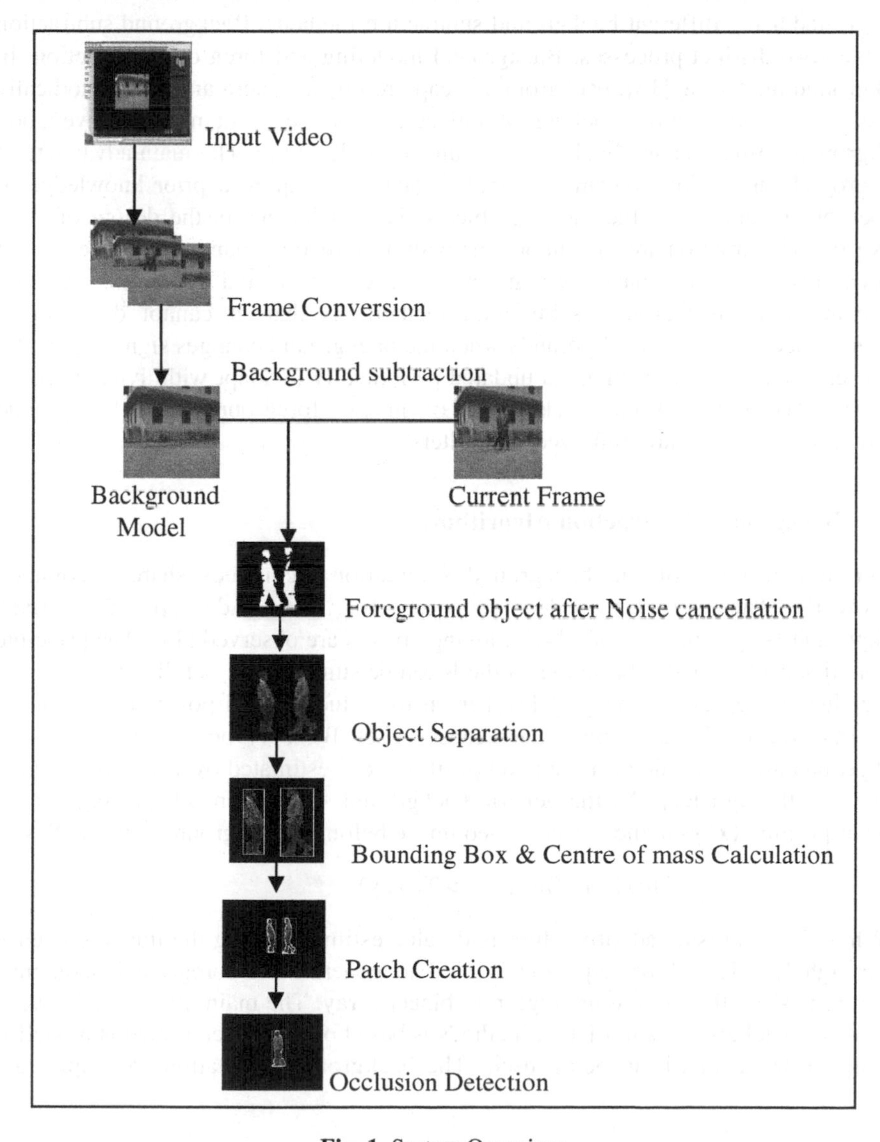

Fig. 1. System Overview

3 Moving Object Extraction

In applications using fixed cameras with respect to the static background, a very common approach is to use background subtraction [12]. In various computer vision applications, the primary task is the moving object separation. One of the most common and efficient approach to focus moving objects is background subtraction. Here the static background model is subtracted from the current frames of video sequence. The process of segmenting the moving objects from the stationary scene is termed as background subtraction. The foreground objects are extracted from the background using different background subtraction methods. Background subtraction involves two distinct processes: Background modeling and foreground extraction. In background modelling [13], background is captured by a camera and it is periodically updated. To obtain good tracking of object in a scene, system must have good background estimation and background subtraction algorithm. The main advantage of background subtraction algorithm is that, it does not require a prior knowledge of shapes or movements of the moving objects. The challenges in the design of good background subtraction are; it cannot cope with illumination changes, the detection of unnecessary non stationary background object pixels and shadows. Another disadvantage of background subtraction methods is that, it cannot discriminate moving objects from the backgrounds when the background changes significantly. So the reference background image is updated periodically to adapt with dynamic scene changes. The noise in the background or in the foreground after background subtraction was efficiently removed with filters.

3.1 Background Subtraction Algorithms

Different but, most of the background subtraction techniques share a common framework. The observed video sequence $It_{(x,y)}$ is made up of a fixed background($Bt_{(x,y)}$), in front of which moving objects are observed [14]. The principle of adaptive background subtraction methods can be summarized as follows.

Let It(x, y) represents the gray level intensity value at pixel position (x,y) and at time instance t of video image sequence I. Let Bt(x, y) be the corresponding background intensity value for the pixel position (x,y) estimated over time from video images B_0 through B_{n-1}. As the generic background subtraction scheme suggests, a pixel at position (x, y) in the current video image belong to foreground if it satisfies:

$$| It(x, y) - Bt(x, y)| > Tt(x, y) \quad (1)$$

Where Tt(x, y) is an adaptive threshold value estimated using the image sequence I_0 through I_{n-1}. The above equation is used to generate the foreground pixel map which represents the foreground region as binary array. The main difference between most of the background subtraction methods is based on how background is modelled and the distance metric is being used. The background adaptation techniques are categorized as: non-recursive and recursive. The commonly used non recursive techniques are, frame differencing and median filtering. These non recursive methods need buffers to store previous 'n' frames, but possible errors visible for longer time in background model. But recursive method does not need buffers. One of the most popular recursive background methods is Gaussian mixture model (GMM). The GMM with multi-model distributions are efficient than single Gaussian model, if the background scene involves large or sudden changes. The mixture of Gaussians technique was first introduced by Stauffer and Grimson in [15].

In the proposed system, GMM is chosen for extracting the foreground object without cast shadow. For the foreground objects detection, each pixel is compared with each Gaussian and is classified according to corresponding Gaussian. In this model each pixel (X_1, X2... Xt) in the scene is modeled by a mixture of K Gaussian distributions. Based on the resolution and the variance of each Gaussian of the mixture, it is determined that which Gaussian may correspond to background colors. If any pixel values, that do not fit the background distributions are considered as foreground. The probability of observing certain pixel has an intensity value.

The probability of observing current pixel has the intensity is considered as,

$$P(X_t) = \sum_{i=1}^{k} \omega_i * \eta\,(X_t, \mu_i, \Sigma_i) \quad . \tag{2}$$

Where K is the number of distributions, ω_i is the estimate of the weight associated to the ith Gaussian in the mixture, μ_i is the mean value of the ith Gaussian , Σ_i is the covariance matrix for the ith distribution, and η is the Gaussian probability density function.

4 Noise Cancellation and Object Separation

The binary object images after background subtraction based on the GMM often have the discrete noises and holes in the object region [16], therefore the proposed system uses mathematical morphological processing operations such as erosion and dilation to remove the isolated noise and fill the hole in the object region. Object separation is performed in the proposed system after moving object extraction. The result of moving object extraction with morphological operation and object separated result are shown in fig 2.

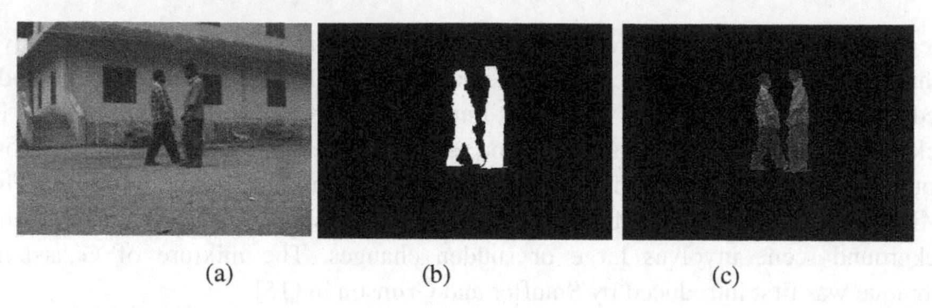

Fig. 2. Moving object extraction based on GMM. (a) Current frame image (b) foreground object with morphological operations(c) Object separated Image

5 Object Tracking

Object tracking is a pre-requisite for the important applications of computer vision. To detect and track people in a video sequence, group the pixels that represent individual people together and calculate the appropriate bounding box for each person. Tracked objects are labeled with numbers and surrounded by bounding box. In the proposed system, to form bounding box, we need four points, starting position(x,y),length and breadth. Fig. 3 shows the object separated image with bounding box.

The algorithm to create bounding box is as follows.

1. Let StX and StY be the Minimum value of row and column minus 0.5 gives starting position(x,y) respectively.
2. Bdth=Max(col)-Min(col)+1;
3. Ln=Max(row)-min(row)+1;
4. Boundingbox BB=[StX StY Bdth Ln];
5. Display(BB);

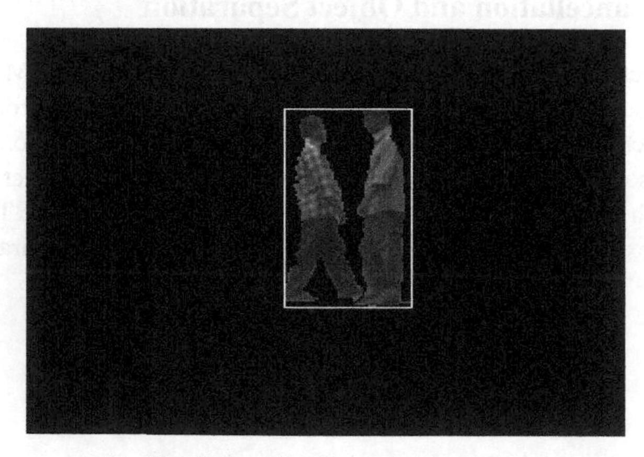

Fig. 3. Bounding box creation result

6 Patch Based Framework

Patch is a small unit with same size and is created around each object .In the proposed system, the patch based frame works were used for identifying the patch which are going to be occluded.

In this novel approach patches are created only at the boundary parts and check the distance (d) between the patch in two objects. If this distance is greater than the predefined threshold, we change the color of the corresponding patches in that objects. To create patches, first we create a mask with its size is equal to the size of the frame. Then, the edges of each object are finding out. From these edges each row and column values are stored in an array and then the size of patch is determined. In our work, the size of patch was selected as 4x4. By using the patch size and number of row values in the edges, we can find the total number of patches that can be created in each object. For creating patches, first we find a suitable location to fit the patch. For that, we search the position in all direction (top, bottom, left and right).If any patch is fit in any of the direction; the patches are placed in that direction one by one. After the creation of patches, we preserve the details of each patch and are called one patch set.

The Patch set Pi has 3 parameters.

$$Pi= \{SI\ (R1, C1), PDI\ ((R1, C1), (R2, C2),..(Ri,Ci)),IL\} \tag{3}$$

Where,

SI (R1, C1) is the starting index of the first patch.PDI ((R1, C1), (R2, C2), ………, (Ri, Ci)) is the patch data index and it represents the coordinates from starting point to the patch size-1. IL is the index label of each patch (i.e., up, down, left or right).
After creating patches at the object boundary, the following steps are used to find the patches that make occlusion.

First, we assume a threshold value of distance between the patches. Then, take the starting index of all objects in the entire frames. After that, the staring index of the first patch in first object is subtracted with all the starting index of all patches in second object. If this difference is less than or equal to the predefined threshold value or greater than or equal to zero, then the direction of that patch (i.e., index label) is compared with the entire direction (Up, Down, Left and Right).At the time of matching, the color of the corresponding patches in all object in that direction is changed shown in fig.4.

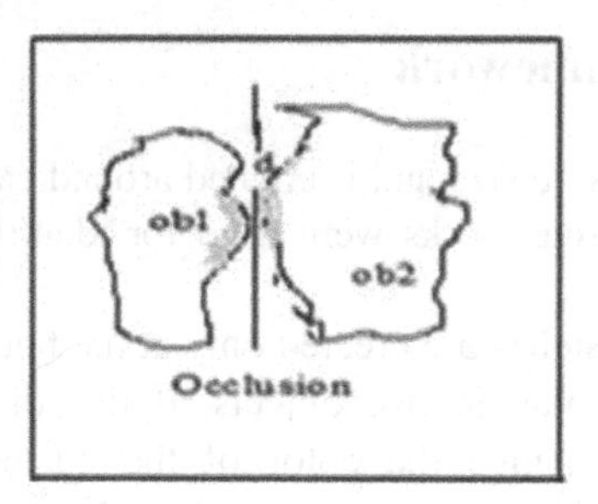

Fig. 4. Occluded patch identification

If the difference between that patches is not matched with the predefined threshold, the next patch in first object is taken and it is compared with all the patches in second object in the corresponding frame .This process is repeated to locate the patches that are going to occlusion. The result of Patch creation is shown in figure 5.

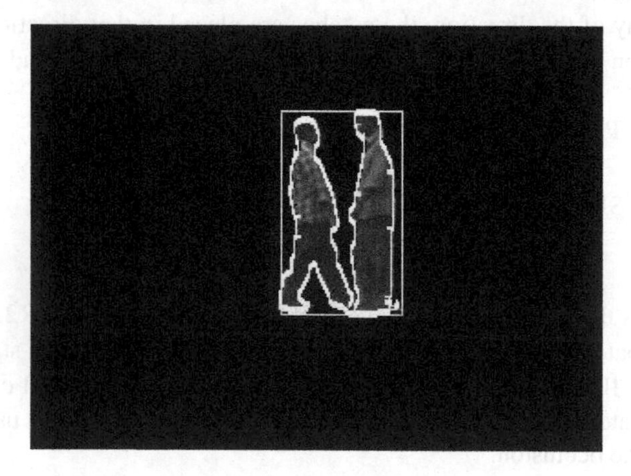

Fig. 5. Patch creation result

7 Occlusion Detection

Centre of mass for all objects are calculated for occlusion detection. Using bounding box parameter in region props algorithm, we find the four extreme points of all object. From these points, select any two diagonal points and by using these values, calculate the centre of mass point using the equation 4.

$$((C(2)+(C(2)+C(4)))/2,C(1)+(C(1)+C(3)))/2\ . \tag{4}$$

Where (C1, C2) is the starting co-ordinates, (C3, C4) is the co-ordinate value of column increment and row increment respectively shown in fig. 6.

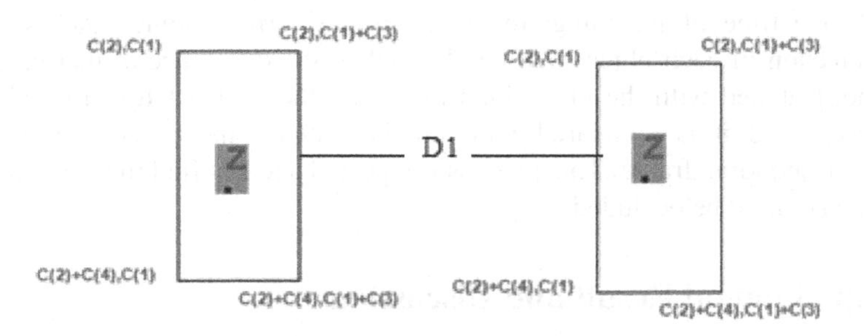

Fig. 6. Occlusion detection by centre of mass

During Occlusion, the distance between two centre of mass point is Zero. The result of centre of mass calculation and patch creation are shown in fig.7.

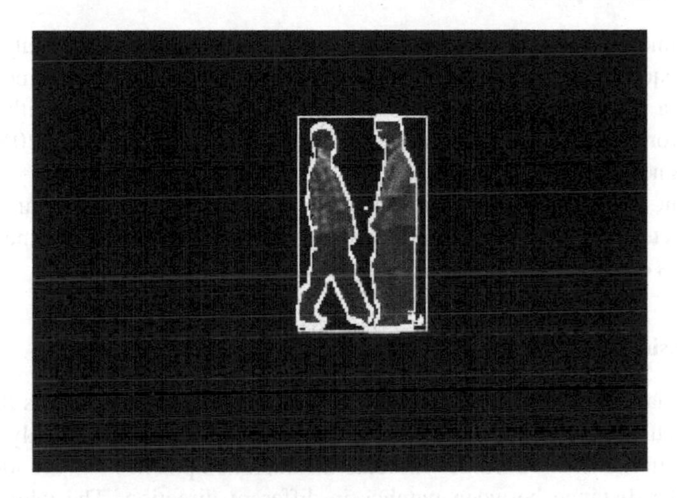

Fig. 7. Centre of mass Calculation

After creating patches at the object boundaries, the following steps are used to find the patches which are going to be occluded.

Algorithm for occluded patch identification

1. First, set a threshold value.
2. Then, take the starting index of each object in all frames.
3. After that, the staring index of the first patch in first object is subtracted with all the starting index of the each patch in second object. If this difference is less than or equal to the predefined threshold value, and greater than or equal to zero, then the direction of that patch (i.e., index label) is compared with the entire direction(Up, Down, Left and Right).

4. At the time of matching, the color of the corresponding patches in that direction of each object will be changed. If the difference of that patches is not matched with the predefined threshold, the next patch in first object is taken and it is compared with all the patches in second object in the corresponding frame .This process is repeated and can find the patches which are going to be occluded.

8 Experimental Result and Discussions

To show the accuracy and robustness of our algorithm, it was tested for many occluded image sequences. The experimental setup and the performance of our system make some assumptions and are reported in the following section.

8.1 The Data Acquisition

In the experiment, we simulate our occlusion detection algorithm in mat lab 8.2.We used video sequences including some videos from PETS Dataset and some videos are recorded on a pleasant morning of March by canon digital camera with some pre-processing works such as compression. The sizes of these videos are 320X240 at the rate of 25 frames per second.

Subsequent to the reading of those videos, processing frame by frame and then moving objects extraction was performed. Based on patch based frame work and centre of mass concepts, occlusion was detected successfully.

8.2 Occlusion Detection

Occlusion is happened when the objects comes closer. So figure 8 shows the result of occluded patch identification in the key frames of the video sequence only. The green colored units in the tracked frames indicate the occluded patches that are identified by calculating the distance between patches in different direction. The white bounding box around each object is used to track the objects and occlusion is determined by using centre of mass. We track the occluded object by calculating the distance between the centre of mass of objects in both direction. For calculating this distance, one predefined threshold value is used. In the proposed system, the threshold value used is 50. During occlusion, the width of the bounding box is increased and in full occlusion, only one centre of mass is present between the occluded objects. But just after occlusion, again the width of bounding box will be increased and the bounding boxes are seemed separately in the frame. Similarly the centre of mass is also seemed as twice and it is shown in figure 8.

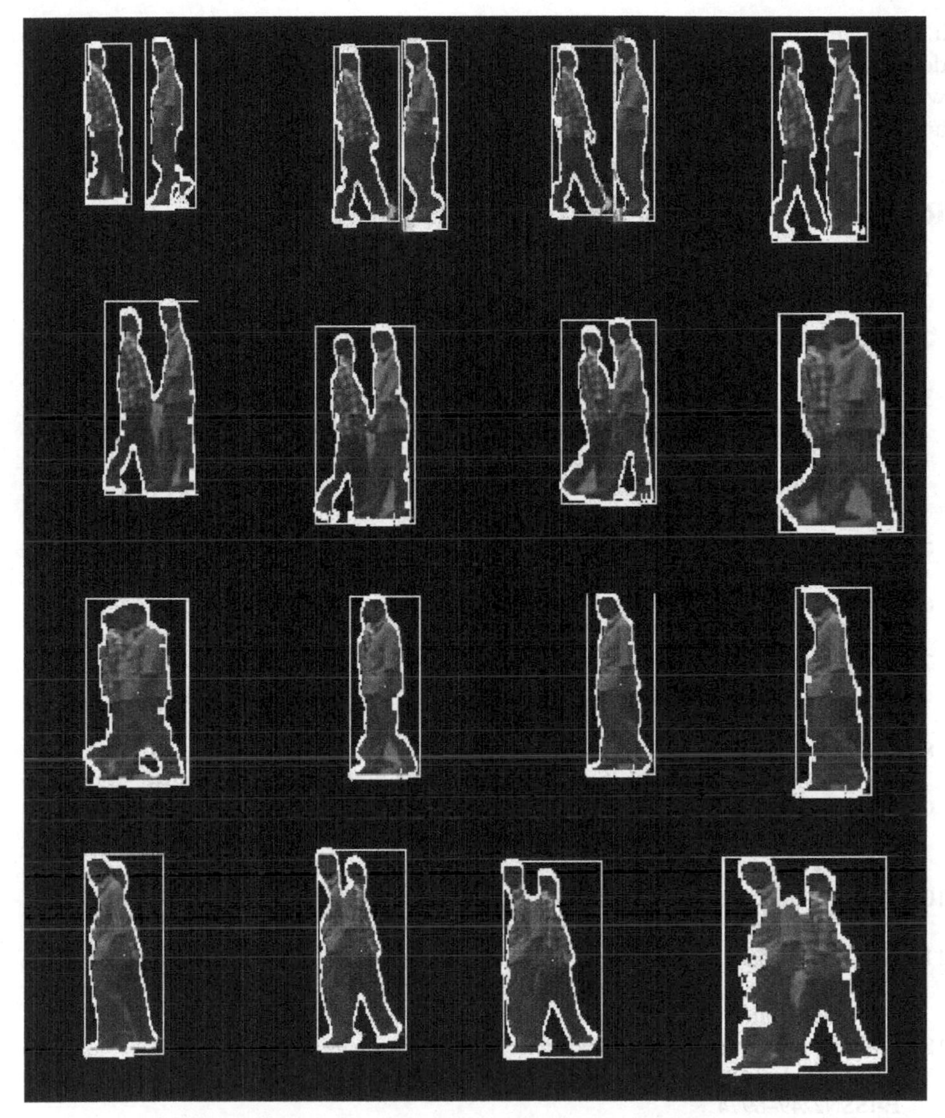

Fig. 8. patch going to be occluded (first row first image), occluded images (remaining images except last image) .During occlusion, only one bounding box and centre of mass and just after occlusion (last image).

9 Conclusion

Today video surveillance system is used in many sites for human detection and identification. In this paper, we used the Gaussian mixture model for motion detection. The detected objects are tracked using bounding boxes. The main innovation of this paper is occlusion detection in human parts. In the proposed work, a robust patch creation idea is used for occluded part identification. The occlusion detection by centre of mass gives best result during human tracking. In the future works, the scope of present work can be extended to develop a simple model to detect and handle occlusion in infrared videos.

References

1. Marques, J.S., Jorge, P.M., Abrantes, A.J., Lemos, J.M.: Tracking groups of pedestrians in video sequences. In: IEEE International Conference on Pattern Recognition, vol. 9 (2003)
2. Chang, T.-H., Gong, S., Ong, E.-J.: Tracking multiple people under occlusion using multiple cameras. In: Proc. 11th British Machine Vision Conference (2000)
3. Dockstader, S.L., Tekalp, A.M.: Multiple camera fusion for multi-object tracking. In: Proc. IEEE Workshop on Multi-Object Tracking, pp. 95–102 (2001)
4. Jepson, A.D., Fleet, D.J., El-Maraghi, T.F.: Robust online appearance model for visual tracking. IEEE. Trans. on Pattern Analysis and Machine Intelligence 25(10), 1296–1311 (2003)
5. Lipton, A., Fuyiyoshi, H., Patil, R.: Moving target classification and tracking from real-time video. In: Proc. Fourth IEEE Workshop on Applications of Computer Vision (1998)
6. Comaniciu, D., Ramesh, V., Meer, P.: Kernel-based object tracking. IEEE Trans. PAMI 25(5), 564–575 (2003)
7. Housni, K., Mammass, D.: Moving Objects Tracking in Video by Graph Cuts and Parameter Motion Model. International Journal of Computer Applications (0975 - 8887) 40(10) (2012)
8. Youssef, S.M., Hamza, M.A., Fayed, A.F.: Detection and Tracking of Multiple Moving Objects with Occlusion in Smart Video Surveillance Systems. In: IEEE (2010)
9. Madhavi, B.S.M., Ganeswara Rao, M.V.: A fast and reliable motion human detection and tracking based on background subtraction. IOSR Journal of Electronics and Communication Engineering 1(1), 29–35 (2012) ISSN: 2278-2834
10. Takala, V., Pietikainen, M.: Multi-Object Tracking Using Color, Texture and Motion. IEEE Trans. on Image Processing (2007)
11. Lin, D.-T., Chang, Y.-H.: Occlusion handling for pedestrian tracking using partial object template-based component particle filter. IADIS International Journal on Computer Science and Information Systems 8(2), 40–50, ISSN: 1646-3692
12. Lakhotiya, S.A.: Robust shadow detection and optimum removal of shadow in video sequences. International Journal of Advanced Engineering Research and Studies 2 (2013) E-ISSN2249–8974
13. Mike McHugh, J., Konrad, J., Saligrama, V., Jodoin, P.-M.: Foreground-Adaptive Background Subtraction. IEEE Signal Proc. Letters (2009)
14. Benezeth, Y., Jodoin, P.M., Emile, B., Laurent, H., Rosenberger, C.: Comparative Study of Background. Subtraction Algorithms. Journal of Electronic Imaging 19 (2010)
15. Stauffer, C., Grimson, W.E.L.: Adaptive Background Mixture Models for Real-Time Tracking. In: IEEE Computer Society Conf. on Computer Vision and Pattern Recognition, CVPR, vol. 2, pp. 246–252 (1999)
16. Ye, Q., Gu, R., Ji, Y.: Human detection based on motion object extraction and head–sholder fearure. Optik 124, 3880–3885 (2013)

Playlist Generation Using Reactive Tabu Search

Sneha Antony and J.N. Jayarajan

Department of Computer Science & Engineering
Rajagiri School of Engineering & Technology
Kochi, India
antony.sneha@yahoo.com, jayarajanjn@rajagiritech.ac.in

Abstract. This paper proposes a reactive tabu search algorithm to solve the automatic playlist generation problem effectively. The conventional tabu search, though produces promising results, has the disadvantage of the tabu size being fixed. The tabu size is fixed by the user and it determines the capacity of the memory. A preset parameter does not yield appropriate results in a heuristic algorithm and so there must be mechanisms to make the tabu list size variable in order to get optimum results. The reactive tabu search, has mechanisms to make tabu size variable and thereby avoids often repeated solutions. This proposed method for the playlist generation problem yields better results than the conventional tabu search. The concrete implementation of the method has been described, which is supported by experimental analysis.

Keywords: Reactive tabu search, prohibition period, Tabu search, playlist generation, music.

1 Introduction

A revolution has taken place in the way we as a society listen to and consume music. The evolution in the distribution of music from having to hear it at the performance place to being able to hear it at one click today, has happened because of the emergence of the internet. This has brought about with it demands for fast and reliable ways to discover music according to the user's preference online. The problem here is, with the million song collection available online, how can a user discover or select a few songs suitable for him, in the most efficient manner. The Automatic Playlist Generation Problem is to generate a playlist that satisfy user constraints, from the songs in the music database [5].

The playlist generation problem is a combinatorial optimization problem. An exhaustive search in this case may not be feasible so a near optimum result that can be computed with minimal cost is desired. The popular meta-heuristic algorithms for the combinatorial optimization problems include ant colony optimization, particle swarm optimization, simulated annealing, genetic algorithms and tabu search. A study done in [4], points out the effectiveness of tabu search when compared to other meta heuristics. In [9], the specific strategies of tabu search is compared and contrasted with strategies based on evolutionary algorithms. It shows how genetic algorithm procedures can be improved by alternate

S.C. Satapathy et al. (eds.), *Emerging ICT for Bridging the Future – Volume 2*,
Advances in Intelligent Systems and Computing 338, DOI: 10.1007/978-3-319-13731-5_12

frameworks like tabu search that use adaptive memory structures. Hence, tabu search has emerged as a promising technique to solve combinatorial optimisation problems like playlist generation.

A tabu search based method to solve this problem is given in [1]. But the problem with the approach is that the internal parameter of tabu size has to be preset manually by the user. It is not appropriate to fix parameter value used in a meta heuristic algorithm. There must be mechanisms to change the value according to different problems and more importantly according to different runs of the same program. This paper proposes a method to solve this problem by using reactive tabu search. This work is more distinguished as it uses adaptive tabu tenure. A reactive tabu search brings in the element of machine learning whereby the size of the tabu tenure is automated, by monitoring its behaviour in each iteration of a run of the program. The search process is monitored, and the algorithm responds to the formation of cycles by adapting the tabu list size.

The paper is organised as follows: Section 2 gives a brief overview on the working of tabu search. Section 3 mentions the problem with manually setting the value of tabu tenure. This is explained with the help of graphs. Section 4 elaborates on the proposed approach with its experimental results in section 5. Section 6 concludes the paper.

2 Tabu Search

Tabu search is a meta-heuristic search method employing local search methods used for mathematical optimization [10]. Local search techniques randomly generate a single solution and then jump from the solution to its neighbouring solution in the hope of finding a better solution. Neighbouring solutions are generated by making small changes to the solution in hand. Tabu search enhances the performance of the local search techniques by using memory structures that keeps track of the visited solutions. If a solution has been previously visited within a specific short-term period, it is marked as "tabu" so that the same solutions are not selected again. Tabu search avoids being stuck at local optimum because of the memory structures and as a result, this becomes an advantageous technique for solving combinatorial optimization problems [1]. In [1], the automatic playlist generation has been solved using tabu search. The results of this approach, given its benefits over evolutionary methods, are surely promising. The steps of tabu search are as follows:

1. Generate a solution randomly and set it as the $Best_{global}$.
2. Evaluate its fitness and insert it into tabu list.
3. Generate neighbourhood of the solution and evaluate fitness of each.
4. Select the best admissible solution, $Best_{admissible}$ and insert it into tabu list.
5. Update tabu list, by popping 1st In value if the size exeeds the limit.
6. If $Best_{admissible} > Best_{global}$, then set $Best_{global} = Best_{admissible}$
7. Check whether stopping criterion is satisfied, if so, get final solution, else repeat the process with the $Best_{admissible}$ solution selected as the current solution.

The two main concepts in tabu search are that of intensification and diversification [7]. Intensification is the process by which the algorithm thoroughly checks the portions of the search space that seem promising, in order to make sure that the best solutions in these areas are found. Diversification is the process by which the algorithm prevents being stuck in a local optimum by forcing the search into previously unexplored areas of the search space.

3 Problem Statement

The major task when applying a heuristic algorithm to a problem is the representation of the problem and the fitness function. Setting these two items will determine the bridge between the algorithm context and the original problem. Determining the internal parameters of the heuristic algorithm will depend on both the representation taken for the particular problem and the fitness function. Due to the element of randomness involved, setting the right parameter values can only be done on a trial and error basis. Even then, getting them right is not possible because these values tend to change from problem to problem and even with each run of the same problem. In case of tabu search, the tabu list size is kept fixed. It is this tabu tenure that determines the occurrences of cycles when searching for the solution. The tabu tenure or prohibition period determines the time for which a previously selected solution is prevented from being selected again. Setting the prohibition period low will hinder the diversification process of tabu search, whereby getting stuck in a local optimum. Setting the value high, will hinder the intensification process whereby missing out on spotting the best solution. Therefore setting the value must strike a balance between the intensification and diversification elements [11].

Previous works on the heuristic algorithms for playlist generation have been implemented by presetting the parameter values based on trial and error method. Fig. 1 shows the output of the playlist generation problem solved using tabu search when using a prohibition period of 3 and 5 respectively. The program was made to run for 15 times and the fitness of the output was plotted. The fitness of a solution (playlist) is the number of constraints satisfied by it. The number of constraints given by user is 20. Both the programs were made to run with the same initial solution in parallel for 50 iterations in each run. In figure 1 it can be seen that while at some runs a prohibition of 5 gives better output, at some points it is tabu search with prohibition period 3 that gives better output. In the 2nd run, prohibition period of 5 gives output with fitness value 19, whereas prohibition period of 3 gives only 17. In case of the 6th run, prohibition period 5 gives 16 whereas prohibition period 3 gives 19. This shows that presetting values for a meta heuristic algorithms is not appropriate. There must be mechanisms to automate this process and dynamically generate the prohibition period in each iteration.

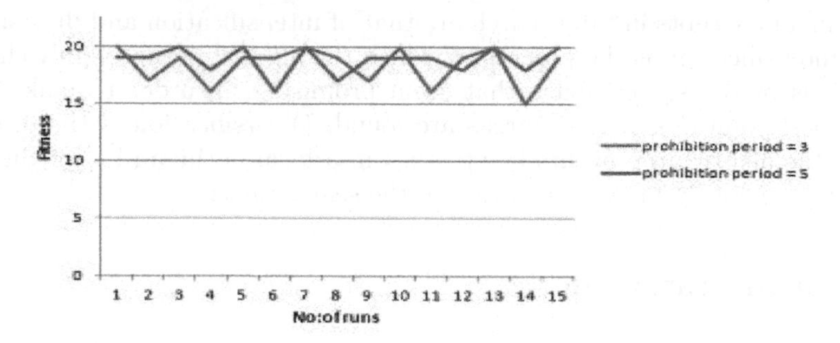

Fig. 1. Graph showing comparison between prohibition period of 3 and 5

Globally there are two methods for setting parameter values: parameter tuning and parameter controlling. Parameter tuning is the age old method in which the value is set before the run of the program. Parameter control is a alternative to parameter tuning as it starts with an initial value which gets changed during the run of the program [8]. The proposed method of reactive tabu search for the automatic playlist generation problem advocates the use of parameter control.

4 The Proposed Method : RTS

Reactive Search Optimisation uses machine learning concepts on local-search heuristics. The Reactive Search framework introduces feedback schemes in heuristics for combinatorial optimisation problems [2][12]. The conventional meta heuristics like genetic algorithm, simulated annealing, tabu search etc, require manual adjustment of search parameters in advance for efficient search. The disadvantage of parameter tuning is corrected by parameter control in reactive search optimisations. Reactive Tabu Search (RTS) is an extension of tabu search and forms the basis for reactive search optimisations. RTS explicitly monitors the search and reacts to the formation of cycles by adapting the prohibition period [6].

RTS uses a feedback-based tuning mechanism of the tabu list size(prohibition period) and strikes a balance between the intensification and diversification mechanisms. Initially the prohibition period is fixed. When a solution previously selected is encountered, the prohibition period increases. If a repetition does not occur for a specific short period of time, the prohibition size is decreased.[3] In order to avoid the long search cycles, RTS uses the escape procedure which is a mechanism for diversification.

4.1 System Architecture

The system architecture is shown in Fig.2. It mainly consists of 4 modules: User Interface module, User Constraint module, Music database and the Playlist Generation module.

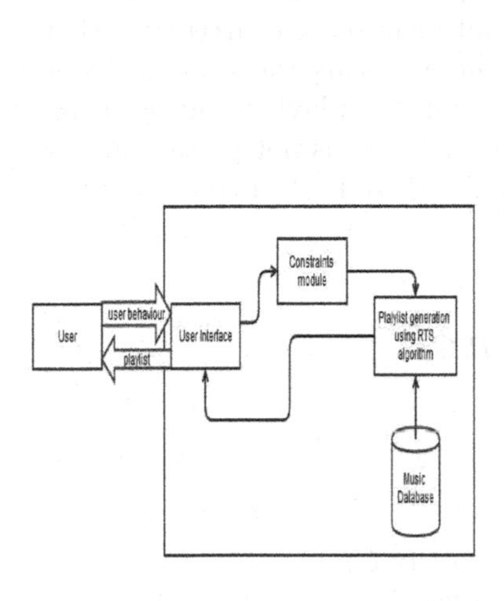

Fig. 2. System Architecture of proposed Playlist Generation System

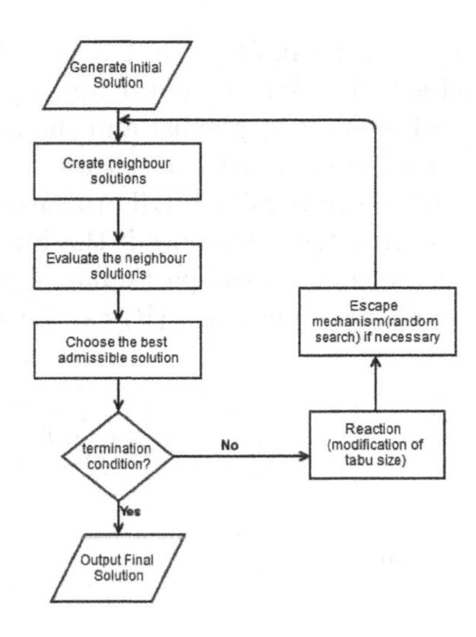

Fig. 3. Flowchart for Reactive Tabu Search

- User Interface module: The user interface gets user behaviour from the user and dispatches commands to the algorithm and displays results of the algorithm. It is this module that interacts with the user.
- Music Database module: The music database is the online or offline collection of songs of the user. It is from this set of about million songs that the playlist is generated.
- User Constraints module: This module contains all the user constraints. In the proposed system all the user constraints have been preset by the user depending upon his taste.
- Playlist Generation module: The playlist generation module is the core algorithm. It uses the Reactive Tabu Search to solve the automatic playlist generation problem.

4.2 Implementation Details

The playlist generation problem is to automatically generate a playlist that satisfies user constraints, from a large music database. A playlist is a set of songs from the music database. Each song is represented by a vector of attributes. We have used 10 attributes in this system, namely: title, album, singer, lyricist, genre, mood, language, duration and year. Each song will have values for each of these attributes, and this will uniquely identify a song. In the proposed system, the set of songs in the playlist is represented by their songIds. The fitness of a solution (playlist) is determined by the number of constraints satisfied by each song in the playlist. For the purpose of comparison between the two approaches,

RTS and Tabu Search, 2 types of constraints have been used: Include and Exclude. The Include constraint mentions the attributes of songs which must be included in the playlist and the Exclude constraint mentions the attributes that must be excluded. The fitness is calculated by incrementing the value for each Include constraint attribute satisfied by each song in the playlist and by incrementing the value once if the Exclude constraint attribute is not present in any of the songs in the particular playlist. For a playlist P, and set of constraints C, the fitness function F(P) can be given as :

$$F(P) = \sum_{C_j \in C} satisfy\ (P, C_j\ ) \tag{1}$$

, where

$$satisfy(P, C_j) = \begin{cases} 1 & \text{if P satisfy } C_j \\ 0 & \text{otherwise} \end{cases}$$

The flowchart of the reactive tabu search algorithm is given in Fig.3. Initially a solution (in this case: a playlist)is generated randomly. The fitness of the solution is evaluated and it is entered into the tabu list. The size of prohibition period is set to be 1 initially. Then neighbours of the solution are generated and evaluated. The best fit among them is selected. The tabu list is updated in a reactive manner. If the solution happens to be the same as previously visited, then the prohibition period is increased or else it is decreased by a percentage. The increase value can be between 1.1 and 1.3. The decrease value can be set between 0.8 and 0.9 [6]. The increase and decrease values are set to be 1.1 and 0.9 in the program. The program terminates when a set no:of iterations is reached or when the maximum fitness value is attained by the playlist.

Two terms that come about in reactive tabu search are: Escape mechanism and Aspiration criterion. Escape mechanism is done so as not to be stuck in a local optimum. The new iteration in this case will start from a random point other than selecting best admissible from the neighbours of the current solution. This is done when the solution selected is found to be in the repeatedly-visited-list. Aspiration is the method by which some tabu moves are disregarded to obtain better local solutions. Generally these are performed when a tabu move leads to an output better than the last best known so far.

Once the best from the neighbourhood has been selected, it is included in the tabu list. The repetition interval of solutions is monitored by maintaining a separate visited list which contains the no:of repetitions and the interval of repetitions of each visited solution. The selected solution is compared with the best known solution so far and then the best is updated accordingly. The process is repeated till the termination condition is reached. The termination condition can be a threshold of fitness level or a limit to the no:of iterations. When the program terminates, it returns the best solution known so far.

5 Experimental Analysis

The Reactive Tabu Search algorithm for playlist generation and Tabu Search were tested on the same set of inputs. The music collection, values for attributes of songs, playlist size, constraint values and even the initial random playlist generated were made sure to be the same for both the programs. Both algorithms were executed in the same program simultaneously. Fig. 4 shows the output obtained on recording values when the program was made to run 20 times. The value of fitness of the returned playlist is plotted for both the algorithms. Reactive tabu search (RTS) is seen to outperform Tabu search (TS). For the experimental setup, 50 constraints were given by the user beforehand. The playlist size of 10 has been used. The program was executed 20 times, each for 100 iterations.

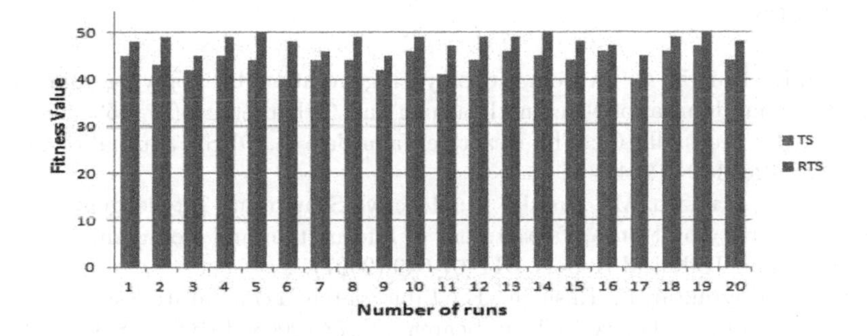

Fig. 4. Graph showing comparative results for TS and RTS

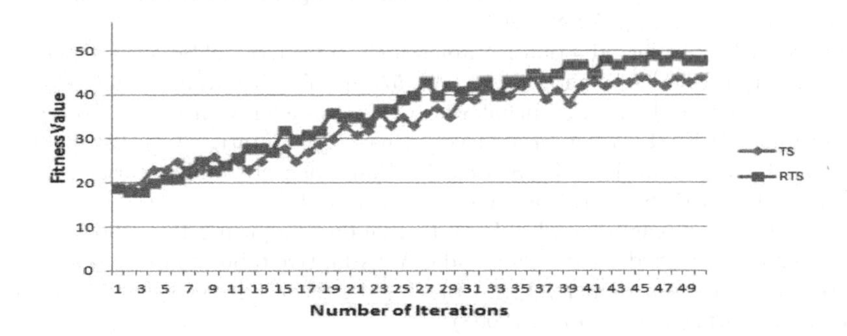

Fig. 5. Graph showing comparative results in progress of a run for TS and RTS

Fig.5 shows the progress in selection of solutions in each iteration of a single run. In this case also both the algorithms are executed with the same set of inputs. The best admissible solution selected from the neighbourhood of the current solution is plotted in each iteration. The program is made to run for 50 iterations. From the graph we see that, the progress of RTS is greater than that of TS. RTS gives an output with fitness 49, whereas TS gives 44.

6 Conclusion

The automatic playlist generation problem being a combinatorial optimisation problem, can be solved best using heuristic or meta-heuristic algorithms, like genetic algorithms, simulated annealing, ant colony optimisation, particle swarm optimisation or tabu search. Tabu search has been shown to give much promising results compared to others in solving combinatorial optimisation problems. But presetting the prohibition period for tabu search will limit the output to local optimum due to repeated cycles caused by inappropriate prohibition period. Reactive tabu search brings in an element of machine learning and changes the value of prohibition period dynamically when the program is executed. The results obtained are proof for the goodness of the algorithm.

References

1. Hsu, J.-L., Lai, Y.-C.: Automatic playlist generation by applying tabu search. International Journal of Machine Learning and Cybernetics 5(4), 553–568 (2014)
2. Battiti, R., Tecchiolli, G.: The Reactive Tabu Search. ORSA Journal of Computing 6(2), 126–140 (1994)
3. Battiti, R., Brunato, M., Mascia, F.: Reactive Search and Intellengent Optimisation. University of Trento, Department of Information and Communication Technology, Italy, Technical Report DIT-07-049 (2007)
4. Pirim, H., Bayraktar, E., Eksioglu, B.: Tabu Search: a comparative study. In: Local Search Techniques: Focus on Tabu Search, p. 278 (2008) ISBN: 978-3-902613-34-9
5. Fields, B.: Contextualize your listening: the playlist as recommendation engine. PhD Thesis, Goldsmiths, University of London (2011)
6. Brownlee, J.: Reactive Tabu Search. In: Clever Algorithms: Nature-Inspired Programming Recipes, pp. 79–86 (2011)
7. Gendreau, M.: An introduction to tabu search. International Series in Operational Reasearch and Management Science, vol. 57, pp. 37–54. Springer US (2003)
8. Eiben, E., Hinterding, R., Michalewicz, Z.: Parameter Control in Evolutionary Algorithms. IEEE Transactions on Evolutionary Computation 3(2), 124–141 (1999)
9. Glover, F., Laguna, M.: Tabu Search. In: Handbook of Combinatorial Optimization, pp. 621–757. Kluwer Academic Publishers (1998)
10. Glover, F.: Tabu search-part I. ORSA Journal on Computing 1(3), 190–206 (1989)
11. Devarenne, I., Mabed, M.H., Caminada, A.: Adaptive tabu tenure computation in local search. In: van Hemert, J., Cotta, C. (eds.) EvoCOP 2008. LNCS, vol. 4972, pp. 1–12. Springer, Heidelberg (2008)
12. Battiti, R., Tecchiolli, G.: Training Neural Nets with the Reactive Tabu Search. IEEE Transactions on Neural Networks 6(5), 1185–1200 (1995)

Design of Fuzzy Logic Controller to Drive Autopilot Altitude in Landing Phase

Adel Rawea and Shabana Urooj

School of Engineering, Gautam Buddha University,
Gr. Noida, UP, India
adelrawea@gmail.com,
shabanaurooj@ieee.org

Abstract. Nowadays the automatic control science is an important filed. Its importance has been raised from equipment evolution and increased control requirements over the many science fields, such as medical and industrial applications. Hence, it was necessary to find ways for studying and working to develop control systems. This paper proposes study of an aircraft as control object in longitudinal channel during the landing phase, and study altitude due to its importance during this phase. This paper introduces the possibility of implementing automated control process using fuzzy controllers. Finally, comparing the results of fuzzy and traditional controllers has been discussed.

Keywords: Fuzzy logic controller, proportional-integral controller, autopilot.

1 Introduction

The Cybernetics Considered from the important science of modern times has come from the development of these important machines used and increase the requirements imposed on it in all industrial fields [4].

As a result of these requirements increased with the time it was necessary to develop the control modes that's used in these areas. So the need arise to find studies and methods of work on the development of this science. Happened in the twentieth century a quantum leap in the field of scientific development in all directions. Has appeared in recent decades, the theory of Fuzzy Logic, which had been widespread in practical applications, especially in the Control Systems has led to the emergence of the Fuzzy Logic Controllers [1].

2 Motivation of Paper

The importance of scientific research to study the possibility of implementing automated control operations in the control systems to replace traditional controllers by Fuzzy Controllers, to move to artificial intelligence in control systems in all fields [2].

S.C. Satapathy et al. (eds.), *Emerging ICT for Bridging the Future – Volume 2*,
Advances in Intelligent Systems and Computing 338, DOI: 10.1007/978-3-319-13731-5_13

3 Research Methods

The research adopted on the method of mathematical modeling to control the aviation altitude in landing phase thus computerized simulation, and design of the proposed fuzzy controller using MATLAB [3].

3.1 The Principle of Building Control System in the Longitudinal Channel by the Center of Gravity of the Aircraft

The process of take-off and landing is considered of the most complex phases of flight due to large change parameters, especially in the final stage of it. We are always searching for the path of landing provides the requirements imposed on control systems, and from these paths that provides this process is the path of exponential landing as shown in the following equation [1]:

$$h = (h_0 + \Delta h_o)e^{-kx} - \Delta h_0 \tag{1}$$

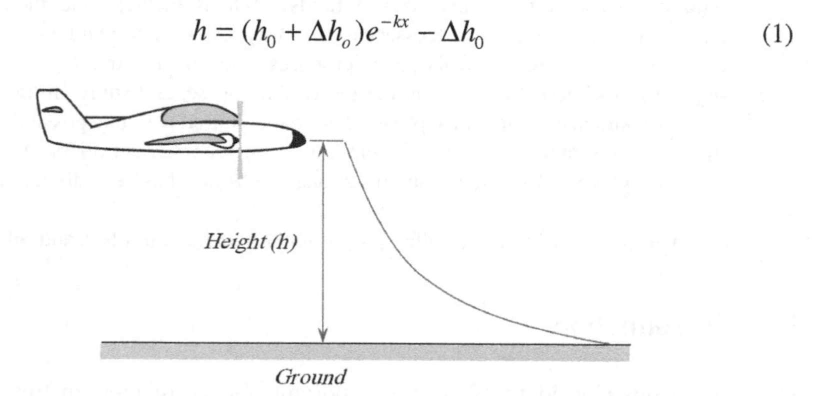

Fig. 1. The suggested path of landing

Based on the above the process of driving control systems in the last stage of the landing becomes control the aircraft altitude. The control of aviation altitudes is considered from the high levels of aviation control, where it is used during the implementation of trip aviation, take-off, landing and aviation at low altitudes. The transfer function for Elevator displacement is [2-13]:

$$\dot{h} + a_h^{\vartheta}\dot{\vartheta} + a_h^{\theta}\theta = 0 \tag{2}$$

The transfer function between θ and h is:

$$W_\theta^h = \frac{\vartheta}{s}$$

After that the function between h and ϑ according to the function between θ and ϑ as shown in the following equation:

$$W_{\vartheta}^{\theta}(s) = \frac{1}{T_{\theta}s+1} \tag{3}$$

In the end the transfer function between pitch angle ϑ and elevator displacement is:

$$W_{\dot{\vartheta}}^{\text{в}}(s) = -\frac{k_{\dot{\vartheta}}^{\text{в}}\omega_{\alpha}(T_{\theta}s+1)}{s^2+2\xi_{\alpha}\omega_{\alpha}s+\omega_{\alpha}^2} \tag{4}$$

We will use the following equation to update the state variable for h:

$$\begin{aligned} \delta_{\text{в}} &= k_{\vartheta}(\vartheta - \vartheta_d) + k_{\omega z}\omega_z \\ \vartheta_d &= \frac{k_h}{k_{\vartheta}}(h - h_d) \end{aligned} \tag{5}$$

Where [2]:

$\delta_{\text{в}}$:elevator displacement. - k_{ϑ}: constant of ϑ angle.

ϑ : current pitch angle. - ϑ_d : desired pitch angle.

$k_{\omega z}$: constant of the angular velocity of pitch.

ω_z : the angular velocity of pitch. - k_h : the constant of altitude.

h: the current altitude. - h_d : the desired altitude.

Based on the above the block diagram of the control system of aviation altitude as shown in the following figure [2]:

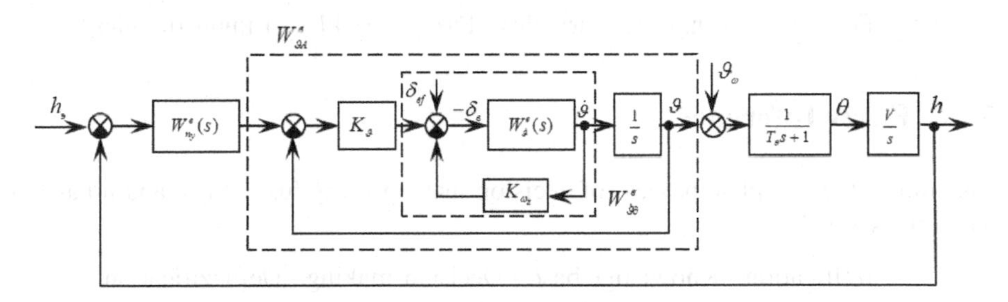

Fig. 2. Block diagram of the control system of aviation altitude

We can model this case using MATLAB according to the aviation system in which specific constants as the following [2]:

kh= K_H =5; knew= K_{ϑ}=5.97;

kwz= $K_{\omega z}$ =2.53; ttheta= T_{θ} =0.3;

walfa= ω_{α} =1.61; psayalfa= ξ_{α} =0.335;

We must pointed here that the design of controllers and control systems to drive the movement of aircraft considered all the elements ideally (linear elements). However, the reality is different, where some parts of the system are nonlinear elements, for example execution units.

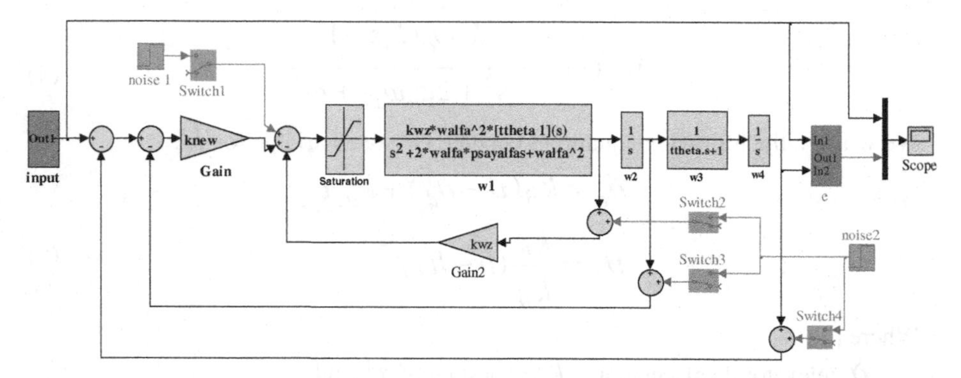

Fig. 3. Block diagram of the control system of aviation altitude in MATLAB with nonlinear elements

We must pointed here that the design Of the advantages of fuzzy logic that unaffected by nonlinear elements, thus the Fuzzy Controllers can be designed to able to deal with the disadvantages of traditional controllers.

4 Fuzzy Logic

This logic is based on some of the most important basic concepts [1]:

Fuzzy sets - Linguistic variables - Fuzzy variables - Linguistic rules.

5 Fuzzy Inference

This unit is the complete process of decision-making using fuzzy logic and it has four basic steps are [1]:

Fuzzification - Knowledge base - Decision-making - Defuzzification.

6 Using of Fuzzy Logic in Control Systems

Control idea is built on the of forming the basis of the control signals (Error Signals) in accordance with the law of a specific model for the effect on the plant and guidance of its processes to achieve the desired goal [1,10].

The main idea in the design of fuzzy control systems is replacing the traditional controller by fuzzy controller, which is built on the basis of Use of the fuzzy

Inference. Then the error signal becomes the input signal of fuzzy controller and processing by fuzzy inference system in fuzzy controller [10].

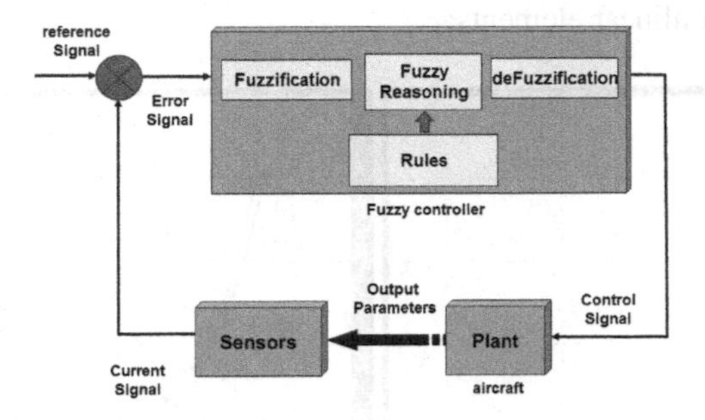

Fig. 4. General scheme of fuzzy control system

7 Design of Fuzzy Logic Controller

For the design of Fuzzy Logic Controller we must initially determine input and output variables for this controller in a case study [9,10]:

7.1. Input variable is the difference between the current altitude and desired altitude (Error signal).
7.2. Output variable is the altitude elevator displacement.

These signals in Fuzzy Logic are linguistic variables and we can describe it by fuzzy variables, thus we can give the input signal (Error) five cases and these cases are:

UL- A Positive Large Error. US- A Positive Small Error.
Z- No Error.
DS- A Negative Small Error. DL- A Negative Large Error.

According to above in design of the Fuzzy Logic Controller the diagram of the control system of aviation altitude with (FLC) becomes as shown in figure 5. [9,10]:

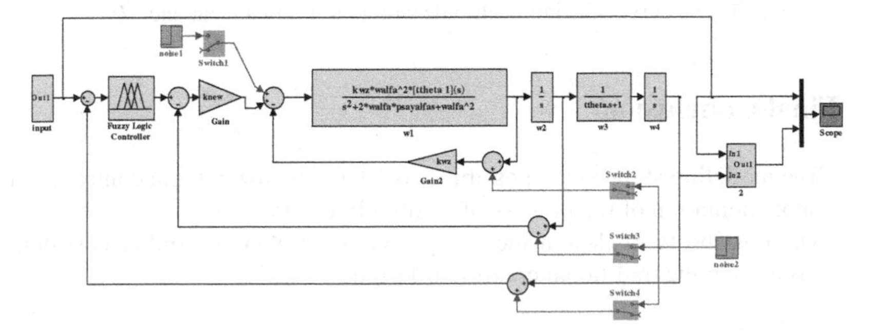

Fig. 5. Block diagram of the control system of aviation altitude with (FLC)

Now the results of comparing between the traditional system and fuzzy system with effect of nonlinearity and measurement error of angular velocity ω_z :

1. Effect of nonlinear elements:

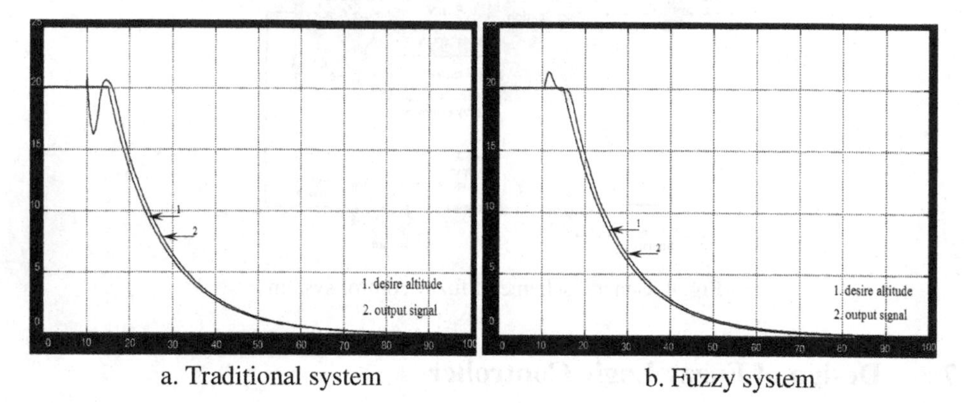

a. Traditional system b. Fuzzy system

Fig. 6. The curve of aviation altitude with nonlinear element

2. Effect of measurement error of angular velocity ω_z :

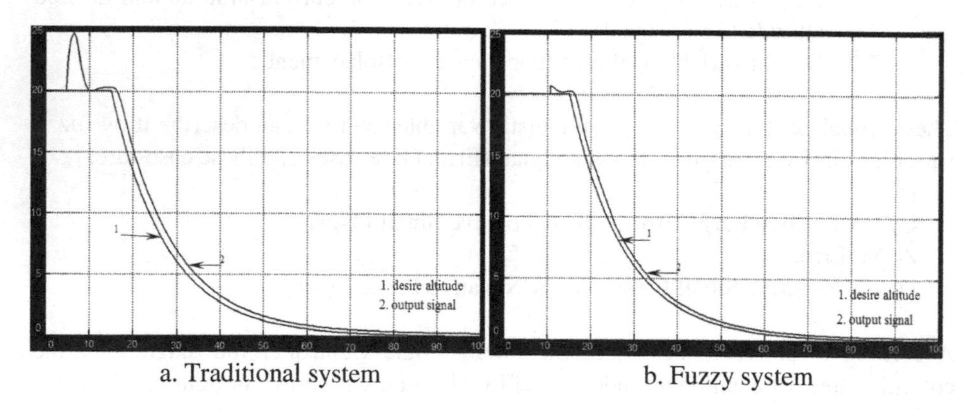

a. Traditional system b. Fuzzy system

Fig. 7. The curve of aviation altitude with nonlinear element and ω_z

8 Final Conclusion

8.1. The modeling showed the possibility of using the fuzzy logic controller in the implementation of the process of landing in theory.

8.2. The possibility of design the fuzzy logic controller in nonlinear systems is easier than the traditional controller design.

9 Results

9.1. The modeling showed that preference of fuzzy logic controller better than the traditional controller when dealing with nonlinear elements in control systems fig. 6. And fig. 7.

9.2. The results showed that the possibility of dealing with some disturbances (ω_z) better than traditional controller fig. 8. And fig. 9

References

[1] Ross, T.J.: Fuzzy Logic with Engineering Applications, 3rd edn. (2010)

[2] Zenh, B.: Aviation control systems – Aleppo University, 2nd edn. (2010)

[3] Moore, H.: MATLAB for Engineers, 3rd edn. Esource/Introductory Engineering and Computing (2011)

[4] Nise, N.S.: Control Systems Engineering, 6th edn. (2011)

[5] Megson, T.H.G.: An Introduction to Aircraft Structural Analysis (2010)

[6] Wahid, N., Rahmat, M.F.: Pitch control system using LQR and Fuzzy Logic Controller. In: 2010 IEEE Symposium on Industrial Electronics & Applications, ISIEA (2010)

[7] Wahid, N., Hassan, N.: Self-Tuning Fuzzy PID Controller Design for Aircraft Pitch Control-Intelligent Systems. In: 2012 Third International Conference on Modeling and Simulation, ISMS (2012)

[8] Fraga, R., Sheng, L.: Non-linear and intelligent controllers for the ship rudder control-Electronics. In: 2011 Saudi International Communications and Photonics Conference, SIECPC (2011)

[9] MathWorks, Fuzzy Logic Toolbox-Design and simulate fuzzy logic systems

[10] Azeem, M.F.: Fuzzy Inference System - Theory and Applications (May 2012)

[11] Attaran, S.M., Yusof, R.: Application of the Fuzzy-logic Controller to the New Full Mathematic Dynamic Model of HVAC System. International Journal of Engineering and Innovative Technology (IJEIT) 2(11) (May 2013)

[12] Dadios, P.: Fuzzy Logic - Controls, Concepts, Theories and Applications (2012)

[13] Lombaerts, T.: Automatic Flight Control Systems - Latest Development (2012)

Video Shot Boundary Detection: A Review

Gautam Pal[1], Dwijen Rudrapaul[2], Suvojit Acharjee[3], Ruben Ray[4], Sayan Chakraborty[5], and Nilanjan Dey[6]

[1] Dept. of CSE, TIT Agartala, India
[2] Dept. of CSE, NIT Agartala, India
[3] Dept. of ECE, NIT Agartala, India
[4] Dept. of IT, GCELT Kolkata, India
[5] Dept. of CSE, BCET, Durgapur
[6] Dept. of ETCE, Jadavpur University Kolkata, India
{goutamtit2003,dwijen.rudrapal,acharjeesuvo,ruben.ray, sayan.cb,neelanjan.dey}@gmail.com

Abstract. Video image processing is a technique to handle the video data in an effective and efficient way. It is one of the most popular aspects in the video and image based technologies such as surveillance. Shot change boundary detection is also one of the major research areas in video signal processing. Previous works have developed various algorithms in this domain. In this paper, a brief literature survey is presented that establishes an overview of the works that has been done previously. In this paper we have discussed few algorithms that were proposed previously which also includes histogram based, DCT based and motion vector based algorithms as well as their advantages and their limitations.

Keywords: Histograms, Transition, Shots, Thresholding, Shot boundary detection.

1 Introduction

Shot boundary detection (SBD) is necessary for automatic video indexing and browsing. It can be applicable for numerous applications like indexing in video database, video compression etc. The basic unit of any video is frame. The structure of a video is shown in Fig. 1. The frame sequences are indexed by frame number. After breaking the video, the obtained frames have identical size. Generally in every one second 25-30 frames are taken. A video shot is a sequence of interrelated consecutive frames taken by a single camera at a stretch. In general, shots are combined to produce a video. Scene may be consist of a single or multiple shots which describe a story unit within a video. Shot boundary detection is based on the identification of visual dissimilarity due to the transitions. The mismatch between two frames generally found while shot change. This dissimilarity appears in different form which is categorized into two types: abrupt (as hard cut) and gradual (dissolve, fade in, fadeout, wipe). An abrupt shot change can be seen in single frame. Fade refers to the slow change of brightness in video that often results in a solid black frame. If the

S.C. Satapathy et al. (eds.), *Emerging ICT for Bridging the Future – Volume 2*,
Advances in Intelligent Systems and Computing 338, DOI: 10.1007/978-3-319-13731-5_14

first shot's pixel are replaced with pixels of second shot in a sequential pattern (for eg. in a pattern from left edge of the video frames to the right), then wipe occurs. A dissolve refers to the process of first shot's images getting dimmed and second shot's images getting brighten, that results into frames within the transition illustrating one image is getting superimposed on the other images.

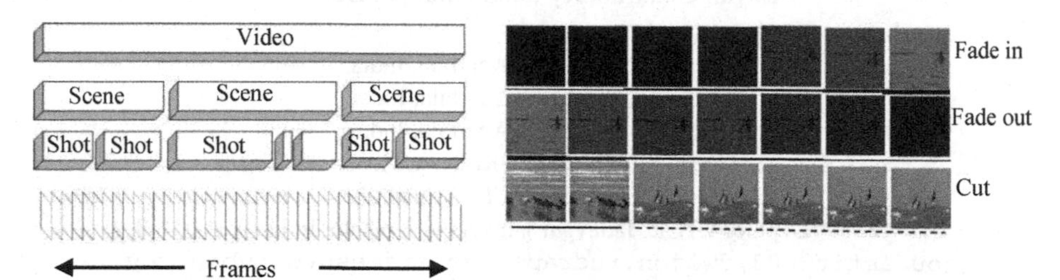

Fig. 1. Structure of a Video

Fig. 2. Gradual and Sudden shot change

1.1 Literature Review

This section focuses on the previous works carried out in the domain of shot boundary detection. A discussion proposed by Boreczky et al. [1] in the year 1996, described different video shot boundary detection techniques and presented a study on the performances of various shot boundary detection algorithms. In the year 1999 Gauch et al. [2] introduced a new technique for Shot Change Detection. This work stated, if two similar shots were joined with a gradual cross fade, then the visual changes might become much smaller than expectation. Shot detection in the VISION (video indexing for searching over networks) system is done by combining three parameter : (1) the average brightness of each video frame, (2) the change in pixel values between two successive frame and (3) the change in color distribution between two consecutive frame. These three quantities are compared with dynamic thresholds to identify the shot change boundaries. In 2001, Heng et al. [3] proposed a shot detection technique based on objects. In this work, finding information with the help of a time stamp transferring mechanism from multiple frames was proposed. Gradual transitions were efficiently handled by this mechanism. This algorithm showed many advantages on the traditional algorithms. In the same year, a simple method of histogram comparison was proposed by Lee et al. [4]. This method ignored spatial information in the frame and was not suitable during highly abrupt luminance changes. Later, Liaoff et al. [5] proposed a smart approach for detecting dissolve in a video by using binomial distribution model. This method determined the threshold needed for discriminating the dissolve due to motions. Liu et al. [6] proposed an algorithm with constant false-alarm ratio (CFAR) for video segmentation. For video cut detection, a theoretical threshold determination strategy using the non-parametric CFAR was developed. It was also capable of finding a controllable precision. Liang et al. [7] used number of edge-based for detecting sudden shot boundaries to avoid the influence of frequent flashlights in many videos. Fuzzy logic was used by Fang et al. [8] to integrate hybrid features that can detect shot boundaries correctly. Shot boundary with gradual shot cuts and sudden shot cuts were detected by different process. Different features were used for fuzzy logic approach based temporal segmentation of videos. Cao et al. [9] proposed a classifier based approach to find the wipes and digital video effects. Six parameters that are causes the formation of feature vector were evaluated for every frame in a temporal window. A supervised SVM classifier based on feature vectors classified the frames in sudden shot change, gradually shot change and no scene change categories. An automatic shot detection technique proposed by Huang, et al. [10] performed very poorly due to high false detection rates caused by camera or object motion. This work tried to overcome the issues in shot detection mechanisms, using local key point matching of video frames that detected both sudden and gradual change efficiently. Changes with long time (i.e fade in, fade out or dissolve) are very hard to locate using low level features. Although, objects compared between two consecutive frames can help to easily identify shot [11] changes. On one hand effect of camera motion and object motion can be easily avoided by detecting objects inside the frames. At the same time it can efficiently detect both abrupt transitions and gradual transitions. This provides a single approach for different shot boundary detection. ASCD can be considered as an automatic operation in real time application. Weighting variance of frame difference and histogram variation extensively used by ASCD. Adaptive threshold can be calculated from the following frame of a last shot change to the previous one of a current frame. It also uses histogram variation of successive frames to set an automatic weighting factor. An automatic SCD algorithm using mean and variance based have a higher detection rate than pixel based methods. K-means clustering was used by Xu et al. [12] in the shot detection algorithm. It first extracted the color feature and then obtained the dissimilarity of video frames using the features. Video frames were grouped by performing graph-theoretical algorithm in Xu et al.'s [13] proposed work. Block based motion was used in order to detect the change in shot which was described by Park et al. [15]. The modified displaced frame difference (DFD) and the block wise motion similarity are combined for the detection of change in shot. Mishra et al. [16] proposed an algorithm that firstly mined structure features from every video frame [17] using dual tree complex wavelet transform and then spatial domain structure similarity was computed between consecutive frames. Later a comparative study was performed [18] between block matching SBD algorithm and dual tree complex wavelet transform based SBD algorithm, in terms of various parameters like false rate, hit rate, miss rate tested on a set of different video sequence. Lu et al. [17] presented a Video Shot Boundary Detection technique that was based on segment selection and singular value decomposition (SVD) with Pattern Matching. In this work shot boundaries' position and gradual transitions' length were calculated using adaptive thresholds and most of the non-boundary frames were discarded at the same time.

2 Methodology

2.1 Features Based Techniques

Features are calculated either from the whole frame or from a portion of it, which is also known as region of interest (ROI). Features include the following parameters: (I)

Luminance and color: The simplest feature which is used to recognize a ROI is its average grayscale luminance. It is capable of illumination changes [19, 20], (II) Histogram: Another feature for ROI is grayscale or color histogram [21]. It is easier to calculate and often insensitive to rotational, translational and zooming motion of the camera, (III) Image edges: This feature is based on edge information of a ROI. Edges can be used to combine objects or to extract ROI statistics. These are independent of illumination changes, camera motion and correspond to the human visual perception. Although, it has high computational cost, high noise sensitivity and high dimensionality, (IV) Transform coefficients (DFT, DCT, and wavelet): These features are used to describe the texture of a ROI.

Merits: Easy to calculate and fast processing
Demerits: These kinds of features are variant to camera zoom.

2.2 Spatial Domain Feature

The region from where, individual features taken, plays a great role in the performance of shot change detection. Features like luminance are extracted from every pixel of a frame and used for shot detection. In next approach each frame is divided into equal-size blocks and a set of features is extracted per block.

Merits: 1) This approach is invariant to small camera and object motion [20].
2) Feature extraction from arbitrarily shaped and sized regions exploits the homogeneous regions, with better detection of discontinuities.
Demerits: 1) This kind of feature is high computational complexity and instability due to the complexity of the algorithms involved.
2) It also has poor performance while measuring the difference [21] among two similar shots.

2.3 Temporal Domain of Continuity Metric

Another choice of shot boundary detection algorithms in which temporal window of frames is used to do shot change detection also referred as temporal domain of continuity metric. Dissimilarity between two successive frames can be measured by looking for discontinuity metric's higher value between two adjacent frames [22]. This approach may fail if different parts' activity varies significantly. To resolve the above problems the dissimilarity can be detected by using the features of all existing frames within that temporal window [21]. This can be done by calculating dynamic threshold and comparing with frame-by-frame discontinuity metric or by measuring discontinuity metric directly on the window. In another method one or more statistical features is calculated for whole shot and compare with the next frame for consistency [21, 22]. If there is existence of variation within shots, statistical features then calculation for an entire shot may not be effective. In different approach the complete video is taken as consideration to measure its characteristics for detecting shot change [28].

Demerits: 1) The system fails if the video has variation within and between the shots. 2) If there is existence of variation within shots, statistical features then calculation for an entire shot may not be effective.

2.4 Shot Change Detection Technique

(I) Thresholding: Calculated feature values are compared with a fixed threshold [21, 22], (II) Adaptive Thresholding: In this type of thresholding the above mentioned problem is solved by taking threshold value which can vary based on average discontinuity within a temporal domain [19], (III)Probabilistic Detection: Shot changes detection can be done by modeling the pattern of specific types of shot transitions and then changing the shot estimation assuming their specific probability distributions [20,23], (IV) Trained Classifier: This technique formulates the shot change detection as a classification problem, with two classes: "Shot change" and "no shot change" [24].

Merits: This is more efficient than previous approaches.
Demerits: This type of techniques comes with a higher computational complexity.

3 Various Shot Difference Measurement Approach

Previous works were done to detect automatic cut within video, but recent research are mainly focused on detecting gradual transitions. Thus, the techniques used recently for these types of works use pixel differences, statistical differences, histogram comparisons, edge differences, compression differences, and motion vectors.

3.1 Pixel Comparison

Pixel difference between two successive video frames or the percentage of pixels that has been changed in two successive frames is compared. This approach is sensitive to fast object and camera movement, camera panning or zooming,

3.2 Statistical Based Difference

Frames are divided into small regions and statistical feature of each pixels within these regions are calculated of each successive frames. Kasturi and Jain [25] calculated standard deviation as well as the mean of the gray levels in various regions of the images. This approach is noise tolerant, but slow due to complex statistical computation.

3.3 Likelihood Ratio

It minimizes the problem of false detection due to camera movements. Without comparing the pixels, likelihood ratio compares statistical features known as likelihood ratio of the corresponding regions or blocks in two successive frames. If the likelihood ratio is larger than the threshold, then it is assumed that region is changed.

3.4 Histogram-Based Difference

Color histogram of each frame is calculated and compared among each other to detect shot boundaries. If the bin-wise difference among the two histograms becomes larger than the preset threshold, then a shot boundary is detected. In order to detect shot boundaries, Ueda *et al.* [26] applied the color histogram change rate.

3.5 Region Based Histogram Differences

Boreczky and Rowe [28] spitted each image into 16 blocks consists of 4x4 pattern. For each image, a 64-bin gray-scale histogram is measured for each region. A Euclidean distance is further calculated to find the difference between the region histograms of two successive images. If the distance is more than a threshold, the region count for that image is incremented. If the value of count is larger than some preset threshold, a shot boundary is detected.

3.6 Edge Based Difference

In this approach the edges of consecutive aligned frames are detected first and then the edge pixels are paired with nearby edge pixels in the other image to find out if any new edges have entered the image or whether some old edges have disappeared. Zabih *et al.* [29] compared color histograms, chromatic scaling and their own technique based on edge detection.

3.7 Motion Vectors

Ueda *et al.* [30] and Zhang *et al.* [31] used motion vectors [32,32,34] on block matching algorithm in order to measure if the shot was a pan or a zoom.

3.8 Adaptive Thresholding

Choosing the proper threshold value is an important criterion in both color histogram comparison and edge change tracking algorithms. An adaptive threshold can be a better option to enhance the shot change detection precision. It uses the local thresholds of the feature or similarity function to be compared, which may be histogram similarity and equivalent contextual region (ECR), respectively.

4 Results and Discussion

Different features and different methods were used to solve the different challenges in the shot boundary detection algorithm. Sometimes multiple features are combined for selection. In this section, a comparative study of popularity of different features and techniques is provided. Though these algorithms are efficient for structured videos but still shot boundary detection is a huge challenge till now for videos with fast changes between shots. This study will help us to find the most popular features and techniques as well as the less explored features and techniques in SBD. The less explored ones can be explored further to solve the challenges faced in SBD in our future work.

Table 1. Utilization popularity of different features in Shot Boundary Detection algorithms

Luminance and Color	Histogram Analysis	Edge Information	Transformation Coefficients (DCT, DFT etc.)	Statistical Measurement	Motion Analysis	Object Detection
4	28	3	7	16	13	2
8%	56%	6%	14%	32%	26%	4%

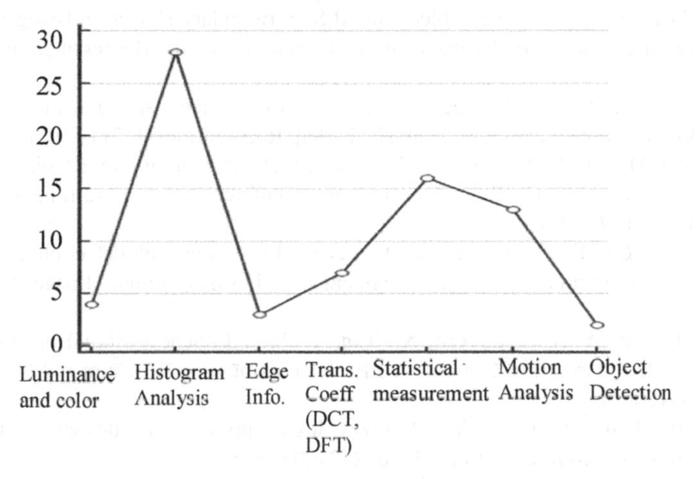

Fig. 3. Utilization popularity of different features in Shot Boundary Detection algorithms

Table 1 and Fig. 3 show the utilization popularity of different features in shot boundary detection algorithm. Fifty research papers were studied from year 1996 to 2014 to calculate the utilization popularity. Hence, the percentage in Table 1 suggests the percentage of usage popularity of every feature in the procedures. The most popular feature to find the change in shot is histogram analysis. Histogram can be of gray images or color images that are used mainly in histogram analysis. Different statistical measurement like variance and mean of intensity and color are also used frequently in different algorithms. Sometimes statistical measurements are combined with histogram and other features for detection of shot boundary. Motion analysis between two consecutive frames is another most popular technique in shot boundary detection.

5 Conclusion

Shot change detection is a very challenging task. For unstructured video, the shot change detection techniques performed were very poor. Only for structured video, these techniques produced better and comprehensive results with a very steady shot change rate. Besides these, the results of the discussed techniques in this work were inefficient for videos with very fast shot change. Hence, there is a scope of further improvement in these techniques. The less explored features like object detection and object analysis can be explored more to improve the techniques.

References

1. Boreczky, J.S., Rowe, L.A.: Comparison of video shot boundary detection Techniques. Journal of Electronic Imaging 5(2), 122–128 (1996)
2. Gauch, J.M., Gauch, S., Bouix, S., Zhu, X.: Real time video scene detection and classification. Information Processing and Management 35, 381–400 (1999)
3. Heng, W.J., Ngan, K.N.: An Object-Based Shot Boundary Detection Using Edge Tracing and Tracking. Journal of Visual Communication and Image Representation 12, 217–239 (2001)
4. Lee, M.-S., Yang, Y.-M., Lee, S.-W.: Automatic video parsing using shot boundary detection and camera operation analysis. Pattern Recognition 34, 711–719 (2001)
5. Liao, H.Y.M., Su, C.W., Tyan, H.R., Chen, L.H.: A motion-tolerant dissolve detection algorithm. In: IEEE 2nd Pacific-Rim Conference on Multimedia, vol. 2195, pp. 1106–1112 (2002)
6. Liu, T.-Y., Lo, K.-T., Zhang, X.-D., Fengc, J.: A new cut detection algorithm with constant false-alarm ratio for video segmentation. J. Vis. Commun. Image R. 15, 132–144 (2004)
7. Liang, L., Liu, Y., Lu, H., Xue, X., Tan, Y.-P.: A Enhanced Shot Boundary Detection Using Video Text Information. IEEE Transactions on Consumer Electronics 51(2), 580–588 (2005)
8. Fang, H., Jiang, J., Feng, Y.: A fuzzy logic approach for detection of video shot boundaries. Pattern Recognition 39, 2092–2100 (2006)
9. Cao, J., Cai, A.: A robust shot transition detection method based on support vector machine in compressed domain. Pattern Recognition Letters 28, 1534–1540 (2007)
10. Huang, C.-R., Lee, H.-P., Chen, C.-S.: Shot Change Detection via Local Key point Matching. IEEE Transactions on Multimedia 10(6), 1097–1108 (2008)
11. Kim, W.-H., Moon, K.-S., Kim, J.-N.: An Automatic Shot Change Detection Algorithm Using Weighting Variance and Histogram Variation. In: ICACT, pp. 1282–1285 (2009)
12. Xu, L., Xu, W.: A Novel Shot Detection Algorithm Based on Clustering. In: 2010 2nd International Conforence on Education Technology and Computer (ICETC), pp. 1570–1572 (2010)
13. Xu, W., Xu, L.: A Novel Shot Detection Algorithm Based on Graph Theory. In: 2010 2nd International Conference on Computer Engineering and Technology (ICCET), pp. 3628–3630 (2010)
14. Donate, A., Liu, X.: Shot Boundary Detection in Videos Using Robust Three-Dimensional Tracking. IEEE Transactions on Consumer Electronics, 64–69 (2010)
15. Park, M.-H., Park, R.-H., Lee, S.-W.: Efficient Shot Boundary Detection for Action Movies Using Blockwise Motion-Based Features. In: Bebis, G., Boyle, R., Koracin, D., Parvin, B. (eds.) ISVC 2005. LNCS, vol. 3804, pp. 478–485. Springer, Heidelberg (2005)
16. Mishra, R., Singhai, S.K., Sharma, M.: Video shot boundary detection using dual-tree complex wavelet transform. In: 2013 IEEE 3rd International Advance Computing Conference (IACC), pp. 1201–1206 (2013)

17. Lu, Z.M., Shi, Y.: Fast Video Shot Boundary Detection Based on SVD and Pattern Matching. IEEE Transactions on Image Processing 22(12), 5136–5145 (2013)
18. Mishra, R., Singhai, S.K., Sharma, M.: Comparative study of block matching algorithm and dual tree complex wavelet transform for shot detection in videos. In: 2014 International Conference on Electronic System, Signal Processing and Computing Technologies (ICESC), pp. 450–455 (2014)
19. Campisi, P., Neri, A., Sorgi, L.: Automatic dissolve and fade detection for video sequences. In: Proc. Int. Conf. on Digital Signal Processing, vol. 2, pp. 567–570 (2002)
20. Lelescu, D., Schonfeld, D.: Statistical sequential analysis for real-time video scene change detection on compressed multimedia bit stream. IEEE Trans. on Multimedia 5(1), 106–117 (2003)
21. Cernekova, Z., Kotropoulos, C., Pitas, I.: Video shot segmentation using singular value decomposition. In: Proc. 2003 IEEE Int. Conf. on Multimedia and Expo, vol. II, pp. 301–302 (2003)
22. Heng, W.J., Ngan, K.N.: An object-based shot boundary detection using edge tracing and tracking. Journal of Visual Communication and Image Representation 12(3), 217–239 (2001)
23. Hanjalicb, L.: Shot-boundary detection: Unraveled and resolved. IEEE Transaction Circuits and Systems for Video Technology 12(2), 90–105 (2002)
24. Lienhart, R.: Reliable dissolve detection. In: Storage and Retrieval for Media Databases. Proc. of SPIE, vol. 4315, pp. 219–230 (2001)
25. Kasturi, R., Jain, R.: Dynamic vision. In: Kasturi, R., Jain, R. (eds.) Computer Vision: Principles, pp. 469–480. IEEE Computer Society (1991)
26. Ueda, H., Miyatake, T., Yoshizawa, S.: IMPACT: an interactive natural-motion-picture dedicated multimedia authoring system. In: Proc. CHI 1991, pp. 343–350 (1991)
27. Nagasaka, A., Tanaka, Y.: Automatic video indexing and fullvideo search for object appearances. In: Visual Database Systems II, pp. 113–127 (1992)
28. Boreczky, J.S., Rowe, L.A.: Comparison of video shot boundary detection techniques. In: Storage and Retrieval for Image and Video Databases (SPIE), pp. 170–179 (1996)
29. Zabih, R., Miller, J., Mai, K.: A feature-based algorithm for detectingand classifying scene breaks. In: Proc. ACM Multimedia 1995, pp. 189–200 (1995)
30. Ueda, H., Miyatake, T., Yoshizawa, S.: IMPACT: an interactive natural-motion-picture dedicated multimedia authoring system. In: Proc. CHI 1991, pp. 343–350 (1991)
31. Zhang, H.J., Kankanhalli, A., Smoliar, S.W.: Automatic partitioning of full-motion video. Multimedia Systems, 10–28 (1993)
32. Acharjee, S., Chaudhuri, S.S.: A New Fast Motion Vector Estimation Algorithm for Video Compression. In: ICIEV 2012, pp. 1216–1219 (2012)
33. Acharjee, S., Dey, N., Biswas, D., Chaudhuri, S.S.: A Novel Block Matching Algorithmic Approach with Smaller Block Size for Motion Vector Estimation in Video Compression. In: ISDA 2012, pp. 668–672 (2012)
34. Acharjee, S., Dey, N., Biswas, D., Chaudhuri, S.S.: An Efficient Motion Estimation Algorithm using Division Mechanism of Low and High Motion Zone. In: i-Mac4s 2013, pp. 169–172 (2013)

User Aided Approach for Shadow and Ghost Removal in Robust Video Analytics

I. Lakshmi Narayana, S. Vasavi, and V. Srinivasa Rao

Dept. of Computer Science and Engineering,
Velagapudi Ramakrishna Siddhartha Engineering College, Vijayawada, India
{ilnarayana1226,vasavi.movva}@gmail.com

Abstract. In almost all computer vision applications moving objects detection is the crucial step for information extraction. Shadows and ghosts will often introduce errors that will certainly effect the performance of computer vision algorithms, such as object detection, tracking and scene understanding. This paper studies various methods for shadows and ghost detection and proposes a novel user-aided approach for texture preserving shadows and ghost removal from surveillance video. The proposed algorithm addresses limitations in uneven shadow and ghost boundary processing and umbra recovery. This approach first identifies an initial shadow/ghost boundary by growing a user specified shadow outline on an illumination-sensitive image. Interval-variable pixel intensity sampling is introduced to eliminate anomalies, raised from unequal boundaries. This approach extracts the initial scale field by applying local group intensity spline fittings around the shadow boundary area. Bad intensity samples are substituted by their nearest intensities based on a log-normal probability distribution of fitting errors. Finally, it uses a gradual colour transfer to correct post-processing anomalies such as gamma correction and lossy compression.

Keywords: Shadow and ghost removal, user-assisted, spline fittings, interval variable.

1 Introduction

A set of connected pixel points detected in motion but not related to any real moving object is called as Ghost. A set of connected background pixel points modified by a shadow cast over them by a moving object is called as Shadow. Shadows and ghosts are ubiquitous in natural scenes, and their removal is an important area of research. Even though there are lot of automatic methods[1,2] available from past work ,they are not matured enough to completely remove shadow/ghost artefacts from images. This paper focuses on user-aided single image shadow and ghost removal. User-aided methods generally provide better shadow/ghost detection and removal at the cost of user given input. According to previous work shadow and ghost effects can be specified as an additive scale fields S_c *and* G_c in the log domain. In this domain, an

S.C. Satapathy et al. (eds.), *Emerging ICT for Bridging the Future – Volume 2*,
Advances in Intelligent Systems and Computing 338, DOI: 10.1007/978-3-319-13731-5_15

image $\hat{H}_c$ with these effects added to an original source image H_c can be represented as follows:

$$\hat{H}_c(x,y) = H_c(x,y) + S_c(x,y) + G_c(x,y) \quad (1)$$

Where c €{R, G, B} indicates each RGB colour space channel, x and y are the pixel coordinates. The darkest area of the shadow region is Umbra, while the wide outer boundary with a nonlinear intensity changes between the umbra and lit area of the image is Penumbra. Most of the user-aided approaches to assist boundary detection require careful highlighting of the boundary area. We propose a method that requires only one rough stroke to mark shadow and ghost sample Major contributions of this approach are as follows:

Easy User Input Specification
Previous work requires precise and exact user-inputs defining the shadow and ghost boundaries. This proposed method only requires a segment highlighted by one rough stroke and grows it on an illumination sensitive image to get initial boundaries.

Interval Variable Pixel Intensity Sampling
Previous work considers only interval-fixed sampling around shadow and ghost regions that causes anomalies near uneven shadow boundaries. To solve this, we proposed an interval-variable pixel intensity sampling according to boundary curvature.

Local Group Optimization for Selected Samples
We propose a local group optimization that balances curve fitness value and local group similarities. Unlike previous work, this approach will filter degraded samples that are replaced with their closest pixel values according to a log-normal probability distribution. This will reduces shadow and ghost removal anomalies.

Gradual Colour Transfer
Post-processing effects will lead to inconsistent shadow and ghost removal compared with the lit areas both in tone and as well as in contrast. We make use of statistics in thin shadow boundaries and the shadow scale field to correct these issues.

2 Related Work

Shadows will come into picture when an object covers the light source, are an ever-present characteristic of our visual experience. If we succeed in detecting ghosts and shadows, then we can easily locate object's shape, and decide where objects contact the ground. These Detected shadows and ghosts also contribute information about illumination constraints and scene geometry.

[3]The main intent of this paper is to build multiple human object tracking based on motion assessment and detection, background subtraction, shadow and ghost removal and occlusion detection. This approach uses morphological operations to identify and remove ghosts and shadows in frames. This proposed strategy can potentially operate even in the presence of occlusions in video streams by detaching them.

Advantages:

- Shadow detection accuracy is enhanced by putting spatial constraint between sub regions of foreground i.e. between shadow and human .
- This approach can effectively remove shadows and ghosts in the presence of occlusions also.

Disdvantages:

- This approach cannot preserve the texture contrast.
- This approach fails to perform dynamic background updation while creating reference background image.

[4]This paper is aimed at detection and removal of shadow in a single image of natural scenes. This approach makes use of region based strategy. Besides taking into account the discrete regions individually, this proposal estimates respective illumination data between segmented regions from their appearances and after that performs pair wise classification based on the collected illumination data. It uses Classification results to build graph of segments and later graph-cut method is employed for discriminating the regions shadow and non-shadow. Finally image matting technique is used to purify the detection results .

Advantages:

- This method can effectively identify and abolish shadows and ghosts from a single still image of natural scenes.
- Every individual pixel's illumination conditions are better recovered by soft matting particularly for those which are at shadow areas boundary.

Disdvantages:

- Shading dissimilarities cannot be discriminated because of cast shadows and surface orientations.
- When there are multiple light sources this algorithm may not work properly.

[5] Shadow and ghost extraction from a single compound natural scene was addressed in this paper. No other streamlined assumption is used other than the Lambertian assumption on camera and on the light source. As it can convert the user supplied rough hints to the potential prior and likelihood functions for Bayesian optimization, this method gains the popularity. Reasonable estimation of the shadowless image is needed by the likelihood function, which can be obtained by solving respective Poisson equation.

Advantages:

- Soft and hard shadows can be removed effectively by preserving the texture under extracted regions.
- It can effectively transform user supplied marks into the productive likelihood and prior functions to the Bayesian optimization.

Disdvantages:

- Shading and shadow effects cannot be distinguished is still remains as a limitation in this single image based approach..

[6]This approach for editing, and rendering shadows and ghost edges in a synthetic image or photograph allows users to isolate image component with the shadow and allows users to make some alterations to its sharpness, position and intensity. For modification of shadows in images tools were developed in this paper. Usually shadows will be distinguished by their soft boundaries and varying sharpness details along the edges.

Advantages:

- This technique can remove the shadows without affecting the underlying image texture.

Disdvantages:

- This technique cannot handle the smooth shadow edges.
- Some shadow edges will not be accurately addressed by this approach.

[7] Only with the very less user support this approach uses Constant illumination distance measure to spot lit and shadowed areas on the same surface of the scene. Affine shadow generation model's parameters can be calculated by these areas. In the end reconstruction process which is of type pyramid-based approach is used to provide shadow-free image with texture preserving and no noise inclusion. At the end image inpainting is applied along the thin boarder to assure smooth evolution between original and recuperated regions.

Advantages:

- Texture and color information at the shadow regions can be well recovered with this approach.
- By reconstructing the affine shadow model's parameters, Shadows with varying intensity can be better handled.

Disdvantages:

- Producing shadow-free images with high quality is still not guaranteed in this approach.
- Even after the shadow is removed some shadow residue is still present in the recovered image.

3 Proposed Method

Given an input image and a user specified umbra segment, we detect the initial shadow and ghost boundaries by expanding the given umbra segment on an illumination image using an active contour. To keep boundary details, we sample pixel intensities along the sampling lines perpendicular to the shadow boundary at variable intervals. We perform a local group optimization task to estimate the illumination change which refines shadow boundary detection and provides an initial scale field. According to an adaptive sample quality threshold, sampling lines with bad samples are replaced by their nearest neighbors and a later local group optimization is applied to them. Finally, this approach relights the shadow area using our scale field and correct post-processing artefacts using gradual colour transfer. Our Proposed method has the following 3 steps as shown in fig. 1:

1.Initial Shadow/Ghost Boundary Detection.

2.Scale Field Estimation.

3.Gradual Colour Transfer.

3.1 Initial Shadow/Ghost Boundary Detection

Determining the initial shadow or ghost boundary is the first step of penumbra detection in most of the previous methods[8,9,10]. we fuse four normalised candidate illumination-sensitive channels from different colour spaces into an illumination image. We measure the confidence values of each candidate channel by using an exponential incentive function φ representing the textureness of each of their umbra sample segments:

$$\varphi(x) = x^{-\lambda} \left(\lambda > 0\right) \tag{2}$$

where x is the pixel intensity, λ(default value 5) specifies the steepness of the incentive function. Lower textureness is preferred as it means higher intensity uniformity of umbra segment. The fused image F is computed as a weighted sum of each normalised candidate channel c_l as follows:

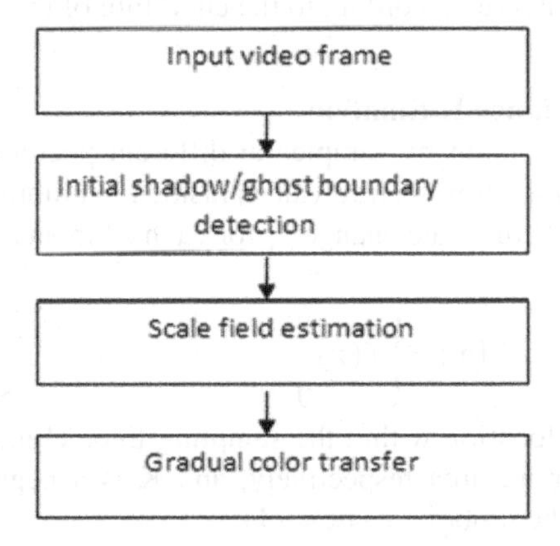

Fig. 1. -Flow diagram of the proposed method

$$F = \left(\sum_{l=1}^{4} c_l \varphi(\sigma_l) \middle/ \left(\sum_{l=1}^{4} c_l \varphi(\sigma_l) \right) \right) \tag{3}$$

Where l is the channel index, σ_l is the standard derivation of the umbra sample intensities of C_l. To avoid texture noise, we apply a bilateral filter [11] to F first. We grow a sparse-field active contour [12] on the fused image to detect the initial shadow boundary.

3.2 Scale Field Estimation

Shadow and ghost affects are represented by varying (or different scale) intensity values. Using a scale field better represents the penumbra and umbra variations, and is used to relight the shadow/ghost area using Eq. 1.we first sample the log domain pixel intensity along the sampling lines perpendicular to the shadow boundary. We adopt a local group spline fitting optimization through the measured sampling line pixel intensities to estimate sparse scales from the initial intensity samples. We replace bad intensity samples with their nearest alternatives and re-optimize for them.

1. Interval-Variable Pixel Intensity Sampling

According to Eq. 1, the logarithms of the original image are supplied for sampling. We sample pixel intensities along the lines perpendicular to the initial shadow/ghost boundary. Uneven boundaries can result in non-smooth vector normal estimations along the shadow boundary. To overcome this, we apply cubic spline smoothing to the initial boundary points before we compute their normals and curvatures. Under-sampling along the boundary neglects sharp details and causes artefacts, while oversampling incurs penumbra removal noise due to texture. More sparse pixel scales are computed for curvy boundary parts for precise in-painting. To avoid texture artefacts, we apply a bilateral filter to the input image before sampling. Our method adjusts the sampling interval according to the curvature of the smoothed boundary.

2. Illumination Variance Estimation

Having obtained sparse intensity samples at different positions along the boundary, our goal is to find illumination scaling values inside the umbra, penumbra and lit area. We model the illumination scale change S_i for each ith intensity sample of each RGB channel as follows:

$$S_i(x) = \begin{cases} K & x < x_1 \\ f(x) & x_1 < x \leq x_2 \\ 0 & x > x_2 \end{cases} \tag{4}$$

Where x is a pixel location within the sampling line, x1and x2 determine the start and end of the penumbra area respectively, and K is a negative scale constant for sample points within the umbra area (x < x1).

3. Gradual Colour Transfer

Image acquisition devices usually apply the concept of post processing, e.g. Gamma correction. Lossy compressions, e.g. JPEG, are also common in images such that compression artefacts (e.g. affecting contrast) in the shadow area become noticeable when removal is applied. To resolve this, we extend the colour transfer with scale field *Sm*. This approach will compute the normalized scale increase hi of the ith sampling line according to Eq. 5 as follows:

$$h_i(x) = (exp(f_i(x)) - exp(K_i))/(1 - exp(K_i)) \quad (5)$$

Here x is the pixel's location of a sampling line, K_i and f_i are respectively the lit constant K and the cubic function piece of the ith sampling line.

Algorithm UASremoval

{

Input:

V: Video

F(1..n) :Frames

U_i: User Input

K(1..f): key frames

S_d: Shadow detection

Output:

Shadow/ghost removed image, S_r.

Method:

```
for input video V
        Convert V into F(1..n)
        For all F1 to Fn
        Compute Shadow detection;
    endfor
  endfor
  for all F1 to Fn do
    select K(1..f)∈ F(1..n)
        for i=1 to f do
        specify U_i on K(1..f) to get S_d and S_r on K(1..f);
        end for
  end for
}
```

This shadow removal algorithm takes video V as input and it converts it into frames F(1..n).This algorithm identifies shadows in all frames F(1..n).Among all frames some of the frames will be considered as key frames k(1..f). For each key frame the user specifies rough sketch U_i on the shadow area .This rough user input will extend up to all shadow region. This algorithm will remove the shadow region from the key frame.

4 Conclusion and Future Work

We have presented a user-friendly texture-preserving shadow and ghost removal method that overcomes some common limitations from the past work. Specifically, our approach retains shadowed texture and performs well on highly-uneven shadow boundaries, non-uniform ghost illumination, and non-white lighting. Our main technical contributions include: (1) highly user-friendly input (2) interval-variable pixel intensity sampling (3) local group optimization and (4) gradual colour transfer. In future work, this approach will focus on more complex cases, such as highly broken shadows, shadowed surfaces with very strong shadow-like textures, and complex reflections in transparent scenes.

References

[1] Weiss, Y.: Deriving intrinsic images from image sequences. In: Proc. Eighth IEEE Int. Conf. Computer Vision 2001, vol. 2, pp. 68–75 (2001)

[2] Finlayson, G.D., Hordley, S.D., Lu, C., Drew, M.S.: On the removal of shadows from images. IEEE Trans. Pattern Analysis and Machine Intelligence 28(1), 59–68 (2006)

[3] Lakhotiya, S.A., Ingole, M.D.: Robust shadow detection and optimum removal of shadow in video sequences. International Journal of Advanced Engineering Research and Studies (2013) E-ISSN2249–8974

[4] Guo, R., Dai, Q., Hoiem, D.: Single-image shadow detection and removal using paired regions. In: Proc. IEEE Conf. Computer Vision and Pattern Recognition, pp. 2033–2040 (2011)

[5] Wu, T.-P., Tang, C.-K.: A Bayesian approach for shadow extraction from a single image. In: Proc. IEEE Int. Conf. Computer Vision, vol. 1, pp. 480–487 (2005)

[6] Mohan, A., Tumblin, J., Choudhury, P.: Editing soft shadows in a digital photograph. IEEE Computer Graphics and Applications 27(2), 23–31 (2007)

[7] Shor, Y., Lischinski, D.: The shadow meets the mask: Pyramid-based shadow removal. Comput. Graph. Forum 27(2), 577–586 (2008)

[8] Liu, F., Gleicher, M.: Texture-consistent shadow removal. In: Forsyth, D., Torr, P., Zisserman, A. (eds.) ECCV 2008, Part IV. LNCS, vol. 5305, pp. 437–450. Springer, Heidelberg (2008)

[9] Arbel, E., Hel-Or, H.: Shadow removal using intensity surfaces and texture anchor points. IEEE Trans. Pattern Analysis and Machine Intelligence 33(6), 1202–1216 (2011)

[10] Arbel, E., Hel-Or, H.: Texture-preserving shadow removal in color images containing curved surfaces. In: Proc. IEEE Conf. Computer Vision and Pattern Recognition, pp. 1–8 (2007)

[11] Paris, S., Durand, F.: A fast approximation of the bilateral filter using a signal processing approach. International Journal of Computer Vision 81(1), 24–52 (2009)

[12] Whitaker, R.T.: A level-set approach to 3d reconstruction from range data. International Journal of Computer Vision 29(3), 203–231 (1998)

A Discrete Particle Swarm Optimization Based Clustering Algorithm for Wireless Sensor Networks

R.K. Yadav, Varun Kumar, and Rahul Kumar

Delhi Technological University, New Delhi, 110042, India
rkyadav.dce.edu, {varuun08,rahul.kmr841}@gmail.com

Abstract. Clustering is a widely used mechanism in wireless sensor networks to reduce the energy consumption by sensor nodes in data transmission. Partitioning the network into optimal number of clusters and selecting an optimal set of nodes as cluster heads is an NP-Hard problem. The NP-Hard nature of clustering problem makes it a suitable candidate for the application of evolutionary algorithm and particle swarm optimization (PSO). In this paper, we shall suggest a PSO based solution to the optimal clustering problem by using residual energy and transmission distance of sensor nodes. Simulation results show a considerable improvement in network lifetime as compared to existing PSO based algorithms and other clustering protocols like LEACH and SEP.

1 Introduction

The use of wireless sensor network has grown enormously in last decade. Wireless sensor networks are used in variety of applications such as monitoring physical and environmental conditions, military surveillance, live stock tracking, home appliance monitoring etc. In most wireless sensor network (WSN) applications nowadays the entire network must have the ability to operate unattended in harsh environments in which pure human access and monitoring cannot be easily scheduled or efficiently managed or it's even not feasible at all [1]. There is a crucial need for scalable and energy efficient routing and data gathering and aggregation protocols in corresponding large-scale environments. In many significant WSN applications the sensor nodes are often deployed randomly in the area of interest by relatively uncontrolled means and they form a network in an ad hoc manner [2, 3]. Sensors in such networks are battery powered and energy constrained and their batteries usually cannot be recharged. Therefore we need energy aware routing and data gathering protocols that offer high scalability and low energy consumption for a long network lifetime. In wireless sensor network hundreds to thousands of sensor nodes are deployed usually randomly. Sensors sense their environment and send their sensed data to a processing centre, called as "Sink" or "Base Station" where all the data is collected and processed [4]. Many routing algorithms have been proposed for efficient transmission of data between base station and sensor nodes. Grouping of sensor nodes into clusters has been widely used by researchers to satisfy the scalability,

S.C. Satapathy et al. (eds.), *Emerging ICT for Bridging the Future – Volume 2*,
Advances in Intelligent Systems and Computing 338, DOI: 10.1007/978-3-319-13731-5_16

high energy efficiency and prolong network lifetime objectives. In clustering the whole sensor network is partitioned into multiple groups of sensor nodes. Each group is called a cluster and each cluster has a leader called cluster head that perform special tasks such as data aggregation and fusion. In addition to supporting network scalability and decreasing energy consumption through data aggregation, clustering has numerous other secondary advantages and corresponding objectives. It can localize the route setup within the cluster and thus reduce the size of the routing table stored at the individual node. It can also conserve communication bandwidth by limiting the scope of inter-cluster communication among CHs and reduce redundant exchange of messages among sensor nodes. Moreover, clustering can stabilize the network topology at the level of sensors and thus cuts on topology maintenance overhead [1] [2] [5].

2 Related Work

Various clustering based routing algorithm has been proposed for efficient utilization of energy of sensor nodes in wireless sensor networks.

LEACH [3] is one of the most popular and widely used probabilistic clustering algorithm that use randomized rotation of CHs and a distributed process of cluster formation based on some priori optimal probability. Many modifications such as LEACH-C [4], LEACH-M [5] etc, to the original LEACH protocol have been proposed.

HEED [6] is another improved and very popular energy efficient protocol. It is a hierarchical, distributed, clustering scheme in which a single-hop communication pattern is retained within each cluster, where as multi-hop communication is allowed among CHs and the base station.. The CH nodes are chosen based on two basic parameters, residual energy and intra-cluster communication cost.

PSO-C [8] is a centralized, PSO based Clustering Algorithm that aims to minimize intra-cluster communication.

EEHC is a probabilistic clustering protocol proposed in [7].The main objective of this algorithm was to address the shortcomings of one-hop random selection algorithms such as LEACH by extending the cluster architecture to multiple hops. It is a distributed, *k*-hop hierarchical clustering algorithm aiming at the maximization of the network lifetime.

In [9] the authors propose such a swarm intelligence-based clustering algorithm based on the ANTCLUST method. ANTCLUST is a model of an ant colonial closure to solve clustering problems. In colonial closure model, when two objects meet together they recognize whether they belong to the same group by exchanging and comparing information about them. In the case of a WSN, initially the sensor nodes with more residual energy become CHs independently. Then, randomly chosen nodes meet each other, exchange information, and clusters are created, merged, and discarded through these local meetings and comparison of their information.

3 Particle Swarm Optimization

Particle swarm optimization (PSO), developed by Dr. Eberhart and Dr. Kennedy in 1995 and inspired by social behaviour of bird flocking or fish schooling is a population based stochastic technique to solve continuous and discreet optimization problems,.

PSO learned from the scenario and used it to solve the optimization problems. In PSO, each single solution is a "bird" in the search space. We call it "particle". All of particles have fitness values which are evaluated by the fitness function to be optimized, and have velocities which direct the flying of the particles. The particles fly through the problem space by following the current optimum particles [10].

PSO is initialized with a group of random particles (solutions) and then searches for optima by updating generations. In each iteration, each particle is updated by following two "best" values. The first one is the best solution (fitness) it has achieved so far. (The fitness value is also stored.) This value is called pbest. Another "best" value that is tracked by the particle swarm optimizer is the best value, obtained so far by any particle in the population. This best value is a global best and called gbest. When a particle takes part of the population as its topological neighbours, the best value is a local best and is called lbest [11].

Suppose, there is a group of K random particles in an n-dimension searching space, the position of the ith particle is $X_i = (x_{i_1}, x_{i_2}, \dots \dots, x_{i_n})$, the personal best value of the particle is $pbest_i = (p_{i_1}, p_{i_2}, \dots \dots, p_{i_n})$, and the velocity of the particle is $V_i = (v_{i_1}, v_{i_2}, \dots \dots, v_{i_n})$. The best value obtained so far by any particle in the population is $gbest = (g_1, g_2, \dots \dots, g_n)$. After finding the two best values, pbest and gbest the particle updates its velocity and positions as follows

$$v_{i_j} = w.v_{i_j} + c_1.r_1\left(p_{i_j} - x_{i_j}\right) + c_2.r_2\left(g_j - x_{i_j}\right) \quad (1)$$

$$x_{i_j} = x_{i_j} + v_{i_j} \quad (2)$$

Where w is inertia and used to control the trade-off between the global and the local exploration ability of the swarm, c1 and c2 are learning factors, r1 and r2 random numbers between 0 and 1.

4 Proposed Algorithm

In this section we describe in detail the working of our proposed algorithm. We assume a wireless sensor network with sensor nodes uniformly distribute across the network. We also assume that location of Base station is fixed inside or outside the sensor network and location of sensor nodes is also known to base station.

4.1 Radio Energy Model

For energy dissipation inside a sensor node for transmitting the data, the first order radio energy model as described in [3] is used.
In order to achieve an acceptable SNR to transmit an L bit message to a node situated at distance d, the energy consumed by radio is given by-

$$E_{Tx}(L,d) = \begin{cases} L.E_{elec} + L.\varepsilon_{fs}.d^2 & \text{if } d \le d_0 \\ L.E_{elec} + L.\varepsilon_{mp}.d^4 & \text{if } d > d_0 \end{cases} \quad (3)$$

Where E_{elec} is the energy, dissipated per bit to run the transmitter or the receiver circuit, ε_{fs} and ε_{mp} depend on the transmitter amplifier model, and d is the distance between the sender and the receiver; By equating the two expressions at $d = d_0$, we get, $d_0 = \sqrt{\frac{\varepsilon_{fs}}{\varepsilon_{mp}}}$.

To receive an L–bit message the radio expends-

$$E_{Rx} = L.E_{elec} \quad (4)$$

4.2 Fitness Function

Success of our proposed algorithm will depend greatly on the formulation of fitness function. So we are defining a fitness function that includes all optimization criteria. Our aim is to minimize the intra-cluster communication energy and energy loss due to cluster head and base station communication, so we can define the fitness of a particle i as

$$F(P_i) = E_1(P_i) + \mu\, E_2(P_i) \quad (3)$$

$$E_1(P_i) = \sum_{k=1}^{K} \sum_{\forall n_{k_j} \in C_k} \frac{f(n_{k_j}, CH_k) - E_{min}}{E_{max} - E_{min}} \quad (4)$$

$$E_2(P_i) = \sum_{k=1}^{K} \frac{g(CH_k, BS) - E_{min}}{E_{max} - E_{min}} \quad (5)$$

$$f(n_{k_j}, CH_k) = \begin{cases} s^2(n_{k_j}, CH_k) & \text{if } s(n_{k_j}, CH_k) \le d_0 \\ s^4(n_{k_j}, CH_k) & \text{if } s(n_{k_j}, CH_k) > d_0 \end{cases} \quad (6)$$

$$g(CH_k, BS) = \begin{cases} d^2_{CH_k,BS} & if\ d_{CH_k,BS} \leq d_0 \\ d^4_{CH_k,BS} & if\ d_{CH_k,BS} > d_0 \end{cases} \tag{7}$$

$$s(n_i, CH_k) = \begin{matrix} min\ (s_{n_i,CH_k}) \\ \forall k = 1,2,\dots,K \end{matrix} \tag{8}$$

Where, $d_{i,j}$ is the distance between node i and node j; s is a function that find the minimum distance cluster head for a given node; f is a function whose value for a given node is proportional to the energy consumed in communication between the node and its cluster head; similarly g signifies the energy loss due to cluster head and base station communication; E_{max} and E_{min} are the maximum and minimum energy loss in the network. C_k is kth cluster in a solution or particle.

E_1 and E_2 are two normalized functions that represent the energy dissipated in intra-cluster communication and due to communication between sink and CHs respectively. F is fitness function and our aim is to minimize this function.

μ is a controlling parameter that control the distance between base station and cluster heads. The higher the value of μ will be the closer will be the CHs from BS. K is the optimal number of cluster heads.

For each particle or solution we will choose k random nodes as cluster heads and remaining nodes will join the cluster whose CH is at minimum distance from it.Then we will evaluate the value of fitness function for each particle and will calculate pbest and gbest. Then we will update the velocity vector and position vector according to equation (1) and (2).

4.3 A New Operator $\oplus_{NW}$

We will define a new operator $\oplus_{NW}$ that when applied on a location with respect to a network, will return a valid sensor node location in the network. In each iteration of our algorithm we will update the location of CHs in each particle or solution. Keeping this into consideration we define $\oplus_{NW}$ as follows:

Suppose $\dot{a} = (a_1, a_2)$ is any location with respect to a sensor network NW then $\oplus_{NW}\ \dot{a}$ will return a valid location in network NW. The operator $\oplus_{NW}$ will first check if $\dot{a}$ a valid location in network or not. If $\dot{a}$ is a valid location than it return $\dot{a}$ as it is; if not then it will return nearest valid location in the network NW toward base station with highest residual energy. After calculating new velocity and position using equation (1) and (2) we will apply our operator to the calculated positions to get valid new positions.

4.4 Clustering Algorithm

```
1. Create and initialize a K-dimensional swarm of P
   particles by choosing K CHs with residual energy
   higher than average energy of network for each
   particle.
2. repeat
3. for each particle i = 1,2,...,P do
4.     if F(X_i) < lbest_i then
5.         lbest_i = X_i
6.     end
7.     if F(X_i) < gbest then
8.        gbest = X_i
9.     end
10. end
11. for each particle i = 1,2,...,P do
12.     update velocity V_i using equation (1)
13.     update position vector X_i using equation(2)
14.     apply ⊕_NW operator to updated position
15. end
16. until the maximum number of iteration reached
```

5 Simulation Results

For simulation we assume a square network field of size 100m X 100m with 100 sensor nodes deployed uniformly in it. We assume that sink is at the centre of the field. We will compare the performance of our proposed algorithm with LEACH and its popular variant LEACH-C. We are using same simulation parameters are described in Table 1. Figure 1, 2 and 3 show the no. of alive nodes in each round of LEACH, LEACH-C and our proposed protocol PSOBC. Simulation results show a considerable improvement in network lifetime.

Table 1. Simulation Parameter

Description	Parameter	Value
Initial energy	E_0	0.5J
Electronic circuitry energy	E_{elec}	50nJ/bit
Multi-path co-efficient	ε_{mp}	10 pJ/bit/m2
Free space co-efficient	ε_{fs}	0.0013 pJ/bit/m4
Data aggregation energy	E_{DA}	5 nJ/bit/signal
Total no. of nodes	N	100
Optimal percentage of CHs	P_{opt}	0.1

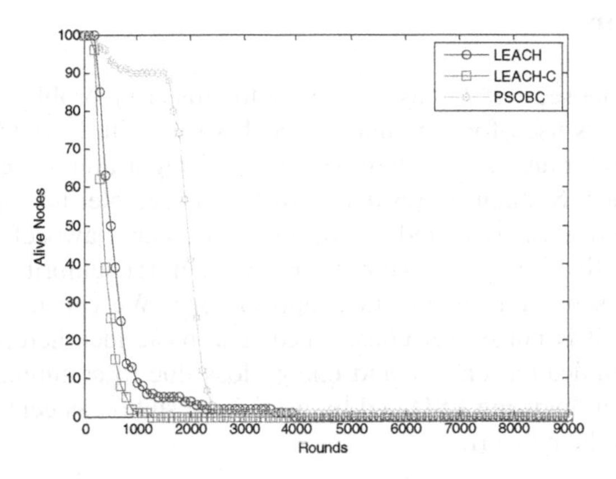

Fig. 1. Alive nodes per round for BS position (50, 0)

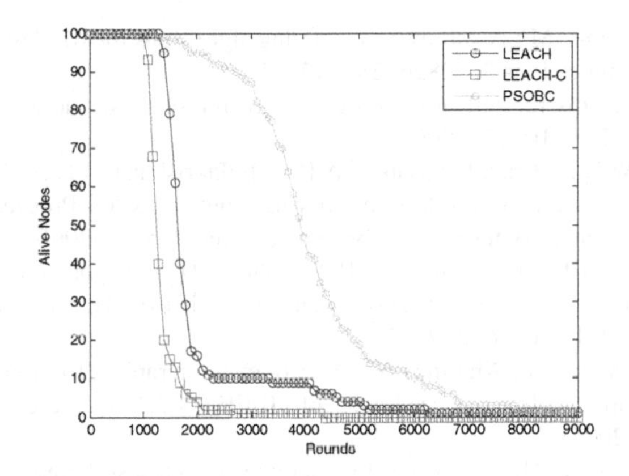

Fig. 2. Alive nodes per round for BS position (0, 50)

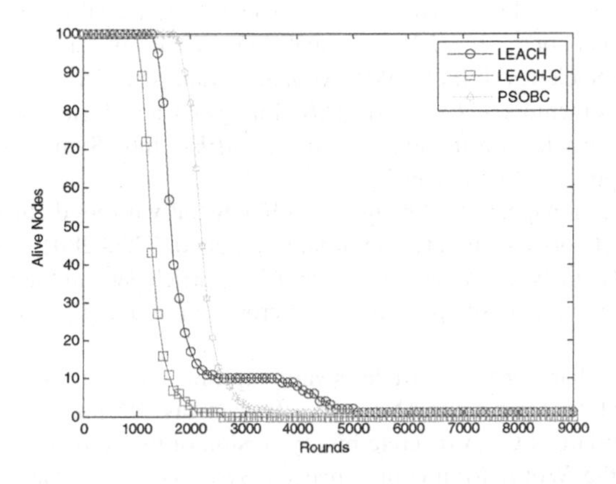

Fig. 3. Alive nodes per round for BS position (200, 200)

6 Conclusion

In this work we proposed a PSO based solution to clustering problem. We used same PSO algorithm that is used for continuous search space with little modification. We defined a new operator and used it with original PSO algorithm to make it work with discrete search space. Simulation results show a considerable increment in Network lifetime as compared to LEACH and LEACH-C. The main drawback of this easy and efficient solution is that it requires the presence of a central authority for cluster setup but it is not always possible in practical applications. We can use base station as central authority if it is not power constrained. The basic idea here was to optimize intra cluster communication energy and energy loss due to communication between CHs and base station by using PSO and by using base station as centralized authority for cluster set up in the network.

References

1. Abbasi, A.A., Younis, M.: A survey on clustering algorithms for wireless sensor networks. Computer Communications 30, 2826–2841 (2007)
2. Sohrabi, K.: Protocols for self-organization of a wireless sensor network. IEEE Personal Communications 7(5), 16–27 (2000)
3. Heinzelman, W.R., Chandrakasan, A.P., Balakrishnan, H.: Energy efficient communication protocol for wireless micro sensor networks. In: Proceedings of the 33rd Hawaiian International Conference on System Sciences (January 2000)
4. Heinzelman, W.B., Chandrakasan, A.P., Balakrishnan, H.: An application specific protocol architecture for wireless microsensor networks. IEEE Transactions on Wireless Communications 1(4), 660–670 (2002)
5. Loscri, V., Morabito, G., Marano, S.: A two-level hierarchy for low-energy adaptive clustering hierarchy. In: Proceedings of IEEE VTC Conference 2005, vol. 3, pp. 1809–1813 (2005)
6. Younis, O., Fahmy, S.: HEED: A hybrid, energy-efficient, distributed clustering approach for Ad Hoc sensor networks. IEEE Transactions on Mobile Computing 3(4), 366–379 (2004)
7. Bandyopadhyay, S., Coyle, E.: An energy efficient hierarchical clustering algorithm for wireless sensor networks. In: 22nd Annual Joint Conf. of the IEEE Computer and Communications Societies (INFOCOM 2003), San Francisco, CA (April 2003)
8. Latiff, N.M.A., Tsimenidis, C.C., Sharif, B.S.: Energy-aware clustering for wireless sensor networks using particle swarm optimization. In: IEEE Intl. Symposium PIMRC 2007, Athens, Greece, pp. 1–5 (September 2007)
9. Selvakennedy, S., Sinnappan, S., Shang, Y.: A biologically inspired clustering protocol for wireless sensor networks. Computer Communications 30, 2786–2801 (2007)
10. Eberhart, R.C., Kennedy, J.: A new optimizer using particle swarm theory. In: Proceedings of the Sixth International Symposium on Micromachine and Human Science, Nagoya, Japan, pp. 39–43 (1995)
11. Kennedy, J., Eberhart, R.C.: Particle swarm optimization. In: Proceedings of IEEE International Conference on Neural Networks, Piscataway, NJ, pp. 1942–1948 (1995)
12. Kennedy, J., Eberhart, R.C.: A discrete binary version of the particle swarm algorithm. In: Proceedings of the World Multiconference on Systemics, Cybernetics and Informatics 1997, Piscataway, NJ, pp. 4104–4109 (1997)

Natural Language Query Refinement Scheme for Indic Literature Information System on Mobiles

Varsha M. Pathak[1] and Manish R. Joshi[2]

[1] Institute of Management and Research affiliated to North Maharashtra University, Jalgaon (MS) India
Pathak.vmpathak.varsha@gmail.com
[2] School of Computer Science , North Maharashtra University, Jalgaon (MS) India
joshmanish@gmail.com

Abstract. The concept and realization of 'Information Pulling' on handheld mobile devices facilitate an easy and effective information access. SMS driven English language information systems for various domains like agriculture, health, education, banking and government are available. However, information systems that support Indic languages are not reported. We have developed an information system that retrieves appropriate transliterated literature documents of Marathi and Hindi languages in response to SMS queries. The proposed system uses Vector Space Model (VSM) to rank the documents according to similarity score. We proposed and used appropriate data structures to enhance the basic VSM for the transliterated document domain. Furthermore we propose to customize the rank of the results obtained in response to user queries. The rank management is based on corresponding relevance feedback given by users.

In this paper we describe the development of the information system for transliterated documents and share our experimental results of relevance feedback mechanism. We have designed a probability relevance model for rank management customization. The performance improvement of the system is demonstrated with the help of standard measures Normalized Discounted Cumulative Gain (NDCG) and Discounted Cumulative Gain (DCG).

Keywords: Information Retrieval, Relevance Feedback Mechanism, Probabilistic Model, Vector Space Model, Discounted Cumulative Gain.

1 Introduction

Mobiles are replacing personal computers with enhanced features. The web based information systems are now accessible on these smart phones. The Information Systems can be benefited with location based information access, customized information access because of the mobile technology. SMS technique is the basic feature of mobile technology mainly used for chatting. If this Human to Human interaction is replaced with Human to Machine interaction, many fruitful applications are possible. Many of such applications are now coming up. GSM based automated remote controlling systems and SMS based information systems are the upcoming

S.C. Satapathy et al. (eds.), *Emerging ICT for Bridging the Future – Volume 2*,
Advances in Intelligent Systems and Computing 338, DOI: 10.1007/978-3-319-13731-5_17

applied research fields. SMS based information system is used for many applications in agriculture, education, health care, tourism, governance, banking sectors [15].

Our literature survey reveals that "SMS based information systems using natural language query" is a dynamic and challenging research extension to the existing Information Retrieval. Transliterated query, SMS style query terms and mixed language terms are the challenging characteristics of the SMS based Information Systems. Also most of this development is related to Information Systems in English language. No significant work is reported in the field of "Literature Information Access in Indic languages on mobiles". Such information system can answer queries like "Who is author of *audu.mbar* poem?" or "What is the title of a song sung by singer *lataa* of a film *meraa naama jokara?"*.

Thus we focus to develop a SMS based Literature Information System for Indic languages. We propose to augment the Vector Space Model with Relevance Feedback Mechanism (RFM) that promises utmost user satisfaction regarding the search results. Literature documents in ITRANS format are collected as the knowledge base at server side. ITRANS is one of the systematic transliteration method specially developed to interconnect Indic languages [20].

We have enhanced the basic Vector Space Model (VSM) in our system. We preferred literature documents in transliterated form with standard encoding like ITRANS. The contents in these documents include self informing tags. We have customized VSM using appropriate data structure to accommodate document term vector for this type of knowledge domain. Further we considered user query comes as free form natural language query. We have designed XML tagging to refine natural language query terms to matching informative tags of our text documents. The relevance rank order produced by VSM module is refined by using relevance feedback mechanism. The probability relevance model is applied for this rank customization. The first call of specific query produces relevance order by applying basic VSM. The system demands for user feedback explicitly. The probability relevance model will train the system using this feedback to produce improved relevance order of literature documents for this specific type of query. The performance of the system is improved with iterative training over the set of queries. This customizes the relevance ranking using relevance feedback.

In second section, we describe methodology used for enhancement in the basic VSM. In third section we describe our enhanced model that refines user query and customizes relevance ranking. The improvement in system performance is demonstrated in fourth section. Analysis and results are presented in the conclusion section of this paper.

2 Methodology

Our information system is designed by applying extension to the conventional information retrieval [1,2, 3] methods. We are using literature of Marathi and Hindi languages as sample knowledge base of our system. We have chosen ITRANS encoding to represent this knowledgebase. This section describes the basic model, the related data structure used for implementation of SMSbLIS system. The enhancement done in the basic model is presented in consecutive subsection.

2.1 Basic Model

We have developed our system using the Vector Space Model (VSM) [6,7]. In this model documents are processed to extract set of keywords of respective document [13]. A key term may relate to more than one document. This results in a vector of terms that germane to the whole document space. Similarly the query submitted by a user is processed to extract query term vector. The angular similarity between the Document Term Vector and Query Term Vector is computed by Cosine Correlation as expressed in Eq. (1). Where doc_i is the ith document of the corpus , $query_j$ is the jth query , $term_{ik}$ is the term weight of kth term in i^{th} document, $qterm_{jk}$ is term weight of kth term of query j. The weights of terms are calculated using Term Frequency/ Inverse Document Frequency (TF-IDF) formulation.

$$Cos(doc_i, query_j) = \frac{\sum_{k=1}^{n}(term_{ik}, qterm_{jk})}{\sqrt{\sum_{k=1}^{n}(term_{ik})^2 \sum_{k=1}^{n}(qterm_{jk})^2}} \quad (1)$$

For a given query each document is assigned a cosine score. The VSM theory says that higher is the score higher is the relevance. Thus documents are arranged in decreasing order of their cosine score to obtain document ranking. We assume that the top ranked documents are precise to obtain required information. Table 1 depicts result of a query “s d burman ne sa.ngeet diya aur lataa ne gaayaa huaa gaanaa”. The query is processed to remove stop words ‘ne’, ‘aur’, ‘huaa’ and query term vector is formed.

2.2 Enhanced Model

As the theory says VSM is keyword searching method. It assigns similarity weights on the basis of angular distance between the query term vector and document term vector. The semantic relation amongst terms is not taken into account in the basic VSM. We found result of VSM based system as quite satisfactory. Relevant documents are top ranked for most of the queries. But as relevance between query and answer given by an IR system is subjective, we found differences in judgments obtained from expert users. In our example as above top ten documents have same similarity score (0.89). When we received judgment of an expert, he has given positive remark to only five documents out of first ten documents. For example Doc#63 (Table-1) at 3rd position has received negative remark.

Thus we explored the documents to understand the possibility of using the tags of these semi-structured literature documents.

Table 1. Sample Query- Top 10 documents

Query Keywords :

1:S 2:D 3:Burman 4:sa.ngeet 5:diya 6:Lata

7:gaayaa 8:gaanaa

Rank#	Document#	Cosine Score
1	19	0.8944271909999159
2	23	0.8944271909999159
3	63	0.8944271909999159
4	133	0.8944271909999159
5	207	0.8944271909999159
6	326	0.8944271909999159
7	341	0.8944271909999159
8	418	0.8944271909999159
9	422	0.8944271909999159
10	447	0.8944271909999159

Self Tagged Documents. We studied different transliterated Marathi and Hindi documents. We found that these documents are self tagged semi-structured documents. These documents are encoded using ITRANS encoding. Tags are used to attribute information about the literature presented in the documents. We propose to standardize these tags to bring uniformity in identification of different type of information related to the literature. As literature category, language, title, writers, editor, date of compilation, date of creation are the few attribute tags. Different parts of the contents like 'Prakaran' (Chapter), 'Pada' (Stranza) are also tagged. All these tagging words are termed as "Tag Term". The actual values and other tokens are called as "Content Term".

Result analysis of VSM module and the feedback obtained from users reveal that there is need to consider the semantic relation between the tag terms and content terms. Like <singer:lataa> is different than <singer:lataa, kishora>. Similarly <music: burman> is different than <singer:burman> or <lyrics:burman>. As expressed in equation 1, our basic VSM model applies unigram term weighting method for cosine correlation calculation. Semantic relation between the words like 'gaayaa' (action, sing) and 'lataa' (person, singer) is not considered. Same applies to words 'sa.ngeet' (synonym, music) and 'burman' (person, music director). Thus unigram *tfidf* weights count same similarity score for documents where the content terms 'burman' and 'lataa' occur. Table-2 demonstrates a few tags and related contents used in Hindi Song documents.

Table 2. Tag-Content Pairs - A Sample

Tag Term	Content Terms
Engtitle	tere mere sapane
Itxtitle	Tere mere sapane
Si.nger	Lataa, Kishora
Lyrics	Yashawa.nta
Music	r d barman

In user queries, terms 'gaayikaa', 'sa.ngeet', 'geetkaara' occur. Whereas, in ITRANS documents, tags like 'si.nger', 'music', 'lyrics' are the synonyms (in English) of these query terms. These tags are associated with the respective contents. For example 'lataa' is name of a singer. The Hindi song document has tagged line '\singer {lataa}'. Here 'singer' is tag and 'lataa' is its value. In user query, word 'gaayikaa' occurs which is Hindi synonym of 'singer'. We need to design a query refinement method to modify original query to add words like 'singer', 'music', 'lyrics' obtained from tags. Similarly remove words like 'gaayikaa', 'sa.ngeet', 'geetkaara' if they are unidentified. The words with lower *tfidf* weights are recognised as unidentified words. In our problem we considered the lower bound value of *tfidf* weights as zero.

Let us see how we modify the basic VSM to accommodate the association of tags and content terms. We explain in next part of this subsection about the document indexing designed for this enhanced VSM model.

Document Indexing. To implement our scheme we used JAVA platform. Let D= $\{d_1, d_2, ...d_m\}$ is the document space and T = $\{ t_1, t_2, ...t_n \}$ is the term vector. A term t_i may occur 0 to k times in any document $d_i \in D$. We used two HASHMAPs as data structure to represent this Document Term Vector Space (DTVS). The tagged lines of literature documents are processed to add both "Tag Term" and "Content Term" in DTVS. In first HASHMAP, each term t_i is mapped to a unique code X. X is generated as next entry number in HASHMAP. Each term t_i thus when occurs first time in document space D, is added in first hash map as < t_i , S >. Where, S is a string. With this a new entry is also added in second HASHMAP as <t_i, X>. Where X is the unique identifier of t_i . The string S associated with term t_i carries information of each occurrence of a term t_i. Each occurrence of the term in any document is represented as a triplet <d,j,l>. Where this triplet points to the document d, tag term j with which the term t_i is associated, and the respective location (line number) l. This triplet is appended in the string S at each occurrence of t_i. This data structure thus produces inverse document index. We use this index to search terms occurring in respective documents, their attached tags and line number where term occurs.

As demonstrated in Figure 3 we assume that term vector includes four terms. These terms 'singer' , 'lataa' , 'starring' and 'kishora' are assigned with identifier as 1,2,3 and 4 respectively in HASHMAP-1. HASHMAP-2 holds the occurrence information of each term. For term 'singer' the first triplet is '1,0,5'. The first value d=1 in this triplet indicates that the term occurs in document number 1. As it is a tag term itself its associated tag term number j=0. And the third value l=5 means that the term occurs on 5th line in that document. The figure shows that the term occurs in three more documents 2, 5, and 8 at 9, 7 and 7 lines in respective documents.

Let us look at other terms. The term ‘lataa’ occurs in documents 1, 5 and 8. It is associated with tag term ‘singer’, in all three occurrences. The term ‘kishora’ occurs in document number 2, location 8 as ‘starring’ (id=3) and in document 8, location 7 as ‘singer’ (id=1).

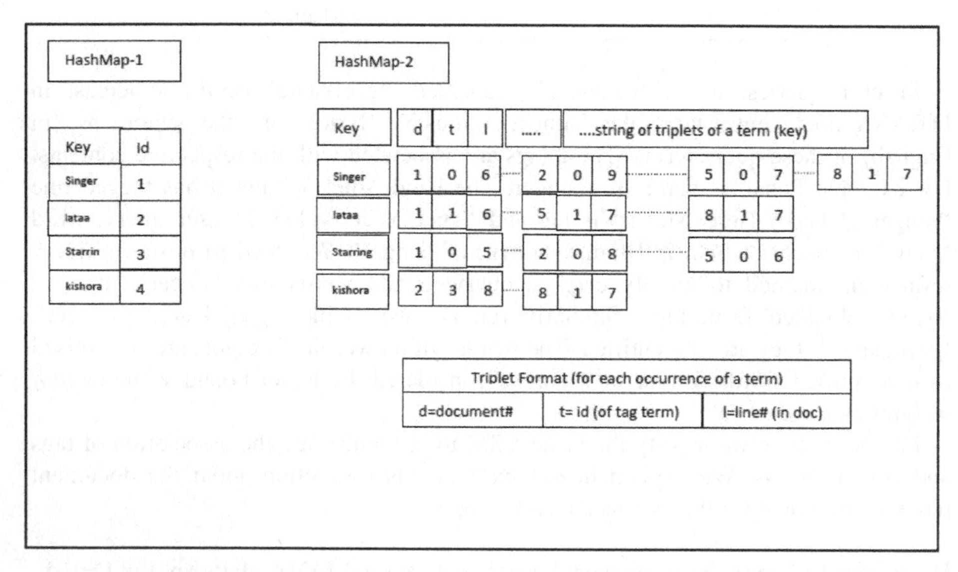

Fig. 1. Structural View of Advanced Model

3 Rank Customization

It is understood by the IR practices that user query gives a vague idea about their information need. To reach up to exact information requirement of the user we need to apply the schemes like query refining, noise filtration, query expansion and relevance feedback mechanism [17]. In our problem we are experimenting on query refining and relevance feedback mechanism [13].

3.1 Query Refinement

Numbers of projects are undertaken by developers and researchers that use XML tagging for indexing the semantically related concepts. Extensible Markup Language (XML) [21] is on the way to concur real world. XML documents are structured documents that can be easily accessed in information systems. Erdmann & Studer [21], express that XML could play important role in the context of knowledge building and dissemination. We have developed XML tagging of document terms to map to their synonyms occurring in user queries. The semantic disambiguation is resolved by adding relative verbs occurred in sample queries. Figure-2 is the sample of the XML tagging we have developed in our system.

```
<Literature>
 <LitSection value="10">
   <SecName>Poetry</SecName>
   <Language>Hindi</Language>
      .
      .
      <Tag>
            <Tname>singer</Tname>
            <Type>person</Type>
         <Synonym>gaayak|gaayeeka|gAyaka|gAyeekA</Synonym>
               <Verb>gaayaa|gAyA</Verb>
               <PreSymbol>slash</PreSymbol>
            <Endswith>hash|percent|dollar</Endswith>
         </Tag>
      .
      .
      .
 </LitSection>
</Literature>
```

Fig. 2. XML Tagging for Literature

Applying this now a query of our example is modified into "s d burman lata music singer title". The scheme removes terms 'sa.ngcct', 'gaayaa' , 'gaanaa' and adds tag terms 'music', 'singer', 'title'. These terms are obtained from 'literature information XML document'. A specific tag set is customized to identify specific class of literature such as poetry, songs, phrases, biographies etc. Thus these terms are useful to sclcct a document set of specific class. Presently we are using collection of only Hindi songs as our domain. Thus the tag set is common to all these documents. There is no effect on Cosine score of respective documents in this domain. We proceed further to customize relevance rank with respect to the *<tag term: content term>* pair values. The method need to have different relevance score to documents with <singer:lataa> than documents with <singer: lataa, kishora>. The documents with <music:burman> should have different relevance score than documents with <lyrics: burman> or <starring: burman>. In next subsection we discuss the Relevance Feedback Mechanism (RFM) [17] that we applied to customize the relevance ranking.

3.2 Probability Relevance Model

Let us consider document is described as the set of attributes and respective values for each attribute, as the literature documents of our system are self documented. The tag terms are the attributes and the content terms are the values of the attributes. In implementation of probabilistic relevance model of our system, we are interested in these <Tag Term , Content Term> pairs. Instead of unigram terms we now consider bigrams in the form $<t_i,t_j>$. where t_i , t_j are two terms occur in same document and are semantically dependent. The semantic relationship between 'tag term' and 'content term' is conveyed by 'XML-tagging'. The system counts occurrences of the bigrams in the documents. To implement relevance feedback mechanism using probability model expert users are asked to judge document D to be relevant or not relevant. As the response has only two values it is known as Binary Relevance Feedback (BRF). Using this relevance feedback for the top ranked documents the probability of relevance of next documents can be estimated applying probability model as discussed below.

If we have some document D and query Q, we have two events:

- L, is the event that D is liked or relevant for a given query q.
- L', is the event that D is not liked or not relevant for a given query q.

As expressed in Ex.2 , let P(L|D) , is the probability that a document with description D is relevant. The description D indicates presence and absence of the $<t_i$, $t_j>$ pairs in the respective document using Vector Space Model of our system.

Given the estimated probability P(D|L) that is the probability that a document is relevant for the given query Q. If prior probability of any document being relevant is P(L) and the probability that a document D is observed independent of its relevance is P(D) then,

The Probability Model follows Naive Bayes' theory [12]. We calculate P(L|D) by applying Bayesian inversion formulation as shown in expression 2 as discussed by Ruthven [13] .

$$P(L \mid D) = \frac{P(D \mid L)(P(L)}{P(D)} \tag{2}$$

Let P_{ik} is $P_q(x_i = 1 \mid L)$ that is the probability of relevance of the document d_k having term x_i for given query q_m . Q_{ik} is $P_q(x_i = 1 \mid L')$ probability of non relevance of the document d_k having term x_i for the query q_m. We derive equation 3 using independence assumption 2 and ordering principle 2 as discussed by Ruthven in [13].

$$C_{ik} = \log \frac{P_{ik}(1 - Q_{ik})}{Q_{ik}(1 - P_{ik})} \tag{3}$$

Where C_{ik} is the weight of individual term $t_i \in T$ calculated applying equation 3. Here T is the term vector. Thus according to equation 4 RSV is calculated as the sum of cost of terms t_i occurring in a document d_k relevant to the query q_m. P_{ik} and Q_{ik} are estimated by applying relevance feedback as described below. One can understand that the probability weight C_{ik} for all term pairs $<t_i ,t_j>$ and documents d_k not only depends on presence of the term but also on absence of the term.

$$RSV = \sum_{t_i \in d_k \cap q_m} C_{ik} \tag{4}$$

When a query asked experts of domain knowledge to give binary (yes for relevant, no for not relevant) judgments for top ten ranked documents of the VSM module. From this judgment remaining documents are judged applying the probability weights of individual <Tag, Value> pair. Documents are ranked based on Retrieval Status Value (RSV) as in equation 4 applying "Probability Ranking Principle" stated by Robertson [11].

4 Results and Analysis

As discussed in this paper we have developed a method that refines given query using XML tags. Also it improves document ranking considering user's feedback. More relevant documents are promoted to higher positions. We used standard measures, Discounted Cumulative Gain (DCG) and Normalized Discounted Cumulative Gain (NDCG) to measure system performance. These measures are useful to understand graded relevance for each rank (r). DCG cumulates the relevance gain of previous r-1 ranks to compute relevance gain of rank r. If a retrieval system has placed a document at position p , DCG_p value indicates as per user's judgments whether it is suitable for specified grade on not. NDCG compares the relevance order of the system with relevance order given by participants.

We applied our relevance improvement model using number of queries of Marathi and Hindi Poems and lyrics of Marathi, Hindi songs. In this paper we present result of performance for a few queries.

For each specified query q, we obtained DCGp for each rank p applying five points grading of the top ten documents given by system. The point 5 denotes highest relevance. That means exact answerable document is observed. Point 0 means no relevance with the query at all. And the intermediate points 4, 3, 2, 1 respectively denote partial relevance with high, moderate, low and very low levels. Also we analyzed results by applying Average Normalized Discounted Cumulative Gain over top ten positions (ranks) for each of sampled queries. Both DCG_p and Average NDCG analysis results are graphically presented in Figure 3 and Figure 4.

From above graphical representation we can observe a definite improvement in the performance of the system. In Figure 3 we can understand the performance improvement for each query individually. Each chart denotes comparison of the DCG for each ranked position for ranks 1 to 10. We could observe that there is no noticeable difference in first 4 to 5 ranks. But next ranks are affected by getting more relevant document at that position. That means for all the sampled queries the relevant documents from top 25 documents have been rearranged by the system at appropriate places as compared to the basic model. The graph in Figure 4 also conveys the same message of improvement in the performance of the system. We can see definite increase in the Average NDCG over all ten ranks for each query.

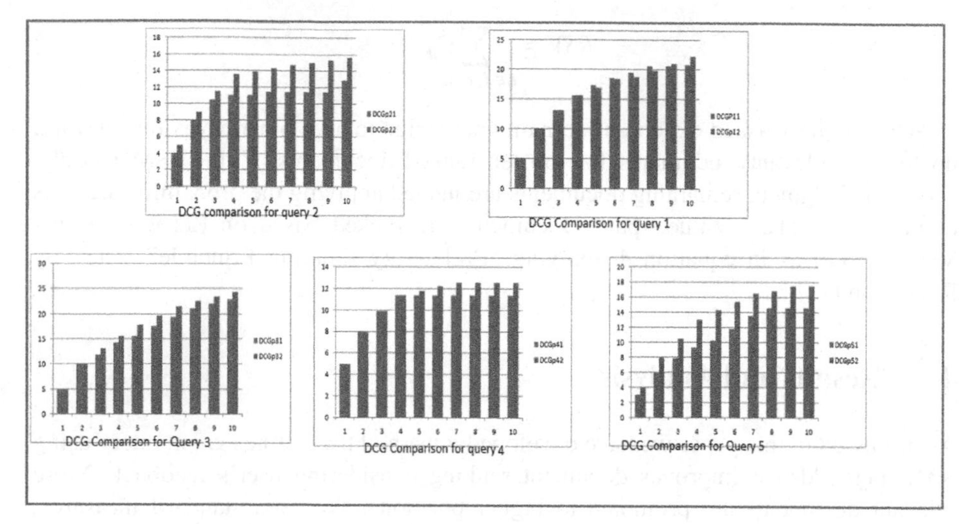

Fig. 3. Discounted Cumulative Gain Comparison of top 10 ranks for 5 selected queries

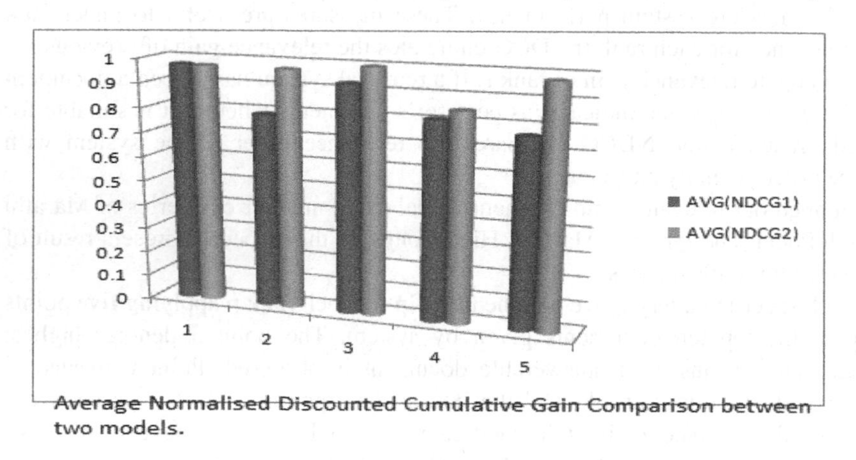

Fig. 4. Average NDCG Comparison

5 Conclusion

This paper is about development of our SMS based Literature Information System (SMSbLIS). The system expects an SMS query from user asking for literature information. The system has to respond with list of relevant document. The system is based on Vector Space Model (VSM) and has support of a suitable knowledgebase built on 'ITRANS formatted Indic Language Literature text documents'. The documents are self documented. We used this feature to refine user query to add more relevant words. We recovered these relevant words by mapping literature attributing words occurring in user query with the synonym words occurring as self documenting tags in literature documents. This mapping is maintained by the system in the form of

'XML tagging'. Query refinement is the first step to enhance relevance of the search result. In second step our VSM based SMSbLIS model progresses to enhance

user satisfaction level. For this the VSM based system is blended with Relevance Feedback Mechanism. We aim to use RFM by indulging benefits of mobile technology. Our efforts are initialized by implementing binary relevance feedback mechanism using Probability Model.

We used DCG and NDCG as the standard measure to demonstrate improvement in the system performance. We found there is definite improvement in the user satisfaction. This result is graphically presented in this paper in Figure 3 and Figure 4.

Encouraged with this result we want to test system performance for more rigorous queries using a large Corpus as our next task.

References

1. Sparck Jones, K.: Automatic Keyword Classification for Information Retrieval. Butterworth's, London (1971)
2. Luhn, H.P.: The automatic creation of literature Abstracts. IBM Journal of Research and Development 2, 159–165 (1958)
3. Schultz, C.K., Luhn, H.P.: Pioneer of Information Science - Selected Works. Macmillan, London (1968)
4. Porter, M.F.: An algorithm for suffix stripping. Program 14(3), 130–137 (1980)
5. Salton, G.: Automatic Information organization and Retrieval. McGraw-Hill, New York (1968)
6. Salton, G.: Automatic text analysis. Science 168, 335–343 (1970)
7. Salton, G. (ed.): The SMART retrieval system – experiments in automatic document processing (1971)
8. Buckley, C., Salton, G., Allan, J., Singhal, A.: Automatic query expansion using SMART TREC-3. In: Harman, D.K. (ed.) Proceedings of the Third Text Retrieval Conference (TREC-3), pp. 69–80. NIST Publication 500-225 (1995)
9. Sparck Jones, K.: A statistical interpretation of term specificity and its application in retrieval. Journal of Documentation 28, 111–121 (1972)
10. Yu, C.T., Salton, G.: Effective information retrieval using term accuracy. Communications of the ACM 20, 135–142 (1977)
11. Robertson, S.E.: The probability ranking principle in IR. Journal of Documentation 33(4), 294–304 (1977)
12. Van Rijsbergen, C.J.: Information retrieval, 2nd edn. Butterworth's (1979)
13. Ruthven, I., Lalmas, M.: A survey on the use of relevance feedback for information access systems. The Knowledge Engineering Review 18(2), 95–145 (2003)
14. Voorhees, E.M., Harman, D.: Overview of the fifth Text REtrieval Conference (TREC- 5). In: Proceedings of the 5th Text Retrieval Conference, pp. 1–29. NIST Special Publication 500-238, Gaitherburg (1996)
15. Joshi, M., Pathak, V.: A Functional Taxonomy of SMSbIR Systems. In: 3rd International Conference on Electronics Computer Technology, Kanyakumari, April 8-10, pp. V6:166–V6:170. IEEE Xplore (2011) ISBN: 978-1-4244-8677-9
16. Ran, A., Lencevicius, R.: Natural Language Query System for RDF Repositories. To appear in Proceedings of the Seventh International Symposium on Natural Language Processing, SNLP (2007)

17. Rocchio, J.: Relevance Feedback in Information Retrieval. In: Salton (ed.) The SMART Retrieval System: Experiments in Automatic Document Processing, ch. 14, pp. 313–323. Prentice-Hall (1971)
18. Hassell, J., Aleman-Meza, B., Arpinar, I.B.: Ontology-driven automatic entity disambiguation in unstructured text. In: Cruz, I., Decker, S., Allemang, D., Preist, C., Schwabe, D., Mika, P., Uschold, M., Aroyo, L.M. (eds.) ISWC 2006. LNCS, vol. 4273, pp. 44–57. Springer, Heidelberg (2006)
19. Karanastasi, A., Christodoulakis, S.: Ontology-driven semantic ranking for natural language disambiguation in the OntoNL framework. In: Franconi, E., Kifer, M., May, W. (eds.) ESWC 2007. LNCS, vol. 4519, pp. 443–457. Springer, Heidelberg (2007)
20. Pathak, V., Joshi, M.: ITRANSed Marathi Literature Retrieval Using SMS based Natural Language Query. Advances in Computational Research 4(1), 125–129 (2012)
21. Erdmann, M., Studer, R.: Ontologies as conceptual models for XML documents. In: Proceedings of the 12th International Workshop on Knowledge Acquisition, Modelling and Mangement (KAW 1999), Banff, Canada (October 1999)

Quality of Service Analysis of Fuzzy Based Resource Allocator for Wimax Networks

Akashdeep

UIET, Panjab University, Chandigarh, India
akash.akashdeepsharma@gmail.com

Abstract. WiMAX is an upcoming technology gaining grounds in recent times that has inherent support for real and non real applications. The rise in number of real time application with popularity of mobile phones always tests scheduler performance of broadband wireless systems like WiMAX. Distribution of resources in such networks has always been a challenging phenomenon. This problem can be solved if scheduling decision is based on traffic conditions of incoming traffic. This paper proposes an application of fuzzy logic by virtue of which an intelligent system for distribution of resources has been defined. The system works as adaptive approach in granting bandwidth to those traffic classes that has relatively higher share of incoming traffic in its queues. The results demonstrate significance of the proposed method.

Keywords: Fuzzy Logic, WiMAX, Quality of service.

1 Introduction

Worldwide Interoperability for microwave access is IEEE 802.16 standard popularized by WIMAX forum under the name WiMAX[1]. It s broadband wireless technology which by virtue of its technical specification is gaining popularity among end users as it provides support for number of real time applications. The popularity of mobile phones has put lot of pressure on today's wireless networks to provide required quality of service. The demand for number of applications is increasing while amount of resources remains limited. Distribution of resources in such networks is always a challenging task as growing quality of service demands of real time applications are always difficult to met. The increase in number of real time applications sometimes makes low priority non real time classes starve for resources.

Traffic in WiMAX network is categorized into five different service classes namely UGS(unsolicited grant service), ertPS(extended real time polling service), rtPS(real time polling service), nrtPS(non real time polling service) and BE(best effoert). IEEE 802.16 standard specifies only priority to these classes and does not specify any fixed mechanism for allocation of resources to these. Equipment manufacturers are free to design and implement their own algorithms [2]. Increasing number of multimedia applications makes resource allocation a very complex and tedious process as real time applications are always hungry for more resources. It becomes very intricate to maintain relatively good quality of service levels for all

S.C. Satapathy et al. (eds.), *Emerging ICT for Bridging the Future – Volume 2*,
Advances in Intelligent Systems and Computing 338, DOI: 10.1007/978-3-319-13731-5_18

sorts of traffic classes and situation gets more complex with rise in number of packets in network. In order to maintain good quality of system performance, allocation of resources shall be immediate and dynamic. This requires scheduling system to be intelligent and powerful so that it can adapt itself to incoming traffic pattern of various applications. This paper discusses performance of one such system developed using fuzzy logic. The fuzzy logic system works according to changes in traffic patterns of incoming traffic and adapts itself to these changes so that appreciable performance level can be maintained for all service classes.

Fuzzy logic is useful where information is vague and unclear and resource allocation process in WiMAX suits application of fuzzy logic to it. Design of intelligent systems for WIMAX networks has started gaining popularity very shortly as number of papers in this direction is still limited. Few of these studies are available at [4]-[10]. Fuzzy logic has been employed by Tarek Bchini et. al. [3] and Jaraiz Simon et. al[4] in handover algorithms. Use of fuzzy logic for implementing inter-class scheduler for 802.16 networks had been done by Yaseer Sadri et al.[6]. Authors had defined fuzzy term sets according to two variables dq_{rt} which means latency for real time applications and tq_{nrt} meaning throughput for non real time applications. Shuaibu et al.[5] has developed intelligent call admission control (CAC) in admitting traffics into WiMAX. Mohammed Alsahag et al.[7] had utilized uncertainty principles of fuzzy logic to modify deficit round robin algorithm to work dynamically on the basis of approaching deadlines. The Fuzzy based scheduler dynamically updates bandwidth requirement by different service classes according to their priorities, latency and throughput by adjusting the weights of respective flows. Similar studies were also given by Hedayati[8], Seo[9] and Akashdeep[10] et al. Authors of current study has already implemented such study that works on fuzzy logic [10]. The study works on two input parameters and outputs a weight value to be used for bandwidth allocation. This paper presents an extension of that approach that utilizes values of instantaneous queue length as an additional variable.

2 Working of System

WiMAX implement a request grant mechanism for resource allocation. Different SS connected to BS request resources from BS and these requests are classified into different queues by classifier of IEEE 8021.6. The scheduler at BS listens to these request and serves these queues by performing two different functions:- allocating resources to different request made by SS and transmission of data to different destinations. BS scheduler component serves these requests on per connection ID basis by taking into consideration available resources and request made by that particular connection. Presently IEEE 802.16 specify priority order for various traffic classes and does not specify any algorithms for resource allocation among these classes. Real time classes have high priority as a result of which low priority non real time classes tend to suffer increased delays in resource allocation process and may sometimes be malnourished. This can be improved by devising strategy that could adapt itself to changing requirements of incoming traffic.

The proposed system is motivated by theories of fuzzy logic where fuzzy logic can work to serves queues belonging to different scheduling services using its uncertainty principles. The designed system works as component of base station and works on three input and one output variables. The input variables are taken as:-.

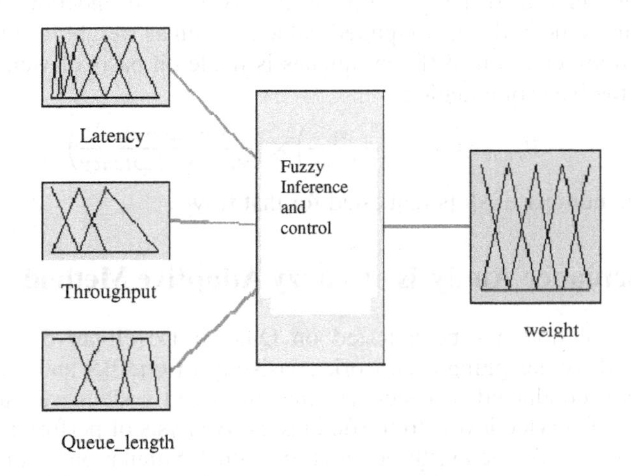

Fig. 1. Structure of fuzzy Logic based system for Resource Allocation

Latency for real time applications, throughput for non real time applications and queue length share of real and non real time applications considered together. The output of fuzzy system is taken as weight of queues serving real time traffic. Membership functions for these variables have been defined utilizing knowledge of domain expert as shown in figure 1. Five different linguistic levels are defined for first input variable and output variables. The membership function are defined as NB(Negative Big), NS(negative small), Z(Zero), PS (Positive small) and PB(positive big). Three membership functions are considered for second and third input variable. The dynamism of variables is taken in range between 0 and 1. The rule base consists of 45 rules which have been framed considering qualities of input variables into consideration. The rule base has been defined considering the nature and dynamism of input traffic and is considered to be sufficiently large.

The initial weight for any flow (i) is calculated from the following equation

$$w_i = \frac{R_{min(i)}}{\sum_{i=0}^{n} R_{\min(i)}} \tag{1}$$

where $R_{min(i)}$ is the minimum reserved rate for flow(i).

$$\sum_{i=0}^{n} w_i = 1 \qquad 0.001 \leq w_i \leq 1 \tag{2}$$

All flows shall satisfy the constraint of equation 2. Equation (2) enables system to allocate minimum value of bandwidth to all flows as weights of queues cannot be zero. Whenever new bandwidth request is received by BS, BS calls fuzzy inference system. The fuzzy system reads values of three input variables, fuzzifies these values and inputs it to the fuzzy scheduler component at BS. Fuzzy reasoning is thereafter applied using fuzzy rulebase and a value in terms of linguistic levels is outputted. At last, de-fuzzification of output value is done to get final crisp value for weight. De-fuzzification is performed using centre of gravity method and inference is applied using Mamdami's method. The outputted value is taken as weight for real time traffic. The bandwidth allocation to different queues is made on basis of weight assigned to that queue on the basis of equation

$$B_{real} = S_i \times \left(\frac{w_i}{\sum_{i=1}^{n} w_i}\right) \times \left(\frac{FrameDuration}{Maximum\ Latency}\right) \quad (3)$$

where S_i is the number of slots requested for that flow.

3 Performance Analysis of Fuzzy Adaptive Method

The proposed solution has been tested on Qualnet Developer 5.2. The proposed scheme is tested by designing a network consisting of one BS and a number of SS. Experiments are conducted to check whether proposed system was able to provide desired quality of service levels to traffic classes. Analysis of performance is done on basis of parameters like delay, throughput and jitter. Simulation is aimed at making sure that proposed scheme is able to provide a relative good QoS levels to all traffic classes. Performance is measured by varying number of SS in ratio of 1:1:1:3:4 for example when total number of SS was 60, the number of UGS, ertPS and rtPS was taken as 6 while number of nrtPS connections was 18 and number of BE connections was 24. Results presented in this section justifies that proposed system was able to provide enough bandwidth opportunities to satisfy increasing requirements of different types of traffic.

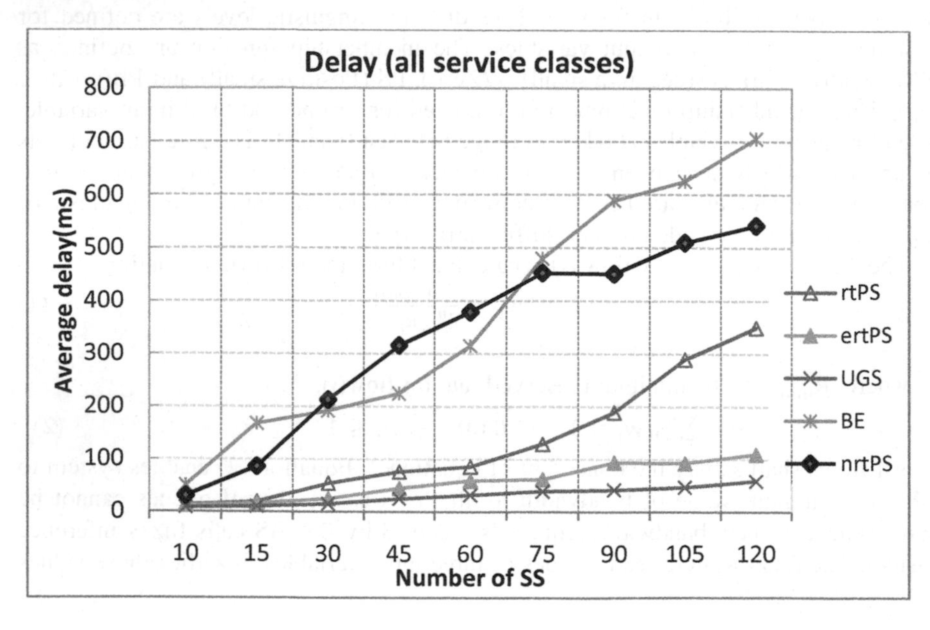

Fig. 2. Delay of various service classes

Figure 2 shows average delay incurred by various services in our fuzzy based inference system. It is evident from figure that delay of UGS and ertPS classes is almost bounded as required by IEEE 802.16 standard. This comes from the fact that scheduler offers higher precedence to UGS and ertPS classes and makes periodic allocations to these classes. Delay of rtPS class shows linear increase till number of SS is about 65 and thereafter growth is almost exponential. This may be attributed to increase in traffic of UGS and ertPS classes which are more prioritized. Delays for nrtPS and BE service classes shows an increasing trend as number of SS increases this is because real time service flows are being offered more share of bandwidth. The delay for BE is better as compared to delay for nrtPS class till a limited number of SS, this is because scheduler was able to provide residual bandwidth opportunities to BE connections. Delay for nrtPS eventually outperforms BE service class as number of connections increases. Nevertheless scheduler was able to avoid starvation of BE flows.

Figure 3 shows throughput of different service classes with an increase in number of SS. Throughput increases as number of connections increase which is expected. Higher throughput for UGS and ertPS is evident as their increasing demand forces system to allocate more amount of bandwidth. Throughput for UGS is almost constant as BS allocates slots to UGS class after fixed interval. Throughput for ertPS shows a small decline as number of SS increase beyond 75, this may be attributed to an increase in number of UGS connections which has high priority. Throughput for rtPS , nrtPS and BE remains almost neck to neck till number of SS are limited(30) as requirements for all classes are getting met thereafter scheduler starts to assign more priority to rtPS as compared to nrtPS and BE in order to satisfy its latency requirements. The throughput for BE class is minimum as there are no QoS requirement for BE class.

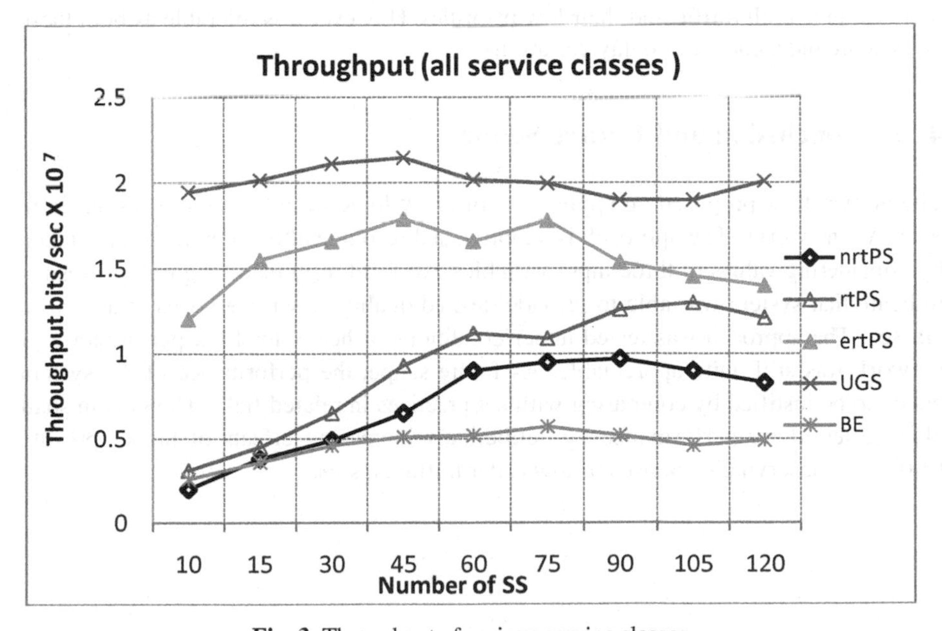

Fig. 3. Throughput of various service classes

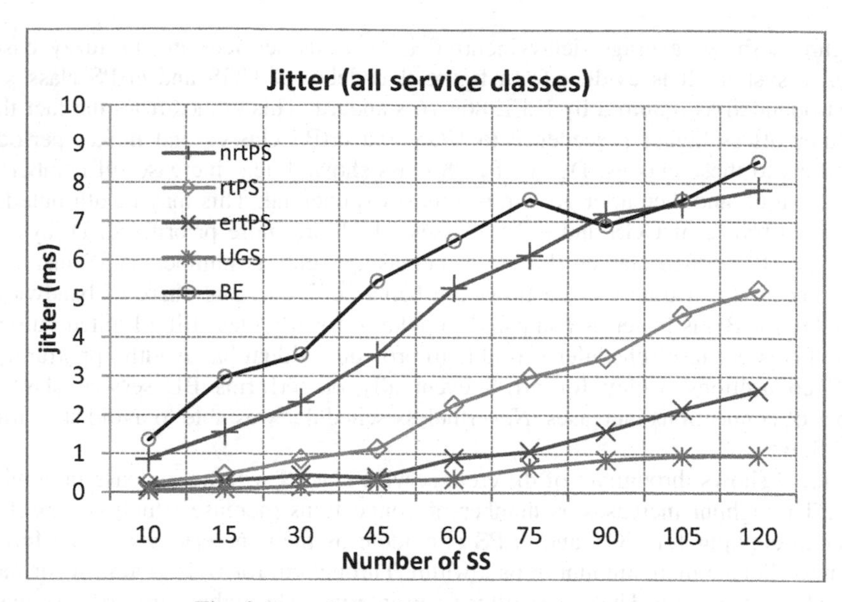

Fig. 4. Jitter observed by various service classes

Figure 4 shows plot of average jitter for our fuzzy based method as function of number of SS. The average jitter for UGS is very small and shows a marginal rise with increase in number of SS. This is because of increase in amount of over all traffic in network and it shows that even UGS class may have packet losses. Jitter of ertPS and rtPS is relatively good considering amount of load being handled by system. Jitter in case of nrtPS and BE is high as expected because of dual reason of increase in overall traffic and their low priorities. However it is tolerable as both these classes are independent of delay variations.

4 Conclusion and Future Scope

The above study proposed an application of fuzzy logic for allocation of resources in WiMAX networks. The approach is adaptive and resource allocation decision is taken by considering values of three input variables extracted from incoming traffic. Results indicate that system was able to provide desired quality of service levels to all traffic classes. The approach was tested under conditions of heavy load but performance of network was still quite appreciable. As future scope, the performance of the system needs to be justified by comparing with set practices in related field. The system shall also be tested for performance by deliberately increasing effects of higher priority traffic and observing responses of low order traffic classes.

References

1. IEEE, Draft: IEEE standard for local and metropolitan area networks. 727 Corrigendum to IEEE standard for local and metropolitan area networks—Part 16: 728 Air interface for fixed broadband wireless access systems (Corigendum to IEEE Std 729 802.16- 2004). IEEE Std P80216/Cor1/D2.730 (2005)
2. IEEE, Draft: IEEE standard for local and metropolitan area networks. 731 Corrigendum to IEEE standard for local and metropolitan area networks – 732 Advanced air interface. IEEE P80216m/D10, 1–1132 (2010)
3. Bchini, T., Tabbane, N., Tabbane, S., Chaput, E., Beylot, A.: Fuzzy logic based layers 2 and 3 handovers in IEEE 802.16e network. J. Com. Comm. 33, 2224–2245 (2010)
4. Simon, J., Maria, D., Juan, A., Gomez, P., Miguel, A., Rodriguez, A.: Embedded intelligence for fast QoS-based vertical handoff in heterogeneous wireless access networks. J. Per. Comp. (2014), http://dx.doi.org/10.1016/j.pmcj.2014.01.009
5. Shuaibu, D.S., Yusof, S.K., Fiscal, N., Ariffin, S.H.S., Rashid, R.A., Latiff, N.M., Baguda, Y.S.: Fuzzy Logic Partition-Based Call Admission Control for Mobile WiMAX. ISRN Comm. and Netw. 171760, 1–9 (2010)
6. Sadri, Y., Mohamadi, S.K.: An intelligent scheduling system using fuzzy logic controller for management of services in WiMAX networks. J. Sup. Com. 64, 849–861 (2013)
7. Alsahag, A.M., Ali, B.M., Noordin, N.K., Mohamad, H.: Fair uplink bandwidth allocation and latency guarantee for mobile WiMAX using fuzzy adaptive deficit round robin. J. Net. Com. Appl. (2013), http://dx.doi.org/10.1016/j.jnca.2013.04.004i
8. Hedayati, F.K., Masoumzadeh, S.S., Khorsandi, S.: SAFS: A self adaptive fuzzy based scheduler for real time services in WiMAX system. In: 2012 9th International Conference on Communications (COMM), June 21-23, pp. 247–250 (2012)
9. Seo, S.S., Kang, J.M., Agoulmine, N., Strassner, J., Hong, J.W. K.: FAST: A fuzzy-based adaptive scheduling technique for IEEE 802.16 networks. In: 2011 IFIP/IEEE International Symposium on Integrated Network Management (IM), May 23-27, pp. 201–208 (2011)
10. Deep, A., Kahlon, K.S.: An Adaptive Weight Calculation based Bandwidth Allocation Scheme for IEEE 802.16 Networks. J. Emer. Tech. Web Inte. 6(1), 142–147 (2014)
11. Deep, A., Kahlon, K.S.: An Embedded Fuzzy Expert System for adaptive WFQ scheduling of IEEE 802.16 networks. J. Expert Sys. With Applications 41(16), 7621–7629 (2014)

An Hybrid Approach for Data Clustering Using K-Means and Teaching Learning Based Optimization

Pavan Kumar Mummareddy and Suresh Chandra Satapaty

Anil Neerukonda Institute of Technology and Sciences,Visakapatnam,India.
{kumarpavan397,sureshsatapathy}@gmail.com

Abstract. A new efficient method for optimization, 'Teaching-Learning Based Optimization (TLBO)', has been proposed very recently for addressing the mechanical design problems and it can also be used for clustering numerical data. In this paper teaching learning based optimization is used along with k-means algorithm for clustering the data into user specified number of clusters. It shows how TLBO can be used to find the centroids of a user specified number of clusters. The hybrid algorithm has been implemented for attaining better results for clustering.

Keywords: K-means, Clustering, TLBO, Optimization.

1 Introduction

Clustering algorithms are divided into two main classes of algorithms namely supervised and unsupervised. The main focus of this paper is on unsupervised clustering. Many unsupervised clustering algorithms have been developed. Most of these algorithms group data into user specified number of clusters. These algorithms include, K-means [7, 8], ISODATA [2], and learning vector quantizers (LVQ) [5].

There have been many population based optimization techniques used for clustering data in data mining literature. Among these PSO,DE,ACO,BF,ABC etc are widely used techniques. Recently a new method for efficient optimization , called ' Teaching Learning Based Optimization(TLBO)' has been introduced for the design optimization of mechanical design problems. This algorithm works based on the influence of a teacher on learners which effects the knowledge level on the learners. Like many other algorithms inspired by nature, TLBO is also a population based stochastic method and uses a population i.e, set of solutions to proceed to the global solution. This paper explores the applicability of TLBO to cluster data vectors. The objective of the paper is to show that the K-Means and TLBO hybrid algorithm can be used to cluster arbitrary data.

S.C. Satapathy et al. (eds.), *Emerging ICT for Bridging the Future – Volume 2*,
Advances in Intelligent Systems and Computing 338, DOI: 10.1007/978-3-319-13731-5_19

2 K-Means Clustering

The clustering algorithms uses, measuring of similarity to determine the closeness of two patterns to each another is the most prominent component. K-means clustering group's data vectors into a user specified number of clusters by considering Euclidean distance as similarity measure. Data vectors with small Euclidean distances are formed as one cluster, and are associated with one centroid vector, represented by the "midpoint" of that cluster. The centroid vector is the mean of the data vectors that belong to the same cluster. For the purpose of this paper, the following symbols are defined as

N_i : denotes the dimension of the input, i.e. the parameter list count of each data vector
N_v : denotes the total number of data vectors to be clustered i.e. dataset
N_k : denotes the number of cluster centroids which were to be specified by the user
d_v : denotes the v-th data vector
c_j : denotes the centroid vector of cluster j
t_j : total number of data vectors in cluster j
S_j : Subset of data vectors that form cluster j.

Using the above notations, the standard K-means algorithm is summarized as

1. Initialize the N_k cluster centroid vectors randomly
2. Repeat
(a) For each data vector, assign the data vector to the cluster with the nearest distance to the centroid vector, i.e. the Euclidian distance to the centroid is obtained using the below formula

$$d_e(d_v, c_j) = \sqrt{\sum_{k=1}^{N_k} (d_{vk} - c_{jk})^2} \tag{1}$$

where k subscripts the input dimension.

(b)For each cluster recalculate the cluster centroid vectors, using

$$c_j = 1/t_j \sum_{\forall D_v \in S_j} D_v \tag{2}$$

until a stopping criterion is satisfied.

The K-means clustering process can be stopped when the user specified maximum number of iterations has been exceeded or when there is small change in the centroid vectors over a number of iterations, or when there are no change in the data vectors of that cluster . For the purposes of this study, when the user-defined number of iterations has been exceeded the algorithm is stopped.

3 Teaching Learning Based Optimization

TLBO algorithm is inspired by a teaching-learning process based on the effect of influence of a teacher on the group of learners in a class. Teacher and learners are the two important components of the algorithm and describes two basic modes of the learning, through teacher (known as teacher phase) and interacting with the other learners (known as learner phase). The TLBO algorithm output is considered in terms of results of the learners which depend on the quality of teaching. So, teacher is usually considered as a highly learned person who trains other learners to obtain better results in terms of their ranks. Moreover, learners can also learn from other learners interacting among themselves which also helps in improving their results. of the entire class in each subject is estimated

TLBO is population based method. In this optimization algorithm, population is the group of learners in that area and different design variables are the different subjects offered to the learners and learners result is considered as the 'fitness' value of the optimization problem. The teacher is the best solution in the entire population. The working of TLBO is categorized into two phases, 'Teacher phase' and 'Learner phase'. Working of the two phases is given below.

3.1 Teacher Phase

This phase of the algorithm simulates the training of the students (i.e. learners) by the teacher. During this phase a teacher exhibits knowledge among the learners and puts maximum effort to increase the average (mean) result of the class. Consider there are 'm' number of subjects (i.e. design variables) offered to 'p' number of learners (i.e. population size, l=1,2,...,p). At any sequential teaching-learning cycle k, $M_{i,k}$ be the mean result of the learners in a particular subject 'i' (i=1,2,...,q). Since a teacher is the most experienced and know ledged person on a subject, so the teacher is the best learner in the entire population in this algorithm. Let $Y_{total\text{-}lbest,k}$ is the result of the best learner among all the subjects, who is considered as a teacher for that cycle. Teacher will have to give maximum effort to increase the knowledge level of the entire class, but the knowledge gained by the learners is according to the talent of teaching delivered by a teacher and the talent of learners present in the class. The knowledge difference between teachers result and mean result as,

$$\text{Mean\_Difference}_{i,k} = r_i * (Y_{i,lbest,k} - T_f Q_{ik}) \quad (3)$$

where, $Y_{i,lbest,k}$ is the result of the teacher (i.e. best learner) in subject i. T_F is the teaching factor which decides to what extent the value of mean to be changed, and random number r_i is considered within the range [0, 1]. Value of T_F can be either 1 or 2. The value of T_F is decided randomly as,

$$T_F = \text{round}\ [1+\text{rand}(0,1)\{2,1\}] \tag{4}$$

T_F is not a parameter of the TLBO algorithm. The T_F value is not taken as an input to the algorithm and its value is randomly decided by the above equation i.e, using Eq. (4).Based on the Mean_Difference $_{i,l,k}$, the existing solution is updated in the teacher phase according to the following expression.

$$Y'_{i,l,k}=Y_{i,l,k} + \text{Mean\_Difference}_{\ i,l,k} \tag{5}$$

where $Y'_{i,l,k}$ is the updated value of $Y_{i,l,k}$. $Y'_{i,l,k}$ is accepted if it gives better function value. All the better function values are maintained at the end of the teacher phase and these values become the input to the learner phase.

3.2 Learner Phase

The learner phase of the algorithm simulates the learning among the students through interaction. The students can be gained knowledge by discussing and interacting among themselves. A learner which has more information will shares his or her knowledge to the other learners so that they will learn new information. The learning phenomenon of this phase is explained below.

Select two random learners M and N such that $Y'_{total-M,k} \neq Y'_{total-N,k}$ (where, $Y'_{total-M,k}$ and $Y'_{total-N,k}$ are the modified values of $Y_{total-M,k}$ and $Y_{total-N,k}$ respectively at the end of teacher phase.

$$Y''_{i,M,k} = Y'_{i,M,k} + r_i * (Y'_{i,M,k} - Y'_{i,N,k}) \quad \text{if } (Y'_{total-M,k} < Y'_{total-N,k}) \tag{6}$$

$$Y''_{i,M,k} = Y'_{i,M,k} + r_i * (Y'_{i,N,k} - Y'_{i,M,k}) \quad \text{if } (Y'_{total-M,k} > Y'_{total-N,k}) \tag{7}$$

$Y''_{i,M,k}$ is accepted if it has a better function value.

4 TLBO Cluster Algorithm

In the clustering concept, a single particle represents the N_k cluster centroid vectors. That is, each particle y_i is constructed as follows:

$$y_i =(c_{i1} \ldots\ldots c_{ij} \ldots\ldots c_{iN_k}) \tag{8}$$

where c_{ij} refers to the j-th cluster centroid vector of the i-th particle in cluster. Therefore, a swarm represents a number of candidate clusterings for the current data vectors. The quantization error is considered as the fitness of particles , is measured using,

$$J_e = \sum_{\forall D \in C_{i,j}} d(D_v, c_j) \quad (9)$$

where d is defined in equation (1), and | S_{ij} | is the number of data vectors belonging to cluster S_{ij} , i.e. the frequency of that cluster.

Using the standard TLBO, data vectors can be clustered as follows:

1. Initialize each particle to contain N_c randomly selected cluster centroids.
2. For t = 1 to t_{max} do
   (a) For each particle i do
   (b) For each data vector d_v
      i. calculate the Euclidean distance d(d_v,c_{ij}) to all cluster centroids S_{ij}
      ii. assign d_v to cluster S_{ij} such that

$$d(d_v, c_{ij}) = min_{\forall c=1.......N_k} \{d(d_v, c_{ik})\}$$

      iii. calculate the fitness using equation (9)
   (c) Update the mean values
   (d) Cluster centroids are updated using equations (5) ,(6) and (7).

where t_{max} is the maximum number of iterations. The population-based search of the TLBO algorithm reduces the effect that initial conditions has, as the K-means algorithm opposed; the search starts in parallel from multiple positions.

5 Hybrid TLBO and K-Means Clustering

The K-means algorithm tends to converge faster (after less function evaluations) than the TLBO, but usually with a less accurate clustering [13]. This section shows that the performance of the TLBO clustering algorithm can further be improved by seeding the initial swarm with the result of the K-means algorithm clusters. The K-means algorithm executes first and then TLBO executes by considering the result of K-means. In this case the K-means clustering is terminated when (1) the maximum number of iterations is exceeded, or when (2) the average change in centroid vectors is less that 0.0001 (a user specified parameter). The K-means algorithm result is then used as one of the particles, while the rest of the particles are initialized randomly.The TLBO algorithm as presented above is then executed.

6 Results

This section compares the results of k-means and hybrid clustering algorithms on three classification problems. The major purpose is to compare the intra-cluster distance of the respective clustering's. Intra-cluster distances, i.e. the distance between data vectors within a cluster, where the fitness function objective is to minimize the intra-cluster distance.

For all the results reported, average values over 10 simulations are taken. The numbers of user defined clusters have to be specified before the algorithm evaluates. All algorithms are run for 100 function evaluations and the hybrid algorithm used 5 particles. The teaching factor for the hybrid algorithm is one. Here no parameters are required to tune the algorithm for better convergence.

The classification datasets used for the purpose of this paper are explained below in detail:

Iris Plant Dataset: This is a well-understand multi-variate dataset with 4 attributes, 3 classes and 150 data vectors. The dataset has no missing values. Clustering is done partially using k-means algorithm whereas complete clustering is done by the hybrid algorithm. Intra-cluster distance is also optimized by using hybrid algorithm. The k-means algorithm traps at local optima because it searches for a solution locally.

Wine Dataset: This is a classification problem with "well-organized" class structures. It consists of 13 attributes, 3 classes and 178 data vectors. The dataset consists of no missing values.

Glass Dataset: This is a "well-behaved" dataset with 10 attributes, 6 classes and 214 data vectors. It consists of no missing values. The k-means algorithm can't cluster the dataset whereas the hybrid algorithm clusters.

Table 1. Comparision of K-means, Hybrid(TLBO) clustering algorithms

Dataset	Algorithm	Intra cluster distance (range)	Intra cluster distance (average)	Best intra Cluster distance	Clustering done	comments
Iris	K-means	104.11-110.46	106.21	104.11	No (partial)	Clustering is done partially
	Hybrid (TLBO)	97.46- 97.75	97.56	97.46	Yes	Clustering is done
Wine	K-means	$1.665 * 10^4$	$1.665 * 10^4$	$1.665 * 10^4$	No (partial)	Clustering is done partially
	Hybrid (TLBO)	$1.635 * 10^4$ - $1.655 * 10^4$	$1.649 * 10^4$	$1.635 * 10^4$	Yes	Clustering is done
Glass	K-means	258–260.89	260.1	258	No	No clustering is done
	Hybrid (TLBO)	339.5-342.6	341.2	339.5	Yes	Clustering is done

The following are the graphs for the wine and iris datasets. The graph consists of the number of iterations on x-axis and the intra-cluster distance on y-axis.

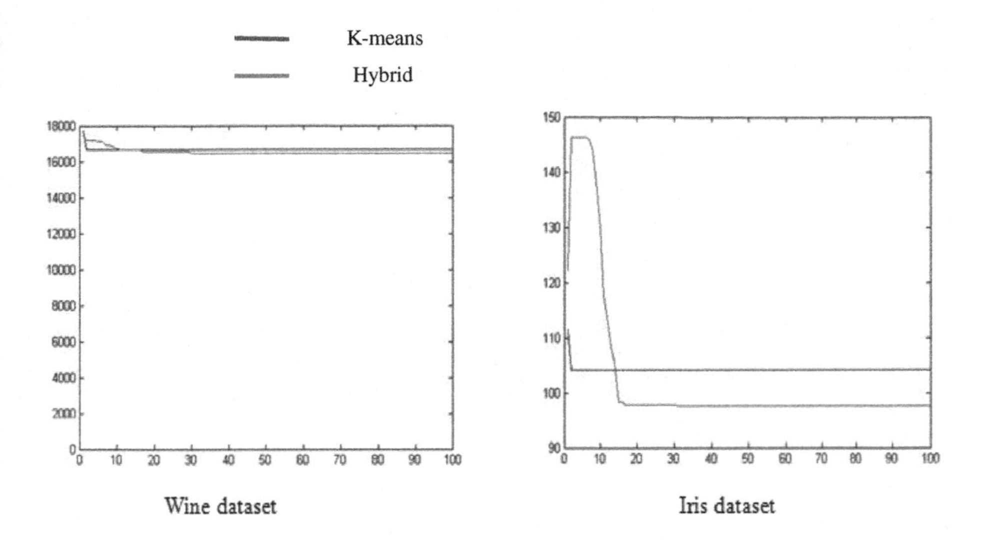

7 Conclusion

By observing the results of k-mean clustering and teaching learning based optimization we found that, the k-mean is sensitive to the initial condition, makes the algorithm to converge to the local optimal solution. On the other side teaching learning based optimization is less sensitive for initial condition due to its population based nature. So teaching learning based optimization is more likely to find near optimal solution.

References

1. van der Merwe, D.W., Engelbrecht, A.P.: Data Clustering using Particle Swarm Optimization
2. Satapathy, S.C., Naik, A.: Data Clustering Based on Teaching-Learning-Based Optimization. In: Panigrahi, B.K., Suganthan, P.N., Das, S., Satapathy, S.C. (eds.) SEMCCO 2011, Part II. LNCS, vol. 7077, pp. 148–156. Springer, Heidelberg (2011)
3. Andrews, H.C.: Introduction to Mathematical Techniques in Pattern Recognition. John Wiley & Sons, New York (1972)
4. Ball, G., Hall, D.: A Clustering Technique for Summarizing Multivariate Data. Behavioral Science 12, 153–155 (1967)
5. Fausett, L.V.: Fundamentals of Neural Networks. Prentice Hall (1994)
6. Fisher, D.: Knowledge Acquisition via Incremental Conceptual Clustering. Machine Learning 2, 139–172 (1987)
7. Forgy, E.: Cluster Analysis of Multivariate Data: Efficiency versus Interpretability of Classification. Biometrics 21, 768–769 (1965)
8. Hartigan, J.A.: Clustering Algorithms. John Wiley & Sons, New York (1975)
9. Kennedy, J., Eberhart, R.C.: Particle Swarm Optimization. In: Proceedings of the IEEE International Joint Conference on Neural Networks, vol. 4, pp. 1942–1948 (1995)
10. Kennedy, J., Eberhart, R.C., Shi, Y.: Swarm Intelligence. Morgan Kaufmann (2002)

A Short Survey on Teaching Learning Based Optimization

M. Sampath Kumar[1] and G.V.Gayathri[2]

[1] Department of Computer Science and Systems Engineering,
College of Engineering, Andhra University, India
sampathau@yahoo.com
[2] Department of Computer Science and Engineering,
Anil Neerukonda Institute of Technology and Sciences, Visakhapatnam, India
gayathri.ganivada@gmail.com

Abstract. Optimization is a process for finding maximum or minimum of a function subject to set of functions. In other words optimization finds the most suitable value for a function within a given domain. Global Optimization means finding minimum value of all local minima and maxima value from all local maxima. The procedure to find out the global maxima or minima point is called a Global Optimization. Teaching-Learning-Based Optimization (TLBO) is recently being used as a new, reliable, accurate and robust optimization technique scheme for global optimization over continuous spaces. In this paper we are comparing the performance of various versions of TLBO.

Keywords: Global Optimization, TLBO, variants of TLBO.

1 Introduction

A wide range of optimization algorithms are there in swarm intelligence and evolutionary computation literature like Genetic Algorithm(GA),Particle Swarm Optimization(PSO), Artificial Bee Colony(ABC),Ant Colony Optimization(ACO), Harmony Search(HS),the Grenade Explosion Method(GEM),etc. are a few of them. GA uses the Darwinian theory of evolution based on the survival of the fittest [1], PSO implements the seeking the behavior of a flock of birds or a school of fish for searching food [2], ABC imitates the scouring behavior of a honeybee[3], ACO works on describing the behavior of an ant searching for a destination from the source [4], HS works on the principle of music improvisation by a group of music players [5] and GEM works on the principle of the explosion of a grenade .

Recently a new optimization algorithm known as Teaching–Learning Based Optimization (TLBO)[6] based on the concept of classroom teaching scenario was proposed by Rao etal. . It soon became a popular tool for solving global optimization problems because of several attractive features like free from the algorithm-specific parameters,(i.e.no algorithm-specific parameters are required for the working of the algorithm),ease in programming, efficiency etc. The basic principle on which TLBO

S.C. Satapathy et al. (eds.), *Emerging ICT for Bridging the Future – Volume 2*,
Advances in Intelligent Systems and Computing 338, DOI: 10.1007/978-3-319-13731-5_20

works lies in the impact of a teacher on the output of learners (students) in a class. The performance of a learner is measured by the results or grades which student (he/she) gets. The teacher plays the role of knowledge supplier[7]. A better teacher produces a better knowledge to the student. It is obvious that a good teacher trains learners such that they can have better results in terms of the marks or grades.

TLBO has been applied to solve different optimization problems effectively and efficiently. Many researchers have proposed few modifications of basic TLBO like Improved Teaching-Learning-Based Optimization (ITLBO), Weighted Teaching-Learning-Based Optimization (WTLBO), Orthogonal Teaching-Learning-Based Optimization (OTLBO), Cooperative Teaching-Learning-Based Optimization (CoTLBO) and Modified Teaching-Learning-Based Optimization (MTLBO). In this paper we compared the better performance of all these algorithms on various real life datasets with various bench mark functions in terms of convergence characteristics and optimum results.

The remaining of the paper is organized as follows: in the Section 2, we give a brief description of TLBO. Section 3 we give description about different variants of TLBO algorithms and their formulae. In Section 4 we compare different variants of TLBO.

2 Teaching Learning Based Optimization (TLBO)

TLBO is a optimization algorithm based on teaching and learning process in a classroom. The searching process consists of two phases, i.e. Teacher Phase and Learner Phase. In teacher phase, learners first get knowledge from a teacher and then from other classmates in learner phase. In the entire population, the best solution is considered as the teacher (*Xteacher*). On the other hand, learners seek knowledge from the teacher in the teacher phase. In this phase, the teacher tries to improvise the results of other individuals (*Xi*) by increasing the mean result of the classroom (*Xmean*) towards his/her position *Xteacher*. In order to maintain uncertain features of the search, two randomly-generated parameters *r* and T_f are applied in update formula for the solution *Xi* as:

$$X_{new} = X_i + \mathrm{r}\,(X_{teacher} - T_f \,.\, X_{mean})$$

where *r* is a randomly selected number in the range of 0 and 1 and T_f is a teaching factor which can be either 1 or 2:

$$T_f^i = round\,[1 + rand(0,1)\{2-1\}]$$

Moreover, X_{new} and X_i are the new and existing solution of *i*.

In the the learner phase, the learners attempt to increase their information by interacting with other learners. Therefore, an individual learns new knowledge if the other individuals have more knowledge compared to the other learner. Throughout this phase, the student *Xi* interacts randomly with another student X_j(where $i \neq j$) in order to improve his/her knowledge. In the case that if X_j is better than X_i (then $f\,(\,X\,)$

$<f(X)$ for minimization problems), Xi is moved toward X_j. Otherwise it is moved away from X_j:

$$X_{new}=X_i+r.(X_j-X_i) \text{ if } f(X_i)>f(X)$$
$$X_{new}=X_i+r.(X_i-X_j) \text{ if } f(X_i)<f(X)$$

If the new solution X_{new} is better, it is accepted in the population. The algorithm will continue until the termination condition is met. The pseudo of TLBO algorithm step by step is as follows:

```
Set k=1
Objective function f(X), X=(x_1,x_2,.........x_d)^T                d=no. of design
variables
Generate initial students of the class room randomly X^i, i=1,2,......,n
n= number of students
Calculate objective function f(X) for whole students of the classroom
WHILE (the termination condition are not met)
    {Teacher Phase}
  Calculate the mean of each design variable X_Mean
  Identify the best solution (teacher)
FOR i=1→n
  Calculate teaching factor T^i_F =round [1+ rand (0, 1) {2-1}]
Modify solution based on best solution (teacher) X^i_new= X^i + rand (0, 1) [X teacher
– (T_F^i . X mean)]
        Calculate objective function for new mapped student f(X^i_new)
IFX^i_new is better than X^i, i.e f(X^i_new) < f(X_i)
        X^i=X^i_new
END IF {End of Teacher Phase}
        {Student Phase}
        Randomly select another leaner X^j , such that j≠i
IF X^i is better than X^j , i.ef(X^i) < f(X^j)
X^i_new =X^i + rand (0,1) (X^i – X^j)
ELSE
X^i_new=X^i + rand (0,1) (X^j - X^i)
END IF
IF X^i_new is better than X^i , i.e f(X^i_new) < f(X^i)
        X^i = ^i_new
END IF {End of Student Phase}
END FOR
  Set k=k+1
END WHILE
```

3 Different Versions of TLBO

3.1 Improved Teaching Learning Based Optimization (iTLBO)

In the traditional TLBO, the teacher phase makes the algorithm proceeds by shifting the mean of the learners towards it teacher where as iTLBO tries to obtain a new set of improved learners by forming a random weighted differential vector the current mean and the desired mean and it is added to the existing population of learners. The learner phase the algorithms proceeds in the similar way by interaction among learners improve his or her knowledge. To obtain a new set of improved learners a random weighted differential vector is formed from a given learner $X_g(i)$ another learner $X_g(r)$ is randomly selected ($i \neq r$) and added to the exiting learner[8].

In this algorithm it is proposed to vary this random weighted differential vector in a random manner in the range (0.5,1) by using the relation

$$0.5*(1+rand(0,1))$$

where rand(0,1) is a uniformly distributed random number within the range[0,1]. So the mean value of this weighted differential scale factor is 0.75. This allows for uncertain variations in the elaboration of the difference vector and thus helps retain the fair chances of obtaining a better location on the multimodal functional surface. Therefore the fitness of the best vector in a population is much less likely to get static until a global optimum is reached.

So the new set of improved learners can be made by using equation in the teacher phase

$$X_{new}{}^{g}{}_{(i)}=X_{(i)}{}^{g} + 0.5 * (1+rand(0,1))*(X_{Teacher}{}^{g}- T_F\, M^g)$$

So the new set of improved learners can be made by using equation in the learner phase

$$X_{new}{}^{g}{}_{(i)}=\begin{cases} X^g_{(i)} + 0.5 * (1 + rand(0,1)) * (X^g_{(i)} - X^g_{(r)}) & if\ f(X^g_{(i)}) < f(X^g_{(i)}) \\ X^g_{(i)} + 0.5 * (1 + rand(0,1)) * (X^g_{(r)} - X^g_{(i)}) & otherwise \end{cases}$$

3.2 Weighted Teaching Learning Based Optimization (wTLBO)

TLBO is based on the principle of teaching-learning approach and a teacher always wishes that his/her student should achieve the knowledge equal to him in fast possible time. Sometimes it becomes difficult for a student to attain the knowledge due to his/her forgetting characteristics. Teaching-learning process is an iterative process where continuous interaction takes place for the transfer of knowledge. Every time a teacher interacts with a student he/she finds that the student is able to recall the part of the lessons learnt from the last session[9].

Sometimes learner may not be in position to recall the knowledge taught in the previous session because this is mainly due to the physiological phenomena of neurons in the brain. Hence a parameter known as "weight" is included to every learner. While computing the new learner value the part of its previous value is considered and that is decided by a weight factor *w*.

In wTLBO, each learner is sampled to different zone of the search space during the early stages of the search. In the later stages adjustments are done to the movements of trial solutions finely so that they can explore the interior of a relatively small space in which the expected global optimum lies. To meet this objective we reduce the value of the weight factor linearly with time from a (predetermined) maximum to a (predetermined) minimum value:

$$w=W_{max}-(\frac{w_{max}-w_{min}}{maxiteration})*i$$

where max and min are the maximum and minimum values of weight factor *w*, *I* iteration is the current iteration number and maxiteration is the maximum number of allowable iterations. W_{max}andW$_{min}$ are selected to be 0.9 and 0.1, respectively.

Hence, in the teacher phase the new set of improved learners can be

$$X_{new}{}^{g}{}_{(i)}= \mathbf{W} * \boldsymbol{X^g_{(i)}} + \boldsymbol{rand} * (\boldsymbol{X^g_{Teacher}} - \boldsymbol{T_f M^g})$$

and a set of improved learners in learner phase as

$$\begin{cases} w * X^g_{(i)} + (rand) * \left(X^g_{(i)} - X^g_{(r)}\right) & if\ f\left(X^g_{(i)}\right) < f(X^g_{(r)}) \\ w * X^g_{(i)} + (rand) * \left(X^g_{(r)} - X^g_{(i)}\right) & otherwise \end{cases}$$

3.3 Orthogonal Teaching-Learning-Based Optimization (OTLBO)

Consider an experiment that involves some factors, each of which have several possible values called levels. Suppose that there are P factors, each factor has Q levels. The number of combinations is Q_p, and for large P and Q it is not practical to evaluate all combinations[10].

Orthogonal design is developed as a mathematical tool to study multi-factor and multi-level problems. It aims to extract an orthogonal array L if M row, where each row represents a combination to be evaluated. The array has three key properties.

1. During the experiment, the array represents a subset of M combinations, from all possible Q^P combinations. Computation is reduced considerably because M<<Q^P.
2. Each column represents a factor. If some columns are deleted from the array, it means a smaller number of factors are considered.
3. The columns of the array are orthogonal to each other. The selected subset is scattered uniformly over the search space to ensure its diversity.

A simple but efficient method is proposed in to generate an orthogonal array L. where

M=Q*Q and P=Q+1. The steps of this method are as follows:

Procedure for generating an orthogonal array L

Input: The number of levels Q
Output: An orthogonal array L
Calculate M= Q*Q and P=Q+1
Initialize an zero matrix L with rows and P columns

for i=1 to M do

$L_{i,1}$=mod ($\lfloor (i-1)/q \rfloor, q$)

$L_{i,2}$=mod (i-1),q)

For j=1 to P-2 do

$L_{i,2+j}$=mod ($L_{i,1}$ * j + $L_{i,2}$,q)

end

end

3.4 Cooperative Teaching-Learning-Based Optimization (CoTLBO)

To improvise the performance of each individual *competition* should be present among the learners. Hence we divide the learners into groups and conduct competitions among them. Instead of optimizing the whole solution vector in one population, the vector is split into its constituent components and assigned to multiple GA (Genetic Algorithm) populations. In this configuration, each population is then optimizing a single component of the solution vector as one-dimensional (1-D) optimization problem. To produce a solution vector for the function being minimized all the populations have to cooperate, as a valid solution vector can only be formed by using information from all the populations. This means that on top of the inherent cooperation in the population itself, a new layer of cooperation between populations are added[11].

The same concept is applied to TLBO, by creating a family of CoTLBO. Instead of having one class (of N learners) trying to find the optimal D-dimensional vector, the vector is split into its components so that classes (of N learners each) are optimizing a 1-D vector.

3.5 Modified Teaching-Learning-Based Optimization (mTLBO)

In this modification is done only to the learner phase of basic TLBO. The Teacher phase remains same as in original. Through the exhaustive analysis of TLBO concept, it is clearly evident that the more the learner learns the better the solution. In a traditional teaching–learning environment the learners output depends on the interaction between learners and the class room delivery by teachers. To further enhance the learning capability of students an extra training through the tutorial helps. Group discussions, presentations, formal communications etc. provide a right platform for the learners to interact with their peers though randomly, and at the same time he or she can discuss elaborately with the teacher who is better knowledgeable person in a tutorial class. A learner always learns something new from the teacher and if the other learner has more knowledge than him or her the n only he or she gets more knowledge. Hence, an extra term has been added in the learner phase equation to modify TLBO[12]. Inspired by the concept to Differential Evolution with Random Scale Factor (DERSF) this term is scaled by scale factor in a random manner in the range(0.5,1) by using

$$0.5*(1+rand(0,1))$$

where rand (0,1) is a uniformly distributed random number within the range[0,1].The mean value of the scale factor is 0.75. This allows for stochastic variations in the available solutions and thus helps retain learner diversity as the search progresses. Even when most of the candidate solutions point to locations clustered near a local optimum, this term gives a fair chance of pointing at an even better location on the multi modal functional surface. Therefore the fitness of the best solution in a population is much less likely to get stagnant until a truly global optimum is reached.

$$X_{new}{}^{g}{}_{(i)}=$$

$$\begin{cases} X^g_{(i)} + rand * (X^g_{(i)} - X^g_{(r)}) + 0.5 * (1 + rand) * (X^g_{(Teacher)} - X^g_{(i)}) & if\ f(X^g_{(i)}) < f(X^g_{(i)}) \\ X^g_{(i)} + rand * (X^g_{(r)} - X^g_{(i)}) + 0.5 * (1 + rand) * (X^g_{(Teacher)} - X^g_{(i)}) & otherwise \end{cases}$$

Now the New Learner modification is expressed a sin place of its corresponding equation of basic TLBO.

The third term in the above equation represents the close interaction between a teacher and learner analogous to tutorial concept. The performance of modified TLBO has been investigated over many common benchmark functions and compared with basic TLBO and few variants of PSO, DE etc. The results reveal the superior performance of mTLBO over other investigated algorithms.

4 Comparison between Variants of TLBO

In this section we have compared variants of TLBO and it is described in two tables. Table 1 contains parameters used in the original TLBO and new parameters included in its variants. Table 2 contains all variants of TLBO and their applications.

Table 1. Parameters involved in variants of TLBO

Name of TLBO Variant	Parameter description
Teaching-Learning-Based Optimization (TLBO)	n – number of students d – number of design variables X_i – initial students of classroom X_{new} - new value of each student $X_{teacher}$ – value of teacher T_f - teaching factor
Improved Teaching-Learning-Based Optimization (ITLBO)	$X_g(i)$ – weight added to performance of one learner $X_g$ri) – weight added to performance of another learner M^g - mean weight of all learners
Weighted Teaching-Learning-Based Optimization (WTLBO)	w- weight factor of each learner w_{max} – maximum weight factor w_{min} - minimum weight factor
Modified Teaching-Learning-Based Optimization (MTLBO)	$X^g_{(Teacher)}$ – weight factor for teacher $X_g(i)$ – weight added to performance of one learner $X_g(r)$ – weight added to performance of another learner

Table 2. Applications of variants of TLBO

Name of TLBO Variant	Applications	Reference Number
Teaching-Learning Based Optimization (TLBO)	Constrained Mechanical Design Optimization Problems Optimization Method for Continuous Non-Linear Large Scale Problems Constrained and Unconstrained Real Parameter Optimization Problems Shape and Size Optimization of Truss Structures With Dynamic Frequency Constraints	13, 14, 15, 16
Improved Teaching Learning Based Optimization (ITLBO)	Global Optimization Global Function Optimization	8
Weighted Teaching Learning Based Optimization (WTLBO)	Global Optimization Global Function Optimization	9

Table 2. *(continued)*

Orthogonal Teaching Learning Based optimization (OTLBO)	Global Optimization Global Function Optimization	10
Cooperative Teaching Learning Based Optimization (CoTLBO)	Stochastic RCPSP	11
Modified Teaching Learning Based Optimization (mTLBO)	Global Numerical Optimization	12
Multi objective Teaching Learning Based Optimization (MO-TLBO)	Economic Emission Load Dispatch Problem	18
Refined Teaching Learning Based Optimization	Dynamic Economic Dispatch of Integrated Multiple Fuel and Wind Power Plants	17
Toward Teaching Learning Based Optimization	Real-Parameter Optimization Problems	16

References

1. Mitchell, M.: An Introduction to Genetic Algorithms. MIT Press (1996)
2. Kennedy, J., Eberhart, R.C.: Particle Swarm Optimization. In: Proceeding of IEEE International Conference on Neural Network. IEEE Press (1995)
3. Karaboga, D., Basturk, B.: On the performance of artificial bee colony algorithm. Applied Soft Computing 8(1), 687–697 (2008)
4. Lhotská, L., Macaš, M., Burša, M.: PSO and ACO in Optimization Problems. In: Corchado, E., Yin, H., Botti, V., Fyfe, C. (eds.) IDEAL 2006. LNCS, vol. 4224, pp. 1390–1398. Springer, Heidelberg (2006)
5. Geem, Z.W., Kim, J.H., Loganathan, G.V.: A new heuristic optimization: harmony search. Simulation 76, 60–70 (2001)
6. Venkata Rao, R., Patel, V.: An elitist teaching-learning-based optimization algorithm for solving complex constrained optimization problems. International Journal of Industrial Engineering Computations 3, 535–560 (2012)
7. Baghlani, A., Makiabadi, M.H.: Teaching-Learning-Based Optimization Algorithm for Shape and Size Optimization of Truss Structures With Dynamic Frequency Constraints. IJST 37(C+), 409–421 (2013)
8. Venkata Rao, R., Patel, V.: An improved teaching-learning-based optimization algorithm for solving unconstrained optimization problems. Scientia Iranica (2013)

9. Satapathy, S.C., Naik, A.: A Weighted teaching learning based optimization for global function optimization. Applied Mathematics 4, 429–439 (2013), doi:10.4236/am.2013.43064
10. Satapathy, S.C., Naik, A.: A A teaching learning based on orthogonal design for solving global optimization problems. Spinger Plus Journal 2, 130 (2013), doi:10.1186/2193-1801-2-130
11. Zheng, H.-Y., Wang, L., Wang, S.-Y.: A Co-evolutionary Teaching-learning based Optimization Algorithm for Stochastic RCPSP. In: IEEE Congress on Evolutionary Computation, CEC (2014)
12. Satapthy, S.C., Naik, A.: A Modified Teaching–Learning-Based Optimization algorithm for global numerical optimization—A comparative study. Swarm and Evolutionary Computation 16, 28–37 (2014)
13. Rao, R.V., Savsani, V.J., Vakharia, D.P.: Teaching–learning-based optimization: A novel method for constrained mechanical design optimization problems 43(3), 303–315 (2011, 2012)
14. Rao, R.V., Savsani, V.J., Vakharia, D.P.: Teaching–learning-based optimization: Optimization Method for Continuous Non-Linear Large Scale Problems 183(1), 1–15 (2011)
15. Baghlani, A., Makiabadi, M.H.: Teaching-Learning-Based Optimization Algorithm for Shape and Size Optimization of Truss Structures With Dynamic Frequency Constraints. IJST, Transactions of Civil Engineering 37(C+), 409–421 (2013)
16. Wang, K.-L., Wang, H.-B., Yu, L.-X., Ma, X.-Y., Xue, Y.-S.: Toward Teaching-Learning-Based Optimization Algorithm for Dealing with Real-Parameter Optimization Problems. In: ICCSEE 2013 (2013)
17. Krishnasamy, U., Nanjundappan, D.: A Refined Teaching-Learning Based Optimization Algorithm for Dynamic Economic Dispatch of Integrated Multiple Fuel and Wind Power Plants. Hindawi Publishing Corporation, Mathematical Problems in Engineering 2014, Article ID 956405, 14 pages (2014)
18. Tekumalla, D.V., Vinod Kumar, D.M.: Multi Objective Economic Emission Load Dispatch Using Teacher-Learning Based Optimization Technique. In: Advances in Control and Optimization of Dynamical Systems, vol. 3, pp. 819–826 (2014)

Linear Array Optimization Using Teaching Learning Based Optimization

V.V.S.S.S. Chakravarthy[1], K. Naveen Babu[1], S. Suresh[1],
P. Chaya Devi[1], and P. Mallikarjuna Rao[2]

[1] Department of ECE, Raghu Institute of Technology, Visakhapatnam, India
[2] Department of ECE, Andhra University, Visakhapatnam, India
sameervedula@gmail.com, naveen.kota@outlook.com

Abstract. Optimization is a serious problem in antenna arrays. Many evolutionary and meta-heuristic algorithms are used in solving such array optimization problems. In this paper a new optimization algorithm known as Teaching Learning Based Optimization technique (TLBO) is applied on side lobe level (SLL) reduction. The Non Uniform Linear Array (NULA) considered has uniform spacing and non-uniform amplitude distribution with symmetry around the axis. The desired SLL is achieved by properly defining the amplitude distribution for the NULA.

Keywords: side lobe level, linear array, TLBO, optimization.

1 Introduction

An Antenna can be termed as an electromagnetic radiator, interceptor, sensor, transducer or an impedance matching device with extensive applications in all communications, radar and in bio-medical systems. The gain of a single antenna element is not sufficient for most applications. Enhancement of the gain and directivity involves in increasing the physical dimensions of the antenna which makes the antenna more bulky forfeiting the ease of installation. Under this circumstance use of arrays provide the solution. An Array is a collection of well-versed elements (antenna). It is a geometrical configuration of antenna elements grouped in such a way as a to direct radiated power towards particular angular direction in space there by increasing the directivity and decreasing the beam width.

The radiation characteristics of an antenna array depends on some input parameters like the relative magnitude and phase of excitation of currents of each radiating element, the geometrical configuration of the array and the spacing between the array elements To control the radiation characteristics of an array the above mentioned parameters should be modified.

The genesis of the problem may be discussed as follows. The Array synthesis issues may be classified into many cases depending up on the requirements like having nulls in the desired direction, low side lobes, narrow beam, sectored beam, beam steering, desired beam area etc. Array synthesis for the desired characteristics

S.C. Satapathy et al. (eds.), *Emerging ICT for Bridging the Future – Volume 2*,
Advances in Intelligent Systems and Computing 338, DOI: 10.1007/978-3-319-13731-5_21

involves in defining the above discussed parameters of the individual elements as well as the geometry of the array. There are several traditional computational techniques and numerical methods to define these parameters. Some of them are Finite Difference Time Domain Method, Finite Element Method, Finite Volume Method and Method of Moments. All the listed techniques are time consuming, requires wide computational time to converge and failed to handle multi modal problems.

Considering the above difficulties associated with mathematical optimization several modern evolutionary and heuristic algorithms have been developed for searching near-optimum solutions to the problems. These algorithms can be classified into different groups depending on the criteria being considered such as population based, iterative based, stochastic, deterministic, etc. Genetic Algorithm, Particle Swarm Optimization, Artificial Bee Colony [1], NSGA-II [2] are some to list which have wide applications in array synthesis and optimization. Keeping up with trend, Teaching Learning based Optimization (TLBO) technique algorithm is considered for antenna array synthesis application. In this paper a linear antenna array synthesis problem is considered and TLBO algorithm is applied to achieve a very low SLL. As TLBO algorithm is said to be algorithm specific parameter free, there is no clear tuning the parameter which makes the algorithm more simple [4, 5].

2 Linear Array Design and Formulation

The design of the Linear Array oriented along the z-axis is as shown in the fig.1. The corresponding array factor formulation is given as follows for the case of non-uniform current excitations with uniform spacing [3].

$$AF = \sum_{n=1}^{N} 2a(\mathrm{n})\, e^{j\phi} \cos(\pi(\mathrm{n}-0.5)(\mathrm{u}-\mathrm{u}_0)) \tag{1}$$

where $u = \cos(\theta)$, $\mathrm{u}_0 = 0$ and a(n) is current excitation of nth element

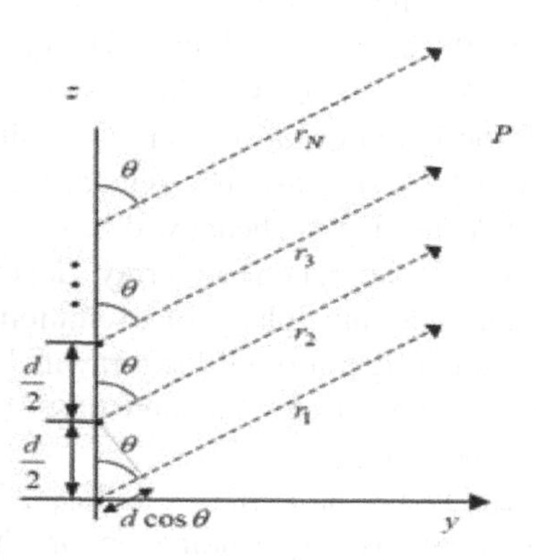

Fig. 1. Linear Array geometry oriented along z-axis

3 Fitness Evaluation

The formulation for fitness evaluation is given as follows.

$$F= \max[AF_{ndB} (\theta_{FN},\ldots\ldots\ldots \theta_{max})] \quad (2)$$

where
AF_{ndB} = normalized array factor
= AF_{dB}-max(AF_{dB})
$AF_{dB} = 20\log_{10}(AF)$
θ_{FN} = First null (towards right hand side of the pattern)
θ_{max} = Maximum value of azimuthal angle

The parameters mentioned in the fitness function are described in Fig.2.

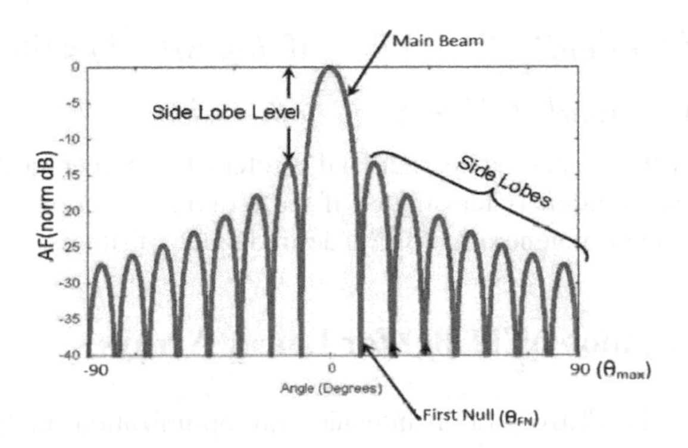

Fig. 2. Template radiation pattern of linear array

The objective function that describes the problem statement is called fitness function. The numerical outcome of the fitness function is fitness value. As the optimization problem mentioned in this work is minimization, the objective of the experiment is to minimize this fitness value. The fitness value is computed according to the eq.(2). The array geometry considered in this work has a symmetric pattern with main positioned in the broadside. Hence the design parameters are computed for nly one half of the array elements and the same are copied to the other half.

4 TLBO Algorithm

TLBO is novel meta-heuristic optimization algorithm based on the exchange of knowledge that possibly takes place in two different ways namely between teacher and learner, and learner and fellow learner in the class room [4,5]. This is classified as Teaching phase and learning phase. The optimization technique has been extensively applied for solving various multi-objective mechanical engineering and clustering problems. It is now verified for application to antenna array optimization problem. Each student is treated as an array and the corresponding subjects in the class are array elements. The best student in the class refers to the array with best fitness function. The understanding of the algorithm is classified in to initialization, teaching phase, learning phase and termination criterion. During the initialization phase the population is generated with the specified upper and lower boundary. During the teaching phase the mean of each of the subjects for all generations is calculated and presented as M^g. In the minimization problems the learner with least mean is considered as teacher for that iteration. All the learners now takes a shift towards the teacher with the following expression [6].

$$Xnew^g = X^g + rand * (X^g_{teacher} - \mathrm{Tf} * \mathrm{M}^g) \quad (3)$$

where X is the learner, g is generation, Tf is Teaching factor which is in the range of 1 to 2.

In the learning phase the learner's knowledge is updated using the following equation.

$$Xnew_i^g = \begin{cases} X_i^g + rand * (X_i^g - X_r^g) & \text{if } fitness(X_i^g) < \text{fitness}(X_r^g) \\ X_i^g + rand * (X_r^g - X_i^g) & otherwise \end{cases} \quad (4)$$

where i refers to considered learner and r refers to another randomly selected learner. Finally the program is terminated if the required criterion is obtained. The criterion can be number of generation or the desired value of fitness.

5 Implementation of TLBO for Linear Arrays

Implementation of the TLBO to our antenna array optimization problem starts with initialization of learners with subjects [7, 8]. Each learner corresponds to an antenna array under trial and each subject of the learner corresponds to an element of the respective antenna array. Marks/score obtained in the subject corresponds to the amplitude of current excitation of the respective element in the array. The range of numerical values that are representing marks scored in a subject is (0, 1). Fitness value computed gives the rank of the student and it corresponds to the respective array's highest SLL which has to be minimized..

6 Results and Discussions

The experimentation is carried on for Number of elements (N) equals to 10,12,14,16,18,20,22 and 24. As symmetry is considered and hence only N/2 elements are generated for every learner during the population generation. The number of iterations in every trial is fixed to 50 and the experimentation is carried on for 5 trials and the weights corresponding to the best trial i.e.; least cost is considered and presented here. From figure 3 through figure 10, the figures (a) shows radiation patterns of Linear Array for uniform excitation (dotted line) and non-uniform excitation (solid line) for N=10,12,14,16,18,20,22,24 and the figures (b) show their respective convergence plots.

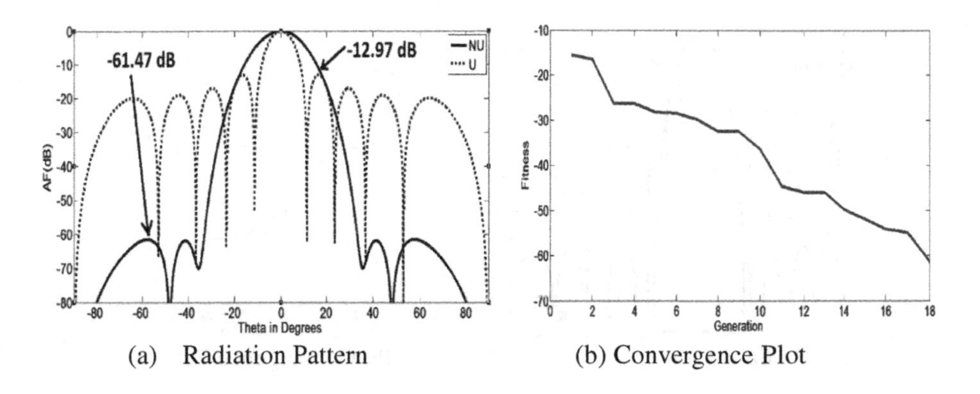

(a) Radiation Pattern (b) Convergence Plot

Fig. 3. Plots corresponding to N=10

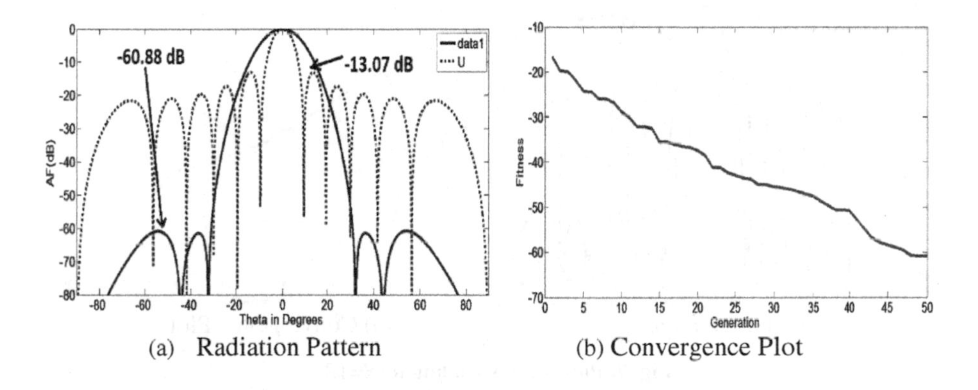

(a) Radiation Pattern (b) Convergence Plot

Fig. 4. Plots corresponding to N=12

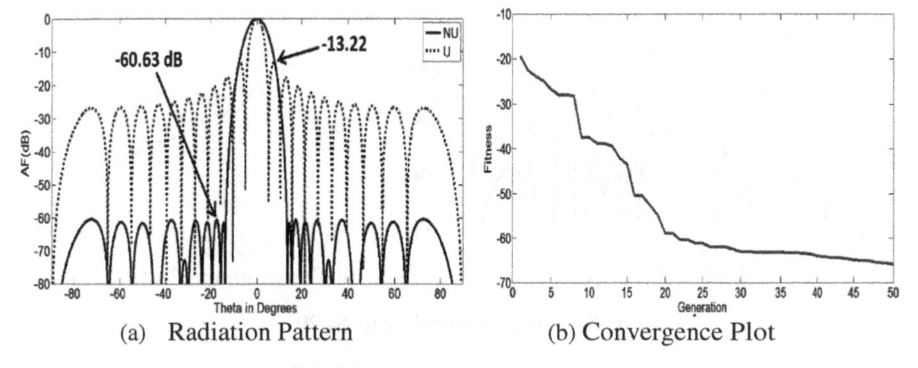

(a) Radiation Pattern (b) Convergence Plot

Fig. 5. Plots corresponding to N=14

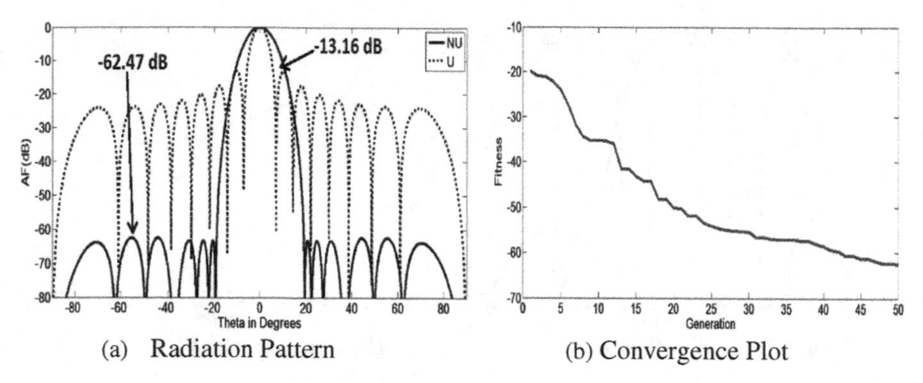

(a) Radiation Pattern (b) Convergence Plot

Fig. 6. Plots corresponding to N=16

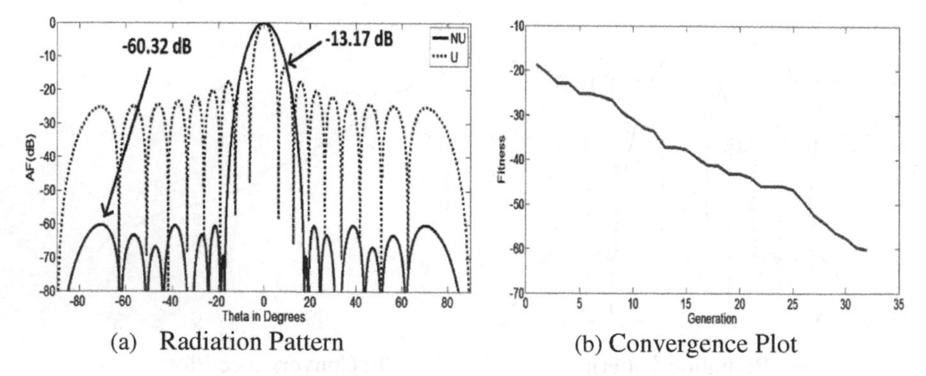

(a) Radiation Pattern (b) Convergence Plot

Fig. 7. Plots corresponding to N=18

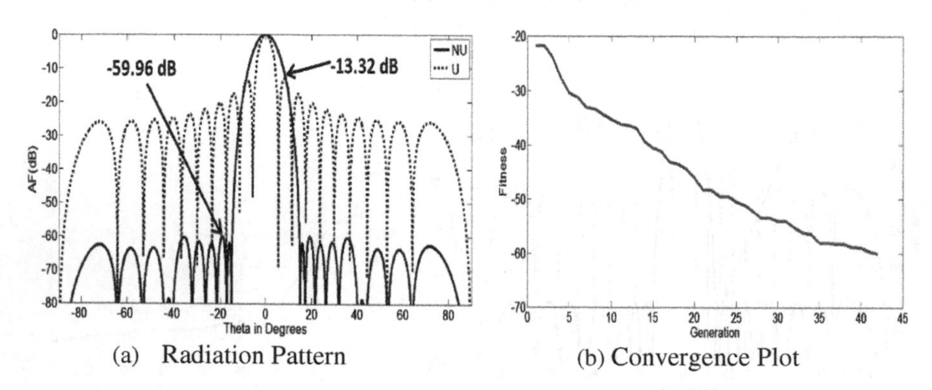

(a) Radiation Pattern (b) Convergence Plot

Fig. 8. Plots corresponding to N=20

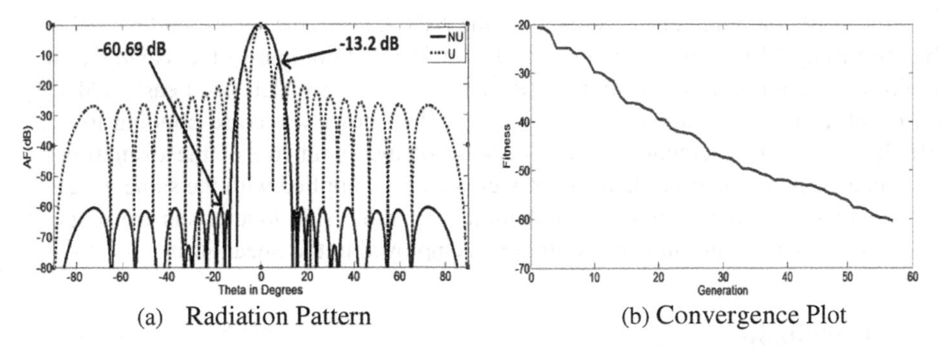

(a) Radiation Pattern (b) Convergence Plot

Fig. 9. Plots corresponding to N=22

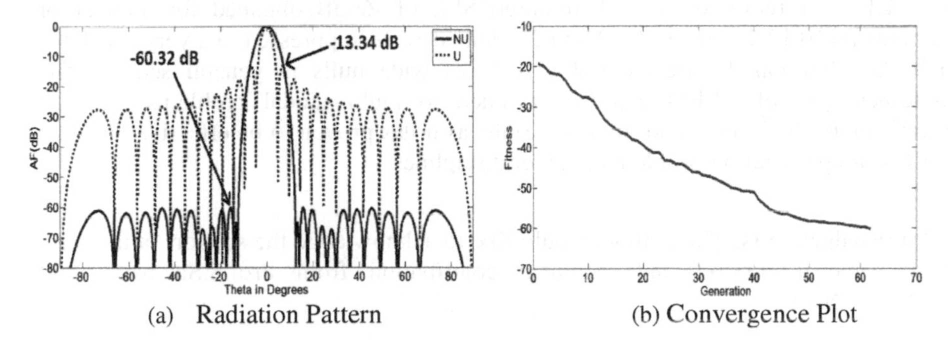

(a) Radiation Pattern (b) Convergence Plot

Fig. 10. Plots corresponding to N=24

Table 1. Weights obtained for non-uniform excitation of NULA

S.No	No. of Elements	Weights
1	10	0.99604 0.78266 0.4708 0.20183 0.050541
2	12	0.98668 0.84304 0.60884 0.36113 0.16556 0.050953
3	14	0.97253 0.86926 0.69049 0.48113 0.28659 0.13863 0.050003
4	16	0.99266 0.91408 0.7711 0.59081 0.40887 0.24762 0.12587 0.050609
5	18	0.97971 0.91987 0.80586 0.66247 0.49759 0.3468 0.21232 0.11769 0.050885
6	20	0.99906 0.95292 0.85552 0.72315 0.57843 0.43336 0.29971 0.18706 0.097634 0.051174
7	22	0.90997 0.87932 0.81452 0.71624 0.59991 0.48509 0.36849 0.26863 0.16883 0.10263 0.059317
8	24	0.52633 0.5084 0.46615 0.45294 0.38749 0.34163 0.2537 0.2173 0.18042 0.11688 0.069198 0.050029

Table.1 lists the number of elements in the NULA and their corresponding weights obtained using TLBO. It can be observed from the 2nd column of the Table that the amplitude distribution is non-uniform to obtain a very low SLL of atleast -60dB. It can be observed in the above Table.1 that weights shown are corresponding to only half the number of elements. This is because of the symmetry of the excitation of antenna elements. The other half of array elements are applied with the same weights. It is obvious from the plots of the radiation pattern the side lobe levels are reduced considerably when non-uniform excitation is employed in the linear array.

7 Conclusion

The TLBO algorithm is effectively applied to the fundamental problem of optimizing the NULA for required SLL. A required SLL of -60dB obtained for number of elements N=10,12,14,16,18,20,22 and 24. The SLL are suppressed to a very low level such that they can be treated collectively as wide nulls in general sense. The implementation of TLBO can be extended to multi modal problems in array optimization which may lead to good results as it has proven to be effective on some non-linear optimization problems in other disciplines.

Acknowledgements. The authors would like to acknowledge the support in the form of valuable discussion and technical contribution from Prof.S.Satapathy and Ms.Anima Naik.

References

1. Basturk, B., Karaboga, D.: An Artificial Bee Colony (ABC) Algorithm for Numeric Function Optimization. In: IEEE Swarm Intelligence Symposium, Indianapolis, Indiana, USA (2006)
2. Deb, K.: An Efficient Constraint Handling Method for Genetic Algorithm. Computer Methods in Applied Mechanics and Engineering 186, 311–338 (2000)
3. Antenna Theory: Analysis and Design. John Wiley & Sons, Publishers, Inc., New York (1997, 1982, 2005)
4. Rao, R.V., Savsani, V.J.: Mechanical Design Optimization using Advanced Optimization Techniques. Springer, London (2012)
5. Rao, R.V., Savsani, V.J., Vakharia, D.P.: Teaching-Learning-Based Optimization: A Novel Method for Constrained Mechanical Design Optimization Problems. Computer-Aided Design 43(3), 303–315 (2011)
6. Satapathy, S., Naik, A.: Improved Teaching Learning Based Optimization for Global Function Optimization. Decision Science Letters 2(1), 23–34 (2013)
7. Dib, N.I.: Synthesis of Thinned Planar Antenna Arrays using Teaching-Learning Based Optimization. International Journal of Microwaves and Wireless Technologies available on CJO (2014), doi:10.1017/S1759078714000798
8. Dib, N., Sharaqa, A.: Synthesis of Thinned Concentric Circular Antenna Arrays using teaching-learning based optimization. International Journal of RF & Microwave Comp. Aided Engg. 24, 443–450 (2014), doi:10.1002/mmce.20784

Partition Based Clustering Using Genetic Algorithm and Teaching Learning Based Optimization: Performance Analysis

Kannuri Lahari[1], M. Ramakrishna Murty[1], and Suresh C. Satapathy[2]

[1] Dept of CSE, GMR Institute of Technology , Rajam,Srikakulam(Dist) A.P. India
{kannurilahari,ramakrishna.malla}@gmail.com
[2] Dept of CSE, ANITS, Visakahapatnam, A.P, India
sureshsatapathy@gmail.com

Abstract. Clustering is useful in several machine learning and data mining applications such as information retrieval, market analysis, web analysis etc. The most popular partitioning clustering algorithms are K-means. The performance of these algorithms converges to local minima depends highly on initial cluster centroids. In order to overcome local minima apply evolutionary and population based methods like Genetic Algorithms and Teaching Learning Based Optimization (TLBO). The GA and TLBO is applied to solve the challenges of partitioning clustering method K-means like initial centroid selection and compare the results with using GA and TLBO. TLBO is latest population based method and uses a population of solutions to proceed to the global solutions and overcome the local minima problem. TLBO method is based on effect of the influence of a teacher on the output of learners in a class.

Keywords: partitioning clustering, Evolutionary, Genetic Algorithm, TLBO.

1 Introduction

The determination of cluster analysis has been in performance asignificant role in cracking many problems in medicine, psychology, biology, sociology, pattern recognition and image processing. Clustering is the jobgrouping a set of objects so that objects in the same group (called a cluster) are more similar (in some sense or another) to each other and different with objects of other groups. Clustering is supposed one of the most difficult and stimulating problems in machine learning, particularly due to its unsupervised nature. [1] Due to its unsupervised nature of the problem implies that its structural characteristics are not known, except if there is some kind of domain information available in advance. Specifically, the spatial distribution of the data in terms of the number, volumes, densities, shapes and orientations of clusters (if any), are unknown. These difficulties may be possible even more by an eventual need for dealing with data objects described by attributes of different natures conditions and scales (ordinal and nominal) Many unsupervised

S.C. Satapathy et al. (eds.), *Emerging ICT for Bridging the Future – Volume 2*,
Advances in Intelligent Systems and Computing 338, DOI: 10.1007/978-3-319-13731-5_22

clustering algorithms have been developed. Data is grouped into clusters independent of the topology of input space in most of these algorithms. Clustering algorithms can be grouped into two main classes of algorithms, called supervised and unsupervised. [4] These algorithms include among others, K-means and Expectation Maximization etc.

In this work used partition based algorithm called K-means with using evolutionary based techniques GA and TLBO. By applying these algorithms on bench mark data sets namely Iris and Wine get certain number of clusters as output. GA and TLBO is applied on the obtained clusters which gives improved and global solution when compared to previous clusters from an optimization perspective, clustering can be considered as a particular kind of NP- hard grouping[11]. This has inspired the search for efficient approximation algorithms, including not only the use of heuristics for particular classes or instances of problems, but also the use of general-purpose meta-heuristics. Particularly, evolutionary algorithms are meta-heuristics widely believed to be actual on NP-hard problems, being able to provide near-optimal solutions to such problems in rational time. [13]Under this hypothesis, a large number of evolutionary algorithms for solving clustering problems have been proposed in the literature.[2] These algorithms are based on the optimization of some objective function (i.e., the so-called fitness function) that guides the evolutionary search. [3][4] The performance of partitioning based clustering technique K-means with evolutionary prospective.

The paper is organized as in section 2 discussed Evolutionary Techniques. In the section 3 experimentation approach, section 4 results discussion and section 5 conclusion.

2 Evolutionary Techniques

Scientists have been give emphasis to using natural evolutionary process in inclosing a computerized optimization algorithm. The belief of targeting good solutions and removing bad solutions seems to dovetail well with preferred properties of good optimization procedure. The nature has been optimizing a plentiful new complicated objective function by means of natural genetics and natural selection than the search and optimization problems. The most popular evolutionary techniques are Genetic Algorithms, Deferential Evolution, Particle Swarm Optimization, Teaching Learning Based Optimization etc.

2.1 Clustering with Using Genetic Algorithms

Genetic Algorithm (GA) is an iterative optimization procedure, first proposed by J.H.Holland in the year 1970s. A genetic algorithm emulates nature and its "Survival of the fittest". First of all pool the population of solutions and calculate their strength regarding the objective function. [5] The strongest members of the population are combined to get offspring, this offspring is added to the population and the procedure is repeated until a certain stopping criterion is met.GA has been used for the purpose of aptly finding a fixed number of clusters (k) of cluster centers in n objects, there by suitably clustering the set of n unlabeled points. The clustering metrics has been adopted in the sum of the Euclidean distances of the points from their respective cluster centers.

GA Operations: The genetic algorithm involves three type of operations namely selection, crossover and mutation.

Selection: This operator is used for select population for reproduction

Crossover: This operator is chooses randomly a position of a chromosome and exchange the subsequences before and after the position between two chromosomes to create two offspring. For example, the strings 10100110 and 11111111 would be crossed over after the third position in each to produce the two offspring 10111111 and 11100110. This operator is approximately impersonators biological recombination between two single –chromosome organisms.

Mutation: This operator randomly handsprings some of the bits in a chromosome. For example, the string 00100100 may be mutated in this second position to get 01100100. Mutation can occur at each bit position in a string with some possibility, usually occurs very small.

```
Begin
  1.  x=0
  2.  initialize population pop(x)
  3.  compute fitness pop(x)
  4.  x = x+1
  5.  if termination criterion
      achieved go to step 10
  6.  select pop(x) from pop(x-1)
  7.  crossover pop(x)
  8.  mutate pop(x)
  9.  go to step 3
  10. output best and stop

End
```

Fig. 1. Basic algorithm steps in GA

2.2 Teaching Learning Based Optimization(TLBO)

TheTLBO algorithm is a newly developed optimization algorithm. This algorithm is explained on the basis of transformation of knowledge to the classroom environment, so that learners will able to acquire knowledge from the teacher and next from their fellow students. The algorithm can be explained in the following manner:

Teacher Phase: In this phase all the solution entrants should be distributed in the available search space randomly. Thus, among all the available solutions the best individual solution should be selected and will discuss the content with other candidates.[6] Since the most experienced person about the subject in the population, influences the student's deed to take part some pre-planned aim. Depending on teacher skill he/she enriches the mean of the class information level. According to the merit of training provided, learners will gain information in the class.

This optimization method is based on the effect of the influence of a teacher on the output of learners in a class. It is a population based method in order to reach the global solution it uses a number of solutions [7] In TLBO a group of learners setup the population. There are number of design variables in any optimization algorithms .In TLBO the various variables are similar to the various subjects that are given to the learners and finally the learners result is comparable to the fitness assign other population-based optimization techniques. As the teacher is considered the most knowledgeable person in the society, the best solution is comparable to Teacher in TLBO. [8][9]The phases of TLBO are divided into two phases. The first phase contains the "Teacher Phase" and the second phase contains the "Learner Phase". The "Teacher Phase" means the learner learns from the teacher and the "Learner Phase" means learning through the interaction between learners. In the sub-sections below briefly discussed the implementation of TLBO [12].

A. Initialization

Following arethe representations used forlabelling theTLBO:

N: number of learners in the class ("class size")

D: numberof offeredcourses to thelearners

The population Xis randomly initialized by a search space bounded by matrix of N rows and M columns. The jth parameter of the ith learner is assigned values randomly using the equation

$$x^{0}_{(i,j)} = x_j^{min} + rand \times (x_j^{max} - x_j^{min}) \quad (1)$$

Where rand representsa uniformlydistributed random variable within the range (0, 1), X_j^{min}andX_j^{max} represent the minimum and maximum value for jth parameter. The parameters of ith learner for the generations are given by

$$X^{g}_{(i)} = [X^{g}_{(i,1)}, X^{g}_{(i,2)}, X^{g}_{(i,3)}, \dots\dots\dots X^{g}_{(i,j)} \dots\dots, X^{g}_{(i,D)}] \quad (2)$$

B. TeacherPhase

The mean parameter M^g of each subject of the learners in the class at generation is given as

$$M^g = [M_1^g, M_2^g, \ldots\ldots\ldots M_j^g, \ldots\ldots M_D^g] \quad (3)$$

The learner with the minimum objective function value is considered as the teacher$X_{Teacher}^g$ for respective iteration.[8] The Teacher phase makes the algorithm proceed by shifting the mean of the learners towards its teacher. To obtain a new set of improved learners a random weighted differential vector is formed from the current mean and the desired mean parameters and added to the existing population of learners [14].

$$Xnew_{(i)}^g = x_{(i)}^g + \text{rand} \times (X_{Teacher}^g - T_F M^g) \quad (4)$$

T_F is the teaching factor which decides the value of mean to be changed. Value of T_F can be either 1 or2. Thevalue of T_F is decided randomly with equal probability as,

$$T_F = \text{round}\,[1 + \text{rand}\,(0\text{-}1)\{2\text{-}1\}] \quad (5)$$

B. Learner Phase

In this phase the communication of learners with one another takes place.[9]The process of mutual communication increases the knowledge of the learner. The casual interaction among learners enhances his or her knowledge. [10]For a given Xnew learner$X_{(i)}^g$, another learner $X_{(r)}^g$ is randomly selected (i$\neq r$).The ith parameter of the matrix Xnew in the learner phase is given as

$$Xnew_i^g = X_{(i)}^g + \text{rand} \times \left(X_{(i)}^g - X_{(r)}^g\right) \qquad \text{if } f\left(X_{(i)}^g\right) < f\left(X_{(r)}^g\right)$$

$$X_{(i)}^g + \text{rand} \times \left(X_{(r)}^g - X_{(i)}^g\right) \qquad \text{otherwise --- (6)}$$

3 Experimentation

The experimentation performs in the Java platform with using NetBeans support. The experimental results comparing with GA and TLBO based clustering with K-means algorithm are provided for two bench mark data sets IRIS flower and Wine data set taken from the UCI machine repository. The details of data are given below.

Data sets:

Iris: This is a best known data set for the clustering applications. It has 150 instances with four attributes namely sepal length, sepal width, petal length and petal width. The data is distributed into three classes. In each class 50 instances

Wine: In this data set there are 13 attributes and 178 instances with 3 classes.

In this interface to browse the data sets by the user as per their requirement. Here we used two kind of benchmark data sets namely Iris and Wine. After select the data set normalizes the data set. Normally the bench mark data set need not normalized but in general real world data set should be normalize due to its abnormal characteristics .In the process of clustering choose a distance measure based on the requirement as shown in the Fig:3.In this project three possible distance metrics namely Euclidean, Manhattan and Cosine similarity are implemented. The results are reported in the form of tables and graphs. In this process the basic clustering k-means is used to get the optimized results so apply it with evolutionary based technique like GA and TLBO.

4 Results and Discussion

The experiment results shows the performance of the evolutionary based clustering is provide better results than traditional clustering methods alone. In the experimentation created user interface to perform clustering operations with using evolutionary methods.

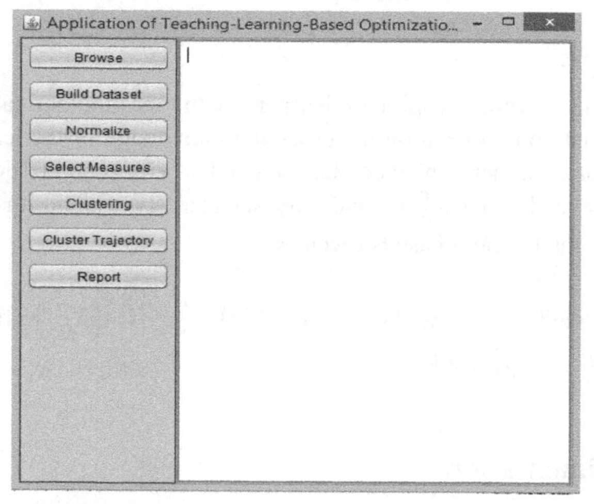

Fig. 2. User interface to perform implementation

In the Fig 2 provide interface for the users to browse the data, which is required for the experiments and also buld their data set size as per requirement. The data normalization process is also provided for dealing with real world data while using by the users and also select the distance measures as shown like the Fig3.

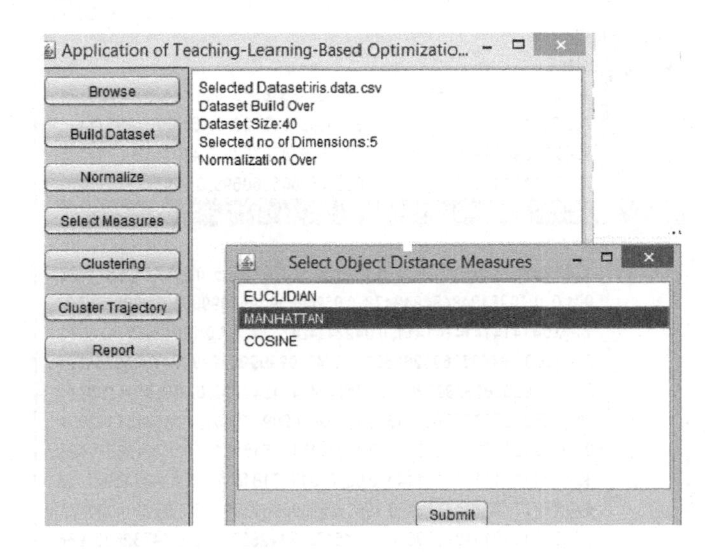

Fig. 3. Select the distance measure as per user requirement

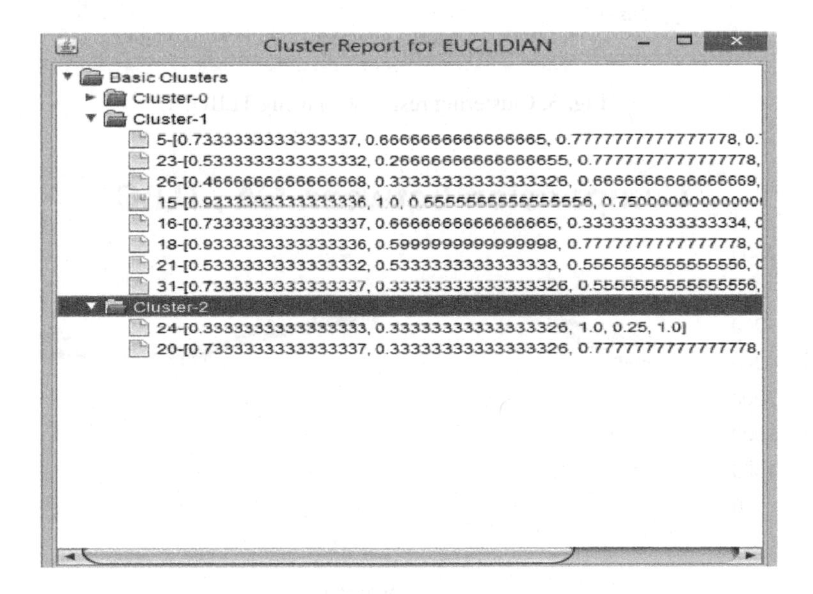

Fig. 4. Clustering results with GA

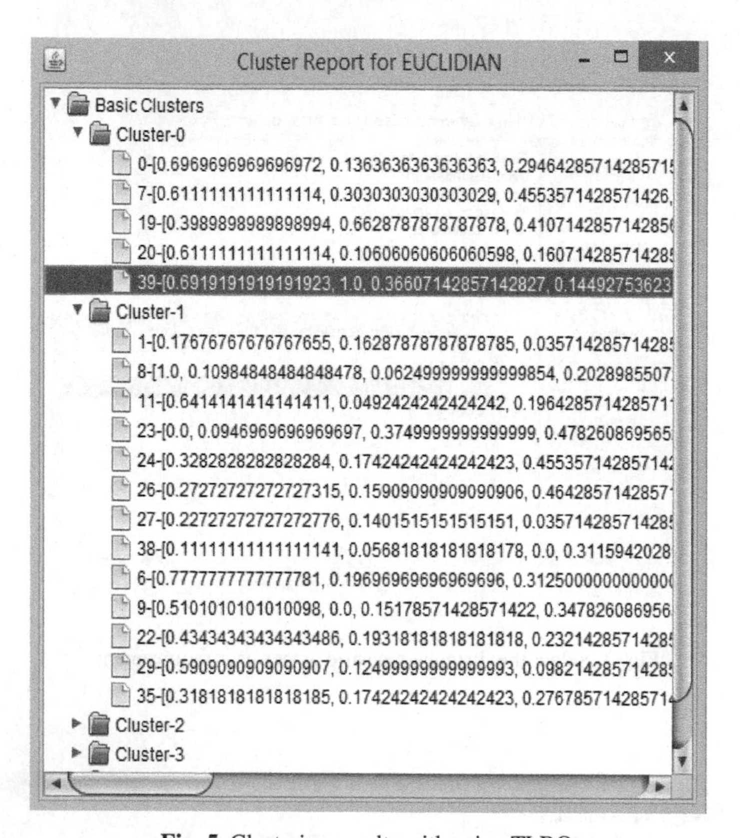

Fig. 5. Clustering results with using TLBO

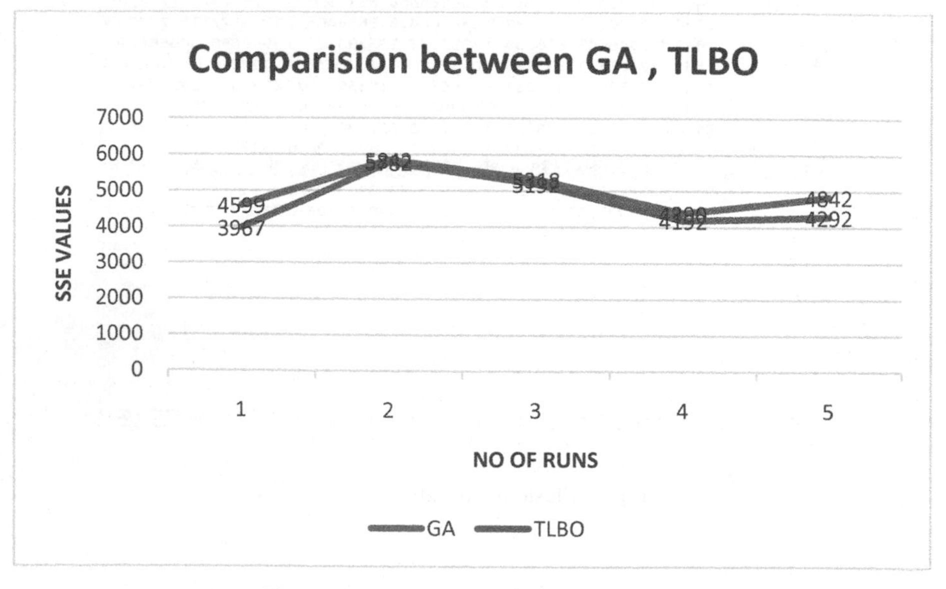

Fig. 6. SSE value with 3 number of clusters

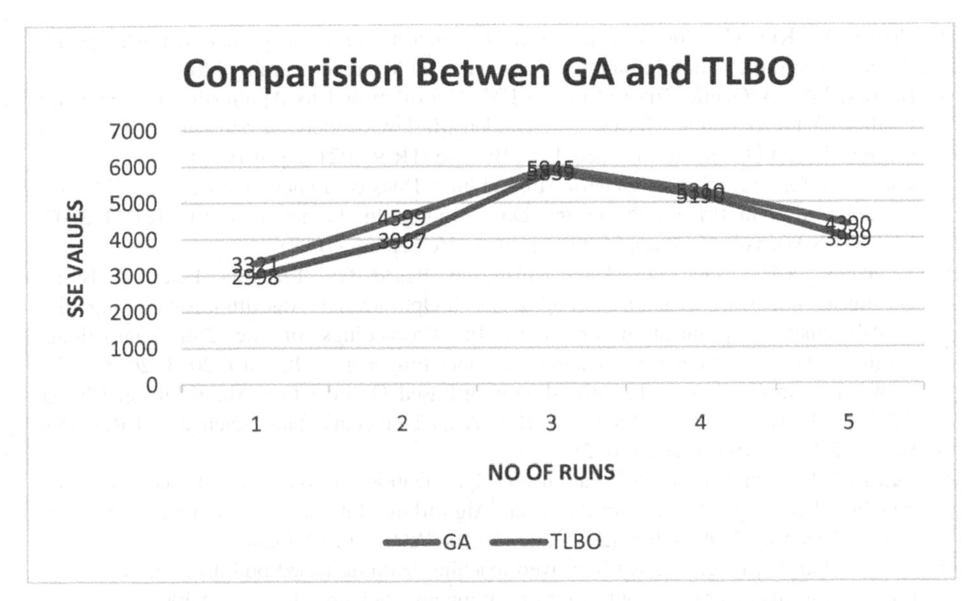

Fig. 7. SSE value with 5 numbers of clusters

The performance of the evolutionary based clustering methods is compared with the results produced by experiments. The TLBO based clustering method performance is better than the GA based cluster as fitness function is SSE (Sum of Squared Error). The compared results are shown in the Fig:6 and Fig: 7 based on the SSE. In the implementation process used few fitness functions for both GA and TLBO based clustering. In the Fig 4 and Fig 5 shows the results with some cluster values are not shown because limited in the report space.

5 Conclusion

The implementation of K-means algorithm is with using evolutionary prospective. Experimented with bench mark data sets namely Iris, Wine. The results compare with GA based clustering and TLBO based clustering with using SSE is fitness function. The results are showing the TLBO based clustering provide better results than GA. The Teaching Learning Based Optimization can be made even more dynamic to give improved results. Modifications can be further made to the original TLBO algorithm.

References

1. Hruschka, E.R., Richard, J., Campbell, G.B., Freitas, A.A., de Carvalho, A.C.P.L.F.: A survey of Evolutionary Algorithms for Clustering. IEEE Transactions 39(2) (March 2009)
2. Soni Madhulatha, T.: An over view on clustering methods. IOSR Journal of Engineering 2(4), 719–725 (2012)

3. Claire, W., Kin, C.: Constrained K-means clustering with background Knowledge. In: Proceedings of the 18th International Conference on Machine Learning (2001)
4. Bilmes, J.A.: A Gentle Tutorial of the EM Algorithm and its Application to Parameter Estimation for Gaussian Mixture and Hidden Markov Models. Department of Electrical Engineering and Computer Science, U.C. BerkeleyTR-97-021 (April 1998)
5. Tan, V.Y.F., Ng, S.-K.: Generic Probability Density Function Reconstruction for Randomization in Privacy-Preserving Data Mining. In: Perner, P. (ed.) MLDM 2007. LNCS (LNAI), vol. 4571, pp. 76–90. Springer, Heidelberg (2007)
6. Satapathy, S.C., Naik, A.: Data Clustering Based On Teaching Learning Based Optimization Toward Teaching-Learning-Based Optimization Algorithm for Dealing with Real-Parameter Optimization Problems. In: Proceedings of the 2nd International Conference on Computer Science and Electronics Engineering, ICCSEE 2013 (2013)
7. Amiri, B.: Application of Teaching-Learning-Based Optimization Algorithm on Cluster Analysis. Journal of Basic Appl. Sci. Res. Azad University, Fars Science and Research Branch 2(11), 11795–11802 (2012)
8. Pawar, P.J., VenkataRao, R.: Parameter Optimization of Machining Processes Using Teaching-Learning Based Optimization Algorithm. International Journal Advanced Manuf. Techno. 67, 995–1006 (2013), doi:10.1007/s00170-012-4524-2
9. VenkataRao, R., Patel, V.: An improved teaching-learning based optimization Algorithm for solving unconstrained optimization problems. Scientia Iranica D 20(3), 710–720 (2013) (received June 11, 2012; revised August 16, 2012; accepted October 9, 2012)
10. Niknam, T., Golestaneh, F., Sadeghi, M.S.: θ-Multi objective Teaching–Learning-Based Optimization for Dynamic Economic Emission Dispatch. IEEE Systems Journal 6(2) (June 2012)
11. Ramakrishna Murty, M., Murthy, J.V.R., Prasad Reddy, P.V.G.D., Naik, A., Satapathy, S.C.: Performance of teaching learning based optimization algorithm with various teaching factor values for solving optimization problems. In: Satapathy, S.C., Udgata, S.K., Biswal, B.N. (eds.) FICTA 2013. AISC, vol. 247, pp. 207–216. Springer, Heidelberg (2014)
12. Satapathy, S.C., Naik, A., Parvathi, K.: A teaching learning based optimization based on orthogonal design for solving global optimization problems, doi:10.1186/2193-1801-2-130
13. Tan, P.-N., Steinbach, M., Kumar, V.: Introduction To Data Mining concepts and techniques, 2nd edn., ch. 8, pp. 496–513. Pearson Publication (2008)
14. Ramakrishna Murty, M., Naik, A., Murthy, J.V.R., Prasad Reddy, P.V.G.D., Satapathy, S.C.: Automatic Clustering Using Teaching Learning Based Optimization. International Journal of Applied Mathematics 5(8), 1202–1211 (2014)

Multilevel Thresholding in Image Segmentation Using Swarm Algorithms

Layak Ali

School of Engineering,
Central University of Karnataka,
Gulbarga, Karnataka - 585311, India
informlayak@gmail.com

Abstract. Image segmentation is the most important part of image processing, and has a large impact on quantitative image analysis. Among many segmentation methods, thresholding based segmentation is widely used. In thresholding method, selection of optimum thresholds has remained a challenge over decades. In order to determine thresholds, most of the methods analyze the histogram of the image. The optimal thresholds are found by optimizing an objective function built around image histogram. The classical segmentation methods often fails to give good result for images whose histograms have multiple peaks. Since Swarm algorithms have shown promising results on multimodal problems, hence the alternative methods for optimal image segmentation. This papers presents the comprehensive analysis of Swarm algorithms for determining the optimal thresholds on standard benchmark images. An exhaustive survey of various Swarm algorithms on multilevel image thresholding was carried out and finally comprehensive performance comparison is presented both in numerical and pictorial form.

1 Introduction

Image segmentation is a process where an image is partitioned into non overlapping regions such that each region is homogeneous and two adjacent regions are heterogeneous. It is one of the most important and difficult image analysis tasks and has attracted many researchers across the globe. The segmentation of images into meaningful objects is important for object classification and recognition. There are various methods of image segmentation, among all, thresholding based segmentation is more popular. For efficient image segmentation, a multilevel thresholding is used, where multiple thresholds need be provided in advance. Thus the selection of optimum thresholds has remained a challenge over decades. The optimal thresholds are found by either minimizing or maximizing an objective function with respect to the values of the thresholds. The objective function for optimization is developed using histogram information of an image. The classical thresholding methods include: P-tile method [1], Maximum entropy method [2], Otsu's method [3], Moment-preserving thresholding [4]. The classical segmentation methods often fails to give good result for images whose histograms have multiple peaks.

S.C. Satapathy et al. (eds.), *Emerging ICT for Bridging the Future – Volume 2*,
Advances in Intelligent Systems and Computing 338, DOI: 10.1007/978-3-319-13731-5_23

Alternative to the classical segmentation methods are Swarm algorithms. There are several Swarm algorithms applied for image segmentation [5,6,7]. The Swarm algorithms found to be superior than classical optimization algorithms when the function to be optimized is discontinuous, non-differentiable and multimodal [8,9,10,11]. Swarm algorithms are nature inspired, population based and stochastic in nature. Most of the Swarm algorithms mimics the target searching behavior of the natural agents, this particular property of the Swarm is harnessed and used for solving optimization problem. These algorithms consists of natural and artificial agents, which are self-organizing and coordinated through decentralized control. Almost all the Swarm algorithms have shown promising results on multimodal optimization problems. There are many Swarm algorithms available, some of the major ones can be found in [8,9,10,11,12,13]. This papers presents a detailed comprehensive analysis of various swarm algorithms for determining the optimal thresholds on standard benchmark images. An exhaustive survey of various Swarm algorithms on multilevel image thresholding is presented both in pictorial and numerical form.

The reminder of the paper is organized as follows. Section 2 describes basic of image segmentation and multilevel thresholds. Section 3 presents Swarm algorithms and variants. Section 4 presents the simulation setup and results. The paper is concluded in Section 5.

2 Image Segmentation

Image segmentation is the process of distinguishing or separating an image into different parts or regions. For human beings it is easy to view and separate various parts in an image, but it is difficult for computing machines to do so. Since computing machines do not have any in-built intelligence, some alternative methods need be designed for automatic segregation and recognition of important regions in an images. The segmentation process is based on various characteristics found in an image like color or pixel information. In general, image segmentation may be divided into different ways like texture analysis based methods, histogram thresholding based methods, clustering based methods and region based split and merging methods [14]. One of the most common methods for segmentation of images is the thresholding method, which is commonly used for segmentation of an image into two or more clusters [15]. Thresholding is a very simple form of segmentation and it converts gray level image into binary image. A threshold will be determined and then every pixel in an image is compared with the selected threshold. If the pixel lies above the threshold it will be marked as foreground, and if it lies below the threshold marked as background. Again thresholding techniques can further be divided into two types: optimal thresholding methods and property-based thresholding methods. The optimal thresholding methods search for the optimal thresholds by optimizing an objective function. Generally this objective function for optimization is built around histogram of an image. However the number of optimal thresholds is hard to determine and needs to be specified in advance.

2.1 Multilevel Thresholds

Multilevel thresholding is one of the thresholding methods used in an image segmentation. It provides multiple thresholds in advance for segmentation and is an efficient way to perform image analysis. The optimal thresholds are found by either maximizing or minimizing a certain criterion. However, automatic selection of a robust and optimum *n-level* thresholds has remained a challenge. The *n-level* threshold problem can be converted to an optimization problem as given in equation (1).

$$T_{C=[RBG]} = max \sum_{j=1}^{n} W_j^C (\mu_j^C - \mu_T^C)^2 \tag{1}$$

Where j represents a specific class. The W_j^C is the probability of occurrence and μ_j^C is the mean of class j, these will be calculated in a standard way for L intensity levels and $n-1$ threshold levels.

3 Swarm Algorithms

Swarm algorithms may also be called as Swarm Intelligence (SI) algorithms. These algorithms are generally composed of natural or artificial agents or individuals, that coordinate among themselves and are self-organizing. Though the individuals here have very simple behaviour their collective behavior resulting from local interactions and with the environment gives extraordinary intelligent behaviour [16,17,18]. There are many SI algorithms including Particle Swarm Optimization (PSO) [8,9], Artificial Bee Colony (ABC) [10,11], Firefly Algorithm (FFA) [12] and Cuckoo Search Algorithm (CSA) [13]. The detailed description of state of-the-art algorithms is presented below.

3.1 Particle Swarm Optimization Algorithm

Particle Swarm Optimization (PSO) is one of the Swarm algorithms introduced in 1995 by Kennedy and Eberhart [8,9]. The PSO imitates the social life of swarm of birds or fish school [19]. The food foraging capability of the swarm is extracted and used for problem solving. The main source of swarm's search capability is their interaction among themselves and with the environment. In PSO terminology, each member of the swarm is called a particle and represents a solution search space. Every particle has memory and remember it's position in the search space. The best position of an individual ever visited is called personal best (*pbest*) and the best position of the whole swarm is called as global best (*gbest*). The information of global best particle is communicated to all the particles and hence movements of the particles are changed in the direction of the swarm's *gbest* and it's own *pbest*. Since the particle's position in a swarm represents the solution, hence is evaluated based on the objective function to be optimized.

Fig. 1. Segmentation with 2 threshold for Airplane (ABC, CSA, FFA and PSO)

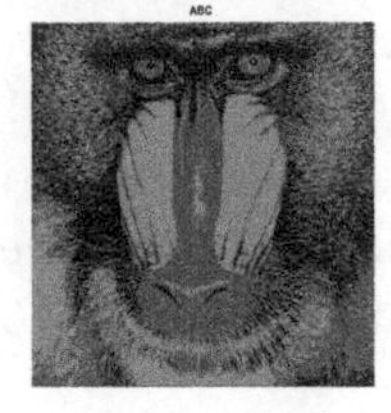

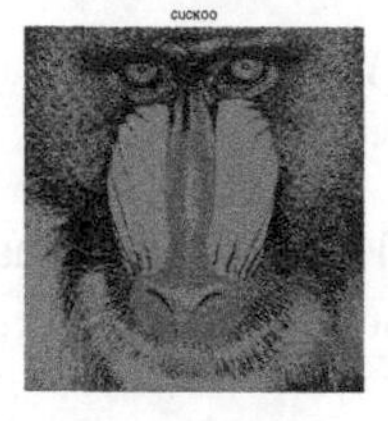

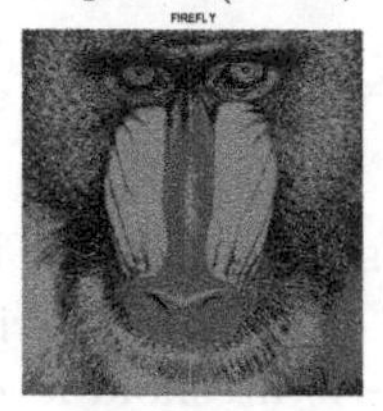

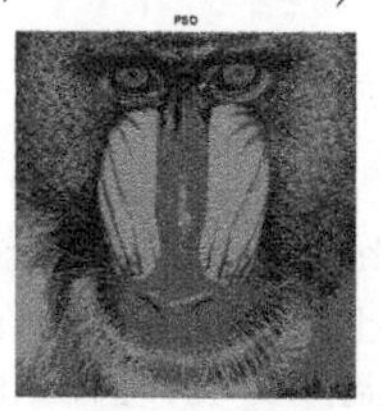

Fig. 2. Segmentation with 2 threshold for Baboon (ABC, CSA, FFA and PSO)

Fig. 3. Segmentation with 2 threshold for Barbara (ABC, CSA, FFA and PSO)

Fig. 4. Segmentation with 2 threshold for Hunter (ABC, CSA, FFA and PSO)

Fig. 5. Segmentation with 2 threshold for Lena (ABC, CSA, FFA and PSO)

3.2 Artificial Bee Colony Algorithm

Karaboga has introduced an Artificial Bee Colony (ABC) algorithm [10,11] in the year 2005. The ABC algorithm is a Swarm algorithm that simulates the honey collecting behavior of the honey bees. The colony of artificial bees can be classified into three groups: employed, onlookers and scouts. The half of the bee colony consists of employed bees and the remaining onlookers. One employed bee is dedicated to one food source and number of employed bees is equal to the number of food sources around the hive. The employed bee whose food source has been exhausted becomes a scout. The information exchange among bees is through typical dance called waggle dance. The dance floor provides current rich source information and more profitable food source will be chosen for further foraging. Employed foragers share their information with a probability, which is proportional to the profitability of the food source, and the sharing of this information through waggle dance is longer in duration. Hence, the recruitment for further foraging is proportional to the profitability of a food source. The nectar searching behavior of honey bees is used for solving optimization problems.

3.3 Firefly Algorithm

Xin-She Yang in 2009 introduced Firefly Algorithm (FFA) [12], it is also population based and stochastic in nature. The FFA algorithm was developed by the inspiration of flashing behavior of fireflies. The main reasons of flash light of firefly insects are, to attract mating partners, attract potential prey and protective warning mechanism. The FFA algorithm follows very simple rules for implementation. It assumes all fireflies are unisex and their attraction is unbiased with respect to the sex. The attraction is proportional to the brightness, and hence the less brighter one will move towards the more brighter one. If there is no brighter firefly then they walk randomly. The attractiveness decrease with increase in the distance. The brightness of a firefly is also affected by the nature of the objective function.

3.4 Cuckoo Search Algorithm

Cuckoo Search algorithm (CSA) is also a population based stochastic algorithm and was developed in 2009 by Xin-she Yang and Suash Deb [13]. It was developed by the inspiration of reproducing mechanism of some cuckoo species. These species lay their eggs in the nests of other birds, especially of other species. The host birds may have the conflicts, like, if they find that the eggs in the nest are not of their own, then either they abandon the eggs or leave the nest and build the new one. In CSA, an egg refers to a solution of the optimization problem and a nest holds all possible solutions. The number of host bird nests is fixed and a nest can hold several eggs. It follows very simple rules, each Cuckoo lays one egg at a time and places randomly in the nest. An egg will be discovered by the host bird with some probability, then it either discards an egg or nest and build the new one. It makes use of random walk for updating and moving to new solution.

Fig. 6. Segmentation with 2, 3, 4, 5 thresholds for Airplane by PSO

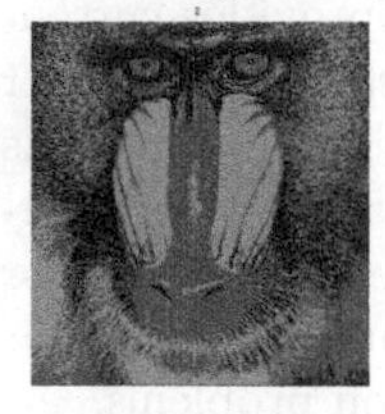 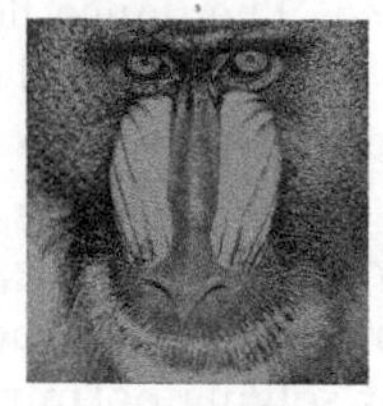  

Fig. 7. Segmentation with 2, 3, 4, 5 thresholds for Baboon by PSO

Fig. 8. Segmentation with 2, 3, 4, 5 thresholds for Barbara by PSO

Fig. 9. Segmentation with 2, 3, 4, 5 thresholds for Hunter by PSO

Fig. 10. Segmentation with 2, 3, 4, 5 thresholds for Lena by PSO

Table 1. Average results obtained for different images

Images	Thresholds	ABC	CSA	FFA	PSO
Airplane	2	1.904388e+3	1.946610e+3	1.944421e+3	1.948009e+3
	3	2.005063e+3	2.006480e+3	1.984747e+3	2.007297e+3
	4	2.033534e+3	2.034668e+3	2.029266e+3	2.051740e+3
	5	2.052734e+3	2.050324e+3	2.041940e+3	2.065112e+3
Baboon	2	1.554512e+3	1.528185e+3	1.540171e+3	1.538312e+3
	3	1.610448e+3	1.607994e+3	1.616532e+3	1.626259e+3
	4	1.646982e+3	1.645987e+3	1.637817e+3	1.663614e+3
	5	1.663242e+3	1.667541e+3	1.662853e+3	1.689206e+3
Barbara	2	2.615840e+3	2.592955e+3	2.568790e+3	2.607399e+3
	3	2.763160e+3	2.744756e+3	2.721833e+3	2.716919e+3
	4	2.783231e+3	2.795635e+3	2.780773e+3	2.828561e+3
	5	2.813618e+3	2.827841e+3	2.822695e+3	2.851628e+3
Hunter	2	2.960069e+3	2.934344e+3	2.923699e+3	2.946005e+3
	3	3.104912e+3	3.099266e+3	3.086077e+3	3.105175e+3
	4	3.167272e+3	3.138026e+3	3.152022e+3	3.171712e+3
	5	3.203339e+3	3.189082e+3	3.171094e+3	3.217985e+3
Lena	2	1.968549e+3	1.947258e+3	1.942799e+3	1.960818e+3
	3	2.092376e+3	2.084348e+3	2.056784e+3	2.105499e+3
	4	2.141112e+3	2.129420e+3	2.122012e+3	2.149109e+3
	5	2.153977e+3	2.154921e+3	2.154124e+3	2.179651e+3

4 Simulation and Results

This section presents the detailed simulation setup used for comprehensive analysis and comparative results.

4.1 Simulation Setup

The simulations were carried out using Pentium Core2Duo, 2GHz with 2GB RAM. Algorithms are coded in Matlab 7.2 on a Windows-XP platform. The Swarm algorithms used for comprehensive analysis are Particle Swarm Optimization [8,9], Artificial Bee Colony (ABC) [10,11] , Firefly Algorithm (FFA) [12] and Cuckoo Search Algorithm (CSA) [13]. The parameter setting of these algorithms are as in ABC [10,11], CSA [13], FFA [12] and PSO [8,9]. The mentioned Swarm algorithms are tested on well known benchmark images like *Airplane*, *Baboon*, *Barbara*, *Hunter* and *Lena*. The results presented are the average of 25 trials over 500 iterations.

4.2 Results and Discussions

Simulation results are presented in two ways: 1) Pictorial representation and 2) Numerical representation. The Fig 1 to Fig 10 depicts the pictorial analysis. The Table 1 and Table 2 presents numerical analysis.

Pictorial Representation. The Fig 1 to Fig 5 gives the segmented images by different state of art Swarm algorithms for two level thresholds. Every figure has four segmented images, the first one is segmented image by ABC algorithm, second one is by CSA, third one is by FFA and the fourth one is by PSO algorithm. From these figures, it can be seen that PSO obtains better segmentation than other. The worst results are shown by ABC, where as the intermediate results are shown by FFA and CSA algorithms. As PSO has shown good results on segmentation, further the segmentation results are presented with PSO on different thresholds.

The Fig 6 to Fig 10 shows the segmented images by PSO with different threshold levels ranging from 2 to 5. The first image is the segmented image by PSO with threshold 2, second, third and fourth are with threshold 3, 4 and 5 respectively. It is evident from these figures, that as threshold levels increases the segmented images will be smoother. This shows that multilevel thresholding in an image segmentation plays an important role.

Numerical Representation. The Table 1 and Table 2 presents the results in numerical form. The Table 1 presents average fitness results over 25 trials and 500 iterations. From Table 1, it can be concluded that PSO obtains good fitness results in almost all the cases. It can again be said from Table 1 that as number of threshold increases from 2 to 5, the complexity of the objective

Table 2. Average thresholds of different segmentation swarm algorithms

Images	Thresholds	ABC	CSA	FFA	PSO
Airplane	2	88, 130	118, 174	114, 174	114, 173
	3	104, 137, 164	88, 138, 183	87, 140, 187	85, 138, 185
	4	142, 128, 134, 147	66, 121, 156, 190	64, 109, 150, 191	81, 118, 166, 199
	5	74, 79, 114, 148, 142	60, 105, 149, 175, 198	50, 72, 110, 150, 193	79, 110, 138, 172, 199
Baboon	2	97, 120	93, 149	98, 149	95, 150
	3	116, 89, 143	80, 120, 160	82, 121, 157	79, 123, 161
	4	92, 78, 112, 102	65, 105, 136, 161	59, 96, 128, 161	71, 99, 136, 168
	5	98, 107, 132, 112, 119	59, 93, 127, 149, 170	38, 69, 101, 131, 163	59, 90, 121, 150, 175
Barbara	2	62,134	84,146	82,146	81,146
	3	74,121,166	74,125,179	74,126,175	69,123,170
	4	84, 76,142,139	65,103,143,178	64,100,136,178	60,100,140,183
	5	99,135,122,127,131	53, 81,119,157,194	51, 78,107,139,180	65, 77,104,140,180
Hunter	2	92,160	56,127	57,124	56,124
	3	46, 62,100	43, 92,149	40, 91,140	38, 90,144
	4	51,117, 72,119	35, 76,120,159	31, 70,109,149	37, 70,115,160
	5	80, 96,108, 68,136	32, 65,106,141,178	23, 46, 76,111,150	28, 62, 97,132,182
Lena	2	93,135	92,153	93,151	92,151
	3	103,119,157	86,126,173	81,126,171	78,128,172
	4	118,122,122,116	75,112,145,180	73,110,140,177	78, 97,135,176
	5	115,118,135,132,121	72, 91,119,150,181	63, 85,113,143,179	67, 98,118,147,183

function increases and performance of other algorithms decreases. The Table 2 gives the average thresholds obtained by different Swarm algorithms.

5 Conclusion

In this papers the comprehensive analysis of Swarm algorithms for determining optimal thresholds is presented. Among many segmentation methods, thresholding is more popular. For efficient segmentation, multilevel threshold is preferred, but it needs to be provided with multiple threshold in advance. The optimum thresholds are selected by optimizing an objective function built by using histogram of the image. Since in most of the cases, histogram of the images have multiple peaks, hence the classical optimization methods often fails to give good results. As swarm algorithms have shown promising results on multimodal problems, they are used here for optimal image segmentation. Simulation results show that the PSO algorithms achieve the better segmentation compared to ABC, FFA and CSA algorithms. The FFA and CSA algorithms shows intermediate results where as ABC shows poor results on almost all the benchmark images.

References

1. Doyle, W.: Operation useful for similarity-invariant pattern recognition. J. Assoc. Comput. 9, 259–267 (1962)
2. Kapur, J.N., Sahoo, P.K., Wong, A.K.C.: A new method for gray-level picture thresholding using the entropy of the histogram. Computer Vision Graphics Image Processing 29, 273–285 (1985)
3. Otsu, N.: A threshold selection method from gray-level histograms. IEEE Transactions on Systems, Man Cybernet. 9, 62–66 (1979)
4. Tsai, W.: Moment-preserving thresholding: a new approach. Computer Vision Graphics Image Processing 29, 377–393 (1985)
5. Lai, C.C., Tseng, D.C.: A hybrid approach using gaussian smoothing and genetic algorithm for multilevel thresholding. International Journal of Hybrid Intelligent Systems 1, 143–152 (2004)
6. Yin, P.Y., Chen, L.H.: A fast scheme for optimal thresholding using genetic algorithms. Signal Processing 72 (1999)
7. Yin, P.Y., Chen, L.H.: New method for multilevel thresholding using the symmetry and duality of the histogram. Journal of Electronics and Imaging 2 (1993)
8. Eberhart, R., Kennedy, J.: Particle swarm optimization. In: Proceedings of IEEE Int. Conference on Neural Networks, Piscataway, NJ, pp. 1114–1121 (1995)
9. Eberhart, R., Kennedy, J.: A new optimizer using particle swarm theory. In: Proceedings of 6th Int. Symp. Micro Machine and Human Science (MHS), Cape Cod, MA, pp. 39–43 (1995)
10. Karaboga, D.: An idea based on honey bee swarm for numerical optimization. Technical Report TR06, Computer Engineering Department, Engineering Faculty, Erciyes University, Turkey (2005)
11. Karaboga, D., Basturk, B.: An artificial bee colony (abc) algorithm for numeric function optimization. In: IEEE Swarm Intelligence Symposium, Indianapolis, Indiana, USA (2006)
12. Yang, X.-S.: Firefly algorithms for multimodal optimization. In: Watanabe, O., Zeugmann, T. (eds.) SAGA 2009. LNCS, vol. 5792, pp. 169–178. Springer, Heidelberg (2009)
13. Yang, X.S., Suash, D.: Cuckoo search via lévy flights. In: Proceedings of the World Congress on Nature and Biologically Inspired Computing, NaBIC 2009 (2009)
14. Brink, A.D.: Minimum spatial entropy threshold selection. IEE Proceedings on Vision Image and Signal Processing 142 (1995)

15. Kulkarni, R.V., Venayagamoorthy, G.K.: Bio-inspired algorithms for autonomous deployment and localization of sensor. IEEE Transactions on Systems 40, 663–675 (2010)
16. Beni, G.: The concept of cellular robotic systems. In: Proceedings of 6th International Symposium on Intelligent Control, pp. 57–62
17. Beni, G., Wang, J.: Swarm intelligence. In: Proceedings of 7th Annual Meeting of the Robotics Society of Japan, Japan, pp. 425–428.
18. White, T., Pagurek, B.: Towards multi-swarm problem solving in networks. In: Proceedings of 3rd International Conference on Multi-agent Systems (ICMAS 1998), pp. 333–340 (1998)
19. Robinson, J., Samii, Y.R.: Particle swarm optimization in electromagnetic. IEEE Transactions on Antenna and Propagation 52, 397–400 (2004)

A Honey Bee Mating Optimization Based Gradient Descent Learning – FLANN (HBMO-GDL-FLANN) for Classification

Bighnaraj Naik, Janmenjoy Nayak, and H.S. Behera

Department of Computer Science Engineering & Information Technology
Veer Surendra Sai University of Technology, Burla, Sambalpur-756018, Odisha, India
{mailtobnaik,mailforjnayak,mailtohsbehera}@gmail.com

Abstract. Motivated from successful use of the Honey Bee Mating Optimization (HMBO) in many applications, in this paper, a HMBO based Gradient Descent Learning for FLANN classifier is proposed and compared with FLANN, GA based FLANN and PSO based FLANN classifiers. The proposed method mimics the iterative mating process of honey bees and strategies to select eligible drones for mating process, for selection of best weights for FLANN classifiers. The classification accuracy of HMBO-GDL-FLANN is investigated and compared with FLANN, GA-based FLANN and PSO-based FLANN. These models have been implemented using MATLAB and results are statistically analyzed under one way ANOVA test. To obtain generalized performance, the proposed method has been tested under 5-fold cross validation.

Keywords: Honey Bee Mating Optimization, Functional Link Artificial Neural Network, Gradient Descent Learning, Machine Learning, Data Mining, Classification.

1 Introduction

Now-a-days, Nature inspired optimization algorithms are frequently used in solving various problems of intelligent computing. These methods are motivated by different theories from nature and biology including advances in structural genomics, gene analysis, modeling of complete cell structures, functional genomics, self-organization of natural systems [1][16]. After the successful use of some popular optimization algorithms like Particle Swarm Optimization (PSO), Ant Colony Optimization (ACO), Clonal Selection Algorithm (CSA), Firefly Algorithm, Invasive Weed Optimization, Monkey Search Algorithm etc., researchers focused into some new swarm optimization techniques based on behavior of bees. These algorithms are broadly classified into two types like: the algorithms based on the **(1) Foraging behavior** (Artificial Bee Colony (ABC) Algorithm [2,3], Virtual Bee Algorithm [4], Bee Colony Optimization Algorithm(BCO) [5], Mutable Smart Bee Algorithm (MSBA) [6,7] the Bee- Hive Algorithm [8], Bee Swarm Optimization Algorithm(BSO) [9] and Bees Algorithm [10,11]) and **(2) Mating behavior** (Honey

S.C. Satapathy et al. (eds.), *Emerging ICT for Bridging the Future – Volume 2*,
Advances in Intelligent Systems and Computing 338, DOI: 10.1007/978-3-319-13731-5_24

Bees Mating Optimization Algorithm (HBMO) [12, 13]). Since its inception , HBMO has been applied in various applications like reservoir operation[14], clustering [15][16], water resource optimization[17], financial classification[18], vehicle routing problems[19,20],Location routing[21], TSP problems[22].

In this paper, a FLANN model with hybrid honey bee mating optimization (HBMO) and Gradient descent learning (GDL) method for classification has been proposed and compared with previously available alternatives(FLANN, GA-FLANN, PSO-FLANN). Prior to this, a good number of FLANN model with different learning algorithm [23-38] has been designed for supervised machine learning tasks like classification, prediction, system identification etc. The remaining part of the paper is organized as follows: Preliminaries, Proposed Method, Experimental Setup, Results and Analysis, Conclusions and References.

2 Preliminaries

2.1 Honey Bee Mating Optimization

In the air, the mating process starts with queen's mating flight during which queen attracts drones (males) and mates with them. Colony of bees is consists of three types of bees: queen, drones & worker and each of them having different responsibility. Mating process of queen is governed by her speed, velocity and probability of mating with each drone. The speed and energy of queen is initialized at beginning of her mating flight. If drone is eligible for mating, queen mates with the drone and stores the sperm (genotype) of the drone in her spermatheca. Mating is not a onetime event with a single drone rather than this process is continued with several eligible drones until spermatheca in not full or speed and energy of queen are reached to threshold level. A crossover operator is used in order to create the broods form sperms in spermatheca. Workers will not participate in mating rather they simply dedicated for the brood care. In the next generation, if a brood is better (fittest) than the queen or any drones in the population then this brood replaces the queen or drones. The complete process of Honey Bee Mating Optimization along with description can be obtained from these papers [12, 13].

2.2 Functional Link Artificial Neural Network Architecture

The Functional Link Neural Network (FLANN) [39, 40] is a class of Higher Order Neural Networks that consume higher combination of its inputs. It comprises of a single-layer network but still is able to handle a non-linear separable classification task compared to the MLP. In FLANN, a single input value of input pattern is extended to 2n+1 number of functionally extended input values for a chosen value of 'n'. This mechanism is known as functional expansion. If 'x' is a dataset with data in a matrix of order 'm x n' then functionally expanded values $\varphi(x_i(j))$ for any input value $x_i(j)$ of a pattern can be generated by equation-1.

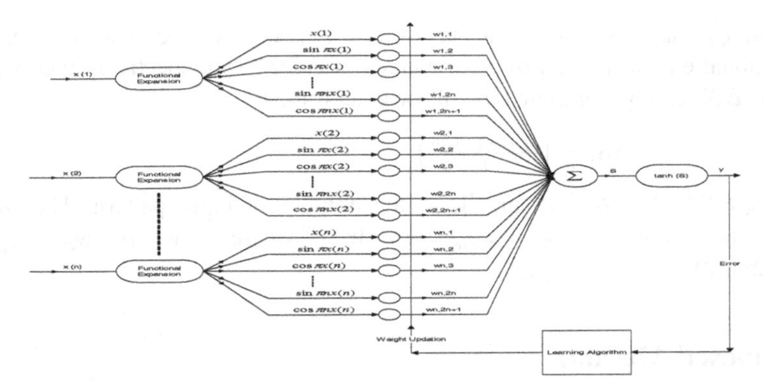

Fig. 1. Basic FLANN Architecture

$$\boldsymbol{\varphi}(\mathbf{x}_i(\mathbf{j})) = \{\, \mathbf{x}_i(\mathbf{j}), \mathbf{cos}\,\boldsymbol{\Pi}\mathbf{x}_i(\mathbf{j}), \mathbf{sin}\,\boldsymbol{\Pi}\mathbf{x}_i(\mathbf{j}), \mathbf{cos}\,\mathbf{2}\boldsymbol{\Pi}\mathbf{x}_i(\mathbf{j}), \mathbf{sin}\,\mathbf{2}\boldsymbol{\Pi}\mathbf{x}_i(\mathbf{j}) \ldots \mathbf{cos}\,\mathbf{n}\boldsymbol{\Pi}\mathbf{x}_i(\mathbf{j}), \mathbf{sin}\,\mathbf{n}\boldsymbol{\Pi}\mathbf{x}_i(\mathbf{j}) \,\} \quad (1)$$

for i=1,2….m and j=1,2…n, Where ‘m’ and ‘n’ are number of input pattern and number of input values of each input pattern respectively except class level (Probably last column of dataset x). For a input pattern, the functionally expanded input vector can be represented as:

$$\boldsymbol{\varphi} = \{\{\boldsymbol{\varphi}(\mathbf{x}_1(\mathbf{1})), \boldsymbol{\varphi}(\mathbf{x}_1(\mathbf{2})) \ldots \boldsymbol{\varphi}(\mathbf{x}_1(\mathbf{n}))\}^{\mathbf{T}}, \{\boldsymbol{\varphi}(\mathbf{x}_2(\mathbf{1})), \boldsymbol{\varphi}(\mathbf{x}_2(\mathbf{2})) \ldots \boldsymbol{\varphi}(\mathbf{x}_2(\mathbf{n}))\}^{\mathbf{T}} \ldots$$
$$\ldots \{\boldsymbol{\varphi}(\mathbf{x}_m(\mathbf{1})), \boldsymbol{\varphi}(\mathbf{x}_m(\mathbf{2})) \ldots \boldsymbol{\varphi}(\mathbf{x}_m(\mathbf{n}))\}^{\mathbf{T}}\}$$

The weight vector W_i is initialized randomly for a single input value of a input pattern as $\mathbf{W}_i = \{\mathbf{w}_{i,1}, \mathbf{w}_{i,2}, \ldots \mathbf{w}_{i,2n+1}\}$, Where i=1,2….n. Hence, for all input values of a single pattern, the weight vector can be visualized as $\mathbf{W} = \{\mathbf{W}_1, \mathbf{W}_2 \ldots \mathbf{W}_m\}^{\mathbf{T}}$. Net output of FLANN network is obtained by euation-2 and equation-3.

$$\mathbf{Y} = \mathbf{tanh}\,(\mathbf{S}) = \{\mathbf{tanh}\,(\mathbf{s}_1), \mathbf{tanh}\,(\mathbf{s}_2) \ldots \mathbf{tanh}\,(\mathbf{s}_m)\} = \{\mathbf{y}_1, \mathbf{y}_2 \ldots \mathbf{y}_m\}, \quad (2)$$

Where S is calculated as

$$\mathbf{S} = \boldsymbol{\varphi}\,\mathbf{X}\,\mathbf{W} = \{\mathbf{s}_1, \mathbf{s}_2 \ldots \mathbf{s}_m\}. \quad (3)$$

2.3 Gradient Descent Learning

The Gradient descent learning is the most frequently used training methods in which weights are changed in such a way that network error is declined as rapidly as possible. The learning process in Gradient descent method with error has been described below.

The error of kth input pattern is generated as $\mathbf{e(k)} = \mathbf{Y(k)} - \mathbf{t(k)}$ which is used to compute error term by equation-4. Where y(k) and t(k) are k^{th} output value and corresponding target value respectively. Then error terms of the network are computed by using equation-4.

$$\boldsymbol{\delta}(\mathbf{k}) = \left(\frac{\mathbf{1} - \mathbf{y_k}^2}{\mathbf{2}}\right) \times \mathbf{e(k)} \quad (4)$$

If $\boldsymbol{\varphi} = (\boldsymbol{\varphi}_1, \boldsymbol{\varphi}_2 \dots \boldsymbol{\varphi}_L)$, $\mathbf{e} = (\mathbf{e}_1, \mathbf{e}_2 \dots \mathbf{e}_L)$ and $\boldsymbol{\delta} = (\boldsymbol{\delta}_1, \boldsymbol{\delta}_2 \dots \boldsymbol{\delta}_L)$ are vector which represent set of functional expansion, set of error and set of error tern respectively then weight factor of w $'\Delta W'$ can be computed as follow(equation-5).

$$\Delta \mathbf{W_q} = \left(\frac{\sum_{i=1}^{L} 2 \times \mu \times \varphi_i \times \delta_i}{L}\right) \tag{5}$$

Where $\mathbf{q} = \mathbf{1}, \mathbf{2} \dots \mathbf{L} \times (\mathbf{2n} + \mathbf{1})$ and $'L'$ is the number of input pattern. The weight updating is done as $\mathbf{w_{new}} = \mathbf{w} + \Delta \mathbf{W}$ where $\mathbf{w} = (\mathbf{w_1}, \mathbf{w_2} \dots \mathbf{w_{L \times (2n+1)}})$ and $\Delta \mathbf{W} = (\Delta \mathbf{W_1}, \Delta \mathbf{W_2} \dots \Delta \mathbf{W_{L \times (2n+1)}})$.

3 Proposed Method

In this section, a HBMO based Gradient descent learning-FLANN classifier has been proposed by using basic problem solving strategies of HBMO.

Algorithm – 1 HBMO-GDL-FLANN for Classification

Weight-set: Set of weights of FLANN in an instance of time.
Generate initial population 'P' of weight-sets (Bees)
Compute fitness of all weight-sets in population P by using Procedure CalculationOfFitness.
F = CalculationOfFitness ($\boldsymbol{\varphi}$, w, t, μ, P);
Select weight-set (Bee) with maximum fitness as queen and rest weight-sets as drones.
Select maximum number of mating flight M (epochs).
i=0;
While (i <=M)
  Set queen's spermatheca to 25% of total number of weight-sets (Bees) in population P. Set time varying parameter energy 'E (t)' and speed 'S (t)' of queen randomly between 1 to 100.
  Set threshold value λ between 1 to 10. Select α in between 0 to 1.
  Set Counter=0.
  While (E > λ && counter != size of speratheca)
    Select a weight-set w1 from population P as a drone.
    Fw= FitnessFromTrain ($\boldsymbol{\varphi}$, w, t, μ).
    Fw'= FitnessFromTrain ($\boldsymbol{\varphi}$, w1, t, μ).
    Compute difference between fitness of w (queen) and w1 (drone) as Δ(f)=|Fw - Fw'|.
    Compute probability of drone to participate in mating as $P(w1) = e^{\frac{-\Delta(f)}{S(t)}}$.
    If (P(w1)>=0.75)
      Spermatheca(Counter,:) = w1;
      Counter = Counter +1;
    EndIf
    S(t+1) = S(t) + α.
    E(t+1) = E(t) + α.
  End While
  For i = 1:1:N, Where N is the size of Spermatheca.
    Select a weight-set w2 (Sperm) from Spermatheca randomly.
    Generate two child weight-sets w3 and w4 (broods) by applying two point crossover operator between w2 & w.
    Improvise fitness of broods until they are better than their current fitness.
    Do
      $w3' = w3 \pm rand(1)$
      $w4' = w4 \pm rand(1)$
      Fw3 = FitnessFromTrain ($\boldsymbol{\varphi}$, w3, t, μ).
      Fw4 = FitnessFromTrain ($\boldsymbol{\varphi}$, w4, t, μ).
      Fw3' = FitnessFromTrain ($\boldsymbol{\varphi}$, w3', t, μ).
      Fw4' = FitnessFromTrain ($\boldsymbol{\varphi}$, w4', t, μ).

While (Fw3' > Fw3 && Fw4' > Fw4);
If any weight-set (brood) among w3' and w4' is having better fitness then queen
Then replace the queen with that brood.
Else
If weight-set (brood) among w3' and w4' is having better fitness then any weight-set (drone) in population P.
Then replace the drone with brood.
EndIf
EndIf
EndFor
i=i+1;
EndWhile

Algorithm – 2 CalculationOfFitness Procedure

function F = *Calculationoffitness* (φ, w, t, μ , P)
for i =1:1:nr, where nr is the number of row in P.
Calculate fitness vector 'F' which keeps fitness of individual weight set w = P(i,:) from population 'P' by using FitnessFromTrain (φ, w, t, μ) *procedure.*
End
end

Algorithm – 3 FitFromTrain Procedure

function F=*fitfromtrain* (φ, w, t, μ)
$S = \varphi X w$
$Y = tanh(S)$;
If $\varphi = (\varphi_1, \varphi_2 \dots \varphi_L)$, $e = (e_1, e_2 \dots e_L)$ *and* $\delta = (\delta_1, \delta_2 \dots \delta_L)$ *are vector which represent set of functional expansion, set of error and set of error tern respectively then weight factor of w* $'\Delta W'$ *is computed as follow* $\Delta W_q = \left(\frac{\sum_{i=1}^{L} 2 \times \mu \times \varphi_i \times \delta_i}{L}\right)$.
compute error term $\delta(k) = \left(\frac{1 - y_k^2}{2}\right) \times e(k)$, *for k=1,2…L where L is the number of pattern.*
e = t - y;
Compute root mean square error (RMSE) from target value and output.
F=1/RMSE, where F is fitness of the network instance of FLANN model.
end

---

4 Experimental Setup

The proposed method has been implemented on a system (Microsoft Windows XP Professional 2002 OS, Intel Core Duo CPU T2300, 2 GB RAM and 1.66 GHz speed) using Matlab 9.0. Dataset are collected from UCI learning repository [41] and processed by KEEL software tool [42] for 5-folds cross validation. Other parameters are set as follows.

4.1 FLANN Parameter Setting

Number of functional expansion term for one input value: 11 (By setting n=5 in 2n+1 input terms to be generated for functional expansion in FLANN) ; Initialization of weight vector: Values between -1 to 1 ; Number of epochs: 1000.

4.2 Gradient Descent Learning Parameter Setting

Learning rate (μ): 0.13

4.3 Honey Bee Mating Optimization Parameter Setting

Maximum number of mating flights (M) is set equal to 1000 ; Number of queen: 1 ; Number of drones: 25 ; Population Size: 100 ; Number of broods: 50 ; Spermatheca Size: 25 ; Speed of queen: A random number between 90 to 100. ; Energy of queen: A random number between 90 to 100. ; $\alpha = 0.9$, Threshold $T = 10^{-5}$; (Small value of α helps in slow decreasing of energy as well as speed of queen and small value of threshold allows maximum number possible mating)

5 Results and Analysis

5.1 Results Obtained

In this section, we represent the comparative study on the efficiency of our proposed method on benchmark datasets and result of proposed HBMO-FLANN model is compared with FLANN, GA-FLANN based on Genetic Algorithm [43] and PSO-FLANN based on Particle Swarm Optimization [44,45]. The datasets information is presented in table-1. The Min-Max normalization is used for normalization and datasets are scaled in the interval -1 to +1. The Classification accuracy of models is calculated by using eq-6 in terms of number of correctly-classified and miss-classified patterns and listed in table 2.

Table 1. Data Set Information

Dataset	Number of Pattern	Number of Features (excluding class label)	Number of classes	Number of Pattern in class-1	Number of Pattern in class-2	Number of Pattern in class-3
Monk 2	256	06	02	121	135	-
Hayesroth	160	04	03	65	64	31
Heart	256	13	02	142	114	-
New Thyroid	215	05	03	150	35	30
Iris	150	04	03	50	50	50
Pima	768	08	02	500	268	-
Wine	178	13	03	71	59	48
Bupa	345	06	02	145	200	-

$$Accuracy = \frac{\sum_{i=1}^{n}\sum_{j=1,\, i==j}^{m} cm_{i,j}}{\sum_{i=1}^{n}\sum_{j=1}^{m} cm_{i,j}} \times 100 \quad (6)$$

Where 'cm' is the confusion matrix found during classification.

Table 2. Performance Comparison In Terms Of Accuracy

Dataset	Accuracy of Classification in Average From Cross Validation							
	FLANN		*GA-FLANN*		*PSO-FLANN*		*HMBO-FLANN*	
	Train	*Test*	*Train*	*Test*	*Train*	*Test*	*Train*	*Test*
Monk 2	93.82813	92.04303	96.54689	93.19913	97.453134	95.46585	98.72549	96.9683
Hayesroth	90.35938	82.3125	91.062504	83.5625	91.265628	83.9375	91.8946	84.8925
Heart	88.962966	78.48149	89.407408	79.07408	89.777762	79.85185	88.98657	78.9387
New Thyroid	93.918596	76.55813	94.197648	77.53487	94.3023	78.79069	94.38364	79.63856
Iris	96.84712	97.36815	97.12973	98.16639	97.352249	98.65	98.27528	99.53782
Pima	78.416	78.7608	78.64	78.80	80.126048	79.47073	80.8529	80.74956
Wine	92.76	93.18595	94.36842	95.53644	97.762	95.6274	98.72536	98.2743
Bupa	72.16	72.76	74.3208	75.5	76.3842	76.75	77.3285	79.3658

5.2 Statistical Analysis

The performance of proposed method are analyzed under one way ANOVA test by using SPSS-16.0 statistical tool to prove the result statistically significant. The test has been carried out using one way ANOVA in Duncan multiple tests by setting 95 % confidence interval level and linear polynomial contrast. The snap shot of the result is listed below (Fig. 2). The result of one way ANOVA test shows that, the proposed method has an average accuracy of 89.2211 % which outperform other alternatives.

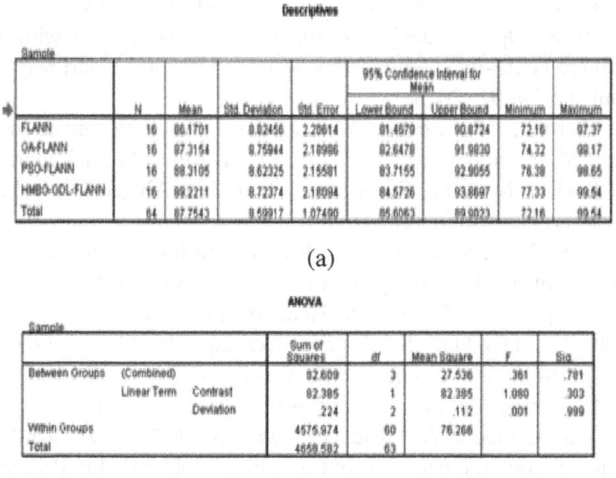

Descriptives

Sample

	N	Mean	Std. Deviation	Std. Error	95% Confidence Interval for Mean		Minimum	Maximum
					Lower Bound	Upper Bound		
FLANN	16	86.1701	8.82456	2.20614	81.4670	90.8724	72.16	97.37
GA-FLANN	16	87.3154	8.75944	2.18986	82.6478	91.9830	74.32	98.17
PSO-FLANN	16	88.3105	8.62325	2.15581	83.7155	92.9055	78.39	98.65
HMBO-GDL-FLANN	16	89.2211	8.72374	2.18094	84.5726	93.8697	77.33	99.54
Total	64	87.7543	8.59917	1.07490	85.6063	89.9023	72.16	99.54

(a)

ANOVA

Sample

			Sum of Squares	df	Mean Square	F	Sig.
Between Groups	(Combined)		82.609	3	27.536	.361	.781
	Linear Term	Contrast	82.385	1	82.385	1.080	.303
		Deviation	.224	2	.112	.001	.999
Within Groups			4575.974	60	76.266		
Total			4658.582	63			

(b)

Fig. 2. (a) (b) One Way ANOVA Test

6 Conclusions

During implementation, the proposed HBMO-GDL-FLANN found to be simple due to fact that HBMO is free from complicated operator like crossover and mutation in GA. Also, it requires less mathematical formulation. The proposed PSO-GA-FLANN model can be computed with a low cost due to less complex architecture of FLANN. The experimental analysis shows that the proposed method performs relatively better than other classifiers. The future work may comprise with the integration of variants of HBMO with other means of higher order neural network for classification in data mining.

References

1. De Castro, L.N., Timmis, J.: Artificial Immune Systems: A New Computational Intelligence Approach. Springer (2002)
2. Karaboga, D., Basturk, B.: A powerful and efficient algorithm for numerical function optimization: artificial bee colony (ABC) algorithm. J. Global Optim. 39, 459–471 (2007)
3. Karaboga, D., Basturk, B.: On the performance of artificial bee colony (ABC) algorithm. Appl. Soft Comput. 8, 687–697 (2008)
4. Yang, X.-S.: Engineering optimizations via nature-inspired virtual bee algorithms. In: Mira, J., Álvarez, J.R. (eds.) IWINAC 2005. LNCS, vol. 3562, pp. 317–323. Springer, Heidelberg (2005)
5. Teodorovic, D., Dell'Orco, M.: Bee colony optimization—a cooperative learning approach to complex transportation problems. In: Proceedings of the 16th Mini – EURO Conference and 10th Meeting of EWGT, pp. 51–60 (2005)
6. Mozaffari, A., et al.: Comprehensive preference optimization of an irreversible thermal engine using pareto based mutable smart bee algorithm and generalized regression neural network, Swarm Evolut. Swarm Evolut. Comput. 9, 90–103 (2013)
7. Mozaffari, A., et al.: Optimal design of classic Atkins on engine with dynamic specific heat using adaptive neuro-fuzzy inference system and mutable smart bee algorithm. Swarm Evolut. Comput. 12, 74–91 (2013)
8. Wedde, H.F., Farooq, M., Zhang, Y.: BeeHive: An efficient fault-tolerant routing algorithm inspired by honey bee behavior. In: Dorigo, M., Birattari, M., Blum, C., Gambardella, L.M., Mondada, F., Stützle, T. (eds.) ANTS 2004. LNCS, vol. 3172, pp. 83–94. Springer, Heidelberg (2004)
9. Drias, H., Sadeg, S., Yahi, S.: Cooperative bees swarm for solving the maximum weighted satisfiability problem. In: Cabestany, J., Prieto, A.G., Sandoval, F. (eds.) IWANN 2005. LNCS, vol. 3512, pp. 318–325. Springer, Heidelberg (2005)
10. Ebrahim Zadeh, A., Mavaddati, S.: A novel technique for blind source separation using bees colony algorithm and efficient cost functions. Swarm Evolut. Comput. (2013), `http://dx.DOI.org/10.1016/j.swevo.2013.08.002`
11. Pham, D.T., Ghanbarzadeh, A., Koc, E., Otri, S., Rahim, S., Zaidi, M.: The bees algorithm—a novel tool for complex optimization problems. In: Proceeding of the 2nd International Virtual Conference on Intelligent Production Machines and Systems, IPROMS 2006. Elsevier, Oxford (2006)
12. Abbass, H.A.: A monogenous MBO approach to satisfiability. In: International Conference on Computational Intelligence for Modelling Control and Automation, CIMCA 2001, Las Vegas, NV, USA (2001)
13. Abbass, H.A.: Marriage in honey-bee optimization (MBO):a haplometrosis polygynous swarming approach. In: The Congress on Evolutionary Computation (CEC 2001), Seoul, Korea, pp. 207–214 (2001)
14. Afshar, A., Haddad, O.B., Marino, M.A., Adams, B.J.: Honey-bee mating optimization (HBMO) algorithm for optimal reservoir operation. J. Frankl. Inst 344, 452–462 (2007)

15. Fathian, M., Amiri, B., Maroosi, A.: Application of honey bee mating optimization algorithm on clustering. Appl. Math. Comput. 190, 1502–1513 (2007)
16. Marinakis, Y., Marinaki, M., Matsatsinis, N.: A Hybrid Clustering Algorithm Based on Honey Bees Mating Optimization and Greedy Randomized Adaptive Search Procedure. In: Maniezzo, V., Battiti, R., Watson, J.-P. (eds.) LION 2007 II. LNCS, vol. 5313, pp. 138–152. Springer, Heidelberg (2008)
17. Haddad, O.B., Afshar, A., Marino, M.A.: Honey-bees mating optimization (HBMO) algorithm: a new heuristic approach for water resources optimization. Water Resour. Manag. 20, 661–680 (2006)
18. Marinaki, M., Marinakis, Y., Zopounidis, C.: Honey bees mating optimization algorithm for financial classification problems. Appl. Soft Comput. 10, 806–812 (2010)
19. Marinakis, Y., Marinaki, M., Dounias, G.: Honey bees mating optimization algorithm for the vehicle routing problem. In: Krasnogor, N., Nicosia, G., Pavone, M., Pelta, D. (eds.) Nature Inspired Cooperative Strategies for Optimization, NICSO 2007. SCI, vol. 129, pp. 139–148. Springer, Heidelberg (2008)
20. Marinakis, Y., Marinaki, M., Dounias, G.: Honey bees mating optimization algorithm for large scale vehicle routing problems. Nat. Comput. 9, 5–27 (2010)
21. Marinakis, Y., Marinaki, M., Matsatsinis, N.: Honey bees mating optimization for the location routing problem. In: IEEE International Engineering Management Conference (IEMC—Europe 2008), Estoril. Portugal (2008)
22. Marinakis, Y., Marinaki, M.: A hybrid honey bees mating optimization algorithm for the probabilistic travelling salesman problem. In: IEEE Congress on Evolutionary Computation (CEC 2009), Trondheim, Norway (2009)
23. Patra, J.C., et al.: Financial Prediction of Major Indices using Computational Efficient Artificial Neural Networks. In: International Joint Conference on Neural Networks, Canada, July 16-21, pp. 2114–2120. IEEE (2006)
24. Mishra, B.B., Dehuri, S.: Functional Link Artificial Neural Network for Classification Task in Data Mining. Journal of Computer Science 3(12), 948–955 (2007)
25. Dehuri, S., Mishra, B.B., Cho, S.-B.: Genetic Feature Selection for Optimal Functional Link Artificial Neural Network in Classification. In: Fyfe, C., Kim, D., Lee, S.-Y., Yin, H. (eds.) IDEAL 2008. LNCS, vol. 5326, pp. 156–163. Springer, Heidelberg (2008)
26. Dehuri, S., Cho, S.: A comprehensive survey on functional link neural networks and an adaptive PSO–BP learning for CFLNN. Springer, London (2009), doi:10.1007/s00521-009-0288-5.
27. Patra, J.C., et al.: Computationally Efficient FLANN-based Intelligent Stock Price Prediction System. In: Proceedings of International Joint Conference on Neural Networks, Atlanta, Georgia, USA, June 14-19, pp. 2431–2438. IEEE (2009)
28. Sun, J., et al.: Functional Link Artificial Neural Network-based Disease Gene Prediction. In: Proceedings of International Joint Conference on Neural Networks, Atlanta, Georgia, USA, June 14-19, pp. 3003–3010. IEEE (2009)
29. Chakravarty, S., Dash, P.K.: Forecasting Stock Market Indices Using Hybrid Network. In: World Congress on Nature & Biologically Inspired Computing, pp. 1225–1230. IEEE (2009)
30. Majhi, R., et al.: Classification of Consumer Behaviour using Functional Link Artificial Neural Network. In: IEEE International Conference on Advances in Computer Engineering, pp. 323–325 (2010)
31. Nayak, S.C., et al.: Index Prediction with Neuro-Genetic Hybrid Network: A Comparative Analysis of Performance. In: IEEE International Conference on Computing, Communication and Applications (ICCCA), pp. 1–6 (2012)

32. Bebarta, D.K., et al.: Forecasting and Classification of Indian Stocks Using Different Polynomial Functional Link Artificial Neural Networks (2012) 978-1-4673-2272-0/12/Crown
33. Mishra, S., et al.: A New Meta-heuristic Bat Inspired Classification Approach for Microarray Data. C3IT, Procedia Technology 4, 802–806 (2012)
34. Mahapatra, R., et al.: Reduced feature based efficient cancer classification using single layer neural network. In: 2nd International Conference on Communication, Computing & Security. Procedia Technology, vol. 6, pp. 180–187 (2012)
35. Mili, F., Hamdi, H.: A comparative study of expansion functions for evolutionary hybrid functional link artificial neural networks for data mining and classification, pp. 1–8. IEEE (2013)
36. Mishra, S.: An enhanced classifier fusion model for classifying biomedical data. Int. J. Computational Vision and Robotics 3(½), 129–137 (2012)
37. Dehuri, S.: An improved swarm optimized functional link artificial neural network (ISO-FLANN) for classification. The Journal of Systems and Software, 1333–1345 (2012)
38. Naik, B., Nayak, J., Behera, H.S.: A Novel FLANN with a Hybrid PSO and GA Based Gradient Descent Learning for Classification. In: Satapathy, S.C., Biswal, B.N., Udgata, S.K., Mandal, J.K. (eds.) Proc. of the 3rd Int. Conf. on Front. of Intell. Comput (FICTA) 2014- Vol. 1. AISC, vol. 327, pp. 745–754. Springer, Heidelberg (2015)
39. Pao, Y.H.: Adaptive pattern recognition and neural networks. Addison-Wesley Pub. (1989)
40. Pao, Y.H., Takefuji, Y.: Functional-link net computing: theory, system architecture and functionalities. Computer 25, 76–79 (1992)
41. Bache, K., Lichman, M.: UCI Machine Learning Repository. University of California, School of Information and Computer Science, Irvine (2013), `http://archive.ics.uci.edu/ml`
42. Alcalá-Fdez, J., et al.: KEEL Data-Mining Software Tool: Data Set Repository, Integration of Algorithms and Experimental Analysis Framework. Journal of Multiple-Valued Logic and Soft Computing 17(2-3), 255–287 (2011)
43. Holland, J.H.: Adaption in Natural and Artificial Systems. MIT Press, Cambridge (1975)
44. Kennedy, J., Eberhart, R.: Swarm Intelligence Morgan Kaufmann, 3rd edn. Academic Press, New Delhi (2001)
45. Kennedy, J., Eberhart, R.C.: Particle swarm optimization. In: IEEE International Conference on Neural Networks, Perth. Australia, pp. 1942–1948 (1995)

Optimal Sensor Nodes Deployment Method Using Bacteria Foraging Algorithm in Wireless Sensor Networks

Pooja Nagchoudhury[1], Saurabh Maheshwari[2], and Kavita Choudhary[1]

[1] JIET Jodhpur, India [2] GWEC Ajmer, India
pooja.28nagchoudhury@gmail.com, saurabh.maheshwari.in@ieee.org, kavity_26@yahoo.co.in

Abstract. The nodes in wireless sensor networks (WSN) need to be deployed optimally to cover whole geographic area and the communication link between them should be optimal. During deployment in remote locations, simulation of the proposed algorithm can be used for optimally deploying the sensor nodes in the area. Bacteria foraging is the nature inspired algorithm which is used to make clusters among various similar entities. Here the nodes to be deployed can be taken as the bacteria that are in search of the food which is depicted by the best possible communication link. This paper presents a novel algorithm for optimal sensor node deployment leading to optimal clustering of WSN nodes. This paper utilizes the fact that an area can be covered fully using regular hexagon. So using a bacteria foraging algorithm for optimization the locations are such adjusted that all the nodes in the network moves to vertices of regular hexagons connected with each other. This leads to complete coverage of the area and all the nodes are equidistant.

Keywords: WSN, sensor node deployment, clustering, bacteria foraging, bio-inspired, nature inspired, hexagon.

1 Introduction

Wireless sensor network is constructed using large number of sensor nodes. Every sensor nodes have sensing unit, energy unit, processing unit, storage unit and transmitting / receiving unit [2, 14]. The various workings elements of wireless sensor networks are sensor nodes, users, base station, clusters and cluster head. All these elements work together to execute the task. Rapid deployment, fault tolerance and self-organize nature makes the WSN to employ in various areas like in military surveillances, tracking application, climatic analysis, health care application commercial applications, agricultural application etc. [14]. While building WSN several issues have to keep in concern: flexibility, easy of deployment, reliability, cost etc. [5]. The WSN is organized in four different ways. They are also known as topologies of WSN, single hop flat model, multi hop flat model, single hop cluster model and multi hop cluster model. Multi hop cluster model is suitable model for wireless sensor network among all models [3].

S.C. Satapathy et al. (eds.), *Emerging ICT for Bridging the Future – Volume 2*,
Advances in Intelligent Systems and Computing 338, DOI: 10.1007/978-3-319-13731-5_25

Data aggregation is essential in WSN, for which the nodes in the field group together to form clusters. Clustering is required because it simplifies the task of data collection, diminish the communication overhead of the network, increases the life of network in WSN, eliminate long distance data sensing and reduces collision in the network hence increases the throughput of the network [3, 19]. The nodes in the field are deployed in arbitrary manner but accurate deployment is essential in the performance of wireless sensor network (WSN). The nodes are deployed using the unmanned aerial vehicle were the manual deployment are not possible [1]. The deployment is easy when the area is accessible by human, the positions of node are registered at the time of deployment and for small network, and otherwise the deployment is complex process [13]. Some of the applications require the information about the nodes. The most common solution of this problem is localization of nodes [4].

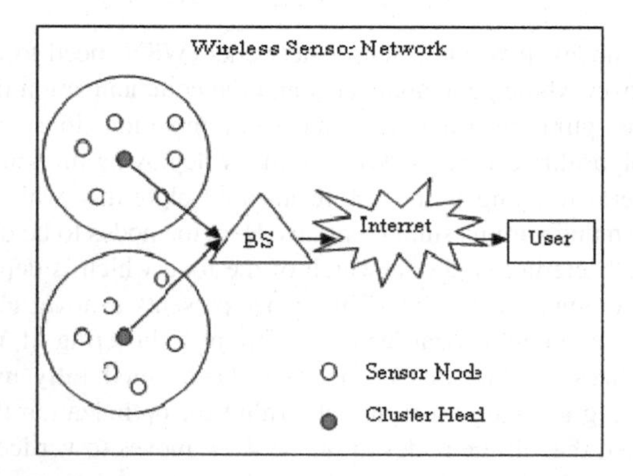

Fig. 1. Block Diagram of Wireless Sensor Network

The grouping of similar type of objects is called clustering. In WSN, the nodes in the network are grouped together to form clusters. Each cluster in the network has a cluster head. The nodes in the cluster sends data to the cluster head, the cluster head performs various operations on the coming data and sends the data to the other cluster heads or to the base station depending upon whether it is directly in communication with the base station or not. If it is in direct communication with the base station, it sends data directly to it and if it is not, then sends data through multi-hop routing to the base station [10]. Cluster size, synchronization of nodes, connectivity etc. are various issues which effects the clustering [15]. To organize a WSN clustering plays an important role. Clustering affects the performance of network [7].

2 Literature Survey

Detail surveys on clustering methods and their performance analysis are given in [3, 11]. Hamid and Masood in [9] concentrate to prolong the network life by proposed a clustering method which extends the LEACH. The method uses gateway nodes for transmission. Gurjot singh et. al. in [11] consider the optimal deployment of sensor nodes as clustering. The Bacterial Foraging Algorithm is used to find the position of nodes in cluster. Thus improves the life of WSN. Authors in [6] presented a comparative study on node clustering using voronoi diagram. The Voronoi diagram of network is constructed to build the clusters. Each cluster contains a single node in it and the small cells in the voronoi diagram are merged to form bigger cluster.

Raghavendra V. Kulkarni et. al. in [14, 16] presented BFA and PSO for localization of deployed nodes. The authors show that BFA finds more accurate position of node as compare to PSO but with slow convergence. Various localization techniques are explained in [15]. The model forms cluster in distributed manner inspired by the behavior of biological model. Various challenges like cluster head rotation, size of cluster, synchronization etc. are faced while developing the clusters which are discussed in [15].

The algorithm prolongs the life of network by MIMO scheme. Ankit Sharma et al. in [5] proposed an idea of clustering by combining the BFA, LEACH and HEED protocols which increase the network life. Qiao Li et al. in [12] presented a Low Energy Intelligent Clustering Protocol (LEICP) which is enhancement of LEACH. It works more efficiently than LEACH. In each round the auxiliary cluster head select the cluster head using BFA and the cluster head decides whom to send the message next using Dijkstra algorithm so that optimal path is used.

Authors in [13] proposed a range free localization algorithm that localizes the node more efficiently which are present at the center of the field or network because it uses three anchor nodes for localization.

3 Proposed Deployment Method

The proposed method is most suitable for the movable sensor nodes in the field to reach an optimum location after initial deployment through aircraft in remote areas. The aim is to move and reach to the optimal position to cover complete geographic area. In this section the proposed method for deployment using bacteria foraging has been discussed.

3.1 Hexagonal Network Architecture

The network consists of several movable nodes equipped with wireless directional antenna, global positioning systems (GPS) for coordinates, memory for storing routing and wireless proximity table. There may be chances that in some area there are multiple sensor nodes whereas in other there may not be any of the nodes fallen. Proper deployment is needed so that complete area has equal number of nodes and no part of the area left un-sensed. Nodes are movable, they can move to the desired locations. The nodes will be thrown to the random locations generally through the airplanes. Algorithm can also be applied for mobile node carrier robots. Nodes move to their optimal locations on the basis of our algorithm.

The aim of the proposed method is to bring the nodes on vertices of a regular hexagon and the cluster head placed on the center. The radius r of the circle circumscribing this hexagon gives the distance between the nodes. The distance of the center from all the vertices is also r. When any two hexagons come in contact with each other no space is left between them (figure 2(b)) as it remains between circles and squares. To cover any area completely without leaving even a smaller portion, hexagonal structure for clusters is favorable where the cluster head will be at the center and nodes taking part in cluster will be at vertices. The nodes on the vertices of any hexagon are centers of other hexagons. This way all the clusters are interconnected and disseminate the sensed information with minimum power consumption.

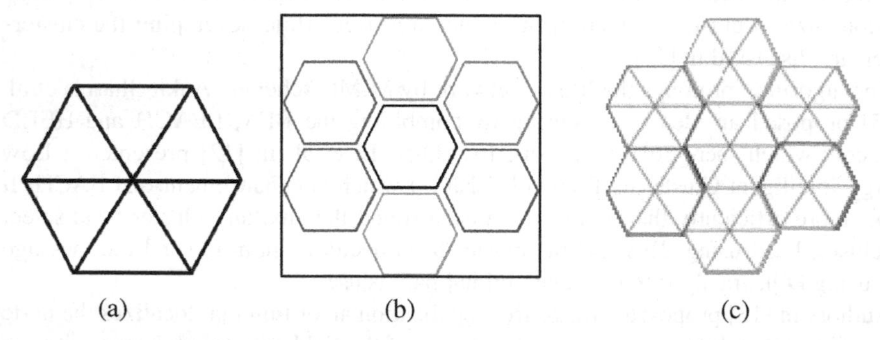

Fig. 2. (a) Hexagonal Structure of cluster (b) Multiple hexagons covering area (c) Vertices of Adjoining hexagons become centers for neighboring clusters

3.2 Bacteria Foraging Algorithm

Bacterial Foraging Algorithm (BFA) copies the searching behavior of Escherichia coli bacteria (E. Coli) which is in the small intestine of human body. E. Coli bacterial finds the nutrient rich location in human intestine. An E. Coli bacterium has two types of movements: swimming and tumbling. In swimming movement, the bacteria moves in a straight line in a given direction, while in tumbling the bacteria randomly changes its direction of movement. BFA has three steps: a) Elimination step, b) Reproduction step and c) Chemotaxis step. The bacteria initially start with any random position and evaluate the objective function to determine the movement in which direction. Then it moves in a particular direction again and evaluates the objective function. If it is better than the last one it continuously moves in the same direction until it gets the bad objective function. This whole step is Chemotaxis step. After a specific number of Chemotaxis steps, the reproduction of the bacteria takes place in which the two bacteria combines to form a healthy bacteria and next is the elimination-dispersal step in which the bacteria splits into two bacteria and the non-healthy bacteria is eliminated in this step. These steps are continuously repeated up to specific number of times until the objective function is reached. In the next section we will see how we have mapped this bacteria foraging optimization algorithm has been mapped for optimal node deployment.

3.3 Optimal Node Deployment Using Bacteria Foraging Algorithm

The proposed approach is for self-deployment, no central intervention is needed. The nodes themselves run the algorithm and calculate optimal location. The aim is to cover whole geographic area with minimum number of nodes and transmit data with minimum power consumption and minimum number of hops between two nodes. The regular hexagonal geometry discussed in previous section is optimal in the entire context. The distance between all the nodes is equal, the nodes equipped with directional antenna has to transmit data in only 6 directions at particular angles. The guarantee that a node will be in the direction path and range is increased since all the vertexes are occupied with nodes. The forwarding node is at center. It forwards to all its neighbors only. This is a cluster. When a location such as a field where the person can visit each and every corner, he may place the nodes explicitly on the calculated positions of regular hexagon vertices. The nodes placed on these vertices will certainly make an optimal directional antenna based WSN. Large remote areas like forests where forest fire detection is to be done and the number of fire sensors will be very large also the area is very vast, the nodes are deployed using airplanes at random locations. They have to be placed such that they cover entire space.

Steps of the proposed algorithm are:

1. Any node X with location (x1,y1) begins the algorithm
2. It creates a virtual regular hexagon around itself by calculating coordinates of its vertices {v11,v12,v13,v14,v15,v16} and broadcast them to all the nodes in its radius r
3. This is the objective function of the bacteria foraging algorithm, that all nodes try to reach to the vertices of the hexagons
4. Any six nodes decided by X moves to the vertices performing swimming step of Chemotaxis
5. Remaining nodes move out of the range r up to a random distance from hexagon performing tumbling step to change the cluster
6. The message will be a directed signal in all six directions separated 60 degrees with each other from the center
7. The vertices of the node X are moved into a queue since they have moved to their final location now they will be new center nodes and will make a cluster around them
8. When all the nodes are in the queue, it means all the nodes have reached to some vertex then the chemotaxis converges
9. If two nodes reach on the same vertex then they reproduce a new node with double battery power but single communication link as per BFO
10. ID of one of the nodes is eliminated so to decrease extra communication without covering extra area in the system
11. Convergence of the algorithm will remove the redundancy of the nodes due to random deployment with their optimal placement positions this will also lead to usage of these nodes as extra battery power or to be locate in the regions which are left without any node.

```
12. This unplaced area will be treated in second pass of the
    algorithm when all the nodes have reached up to some
    vertices.
13. So new nodes will be created using reproduction with new
    locations.
```

4 Simulation Results

The simulation of the proposed node deployment scenario has been done in MATLAB. The area is 100 by 100 Km.

4.1 Deployment Plots

The nodes deployed randomly are represented as blue * in the figure 3 (a). The nodes create virtual hexagon around them. These hexagons are initially overlapping (figure 3(b)) on each other it means the deployment is not proper. After the run of the proposed algorithm the intermediate and final results are shown in figure 3 (c) and (d).

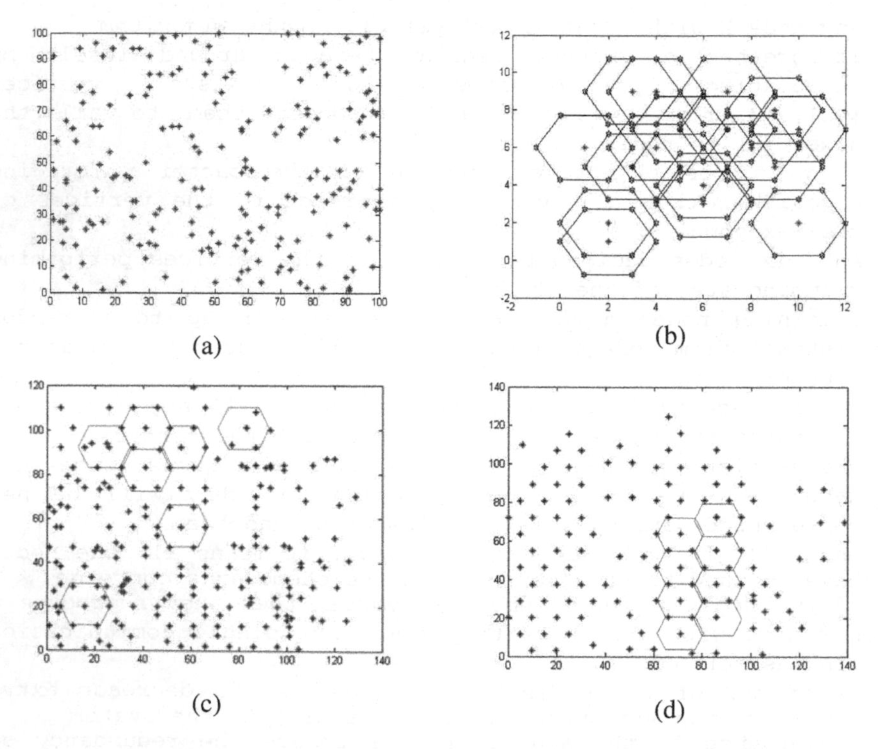

Fig. 3. (a) Random Nodes deployment (b) Overlapping Hexagonal Architecture (c) Intermediate hexagonal structure of clusters (d) Final Hexagonal equidistant nodes

The total number nodes n in simulation is 200, but in final architecture the number of final nodes are less. This is due to elimination of repetitive nodes. Some area is left where no nodes exist it means no node could reach there, in II phase of the method the repetitive nodes move to blank vertices on empty areas making the compelte field covered with sensor nodes. The phase II is currently in development so simulation figure is not available.

Another analysis of the proposed method can be done on the basis of optimal number of nodes is a geographical area. In figure 3(d) it can be seen that all the nodes are arranged at vertices of some node. Complete field is now covered with hexagons only. The optimal number of nodes N needed to cover the area is given by the nodes that are plotted in the figure 3(d).

4.2 Average Distance Moved per Node

This is the parameter that tells the amount of distance particular node has moved to reach to optimal location. Since the movement also needs the power so it has to be minimized. The average distance moved by a node for n=200 and r=10m is less than 10m which is very good.

5 Conclusion and Future Work

The method for optimal deployment of the sensor nodes has been formalized in this paper. The method can be proved in many ways for the optimality since the hexagonal architecture of cluster followed here is optimal in context of power utilization and minimum number of nodes to cover complete area. The node deployment in remote unmanned areas is target application. The hexagonal clustering architecture can be viewed in simulation result section. The movement of nodes needed is very less thus tradeoff between the movement and power saving can be established. The movement is needed only once and the power saving is done for entire lifetime. The proposed method is also useful for offline calculation of node placement coordinates to place the nodes manually in the field, thus increasing the scope of the proposed algorithm. The proposed method needs to be formalized in other manners to compare all the parameters with existing deployment methods. Due to limitation of space we are not able to present them here.

References

1. Sreedevi, I., Mankhand, S., Chaudhury, S., Bhattacharyya, A.: Bio-Inspired Distributed Sensing Using a Self-Organizing Sensor Network. Journal of Engineering (2013)
2. Sribala, S.: Energy Efficient Routing in Wireless Sensor Networks Using Modified Bacterial Foraging Algorithm. IJREAT International Journal of Research in engineering and Advanced Technology 1(1) (March 2013) ISSN: 2320-8791

3. Nithyakalyani, S., Kumar, S.S.: Data Aggregation In Wireless Sensor Network Using Node Clustering Algorithms — A Comparative Study. In: IEEE Conference on Information & Communication Technologies (ICT), April 11-12, pp. 508–513 (2013)
4. Aruna, Gupta, V.: Soft Computing Implementation for Mobile Ad-hoc Network Optimization Using Bacteria Foraging Optimization Algorithm. International Journal of Computer Science and Communication Engineering 2(2) (May 2013) ISSN 2319-7080
5. Sharma, A., Thakur, J.: An Energy Efficient Network Life Time Enhancement Proposed Clustering Algorithm for Wireless Sensor Networks. International Journal of Enhanced Research in Management and Computer Application 2(7), 1–4 (2013)
6. Kumar, A., Khosla, A., Saini, J.S., Singh, S.: Computational Intelligence Based Algorithm for Node Localization in Wireless Sensor Networks. In: 6th IEEE International Conference Intelligent Systems (IS), Sofia, September 6-8, pp. 431–438 (2012)
7. Bhuvaneswari, P.T.V., Karthikeyan, S., Jeeva, B., Prasath, M.A.: An Efficient Mobility Based Localization in Underwater Sensor Networks. In: 2012 Fourth International Conference on Computational Intelligence and Communication Networks (CICN), November 3-5, pp. 90–94 (2012)
8. Kulkarni, R.V., Venayagamoorthy, G.K.: Particle Swarm Optimization in Wireless-Sensor Networks: A Brief Survey. IEEE Transactions on Systems, Man, and Cybernetics, Part C: Applications and Reviews 41(2), 262–267 (2011)
9. Tarigh, H.D., Sabaei, M.: A New Clustering Method To Prolong The Lifetime of WSN. In: 3rd International Conference on Computer Research and Development (ICCRD), March 11-13, vol. 1, pp. 143–148 (2011)
10. Wei, C., Yang, J., Gao, Y., Zhang, Z.: Cluster-Based Routing Protocols in Wireless Sensor Networks: A Survey. In: 2011 International Conference on Computer Science and Network Technology (ICCSNT), December 24-26, vol. 3, pp. 1659–1663 (2011)
11. Gaba, G.S., Singh, K., Dhaliwal, B.S.: Sensor Node Deployment Using Bacterial Foraging Optimization. In: 2011 International Conference on Recent Trends in Information Systems (ReTIS), December 21-23, pp. 73–76 (2011)
12. Li, Q., Cui, L., Zhang, B., Fan, Z.: A Low Energy Intelligent Clustering Protocol for Wireless Sensor Networks. In: 2010 IEEE International Conference on Industrial Technology (ICIT), March 14-17, pp. 1675–1682 (2010)
13. Yang, G., Yi, Z., Tianquan, N., Keke, Y., Tongtong, X.: An Improved Genetic Algorithm for Wireless Sensor Networks Localization. In: 2010 IEEE Fifth International Conference on Bio-Inspired Computing: Theories and Applications (BIC-TA), September 23-26, pp. 439–443 (2010)
14. Kulkarni, R.V., Venayagamoorthy, G.K.: Bio-inspired Algorithms for Autonomous Deployment and Localization of Sensor Nodes. IEEE Transactions on Systems, Man, and Cybernetics, Part C: Applications and Reviews 40(6), 663–675 (2010)
15. Zein-Sabatto, S., Elangovan, V., Chen, W., Mgaya, R.: Localization Strategies for Large-Scale Airborne Deployed Wireless Sensors. In: IEEE Symposium on Computational Intelligence in Miulti-Criteria Decision-Making, MCDM 2009, March 30-April 2, pp. 9–15 (2009)
16. Kulkarni, R.V., Venayagamoorthy, G.K., Cheng, M.X.: Bio-inspired Node Localization in Wireless Sensor Networks. In: IEEE International Conference on Systems, Man and Cybernetics, SMC 2009, October 11-14, pp. 205–210 (2009)

Efficient Data Aggregation Approaches over Cloud in Wireless Sensor Networks

Pritee Parwekar[1], Veni Goel[2], Aditi Gupta[2], and Rajul Kukreja[2]

[1] Anil Neerukonda Institute of Technology & Sciences, Visakhapatnam
[2] Jaypee Institute of Information Technology, Noida-62, India
pritee.cse@anits.edu.in,
{venigoel16,adi.aditig22}@gmail.com, rajul.kukreja@live.com

Abstract. In future wireless sensor network (WSNs) scenarios, the mobility is emerging as an important feature with increased number of sensors. Multifarous obstacles in this research are being encountered as the deployments in sensor networks are growing. However, these issues can be shielded from the software developer in by integrating the solutions into a layer of software services. The Data is ever growing which demands efficient data handling algorithms. In this paper, we propose a technique in which sensed data will be stored over cloud and different data aggregation techniques like clustering and classification will be used to process such big data on the cloud. This will reduce the computation overload on the base station as the data is stored and processed on cloud itself. Clustering is used to omit the abrupt values and cluster the similar data together. Classification algorithms are used for reaching to a final conclusion. A predictive Markov chain model was also developed for the prediction of overall weather outlook. Then a concept of weather forecasting, called Long Range Forecasting was used to predict the exact numeric values of the future weather parameters.

Keywords: Wireless sensor networks, Data Aggregation, Cloud Computing, K-means, Naïve Bayes, Weather Forecasting.

1 Introduction

In today's world, wired networks are progressively being replaced by the wireless networks due to many advantages of wireless over wired networks. No doubt wired networks are fast and cost effective, but wired networks have limitations when it comes to scalability and reachability, especially in remote areas. We cannot take wires everywhere that's why wireless networks have got an edge over wired networks. The concept of WSNs is becoming more and more advanced and sensors are being used in many applications in the area of security (monitoring buildings), environmental monitoring and e-health (e.g., heart rate monitoring of patients). WSNs are generally restricted to a building or a local network as data is being aggregated over local base station, but now there is a need to extend the reachability of wireless networks to wide area networks using Cloud platforms. Cloud provides a storage as

S.C. Satapathy et al. (eds.), *Emerging ICT for Bridging the Future – Volume 2*,
Advances in Intelligent Systems and Computing 338, DOI: 10.1007/978-3-319-13731-5_26

well as a service platform, where complex computations can be done on the data and the final result can be sent to the user.

Cloud computing is a technique that allows users to share resources for parallel computations like managing, storing, processing etc on uthe remote server hosted over internet. This reduces the load on the local server or the personal computer. It provides various service models like IaaS, Saas, Paas. Here, we intend to use cloud's storage platform to store data from sensors and it's Software as a service model to develop and deploy an integrated application to monitor the changes in the sensor environment.

As the data size increases, the load on base station also increases. As a result, the delay involved in giving the final result also increases. In order to solve this problem we need efficient Data Aggregation algorithms.

In this paper two Data Aggregation concepts have been used- 1) Classification and 2) Clustering. Classification methods are used to classify current data with the help of past training data .Clustering is used to eliminate redundant data and abrupt values and to cluster similar data together. Markov chain model is used to predict the future scenario. This method can be used in a number of applications like sales forecast, weather prediction, CRM models etc.

2 Related Work

K-means algorithm can be extensively used to do cluster analysis of many data types texts, images etc. However, there are certain disadvantages in this algorithm. One of the disadvantages is that initial input parameters like number of clusters and the size of cluster have to pre-determined. Reference[1] proposes an algorithm to improve the principal of K-means algorithms and its cluster seed selection method with an aim to make text clustering more stability. Reference[2] proposes a Stepwise Automatic Rival-penalized algorithm that performs clustering without predetermining the correct cluster number. [3] presents a mid-point algorithm to avoid the problem of random selection of the initial centroids for clustering. [4] mitigates the assumption of naïve bayes algorithm that none of the features can be discounted to classify a data instance into a class. This paper proposes a feature weighting method in which, the amount of information a certain feature gives to the target feature is defined as the importance of the feature, which is measured by Kullback-Leibler measure. As far as weather forecasting is concerned a lot of work has been done in this area as a result of its use in different real world applications. Reference [5] presents a simple mathematical model, Multiple Linear Regression (MLR) which involves statistical indicators like moving average (MA), exponential moving average (EMA), rate of change (ROC), oscillator (OSC), moments ($\mu2$, $\mu3$ and $\mu4$) and coefficients of skewness and kurtosis to predict the weather parameters. The authors have used this to record the weather data at a particular station. The main weather parameters like temperature and relative humidity have been predicted based on historic data equipped with correlation values in the weather data series over different periods.

In reference [6] a comparative study of different cloud service providers is given. It shows that none of the cloud service provider is best or weak. A cloud provider must be chosen according to the convenience of the user. Further, [7] evaluates the Microsoft Windows Azure for deploying HPC applications. A comparison between Amazon EC2 and Windows Azure, has also been done in terms of performance and efficiency. The authors have identified that the performance of Azure is the best simulation to real machines, and that it is the most viable alternative for running HPC applications amongst the Cloud Computing solutions.

3 Proposed Approach

Clustering is the way of grouping similar sets of data together such that data in same cluster are more related to each other than the data in other clusters. In centroids based clustering method, clusters are represented by a central point, which may not necessarily be a member of the data set. In k-means clustering, k clusters are defined and each data object is assigned to the cluster whose centroids is the nearest to the data object. The distance can be measured in terms of Euclidean distance or Manhattan distance. The central concept in the k-means algorithm is the centroids. In data clustering, centroid will be the representative value of that cluster.

Naive Bayes Classification methods are based on applying the probabilistic Bayes' theorem as a set of supervised learning algorithms considering the every pair of features are independent. Bayes' theorem states the following relationship

$$P(y \mid x_1 \ldots \ldots x_n) = \frac{P(y)P(x_1 \ldots \ldots x_n \mid y)}{P(x_1 \ldots \ldots x_n)}$$

Where y is the class variable and $x_1$through x_n are the dependent feature vector , And for all i

$$P(y \mid x_1 \ldots \ldots x_n) = \frac{P(y)\prod_{i=1}^{n} P(x_i \mid y)}{P(x_1 \ldots \ldots x_n)}$$

If $P(x_1 \ldots \ldots x_n)$ is taken as a constant then it describes the conditional probability of y occurring for the classes x1,x2...xn. P(xi|y) is the measure how much xi contributes to y. P(y)is called the prior probability. If this does not provide the correct result of one class versus another, then we choose the one between x1,x2...xn having higher probabilities. This is called maximum posterior (MAP) where final value is on the basis of higher probability of the class.

Despite these simplified assumptions, naive Bayes classifiers have found application in many real-world situations. It only requires a small amount of training data to estimate the parameters (means and variances of the variables) necessary for classification. Because independent variables are assumed, only the variances of the variables for each class need to be determined and not the entire covariance matrix.

A **Markov Chain** is a mathematical concept used in predicting the distant future. It assumes that the future events are dependent on the current state of events.A Markov chain is defined as follows: There is a set of states, S = $\{S_1,S_2,\ldots,S_n\}$. The process starts in one of these states and moves successively from one state to another.

Each move is called a step. If the chain is currently in state si, then it moves to state sj at the next step with a probability denoted by pij, and this probability does not depend upon which states the chain was in before the current state. The probabilities pij are called transition probabilities.

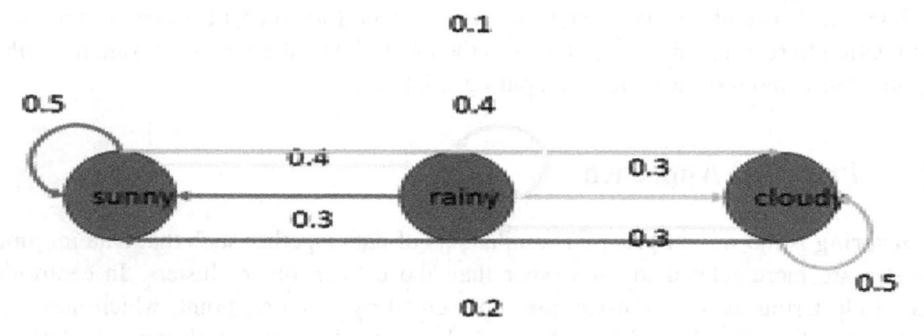

Fig. 1. State diagram representing transition in Markov's Chain

Here the matrix developed from this state diagram is called Transition Matrix.

$$P=\begin{pmatrix} 0.5 & 0.4 & 0.1 \\ 0.3 & 0.4 & 0.3 \\ 0.2 & 0.3 & 0.5 \end{pmatrix}$$

The following equation, with P as the transition matrix of a Markov chain, and u as the probability vector representing the starting distribution the probability of the chain remaining in state S_i after n steps is represented.

$$u(n) = uP^n .$$

Today being sunny, the probability that it will be rainy 4 days later is solved as follows:-

Then the two day forecast would be:-

$$\begin{pmatrix} 0.5 & 0.4 & 0.1 \\ 0.3 & 0.4 & 0.3 \\ 0.2 & 0.3 & 0.5 \end{pmatrix}\begin{pmatrix} 0.5 & 0.4 & 0.1 \\ 0.3 & 0.4 & 0.3 \\ 0.2 & 0.3 & 0.5 \end{pmatrix} = \begin{pmatrix} 0.39 & 0.39 & 0.22 \\ 0.33 & 0.37 & 0.30 \\ 0.29 & 0.35 & 0.36 \end{pmatrix}$$

And the four day forecast would be:-

$$\begin{pmatrix} 0.5 & 0.4 & 0.1 \\ 0.3 & 0.4 & 0.3 \\ 0.2 & 0.3 & 0.5 \end{pmatrix}^2\begin{pmatrix} 0.5 & 0.4 & 0.1 \\ 0.3 & 0.4 & 0.3 \\ 0.2 & 0.3 & 0.5 \end{pmatrix}^2 = \begin{pmatrix} 0.3446 & 0.3734 & 0.2820 \\ 0.3378 & 0.3706 & 0.2916 \\ 0.3330 & 0.3686 & 0.2984 \end{pmatrix}$$

4 Implementation

Wireless sensor networks have variety of implementation scenarios like monitoring for personnel and vehicles, intrusion detection, monitoring and denial, monitoring the environment, vehicular traffic monitoring & control, power substation generation and distribution monitoring, home patient medical monitoring etc. With so many application WSNs also have certain limitations like limited processing power, limited memory, limited range and other problems related to data distribution. In this paper, the main focus has been on these issues. With the help of cloud resources, the data processing and distribution becomes much easier. Since WSNs have limited memory, using cloud as a storage platform can be a great help. Efficient data aggregation techniques will lead to faster computation on cloud, reducing the load of computation on the base station.

In this paper we have implemented a weather monitoring application using wireless sensor nodes that sense different weather parameters like temperature, humidity and pressure. The purpose of this application is to sense the weather, send the data to the cloud for storage and computation and further send the results to the users via email. The application solves the problem of mobility as now the monitoring of the sensor environment is not restricted to the local network. Using different sensors the application can prove to be useful in many scenarios other than just weather monitoring. It can be used for agriculture purposes to monitor soil fertility etc. Alarm systems can also be developed to warn the public in case of floods, cyclones etc.

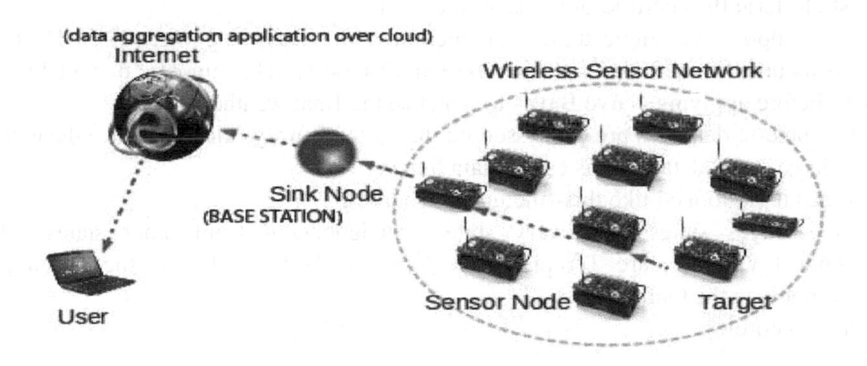

Fig. 2. Secure data aggregation in wireless sensor network

The implementation involves these steps-

Firstly, wireless sensor nodes were deployed within the range of 40 meters. The data sensed via wireless sensor nodes was sent to the base station. The Base station further sends the data to the local server where it is stored temporarily. This data is then sent to the cloud for the purpose of computation and analysis. The cloud used is Windows Azure.

Once the data is stored on the cloud, results are extracted from this raw data using data aggregation techniques. Finally the information is sent to the user via SMTP Server (email).

The hardware used in wireless sensor networks-

- Sensors – LM35 to sense the temperature.
- Xbees modules with shields-for wireless communication between the nodes and base station.
- Arduino (Atmega 328) boards-To power up the sensor node and for serialcommunication.
- Breadboards
- Connecting cables

A typical sensor node contained an Arduino board, sensors and xbees. The data sensed by all the nodes was received by the base station which was further sent to a local server for temporary storage. To read the data from the serial com port a matlab code was used. The data was stored in a .csv file.

The raw data sent to the cloud looked like this- (25,45,1) ie temperature as 25 degree Celsius and humidity as 45% and windy status as 1 if it is windy else 0.

K-means clustering is done on this data to eliminate redundant values and abrupt values and to cluster similar data together. Clustering helped to refine the data and provide a single weather scenario. This value was the most representative value of the largest clusterie the centroid of the largest cluster.

Let's suppose the single tuple generated after the clustering was (27,50,1). This tuple is assumed to be today's temperature and humidity. This numeric data had to be binned before applying Naïve Bayes to generate the final weather report.

The numeric data was pre-processed on the basis of fuzzy rules to reduce the errors due to large set of data. This is called data binning.

Binned tuple looked like this-(medium,high,true).

Naïve Bayes states that every state is independent from other states. The independent variables are Temperature, Humidity, Windy whereas the dependent class variable is the Outlook.

Bayes theorem:

$$P(C|X) = P(X|C) \cdot P(C) / P(X)$$

P(X) is constant for all classes

P(C) = relative frequency of class C samples

The idea is explained in following example: -

Temperature	**Humidity**	**Windy**	**Outlook**
High	High	True	Sunny
Medium	Medium	False	Sunny
Low	Medium	False	Overcast
Low	High	True	Rainy
High	Low	True	Sunny
Medium	Low	False	Overcast
High	Medium	True	Rainy
High	High	True	Sunny
Medium	Medium	False	Sunny
Low	High	True	Overcast
High	High	True	Rainy
High	Low	True	Sunny
High	Low	True	Overcast
Medium	Medium	True	Rainy

Fig. 3. Training data set chosen randomly to compute the class of the current weather scenarios

Let today's weather scenario be high,hightrue.

To find the class of the tuple X=(High,high,true) we find the probabilities as follows-

P(sunny) = 6/14

P(rainy) = 4/14

P(overcast)=4/14

P(sunny|X)= P(high|sunny)*P(high|sunny)*P(true|sunny)

P(sunny|X)= (4/6)*(2/6)*(4/6)*(6/14)

= 0.06

Similarly,

P(rainy|X) = P(high|rainy)*P(high|rainy)*P(true|rainy)

P(rainy|X)= (2/4)*(2/4)*(4/4)*(4/14)

=0.07

P(overcast|X)=P(high|overcast)*P(high|overcast)*P(true|overcast)P(overcast|X)= 1/4. 1/4. 2/4. 4/14

=0.01

Comparing probabilities of above 2 classes it is most likely to be a sunny day.

After giving the result for today's weather next task was to forecast the weather of the coming days. For this Markov chain model was used. To define the Transition probability matrix we used past 30 days data. The final was the one day and two day forecast.

Then the two day forecast would be represented by P^2. Similarly the four day forecast would be P^4.

Then a concept of weather forecasting, called Long Range Forecasting was used to predict the numeric values of the weather parameters.In this technique past data is analyzed to search a day which had similar weather observations as that of today's weather. That is if today is 25 august 2013 we search the previous year's august weather reports and look for a day which was most similar to today's weather. The one day forecast for 26 august 2013 would be the day next to the matched record.

The final results were sent to the users via email.The SMTP server was responsible to multicast the information to the user from the cloud.

5 Result

The temperature sensed in Arduino-

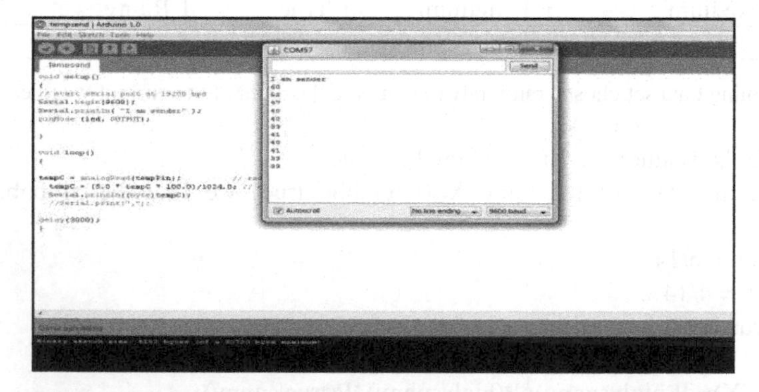

Fig. 4. Arduino monitor showing received values from mobile sensors

Fig. 5. Complete circuit of WSNs

```
Using Naive Bayes with Laplacian smoothing to classify when:
 temperature = cold
 humidity = normal
 windy = true

Probability of sunny  = 0.2345
Probability of rainy = 0.1450
Probability of overcast = 0.6205

Data case is most likely overcast day
```

Fig. 6. Implementation shows higher probability of overcast today

The Wireless sensor nodes were successfully deployed in a room or an open area. Weather parameters were sensed by the sensor nodes and data was sent to the cloud(Windows azure) for computation and storage. Final results were sent to the users via email. K-means Clustering and Naïve Bayes classification methods helped to determine current weather reports while Markov chain model was used for weather forecasting. Integrating the concept of cloud computing with WSNS helped to make WSNs more efficient and easy to handle.

The application provides the user an authentic and easy way of determine the weather. The base stations can receive data from the sensor nodes as long as they are in its range. The processing time should be 4-5 sec of the total application. But, the run time is also affected by the internet speed since it includes cloud.

6 Conclusion

Integrating the concept of cloud with WSNs can take the wireless sensor networks to the next level. Cloud computing allows us to do parallel computations on large number of computers simultaneously over internet. With so many application WSNs also have certain limitations like limited processing power, limited memory, limited range and other problems related to data distribution. In this paper, the main focus has been on these issues. With the help of cloud resources, the data processing and distribution becomes much easier. Since WSNs have limited memory, using cloud as a storage platform can be a great help. Efficient data aggregation techniques lead to faster computation on cloud, reducing the load of computations on the base station.

References

1. Li, X.: Research on Text Clustering Algorithm Based on Improved K-means. In: 2010 International Conference on Computer Design and Appliations, ICCDA 2010 (2010)
2. Cheung, Y.-M.: K*-Means: A new generalized k-means clustering algorithm. Pattern Recognition Letters 24, 2883–2893 (2003)
3. Aggarwal, N., Aggarwal, K.: A mid-point k-mean clustering algorithm for data mining. International Journal on Computer Science and Engineering, IJCSE (2009)
4. Lee, C.-H., Dou, F.G.: Calculating Feature weights in Naïve Bayes with Kullback-Leibler Measure. In: 11th International Conference on Data Mining (2011)

5. Paras, Mathur, S.: A Simple weather forecasting Model using Mathematical regression and trend analysis. Indian Research Journal of Extension Education Special Issue I (January 2012)
6. Goyal, S.: A comparative study of cloud computing service providers. International Journal of Advanced Research in Computer Science and Software Engineering 2(2) (February 2012)
7. Roloff, E., Birck, F., Deoner, M.: Evaluating high performance computing on Azure. In: IEEE Fifth International Conference on Cloud Computing (2012)

Performance Analysis of Modified Bellman Ford Algorithm Using Cooperative Relays

Rama Devi Boddu[1,*], K. Kishan Rao[2], and M. Asha Rani[3]

[1] Department of E.C.E., Kakatiya Institute of Technology and Science, Warangal, Telangana, India
[2] Department of E.C.E., Vaagdevi College of Engineering, Warangal, Telangana, India
[3] Department of E.C.E., Jawaharlal Nehru Technological University, Hyderabad, Telangana, India

ramadevikitsw@gmail.com, prof_kkr@rediffmail.com, ashajntu1@yahoo.com

Abstract. Cooperative communication plays a vital role in modern wireless communication. It improves the performance of the system and supports higher data rates. In this paper, the nodes in the network are classified into amplify-and-forward (AF) and decode-and-forward (DF) relays based on the Signal to Noise Ratio (SNR). The node who's received signal SNR is greater than threshold SNR is considered as DF relay, otherwise it considered as an AF relay. Network lifetime can be increased by operating nodes in the power save mode (PSM). The energy efficient Modified Bellman Ford Algorithm (MBFA) using the power saving mechanism of 802.11g is employed for routing. All the nodes acting as a DF relay will be considered for routing. The performance of MBFA can be improved by cooperative relays. Residual Energy is considered as a routing metric and optimal path is selected using MBFA. In a grid network, nodes are placed in a uniform distance and grid topology gives better performance over random topology. Mobility of nodes cause packet loss and the performance of a mobile network degrades further with PSM. Performance of MBFA with PSM for a static network using grid topology is better compare with mobile networks and random topology. For various node densities, the residual battery capacity of static and mobile networks for grid and random topologies are investigated using MBFA. The bit-error-rate performance of the MBFA using cooperative relay for different modulation schemes BPSK, QPSK and 64-QAM were analyzed.

Keywords: amplify-and-forward (AF), bit error rate (BER), decode-and-forward (DF), Modified Bellman Ford Algorithm(MBFA), power saving mode (PSM), residual energy (RE), signal to noise ratio (SNR).

1 Introduction

The lifetime of an ad-hoc network can be improved by prolonging the lifetime of each node. One-way to achieve this is to minimize power consumption of the network by

[*] Corresponding author.

S.C. Satapathy et al. (eds.), *Emerging ICT for Bridging the Future – Volume 2,*
Advances in Intelligent Systems and Computing 338, DOI: 10.1007/978-3-319-13731-5_27

employing energy efficient routing protocols [1]. The lifetime of a network can be increased by operating the nodes in the power save mode (PSM) [2]. The power saving mechanism of IEEE 802.11g [3] allows the nodes to operate in active mode or in doze state.

Energy consumption can be reduced by forwarding the packet through the shortest path. A proactive routing protocol, Distance Vector (DV) routing is used to find the shortest path. The two basic algorithms of DV can be used to find the shortest path [4]. They are: (i) Bellman Ford Algorithm and (ii) Dijkstra Algorithm.

The node deployment determines the network topology, effects routing and the performance of the system [5]. The grid and random topologies were considered and the performance of a static network without and with PSM was investigated using Bellman Ford Algorithm [6].

Cooperative communication allows single transmitting or receiving antenna and provides cooperative diversity [7] which increase robustness against the channel fading channel. The energy consumed by amplify-and-forward and decoded-and-forward cooperative protocols using OFDM physical layer with PSM was analyzed [8]. Cooperative shortest path relay selection for Multihop MANETs was investigated [9].

An ad-hoc network requires energy efficient routing technique to save battery energy and to increase the network lifetime. In this paper, energy efficient Modified Bellman Ford Algorithm is proposed. MBFA considers the residual energy of the node as a metric. The performance of a static and mobile network with grid and random topologies without and with PSM was analyzed. The battery capacity of static and mobile networks for grid and random topologies without and with PSM were analyzed for different node densities of 15, 30, 45 and 60. The battery capacity of a static network is more compare with a mobile network. Further, a static grid network is considered and all nodes operate in the power save mode. The nodes in the network are classified into amplify-and-forward and decode-and-forward relays based on the received SNR. The received signal SNR of the relay is greater than the threshold, the node acts DF otherwise AF relay. All DF relays consider for routing. The optimal path is calculated using MBFA and the packet is forwarded to the destination. The performance of MBFA can be improved by using cooperative relays. The BER performance of the MBFA using co-operation is compared for different modulation schemes.

This paper is organized as follows: Section 2 describe Modified Bellman Ford Algorithm with PSM. Section 3 describes MBFA using Cooperative Relays. Numerical Results were analyzed in Section 4. Finally, we outline some conclusions in Section 5.

2 Modified Bellman Ford Algorithm (MBFA) with PSM

Let us assume that $N = (n_1, n_2, n_3, ..n_n)$ represents the set of nodes in a wireless ad-hoc network which consists of 'n' number of nodes. $G = (\mathrm{N}, \mathrm{E})$ represent the graph of a network and E represent the link set. The node n_j is considered to be within the transmission range 'r' of node n_i , if dist$(n_i, n_j) \leq r$. The edge $e_{i,j} = (n_i, n_j) \in E$, edge cost is represented by $C_{i,j}$ and D_i^h is optimal cost between node n_i to n_0 (sink) where $h = 0,1,2..$ represents iteration value. The route between n_i and is given by $R = (n_{i_n}, n_{i_{n-1}}, ..., n_{i_1}, \mathrm{n}_{i_0} = \mathrm{n}_0)$.

The shortest path [10] generated by Bellman Ford Algorithm (BFA) can be written as

a. Initialization step:
   $D_1^h = 0$ for all h, $D_i^0 = \infty$ for $i = 2,3,..n$.
b. Evaluate $D_i^{h+1} = \min_{j \in N}[C_{ij} + D_j^h]$ for all $i \neq 1$
   Let h=h+1
c. The algorithm terminates if $D_i^h = D_i^{h-1}$

2.1 Modified Bellman Ford Algorithm

In MBFA, residual energy is considered as a routing metric and included in the routing table. Initially, all nodes are configured in BFA and they start broadcasting the route advertisement packet. MBFA considers residual energy as a metric in the routing table of Bellman Ford Algorithm, in addition to the destination sequence which is standard metric. When the new route advertisement packet arrives, the energy in the incoming route advertisement packet with the energy of the node in the route table for which the packet is being advertised was compared. The route table update with new node energy if certain offset change in its energy. Finally, the node energy is compared with the preconfigured threshold and if that energy is below the threshold, the node will be inactive from the route. The packet is forward through the optimal path using BFA. During the packet forwarding, the energy of the route 'R' can be evaluated by summing the individual energies of all the nodes in the route and all nodes in the route are aware of the total energy utilized by the route. In MBFA, nodes update their residual energies periodically and all nodes aware of all other node energies.

Let us assume that node n_{i_k} is a part of the route $R = (n_{i_n}, n_{i_{n-1}}, ..., n_{i_1}, \mathrm{n}_{i_0} = \mathrm{n}_0)$.The energy consumed by a node n_{i_k} at transceiver is E_{trans}, energy consumed by power amplifier during transmission is E_{tr_amp} and E_{sleep} be the energy consumed by the node during sleep mode if it is in PSM. The initial residual energy of node n_{i_k} is E_k^i . The link between $n_{i_k}, n_{i_{k+1}}$ varies dynamically. It depends on channel variations, power of transmitting antenna and other factors. The transmission power depends on the distance $d_{k,k+1}$ between nodes n_{i_k} to $n_{i_{k+1}}$ and proportional to $d_{k,k+1}^{\alpha}$ where α represents the propagation constant (usually $2 \leq \alpha \leq 4$).

The energy consumed by the node n_{i_k} for transmitting 1-bit data to the next hop node $n_{i_{k+1}}$ can be expressed as

$$E_k^{Tx} = E_{trans} + E_{tr\_amp} d_{k,k+1}^{\alpha} \tag{1}$$

The energy consumed by the node n_{i_k} to receive 1-bit data from the previous hop node $n_{i_{k-1}}$ can be expressed as

$$E_k^{Rx} = E_{trans} \tag{2}$$

If the node is in PSM mode, the energy consumed during sleep mode is

$$E_k^{SM} = E_{sleep} \tag{3}$$

Residual energy at node n_{i_k} after transmission/receiving 1-bit data or after sleep mode can be written as

$$E_k^R = \begin{cases} E_k^i - E_k^{Tx} & \text{for } 1-\text{bit transmission} \\ E_k^i - E_k^{Rx} & \text{for receiving } 1-\text{bit} \\ E_k^i - E_k^{SM} & \text{during sleep } (PSM) \end{cases} \tag{4}$$

The total energy consumed by the route having 'z=n' hops can be expressed as

$$E_T = \sum_{k=0}^{z-1} [E_k^{Tx} + E_{k+1}^{Rx} + E_k^{SM}] \tag{5}$$

3 MBFA Using Cooperative Relays

Two popular cooperative protocols are: (i) amplify and forward (AF) (ii) decode and forward (DF). Consider a conventional cooperative model which consists of source (S), destination (D) and relay (R). The signal received from S is amplified at AF relay and transmitted to D. DF relay decodes the signal received from S and re-encodes and transmit it to the D. AF relay is simple, consumes less power and has less delay. DF relay is complex and has more processing delay but gives better BER for a dynamic channel. In this work, cooperative relay employs two-phase half duplex transmission. During phase-1, S broadcast data to D and R, during phase-2 relay transmits the signal to D. S and R have equal transmitting powers, i.e. $P_S = P_R$ where P_S, P_R represents transmission powers of S and R respectively.

During phase-1, the signal received at R and D from S can be written as

$$y_{SD} = \sqrt{P_S} h_{SD} x + n_{SD} \tag{6}$$

$$y_{SR} = \sqrt{P_S} h_{SR} x + n_{SR} \tag{7}$$

During phase-2, the received signal at D from DF relay can be expressed as

$$y_{RD} = \sqrt{P_R} h_{RD} \hat{x} + n_{RD} \tag{8}$$

where h_{xy} represent channel gain between x-y. The noise between x-y is represented by n_{xy} which follows the Gaussian distribution and has $CN(0, N_o)$.

3.1 AF and DF Relay Classification Based on the SNR Threshold

Consider an ad-hoc network consists of node set $N = (n_1, n_2, n_3, ..n_n)$ which employs multi-hop transmission. The nodes in the ad-hoc network can be divided into amplify-and-forward (AF) and decoded-and-forward (DF) relays depending on the SNR threshold. If the SNR at the relay or node (R_{n_k}) is greater than the threshold SNR value, the relay acts as DF relay, otherwise it acts as a AF relay.

The SNR at the relay R_{n_k} can be calculated by

$$\gamma_{SR_{n_k}} = \frac{P_S \left| h_{SR_{n_k}} \right|^2}{N_0} \tag{9}$$

$$\begin{aligned} \gamma_{SR_{n_k}} &> \gamma_{th} \quad DF\ relay \\ \gamma_{SR_{n_k}} &< \gamma_{th} \quad AF\ relay \end{aligned} \tag{10}$$

The AF relay whose residual energy below the predefined battery capacity value will be discarded. All the DF relays will be considered as a part of the route for cooperation. Once the selection of DF relay process completes, the packet is transmitted using MBFA as explained in the section 2.

4 Numerical Results

Using QualNet simulator, nodes are configured in $1000\,m^2$ area with different node densities of 15, 40 and 60.The OFDM technique using 802.11a/g standards are considered for implementation and simulation parameters used are given in Table 1. The proposed work considers the static and mobile networks using grid and random topologies. The performance of different networks using MBFA with and without PSM was investigated. The screen shots of MBFA for mobile networks using grid and random topologies are shown in Fig. 1.

Table 1. Simulation Parameters

Heading level	Font size and style
Terrain Size	1000 x 1000 meters
Node Density	15,30,45,60
Topology	Grid Placement
Mobility	Static
Routing Protocol	Modified Bellman Ford Algorithm
Energy Configuration	General (As in QualNet)
Battery Capacity	120mAh
Simulation Time	3000 seconds
MAC Layer	802.11g
Transmit Power	100mW
Receive Power	130mW
Idle Power	120mW
Initial Node Energy	6480 Joules, 1.2Ahr with 1.5V battery
Radio Range	Max. 402m, Min. 217m
Data Packet	512bytes
Path Loss Coefficient	2.3 (As per Two RAY MODEL)
Radio Frequency	2.4GHz
ATIM Interval	20 Tus (1Tus = 1024 μs)
Beacon Interval	200 Tus
Bandwidth	20 MHz
Pre-defined SNR threshold for Co-operation	30dB
Threshold Battery Capacity	700mAh
Speed	2m/s

MBFA considers residual energy as a metric; the residual battery capacities of a static network for grid, random topologies without and with PSM are plotted in Fig. 2. For a mobile network, the node with group mobility of 2m/s is considered. In Fig. 3, the residual battery capacity of a mobile network for grid and random topologies without and with PSM are compared. In grid topology nodes are placed with uniform distance and the performance of a grid network is better compared with random topology. In a mobile network, some packets are lost due mobility of the node and PSM degrades the performance further.

The residual battery capacity of MBFA for static and mobile networks with PSM is compared in Fig. 4. The simulation after routing with MBFA, the static network has more battery capacity compared to a mobile network. The distance between nodes may increase due to mobility which increases the energy consumption of a node during transmission, since it depends on the distance between nodes. As the residual battery capacity of a static network is more compare with a mobile network and the performance of grid topology is better compare with random topology, a static grid network is considered for cooperative MBFA.

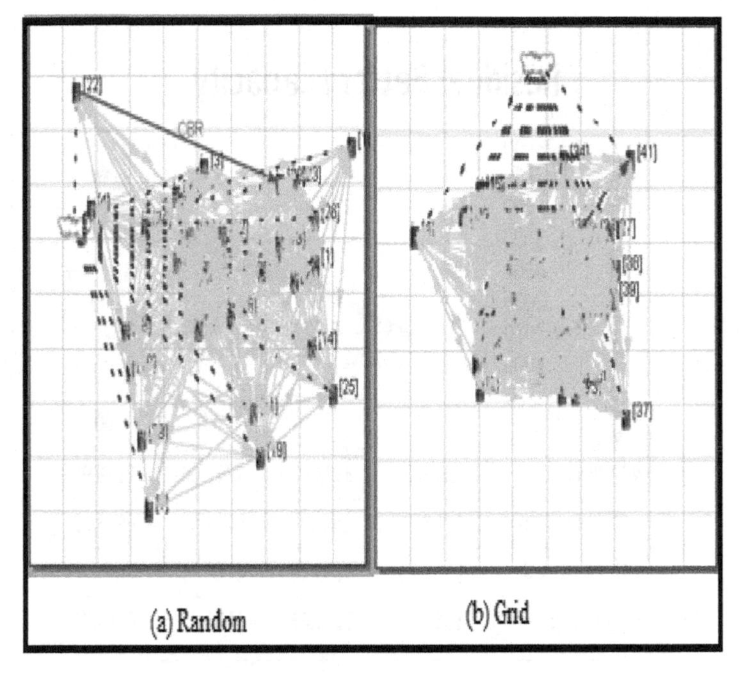

Fig. 1. Screen shot of MBFA for a mobile network with 30 nodes

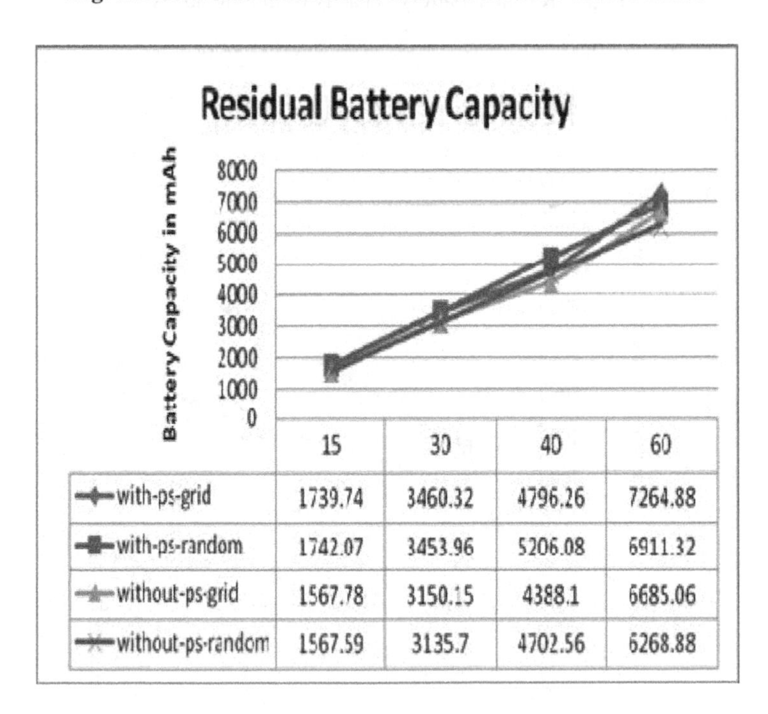

	15	30	40	60
with-ps-grid	1739.74	3460.32	4796.26	7264.88
with-ps-random	1742.07	3453.96	5206.08	6911.32
without-ps-grid	1567.78	3150.15	4388.1	6685.06
without-ps-random	1567.59	3135.7	4702.56	6268.88

Fig. 2. Residual battery capacity static nodes

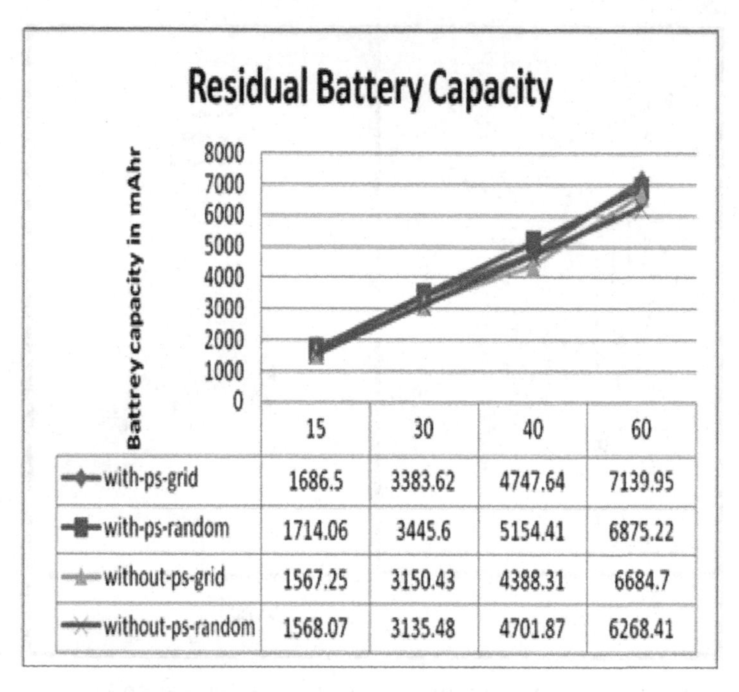

Fig. 3. Residual battery capacity of mobile nodes

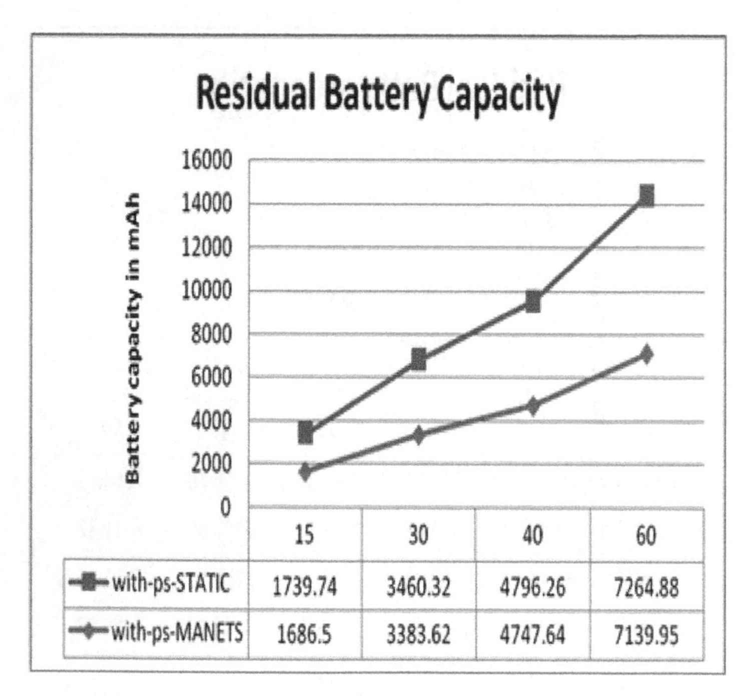

Fig. 4. Statistic comparison for static and mobile networks for grid topology

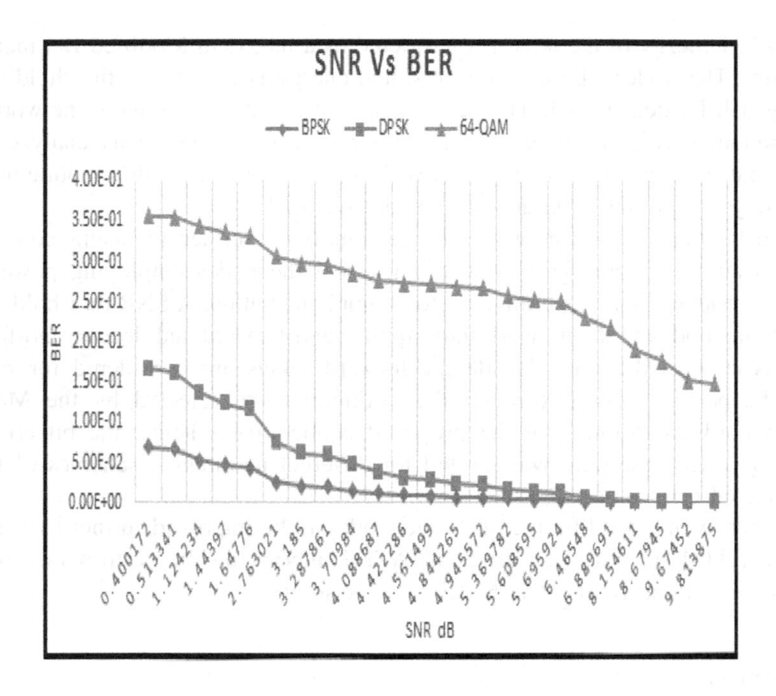

Fig. 5. SNR verses BER for Cooperative MBFA

BER verses SNR for Cooperative MBFA: A static grid network is considered for MBFA with cooperative relays. A threshold SNR of 30dB is considered for the classification of AF and DF relays. Further, all AF nodes whose current battery capacity is less than 700mAh threshold energy (a minimum battery capacity in QualNet) is deactivated. All DF nodes are considered for co-operation and optimum path for packet transmission is selected further using MBFA.

The proposed work employs equal power allocation scheme at source and relay with 100mW transmission power and transmission range of 217-402m.The performance of SNR Verses BER for Cooperative MBFA using different modulation schemes of BPSK, QPSK and 64-QAM were compared and are shown in Fig. 5. Higher order modulation schemes are preferable at high SNR and lower order modulation schemes are suitable at low SNR value.

5 Conclusions

Lifetime of ad-hoc network limited by battery power. There is a requirement of developing energy efficient routing schemes for ad-hoc networks which consumes less power and saves battery energy. The lifetime of the network can be increased by operating all nodes in the power save mode of 802.11g.The network topology, end-to-end delay, lifetime, and performance of the network depend on node deployment. In this paper, energy efficient Modified Bellman Ford Algorithm is proposed.

The residual energy of the node is considered as a metric in Modified Bellman Ford Algorithm. The nodes whose current residual energy is less than a threshold battery capacity will be deactivated. The performance of a static and random network with grid, random topologies without and with power save mode was analyzed. The residual battery capacity of a static network is more compare with a mobile network and static grid topology is suitable for power save mode.

The performance of a network can be improved further by using cooperative communication. A static grid network in which the nodes employing power save mode is considered for cooperative protocol implementation. A SNR threshold is used to divide the nodes in the network into amplify-and-forward and decode-and-forward relays. After classification, decode-and-forward relays are considered for routing. Later, the packet forward through the optimum route selected by the Modified Bellman Ford Algorithm. For the proposed cooperative scheme, the bit error rate verses signal-to-noise ratio was plotted for different modulation schemes of BPSK, DPSK and 64-QAM.

For future work, the lifetime of the network can be improved further by selecting optimum ATIM interval. The effect of ATIM interval on the performance of both cooperative and non-cooperative MBFA will be studied.

References

1. Toh, C.K.: Maximum Battery Life Routing to Support Ubiquitous Mobile Computing in Wireless Ad-hoc Networks. IEEE Communications Magazine 39(6), 138–147 (2001), doi:10.1109/35.925682
2. Namboodiri, V., Gao, L.: Energy-Efficient VoIP over Wireless LANs. IEEE Transactions on Mobile Computing 9(4), 566–581 (2010), doi:10.1109/TMC.2009.150, ISSN: 1536-1233
3. IEEE 802.11-2012 standard, `http://standards.ieee.org/getieee802/download/802.11-2012.pdf`
4. Tekbiyik, N., Uysal-Biyikoglu, E.: Energy efficient wireless unicast routing alternatives for machine-to-machine networks. J. of Network and Computer Applications 34(5), 1587–1614 (2011), doi:10.1016/j.jnca.2011.02.005
5. Sergiou, C., Vassiliou, V.: Energy utilization of HTAP under specific node placements in Wireless Sensor Networks. In: 2010 European Wireless Conference (EW), pp. 482–487 (2010), doi:10.1109/EW.2010.5483490
6. Rama Devi, B., Asha Rani, M., Kishan Rao, K.: Performance of Static Networks for Power Saving Mode. Int. J. Advanced Engineering and Global Technology 2(6) (2014) ISSN No: 2309-4893
7. Nosratinia, A., Hunter, T.E.: Cooperative Communication in Wireless Networks. IEEE Communications Magazine, 74–80 (October 2004)
8. Rama Devi, B., Asha Rani, M., Kishan Rao, K.: Energy Efficient Cooperative Node Selection for OFDM Systems Based on SNR Estimation. Int. Journal of Advances in Computer, Electrical & Electronics Engineering 3(1), 32–36 (2014); Special Issue of IC3T 2014

9. Rama Devi, B., Kishan Rao, K., Asha Rani, M.: Co-operative Shortest Path Relay Selection for Multihop MANETs. In: Satapathy, S.C., Biswal, B.N., Udgata, S.K., Mandal, J.K. (eds.) Proc. of the 3rd Int. Conf. on Front. of Intell. Comput (FICTA) 2014. AISC, vol. 328, pp. 697–706. Springer, Heidelberg (2015)
10. Bechkit, W., Challal, Y., Bouabdallah, A., Koudil, M., Souici, B., Benatchba, K.: A new weighted shortest path tree for convergecast traffic routing in WSN. In: 2012 IEEE Symposium on Computers and Communications (ISCC), pp. 187–192 (2012), doi:10.1109/ISCC.2012.6249291, ISSN: 1530-1346

Selection of Reliable Channel by CRAODV-RC Routing Protocol in Cognitive Radio Ad Hoc Network

Sangita Pal and Srinivas Sethi

Department of CSEA, IGIT, Saranag, Orissa, India
sangitapalmtech.cet@gmail.com, Srinivas_sethi@igitsarang.ac.in

Abstract. The Cognitive Radio Ad Hoc NETwork (CRAHN) technology is a burning novel technology in the wireless communication era and it meets the requirements like self-configuring, self-healing, and robustness with low deployment cost. In this communication environment routing has vital role to establish the path from source to destination and transmit the data between them. The efficient and effective route can be formed by selecting the reliable channel for transmitting the data. In this paper, it has been established the efficient and effective route, using threshold values of TIME-ON of primary users and channel capacity which are important constraints for routing the packets.

Keywords: CRAHN, Routing Protocol, Reliable, Channel Utilization.

1 Introduction

Cognitive radio (CR) knowledge has opened new doors to evolving claims. This has been broadly used in numerous application circumstances including military, public safety networks, post-disaster situations and wireless medical networks. This technology can be used to provide opportunistic entrance to large parts of the less utilized spectrum in networks.

CR technology can help Delay Tolerant Networks (DTNs) to provide reliable, delay-sensitive opportunities for communication. For instance, DTNs and CR technology could be used in urban scenario, where high density of wireless devices causes delay in communication due to contention over the link. Cognitive radios thus help in finding free channels for opportunistic use and ensure timely delivery of messages.

CR technology attains Dynamic Spectrum Access (DSA) of authorized bands of the spectrum by taking the benefit of utilization of the spectrum with an access to the unauthorized users [1]. Initially CR denoted to software radios (SR) enhanced with self-awareness about the features and necessities, in order to decide suitable radio protocol to be used. According to the spectrum bands, the secondary users (SUs) are accessing and sharing the spectrums and the allocation schemes can be categorized into two kinds they are: open spectrum sharing and licensed spectrum sharing [2][3].

S.C. Satapathy et al. (eds.), *Emerging ICT for Bridging the Future – Volume 2*,
Advances in Intelligent Systems and Computing 338, DOI: 10.1007/978-3-319-13731-5_28

The present scenario envisages that the SUs communicate over certain bandwidth initially assigned to a primary network, have drawn research interests. Specifically, the primary users (PUs) are the licensed/authorized users in the primary network, who have every right to access the assigned spectrum bands for them, and ready to share the spectrum band with the SUs, who have no licensed spectrum band [4]. The SUs in CRAHN are acceptable to access the PU assigned bands, as long as the SUs do not disturb or create harmful interference to the PUs.

The available spectrum differs noticeably and depends deeply on how the PUs have to bear interference on their spectrum and the new problem locations get up for the SUs. From the fact on the unpredictable resources may be imperfect, or finding the information may cause major delays. So reliable and efficient spectrum or channel management has important role in CRAHN, which is our objective for better route formation.

The rest of the paper describe as follow. Related works discussed in section 2 and basic CRAODV has been discussed in section 3. Section 4 gives the description of proposed protocol in CRAHN. Results and discussions have been placed in section 5. Section 6 concludes the paper.

2 Related Works

The use of free spectrum in TV bands proposes a centralized and distributed Cognitive Radio Protocol for dynamic spectrum allocation in [5], which enables the nodes to share the free portion of spectrum in optimal way. The authors of it have introduced the concept of using a Time-Spectrum block which represents the time reserved for use of portion of spectrum by the cognitive user. This minimizes spectrum allocation problem.

In [6], the authors describe a carrier sense multiple access (CSMA) based Medium Access Control (MAC) protocols for the Cognitive Radio Network, but also simultaneously transmit packet for a CR during the transmission of MS of the Primary System (PS). The PS and cognitive radio network (CRN) are all carrier sensing based systems. Before transmission, each MS of the PS will sense the carrier for a period of time. If the channel is free then the MS of the PS can transmit packets. When transmissions of a CR or other MSs are taking place, the MS of the PS transmit no packet because the channel is busy.

In [7], the authors discuss a negotiation mechanism which enable to exchange game information among game players (nodes).Then the nodes updates their information in the game timing. A MAC protocol is a multiple access protocol which able to communicate multiple nodes within the same transmission medium. Dynamic Open Spectrum Sharing (DOSS) protocol is a MAC protocol for wireless ad hoc networks which utilizes more spectrums by allowing to use dynamic control channels and arbitrary data channels [8]. The control channels are dynamic in both center frequency and bandwidth. It also allots to interference like jamming, primary user's activities from outside of networks and traffic loads from inside of the networks. The data channels are used for flexible utilization of the radio spectrum which is achieved by performing coordination over the control channels.

A MAC Protocol called Hardware-Constrained cognitive Medium Access Control (HC-MAC) utilizes multiple channels to increase the CRN throughput and complete spectrum utilization [9]. The cognitive radio used by SUs takes two types of hardware constraints that are sensing and transmission constraints. The protocol identifies the problem of spectrum sensing and access decision tradeoff and formulates an optimal stopping problem. The PU has certain terms and conditions **for** the maximum acceptable interference from the SUs.

3 Cognitive Radio Ad Hoc On-Demand Distance Vector (CRAODV)

A CRAODV is based on designs of the original AODV [14]-[17] routing protocol, which is presented by the implementation of the cognitive model. Enhancement of the route request process of AODV has not been affected by primary users (PUs) activities. CRAODV generates route request (RREQ) message and broadcasted by the source node to neighbor node on each channel which is not affected by a PUs activity during route discovery process. Similarly it finishes with a route establishment after the response of a route reply (RREP) from the sink node. An intermediate secondary users (SUs) is supposed to receive and deal RREQs and RREPs among a subgroup of the n channels.

It will verify a channel l is free from PU activities or not, after receiving a RREQ message through channel l. It drops the RREQ packet if the channel is not free from PU. Otherwise, create a reverse route through channel l. A node can keeps numerous paths, one for each available vacant channel, and the paths can be composed by various intermediate nodes. Nodes must verify the existence of a PU before transfer a packet through a channel. It constructed an opposite path toward the source SU using the same channel when an intermediate SU obtains the first RREQ using a channel free from PU event. It sends a RREP to the sender node using the same channel l if the reception SU supplies a valid path for the sink node. Otherwise, it broadcasts a copy of the RREQ packet through the channel l. It provides to check the duplicate RREQ packet and discard it if found. When an intermediate SU gets the RREP using an idle channel, it makes a forward path using the same idle channel to the RREP source node then forwards the same message along the opposite path using channel l. If another new RREP message will be received with a better forward route using channel l, the SU will update the forward path.

The route maintenance process is another important component through which the route will be restored in such dynamic ad hoc network. In addition to mobility the route error can be occurred due to a PU activity which starts to use a previously available channel. So, CRAODV performs two classes of route error (RERR) messages. One of them is RERRs for handling topology modifications in dynamic network as it is ad hoc in nature and another one is PU-RERRs for handling activity of the PU. The node cancels all the routing entries through an accessed channel and it notifies the neighbor SUs that channel l is now unavailable with a PU-RERR packet when a PU presence is sensed through a node in a channel l. The SUs get a packet which cancels the channel l route entries whose succeeding hop is the PU-RERR. In such a way, only the nodes in the network are affected by the PU activity and it must discharge the used channel. The PU-RERR packets are locally used and they are not promoting new route requests. When a node in the network senses a earlier engaged channel becomes free, it re-validates the routing entries using channel l, so that received data may be progressed along with such channel.

4 Proposed Protocol

The proposed routing protocol CRAODV-RC is similar with CRAODV in CRAHN. In this segment we deliver the description of the proposed routing protocol, by describing the route finding or discovery and route maintenance process with reliable channel. The proposed reliable protocol Cognitive Radio Ad hoc On-demand Distance Vector using Reliable Channel (CRAODV-RC) is designed for efficient route discovery using sets of reliable channel. It has been deployed the vital functionality of control packets like RREQ, RREP, SU_RREQ, SU_RREP and HELLO, RERR. Apart from these, it has been also implemented the PU-RERR and proper utilization of the spectrum by protocol like CRAODV in CRAHN. In MAC, multiple sender nodes transmit multiple possible different data over a shared physical medium to one or more destination nodes, which require a channel access scheme, combined with a multiplexing scheme and have applications in the uplink of the network systems.

The HELLO messages like in AODV are advertised periodically from each node which is used to inform about the presence of node in the network to all its neighbor nodes. In the other hand each node in the network system is expected to receive messages regularly from each of its outgoing nodes. The node is supposed to be no longer accessible, if a node has not acknowledged message from outgoing nodes for a certain time period in the network, it eliminates all pretentious route entries.

The RREQ is announced by flooding and transmitted from intermediate nodes to other nodes in the network to discovery the route information during the route discovery process. The initiating node spreads a RREQ message to its adjacent nodes, which communicates the message to them and this process continues till it reaches to destination node in the network. The message is originated with defined time to live (TTL) and a stated hop distance by the originating node. If none of intermediate nodes has the route information of sink node, they retransmit the RREQ with a next hop number. This procedure continues till it reaches to the destination node. Then it sends a route reply (RREP) messages to the originated node with the complete stored information of the route in it, which establish the route from source to destination node and transfer the data through the established path. Nodes along with the route information can be updated their routing table entries.

Now it has been discussed the route request by secondary users (SU_RREQ), which is important for CRAHN. It will verify whether a licensed channel is free from PU activities or not. After receiving a SU_RREQ message through the predefined channel it drop the RREQ packet if the registered/ licensed channel is not free from PU activity. Otherwise, it generates a opposite route through same channel. Similar to CRAODV, a node can keeps numerous routes in CRAODV-RC. The secondary user verifies the existence of a PU activity before transfer a data packet using a channel. It sends a secondary user route reply (SU_RREP) to the source node using the same channel. Otherwise, it transmits a copy of the SU_RREQ message through the channel and it provides to check the duplicate RREQ packet and discard it if found.

Similar to SU_RREQ secondary user route reply (SU_RREP) is another control packet used in the CRAHN to reply the destination information regarding PU activity on a channel with statically information of TIME-ON of PU on a channel and its capacity. The node updates the opposite route and sends SU_RREP through intermediate nodes in the CRAHN. When intermediate nodes receive the SU_RREP over a free channel, it analyzes the reliability of the channel using threshold values of ON time of PU on specified channel and its capacity. It built up a path through the same free channel to the source node if the statistical average ON time periods of channel for PU is less than threshold value and better channel capacity. Then it forwards the SU_RREP message along the opposite path using specified channel. The SU will update the onward path only if the new SU_RREP message is better forward route received using the same channel,.

Like to CRAODV, the route maintenance procedure can be performed as it is a vital role in routing protocol. So, in CRAODV-RC, there are two classes of route error (RERR) message. RERRs is one of them for topology alterations in such dynamic network as it is ad hoc in nature and other one is primary user (PU-RERRs) for managing the activity of the PU. The node in the network cancels all the route records using an accessed channel and it notices the neighboring SUs that used authorized or licensed channel is inaccessible with a PU-RERR message when a PU activity is identified through a node in a same channel. The SUs get a packet nullify the used channel route entries whose succeeding hop is the PU-RERR sender. In this method, the SU nodes in the network pretentious by the primary user activity must withdraw from the eventful channel. The PU-RERR packets are locally advertised and they will not allow to new RREQs. When a node in the network detects a previously engaged channel converts free, it re-evaluates the routing entries using the channel, so that received data may be forwarded along with such channel. In this way the route discovery and maintenance can be performed using reliable channel which make it more efficient and effective.

The simulations have been carried out by using simulations through NS-2 [18], based on the CRCN integrated simulator [19] and evaluate the performance of protocol through throughput, packet delivery ratio (PDR), end-to-end delay and normalized routing load (NRL) as per the following table.

The performance evaluations parameters can be obtained through the NS2 Trace Analyzer are as follows.

The NRL is the total amount of routing packets per total delivery packets. The End-to-End delay shortly states as delay and refers as the average time period taken for a packet to be transmitted across a network from source to destination. The PDR is used to calculate the reliability of a routing protocol, and this can be measured the percentage of the ratio between the number of received packets at destination and the number of packets sent by sources. Further, the throughput another evaluation parameter that is the rate receives data in the network.

Table 1. Simulation Parameters for CRAODV-RC and CRAODV

S.No	Parameters	Values
1	Area size	**500m x500 m.**
2	PU Transmission range	**250 m.**
3	Simulation time	**500 s.**
4	Nodes speed	**5 m/s**
5	Pause times	**5 s.**
6	Data rate	**5 Kbps**
7	Mobility model	**Random any point.**
8	Interface	**1**
9	No. of channel	**5**
10	Numbers of SU Node	**10,20,30,40,50,60,70,80,90,100**
11	Numbers of PU Node	**10**
12	Threshold Value	**60%**
13	No. of Simulation	**5**

5 Results and Discussions

In this paper, reliability of the node in the CRAHN has been calculated by using TIME-ON of PU and channel capacity to make the routing protocol more effective and efficient. At the same time it has been also tried to improve the packet delivery ratio and throughput which denotes the efficiency, effectiveness, and reliability of routing protocol and good channel efficiency.

As per fig. 1, the proposed routing protocol (CRAODV-RC) has less NRL as compare to CRAODV. This show overhead is less as compared to conventional routing protocol CRAODV since some licensed channel has not been consider. This implies the collision is less in our proposed model in CRAHN. This is better significant of protocols for transmission of the data from sender node to receiver node via intermediate nodes between them including SUs.

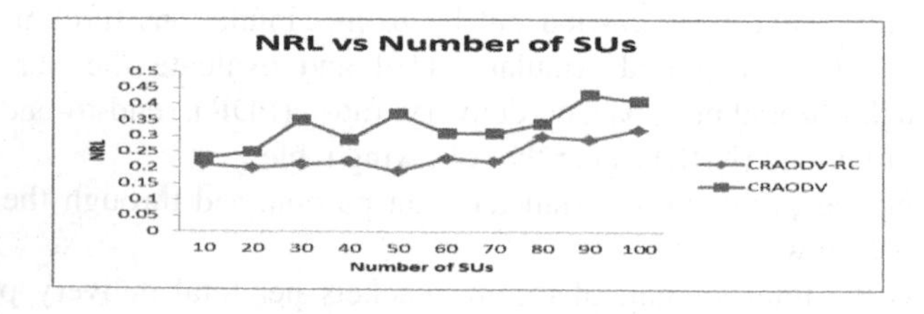

Fig. 1. NRL vs Number of SUs

As per fig. 2, the proposed routing protocol consume less time to transmit the data through intermediate nodes as compared to conventional routing protocol CRAODV, from sender to receiver. This is because of formation of better route using better channel selection process.

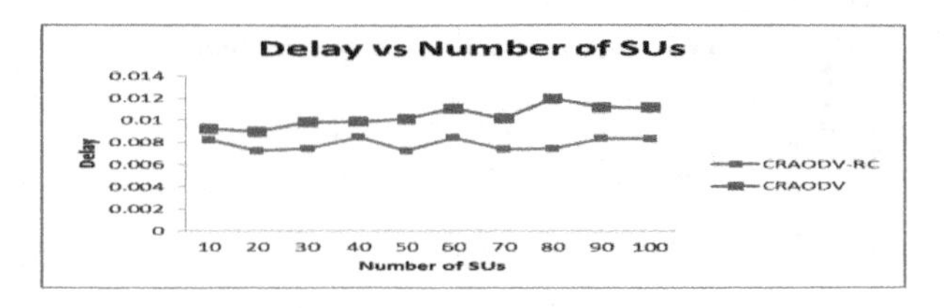

Fig. 2. Delay vs. Number of SUs

The proposed routing model CRAODV-RC shows better results in term of PDR as compared to CRAODV which is displayed in fig. 3. Since end-to-end delay minimize with less NRL the efficiency and effectiveness of our proposed routing model is better as compare to CRAODV.

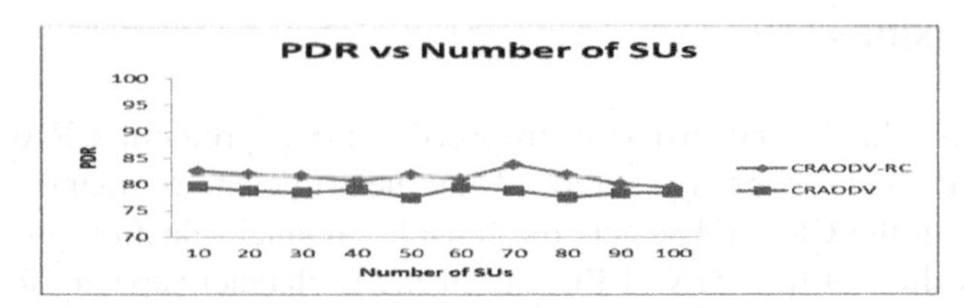

Fig. 3. PDR vs. Number of SUs

The system throughput has been also considered for better channel utilization for evaluation of routing protocols. At a particular time, it can't be compared the resultant data, based on throughput for different SUs node size of CRAODV which is shown as per fig. 4. Further it can be observed that, the average throughput values may get within the range of 28 to 45 bytes/sec. In this paper, it has been used the average throughput to compare between CRAODV and CRAODV-RC. It has been analyzed through the figure 5 and found the performance in terms of average throughput of CRAODV-RC is better as compared to CRAODV.

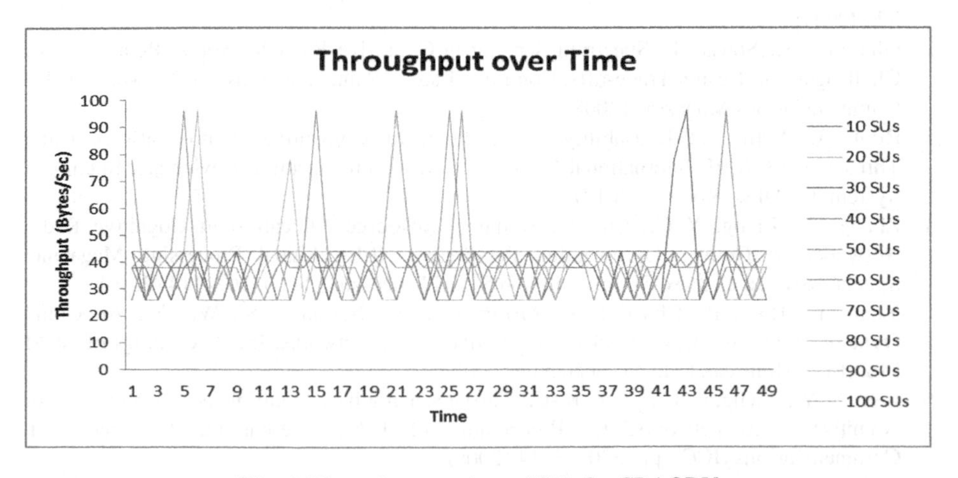

Fig. 4. Throughput vs. time of SUs for CRAODV

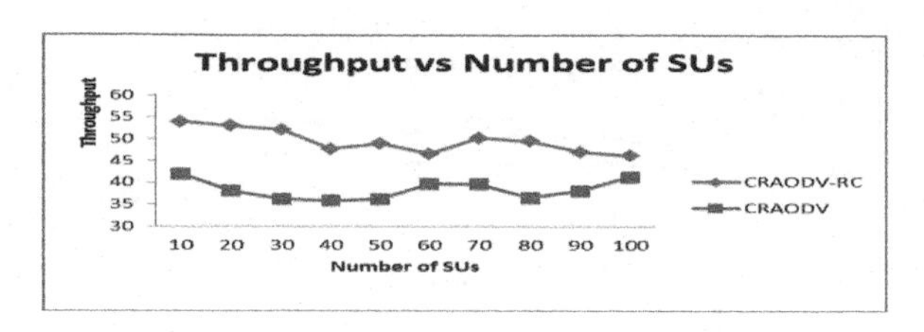

Fig. 5. Throughput vs. Number of SUs

So we observed through different figures from 1 to 3 and 5, that the proposed routing protocol CRAODV-RC is better than conventional routing protocol i.e., CRAODV.

6 Conclusions

In this paper a reliable and efficient proposed routing protocol CRAODV-RC has been developed in the cognitive radio ad hoc network environment. The proposed CRAODV-RC in the CRAHN selects the reliable channel selection process by using the threshold values of time ON of PU and licensed channel capacity for better route construction. Due to this reliable channel selection the route formation by using proposed model CRAODV-RC is better and more numbers of packet are transmitted as compared to conventional one. It improves the packet delivery ratio and throughput with minimize the end-to-end delay and normalized routing load as compared to an existing conventional routing protocol.

References

1. Federal Communications Commission: Facilitating opportunities for flexible, efficient, and reliable spectrum use employing cognitive radio technologies, FCC Report, ET Docket 03-322 (2003)
2. Ghasemi, A., Sousa, S.: Spectrum Sensing in Cognitive RadioNetworks: Requirements, Challenges and Design Trade-offs, Cognitive Radio Communications and Networks. IEEE Communications Magazine (2008)
3. Rout, A., Sethi, S.: Reusability of Spectrum in Congnitive Radio Network using Threshold. In: IEEE International Conference on Communication, Device and Intelligent System (CODIS), Kolkota (2012)
4. Zhang, R., Liang, Y.-C., Cui, S.: Dynamic Resource Allocation in Cognitive Radio Networks: A Convex Optimization Perspective. IEEE Signal Processing Magazine: Special Issue on Convex Optimization on Signal Processing 27(3), 102–114 (2010)
5. Yuan, Y., Bahl, P., Chandra, R., Moscibroda, T., Narlanka, S., Wu, Y.: Allocating Dynamic Time-Spectrum Blocks in Cognitive Radio Networks. In: Proceedings of ACM MobiHoc, Montreal, Canada (September 2007)
6. Lien, S.-Y., Tseng, C.-C., Chen, K.-C.: Carrier sensing based multiple access protocols for cognitive radio networks. In: Proceedings of IEEE International Conference on Communications (ICC), pp. 3208–3214 (2008)

7. Zhou, C., Chigan, C.: A game theoretic DSA-driven MAC framework for cognitive radio networks. In: Proceedings of IEEE International Conference on Communications (ICC), pp. 4165–4169 (2008)
8. Ma, L., Han, X., Shen, C.-C.: Dynamic open spectrum sharing for wireless adhoc networks. In: Proceedings of IEEE DySPAN, pp. 203–213 (2013)
9. Jia, J., Zhang, Q., Shen, X.: HC-MAC: a hardware-constrained cognitive MAC for efficient spectrum management. IEEE J. Selected Areas Commun. 26, 106–117 (2008)
10. Wong, Y.F., Wong, W.C.: A fuzzy-decision-based routing protocol for mobile ad hoc networks. In: 10th IEEE International Conference on Network, pp. 317–322 (2002)
11. Raju, G.V.S., Hernandez, G., Zou, Q.: Quality of service routing in ad hoc networks. In: IEEE Wireless Communications and Networking Conference, vol. 1, pp. 263–265 (2000)
12. Pedrycz, W., Gomide, F.: An introduction to fuzzy sets: analysis and design (complex adaptive systems). MIT Press, Cambridge (1998)
13. Buckley, J.J., Eslami, E., Esfandiar, E.: An introduction to fuzzy logic and fuzzy sets (advances in soft computing). Physica Verlag (2002)
14. Perkins, C.E., Belding-Royer, E.M., Das, S.R.: Ad Hoc On-Demand Distance Vector (AODV) Routing, Internet Draft, draft-ietf-manet-aodv-13.txt (February 2003)
15. Cacciapuoti, A.S., Calcagno, C., Caleffi, M., Paura, L.: CAODV: Routing in Mobile Ad-hoc Cognitive Radio Networks. Wireless Days, 1–5 (2010)
16. Cacciapuotia, A.S., Caleffia, M., Paura, L.: Reactive routing for mobile cognitive radio ad hoc networks. Ad Hoc Networks 10(5), 803–815 (2012)
17. Salim, S., Moh, S.: On-demand routing protocols for cognitive radio ad hoc networks. EURASIP Journal on Wireless Communications and Networking Regular submissions, 1499–2013 (2013)
18. Ns-2 Manual, internet draft (2009), http://www.isi.edu/nsnam/ns/nsdocumentation.html
19. CRCN integration, http://stuweb.ee.mtu.edu/~ljialian/

Sensor Controlled Sanitizer Door Knob with Scan Technique

Tamish Verma[1], Palak Gupta[2], and Pawan Whig[1]

[1] Electronics and Communication Engineering Department,
Guru Premsukh memorial College of Engineering (GGSIPU),
Bhagwan Parshuram Institute of technology (GGSIPU), India
{tamish60,pawanwhig}@gmail.com
[2] Guru Premsukh memorial College of Engineering (GGSIPU), India
palakgupta2207@gmail.com

Abstract. Hand sanitizers were developed for use of cleansing of hands without soap and water. These are made up of gels that contain alcohol in order to kill germs present on the skin. The alcohol works immediately and effectively in order to kill bacteria and many viruses. Sanitizer is low of cost cheaply available to use for cleansing of hands. They are convenient, portable, easy to use and less time consuming. So in this project a hand sanitizer is attached to the door handle and releases automatically when the person touches the door handle. It saves time in our fast life by removing the use of washbasins. Although various other options are already present in which a wet tissue is attached to the door handle. But the drawback of this system is tissues get dried up and hence become of no use. It causes wastage. The sanitizers on the other hand are liquid based. The project has been implemented using FPGA. The FPGA has many advantages as speed, number of input/output ports and performance. This system has been successfully implemented and tested in hardware using Xylinx 14.5 software packages using Very High Speed Integrated circuit hardware description language(VHDL). RTL and technology schematic are included to validate simulation results

Keywords: FPGA, Xylinx, VHDL, RTL.

1 Introduction

Health problems are the serious problems these days. Health problems [1] has been causing many critical problems and challenges in the major and most populated cities. Due to problems related to health, people lose time, miss opportunities, and get frustrated. Health issues directly impacts the companies. Due to bad health there is a loss in productivity from workers, trade opportunities are lost, delivery gets delayed, and thereby the costs goes on increasing.

To solve these health problems, we have to build new facilities & technology but at the same time make it smart. To decrease health related problems we need more

S.C. Satapathy et al. (eds.), *Emerging ICT for Bridging the Future – Volume 2*,
Advances in Intelligent Systems and Computing 338, DOI: 10.1007/978-3-319-13731-5_29

cleansing techniques, more washbasins etc. But for all these things to operate we need a constant running water[2] which is

Slightly a difficult task. This project introduces a technique which will reduce usage of water, saves time and hence decreases health related issues also. This device is applied on the door handle. Sanitizer's [3] release is based on the conductivity [4] through hand. Sanitizer's inlet through the door handle is controlled by the motor. The motor controls the pore of the only inlet. In this device we have used a motor and a sensor [5] which checks the conductivity. We have used a d-flip flop because it gives the same output as the input is given. Motor works according to the sensor. The sensor panel checks the conductivity and makes the motor work accordingly.

If the hand is wet then it is in conductive state[6] and if it is dry it is in non-conductive state. Conductive state needs less amount of sanitizer whereas nonconductive state needs more amount of hand sanitizer. Motor works accordingly.

2 Flowchart

In fig. 1 a flowchart explains the actions taken by the machine of how it will work accordingly. Initially the hole will be closed. Sensor will detect the hand and sends the signal to motor according to condition. This program will allow us to do the modification according to the conditions and need of the market. Like the timer for the conditions can be changed. We have also included the technique of checking if the motor is working accordingly or not. If not it can be replaced. For future development in this project we can add some more conditions to the sensor.

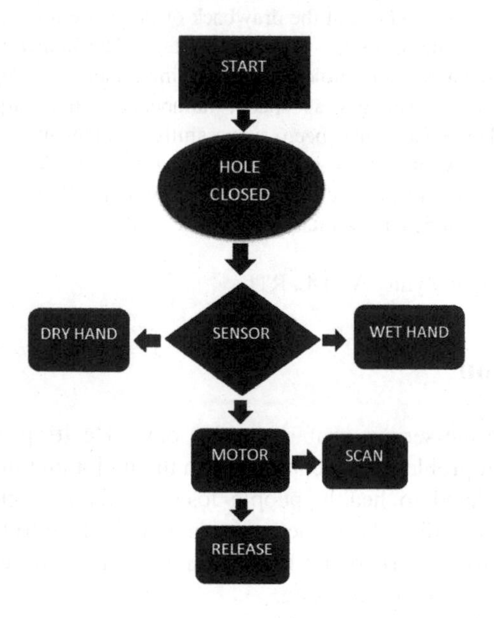

Fig. 1. Flowchart of the project

3 Block Diagram

Block diagram in fig.2 explains that if the hand is touched by the sensor, the sensor panel activates. Sensor panel gives the signal to the motor according the exact specifications. It makes the motor work accordingly. If the hand is wet motor operates for 4sec. and if the hand is dry the motor works for 7sec. and comes to reset position

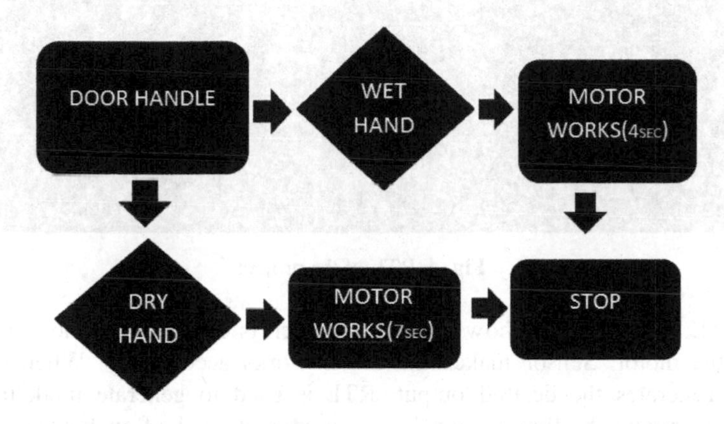

Fig. 2. Block diagram

Fig.3 shows the RTL of the motor used in the developing of the project. It also shows the connections and working of the motor according.

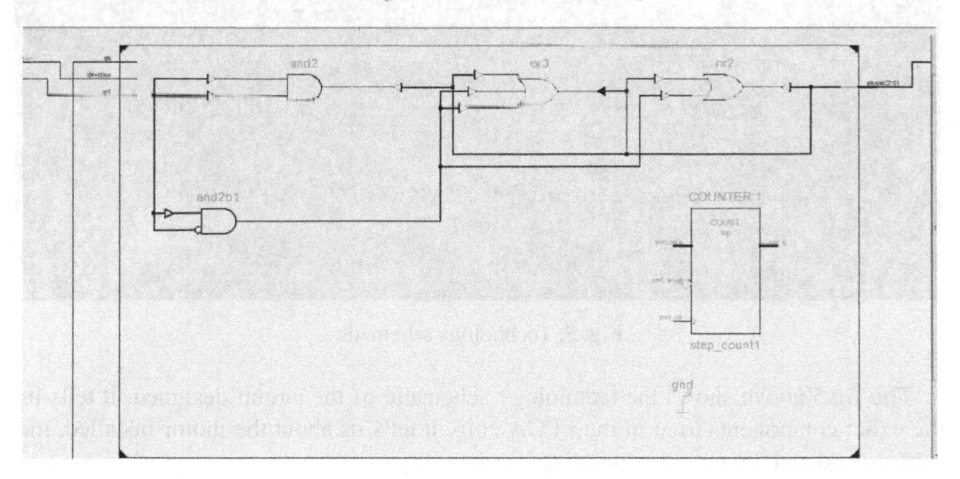

Fig. 3. RTL of the motor used

4 Hardware Description

Hardware description of the project is shown in the following schematic diagrams and images of RTL and Technology schematic. Then simulation of the project is also shown in the following figures.

The corresponding RTL schematic and technology schematic has been shown in the following figures

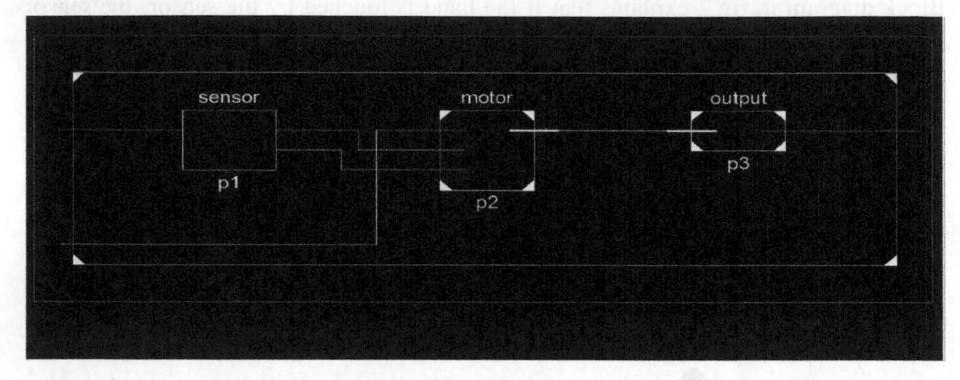

Fig. 4. RTL of the project

The RTL in fig.4 above shows the motor which is connected to the sensor which controls the motor. Sensor makes the motor works accordingly. When the motor works it generates the desired output. RTL is used to generate modeling of the synchronous circuits by flow of signals. It is used to study the flow between hardware registers and logical operations.

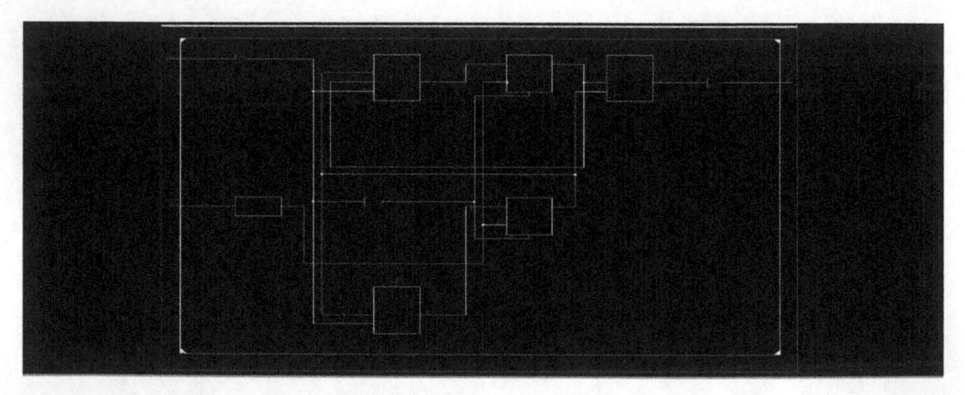

Fig. 5. Technology schematic

The fig.5 above shows the technology schematic of the circuit designed. It tells us the exact components used in the FPGA chip. It tells us about the motor installed, the sensor used and the output generating block.

5 Stimulation and FPGA Elements

The fig.6 tells us the stimulation of the elements which reveals the execution of motor and sensor according the conditions and how the code operates under certain conditions.

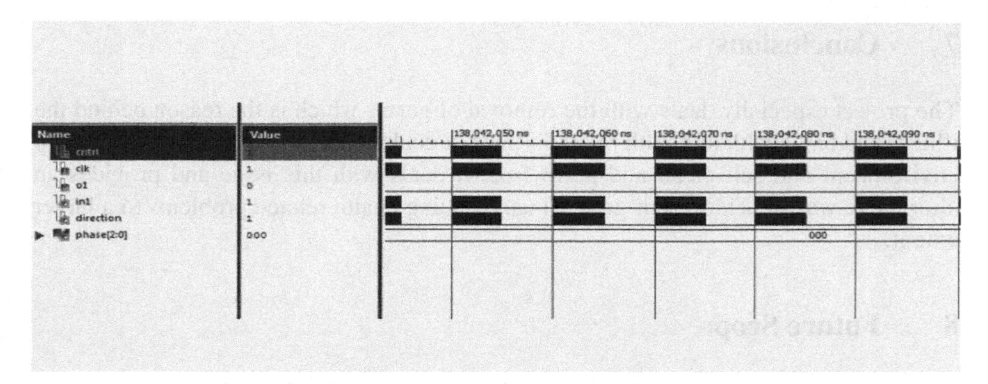

Fig. 6. Simulated Waveform

6 Scan Path Technique

Scan path technique is used to investigate if the motor is broken or not working properly it'll directly initiate the problem without opening the whole system. It has been shown in the block diagram.

The simulated waveform is provided below where by giving the scanning input '0' or '1' the system will switch on and off respectively.

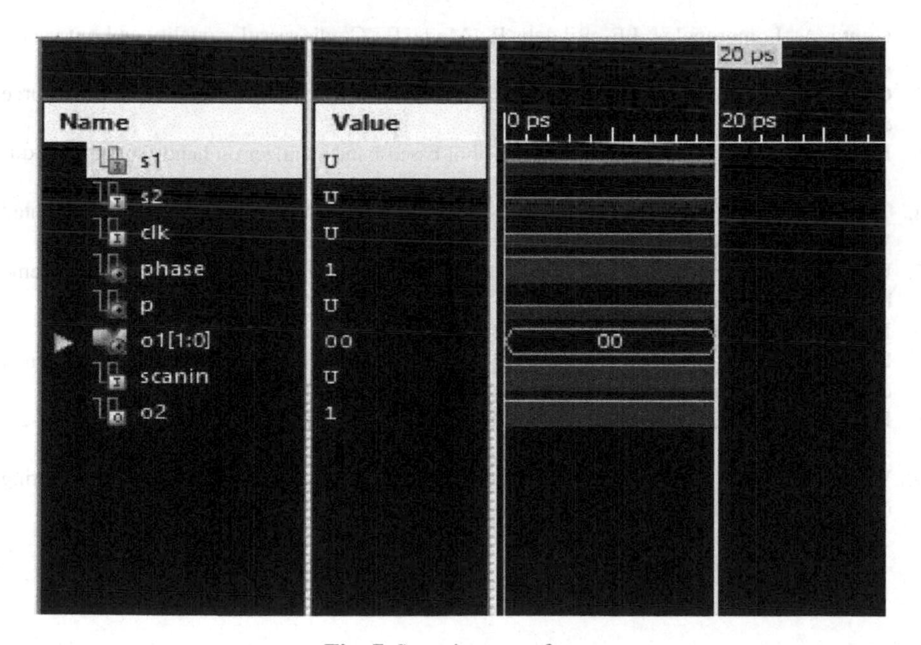

Fig. 7. Scanning waveform

7 Conclusions

The project especially deals with the removal of germs which is the reason behind the illness and bad health or health related issues. It deals with a very small aspect to keep environment and self clean and germ free. It deals with this issue and provides an alternative way to achieve our goal. Thus, reducing health related problems to a larger extent.

8 Future Scope

A significant advantage of the proposed design is its simple architecture and low component count. Therefore, it is very suitable for washrooms in hospitals and malls. This study may be extended for further improvements in terms of power and size, besides the wiring and layout characteristics level.

Acknowledgment. I would like to express my gratitude towards Prof. Pawan Whig without whose efforts time and contribution the project would not be possible.

References

1. Gupta, A., Lampropulos, J.F., Bikdeli, B., Mody, P.: Cardiovascular quality and outcomes. Editorial office, 560 Harrison Ave, Suite 502, Boston
2. Guerin, Thomas Canadian.org Universities of Alberta libraries. Book has an open source editable web page
3. Davis, J., Bohem, A.B.: Efficiency of alcohol based hand sanitizer on hands soiled with dirt and cooking oil ....Amy j.pickering
4. Tentzeris, M.M., Yang, L., Rida, A., Vyas, R., Krussei, C.: Conductive Inkjet-Printed Wireless Sensor Nodes on Flexible low cost-Cost Paper-Based Substrates
5. Kou, W.C., Li, Y.H., Wuu, L.C.: Digitally signed documentation flexible sanitizing scheme based on bilinear maps (June 2011)
6. Kurnia, G.L., De Larminant: Position sensor error analysis for EPS (June 2003)
7. Lopatkkiewicz, R., Nadolony, Z., Przybylek, P.: The influence of water content on thermal conductivity of paper used as transformer winding insulation, July 24-28 (2012)
8. Dai, S., Song, Z., Jia, R.: Web based fluid mechanics. Experimental System (June 25-27, 2010)
9. Whig, P., Ahmed, S.N.: PMOS Integrated ISEFET Device for water Quality Monitoring (August 25, 2013)

Impact of Propagation Models on QoS Issues of Routing Protocols in Mobile Ad Hoc Networks

Attada Venkataramana[1], J. Vasudeva Rao[1], and S. Pallam Setty[2]

[1] Department of Computer Science & Engineering,
GMR Institute of Technology, Rajam, Andhra Pradesh, India
{venkataramana.a,vasudevarao.j}@gmrit.org
[2] Department of CS&SE, Andhra University, Visakhapatnam, India
drspsetty@gmail.com

Abstract. Wireless and mobile communication networks are becoming more and more fashionable in today's mobile world. In wireless communications, Mobile Ad hoc Networks play crucial role. Ad hoc network is an infrastructure-less and self -organized collection of autonomous nodes. In mobile ad hoc network, routing is one of the thrust areas. There are no committed routers in MANETs; every network node should contribute to routing. Simulation tools are used to analyze the performance of routing protocols in MANETs. Propagation models show strong impact on the performance of routing protocols during simulations. This paper investigates the impact of familiar propagation models Two-ray and Shadowing on the performance of AODV and DSR routing protocols using NS-2. Experimental results found that both AODV and DSR show better throughput in two-ray ground model.

Keywords: Mobile Ad hoc Network, Propagation models, AODV, DSR, NS-2.

1 Introduction

Communication is one of the primary factors of science that has constantly been a focal point for exchanging information between communication parties at distant locations. Wireless and mobile communication networks are becoming more and more fashionable in today's mobile world. Wireless communication is an emerging technology since last two decades and it is showing rapid growth worldwide compared to its traditional wired counterpart. This growth is mainly due to trends in mobile equipment like Apple iPhone, Research in Motion's Blackberry, and consequently Android-based phones such as smart phones, tablets, notebooks, laptops, etc., with an increase of connection speed. Wireless communication entails the transmission of information over a distance without help of wires or any other forms of electrical conductors.

Ad hoc networks are on-demand and infrastructure-less networks and these networks are created for temporary purpose, where it is not possible to establish infrastructure wireless networks. Mobile Ad hoc Network is a self-generated, self-organized and self-handled network consisting of collection of independent nodes

S.C. Satapathy et al. (eds.), *Emerging ICT for Bridging the Future – Volume 2*,
Advances in Intelligent Systems and Computing 338, DOI: 10.1007/978-3-319-13731-5_30

communicating one another without aid of any centralized infrastructure. In this network each node plays dual role i.e., acts as both host as well as router. All the nodes are frequently moving around the network region. Due to the dynamic topology, routing is a major issue in Ad hoc network. An efficient routing protocol is required, which provides QoS by minimizing delay and power consumption while maximizing throughput [1]. In Mobile ad hoc networks all the nodes are shared the available resources. Therefore optimal way of utilizing resources is another challenging issue in MANETS.

Mobile Ad-Hoc Network is the rapid rising technology since last two decades. Due to high mobility and dynamic topology, routing is a one of the major challenging area in MANET. Researchers are giving spectacular amount of attention to routing. Routing is the act of moving information from a source to a destination in an internetwork. At least one intermediate node within the internetwork is encountered during the transfer of information [7]. Routing Protocols play crucial role in MANETs. Routing protocols for ad-hoc network can be categorized in to three categories. The three classifications of routing algorithms are Reactive, Proactive and Hybrid routing protocols [2]. This paper evaluates the performance of routing protocols under propagation models using NS-2.

1.1 Ad hoc On-Demand Distance Vector Protocol (AODV)

The AODV is a reactive routing protocol. i.e. Route is established only when it is required by a source node and it reacts to the changes. It maintains only the active routes in the tables for a pre-specified expiration time. These routes are found and are expected to be available at a given instant. It also performs unicast routing. It employs destination sequence numbers to identify the most recent path. AODV adopts the destination sequence number technique [6]. AODV defines three types of control messages for route maintenance. RREQ - A route request message is transmitted by a node requiring a route to a node. As an optimization AODV uses an expanding ring technique when flooding these messages. Every RREQ carries a time to live (TTL) value that states for how many hops this message should be forwarded. This value is set to a predefined value at the first transmission and increased at retransmissions [5]. RREP - A route reply message is unicast back to the originator of a RREQ if the receiver is either the node using the requested address, or it has a valid route to the requested address. When a link breakage in an active route is detected, a RERR message is used to notify other nodes of the loss of the link.

1.2 Dynamic Source Routing Protocol (DSR)

Dynamic source routing protocol (DSR) is an on-demand protocol. It deploys source routing. It is designed to restrict the bandwidth consumed by control packets in ad hoc wireless networks by eliminating the periodic table-update messages required in the table-driven approach [4]. In DSR each node catches the specified route to destination during source routing of a packet through that node. This enables the node to provide route specification when a packet source routes from that node. The error packet is sent by reverse path in case it is observed by a router. DSR also uses three types of packets like AODV [8]. DSR ensures that each packet includes the routing-node address also. The major difference between DSR and the other on-demand routing protocols is that it is beacon-less and hence does not require periodic hello packet (beacon) transmissions, which are used by a node to inform its neighbors of its presence.

2 Related Work

Large numbers of studies have analyzed the performance of Mobile ad hoc networks in terms of routing, provision of QoS, congestion control, impact of MAC layer on routing in MANETs etc. Maan, F et al. presented the performance evaluation of various routing protocols by using various mobility models like random-way point, uniform, grid etc. They concluded that routing protocols shows better performance in Random way point mobility model [1]. Muhammad Ibrahim et al. [2] presented certain QoS models that provide architecture for providing QoS to the mobile nodes in the MANETs. Hanzo, L. and Tafazolli et al. [3] presented problems that may occur in providing QoS to Mobile nodes in Mobile Ad-hoc networks and solution for managing those problems, like dynamic topologies, that change continuously and unpredictable at any time. Hinatariq, et al [4] presented the performance of various routing protocols in MANETs and they concluded that AODV is better in-terms of throughput and DSDV is better in-terms of delay. A.Venkataramana et al. [5] presented the impact of MAC layer protocol on the performance routing protocols in MANETs. Deepali Arora et al. [6] presented the performance evaluation of various reactive routing protocols in mobile ad hoc networks. Devi M. et al [7] presented that, fuzzy based route recovery technique increases packet delivery ratio while decreasing the delay, packet drop and energy consumption when compared to existing route recovery technique. Lajos Hanzo et al [8] presented various QoS issues for mobile ad hoc networks. Santosh Kumar et al [9] presented the impact of mobility models on the performance of routing protocols. Sujata V et al [10] presented the overview of various simulation tools for Ad hoc networks.

3 Propagation Models

Propagation models play crucial role in evaluating the performance of routing protocols in MANETs. In a propagation model, we use a set of mathematical models which are supposed to provide an increasing precision. Different propagation models are: path loss, shadowing, two-ray and fading.

The path loss model can be described as the power loss during the signal propagation in the free space. The shadowing model is characterized by fixed obstacles on the path of the radio signal propagation. The fading category is composed of multiple propagation distances, the fast movements of transmitters and receivers units and finally the reflectors [9]. During the simulation experiment selection of propagation model shows strong impact on the simulation results.

The familiar simulation models used in NS-2 are Two-ray ground, shadowing models. The Two Ray Ground model is also a large scale model. It is assumed that the received energy is the sum of the direct line of sight path and the path including one reflection on the ground between the sender and the receiver. A limitation in NS-2 is that sender and receiver have to be on the same height. It is shown that this model gives more accurate prediction at a long distance than the free space model [3]. The shadowing model of NS-2 realizes the lognormal shadowing model. It is assumed that the average received signal power decreases logarithmically with distance. A Gaussian random variable is added to this path-loss to account for environmental influences at the sender and the receiver.

4 Simulation Environment

The main purpose of simulation environments is to provide a virtual platform to the distributed applications with the required components. Simulation is an important tool in design development and evaluation of mobile ad hoc networks and their protocols. Simulation is an attractive and economical tool to evaluate the performance of wireless and wired networks, because real time performance assessment is very cost effective and difficult. The accuracy of the simulation depends on the quality of the models used. The different types of models used in simulators are mobility models, propagation models, energy models etc. This paper evaluates the performance of routing protocol using Network Simulator-2 (NS 2). It is widely recognized and improved network simulator for MANETs [10]. In Simulation environment the propagation models used are Two-ray and Shadowing models. Simulation environment is shown in Table 1.

Table 1. Parameters used during simulations

Parameters	Description
Routing Protocol	AODV, DSR
Simulation time	300 sec.
Node placement	randomly distributed
Propagation model	Two-Ray ground and Shadowing
Mobility Model	Random way point
Simulation Area	1,000 × 1,000 sq.m.
Network size (nodes)	5, 11, 15, 20
Bandwidth	1Mbps
Simulation time	300 s
Traffic type	Constant Bit Rate (CBR)
Packet size	1024 MB
Antenna Type	Omni directional

5 Results and Discussions

In this section, we display the experimental results and findings about the impact of propagation models on the performance of routing protocols. The QoS parameters used are throughput, end-to-end delay and packet delivery ratio. Values of QoS parameters are shown in Table 2.

5.1 Impact of Two Ray and Shadowing Models on the Performance of AODV

Performance of AODV is measured under two-ray ground model using NS-2 simulator. We have created networks with four different sizes, i.e. with 5, 11, 15 and 20 nodes. The throughput for AODV is Maximum i.e., 1953 b/s when the network contains 5 nodes. Throughput is Minimum i.e. 1943.39 b/s, when the network contains 15 nodes. The average end-to-end delay is minimum i.e., 86.8769 mille sec, when the network size is 5, the delay is high i.e., 168.416 mille sec. Packet delivery ratios are around 99% in all four different sizes of networks. From the overall results we observed that in two-ray ground model AODV performs better for smaller networks when compared with large networks.

Performance of AODV is measured under shadowing model using NS-2 simulator. The throughput for AODV is Maximum i.e., 1912 b/s when the network contains 15 nodes. Throughput is Minimum i.e. 1839.29 b/s, when the network contains 11 nodes. The average end-to-end delay is minimum i.e., 76.4521 ms when the network size is 5; the delay is high i.e., 101.135 milliseconds. Packet delivery ratios are around 99% in all four different sizes of networks. From the overall results we observed that in shadowing model AODV is suitable for small and medium networks. Performance of throughput and delay for AODV is shown in Fig.1.

Table 2. Values of QoS Parameters for AODV and DSR routing protocols.

Name of Routing Protocol	Network Size	Propagation Models			
		Two-Ray		Shadowing	
		Throughput	Delay	Throughput	Delay
AODV	5	1953.81	86.8769	1865.64	101.135
	11	1951.85	87.1249	1839.29	76.4251
	15	1943.39	120.31	1912.35	88.3909
	20	1944.00	168.416	1887.40	92.3898
DSR	5	3807.78	116.024	3041.87	209.506
	11	2907.30	183.052	3021.30	193.596
	15	3169.82	126.437	3169.82	137.602
	20	3622.29	87.541	3518.12	100.183

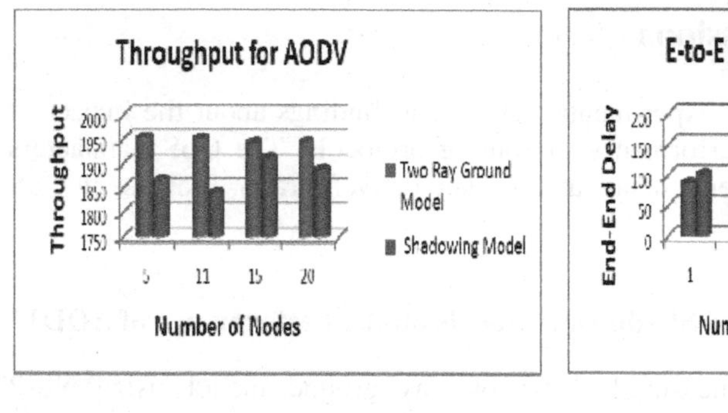

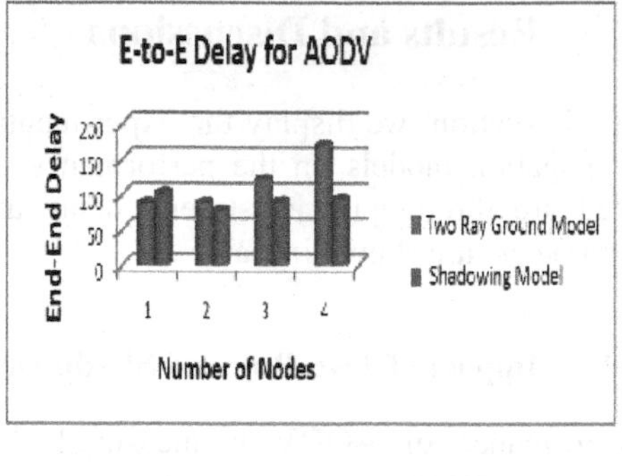

(a) (b)

Fig. 1. Peformance of (a) Throughput and (b) Delay for AODV under Two-ray and shadowing Models

5.2 Impact of Two Ray and Shadowing on the Performance of DSR

Performance of DSR is measured under two-ray ground model using NS-2 simulator. We have created networks with four different sizes, i.e. with 5, 11, 15 and 20 nodes. The throughput for DSR is Maximum i.e., 3807.78 b/s when the network contains 5 nodes. Minimum Throughput is 2907.30 bps, when the network contains 11 nodes. The throughput is increased with increased network size, i.e., when the network contains 20 nodes the throughput values 3622.29 b/s The average end-to-end delay is minimum i.e., 87.541 ms when the network size is 20, the delay is high i.e., 183.052 mille sec. Packet delivery ratios are around 100% in all four different sizes of networks. From the overall results we observed that in two-ray ground model DSR performs better for small and large networks.

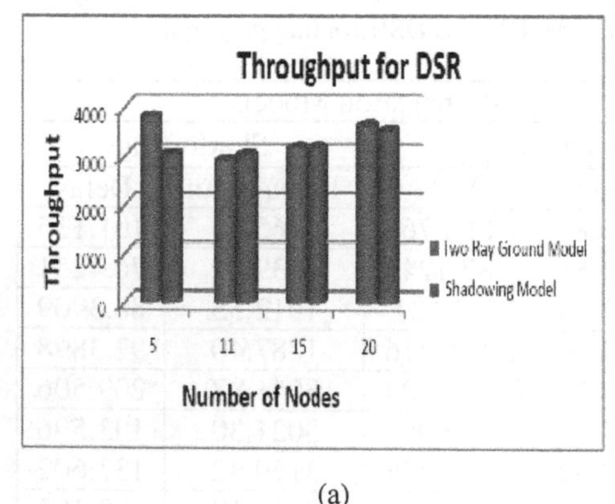

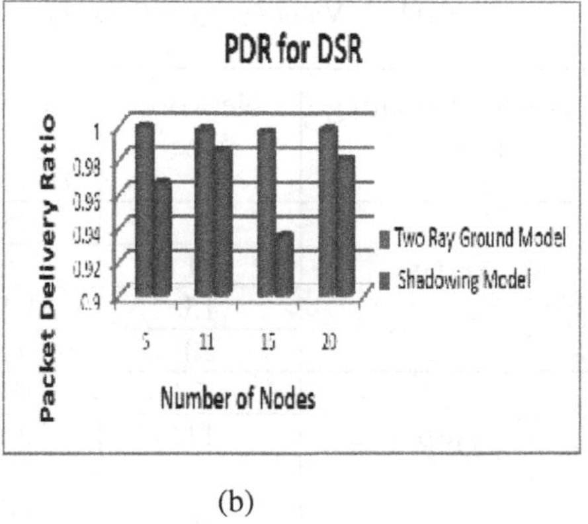

(a) (b)

Fig. 2. Performance of (a) Throughput and (b) PDR for DSR under Two-ray and shadowing Models

Performance of DSR is measured under Shadowing model using NS-2 simulator. We have created networks with four different sizes, i.e. with 5, 11, 15 and 20 nodes. The throughput for DSR is Maximum i.e., 3518.12 b/s when the network contains 20 nodes. Throughput is Minimum i.e. 3021.30 b/s, when the network contains 11 nodes. The throughput is increased with increased network size. The average end-to-end delay is low i.e., 100.183 ms when the network size is 20; the delay is high i.e., 209.506 ms when the network size is 5. Packet delivery ratios are around 100% in all four different sizes of networks. From the overall results we observed that under shadowing model DSR performs better for medium and large networks. Performance of throughput and PDR for DSR is shown in Fig.2.

6 Conclusions

In this paper simulation results reveal that propagation models shows strong impact on the performance of the routing protocols in MANETs. The performance of AODV and DSR routing protocols are measured using NS-2 under propagation models two-ray and shadowing. Performance evaluation is done based on QoS parameters like throughput, end-to-end delay and packet delivery ratio. Experimental results show that both AODV and DSR shows better throughput in two-ray ground model. AODV is performed well for smaller network where as DSR is performed well for large networks. Delay is less in AODV for small networks in two-ray model, where as in shadowing model delay is more for AODV for small networks. Shadowing model gives less delay for AODV in large networks. For example when the network size is 15, 20 the delay for AODV in two-ray model is 120.31 and 168.416 ms respectively. This work can be extended to other routing protocols by varying simulation areas and simulation time.

References

1. Maan, F.: MANET routing protocols vs mobility models: A performance evaluation. In: Third International Conference on Ubiquitous and Future Networks (ICUFN), pp. 179–184 (2011)
2. Venkataramana, A., PallamSetty, S.: Impact of MAC layer on AODV and LAR routing protocols in MANETs. International Journal of Computer Applications (IJCA) 84(4) (2013)
3. Ibrahim, M., Mehmood, T., Ullah, F.: QoS providence and Management in Mobile Ad-hoc networks. In: International Conference on Computer Engineering and Applications (IPCSIT), vol. 2. IACSIT Press, Singapore (2009)
4. Hanzo, L., Tafazolli, R.: A survey of QoS routing solutions for mobile ad hoc networks. IEEE Communications Surveys 9(2), 50–70 (2007)
5. Tariq, H., Khan, U.S., Zafar, S.: Performance Evaluation of Routing Protocols for Mobile Ad Hoc Networks. J. Basic. Appl. Sci. Res. 475, 471–475 (2013)
6. Arora, D., Millman, E., Neville, S.W.: Assessing the Performance of AODV, DYMO and OLSR Routing Protocols in the Context of Larger-scale Denser MANETs. IEEE Press (2011)

7. Devi, M., Rhymend Uthariaraj, V.: Fuzzy Based Route Recovery Technique for Mobile Ad Hoc Networks. European Journal of Scientific Research 83(1), 129–143 (2012)
8. Hanzo, L., Tafazolli, R.: A Survey of QoS routing solutions for mobile ad hoc networks. IEEE Communications 9(2) (2007)
9. Kumar, S., Sharma, S.C.: Bhupendra Suman: Impact of Mobility Models with Different Scalability of Networks on MANET Routing Protocols. International Journal of Scientific & Engineering Research (IJSER) 2(7) (2011)
10. Mallapur, S.V., Patil, S.R.: Survey on Simulation Tools for Mobile Ad-Hoc Networks. International Journal of Computer Networks and Wireless Communications (IJCNWC) 2(2) (2012)

Implementation of Scatternet in an Intelligent IoT Gateway

Lakshmi Mohan, M.K. Jinesh, K. Bipin, P. Harikrishnan, and Shiju Sathyadevan

Amrita Centre for Cyber Security Systems and Networks, Amrita Vishwa Vidyapeetham, Amritapuri, Kollam, Kerala, 691572, India
{lakshmimohan,jinesh,bipink,harikrishnanp,shiju.s}@am.amrita.edu

Abstract. Anything and everything will be connected in the world of IoT. This allows a ubiquitous communication around the world. The communication can be a sensed data from the physical world, or control signal for a device or else a usual internet data communication. There can be several ways with which the real world device can communicate to the IoT platform. Bluetooth is one such technology which allows communication between a device in the real world and IoT network. Bluetooth Scatternet concept, gives a bluetooth network the capability to support multiple concurrent bluetooth device communication over a wide area.. This paper discuss about a custom build intra and inter piconet scheduling model and its real-time validation on our own IoT platform which includes the Gateway, AGway (Amrita Gateway) and Middleware, AIoTm (Amrita Internet of Things Middleware). The paper emphasis on scatternet building and scatternet maintenance. The entire implementation is based on the linux bluetooth stack Blue-Z.

Keywords: Internet of Things,AGway,AIoTm, Bluetooth, Piconet, Scatternet, Bluetooth Stack, L2CAP.

1 Introduction

Internet of things refers to interconnection of heterogeneous devices or objects, which has the ability to interact with the physical world, through the internet, where every physical and virtual "objects" will have identities, physical attributes, and virtual personalities. This implies that every device in the network is unique. Such a uniqueness paved the way for using RFID (Radio Frequency Identifiers), or similar uniquely identifying technologies like MAC IDs, IPs etc in IoT devices [1]. But as with the advance in technology and human necessity, the concept of Internet of Things also got wider, with the term recompiled as Internet of Everything : defined as a dynamic global network infrastructure with self-configuring capabilities based on standard and interoperable communication protocols there by resulting in the advanced interconnection of devices, systems and services which can support varied protocols, domains and applications [2]. This itself alleviates the traditional machine- to-machine communication concept

S.C. Satapathy et al. (eds.), *Emerging ICT for Bridging the Future – Volume 2*, Advances in Intelligent Systems and Computing 338, DOI: 10.1007/978-3-319-13731-5_31

thereby highlighting an intelligent network and that could be seamlessly integrated into the information network. IoT platform supporting protocols can be wireless communication protocols or wired communication protocols, that could be used for a variety of applications like automation, real-time data sensing of varied domains like industries, environment etc, local generation, storing, transportation and distribution of energy, human and commodity tracking and like wise many[3] .One such protocol which supports the IoT platform is bluetooth.

Bluetooth is a widely used wireless communication technology which provides a robust, low power and low cost data transfer mechanism, within a short range say one metre or upto 100 metres. The Bluetooth transceiver operates in the globally available unlicensed 2.4 to 2.485 GHz ISM band, which makes its usage globally for free [4].Adopting bluetooth as a communication protocol offers a range of advantages to todays resource contrained world. This include, reduced power consumption which will be in milliWatts, wireless mode of communication, reduced interference with devices operating in the same frequency range like WiFi or between bluetooth devices itself, noise-resistance ,its ad-hoc networking capability and the 128-bit encryption.

In bluetooth wireless technology the radio signals are transmitted via a rapid signal transmitting method called Frequency-Hopping Spread Spectrum (FHSS). Bluetooth allows transmission of both voice and data , where data transmission occurs via packet switching which are always via Asynchronous-Connection-Less (ACL) links where as voice traffic via Synchronous -Connection-Oriented (SCO) links. This wireless technology basically follows a packet-based communication method which always maintains a master-slave relationship, when it comes to communication. The smallest unit of a bluetooth network is called a piconet and several piconets overlaps to form a scatternet.

The basic concepts which could weave this prposed model includes the Bluetooth technology, Piconet, Scatternet, Master-Slave concept, Bluetooth Protocol Stack, the Client-Server architecture, and time sharing. This entire concept in implemented on our IoT Gateway called the AGway (Amrita Internet of Things Gateway), which communicates the data received from bluetooth network to our own IoT middleware AIoTm (Amrita Internet of Things Middleware) [23,24].

The remaining of the paper is organized as follows.We discuss few of the earlier works done as part of Scatternet Formation Algorithm, then the concept of piconet and scatternet formation , which is followed by the customized bluetoooth protocol stack for our model, then the proposed model algorithm. The real-time implementation and testing follows the algorithm, after which we concluded with our test results and few points to make the model more efficient.

2 Related Works

Bluetooth technology was developed in order to address various issues faced by wired networks and also wireless ad-hoc networks. But the technology itself is not exempted from problems. Bluetooth specification explains about piconets and scatternets, but no where the specification has addressed the issues related to inter-piconet scheduling or intra piconet scheduling.So, many factors are to be taken into account while forming a Bluetooth Scatternet Algorithm.

In [5]Ahmed Jedda put forward different Bluetooth Scatternet Formation (BSF) algorithms to deal with various issues introduced, just during the procedure of device discovery. Jedda has explained BSF algorithms giving priority to time needed to form the scatternet, connectivity of the scatternet as long as its nodes originally form a connected network and out-degree constraint (A piconet should not have more than seven slaves). Also the number of piconets, number of master/slave bridges and number of bridge devices were also taken as a matter of concern.

A scatterenet formation algorithms can be divided into various categories depending on a few criteria. Depending up-on the communication between nodes in a network, we can distinguish algorithms as Single-hop algorithms [6],[7],[8]and multi-hop algorithms [9],[10], [11],[12]. These algorithms can be divided into self-healing algorithms [13]and non-self-healing algorithms [11],[14]depending on the ease with which a particular device can leave from the scatternet. There is one more classification based on selection of leader for a scatternet as centralized algorithms [15]and distributed algorithms[16], [17]. Another classification is dynamic and static algorithms based on how much knowledge a particular node has got about its neighbours [22],[13].

In [18] Albert Huang and Larry Rudolph explains a socket based approach on bluetooth piconet and scatternet formation. Adopting this approach and taking into consideration of below listed scatternet formation factors we have implemented and tested bluetooth scatternet on AGway.

Also our model have taken into consideration few of the above discussed factors, with the implementation and testing done mainly on performance level,with the following objectives.

- Allowable number of slaves for the proposed piconet concept
- Time delay in communication with the increase in number of devices participating in communication
- The possible communication range that could be practically achieved in a piconet with device of employed specifications
- Feasibility of proposed concept in real time application

3 Formation of Piconets and Scatternet

The elementary unit of a Bluetooth network is called a piconet, which could be formed just with two Bluetooth devices, where one among them, who initiates the connection will be acting as a master and the other as a slave.A Bluetooth piconet can accomodate to a maximum of one master and seven slaves [19]. The slave restriction came as part of the Bluetooth packet structure, in which the header part is allowed with three bits to identify a slave. For a communication to start, everytime the master will have to poll the slaves. The communication from the slave to master happens only when the slave is addressed by the master, and also only immediately after being addressed by the master.

In-order to have a wider coverage multiple piconets can share their area of coverage, there by forming a network of bluetooth devices called scatternet. While the piconets overlap, they can have an inter-piconet communication. This inter-piconet communication basically aims at sharing of data between the piconets , by means of making a single or multiple bluetooth devices available to two or more piconets. The shared bluetooth device can either be a master or a slave. That is a particular bluetooth master device can be master in one piconet and at the same time can be a slave in other piconet. This allows sharing of whatever data the master gathered from one piconent, with other piconent where it is a slave. Or a single slave can be accessed by two different piconet masters. This method of slave sharing can be employed in particular locations which requires continuous monitoring. In this model we have adopted the method of slave sharing since few of our gateways are deployed in areas that demands time stipulate surveillance.

This traditional concept of Master-Slave communication is replaced with Client-Server model where multiple clients can access a single server, in this proposed model.Fig.1 shows a simple scatternet with two piconets sharing a single slave.Each of the bluetooth devices will be connected to AGway,our IoT Gateway. AGway is a single board computer which runs on a Linux distribution. The bluetooth devices are supposed to share between the AGways the real-time sensor data, that the AGway receives from various sensors under its control.

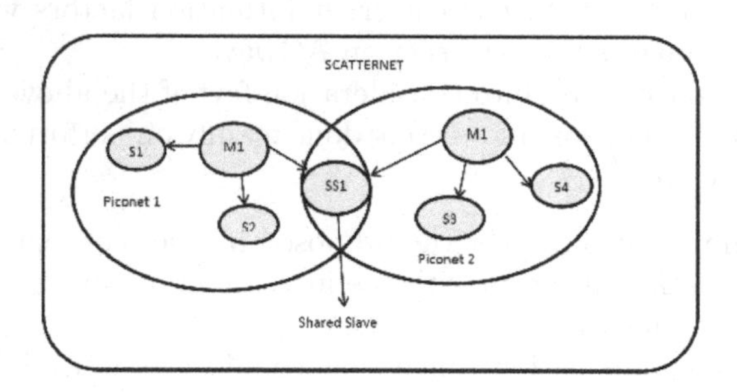

Fig. 1. A Bluetooth Piconet and Scatternt

4 Bluetooth Stack for Proposed Model

The Bluetooth Protocol stack for the proposed model follows a structure as shown in Fig.2

The Bluetooth protocol stack is a combination of lower layer firmware stack or controller stack and the higher layer software stack or host stack. Since we employed linux distribution as our development environment, the entire work is based on open source linux bluetooth stack Blue-Z. The entire software stack portion is included in the Blue-Z stack. The Blue-Z stack comes with every linux kernel version from 3.2 [21], [20].

Above the firmware, from the Host Controller Interface, starts the software stack, with the Logical Link Control and Adaptation Protocol (L2CAP), and by Radio Frequency Communication layer (RFCOMM)following the L2CAP. RFCOMM is equivalent to TCP layer of socket based communication in the matter of services and reliability, whereas L2CAP is equivalent to UDP. These two layers form the transport layer in Bluetooth stack. Bluetooth was developed as a wireless replacement for RS-232 data cables whose function is done by the transport layer RFCOMM. The decision to use one among the two transport layer protocols was a tough part in proposed model. The biggest difference between TCP and RFCOMM is in the number of ports available for communication, which is 65535 for TCP where in for RFCOMM the count is 30. Thus L2CAP over powers RFCOMM in the matter of number of ports. L2CAP offers odd numbered port values from 1 to 32767. Since L2CAP could communicate with the application directly, choosing L2CAP instead of RFCOMM reduces the over burden of managing two layers. Also L2CAP follows a best-effort simple datagram semantics. These advantages that L2CAP offered paved the way for adapting L2CAP as the transport layer in this model [18].

With this socket based approach we have the ability to add more than 8 devices into a piconet. This itself will scale the Bluetooth network. A particular piconet will communicate in one single port or a single channel. The use of L2CAP as transport layer provides with huge number of ports that can make available more number of piconets. Here we have bypassed the entire Bluetooth host controller, which includes the Link Controller, Link Manager and the Baseband Radio. The entire implementation is based on Bluetooth software stack. This was possible since all our Bluetooth devices are in-house products and so we have the MAC of every device known. The entire process involved in this Client-Server based bluetooth communication could be broken up into a series of actions performed by each of the individual Bluetooth Client-Server pair involved in the communication.

Following are the major events

- Choosing a device with the same protocol with which this device can communicate (Inquiry/Paging)
- Trace out a means by which it can communicate
- Initiate an outgoing connection
- Accept an incoming connection
- Send data that need to be shared
- Receive the data

From the above series of events , “Choosing a device for communication” will be trivial in our model. This is because the MAC Id of authorized bluetooth devices are already known, since all Bluetooth device were in-house products. This avoids the initial inquiry/paging which used to delay the time required for establishing communication. Thus we have eliminated the issues related to device discovery. Anyway since we have the MAC of our-own product, wasting device power with unnecessary inquiry and paging is undesirable. Also there is

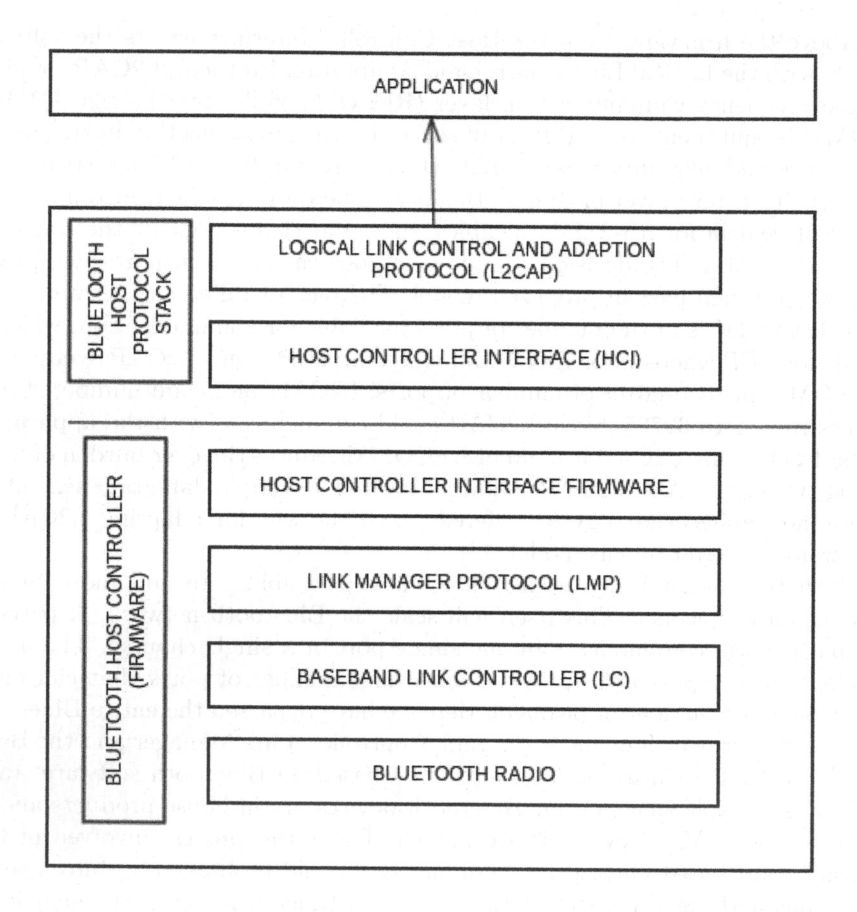

Fig. 2. Proposed Model Bluetooth Protocol Stack

no surety that an inquiry request from a master device will receive reply from the entire Bluetooth devices in range.

The local database involved in this system is MySQL Database which stores the MAC Id of authorized bluetooth devices. This local storage substituted the requirement of inquiry stage by which the need for repeated scanning and connection establishment is avoided. Also the user can add MAC Id to the database from anywhere since the AGway is network enabled. From the different links that is established between a master and a slave device in a piconet, our implementation employs Asynchronous Connection Less links (ACL-links). ACL-links allow packet-switched point-to-multipoint connections. Such links are typically used for data transmission.

5 Algorithm

We have employed MySQL database which is maintained locally in all the AG-ways. The database stores the MAC id of authorized bluetooth devices which alone can participate in piconet formation and there after the scatternet. A particular Gateway user has got the privilege to manipulate and maintain the local database. Before formation of piconet the MAC ids of those bluetooth devices which he wishes to add into his piconet will be inserted into the local database by the user. Here the Master bluetooth device will act as the client and the slave Bluetooth device will be the server. The slave device will be kept in the listening mode as the first stage of piconet formation and will continue this state as long as part of the network. The communication happens only between master and slave. There is no slave to slave communication. The algorithm follows a star topology with communication happening in a round-robin fashion. All the master and slave devices are static.

The intra-piconet scheduling algorithm can be divided into two separate major phases as below

- Authentication phase, where master polls for slave existence:

Once the slaves move into listening mode, master will start polling for the slave existence. After establishing the connection, the Master bluetooth device will send "existence check" request to each of the slave device one after the other, whose MAC is inserted into the local database. The Slave devices which are alive will reply back with an exist acknowledgement. Further communication will happen only with those slave devices which had sent the "exist ack". Here two level of authentication is done. One level is the reception of "exist ack"and the other is the secret key which is passed confirming themselves as authorized to participate in the piconet formation. If either of these authentication level is uncleared by a particular device, that device will be automatically removed from the local database after a particuar wait time. If this slave device again wants to join the piconet, the Gateway user has to insert the MAC id of that device into the local database. Before the next phase, master will authenticate all devices listed in the local database.

- Data exchange phase, where master receives data from authorized slave devices:

This phase follows the authentication phase. In this stage the master bluetooth device will start the actual sensor data excahnge between the master device and the slave device. Consider there is only one slave device in the piconet, the communication bill be between this slave and the master. If more than one slave comes into picture, the network will expand and the number of slave devices also. Time diviion multiplexing is used to access these slaves. The slaves are accessed in a round-robin fashion where each will get a fixed duration of time to communicate with the master. A particular slave is removed from the piconet if that device is not responding to the master in three consecutive cycle. The

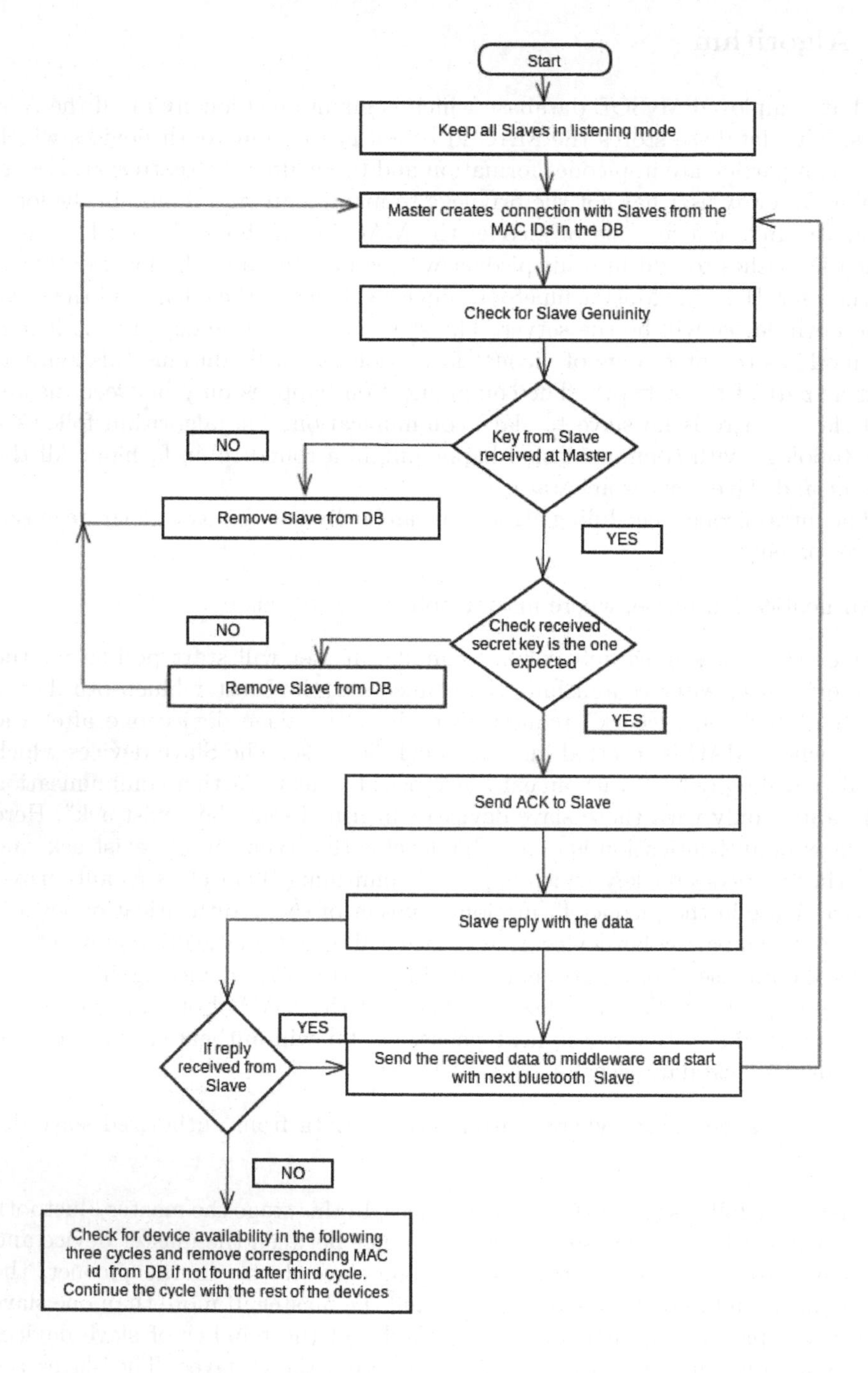

Fig. 3. Intra Piconet Scheduling Flow Chart

removed device should be added further by the AGway user to continue with the network.

The inter-piconet scheduling algorithm follows a fashion as below.

During the formation of scatternet, the communication with in the piconet will follow the rules for intra-piconet scheduling algorithm. Among those slaves in the piconet,a particular slave will be shared between the multiple piconets to form the scatternet. This particular slave will be accessed by the Master of individual piocnets in a time sharing manner. Consider a time interval of T seconds in a minute and two piconets forming the scatternet. In the first t1 seconds the Shared Slave will be accessed by one piconet, and in the next t2 seconds of the minute the Shared Slave will be accessed by the other piconet. This will happen in a cyclic manner as long as the slave is alive in the piconet. Entire communication time can be divided into intervals of T seconds each. Where T is divided into t1, t2. The division of Ts interval depends on the number of piconets accessing a particular slave. Say the total time take for a communication is of N seconds duration. Then 'N' can be represented as

$$N = T + T + T + T..... + T = nT$$

n = number of cycles in round-robin

Suppose there are four piconets sharing a particular slave in this 'T' time , then 'T' can be represented as

$$T = t1 + t2 + t3 + t4$$

where t1=t2=t3=t4

These t1, t2, t3 , t4 can be again further sliced into small intervals where each interval is dedicated to a particluar slave in a piconet. That is, if a piconet has got 8 slaves then,

$$t1 = T1 + T2 + T3 + T4 + T5 + T6 + T7 + T8$$

where the difference between T1 and T2,T2 and T3, T3 and T4 , T4 and T5, T5 and T6, T6 and T7, T7 and T8 are equal. Also 'T1' or 'T2' or 'T3' are the time allotted for particular slaves to communicate with its corresponding master

The entire algorithm for intra-piconet formation is given as a flow chart in Fig.3

6 Implementation and Testing

The implementation has bypassed the Bluetooth Host Controller firmware part. At the protocol software stack level, this proposed model has incorporated the Host Controller Interface and the Logical Link Control and Adaptation Protocol. We employed Class-2 Bluetooth radio modules which deliver a maximum of around 2.5 mW and with a range of 9-10 m. The Bluetooth version with which the implementation was tested is version 2.0 + EDR (Enhanced Data Rate).

Our-own IoT gateway AGway, housed the bluetooth radio modules. The AGway runs on a Linux based platform and hence supports BlueZ protocol stack. The AGway collect sensor data from the environment and has to bypass it to the middleware. There are few AGways which cannot communicate directly with the middleware. In such situations those AGways will pass the sensor data to the near-by AGways which will further make the data available to the middleware via MQTT publish/subscribe method. The middleware then process the data and make it available to the user citeshiju.

Initially individual piconets were tested for the number of bluetooth modules it can support. The running 'piconet 1' used L2CAP port 1 for establishing connection and communication. The very first phase was with just a single master and a single slave bluetooth device. Later on the number of slave bluetooth devices were increased from one to many. As the number of slave devices increased beyond four, communication delay of one to two seconds were observed. This is the delay for esablishing a connection. Beyond six, the delay was further going up to three or four seconds. The setup could support more than seven slave bluetooth devices which was the number of slave support for a conventional Bluetooth piconet. The optimal number of slaves for the best performance was found out to be four to five.The setup was tested for the range with and without obstacles in and around the communication area. The maximum range obtained was upto eight to nine metre without any hindrance on the path and with obstacles the range was six to eight metres. Also packet loss were also observed once in every 15 packets.

Fig.4 and Fig.5 shows the response of slaves to master in a single piconet.

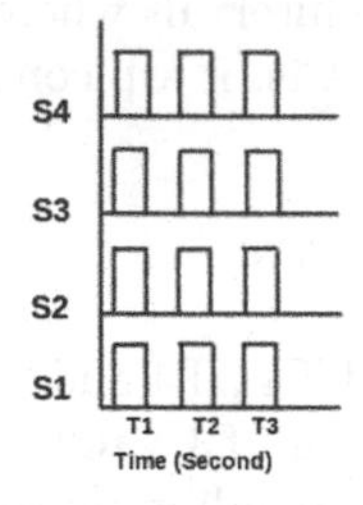

Fig. 4. A Single Piconet with four bluetooth devices

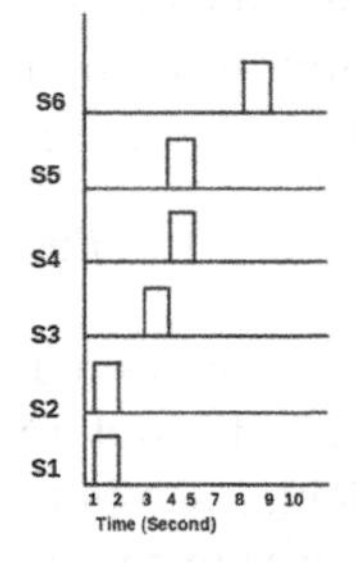

Fig. 5. A Single Piconet with more than four slaves

Fig.4 shows the response time of four slaves in a piconet without obstecles. All the slaves responded at the same time. That is if S1's data was received at T1 time, then S2,S3,S4 also responded at the same time (Note : T1 was in seconds).Fig.5 shows the response time of a piconet with six slaves. S1 and S2 were able to respond within 1 second, where as S3 took one second delay, S4 and S5 took two seconds delay and S6 took five seconds delay. It was observed, that since the algorithm works in a round-robin fashion, the delay of one slave will add to delay of next slave coming in the queue.Fig. 6 shows that as the number of slaves increased to nine, the delay in response was increased.Eventhough delay is there, all the devices were able to respond with in half of a minute.

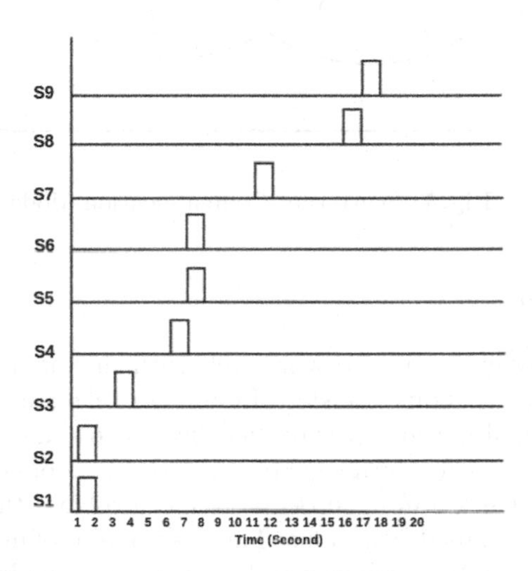

Fig. 6. Piconet with nine bluetooth devices

Scatternet was tested with two piconets each having three slaves and sharing one slave in common. The very same hardware devices which were used for piconet, was used for testing scatternet also. The scatternet was implemented based on the time sharing mechanism. The shared slave will be accessed by a particular piconet at a particular assigned time interval.S1,S2,S3,S4 are the slaves, where S1 and S2 are members of piconet 1 and S3 and S4 are members of piconet 2. SS1 is the shared slave between the two piconets.

That is if we have two time duration T1 and T2,

T1 Time Duration :

Piconet 1 in T1 time duration : S1-S2-SS1

Piconet 2 in T1 time duration : S3-S4

T2 Time Duration :

Piconet 1 in T2 time duration : S1-S2

Piconet 2 in T2 time duration : S4-S3-SS1

This makes sure that the Shared Slave is accessed by only one piconet at a particular time. This method worked well with two piconets and total five slaves and two master devices. Fig. 7 shows a scatternet with a single slave shared. The running piconets 1 and 2 have used L2CAP port 1 and 3 respectively.

The Scatternet was tested with each piconent with more than seven slaves. The delay observed in establishing connection and start communication reached a maximum of 20 seconds between the first slave response and the last slave in the piconet.

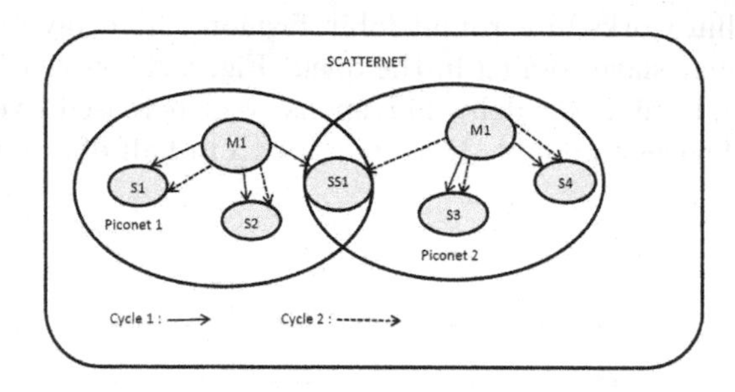

Fig. 7. Scatternet Communication Model

7 Conclusion

This particular setup seemed to work well with maximum of five members in a piconet, where continuous transfer of data can take place from the master to different slaves and also from one slave in a piconet to a master in another piconet which thereby forms a scatternet, expanding the area of coverage of the sensor network. Since much scanning process is not involved, this concept could be used in energy constrained situations. There is no need of manual intervention to connect the master and slave device and start the communication between them, since there is no inquiry/paging. Also the proposed method could successfully connect and communicate with more than seven slave devices in a single piconet, on a time sharing basis.

This particular concept can be developed as an android app since the java supporting Bluetooth stack JBlueZ is readily available as open source, thus can serve a real-time application. Replacing MySQL with SQLite will reduce the database accessing time since SQLite is light weight.

References

1. van Kranenburg, R.: The Internet of Things, A critique of ambient technology and the all-seeing network of RFID
2. River Publishers Series in Communication: Internet of Things-GLobal Technological and Societal Trends
3. `http://www.internet-of-things-research.eu`
4. `http://www.imd.uni-rostock.de/ma/gol/lectures/wirlec/`
5. Jedda, A.: The Device Discovery in Bluetooth Scatternet Formation Algorithms, Master of Science, thesis, Ottawa-Carleton Institute for Computer Science, School of Information Technology and Engineering, University of Ottawa (2009)
6. Barriere, L., Fraigniaud, P., Narajanan, L., Opatrny, J.: Dynamic construction of Bluetooth scatternets of fixed degree and low diameter. In: Proceedings of the ACM Symposium on Discrete Algorithms, SODA (2003)
7. Wang, Y., Stojmenovic, I., Li, X.-Y.: Bluetooth scatternet formation for single-hop ad hoc networks based on virtual positions. In: Proceedings of the IEEE Symposium on Computers and Communications, Alexandria, Egypt (2004)

8. Tan, G., Miu, A., Guttag, J., Balakrishnan, H.: An Efficient Scatternet Formation Algorithm for Dynamic Environments. In: Proceedings of IASTED CCN, Cambridge, MA (2002)
9. Perkins, C.E., Royer, E.M.: Ad Hoc On-demand Distance Vector Routing. In: Proceeding of the 2nd IEEE workshop of Mobile Computing Systems and Applications, pp. 90–100 (1999)
10. Zaguia, N., Daadaa, Y., Stojmenovic, I.: Simplified Bluetooth scatternet formation using maximal independent sets. Integrated Computer-Aided Engineering 15(3), 229–239 (2008)
11. Cuomo, F., Melodia, T., Akyildiz, I.F.: Distributed self-healing and variable topology optimization algorithms for QoS provisioning in scatternets. IEEE Selected Areas in Communication 22(7), 1220–1236 (2004)
12. Dubhashi, D., Häggström, O., Mambrini, G., Panconesi, A., Petrioli, C.: Blue Pleiades, a new solution for device discovery and scatternet formation in multihop Bluetooth networks. Wireless Networks 13(1), 107–125 (2007)
13. Law, C., Mehta, A.K., Siu, K.: A new Bluetooth scatternet formation protocol. Mobile Networks and Applications 8(5), 485–498 (2003)
14. Pamuk, C., Karasan, E.: SF-DeviL:An algorithm for energy-efficient Bluetooth scatternet formation and maintenance. Computer Communications 28(10), 1276–1291 (2005)
15. Mehta, V., El Zarki, M.: A Fixed Sensor Networks for Civil Infrastructure Monitoring. Wireless Networks 10(4), 401–412 (2004)
16. Salonidis, T., Bhagwat, P., Tassiulas, L., LaMaire, R.: Distributed topology construction of Bluetooth personal area networks. In: Proeedings of the IEEE INFOCOM (2001)
17. Ramachandran, L., Kapoor, M., Sarkar, A., Aggarwal, A.: Clustering algorithms for ad hoc wireless networks. In: Proceedings of the ACM DIALM Workshop, pp. 54–63 (2000)
18. Huang, A., Rudolph, L.: Bluetooth for Programmers. Massachusetts Institute of Technology, Cambridge (2005)
19. http://www.epubbud.com
20. http://www.bluez.org/
21. http://www.bluetooth.org
22. Perkins, C.E., Royer, E.M.: Ad Hoc On-demand Distance Vector Routing. In: Proceeding of the 2nd IEEE Workshop of Mobile Computing Systems and Applications, vol. (1), pp. 16–28 (1999)
23. Sathyadevan, S., Achuthan, K., Poroor, J.: Architectural Recommendations in Building a Network Based Secure, Scalable and Interoperable Internet of Things Middleware. In: Satapathy, S.C., Biswal, B.N., Udgata, S.K., Mandal, J.K. (eds.) Proc. of the 3rd Int. Conf. on Front. of Intell. Comput (FICTA) 2014- Vol. 1. AISC, vol. 327, pp. 429–440. Springer, Heidelberg (2015)
24. http://aiotm.in

Reactive Energy Aware Routing Selection Based on Knapsack Algorithm (RER-SK)

Arshad Ahmad Khan Mohammad[1], Ali Mirza[2], and Mohammed Abdul Razzak[3]

[1] KL University, Guntur, India
[2] DMS.SVH College, Machlipatnam, India
[3] Innovative Technologies, India
{Ibnepathan,alimirza.md,connect2razzak}@gmail.com

Abstract. Wireless mobile ad-hoc network technology is designed for the establishment of a network anywhere and anytime, characterized by lack of infrastructure, clients in a network free to move and organize themselves in an arbiter fashion, Communication may have multiple links and heterogeneous radio, can operate in a stand-alone manner, which needs a efficient dynamic routing protocol in terms of energy due to its energy constraint characteristic. We propose a novel energy-aware routing protocol for mobile ad-hoc networks, called reactive energy aware routing selection based on knapsack: RER-SK. It addresses the two important characteristics of MANETs: energy-efficiency and improving life time of network. It considers the node's optimistic information processing capacity with respect to current energy & traffic conditions with the help of knapsack algorithm. The simulation result shows that this work is better than the existing MBCR & MRPC interns of network lifetime by using NS2.

Keywords: Energy, MANETs, Knapsack, lifetime, reactive, optimization.

1 Introduction

The development aim of Mobile ad hoc networks (MANETs) is to support internet access anywhere and anytime, characterized by lack of infrastructure with self configuration & self maintenance capacities .MANETs are also one type of wireless networks (Infrastructure less) composed of a collection of heterogeneous (in terms of energy, computation capacity) mobile devises (nodes) which are connected by a dynamically varying network topology without pre define infrastructure and absence of central coordinator or base station. Where network intelligence placeless inside every mobile node in a network thus nodes in a network act as a router as well as host which means MANETs act as a peer to peer network. The communication between nodes may have a more than one links and heterogeneous radio and can operate in a standalone passion. This network well suited a situation where infrastructure is hard to setup and cost and/or time effective.

The development, design, and performance of MANETs extremely include in routing, Security, quality of service, multicasting, scalability, service discovery, & Resource management (energy, bandwidth, delay, jitter and battery power). Routing protocol design issue of MANETs is inherently related with its applications. The primary purpose of

S.C. Satapathy et al. (eds.), *Emerging ICT for Bridging the Future – Volume 2*,
Advances in Intelligent Systems and Computing 338, DOI: 10.1007/978-3-319-13731-5_32

routing protocol design is to find path discovery but in MANETs it should include Quality of service, resource management with security. Design of Routing protocol in MANETs is an active research area in recent years; number of routing protocols has been developed. Routing protocols are useful when they offer acceptable communication services like route discovery time, communication throughput, end to end delay, and packet loss. Quality of Service is the performance level of service which is offered by the network to its user. But in case of Quality of Service routing process, which has to provide end to end loop free path with guarantee the necessary Quality of Service parameters like bandwidth, delay, energy & availability. Depending up on the application Quality of Service parameters are varies.

Energy-Efficient routing is another powerful factor for MANETs routing due to its energy limitation characteristic so as to reducing the energy cost during data communication. Routing protocol aim is to just considering energy consumption during end to end packet travelling is not reliable routing but it also consider reliable links and residual energy of nodes which not only improve QoS but also improve life time of the network. Various routing protocols have been proposed which aim to improve reliability, energy efficiency and life time of network.

Routing, QoS & Energy efficiency is challenging in MANETs compared to infrastructure network due to its characteristics like dynamic network topology, lack of infrastructure, absence of central coordinator, mobility, resource constraint, error prone channels and hidden & expose node problem. Hence it is a requirement of developing routing protocol which should address the MANETs challenges & must be fully distributed and adaptive to network dynamics, loop free path with minimum route establishment's effort must be in terms of time, energy, power and memory. Finally it must support certain level of QoS parameters with energy efficiency.

In this paper we present a novel mechanism to achieve energy efficient routing protocol named as reactive energy aware routing selection based on knapsack, where route selection is based on node's Optimized information process capacity within remaining battery power. The remaining paper organized as follows: Next section describe about Energy efficient routing in MANETs. In section 3, we discuss a process of finding optimization of information processing capacity of node with respect to energy. Section 4 describes the proposed model. Section 5 discuss related work & our work end with simulation and conclusion.

2 Energy Efficient Routing in MANETs

Energy-Efficient routing is a powerful factor in MANETs to reducing energy cost of data communication due to its energy limitation characteristic. Routing protocol aim is to just finding energy consumption during end to end packet travelling is not reliable routing but it also consider reliable links and residual energy of nodes which not only improve Quality of service but also improve life time of the network. Various routing protocols have been proposed which aim to improve reliability, energy efficiency and life time of network. e.g., [1], [2], [3], [4], [5], [6], [7], [8], [9], [10], [11], [12], [13],[14], [15]. These routing protocols are majorly sub divided into three categories

1. Algorithms that aim to finding reliability of links to find more reliable routes.
2. Algorithms that aim to finding energy-efficient routes
3. Algorithms that aims to finding the routes with higher energy nodes to Improve the network lifetime

For instance, De Couto et al. [10] introduced to find the reliable routes based on expected transmission count, where links in a route require less number of retransmissions to recover from packet loss, which reduce energy consumption of routes but not end to end transmission of packets. The proposed algorithms in [13], [9], [12], [6], [1], [5] aims to find routes based on the energy-efficiency but not consider the remaining battery energy of nodes. These algorithms find the route based on transmission power of nodes but not energy consumption of nodes. In these [12], [6], [1], [5], address energy-efficiency along with reliability. In [15], [14], [8], [7], [4], [2], [3], [11] aims to find the routes with higher energy nodes to Improve the network lifetime but not consider the reliability and energy efficiency which may cause the overall increment of energy consumption in network.

Reference [17], discuss about minimum power routing algorithm based on minimization of power requirement to transmit pocket from source to destination given by

$$Minimize \sum_{i \in path} P(i, i+1)$$

P (i, i+1) = power required to transmit packet between two consecutive nodes. In reference [17] discuss about minimum power routing. Which derived equation for transmits, receive, broadcast and discard of pockets given by

$$E(packet) = b \times packet\_size + c$$

b=packet size dependent energy consumption, c= fixed cost that accounts for acquiring the channel and for MAC layer control negotiation. These approaches selects route depends on packet size, which always selects least power cot routs. At the same time nodes along these routes may die soon as battery energy exhaustion. Another issue about energy efficient node is, it is accepting all the traffic due to enough residual battery capacity, much traffic load be injected on it.

In this paper we design a routing protocol which aims to improve network life time and find reliable links so as to improve the QoS based on residual energy & current traffic of node. In our work we are considering the impact of multiple transmissions over multi hop MANETs on an energy constraint intermediate node. We take consideration of node's optimized information processing capacity within an available energy, and we assign the priority of nodes based on information processing capacity. According to node priority application is initiated on that node.

Our contribution of work mainly as follows

1. Calculation of energy required by a node to process the packet of [x] bit data
2. Optimization of information processing capacity of node with respect to energy
3. Prioritization of nodes
4. Routing based on prioritization

3 Optimization of Information Process Capacity of Node with Respect to Current Energy and Traffic of Node (SK)

In order to optimize information processing capacity say 'SK' of node let 'Z' with energy capacity of 'C_e' Joules, we are considering 'n' information to be process through it. Where information1 say I_1 has [x]bits of data and takes 'e(x)' energy to transmit, process and receive. We need to process as much as information from node 'Z' within a available energy of 'C_e' joules, which in turn need to find the subset of information such that in below conditions

1. All processing information have combined with size (bytes) at most of 'C_e'joules
2. The total amount of information process by node as large as possible
3. Node cannot process a part of information (which should either hole/nothing)

Table 1. Notation Used in Work

Notation	Description
$\varepsilon_{u,v}(x)$	Energy of node when it maximum battery Charged
$\omega_{u,v}(x)$	Energy required by a node 'u' to transmit x [bit] information to node 'v'
	Energy required by a node 'u' to receive x [bit] information coming from node 'v'
e (x)	Energy required by any node to process x[bit] of information
Th_{max}	Maximum Threshold value
Th_{min}	Minimum Threshold Value
Sk	Optimized Information processing capacity of given node with respect to current energy & traffic
r [bits/s]	Bit rates of physical link
A_u	Power required to run the processing circuit of the packet transmitting node 'U'
B_v	Power required to run the processing circuit of the packet Receiving node 'V'

We firstly derive an equation for the energy required by a node to process the information within a multi hop MANETs, which is required to calculate the optimization. Table 1 show the notation used in our work. We consider a multi hop network, where multi sources are need to transmit the information through an intermediate node say 'Z', „which has energy capacity $\underline{C}_e$, which contain B bytes of buffer capacity and can process up to 'n' packets of different source nodes information let '$l_1, l_2, \ldots l_n$' within a given energy capacity. We assume that in given number of Information which needs to be process through 'Z' node in a network, the packets of that information processed completely, and partial packets flow is not possible. We consider n- Topples of positive values as

1. Number of Information which need to process from the given node let (l1, l2, l3…ln) which contain the number of packets with size of x [bits]
2. Energy consumed by information to process by node, include transmit, process and receive is given as 'e'

$$e(x) = \varepsilon_{u,v}(x) + \omega_{u,v}(x) + P_{u,v}(x) \quad (1)$$

Where $\varepsilon_{u,v}(x)$, $\omega_{u,v}(x)$ are the Energy required by a node 'u' to transmit x [bit] information to node 'v' and Energy required by a node 'u' to receive x [bit] information coming from node 'v' respectively and mathematically given by the work present in [16]. Where $P_{u,v}(x)$ the energy required by node to process the x[bit] information within a node, which we are considering negligible in our work.

$$\varepsilon_{u,v}(x) = \left(A_u + \frac{P_{u,v}}{K_u}\right)\frac{x}{r}, \quad \forall x \geq 0, \ \forall (u,v) \in \mathbb{E}, \quad (2)$$

$$\omega_{u,v}(x) = \frac{B_v}{r}x, \quad \forall x \geq 0, \ \forall (u,v) \in \mathbb{E}. \quad (3)$$

Where Au = power required to run the processing circuit of the transmitter of node u, $P_{u,v}$ be the transmission power from node u to node v, B_v be the power required to run the receiving circuit of the wireless interface at node v, and r [bit/s] be the data rate of the physical link.

We want to determine the energy consumption of information in bits $T \in \{x_1, x_2, x_3 \ldots x_n\}$ so as to

Maximize $\sum I(x_i)$ where $I \in T$ Subject to $\sum e_i(x) \leq C_e$

To get maximum information processed by an intermediate node for given energy capacity 'C_e', one possibility is to try for all 2n possible subsets of 'T'. Here we apply Knapsack algorithm [19] to get the optimizing solution, where it construct two dimensional arrays

$$V[0\ldots..n, 0\ldots.C_e] \quad \forall\ 1 \leq i \leq n \ \& \ 0 \leq e(x) \leq C_e$$

Such that V[I, C_e] will process maximum information of any subset of flows with x bits of data {1,2,3,.....i} of energy required to process at most 'e'. Where array entries V[n, C_e] will contain maximum flows to process from given intermediate node. At the same time array entries are not consider in below conditions.

1. V[0, e]=0 $\forall\ 0 \leq e \leq C_e$ no information process from node
2. V[i, e]= $-\infty$ $\forall\ e < 0$, illegal.

Optimization solution is as follows

$$V[i, e] = \max(V[i-1, e], V_i + V[i-1, e-e_i]) \quad \forall \quad 1 \leq i \leq n \quad \text{and} \quad 0 \leq e \leq C_e$$

To compute the actual subset, Knapsack adds an auxiliary Boolean array Keep [i, e] which becomes 1 if node decide to process the 'i' th flow of x bits in V [I, e] and it becomes 0 otherwise. The algorithm to calculate maximum information Process Capacity of a Node with an available energy capacity of nodes is shown in algorithm 1.

4 Proposed Model

Depending on the optimized processing capacity of nodes with respect to current energy and traffic present in a node is used as a metric to find the optimistic path between sources to destination. In this work, each node 'n_i' (where i=1, 2, 3...) computes its optimized Process Capacity say 'sk' in an available energy and current traffic by using the proposed algorithm. Every node in a network during route discovery process needs to maintain the table with entry of its maximum Process Capacity 'sk'. We are considering the two threshold values Th_{min}, Th_{max} depending on the optimized process capacity of node.

1. If nodes optimized Process Capacity 'sk' is greater or equal to Th_{max} , this types of nodes have higher priority and used for any types of application in MANETs environment.
2. If nodes optimized Process Capacity 'sk' is lie between Th_{min} & Th_{max} , This types of nodes has medium priority and used for backup for higher priority nodes.
3. If nodes with optimized Process Capacity 'sk' is less than Th_{min} are not eligible for processes the communication.

Where Th_{max} value is optimized information processing capacity of node under the current energy= $Ce/2$ and minimum traffic condition. Th_{min} value is optimized information processing capacity of node under the current energy= $\frac{Ce}{4}$, where C is the Maximum energy capacity of node. At the same time one can guaranty that a node present in a route have the maximum processing capacity will take account of node's present traffic condition, as a node with high remaining battery power may not survey in heavy traffic load. There may be vocationally status of the selected path can change over time due to switching of node's priority due to variations of their 'sk' values as they go through heavy load and lack of remaining energy, the activation of new path selecting depends on underlying routing protocol. In our proposed model it is based on reactive routing which is on top of already available AODV, instead of hop count as optimistic route we are using priority of node which we discussed above. Every source node has to periodically obtain the status information.

4.1 Algorithm

Algorithm 1. Algorithm is to find the optimistic energy processing capacity of node

```
[1]     Knapsack(l, e, n, Ce)   {
[2]     for (e=0 to Ce )  V [0, e] =0;
[3]     for (i= 1 to n)
[4]     for (e= 0 to Ce)
[5]     If ((e[ i] ≤ e) and (l[i]+V[i-1, e- e[i]] > V[i-1,e]))   {
[6]     V [ i, e] =  l[i] + V [i-1, e- e[i]];
[7]     Keep [i,e]=1;      }
[8]     else      {
[9]     V [i, e] = V [i-1, e];
[10]    Keep [i, e] =0;     }
[11]    K= Ce    ;
[12]    For (I = n down to 1)
[13]    If (keep [i, K]==1)    {
[14]    Output i;
[15]    K= K- e[i];     }
[16]    Return V[n, Ce ];
```

5 Performance Evaluation

We investigate the performance of proposed RER-SK model using the ns-2 simulator with the necessary extension and compare it with MBCR [8] &MRPC [18] in a same condition with respect to network life time. In our simulations, we use a fixed transmission range of 250 meters, which is supported by most of real time and current network interface cards. We used the "random waypoint" with speed of nodes is uniformly distributed between 0 to maximum speed of 15 m/s with pause time value of 30 sec. All mobile nodes to be equipped with IEEE 802.11 network interface card and data rates of 2 Mbps. The initial energy of all the nodes is 10J. The transmission power is 600mW and the receiving power is 300mW. Finally, source nodes generate CBR (constant bit rate) traffic. Traffic sessions are generated randomly on selected different source- destinations with a packet size of 512 bytes. Every node in a network has to run our algorithm whenever it becomes an intermediate node to forward the information of source nodes. Each simulation was run for the duration of 200 seconds and sampled data we collected from simulation is average of 3 times.

Our aim is mainly investigate the node' s optimized information processing capacity with respect to current battery as well as traffic condition, and according to it provide priority to nodes for routing metric. We therefore measure the network lifetime, reliability of link with respect to packet delivery ratio and compare it to existing MBCR [8], MRPC [18]. The network life time in our work is considered as a time that the first node failure occurs in a network due to exhausting of battery. Delay in the failure of first node impacts on other node to be delayed. Fig.1 shows the performance of RER-SK, MBCR & MRPC as function of the packet delivery ratio with respect to network life time. This fig1 clearly shows RER-SK can significantly delay the first node failure compare to other algorithms. In fig2 shows the important characteristic of RER-SK is finding reliable route .In fig3 shows the finding energy efficient route compare to other, where less amount of energy consumed to route a packet. In conclusion, RER-SK increases the lifetime of network with reliable link.

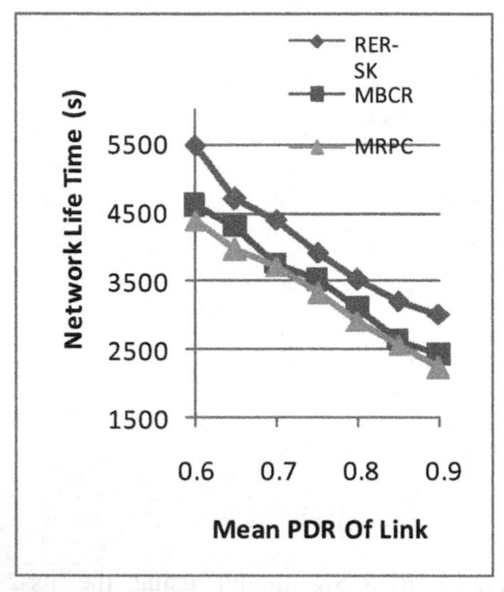

Fig. 1. Average Number of packets delivered to destination with respect to node failure in a network

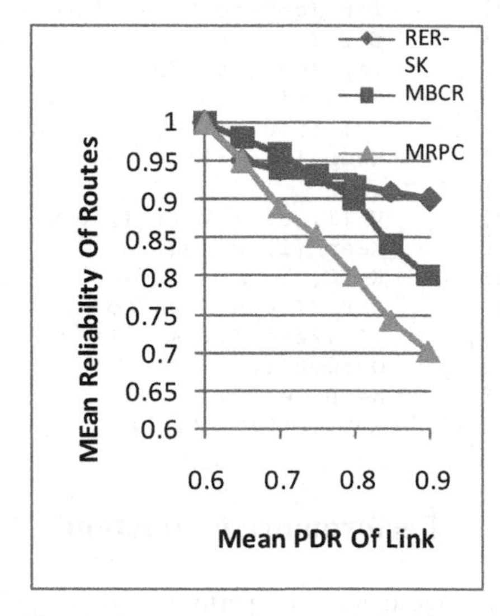

Fig. 2. Average end to end Reliability of selected Paths

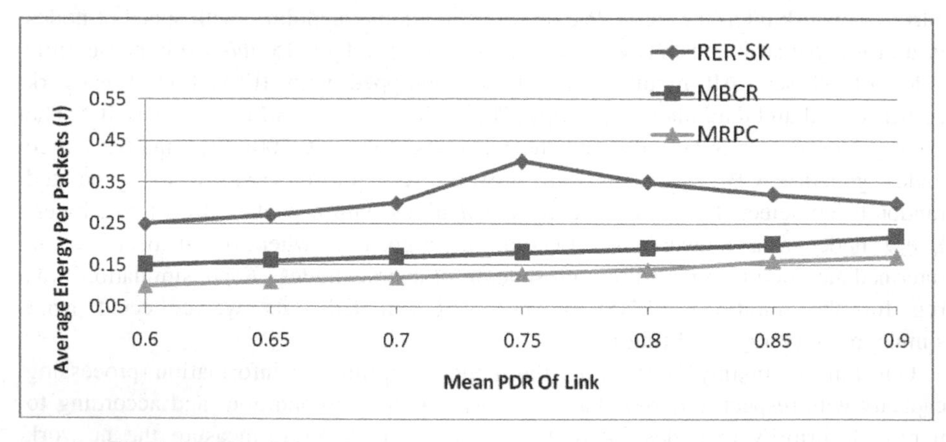

Fig. 3. Average Energy consumed to route packet from Source to Destination

6 Conclusion

In this work we proposed a new metric, "Optimized Information Process Capacity of node with respect to current energy of node", which is used to protect the life time of nodes. This metric Combines with current traffic condition of node to decide whether node to be a part of route or not. We establish the method called the "Optimized information process capacity" which can be used for any reactive routing protocols in MANETs. This metric provide good way to know the current capacity of node with respect to energy as well as current traffic. The main goal of „RER-SK" is not only to prolong the life time of each node but also to improve the life time of link, so as to maximize the network life time.

References

[1] Li, S.X.-Y., Wang, Y., Chen, H., Chu, X., Wu, Y., Qi, Y.: Reliable and Energy-Efficient Routing for Static Wireless Ad Hoc Net-works with Unreliable Links. IEEE Trans. Parallel and Distributed Systems 20(10), 1408–1421 (2009)

[2] Mohanoor, A.B., Radhakrishnan, S., Sarangan, V.: Online Energy Aware Routing in Wireless Networks. Ad Hoc Networks 7(5), 918–931 (2009)

[3] Vergados, D.J., Pantazis, N.A., Vergados, D.D.: Energy-Efficient Route Selection Strategies for Wireless Sensor Net-works. Mobile Networks and Applications 13(3-4), 285–296 (2008)

[4] Nagy, A., El-Kadi, A., Mikhail, M.: Swarm Congestion and Power Aware Routing Protocol for Manets. In: Proc. Sixth Ann. Comm. Networks and Services Research Conf. (May 2008)

[5] Li, X., Chen, H., Shu, Y., Chu, X., Wu, Y.-W.: Energy Efficient Routing with Unreliable Links in Wireless Networks. In: Proc. IEEE Int'l Conf. Mobile Adhoc and Sensor Systems (MASS 2006), pp. 160–169 (2006)

[6] Li, X.-Y., Wang, Y., Chen, H., Chu, X., Wu, Y., Qi, Y.: Reliable and Energy-Efficient Routing for Static Wireless Ad Hoc Net-works with Unreliable Links. IEEE Trans. Parallel and Distributed Systems 20(10), 1408–1421 (2009)

[7] Chang, J.-H., Tassiulas, L.: Maximum Lifetime Routing in Wireless Sensor Networks. IEEE/ACM Trans. Networking 12(4), 609–619 (2004)

[8] Kim, D., Luna Aceves, J.J.G., Obraczka, K., Carlos Cano, J., Manzoni, P.: Routing Mechanisms for Mobile Ad Hoc Networks Based on the Energy Drain Rate. IEEE Trans. Mobile Computing 2(2), 161–173 (2003)

[9] Gomez, J., Campbell, A.T., Naghshineh, M., Bisdikian, C.: PARO: Supporting Dynamic Power Controlled Routing in Wireless Ad Hoc Networks. Wireless Networks 9(5), 443–460 (2003)

[10] De Couto, D.S.J., Aguayo, D., Bicket, J., Morris, R.: A High-Throughput Path Metric for Multi-Hop Wireless Routing. In: Proc. ACM MobiCom, pp. 134–146 (2003)

[11] Misra, A., Banerjee, S.: MRPC: Maximizing Network Lifetime for Reliable Routing in Wireless Environments. In: Proc. IEEE Wireless Comm. and Networking Conf. (WCNC 2002), pp. 800–806 (2002)

[12] Banerjee, S., Misra, A.: Minimum Energy Paths for Reliable Communication in Multi-Hop Wireless Networks. In: Proc. ACM MobiHoc, pp. 146–156 (June 2002)

[13] Singh, S., Raghavendra, C.: PAMAS—Power Aware Multi-Access Protocol with Signalling for Ad Hoc Networks. ACM Computer Comm. Rev. 28, 5–26 (1999)

[14] Toh, C.: Maximum Battery Life Routing to Support Ubiquitous Mobile Computing in Wireless Ad Hoc Networks. IEEE Comm. Magazine 39(6), 138–147 (2001)

[15] Singh, S., Woo, M., Raghavendra, C.S.: Power-Aware Routing in Mobile Ad Hoc Networks. In: Proc. ACM MobiCom (October 1998)
[16] Vazifedan, J., Prasad, R.V., Niemegeers, L.: Energy- Efficient Reliable Routing Considering Residual Energy in wireless Ad Hoc Networks. IEEE Trans. 13(2), 434–447 (2014)
[17] Maleki, M., Dantu, K., Pedram, M.: Power-aware Source Routing Protocol for Mobile Ad HocNetworks. In: Proceedings of the 2002 International Symposium on Low Power Electronics and Design, ISLPED 2002 (2002)
[18] Shih, K.-P., Chang, C.-C., Chen, Y.-D.: MRPC: A Multi-Rate Supported Power Control MAC Protocol for Wireless Ad Hoc Networks. In: Wireless Communications and Networking Conference, WCNC 2009. IEEE (2009)
[19] Martello, S., Toth, P.: Knapsack problems, catalog.enu.kz (1990)

A Novel Approach for Enhancing the Security of User Authentication in VANET Using Biometrics

P. Remyakrishnan and C. Tripti

Department of Computer Science & Engineering
Rajagiri School of Engineering & Technology
Kochi, India
remyakrishnan.p@gmail.com, triptic@rajagiritech.ac.in

Abstract. Vehicular Ad Hoc Network (VANET) offers various services to users. Misusing such network could cause destructive consequences. A perfect user authentication scheme is necessary to secure the VANET system. Use of biometrics in authentication can overcome the limitations of existing random key based authentication techniques. A combination of face and finger print biometrics provide more accurate recognition of users. Here we propose, A novel approach for enhancing the security of User Authentication in VANETs based on biometrics. It concentrates on enhancing the security of Vehicle-to-Infrastructure (V2I) communication in VANET.

Keywords: Authentication, Layered Biometric Encryption, Lightweight, vehicular ad hoc networks (VANETs).

1 Introduction

VANET improves road safety and offer comfort for the travelers. Where, vehicles will be capable of storing and processing great amounts of information, including a drivers personal data and geo-location information, emergency messages etc. VANETs enable Vehicle-to-Vehicle (V2V) and Vehicle-to-Infrastructure (V2I) communication, information ex-change and authentication of Vehicles is significant for ensuring the security. The major components of a VANET are the wireless on-board unit (OBU), the roadside unit (RSU), and the authentication server (AS). OBUs provide wireless communication capability which will be installed with in the vehicles,whereas RSUs are the infrastructures present in the intersections to provide information access for the vehicles within its radio coverage. The AS is responsible for saving the secure parameters to authenticate the user.

The increasing mobility of vehicles has caused a high cost for societies as consequence of the increasing number of traffic congestion, fatalities and injuries. Hence, the users with high degree of mobility in vehicles demands a scheme for authenticating themselves and providing protection for their data.

S.C. Satapathy et al. (eds.), *Emerging ICT for Bridging the Future – Volume 2*,
Advances in Intelligent Systems and Computing 338, DOI: 10.1007/978-3-319-13731-5_33

An excellent survey on securing user authentication in vehicular ad hoc networks is presented in[1]. A detailed survey of VANETs Authentication is presented in[2].Ming-Chin Chuang et al.[3] proposed a decentralized lightweight authentication scheme called Trust-Extended Authentication Mechanism (TEAM) for vehicle-to-vehicle communication networks. The proposed system TEAM adopts the concept of transitive trust relationships to improve the performance of the authentication procedure and only needs a few storage spaces. The Chameleon Hashing for Secure and Privacy-Preserving Vehicular Communications reported by Song Guo, et al.[4] is a new privacy-preserving authentication protocol with Authority.Traceability using elliptic curve based chameleon hashing .An Infrastructure based Authentication in VANETs[5]. In [6], Biometrics-based Data Link Layer Anonymous Authentication in VANETs was proposed with provable link-layer location privacy preservation but make use of only a single biometric feature.

Most of the authentication techniques mentioned are based on cryptographic key generation schemes. Even though some biometric based authentication techniques have been proposed, they were based on a single biometric feature and a remarkable achievement has not been obtained till now. This motivated us to investigate the effect of combining face and finger print biometrics for the authentication of users in VANET.Another major challenge involved is protect the biometric data transferred. Problem is mainly due to enormously large number of on road vehicles, with varying characters of users and high similarity between the data transferred. If the biometric templates are in plaintext, it is easy to be compromised [7]. So, encryption must be performed in order to secure the users biometric templates.

In this paper, we propose a layered biometric approach which make use of the face and finger print images of the owners of the vehicle in order to ensure their identity. And also to secure the users biometric templates, Biometric Encryption is proposed.

This paper is structured as follows: Section 2 introduces the proposed system. Section 3 presents the experimental setup and analysis and finally, Section 4 concludes the paper.

2 Proposed System

The proposed system consists of mainly 4 stages: Registration, Authentication, Session Key generation and Secure communication. During the registration phase the user has to register his details with an Authentication Server. And during the Authentication phase, the user input will be compared against the saved details, and once it matches the user is marked as trusted and the Authentication server will proceed to Key generation phase, otherwise denies the access to that particular user. In the Key generation phase, initially a session key generated for each trusted user based on his biometrics. Once obtained the session key, if a user want to communicate with another trusted user in the range of that Authentication Server, he can make a request to the AS using the session key and once granted the AS will generate a one-time key to secure the communication between those two vehicles. Message transfer between these vehicles will be encrypted using this one-time key.

2.1 Registration

The vehicles in VANET has to perform a registration procedure with the AS.It can be done either through the manufacturer or a secure entity. This initial registration procedure is performed only once. During the registration, the user has to enroll his face and finger print images which will be saved along with vehicle details in the AS.

2.2 Authentication

Authentication usually happens in an on-road scenario. The User who wants to gain access to the VANET has to authenticate himself. The OBU performs the authentication procedure with the AS, For that the user has to enter biometric images which will be encrypted and send to the AS. The AS will decrypt and compare the images with the saved details corresponding to that user, and mark the user a Authorized or Unauthorized. Face recognition is done by using an existing approach called Eigen face approach which is one of the most successful and fast approaches for face recognition [8].And the fingerprint matching is carried out using a minutiae feature extraction method [9]. The steps in authentication procedure are:

1. User i enter his ID, Image(Xi) and fingerprint (Fi).
2. OBUi computes Ci = Xi XOR Fi.
3. OBUi sends the cipher text Ci and user id IDi to AS.
   After receiving the users ID and cipher text, the AS compares the ID with the saved details to obtain the corresponding fingerprint Fi.
4. AS computes Xi = Ci XOR Fi.
5. AS compares image Xi with the saved images of the user.
6. If a match is found the user is marked as Authorized otherwise Unauthorized

Once a user got authenticated, the AS will proceed to the key generation phase, if not any further communications from that user will be denied by the AS. Figure below shows the authentication procedure.

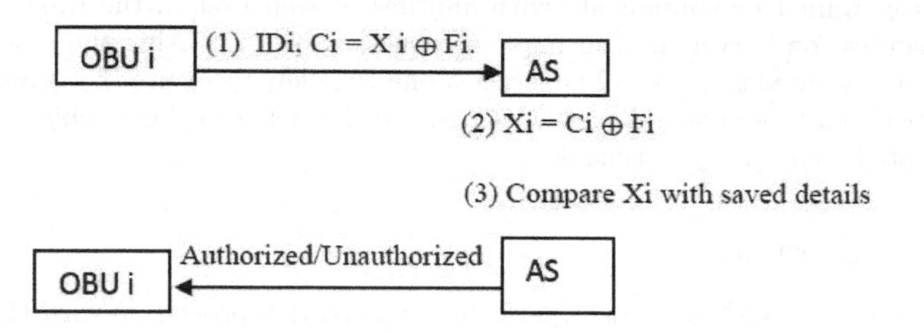

Fig. 1. Authentication Procedure

2.3 Key Generation

Once a user got authenticated the AS will generate a session key corresponding to that user based on his biometrics. This key is used to encrypt any further communication between that user and AS. In this process, if an entire finger print image is taken, it takes several blocks to keep in rounds to generate the session key. So, in order to avoid that, corresponding to each matched fingerprint image, a unique random value is generated and used in the first step of key generation only.The steps are:

1. AS generate a random number Pi and compute session key Ci as: Ai =h(Xi XOR Pi) , Bi = h(Ai) , Ci= h(Fi) XOR Bi.
   Here, Ci is the session key generated. Now, AS will send the session key Ci to OBUi by encrypting it with the fingerprint Fi, so that only the corresponding user will be able to decrypt it.

2. AS sends Di = Ci XOR Fi to OBUi
3. The OBUi computes Ci = Di XOR Fi, and obtains the session key Ci

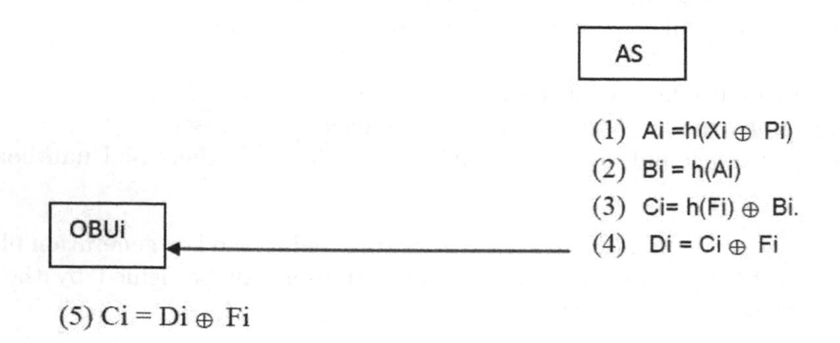

Fig. 2. Key Generation Procedure

2.4 Secure Communication

Once authenticated,the vehicles has to perform a secure communication procedure when they want to communicate with each other. The communication takes place through the AS. Suppose, the OBUi wants to communicate with OBUj ,it will send a request containing the ID of OBUj XORed with its session key Ci to the AS. Now, AS checks whether OBUj is authenticated and in its range. If yes, the AS will send a query containing the ID of OBUi XORed with its session key Cj to OBUj to know whether it is willing to communicate with OBUi. The OBUj will send OK message to AS if it is ready to communicate with OBUi, otherwise send a NOT OK message

1. OBUi sends REQi=IDj XOR Ci and IDi to AS
2. AS obtains Ci corresponding to IDi and computes IDj = REQi XOR Ci
3. AS send QUERYj= IDi XOR Cj to OBUj.
4. OBUj obtains IDi by computing QUERYj XOR Cj.
5. OBUj responds with an OK or NOT OK message to AS.
   Up on receiving an OK message from OBUj, AS will generate a one-time key Kij for communication between the OBUi and OBUj.The key Kij is computed by finding XOR of the session keys Ci or Cj alternatively with a random number Nij and send this key to both OBUs by encrypting it with their corresponding finger prints, so that an attacker will not be able to retrieve it. The steps are:
6. If OK, then AS computes the one-time key Kij = (Ci or Cj) XOR Nij
7. AS sends Kij XOR Fi to OBUi.
8. AS sends Kij XOR Fj to OBUj.
   Once obtained the key Kij, both vehicles can communicate with each other by XORing the messages send between them using Kij. Thus, the communication is secured.The steps are:

9. OBUi send M = MSG XOR Kij to OBUj
10. OBUj obtains the message by computing MSG= M XOR Kij.

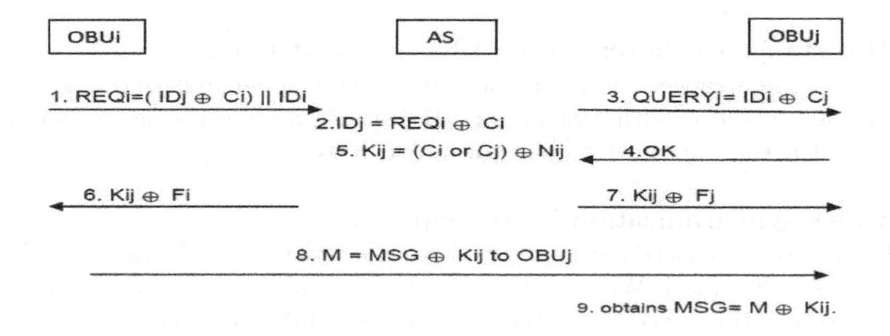

Fig. 3. Secure Communication Procedure

3 Simulation and Analysis

The simulation was done using Java J2SE.An on-road scenario was created with moving vehicles, Road side Units and Authentication Servers. The face and fingerprint image was taken in real-time and working of the algorithm was checked perfectly with varying the input data. The results obtained has been analyzed in terms of security, computational cost, and performance below.

3.1 Security Analysis

The proposed method satisfies the following security requirements:

- **Resistance to replay attacks:**
  The proposed scheme is protected from replay attacks, by using the biometrics to encrypt the authentication message.

- **Session key agreement:**
  The session key is generated by performing XOR of a random number with the finger print image so that it is difficult for the adversary to derive the session key from the intercepted messages.

- **Resistance to modification attack:**
  Use of hash function ensures that the information cannot be modified during transfer.

- **Fast error detection:**
  In the login procedures, the AS can easily detect the error when the attacker enters a wrong user ID or password.

- **Resistance to man-in-the-middle attack:**
  The password and the session key of the system are used to prevent the man-in-middle attack. The session key is encrypted with fingerprint of the user, hence the attacker cannot obtain this.

- **Resistance to chosen-cipher text attack (CCA):**
  Even if the attacker enters one or more known cipher texts into the system and obtain the resulting plaintexts. He cannot recover the secret key used, since it is the finger print image of intended user.

- **Clock synchronization is not required.**
  The timestamp based authentication scheme requires the clocks of all vehicles to be synchronized. Whereas the proposed technique does not require clock synchronization since the authentication mechanism is nonce based.

- **Use of XOR in Key Generation**
  Even though the XOR is reversible, its use in the key generation process will not weaken the security aspect of the algorithm even if the analyst is able to catch it. Because, the re-XOR has to be done with the finger print of the intended user which is not available to the analyst.

3.2 Analysis of Computational Cost

In order to analyze the computational cost, we use the following notations: n represents the number of OBUs in VANET.0 means there is no cost involved for computation in that phase; Chash denotes the cost involved in perfoming the hash operation; CXOR indicates cost of the XOR operation; and Cran is cost for generating the random number. The computational cost of the approach is shown in Table 1 below:

Table 1. Computational Cost of Proposed Scheme

Modules	OBUi	Authentication Server	OBUj
Initial Registration	0	0	0
Authentication	CXOR	CXOR	CXOR
Key Generation	CXOR	Cran + 3Chash +3CXOR	CXOR
Secure Communication	3 CXOR	4CXOR + Cran	3 CXOR

3.3 Performance Analysis

A performance analysis has been conducted by considering a set of 100 vehicles and authentication time required for them. The result obtained is shown in the graph below. The analysis can be checked on large scale of vehicles.

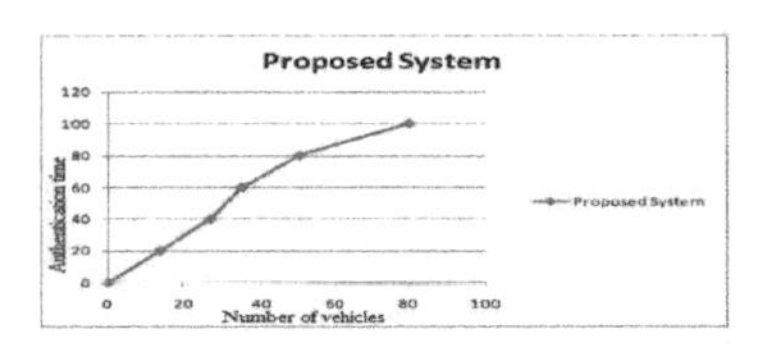

Fig. 4. Performance Analysis

4 Conclusion

In this paper, we have presented a secure method for User Authentication in VANETS using biometrics. The computational cost analysis shows that it is a lightweight scheme because the amount of cryptographic calculations under scheme fewer than in existing schemes. it only uses an XOR operation and a hash function. Moreover, it is based on the concept of biometric encryption, which ensures authentication convenience, as well as secures the biometric template. Security analysis proves that our scheme can resist multiple attacks. In the future works we concentrate on further enhancing the encryption algorithm proposed.

References

1. Dahiya, A., Sharma, V.: A survey on securing user authentication vehicular ad hoc networks. International Journal of Information Security 1, 164–171 (2001)
2. Remyakrishnan, P., Tripti, C.: A review of various Authentication techniques in VANET. International Journal of Advanced Research in Computer Science 5(4) (special issue, 2014)
3. Chuang, M.-C., Lee, J.-F.: TEAM: Trust-Extended Authentication Mechanism for Vehicular Ad Hoc Networks. IEEE Systems Journal 8(3), 749–758 (2013)
4. Guo, S., Zeng, D., Xiang, Y.: Chameleon Hashing for Secure and Privacy Preserving Vehicular Communications. IEEE Transactions on Parallel and Distributed Systems 99(1), 1 (2013)
5. Chaurasia, B.K., Verma, S.: Infrastructure based Authentication in VANETs. International Journal of Multimedia and Ubiquitous Engineering 6(2) (2011)
6. Yao, L., Lin, C., Dengy, J.: Biometrics-based Data Link Layer Anonymous Authentication in VANETs. In: Seventh International Conference on Innovative Mobile and Internet Services in Ubiquitous Computing (IMIS), pp. 182–187 (2013)
7. Anil, K.J., Karthik, N., Abhishek, N., et al.: Biometric template security. EURASIP Journal on Advances in Signal Processing 2008, Article No. 113 (2008)
8. Savvides, M., Vijaya Kumar, B.V.K., Khosla, P.K.: Eigenphases vs. Eigenfaces. In: Proceedings of the 17th International Conference on Pattern Recognition, vol. 3, pp. 810–813 (2004)
9. Jiang, X., Yau, W.Y.: Fingerprint minutiae matching based on the local and global structures. In: International Conference on Pattern Recognition, pp. 1038–1041 (2000)

Detection of Black Hole Attack Using Code Division Security Method

Syed Jalal Ahmad[1], V.S.K. Reddy[2], A. Damodaram[3], and P. Radha Krishna[4]

[1] J B Institute of Engineering & Technology, Hyderabad, India
Jalal0000@yahoo.com
[2] Malla Reddy College of Engineering & Technology, Hyderabad, India
vskreddy2003@gmail.com
[3] Jawaharlal Nehru Technological University, Hyderabad, India
damodarama@rediffmail.com
[4] Infosys Labs , Infosys Limited, Hyderabad, India
radhakrishna_p@infosys.com

Abstract. A mobile Adhoc network (MANET) is a collection of mobile nodes to form a temporary network without use of any predefined infrastructure. Direct communication is possible only when two nodes lie within their sensing range; otherwise communication is made through intermediate nodes till the destination is reached. Such type of networks can allow any node to join in the network or leave the network at any instant of time. So any node can act as a host or the router in the network, which results in security issues in MANETs. A well known attack in MANETs is a Black hole attack. In this paper, we present a simple but effective method called Code Division Security Method (CDSM) for security in order to prevent Black hole attack in MANETs. Black hole node is a malicious node which can mislead a normal node to forward the data through it and corrupt the data so that it can degrade the performance of the network. We validate our approach using network simulator with an example.

Keywords: MANET, security, Black hole, Authentication, Hop code.

1 Introduction

Mobile Ad-hoc Network (MANET) is a network of mobile nodes without any predefined infrastructure. If the two communicating nodes lie within the same sensing range they can directly communicate to each other, otherwise they can communicate through multiple hops where nodes act as intermediate routers. In this type of network, nodes can enter in the network or leave the network at any time without informing the other nodes in the network. Thus, security is a challenging task due to dynamic topology in such type of networks. In MANETs, there is no guarantee that a path from source to destination is free from malicious nodes. Due to lack of central coordination, there are several attacks in MANETs. Spoofing attack occurs when an attacker tries to phishing the node that exists within the route where data packets are transmitting [7]. In the Sybil attack [5], the attacker not only embodies the node but

S.C. Satapathy et al. (eds.), *Emerging ICT for Bridging the Future – Volume 2*,
Advances in Intelligent Systems and Computing 338, DOI: 10.1007/978-3-319-13731-5_34

also assumes the identity of several nodes and thus fails to find the redundancy of the number of protocol. Among all the attacks, a well known attack is a Black hole attack, which is created by a malicious node by sending a very quick reply with highest destination sequence number and shortest path. Black hole node can easily corrupt the information. To avoid such type of attacks, research community pays much attention towards the security of MANETs. The Black hole attack is addressed in the literature either by considering the energies of the nodes [2] or by giving the certification to all the nodes in the network [10]. The first approach does not provide an effective solution to the Black Hole attack as the malicious nodes can be available in the network with different energies. In the second approach, providing certifications to the nodes that are mobile is difficult.

In this paper, we propose a simple and effective method of security called ***Code Division Security Method*** (CDSM). For this purpose, we consider an additional field of one byte with the packet header to represent the code of source node. In our approach, the starting code of source node is '0' (i.e. 00000000). In the first hop, a decimal number '01' (i.e. 00000001) is added to the source code word. In the next hop, '02' (i.e. 00000010) is added to the first hop code word and this process continues till destination node. The presented method not only authenticates the nodes between source and destination, but also reduces the power as well as the time to authenticate the node.

The rest of the paper is organized as follows. Section 2 presents the related work. In section 3, we describe our proposed approach. Section 4 presents the simulation results. We conclude the paper in section 5.

2 Related Work

Mirchiardi and Molva [9] presented a scheme for the misbehavior detection and reaction of a node in Adhoc networks. However, this scheme is having a poor response when collision occurs in the route where data packets are transmitted. Also this approach is imperfect due to less transmission power. Sanjeev and Manpreet [10] described a two hop authenticated model that gives a secure communication between the end users. This approach requires more time and processing power, which leads to problem for multimedia applications. Hu et al [6] presented a secure on demand routing protocol that deals with the life time and the control messages. However, this scheme does not have any authentication for the intermediate nodes. So, malicious node can easily enter in the network and can take part in the route and creates interference in the routing process.

Sharma et al [11] described a solution to the Black hole attack by setting the waiting time of the source node to receive the repeat request from other nodes. The authors assume that the waiting time is exactly equal to half of the route reply (RREP). However, this may not be true for multi hop network when the two routes from source to destination have enough time difference to receive request (due to large propagation time and queuing delay of route 1 in comparison to route 2). Deng et al [3] presented a routing protocol in which every intermediate node requires to send a reply request message to the source node. However, this approach increases the routing overhead and also increases the delay from source to destination. Chandrakant [2] described an approach for protection to MANETs based on energy consumption of a node. This may not be always true when number of malicious nodes will enter with different energies in the network.

Lu et al [8] proposed a Secure Ad-hoc on-demand Distance Vector (SAODV) routing protocol to prevent Black hole attack in MANETs. This approach addresses only few of the security weaknesses of AODV, and thus Black hole attack cannot be removed completely. Deswal and Sing [4] proposed an enhanced version of SAODV protocol by giving password to each of the routing node. This approach may not be valid when new nodes enter in the network, which degrades the performance of the network. In this paper, we propose a model that improves the performance of the network by giving the security at every hop based on the source code.

3 Proposed Model

In this section, we present our proposed CDSM method to find the Black hole attack in MANETs. We assume that all active nodes in the network have the same behavior, so that all nodes know the hop operation as well as recognition property of the code at any hop. That is, our method does not require any certification and hence effective in MANETs due to its dynamic topology. In our approach, we consider an additional field of one byte with the packet header to represent the code of source node and hop count. When the source node is ready to communicate with destination node through intermediate nodes, it appends an eight bit data '00000000' in the packet header to indicate the starting code and hop count. Then, the source node sends the data along with this code word to all its neighbors. All neighbor nodes will receive the data, however only those nodes can access it which knows the one hop operation (i.e., how to operate the code at one hop). In this way, Black hole attack can easily be detected and data can be protected from such attacks. At the next hop, a decimal number '01' (i.e. 00000001) is added to the source code word. Similarly, '02' (i.e. 00000010) is

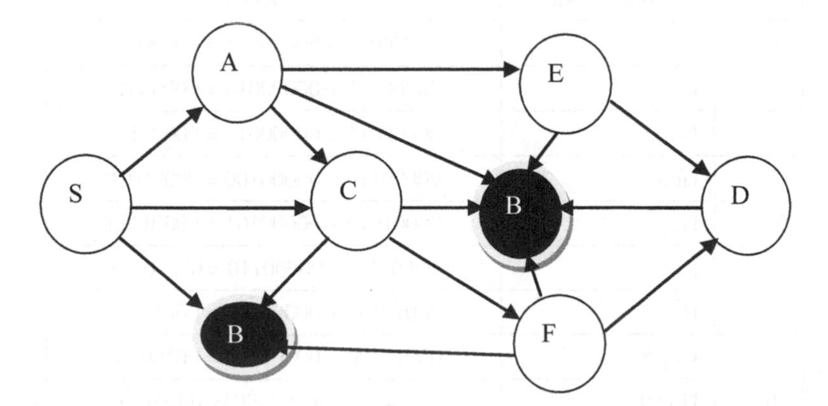

Fig. 1. Three hop MANET with Malicious Nodes

added to the next hop code word. This process continues till it reaches the destination. This approach not only improves the performance of Packet Delivery Ratio (PDR) but also saves the time and overhead. We used the approach described in [1] for route reservation in order to establish optimum route from source to destination.

Figure 1 shows the operation of three hop MANETs. In the figure, **B** represents Black Hole (Malicious Node), **S** represents source node, **A, C, E, F** represent intermediate nodes and **D** represents destination node. Source node sends the route request to all its neighbors, i.e., A, C and B with source code word (i.e. 00000000), all the nodes will receive the request. However, only node 'A' and node 'C' can process the data by adding the one hop code word with the source node code word (see Table 1). Malicious node B cannot process it because it cannot understand the one hop operation. So Black hole attack can be easily detected and data can be saved from such type of attacks.

In the next hop, node 'A' and node 'C' will forward the data along with one hop code to nodes 'E', 'B' and 'F'. Again node 'B' cannot access the data because it does not operate the code word as per the second hop operation. So nodes 'E' and 'F' can only access the data and process it towards the next anchoring node. This process continues till the data reaches the destination. Suppose any malicious node manages to know the hop count, still it cannot construct the code word at that hop stage as it does not know the operation of code word at that stage of hop count. So malicious node can be easily detected and thus it will not drop the packets.

Let the source code be represented by S, and i^{st}, 2nd , 3rd, etc. hop codes can be represented by H_1, H_2, H_3 ,etc. Here,

S = 00000000

H_1= 00000001

H_2= 00000010

H_3= 00000011 and so on till up to last hop.

Table 1. Code word of 10-hops

S. No	Hop Number	Code word
01	O (Source node)	00000000
02	Hop 1	00000000 + 00000001 = 00000001
03	Hop 2	00000001 + 00000010 = 00000011
04	Hop 3	00000011 + 00000011 = 00000110
05	Hop 4	00000110 + 00000100 = 00001010
06	Hop 5	00001010 + 00000101 = 00001111
07	Hop 6	00001111 +00000110 = 00010101
08	Hop 7	00010101 + 00000111 = 00011100
09	Hop 8	00011100 + 00001000 = 00100100
10	Hop 9	00100100 + 00001001= 00101101
11	Hop 10	00101101 + 00001011 = 00111000

The complete code word for the individual hop can be represented as

Source code word $= S$

First hop code word $= S + H_1 = H_{CW1}$

where H_{CW1} is the hop code at first hop and is equal to $S + H_1$.

Second hop code word $= H_{CW1} + H_2 = H_{CW2}$

where H_{CW2} is the hop code at second hop and is equal to $H_{CW1} + H_2$. Similarly, the 'Nth' hop code can be represented as,

$$H_{CWN} = H_{CW(N-1)} + H_N$$

where H_{CWN} is the hop code at Nth hop and H_N is the Nth hop code. So the code word at any hop can be calculated as,

$$\text{Hop code word } (H_{CW}) = \text{Source code word} + \sum_{i=1}^{N} i\,, \quad I = 1, 2, 3....N$$

where i is the hop count represented in terms of 8 bits as shown in Table 1.

4 Simulation Results

In this section, we describe the results of our approach and compare with the two existing approaches: Ad-hoc on-demand Distance Vector (AODV) and Chandrakant [3] approach by using NS-2 Simulator. Figure 2 shows the variation of Packet Delivery Ratio (PDR) versus the simulation time of the source- destination pair. It can

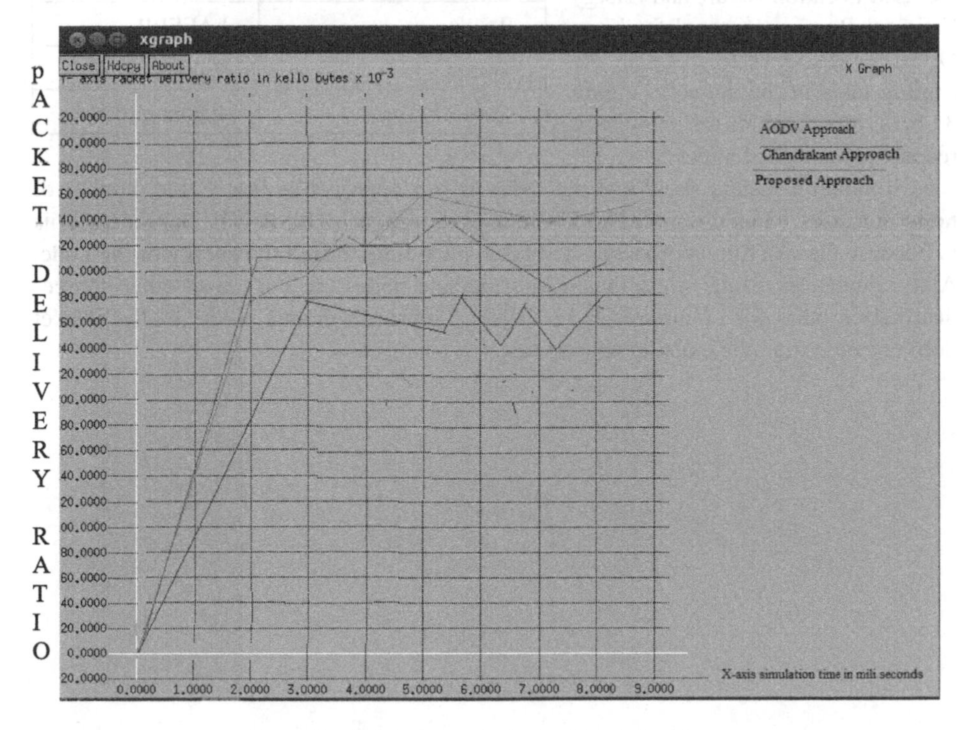

Fig. 2. Packet Delivery Ratio Vs Simulation time

be observed that if the source- destination pair is far away from each other (i.e., source to destination consists of multiple hops), then more malicious nodes can interact in between source and destination. However, our approach produces higher PDR values than other two approaches. Initially, the PDR of Chandrakant approach is better when compared to our approach. This is because the malicious nodes initially have very much energy difference with the actual nodes. After elapse of time, the malicious nodes enter in the network with different batteries (i.e., different energies). Hence energy of malicious node cannot be judged by Chandrakant approach, so it degrades the performance of packet delivery ratio in multi hop network.

Table 2 shows the simulation environment and the various parametric values considered. Total simulation time is 50 ms. Multi-hop network consists of maximum 200 nodes with random distribution in the network. Also for the simulation, we used Location Aware and Energy Efficient Routing Protocol for Long Distance MANETs to maintain the routing table in the network. Figure 3 shows the variation of delay with respect to number of nodes.

Table 2. Simulation parameters

Network Parameters	**Values**
Simulation Time	50 seconds
Number of nodes	2 to 200
Link Layer Type	Logical Link (LL)
MAC type	802.11
Radio Propagation Model	Two-Ray Ground
Queue Type	Drop-Tail
Antenna	Omni antenna
Routing	LAEERP
Traffic	Video
Network Area	1000m x 1000m

Simulation results show that the delay in our approach is less as the number of hops increases when compared to Chandrakant approach and AODV approach. This is because the existing approaches will take much time to find the energy of the node. Also simulation results show that if intermediate nodes are busy with other source-destination pairs for communication, still our approach maintains the higher Packet delivery ratio than the existing approaches.

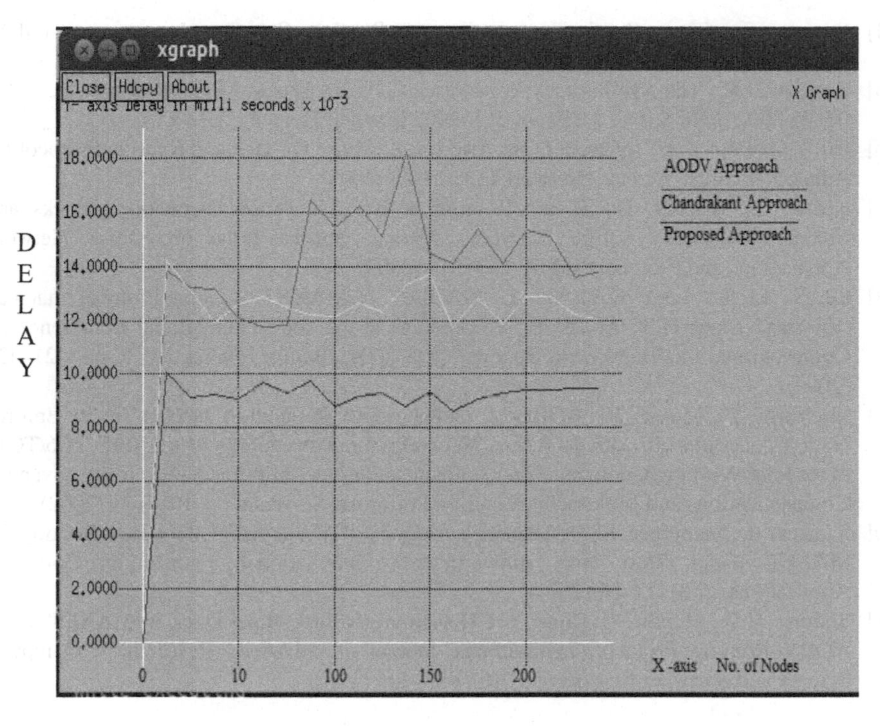

Fig. 3. Delay Vs Number of nodes

5 Conclusion

In this paper, we use a new concept of detecting Black hole attack in MANETs. The presented CDSM approach can easily detect the Black hole attack and also saves the energy and time by reducing the overhead. Our approach guarantees the security against the Black Hole attack in MANETs. Our method performs well when compared to AODV and Chandrakant's approaches. CDSM is very simple and effective in MANETs particularly when number of malicious nodes in the network is more and create number of Black hole attacks with different energies of the nodes.

References

[1] Ahmad, S.J., Reddy, V.S.K., Damodaram, A., Krishna, P.R.: Location Aware and Energy Efficient Routing Protocol for Long Distance MANETs. International Journal of Networking and Virtual Organizations (IJNVO), Inderscience 13(4), 327–350 (2013)

[2] Chandrakant, N.: Self Protecting Nodes for Secured Data Transmission in Energy Efficient MANETs. International Journal of Advanced Research in Computer Science and Software Engineering 3(6), 673–675 (2013)

[3] Deng, H., Li, W., Agrawal, D.P.: Routing Security in Wireless Adhoc Networks. IEEE Communication Magazine 40(10), 70–75 (2002)

[4] Deswal, S., Sing, S.: Implementation of Routing Security Aspects in AODV. International Journal of Computer Theory and Engineering 2(1), 135–138 (2010)

[5] Douceur, J.R.: The Sybil Attack. In: Druschel, P., Kaashoek, M.F., Rowstron, A. (eds.) IPTPS 2002. LNCS, vol. 2429, pp. 251–260. Springer, Heidelberg (2002)

[6] Hu, Y.C., Perrig, A., Johnson, D.B.: Aridane: A Secure On-Demand Routing Protocol for Adhoc Networks. Wireless Networks 11, 21–38 (2005)

[7] Karlof, C., Wagner, D.: Secure Routing in Wireless Sensor Networks: Attacks and Counter Measures. AdHoc Networks Journal, Special Issue on Sensor Network Applications and Protocols 1(2-3), 293–315 (2003)

[8] Lu, S., Li, L., Lem, K.Y., Jia, L.: SAODV: A MANET Routing Protocol that can Withstand Black Hole Attack. In: Proceedings of the 2009 International Conference on Computational Intelligence and Security (CIS 2009), Beging, China, vol. 2, pp. 421–425 (2009)

[9] Michiardi, P., Molva, R.: CORE: A Collaborative Reputation Mechanism to Enforce Node Cooperation in Mobile Adhoc Networks. In: Proceedings of the IFIP TC6/TC11 Sixth Joint Working Conference on Communications and Multimedia Security: Advanced Communications and Multimedia Security, Portorosz, Slovenia, pp. 107–121 (2002)

[10] Sanjeev, R., Manpreet, S.: Performance Analysis of Malicious Node Aware Routing for MANET using Two Hop Authentication. International Journal of Computer Applications 25(3), 17–24 (2011)

[11] Sharma, V.C., Gupta, A., Dimri, V.: Detection of Black Hole attack in MANET under AODV Routing Protocol. International Journal of Advanced Research in Computer Science and Software Engineering 3(6), 438–443 (2013)

Cluster Based Data Replication Technique Based on Mobility Prediction in Mobile Ad Hoc Networks

Mohammed Qayyum[1], Khaleel Ur Rahman Khan[2], and Mohammed Nazeer[3]

[1] Dept. of Computer Engg, King Khalid University, Abha, Saudi Arabia
mdqayyum.se@gmail.com
[2] Dept. of CSE, ACE Engineering College, Hyderabad, India
khaleelrkhan@aceec.ac.in
[3] Dept. of CSE, Muffakham Jha College of Engineering and Technologhy, Hyderabad, India
nazeer857@yahoo.co.in

Abstract. The mobile database system in a MANET is a dynamic distributed database system, which is composed of some mobile MHs. The key issues in MANETs for mobile database are: How to optimize mobile queries, cache and replicate data, manage transactions and routing. In this proposal, we wish to take the problems of data replication in solving the mobile database issues. Replication of data in a MANET environment focuses on to improve reliability and availability of data to the mobile clients (node). There are many issues revolving around replication of data in such a scenario like power, server and node mobility, networking partition and frequent disconnection. We propose a cluster based data replication technique for replication of data and to overcome the issues related to node mobility or disconnection problem in MANET environment. Our approach has two phases; initial phase consists of formation of cluster and cluster head and in the second phase, the distributions of data (replicated data) to the respective cluster head. By NS2 simulation, we will show that our proposed technique attains better data consistency and accuracy with reduced delay and overhead.

1 Introduction

1.1 Mobile Ad-Hoc Network (MANET)

A mobile ad hoc network (MANET) is a compilation of autonomous, mobile, wireless devices which forms a communications network even in the absence of fixed infrastructure. The primary goal of MANET network designers is to provide a self-protecting, "dynamic, self-forming, and self-healing network" for nodes on the move. [1] Each MANET node may move arbitrary and dynamically there by connected to form network depending on their positions and transmission range and as well as can act as a self router. The topology of the ad hoc network depends on the transmission power of the nodes and the location of the Mobile Nodes, which may change with time [2].

S.C. Satapathy et al. (eds.), *Emerging ICT for Bridging the Future – Volume 2,*
Advances in Intelligent Systems and Computing 338, DOI: 10.1007/978-3-319-13731-5_35

The feature of Ad hoc networks has an added advantage with respect to quick deployment and easy reconfiguration, which makes these system an ideal in situations where installing an infrastructure is too expensive or too susceptible.[5] MANETs have applicability in several areas like [4];

- Soldiers relaying information for situational awareness on the battlefield.
- Emergency disaster relief personnel coordinating efforts after a fire, hurricane, or earthquake.
- Personal area and home networking
- Business associates sharing information during a meeting.
- Attendees using laptop computers to participate in an interactive conference.
- Location-based services and sensor networks.

The three major drawback related to the quality of service in MANET are bandwidth constraints, dynamic topology of MANET and the limited processing and storing capacity of mobile nodes.[3]

1.2 Mobile Databases in MANET

The mobile database system in a MANET is a dynamic distributed database system, which is composed of some Mobile Heads (MHs). Each MH comprises of local database system. [8]. In storing capacity of mobile nodes, Transaction Manager (TM) of a mobile multi database management system is accountable for providing dependable and steady units of computing to all its users. [6]

Nodes in a MANET can be classified by there capabilities that is as a client or a server.

1. A Client or *Small Mobile Host (SMH)* is a node with reduced processing, storage, communication and power resources.
2. A Server or *Large Mobile Host (LMH)* is a node having a larger share of resources.

Servers, due to their larger capacity contain the complete DBMS and abide primary responsibility for data transmission and satisfying client queries. Clients typically have enough resources to cache portions of the database as well as some DBMS query and processing modules. [7]

There are three layers in a mobile distributed database system; [8]

- **The application layer** – In this layer the user queries are accepted.
- **The middleware layer** – In this layer, Queries are processed and transmitted to the middleware of other MHs in the network. The middleware of a MH sends queries to the local database system. After the database finishes executing a query, the results are transmitted from the middleware layer to the application layer and then the results are returned to the user.
- **The database layer** – In this layer all the information and data are stored.

The middleware layer is the core of the mobile distributed database system and it is divided into three sub-layers [8];

- **The network layer** – It manages location information of nodes, divides nodes into groups, routs data packets between the query layer, and the cache layer.
- **The cache layer** – It stores the data which are accessed frequently by the query nodes or their neighbors.
- **The query layer** – It parses the syntax of user queries and determines the query types

1.3 Issues of Mobile Databases in MANET

The key issues in MANETs for mobile database are: How to optimize mobile queries, cache and replicate data, manage transactions and routing. [8] Issues related to mobile database in MANET are as; [7,9,10,11]

- **Power** - All mobile devices in the MANET are battery powered. In traditional mobile networks, only the power needs of the clients are considered. But in the present scenario's, the power of the server, which provides DBMS data services, is perhaps more important as it provides DBMS services to potentially many clients.
- **Mobility of the nodes** - Due to the dynamic nature of a MANET, it exhibits frequent and unpredictable topology changes. The MANET not only operates within the ad-hoc network, but may also require access to a public fixed network. MANETs therefore should be able to adapt the traffic and propagation conditions to the mobility patterns of the nodes.
- **Resource availability** – A node should supply mechanisms for proficient use of processing, memory and communication resources, while maintaining low power consumption. A node should bring about its basic operations without resources exhaustion.
- **Response Time** - Regardless of the method of communication used, access time and tuning time must be considered. *Tuning time* is the measure of the amount of time each node spends in Active Mode. This is the time of maximum power consumption for a client. *Access time* measures the sensitivity of the algorithm. It refers to the amount of time a client must wait to receive an answer to a database query.
- **Quality of Service** – Nodes become disengaged for a variety of reasons. This may be due to location or lack of power, dynamic nature and redundancy. The accuracy of information stored at each node: server and client are alike. When portions of the network become separated for a time, data accuracy may become impossible.
- **Data Broadcast** - The size and contents of a broadcast have an effect on power consumption and the frequency of data queries. If the broadcast is too outsized, unnecessary information may be broadcast. If too little information, then wrong information is broadcasted. Thus increasing the on-demand requests. Also if several servers attempt to broadcast simultaneously, there will be a collision and the broadcast of all will be jumbled.

The issues mentioned are some of the major issues which we come across in the mobile database in MANET. Based on the above issues, it is necessary to provide a solution for the mobile database management. Below are some recent literature works which throw a light on the problems and its major solutions.

2 Related Work

Jin-Woo Song et.al. [12] have proposed a lucrative replica server allocation algorithms which induces a present algorithm with more careful analysis of the moving patterns of mobile device users. Here existing four algorithms are studied and customized. First a modified "vertex occurrence counts (VOC) -neighbor reduction" algorithm is introduced for lowering the vertex occurrence counts of the neighbors of the cells selected in VOC. A similar algorithm, called the greedy set-cover algorithm, which designate, replicated servers so that they are not allocated to the cells adjacent to each other. The replicated server clustering algorithm exploits the k-means clustering. The algorithm returns a set of clusters; the center of each cluster becomes the location for a replicated server.

Anita Vallur et.al [13] have proposed a data replication technique called REALM (REplication of data for A Logical group based MANET database) for logically group based MANET real-time databases. REALM, groups mobile owner based on the data items which they need to access. Mobile hosts that access the same set of data items are grouped into the same logical group. Group membership of mobile owner helps in identifying the data items that any mobile owner in the network will need to access, as well as to identify the most frequently accessed data item on every server. REALM tries to increase the percentage of successful transactions in real-time database of MANET.

Deniz Altınbuken and Oznur Ozkasap [14] have come upon with an comprehensive SCALAR (Scalable Data Lookup and Reactive Replication) framework for updated data, named as SCALARUP. It presents a replication service for updated data items in a fully distributed approach. Here the data are updated randomly by the owner of a data item and a new write frequency value is transmitted to the system. When a node acquires a new write frequency value for a data item, it restores its write frequencies table and invalidates old replicas of the updated item. During this restoration process the message overhead is altered only in terms of the write frequency broadcast messages. This dynamic backbone construction algorithm used in this paper, minimizes the time required to search and retrieve data for replication.

Thomas Plagemann et.al. [15] have proposed an approach for reliability management in shared data spaces for emergency and rescue operations. The work shows that the application specific requirement tries to solve the problem of consistency management for optimistic replication in Sparse MANETs. The problem of consistency management during replica synchronization is labeled by showing the need of data deleted at the crisis and rescue environment.

Prasanna Padmanabhan and Le Gruenwald [16] have proposed a data replication technique called DREAM for real time mobile ad-hoc network database systems. It focuses on data convenience while addressing the issue of power restriction by replicating hot data items before cold data items at the servers that have high remaining power. It handles the real-time transaction issue by giving a superior priority for replicating data items that are accessed frequently by firm transactions than those accessed frequently by soft transactions. It addresses disconnection and network partitioning by introducing new data and transaction types and by determining the stability of wireless links connecting servers. The remaining energy of connecting servers is also used to measure their link stability.

Matiasko K and Zabovskı M [17] have proposed an algorithm for dynamic re-allocation of data with a mobile computers incorporated in replication schema. The replication schema used here consists of two steps; Step 1: Test of expansion, implementation is done after the particular number of transactions to develop replication schema expansion when changes improve solution. Step 1: Test of contraction solves the problem with wired nodes included in the replication schema. Thc motivation is given by assumption that is easier to prevent site failure due to the communication network problem than to solve its failure.

Hao Yu et.al [18] have proposed a cluster-based optimistic replication management system for large-scale MANETs named as Distributed Hash Table Replication (DHTR) It uses a dispersed hash table technique and a dispersed replica information directory to enhance the efficiency of update propagation. Distributed clustering offers a method of maintaining hierarchical structures in ad-hoc networks to simplify the system control and decrease overhead messages and Distributed hash table technology is normally applied in peer-to-peer networking environments to help the uscr locatc thc rcsourccs quickly.

3 Proposed Scheme

Replication of data in a MANET environment focuses on to enhance dependability and availability of data to the mobile clients (node). There are many issues revolve around replication of data in such a scenario like power, server and node mobility, networking partition and frequent disconnection [19].We propose an approach for replication of data and to overcome the issues related to node mobility or disconnection problem in MANET environment.

Our approach has two phases; initial phase consists of formation of cluster and cluster head and in the second phase, the distributions of data (replicated data) to the respective cluster head.

3.1 Initial phase – Creating "Basket Node"

In the initial phase, nodes are clustered based on two factors; received signal strength of nodes and distance from past movements. The node which is most stable is elected as the cluster head. These cluster head act ultimately as "basket nodes" (those cluster heads which posses the replicated data) and share with the member nodes (according to the queries). We will discuss about the calculation of received signal strength of nodes and distance from past movements. We compare the two factors and select the best stable node as cluster head.

3.1.1 Received Signal Strength of Nodes

In the Received signal strength [20], we calculate the most stable node with respect to the signal strength received. In the signal strength scheme, each node communicate with there one hop neighbour with sending/receiving an "alive" message. Each node calculates the pairwise relative mobility metrics (RM) by reception of two successive "alive" message. The pairwise relative mobility metrics is calculated by;

$$\text{RM (a,b)} = 10\log_{10}\frac{Rg(b)\times P}{Rg(b)\times P} \quad (1)$$

Where, RM(a,b) denotes the relative mobility metric of node "a" with respect to node "b". $\frac{Rg(b)\times P}{Rg(b)\times P}$ is the ratio of the new and old power level product of received signal strength detected at the receiving node "a". Before sending the next broadcast packet to its neighbors, a node computes the aggregate relative mobility metric (RMAGG), which is calculated by the expected value of the squares of the relative mobility samples from neighbors given by;

$$\text{RMAGG (a)} = \text{E}\,[\,\text{RM(a, i)}^2\,] \quad (2)$$

The mobility of the nodes can be calculated by ratio of the new message and old message. If the power of new message is less than old message, then RM value will be negative, which indicates that the two nodes are moving away with respect to each other. On the other hand, if the power of new message is more than old message, then RM value will be positive, which indicates that the two nodes are moving away with respect to each other.

3.1.2 Distance from Past Movements

The distance from past movements is calculated by "Average Movement (AM) and Range time (RT)" [21] equation. Average Movement represents the average mobility of a resource and/or user based on user and resource mobility. It is calculated based on two recent communications between user/initiator and resource with respect to the user/initiator. AM is calculated by;

$$\text{AM} = \left|\,\text{old location - new location - user average movement}\,\right| \quad (3)$$

The location provides the distance between user and resource, where user average movement is the location of users past location history. This provide us with the Range time (RT), which is given by

$$\mathrm{RT} = \frac{(User\ Range - Dis\ \tan ce)}{Average\ Movement} \tag{4}$$

3.1.3 Combined Scheme

The initial period of all nodes is in the un-clustered random form. During a specific time period every node broadcasts two successive "alive" messages. During its reception, each node calculates the RMAGG and RT. These factors are stored in the neighbor table of each neighbor along with a time-out period (TO) set. A node receives the aggregate mobility values from its neighboring nodes, and then compares its own mobility value with those of its neighbors. If a node has the lowest value of RMAGG as well as, high RT amongst all its neighbors it assumes the status of a Cluster Head. The range leads to the formation of cluster boundaries. If a node is a neighbor of two cluster heads, then it becomes a "gateway" node.

If two neighboring nodes in an un-clustered state have the same value of RMAGG and RT, we resort to comparison of IDs and follow the Lowest-ID algorithm. That is the mobility metric of two cluster-head nodes is the same, and they are in contention for retaining the Cluster Head status, then the selection of the cluster-head is based on the Lowest-ID algorithm wherein the node with the lowest ID gets the status of the Cluster Head. In a mobile scenario, if a node with Cluster Member status with a low mobility moves into the range of another Cluster Head node with higher mobility, re-clustering is not triggered, but a Cluster conflict (CC) period is allowed for incidental contacts between passing nodes. If the nodes are in transmission range of each other even after the CC period has expired, re-clustering is triggered, and the node with the lower mobility metric assumes the status of Cluster Head.

These cluster head act ultimately as "basket nodes" which posses the replicated data and share with the member nodes according to the queries. We assume an offline server to collect queries from the nodes, after the cluster formation. The server maintains two tabular columns regarding node location and cluster details. During a query request, the node sends request message to the server. The message comprises of the query id (QID), the query (QRY), the node id (NID) and the cluster head (CH). After analyzing the query, the server cross checks both the table and locate the node and its cluster head position. After the node and cluster head positioning the server distributes the respective data according to the queries.

QID	QRY	NID	CH

Fig. 1. Query Request

3.2 Second Phase – Distribution of Data to Basket Nodes

In the second phase, the distribution of data to the "basket node" from the main server is done. This distribution (replication data's) depends on the queries from the nodes. According to the queries, the replicated data's are sent to the nearest cluster heads from which uninterrupted data services can be maintained.

Replication of data into a single cluster head increases the overhead of the cluster head as well as the power consumptions. In our approach, the offline server acts as a centralized distributor. The server intakes the queries and based on the clusters formed, the server can distribute the replicated data to nearest cluster heads according to the queries.

We also propose a distributed data replication, where the nearer cluster heads can distribute the data if the queries are similar. For example, when members of two nearer clusters (A and B) access data of similar queries, the server distributes the replicated data proportionally for both the cluster heads A and B. When cluster member of A seeks data which is stored in B, the cluster head of A seeks for the information through cluster head of B. In such a scenario, the overhead on cluster heads decreases due to the distribution of replicated data.

Also in our approach, we allocate a time period to each cluster head to analyze the consistency of data on cluster heads. If a cluster head is idle for allocated time period, then the corresponding replicated data is deleted. Otherwise it is updated according to the queries.

Algorithm

We describe our algorithm in the following step;

1. Creating "Basket Node"

1.1 Node sends successive "alive" message to form cluster.

1.2 Nodes are clustered based on two factors; received signal strength of nodes and distance from past movements.

1.3 With successive messages, nodes calculates RMAGG and RT

1.4 With the values of RMAGG and RT, Cluster head is elected. (if RMAGG is low and RT is high)

1.5 The elected cluster head maintains its members detail and transmit it to the server.

1.6 The server maintains table for node location and cluster details.

2. Distribution of data to basket nodes (refer figure 2)

2.1 The nodes send the query request to the server.

2.2 From the table in server, location of the nearest cluster head is determined. The data's related to the query details are sent to the cluster head.

2.3 If similar data are sent to nearer cluster heads, data can be distributed proportionally between the two cluster heads.

2.4 The query data are sent to the respective cluster member through cluster heads.

2.5 Queried data remain in the cluster head for the certain data time and are deleted if idle for allocated time period

Thus our approach tries to predict the node mobility to replicate the data, by determining the received signal strength of nodes and distance from past movements of nodes. We also reduce the cluster head overhead by timely refreshing or deleting the nodes and by distributing the replicated data proportionally to nearer cluster heads. Distribution of replicated data among the nearer cluster heads reduce the cluster head consumption of memory and power.

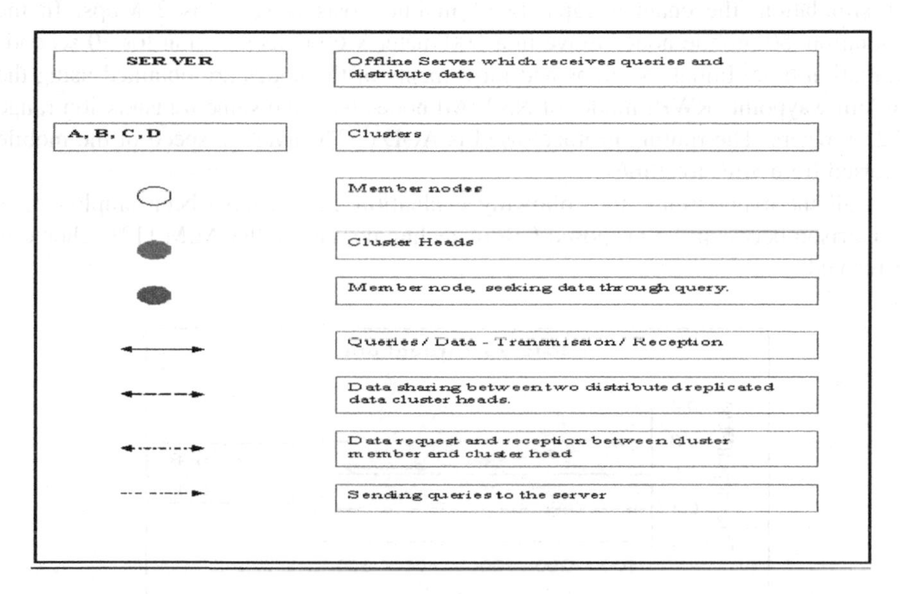

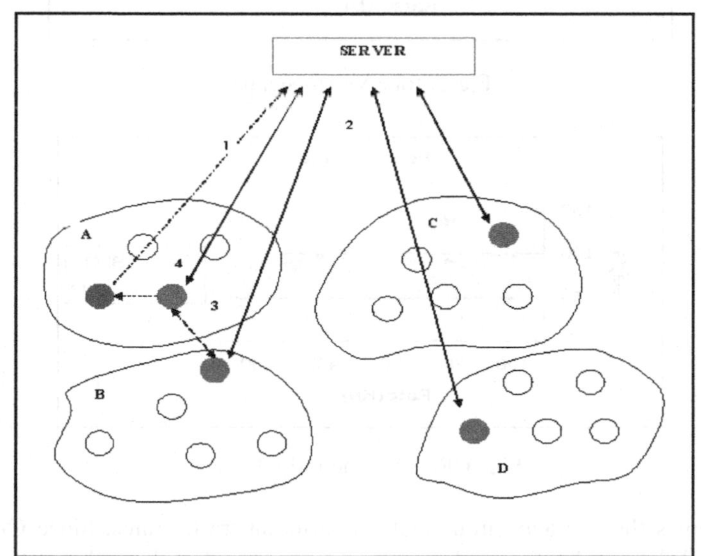

Fig. 2. Data Distribution

4 Simulation Results

4.1 Simulation Setup

This section deals with the experimental performance evaluation of our algorithms through simulations. In order to test our protocol, The NS2 simulation [22] is used. In our simulation, the channel capacity of mobile hosts is set as 2 Mbps. In the simulation, 50 mobile nodes move in a 600 meter x 600 meter region for 50 seconds simulation time. Initial locations and movements of the nodes are obtained using the random waypoint (RWP) model of NS2. All nodes have the same transmission range of 250 meters. The routing protocol used is AODV. The average speed of the mobile is varied from 5m/s to 20m/s.

In all the experiments, the following evaluation criteria have been employed. A comparison between the proposed CBDR technique and the REALM [13] scheme is performed.

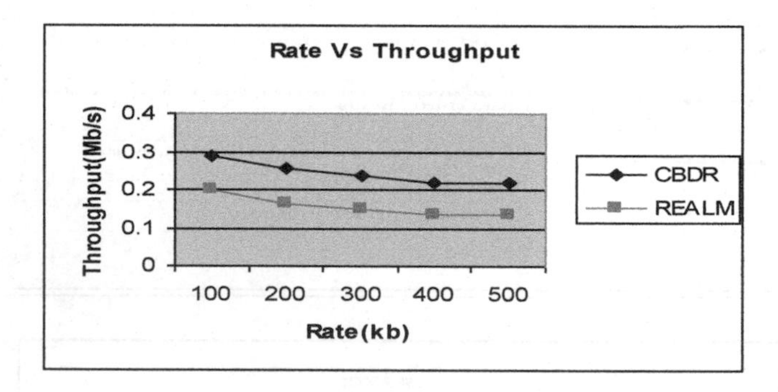

Fig. 3. Rate Vs Throughput

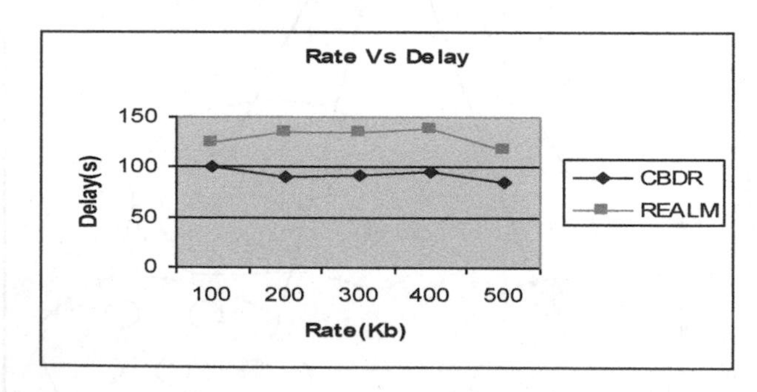

Fig. 4. Rate Vs End-to-End Delay

Fig 3 shows the average throughput for different traffic rates. Since CBDR uses cluster based data replication technique, throughput will be high. From the figure it can be observed that CBDR has higher throughput, when compared with REALM. Fig 4 shows average end-to-end delay for different traffic rates. Since the queries are processed by the resource efficient cluster heads, the end-to-end delay will be significantly less. From figure it can be observed that CBDR has less delay, when compared with REALM.

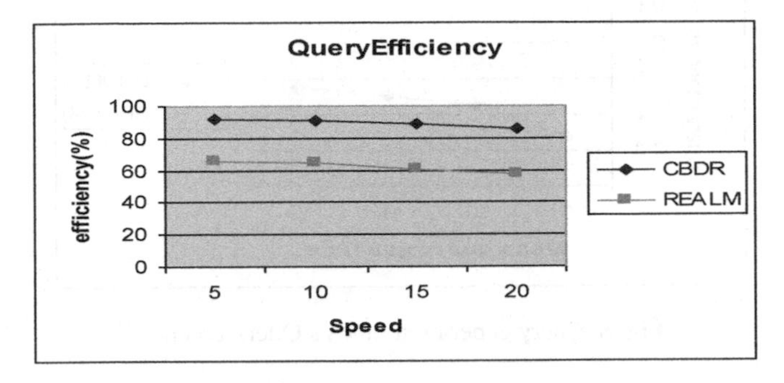

Fig. 5. Mobile Speed Vs Query Efficiency

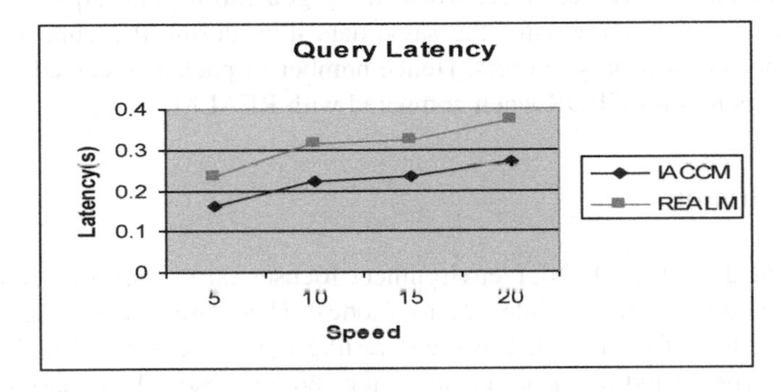

Fig. 6. Mobile Speed Vs End-to-End Delay

Fig 5 and 6 show the average query efficiency and latency for varying the mobile speed. As the mobile speed increases, it results in network disconnection or portioning, thereby reducing the query efficiency and increasing the latency. Since CBDR uses the cluster based data replication technique, the efficiency is high with low latency when compared with REALM.

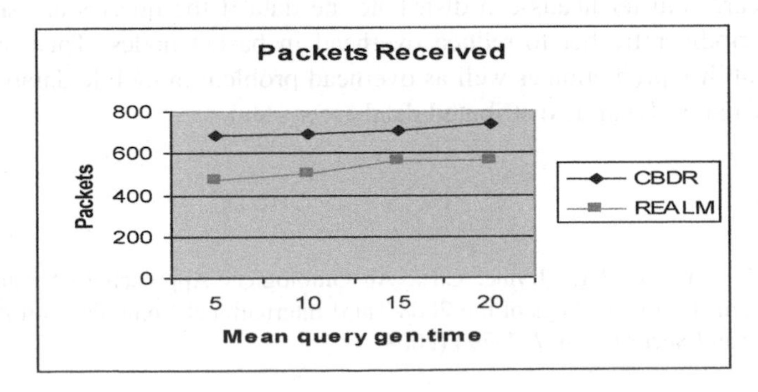

Fig. 7. Query generation timeVs Packets Received

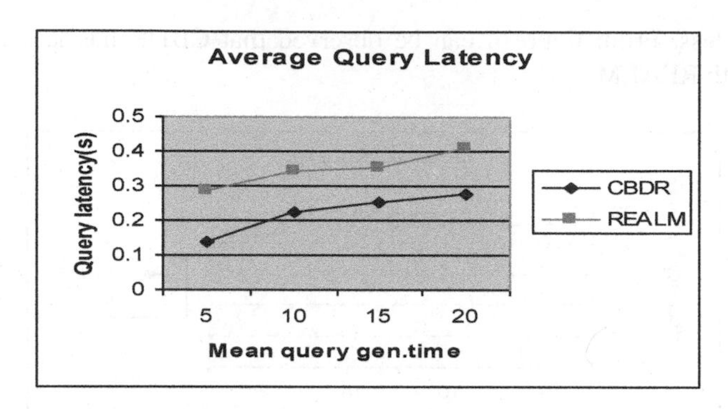

Fig. 8. Query generation timeVs Query Latency

Fig. 7 and Fig 8 show the results for the number of packets received and the average query latency, respectively, when query generation time *T*q is varied. Note that if several mobiles request for the same data item during the same interval, the data are replicated in nearby clusters. Hence number of packets received is more and query latency is less for CBDR when compared with REALM.

5 Conclusion

Replication of data in a MANET environment focuses on to improve reliability and availability of data to the mobile clients (node). There are many issues revolving around replication of data in such a scenario like power, server and node mobility, networking partition and frequent disconnection. We propose a dual phase scheme for replication of data and to overcome the issues related to node mobility or disconnection problem in MANET environment. The initial phase is used to form a cluster and a stable cluster head known as "basket nodes" using the two factors; received signal strength of nodes and distance from past movements. The offline server in our scheme acts as a centralized distributor which intakes the queries and based on the clusters formed will distribute the replicated data to nearest cluster heads according to the queries. Our scheme also proposes a distributed replicated method in which the nearer cluster heads can distribute the data if the queries are similar. We provide a periodic refresher to reduce overhead in basket nodes. Thus our scheme solves the mobility prediction as well as overhead problem in mobile database system in a MANET under dynamic distributed database system.

References

1. Orwat, M.E., Levin, T.E., Irvine, C.E.: An Ontological Approach to Secure MANET Management. In: Proceedings of the 2008 Third International Conference on Availability, Reliability and Security, pp. 787–794 (2008)

2. Saad, M.I.M., Ahmadnn, Z.: Performance Analysis of Random-Based Mobility Models in MANET Routing Protocol. European Journal of Scientific Research 32(4), 444–454 (2009)
3. Uma, M., Padmavathi, G.: A comparative study and performance evaluation of reactive quality of service routing protocols in Mobile Adhoc networks. Journal of Theoretical and Applied Information Technology 6(2) (2009)
4. Wu, B., Chen, J., Wu, J., Cardei, M.: A Survey on Attacks and Countermeasures in Mobile Ad Hoc Networks. In: Xiao, Y., Shen, X., Du, D.-Z. (eds.) Wireless/Mobile Network Security. Springer (2006)
5. Huang, Y., Jin, B., Cao, J., Sun, G., Feng, Y.: A Selective Push Algorithm for Cooperative Cache Consistency Maintenance over MANETs. In: Kuo, T.-W., Sha, E., Guo, M., Yang, L.T., Shao, Z. (eds.) EUC 2007. LNCS, vol. 4808, pp. 650–660. Springer, Heidelberg (2007)
6. Gruenwald, L., Banik, S.M.: A Power-Aware Technique to Manage Real-Time Database Transactions in Mobile Ad-Hoc Networks. In: Proceedings of the 12th International Workshop on Database and Expert Systems Applications, pp. 570–574 (2001)
7. Fife, L.D., Gruenwald, L.: Research Issues for Data Communication in Mobile Ad-Hoc Network Database Systems. ACM SIGMOD Record 32(2), 42–47 (2003)
8. Li, J., Li, Y., Thai, M.T., Li, J.: Data Caching and Query Processing in MANETs. JPCC 1(3), 169–178 (2005)
9. Hadim, S., Al-Jaroodi, J., Mohamed, N.: Middleware Issues and Approaches for Mobile Ad hoc Networks. In: Proceeding of IEEE Consumer Communications and Networking Conference (CCNC 2006), Las Vegas, Nevada (January 2006)
10. Artail, H., Safa, H., Pierre, S.: Database Caching in MANETs Based on Separation of Queries and Responses. In: Proceeding of WiMob 2005, Montreal, Canada (August 2005)
11. Pushpalatha, M., Venkatraman, R., Ramarao, T.: An Approach to Design an Efficient Data Replication Algorithm in Mobile Ad hoc Networks. IJRTE 1(2) (May 2009)
12. Song, J.-W., Lee, K.-J., Kim, T.-H., Yang, S.-B.: Effective Replicated Server Allocation Algorithms in Mobile computing Systems. IJWMN 1(2) (November 2009)
13. Vallur, A., Gruenwald, L., Hunter, N.: REALM: Replication of Data for a Logical Group Based MANET Database. In: Bhowmick, S.S., Küng, J., Wagner, R. (eds.) DEXA 2008. LNCS, vol. 5181, pp. 172–185. Springer, Heidelberg (2008)
14. Altınbuken, D., Ozkasap, O.: ScalarUp: Scalable Data Lookup and Replication Framework for Updated Data. In: Proceeding of 5th International Advanced Technologies Symposium (IATS 2009), Karabuk, Turkey, May 13-15 (2009)
15. Plagemann, T., Munthe-Kaas, E., Goebel, V.: Reconsidering Consistency Management in Shared Data Spaces for Emergency and Rescue Applications. In: Model Management und Metadaten-Verwaltung (BTW-MDM 2007), Aachen (March 2007)
16. Padmanabhan, P., Gruenwald, L.: Managing Data Replication in Mobile Ad-Hoc Network Databases. In: International Conference on Collaborative Computing: Networking, Applications and Worksharing (November 2006)
17. Matiasko, K., Zabovskı, M.: Adaptive replication algorithm for mobile computers. Acta Electrotechnica et Informatica 8(2), 18–21 (2008)
18. Yu, H., Martin, P., Hassanein, H.: Cluster-based Replication for Large-scale Mobile Ad-hoc Networks. In: Proceedings of the Seventh ACM International Workshop on Data Engineering for Wireless and Mobile Access International Workshop on Data Engineering for Wireless and Mobile Access, Vancouver, Canada, pp. 39–46 (2008)

19. Padmanabhan, P., Gruenwald, L., Vallur, A., Atiquzzaman, M.: A survey of data replication techniques for mobile ad hoc network databases. The International Journal on Very Large Data Bases 17(5), 1143–1164 (2008)
20. Basu, P., Khan, N., Little, T.D.C.: A Mobility Based Metric for Clustering in Mobile Ad Hoc Networks. In: Proceedings of the 21st ICDCS, April 16-19, p. 413 (2001)
21. Farooq, U., Mahfooz, S., Khalil, W.: An Efficient Resource Prediction Model for Mobile Grid Environments. In: Proceeding of 1st Annual Post Graduate Symposium on the Convergence of Telecommunication Networking and Broadcasting, IEE/EPSRC. Liverpool John Moores University, UK, June 26-27 (2006)
22. Network Simulator, http://www.isi.edu/nsnam/ns

Mitigating FRI-Attack Using Virtual Co-ordinate System in Multi-hop Wireless Mesh Networks

P. Subhash[1] and S. Ramachandram[2]

[1] Department of CSE, JITS, Karimnagar, India
subhash.parimalla@gmail.com
[2] Department of CSE, UCE, Osmania University, India
schandram@osmania.ac.in

Abstract. Routing in Wireless Mesh Network has become an active field of research in recent days. Most of the existing protocols are vulnerable to many attacks. One such approach that causes serious impact on a WMN is a Fraudulent Routing Information attack known as FRI –Attack, with which an external attacker can drop all the data packets by using a single compromised mesh node. In this paper, we present a Virtual Coordinates based solution to mitigate FRI attack in WMN. Virtual Coordinate System uses the topological structure of a network to get virtual coordinates rather than getting real coordinates using GPS. The proposed mechanism is designed for an on demand routing protocols such as HWMP and it relies on digital signature to mitigate FRI attack in WMNs during route discovery. This is a software based solution and it does not require each node to be equipped with specialized hardware like GPS.

Keywords: Wireless Mesh Networks, FRI Attack, Virtual Coordinate System, Digital signature, HWMP.

1 Introduction

Wireless Mesh Networks (WMNs) have become an emerging technology to meet challenges of next –generation networks and offers cost-effective solutions to the service providers. WMN is a poplar replacement technology for last-mile connectivity to the home and community networking.

A typical WMNs consist of mesh routers and mesh clients. In WMNs, mesh routers are static (or having minimum mobility) and mesh clients are either static or highly mobile. Mesh routers can form mesh backbone network that can be connected to the internet via mesh gateway routers. WMN has many advantages such as low-setup cost, extended coverage and also offer flexible and reliable services, meeting all these requirements and providing security is one of the big challenging task because the open nature of the wireless medium itself is susceptible to various types of attacks [4]. The numerous application scenarios [1] of WMNs include home networking, community networks, enterprise networks, backhaul support for cellular networks, etc. WMNs can be classified into three groups based on their architecture and design:

S.C. Satapathy et al. (eds.), *Emerging ICT for Bridging the Future – Volume 2*,
Advances in Intelligent Systems and Computing 338, DOI: 10.1007/978-3-319-13731-5_36

One-tier mesh networks, Two-tier mesh networks and Hybrid wireless mesh networks. One-tier mesh networks are No-Infrastructure mesh networks, Two-tier mesh networks are infrastructure mesh networks consist of mesh backbone routers, which can perform the routing functionality and communicate with clients, finally, Hybrid mesh networks offer functionalities of both one-tier and two-tier mesh networks. Fig.1 shows a typical architecture of wireless mesh networks.

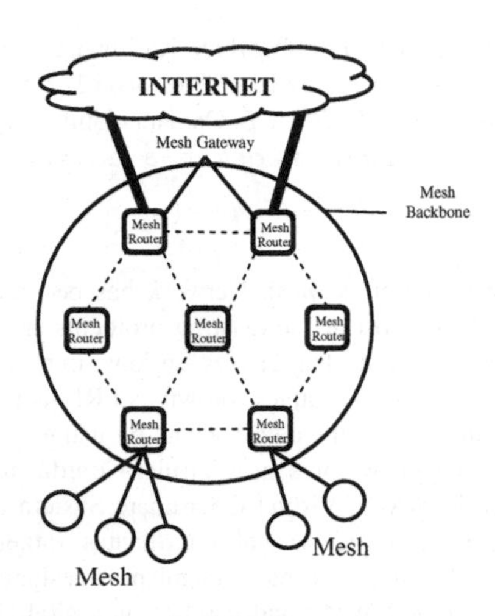

Fig. 1. A typical WMN architecture

Several classes of variant secure routing techniques in the field of WMNs have been investigated. Secure routing is considered to be one of the most challenging issues in WMNs as they are suffering from various types of vulnerabilities. The proposed solution is based on virtual coordinate system to defend against FRI-Attack in WMNs for on-demand hop by hop routing protocols like HWMP (Hybrid Wireless Mesh Routing) protocol. The HWMP is the default routing protocol for path selection in WMNs according to IEEE 802.11s [9] standard to provide interoperability between devices of different vendors. HWMP works on layer 2 with Mac addresses and uses airtime metric [18] for the path selection.

In this paper, we present a solution to defend against FRI-Attack based on Virtual Co-ordinate System (VCS) as these co-ordinates are used for measuring hop distance between two nodes in the network. The rest of the paper is arranged as follows. In Section 2, the related works are shown. The FRI-Attack with example network is discussed in section 3. The Virtual co-ordinate system is presented in section 4. The network and Threat model are presented in section 5. In section 6, the proposed method to defend against FRI-Attack is discussed, discussion on proposed solution is addressed in section 7 and section 8 concludes the paper.

2 Related Work

Many routing protocols have been designed for WMNs to achieve high throughput, all the nodes in the network are assumed to be honest. In many scenarios, this assumption does not hold. Cryptographic solutions is not enough to address various kinds of packet dropping attacks. Keeping security in mind various trust based routing protocols have been investigated [5], [6], [7], [8]. Unfortunately, all are assumed to operate in promiscuous mode to build a trust model. The malicious behavior of the compromised node in FRI-Attack is not detected in promiscuous mode as this attack is holding the capability of the attacker to hide its behavior.

The nature of FRI-Attack is similar to Node clone attack, in which a single node is compromised to access the inside information and an adversary replicate compromised nodes and deploy them in the network to launch a variety of attacks. Since a cloned node has legitimate information, it may involve in the network as non-compromised node.

Few works have been addressed node clone attacks [13], [15], [16], [17] for wireless sensor networks to improve their performance by keeping the limitations of WSNs in mind. To the best of our knowledge, no security frameworks have been proposed to address FRI-Attack in multi-hop wireless mesh networks.

3 FRI-Attack Description

In this section, we address FRI-Attack (Fraudulent Routing Information–Attack) a newly identified attack [3], with which an external adversary node can disturb data communication by using a single compromised mesh node (Mesh Router). The FRI – Attack scenario is shown in Fig. 2, an adversary places an external attacker node X to launch FRI-Attack in conjunction with a single compromised node C. The main aim of an attacker is to launch various types of packet dropping attacks such as greyhole and blackhole. As expressed in Fig. 2, mesh router X (External Node) extends the neighborhood of C to enhance the range of packet dropping attacks. The FRI-Attack on routing in wireless mesh networks is to be considered as main threat against neighborhood discovery protocol.

As expressed in Fig. 2, source node S initiates route discovery to target node T by broadcasting a PREQ packet, initializing metric and hop count values to 0. Up on receiving a PREQ message, an intermediate node updates these fields accordingly and rebroadcasts the PREQ message. Node C (Compromised node) also broadcasts the PREQ to its neighbors as well as to node X via an out-of-band channel. Up on receiving a PREQ from node C, node X broadcasts it to its neighbor nodes J and M. Node J and node M accept this PREQ packet as it seems to be legitimate (Being X impersonated as C). Nodes J and M again rebroadcast the PREQ towards target node T after updating the metric.

The target node T generates PREP (Path Reply) and unicasts along the way that offer best metric. In this example, a path via node M offers better metric. Thus, target node T unicasts a PREP message to node M and further forwarded to node X (Being impersonated as C) by node M. Up on receiving a PREP, node X relays it to node C and further it is transported to source node S via node A. In this way, the compromised node C involves in the selected path to get full control over the data communication between source and destination. In the next stage, it launches a DoS attack without being detected.

Most of the existing security frameworks to defend against packet dropping attacks depends on the observation of forwarding behavior of nodes in promiscuous mode. In this scenario, node C (compromised node) forwards all the data packets addressing to the next node (M). Unfortunately, node M is not the immediate neighbor of node C and therefore no other neighbors receives it. The malicious behavior of a compromised node C is not being detected even in the promiscuous mode as it forwards all the data packets.

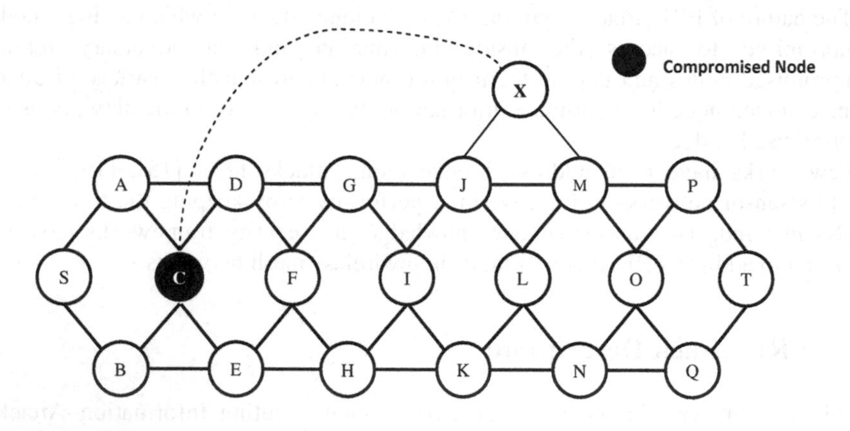

Fig. 2. FRI-Attack Scenario

4 Virtual Coordinate System

Virtual Coordinate System (VCS) uses the topological structure of a network to get virtual coordinates or logical coordinates rather than getting real coordinates using GPS. Virtual Coordinates were introduced recently for geographic routing without the use of Global Positioning System (GPS).Geographic routing does not follow the explicit route discovery, it only requires knowing the position of neighbors and the destination. Many routing protocols have been proposed for geographic routing using virtual coordinate systems [2], [10], [11], [12], [14].

4.1 Hop ID System

Hop -ID system [2] constructs a multidimensional virtual co-ordinates and that are used to calculate the Hop-ID distance between a pair of nodes by using distance function. In Hop-ID vector, each dimension is the hop distance from the node to a pre-chosen landmark nodes. The main idea of this Hop-ID is for each node to maintain hop count to a number of landmark nodes. This Hop-ID vector is the logical coordinates of a node. The Hop-ID distance between two nodes M1 and M2 in the network can be defined as L_h. Assuming there are n landmark nodes. The Hop-ID of node M1 is $H^{(1)}$ i.e. ($H_1^{(1)}$, $H_2^{(1)}$, - - - - ,$H_n^{(1)}$) and the Hop-ID of node M2 is $H^{(2)}$ i.e. ($H_1^{(2)}$, $H_2^{(2)}$, - - - - ,$H_n^{(2)}$), the following triangulation inequality holds:

$$\underset{i}{Max}\left(\left|H_i^{(1)} - H_i^{(2)}\right|\right) \leq L_h \leq \underset{i}{Min}\left(\left|H_i^{(1)} + H_i^{(2)}\right|\right) \quad (1)$$

For each i from 1 to n, L_h is no more than the sum of $H_i^{(1)}$ and $H_i^{(2)}$ since there exist a path from M1 to M2 via landmark i and the total hop count of this path is $H_i^{(1)}$ + $H_i^{(2)}$. One of the key concepts is a calculation of hop distance between two nodes having Hop-IDs. Power distance D_p [2] is used to obtain hop distance between two nodes.

$$D_p = \sqrt[p]{\sum_{i=1}^{n} \mid H_i^{(1)} - H_i^{(2)} \mid^p} \quad (2)$$

Where each i is a landmark node and n is no. of landmark nodes.

5 Network and Threat Model

5.1 Network Model

We consider a typical Wireless Mesh Network architecture consisting mesh routers and mesh clients, where mesh routers can form a wireless mesh backbone among which few mesh routers are designed with an additional gateway functionality to be connected to the internet. We assume all the communication links are bi-directional.

5.2 Threat Model

We assume that an adversary can compromise a single mesh router in the wireless mesh backbone network and can obtain all its security credentials to make an external node as legitimate node in the network. An out of band channel is established between a compromised mesh router and external mesh router as it is used to bypass all the traffic in the network from one location to another.

6 Proposed Solution to Mitigate FRI Attack

In this section, we propose an efficient mechanism to mitigate FRI attack by observing large discrepancies in hop distance between Hop IDs of neighboring nodes presented in the PREQ packet during the route discovery process. Hop distance of a node to all the pre-selected landmark nodes are combined into a vector, assuming every node in the network maintains neighborhood relations with all its 2-hop neighbors. Landmark or Reference nodes are chosen randomly in the example network which are used by each node to obtain virtual coordinates through a virtual coordinate establishment mechanism. The landmark selection algorithms were presented in [2], an user can choose a suitable landmark selection algorithm based on their applications. In the example network , the node D virtual co-ordinates are (1,2,3) as it is 1-hop away from landmark node A, 2-hops way from landmark node H and 3-hops away from landmark node M. In on demand routing, route discovery phase is initiated, whenever a source mesh node (or mesh router) S wants to discover a path to a new target node (T) or when the lifetime of an existing path to a Target node has expired. During route discovery, a source node broadcasts a PREQ packet, which in turn rebroadcasted by the neighbor nodes, this operation will iterate until the sought path is found. Up on receiving a PREQ, an intermediate node update PREQ packet accordingly. The PREQ packet is modified to include the address and Hop ID values of the pre-cursor mesh node that the PREQ has traversed.

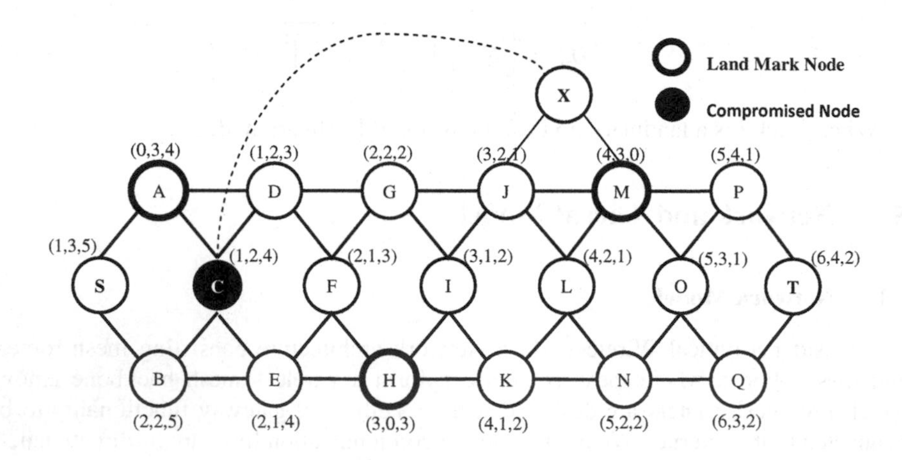

Fig. 3. An Example Network with Virtual Coordinates

For example in Fig. 3, when node C receives PREQ from node A, this is directly bypassed to an external attacker node X using an out of band channel, There the node X will rebroadcast the received PREQ, the node J and node M receives it. Up on receiving PREQ, Node M calculates the Hop ID distance between two pairs of nodes i.e (node M, node C as node X is impersonated as C) and (node C, node A) as Hop ID values of node C and node A is taken from receiving PREQ. If the hop distance value of any of these pairs is abnormal (not an accepted distance), then the PREQ received from node C is not further forwarded and simply discarded by node M. Thus, avoiding a routing path having malicious nodes from being chosen during the route discovery. The Hop distance is calculated using power distance equation (D_p). The large discrepancies in the hop distance in either of node pairs inform node M that the received PREQ has been skipped many hops. The Node C could not able to modify the Hop ID value of its pre-cursor node A since the Hop ID value of each node is digitally signed. If node X modifies the Hop ID value of node C, then the hop distance between node C and node A results in to abnormal

7 Discussion

The proposed solution to mitigate FRI-Attack uses the large discrepancies in hop distance between Hop-IDs of neighboring nodes presented in PREQ packet during the route discovery phase. This mechanism works efficiently to mitigate the impact of FRI-Attack as the power distance D_p is used to accurately measure hop distance between two Hop-IDs in wireless mesh networks. Hop-ID system [2] constructs a multidimensional virtual co-ordinates or logical co-ordinates rather than getting real co-ordinates using GPS.

8 Conclusion and Future Work

In this paper, we propose a Virtual Coordinates based solution to mitigate FRI attack impact on HWMP path selection protocol in Wireless Mesh Networks. Virtual Coordinate System (VCS) uses the topological structure of a network to get virtual coordinates or logical coordinates rather than getting real coordinates using GPS. Virtual coordinate systems were proposed as an alternative method for localization. We use large discrepancies in hop distance between virtual coordinates of two neighbor nodes to defend against FRI-Attack. In future, this can be simulated to evaluate the performance of the protocol under various factors. The virtual coordinate establishment and landmark selection techniques are need to be further investigated extensively.

References

1. Akyildiz, I.F., Wang, X., Wang, W.: Wireless mesh networks: A survey. Computer Networks and ISDN Systems (2005)
2. Zhao, Y., Chen, Y., Li, B., Zhang, Q.: Hop ID: A virtual coordinate based routing for sparse mobile ad hoc networks. IEEE Transactions on Mobile Computing 6, 1075–1089 (2007)
3. Matam, R., Tripathy, S.: FRI-attack: Fraduelent Routing Information- Attack on wireless mesh network. In: Proceedings of the 1st International Conference on Wireless Technologies for Humanitarian Relief. ACM (2011)
4. Zhang, W., Wang, Z., Das, S.K., Hassan, M.: Security Issues in Wireless Mesh Networks. In: Wireless Mesh Networks: Architectures and Protocols. Springer (2008)
5. Ghosh, T., Pissinou, N., Makki, K.: Collaborative Trust Based Secure Routing against colluding malicious nodes in Multihop Ad Hoc Networks. In: Proc. of 29th Annual IEEE International Conference on Local Computer Networks, LCN 2004 (2004)
6. Yau, P., Mitchell, C.: 2HARP: A Secure Routing Protocol for Mobile Ad Hoc Networks. In: Proc. of the 5th World Wireless Congress, WWC 2004 (2004)
7. Oliviero, F., Romano, S.: A reputation-based metric for secure routing in wireless mesh networks. In: Proc. of IEEE GLOBECOM 2008, New Orleans, LA (December 2008)
8. Paris, S., Rotaru, C.N., Martignon, F., Capone, A.: EFW: A Cross-Layer Metric for Reliable Routing in Wireless Mesh Networks with Selfish Participants. In: Proc. of IEEE INFOCOMM, Shanghai, China (2001)

9. IEEE P802.11s/D5.0 Part 11: Wireless LAN Medium Access Control (MAC) and Physical Layer(PHY)Specifications, Amendment 10: Mesh Networking
10. Caruso, A., et al.: GPS free coordinate assignment and routing in wireless sensor networks. In: Proceedings of the IEEE 24th Annual Joint Conference of the IEEE Computer and Communications Societies, INFOCOM 2005, vol. 1. IEEE (2005)
11. Awad, A., German, R., Dressler, F.: Exploiting virtual coordinates for improved routing performance in sensor networks. IEEE Transactions on Mobile Computing 10(9), 1214–1226 (2011)
12. Huang, P., Wang, C., Xiao, L.: Improving end-to-end routing performance of greedy forwarding in sensor networks. IEEE Transactions on Parallel and Distributed Systems 23(3), 556–563 (2012)
13. Choi, H., Zhu, S., La Porta, T.F.: SET: Detecting node clones in sensor networks. In: Third International Conference on Security and Privacy in Communications Networks and the Workshops, SecureComm 2007. IEEE (2007)
14. Tsai, C.-H., Chang, K.-T., Tsai, C.-H., Wang, Y.-H.: A Routing Mechanism Using Virtual Coordination Anchor Node Apply to Wireless Sensor Networks. In: Park, J.J.(J.H.), Arabnia, H.R., Kim, C., Shi, W., Gil, J.-M. (eds.) GPC 2013. LNCS, vol. 7861, pp. 546–555. Springer, Heidelberg (2013)
15. Xing, K., et al.: Real-time detection of clone attacks in wireless sensor networks. In: The 28th International Conference on Distributed Computing Systems, ICDCS 2008. IEEE (2008)
16. Conti, M., et al.: Distributed detection of clone attacks in wireless sensor networks. IEEE Transactions on Dependable and Secure Computing 8(5), 685–698 (2011)
17. Li, Z., Gong, G.: On the node clone detection in wireless sensor networks. IEEE/ACM Transactions on Networking (TON) 21(6), 1799–1811 (2013)
18. Bahr, M.: Proposed routing for IEEE 802.11 s WLAN mesh networks. In: Proceedings of the 2nd Annual International Workshop on Wireless Internet. ACM (2006)

Cluster Oriented ID Based Multi-signature Scheme for Traffic Congestion Warning in Vehicular Ad Hoc Networks

Bevish Jinila[1] and Komathy[2]

[1] Faculty of Computing,
Sathyabama University, Chennai, India – 600119
bevishjinila.it@sathyabamauniversity.ac.in
[2] School of Computing Sciences, Hindustan University, Chennai, India – 603103
gomes1960@gmail.com

Abstract. To report safety messages like traffic congestion warning in a Vehicular Ad hoc Network (VANET) vehicles communicate with each other and with the fixed Road Side Units (RSU). To ensure the messages communicated are true, message authentication plays a vital role. Since the number of incoming messages received by the RSU grows exponentially with time, delay in authentication increases. Existing batch and priority based verification schemes addresses issues like re-verification when there is a single false message. This leads to delay in authentication. In this paper, a cluster oriented ID based multi-signature scheme is proposed to overcome the delay in authentication. In this scheme, the network is clustered and each cluster holds a cluster head which is responsible for generating a multi-signature. Experimental analysis shows that this scheme incurs less delay in authentication, communication overhead and loss ratio when compared to existing approaches.

Keywords: Authentication, Multi-signature, Cluster, Vehicular Ad hoc Networks.

1 Introduction

Safety and comfort on travel has become an important concern in human life. Wireless innovations help public to access internet services everywhere on travel. The present day vehicles are equipped with certain wireless devices known as On Board Unit (OBU) that enable communication between vehicles and the RSUs deployed for every one kilometer and thereby organizing a network known as Vehicular Ad hoc Network (VANET). Such feature helps the vehicles and RSU to report events like traffic congestion warning to other vehicles and thereby making them to take an alternate path to avoid congestion. Such events when not addressed immediately can lead to a great havoc in the network.

Since the network is subject to varied attacks, it is mandatory to verify the authenticity of the messages received and take necessary action. When traffic

S.C. Satapathy et al. (eds.), *Emerging ICT for Bridging the Future – Volume 2*,
Advances in Intelligent Systems and Computing 338, DOI: 10.1007/978-3-319-13731-5_37

congestion is reported, all the vehicles involved in the event will report this message to the nearby RSU for every 100 to 300 milliseconds. The number of incoming messages in the RSU goes high when the number of vehicles reporting the event increases. This leads to an increased authentication overhead and delay.

Existing approaches introduced pseudonymous public key certificates to generate and verify the signatures where each message send from a particular vehicle was verified individually. To overcome the delay, batch verification was introduced by Zhang et.al [1] where incoming messages are grouped into batches and each batch is verified for its authenticity. Since the applications in VANET are time critical applications, this scheme lead to the loss of some important messages thereby degrading the efficiency of the network. In priority based verification proposed by Subir et.al.[2] each message is assigned a priority and verification of the incoming messages is done based on the priority of the messages. This scheme of prioritization leads to the drop of more incoming messages leading to the degradation of the network.

In this paper, we propose a cluster oriented ID based multi-signature scheme to reduce the authentication delay and signature size. This scheme is cluster oriented where all the vehicles travelling in the communication range of an RSU are grouped together into a single cluster and a cluster head is elected and this cluster head generates a multi signature representing all the other vehicles which is ready to report the same event. This scheme eventually reduces the delay in authentication and the signature size.

The rest of the paper is organized as follows. Section 2 describes the related work of signature generation and verification. Section 3 details about the system model and the preliminaries. Section 4 describes the proposed scheme of cluster oriented ID based multi-signature scheme. Section 5 analyses the performance of the proposed scheme. Section 6 concludes the work and proposes the future ideas to enhance the work.

2 Related Work

Problems on security and privacy enhancing mechanisms in a Vehicular Ad hoc Network plays a vital role. Recently many schemes are proposed to enhance the security and privacy of communication in this network. The IEEE 1609.2 standard for VANET [3] has included the public key algorithm Elliptic Curve Digital Signature Algorithm (ECDSA) for authentication. Since the processing time is more, during heavy traffic conditions this scheme leads to increased packet loss.

For applications like traffic congestion warning, where the density of the traffic is very high it is better to verify the messages by the nearby RSU than verifying by the nearby vehicle, since OBU cannot verify all the incoming messages within a short span of time. Lin et. al.[4] proposed a centralized group signature protocol which combines the features of ID based signatures. In this scheme, the time for signature generation grows linearly based on the number of revoked vehicles and this increases the delay in authentication. To overcome this limitation, Lei et. al.[5] proposed a decentralized group signature protocol. This scheme doesn't rely on a tamper proof device and the RSUs are responsible to maintain on the fly groups in its communication range. Since the RSU is completely involved for group formation, a single point of failure may cause a great havoc in the network. And, since the group manager is given the authority to verify the signatures there is a possibility that the manager can be compromised. In [6], a privacy preserved authentication scheme is proposed. In both the categories, sender generates the signatures and receiver verifies it.

In [7],[8]the authors proposed an RSU aided message authentication scheme where RSUs are used to verify the authenticity of all the incoming messages and to reply back. In this scheme, incoming messages are buffered by the vehicle in its local database and once when the signed packet is received from the RSU, the messages are verified. Since the incoming messages are buffered and verified, this increases the delay in authentication.

To reduce the delay in authentication, Zhang et. al [1] proposed a scheme where all the incoming messages are verified in batch by the RSU. This reduces the overhead of the RSU to a greater extent but, this verification fails for Denial of Service (DoS) attacks. To overcome this limitation, Subir et.al.[2] has proposed a priority based verification scheme where messages are assigned priorities and first the verification is done for the high priority messages. This may lead to the drop of more number of medium and low prioritized messages which too can lead to a great havoc in the network. To summarize as in [9], these limitations can be overcome by a multi-signature method.

3 System Model and Preliminaries

3.1 System Model

In this section, we formalize the network and the adversarial model to define the problem.

a) Network Model

The VANET in the metropolitan area is considered. It consists of many number of vehicles, RSUs deployed at various locations which supports the communication and TAs are assumed to be centralized. Fig 1 shows the network model considered for the proposed scheme.

- *Trusted Authorities*: The centralized TA is responsible for vehicle registration. During vehicle registration process, TA is responsible for generating and issuing secret passwords for the vehicle user. In addition, it is responsible for storing all the information regarding the vehicle user in its database for further reference.
- *Traffic Management System (TMS)*:These management systems are distributed across three different zones namely south, north and central and are responsible for any occurrence of events in its zone.

- *Vehicles*: In order to use the applications of the network, each vehicle should be a registered user. All registered vehicles are equipped with an OBU which can communicate with other vehicles and RSU.
- *Road Side Units*: These are deployed across various locations and are responsible for communicating with the vehicles and other RSUs.

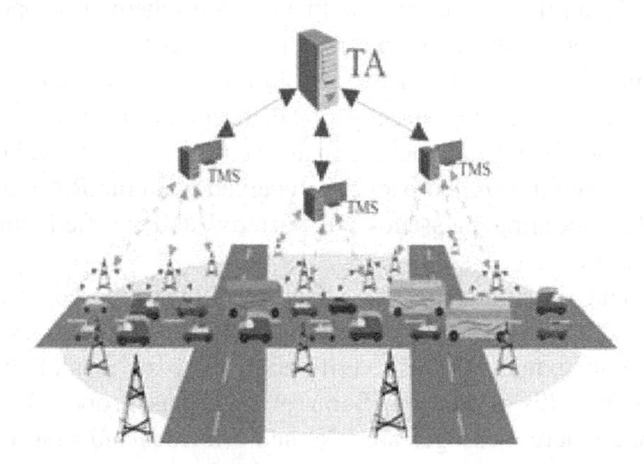

Fig. 1. Network Model

b) Attack Model

The communication channel in VANET is insecure and is susceptible to various attacks. The safety applications liketraffic congestion warning are susceptible to the attacks listed below.

Forging Attack:
The messages communicated during regular intervals or for reporting an event can be forged by an adversary. There is also a possibility where a valid signature is generated even if the message is altered.

False Signature Attack:
In case of batch verification, even if one of the signatures is a false signature there is a possibility for the drop of the entire batch. This attack is difficult to be traced where if there is any problem encountered verification has to be done for the entire batch again which consumes a lot of time and resource.

3.2 Assumptions Made

- The task of traceability is distributed to the Traffic Management Server (TMS) distributed across various zones (North, South, and Central).
- During registration, the TA generates passwords for driver authentication and issues it to the vehicle user.
- Cluster formation and cluster head selection is based on rough a fuzzy set which is out of the scope of the paper.

4 Cluster Oriented ID Based Multi-signature Scheme

This section briefly describes the proposed cluster oriented ID based multi-signature scheme. Initially, the network is grouped into clusters. For a perfect formation of clusters and cluster head selection, it is assumed that rough fuzzy sets are used. The cluster formation and cluster head selection is out of the scope of the paper.

If the same event is to be reported by all the cluster members, the cluster center coordinates and generates a multi-signature representing all the cluster members. This signature is delivered to the RSU and the verification is done for this multi-signature. This greatly reduces the number of incoming messages and thus reduces the authentication delay. In the proposed scheme, the improved multi-signature scheme proposed by Jun [10] is adopted for signature generation and verification.This scheme is an improvement to the basic Shamir's ID based scheme and Harari's scheme. Table 1 show the notations used in the proposed scheme.

Table 1. Notations Used

Notation	Description
TA	Trusted Authority
TMS	Traffic Management System
RSU	Road Side Unit
p,q	Prime numbers
ID	Pseudo ID
H(.)	Hash function SHA-1
M	Message
S_i	Signing Key

The multi-signature scheme consists of three sections.

a) Setup

Let (n,e) be the public key and 'd' be the private key. Let H(.) be a secured one way hash function. The cluster head selects a random integer r_i corresponding to signer 'i' and calculates as shown in equation (1),(2) and (3).

$$r = \prod_i r_i \bmod n \tag{1}$$

$$R = r^e mod\ n \tag{2}$$

$$T = H(m||R) \tag{3}$$

Then <r_i,T> is send to signer i. The cluster head then calculates the signing key $s_i$as shown in equation (4) and sends to the signer through a secure channel.

$$S_i = H(ID_i)^{\phi(l)-1)d} mod\ n \tag{4}$$

b) Signature Generation

Initially, sub signatures are generated for random signer 'i', as shown in equation (5).

$$S_i = r_i m S_i^T mod\ n \tag{5}$$

All the sub signatures are then send to the cluster head. The cluster head computes the multi-signature as shown in equation (6),

$$S = \prod_i S_i \bmod n \tag{6}$$

Then the multi-signature computed, MS=(S,T) is send to the RSU for verification.

c) Signature Verification

Given the multi-signature MS=(S,T) for a message 'm', the verification is done as shown in equation (7).

$$R^* = S^e m^{-el} \left[\prod_{i=1}^{l} H(ID_i)\right]^{-T} \bmod n \tag{7}$$

From the equation (7), if $H(R^*) = H(R)$ then the signature is valid.

5 Performance Evaluation

5.1 Scenario Characteristics

The mobility of the vehicles is generated using the traffic simulator SUMO. This is provided as an input to OMNET++. The scenario includes the creation of a vehicular network which varies from 10 to 100 nodes. The acceleration is taken as 10% of the maximum velocity. The minimum velocity Vmin is fixed to 20 km/hr and the maximum velocity is varied from 20 km/hr to 100 km/hr. The road structure is created with 6 junctions. The radio model used in the simulation is LAN 802.11p which provides a transmission rate of 2 Mbps and a transmission range of 1000m. The update interval for safety messages is fixed to 300 milliseconds. The total simulation time is 1000s.

5.2 Evaluation Criteria

The performance of the proposed scheme is compared with the existing schemes of batch verification and priority based verification. Analysis is done based on the authentication delay, loss ratio and communication overhead.

5.2.1 Message Authentication Delay

The average message authentication delay is computed by the equation (8). In equation (8), N represents the total number of vehicles in the simulation; M is the number of messages sent by the ith vehicle.

$$AD = \frac{1}{N}\sum_{i=1}^{N}\frac{1}{MK}\sum_{m=1}^{M}\sum_{k=1}^{K}(T_R - T_S) \tag{8}$$

Fig. 2 shows the performance of the authentication delay for batch verification, priority verification and the proposed multi-signature scheme. From Fig. 2 it can be inferred that, compared to the existing batch and priority based verification schemes the proposed multi-signature scheme has less authentication delay. Even though, the number of vehicles increases, the delay in verification is less by 10% to 15% when compared to existing approaches.

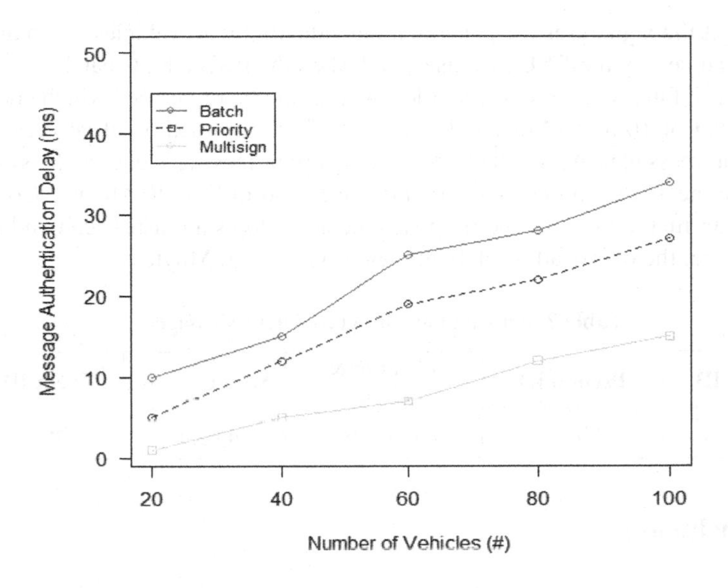

Fig. 2. Number of vehicles vs Message Authentication Delay

5.2.2 Communication Overhead

The overhead in communication is computed based on signature generation.

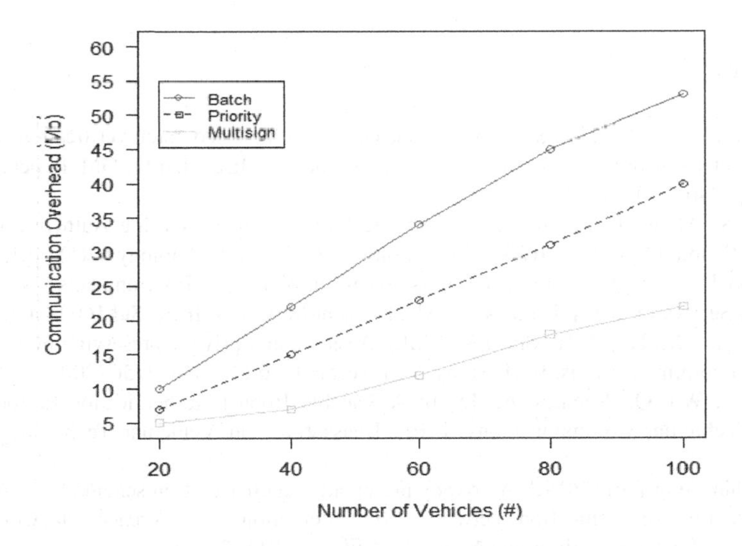

Fig. 3. Number of Vehicles vs Communication Overhead

All the signature generation approaches are assumed to be ID based. The safety message format adopted for the simulation is shown in Table 2.From Fig. 3 it is evident that if the signatures are generated individually by all vehicles, the number of messages received by the RSU increases, and when the RSU adopts batch verification and if there is a false signature, the whole batch should be re-verified which creates an extra overhead of 10 to 50 Mbytes. In the case of priority method of verification, all the important messages are verified and loss of certain messages causes the system to re-verify the messages and creates a relative overhead of 7 to 40 Mbytes. Compared to the existing methods, since the proposed method adopts a cluster head and creates multi-signature, the overhead is relatively less from 5 to 20 Mbytes.

Table 2. Format of an Unsigned Safety Message

Type ID	Pseudo ID	Location& Time	Speed	Event ID
2 bytes	2 bytes	10 bytes	4 bytes	2bytes

6 Conclusion

In this paper, the concept of multi-signatures is adopted for authentication in traffic warning application of VANET. From the analysis, it can be concluded that cluster oriented ID based multi-signature scheme has less authentication delay and less communication overhead when compared to the existing approaches. In future, the multi-signatures can be tested for other applications in VANET and an improved scheme can be proposed to increase the efficiency of the network.

References

1. Zhang, C., Lu, R., Lin, X., Shen, X.: An efficient identity based batch verification scheme for vehicular sensor networks. In: Proceedings of the 27th IEEE INFOCOM, Phoenix, AZ, USA, pp. 246–250 (2008)
2. Biswas, S., Misic, J.: A Cross-Layer Approach to Privacy Preserving Authentication in WAVE-Enabled VANETs. IEEE Transactions on Vehicular Technology 62(5) (2013)
3. IEEE Trial – Use Standard for Wireless Access in Vehicular Environments (WAVE) – Security Services for Applications and Management Messages, IEEE Std.1609.2 (2006)
4. Lin, X., Sun, X., Ho, P.-H., Shen, X.: GSIS: A secure and privacy preserving protocol for vehicular communications. IEEE Trans. Veh. Technol. 56(6), 3442–3456 (2007)
5. Zhang, L., Wu, Q., Solanas, A., Josep: A Scalable Robust Authentication Protocol for Secure Vehicular Communications. IEEE Transactions on Vehicular Technology 59(4) (2010)
6. Ming-Chin, Jeng-Farn: PPAS: A privacy preservation authentication scheme for vehicle to infrastructure communication networks. In: International Conference on Consumer Electronics, Communications and Networks, CECNet 2011 (2011)
7. Zhang, C., Lin, X., Lu, R., Ho, P.-H.: RAISE: An efficient RSU-aided message authentication scheme in vehicular communication networks. In: Proc. IEEE ICC, Beijing, China, pp. 1451–1457 (2008)

8. Hsin-Te, Wei-Shuo, Tung-Shih, Wen: A novel RSU based message authentication scheme for VANET. In: Proceedings of the Fifth International Conference on Systems and Network Communications (2010)
9. BevishJinila, Y., Komathy, K.: A privacy preserving authentication framework for safety messages in VANET. In: Proceedings of IET Chennai 4th International Conference on Sustainable Energy and Intelligent System (2013)
10. Jun, L.V.: Improvement of an Identity Based Multi-signature Scheme. In: 4th International Conference on Computational and Information Sciences (2012)

Source Node Position Confidentiality (SNPC) Conserving Position Monitoring System for Wireless Networks

Darshan V. Medhane[1] and Arun Kumar Sangaiah[2]

[1] Department of IT, Sinhgad College of Engineering, Pune, Maharashtra, India
dvmedhane.scoe@sinhgad.edu
[2] School of Computing Science and Engineering, VIT University, Vellore, Tamil Nadu, India
sarunkumar@vit.ac.in

Abstract. Position Monitoring System (PMS) is one of the applications of wireless networking that aims for monitoring and tracking objects' current position in the wireless area. For example, in order to keep an eye on the movement of an enemy in military applications the position monitoring system in wireless network can be set up. Battlefield surveillance is the classic example of this category in which the soldier needs protection from attackers. Consequently the issue is: how to conceal the position of the wireless node from the attacker? In this paper, we present a novel scheme for preserving confidentiality of the source node position in wireless network. The main objective behind presenting this scheme is to maximize the accuracy of the aggregate position information, to minimize the communication and computational cost and to offer source node position confidentiality by achieving numerous aspects of security, anonymity, traceability, revocation, data unlinkability and with the help of fake data resources. At the end, the experimental work and simulation results depicts the effectiveness of the proposed scheme.

Keywords: Position confidentiality, Source node position confidentiality, Position Monitoring System, Anonymity, Traceability, Revocation, Unlinkability.

1 Introduction

A wireless network is a kind of network committed for exchange of data between two or more nodes by means of radio signals. It can consist of different types of physical devices and offer different type of applications. The wireless nodes or stations are positioned into an area in which they want to examine real time happenings. In this paper, we are paying attention to monitoring and tracking mechanisms or applications that are available in the literature in today's date and proposing a novel source node position confidentiality conserving position monitoring system for wireless networks. Battlefield supervision, patient monitoring in a hospital and wildlife tracking are some of the applications where source node position confidentiality is one of the immense issue. The structural design of a wireless network for any of these applications comes with following scenario: source nodes inside a network monitor an area and search for the occurrence of a certain kind of entity which is known as an object. This object can be anything, for instance a vehicle, an animal, or a human depending on the type of

S.C. Satapathy et al. (eds.), *Emerging ICT for Bridging the Future – Volume 2*,
Advances in Intelligent Systems and Computing 338, DOI: 10.1007/978-3-319-13731-5_38

application. A sink node is a node that has more computation power and storage space in comparison with a normal source node and it is capable of performing a great deal of computing tasks that normal wireless nodes are not skilled and practiced. In addition, the sink node is responsible for collection of all the data of a wireless network.

Let us take the battlefield surveillance [1], [2] as an instance where the soldier needs protection from attackers: soldiers track the position of attackers in an area. A soldier that gets information about the attacker notifies the sink node through message passing mechanism by means of routing messages through intermediary nodes to the sink. In the meantime, the attacker tries to find out the soldier in the battlefield. The attacker traces the messages from the wireless network all the way to the source node that sensed the soldier, in order to kill the soldier. In this scenario, how to protect the soldier from the attacker? A same dilemma arises in other applications, for instance: monitoring the patients and doctors in a hospital [3], and the panda hunter game [4], [5], [6]. So as to protect an object from the attacker, we have to hide the position of the source that tracks the object. For this reason, we aim at the *source node position confidentiality* (SNPC). SNPC needs more than secrecy of the messages exchanged between nodes. SNPC requires that the flow of the messages does not give away the position of a source node. Actually, the privacy of a message is part of another privacy category, known as content privacy. Content privacy deals with providing integrity, secrecy, non-repudiation, and confidentiality of the messages exchanged in a wireless network. SNPC is a fraction of context privacy that focuses on hiding the contextual information of a wireless network [7], [8], [10]. The variety of the scenarios and their complexity are the motivation for a large set of studies in the direction of solutions that provide SNPC.

1.1 Contribution of the Proposed Work

In this paper, we present a source node position confidentiality conserving position monitoring system. The foremost contributions of this paper can be summarized as:

1. We propose SNPC preserving position monitoring system for wireless networks with the help of fake data resources and limited data unlinkability.
2. SNPC provides position anonymity of source, destination and roaming nodes.
3. The proposed SNPC scheme mainly makes use of broadcasting messages within the legitimate and authorized objects inside the wireless cloaked area.
4. SNPC maximizes the accuracy of the aggregate position information and minimizes the communication and computational cost.
5. We have implemented a simulation model for the proposed system and conducted comprehensive experiments in order to evaluate the performance of SNPC scheme in comparison with the existing system in the literature.

2 Related Works

In this section, the different concepts used for describing source node position confidentiality are discussed.

2.1 Source Node Position Confidentiality

P. Kamat et al. [4] were the first to formalize the node position problem on the basis of the panda hunter game. The concept of source node position confidentiality (SNPC) corresponding to location privacy was first described by C. Ozturk et al. [3] based on the panda hunter game. The essence of the panda hunter game is that the hunter can only make use of the traffic flow in order to track the panda. Both P. Kamat and C. Ozturk argue that it is up to the routing scheme to hide the position of the object, with respect to an opponent that only traces packets through the wireless network. P. Kamat make use of two metrics for calculating the confidentiality delivered by means of the routing algorithms. C. Ozturk shared the safety period of a routing protocol as a one of the metric. The safety period is the number of messages a wireless node sends out before the attacker traces the object in the wireless cloakless area, given the movement strategy of the attacker. Provision of node position confidentiality means one has to counteract the traffic analysis. The dilemma of countering traffic analysis is much older than the problem of SNPC in a wireless network. Countering traffic analysis brings us to the context of the Internet. Anonymity is defined by A. Pfitzmann et al. [6] as "*the state of being not identifiable within a set of subjects, the anonymity set*". Anonymity masks the identity of a wireless node as the identity could belong to any of the wireless nodes within the set [9], [10].

3 Research Gaps from Existing Work

In this section, we introduce the categories that we used to classify the solutions of source node position confidentiality.

3.1 Categories for the Solutions

Following is a set of categories on the basis of the core techniques used within each of the solutions for source node position confidentiality.

1) ***In Network Position Anonymization***: The following are the solutions under this category: the simple anonymity scheme, the cryptographic anonymity scheme, the anonymous communication scheme, destination controlled anonymous routing protocol for sensornets and phantom ID. Solutions in this category hide either the position or the identity of a source node.
2) ***Using Fake Data Sources***: The main objective of solutions under this category is that an attacker should no longer be able to see which part of the traffic is dummy, and which part is genuine or authentic. We found the following available solutions in this category: the group algorithm for fake-traffic generation, aggregation-based source location protection scheme and the cloud-based scheme for protecting source location privacy.
3) ***Delay***: This category comprises of solutions that change the flow of the traffic. In this type, we identified the following two solutions: rate controlled adaptive delaying and probabilistic reshaping.
4) ***Limiting Node Detectability***: We have recognized the following solutions in this category: anti localization by silencing, context-aware location privacy and multi

co-operator power control. This category consists of solutions that limit the transmission power of the nodes to make them harder to detect.
5) ***Network Coding***: In network coding, each node divides its message and forwards it in smaller pieces. These pieces are then forwarded through different routes towards the sink node. In this way, we have introduced in short each of the categories of available solutions for SNPC.

4 System Architecture

In this section, we depict the system architecture along with the introduction to basic scenario where there are three major entities: wireless nodes, position-based services (PBS) server and system users. Fig. 1 represents the architectural model of the proposed system. We will define the problem addressed by our system which is then followed by details of each entity involved in the architecture, data flow inside the system and the privacy model of our system.

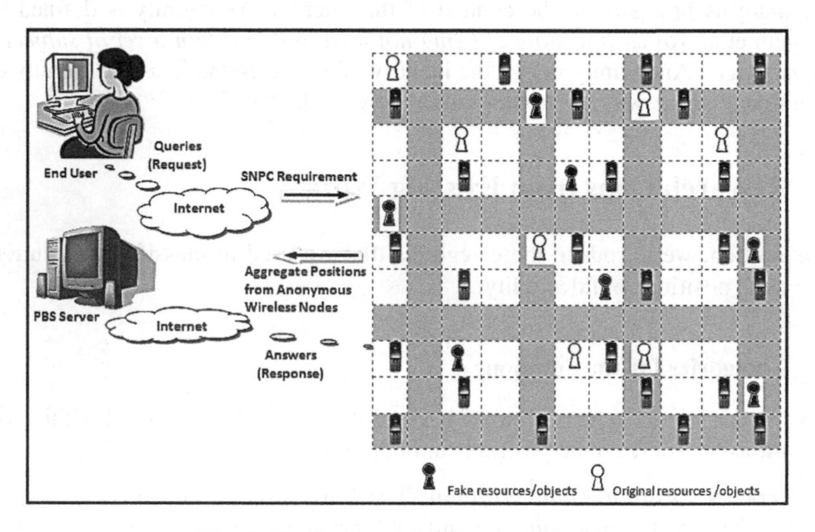

Fig. 1. The Proposed Architectural Model

a) Problem Statement: To design source node position confidentiality (SNPC) conserving position monitoring system for wireless networks.
b) Problem Definition: Given a set of wireless nodes $w_1, w_2, \ldots, w_n$ with wireless range areas $a_1, a_2, \ldots, a_n$ respectively, a set of moving objects $o_1, o_2, \ldots, o_n$ and a required anonymity level k; we find a cumulative position for each wireless node w_i, in a form of $R_i = (Area_i, N_i)$, where $Area_i$, is a rectangular area containing the wireless range area of a set of wireless nodes w_i and N_i is the number of objects residing in the wireless range areas of the wireless nodes in w_i such that $N_i \geq k$. An aggregate query Q that asks about the number of objects in a certain area $Area_i$ is answered on the basis of the aggregate positions reported from the wireless nodes.
c) System Elements: various wireless nodes are present in a trusted zone. All wireless nodes are responsible for calculating moving objects in their respective areas on their

own and are anonymous in nature. Wireless nodes communicate with the other wireless nodes inside the network to prepare a peer list by broadcasting a message. After formation of a peer-list, all valid wireless nodes form a cloaked area in which at least k no. of objects should be present. The cloaked area is the masked area which cannot be seen by other wireless nodes or objects inside the network. PBS server can be called a central node since every wireless node involved in the network is connected to it. PBS server keeps information of all wireless nodes inside the network. Trusted zone consists of several wireless objects and this zone is known as trusted since the anonymous wireless objects are present in it. Anonymous nature of wireless nodes helps hiding from other wireless nodes inside the network. In our system, the trusted zone is constituted by numerous wireless nodes where they communicate with each other through a secure network channel in order to avoid local as well as global attacks.

5 Proposed Methodology

In this section, we show the proposed five step procedure along with the algorithm to achieve source node position confidentiality in wireless networks.

A. Group Formation & Message Multicasting: This is the first step where all wireless nodes present in a wireless range area transmit a message to closer wireless nodes can be called as neighbor nodes. This message comprises of three attributes that are wireless node id for unique identification, its wireless range area and count of nodes involved in its wireless range area. All wireless nodes should make their own neighbor node list.

B. Determination of a search space: It is expensive for any wireless node w to congregate the information of all the wireless neighbor nodes w in order to calculate and figure out its relative minimal cloaked area *MCA* in view of the fact that a classic wireless network has a huge number of wireless nodes n.

C. Minimal cloaked area calculation: A set of wireless nodes positioned in the search space S, are considered as an input in this third step and this step plays an active role in computation of the minimal cloaked area *MCA* for the wireless node w. In this step, it is not necessary to scrutinize or inspect all of the permutations of the wireless nodes in the search space S. As a substitute, it only necessitates the grouping of at least four nodes for the reason that at least two wireless nodes describe width of *MCA* and at least two wireless nodes describe height of *MCA*. One more thing is important in this regard is with respect to the minimal cloaked area *MCA* of a set of wireless nodes N that has the monotonicity property, for the reason that adding wireless nodes n to the search space S should not decrease the area of the *MCA* of search space S or the number of objects contained by the *MCA* of the search space S.

D. Cloaked area configuration: The fundamental notion of this step is that every wireless node w shapes its wireless range area into a cloaked area which comprises at least k objects, in order to satisfy the k-anonymity privacy requirement. For every wireless node w, w initializes a search space S and then determines a value for each node in its respective neighbor list. The value is defined as a ratio of the object count of the node n to the distance between the node n and wireless node w. Finally, wireless node w establishes the cloaked area that is a minimum cloaked area *MCA* which covers the wireless range area of the wireless nodes in search space *S*, and the total number of objects in neighbor list.

E. Verification & Confirmation: This is the final step in order to avoid reporting aggregate positions with a relationship to the PBS server. We do not agree to the wireless nodes inside the network to report their aggregate positions with the relationship to the PBS server, for the reason that merging these aggregate positions may possibly pose confidentiality leakage. The detailed algorithm to be followed by source or intermediary nodes in order to achieve confidentiality can be explained as follows:

Algorithm SNPC (WirelessNode *w*, NeighborNode *n*, AnonymityThresholdValue *k*, NodeList *N*, SearchSpace *S*, WirelessRangeArea *Area*, WirelessNodeCount *Count*)
// Step 1: Group Formation & Message Multicasting
Each wireless node ***w*** is supposed to forward a message to all of the fellow citizens of the respective wireless range area and the message comprises of *w*'s identity*w.ID*, wireless range area *w.Area*, and total count of wireless nodes in the respective wireless range area *w.Count*.
IF node *w* gets a message from a neighbor node *n*, i.e.*(n: ID, n: Area, n: Count)* THEN wireless node *w* will merely add the neighbor node *n* to its neighbor-list.
// Step 2: Determination of search space step
A search space *S* should be decided in association with node list *N* on the basis of group formation and message multicasting which is then followed by detailed information gathering of the wireless nodes *w* and their respective neighbor nodes *n* situated in search space *S*.
// Step 3: Minimal cloaked area calculation
Append each node *w* or *n* positioned in search space *S* to *MCA [1]* like an object. Add *w* in all object-sets in *MCA [1]* as the first object

```
FOR x = 1; x • 4; x + +DO
FOR each object-set A = a1.   . . ai+1 MCA [i] do
IF Area (MCA (A)) < Area (existing MCA) THEN
IF n (MCA (A)) • k THEN existingMCA • A
Take out A from MCA [i] ELSE
Take out n from MCA [i]END FOR
IF x < 4 then
FOR each object-set A = a1.., ai+1, B = b1.., bi+1
IF a1 = b1.., ai = bi and ai+1 = bi+1THEN
Insert an object-set a1...ai+1, bi+1 to MCA [i+1]
```

// Step 4: Cloaked area configuration
Now, it is necessary to calculate a score of all nodes in neighbor list of the respective wireless node *w*. Choose the node *n* with the maximum peak value from neighbor list in search space *S* again and again in anticipation of the total count of objects in search space *S* is no less than *k*.
// Step 5: Verification and confirmation
Send (Area, n) message to the nodes within Area as well as to the PBS server IF there is no relationship with Area in addition to n ε N. ELSE IF wireless range area of *w*'s is restricted by several nodes *n* such that n ε N THEN Pick *N ε n* at random in such a

way that *N.Area* comprises wireless range area of *w* and forward *N* to the nodes surrounded by *N.Area* as well as the PBS server ELSE convey *Area* with a masked *n* to the nodes within *Area* and the PBS server.

6 System Evaluation and Result Analysis

This section explains our simulation setup and some of the evaluation results in comparison with existing protocols. Different parameters described in SNPC system are anonymity threshold (k), user mobility method, No. of query (Q) and time interval (T).

6.1 Simulation Setup

We develop a java program to simulate the performance of our system and compare it with the existing baseline system. All these work are run in a Windows 8 laptop. To simulate our work, 500 mobile users are considered within the simulation area of 2000 M X 2000 M. Following Table 1 describes parameters and their respective substantial values. To make the obtained results more precisely and accurately, each result in our experiment is an average of 1000 times running.

Table 1. Parameters and their respective values

Sr. No.	Name of Parameter	Parameter value
1	Simulation Area	2000 M * 2000 M
2	User Mobility Method	Plus Mobility
3	No. of Users	500
4	Anonymity Threshold	Low (5), Medium (7) and High (10)
5	Time Interval	2 Hours
6	No. of Queries	In a range of 15 – 300

6.2 Results

The system is tested for a set of specific datasets with parameters as mentioned in Table 1. The final outcome is in the form of graph G with total no. of users on X-axis

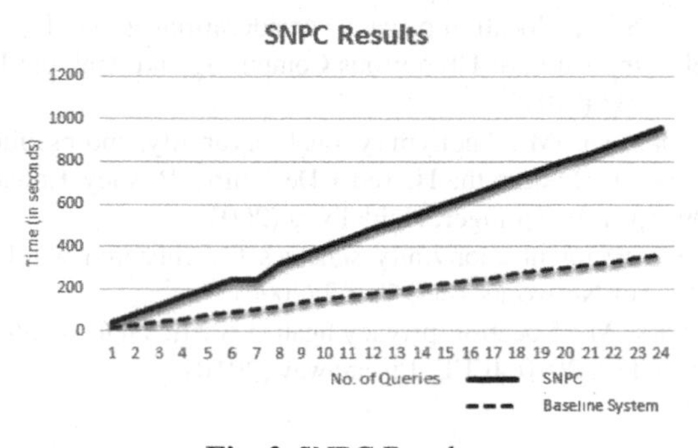

Fig. 2. SNPC Results

and time taken by the system to respond to the user's query on Y-axis as shown in Fig. 2. The graph shows that more confidentiality is preserved with proposed mechanism in comparison with the existing system.

7 Conclusion and Future Work

In this paper, we have given a novel source node position confidentiality (SNPC) conserving position monitoring system for wireless networks. Based on literature survey, we have the following concluding remarks. At the outset, anonymization techniques alone are not enough to prevent traffic analysis. Hiding an identity of a node alone will not provide SNPC by itself. Hence, we have introduced the system for maintaining confidentiality of source nodes in mobility. Secondly, we observed that most of the fake data source solutions that try to confuse real traffic frequently consume a lot more energy than solutions that try to imitate the presence of an object. We have overcome with this problem in our system. Ultimately, we have given the source node position confidentiality conserving position monitoring system for wireless networks in order to address the data confidentiality issue in existing anonymity, traceability and limiting node detectability approaches. For the future work, we will study the potential performance of internal adversaries and extend the SNPC system to oppose different adversaries in more effective manner. Furthermore, we will design the extended version of the proposed system for supporting very large group of wireless nodes in wireless area and detailed mathematical model of this work will be discussed in our future work.

References

1. Chow, C., Mokbel, M., He, T.: A privacy-preserving location monitoring system for wireless sensor networks. IEEE Trans. Mobile Computing 10(1), 94–107 (2011)
2. Shaikh, I., Jameel, H., Auriol, B., Lee, H., Lee, S., Song, Y.-J.: Achieving network level privacy in wireless sensor networks. Sensors 10(3), 1447–1472 (2010)
3. Ozturk, C., Zhang, Y., Trappe, W.: Source-location privacy in energy-constrained sensor network routing. In: SASN 2004, pp. 88–93. ACM, New York (2004)
4. Kamat, P., Zhang, Y., Trappe, W., Ozturk, C.: Enhancing source location privacy in sensor network routing. In: ICDCS 2005, pp. 599–608. IEEE Computer Society (2005)
5. Rios, R., Lopez, J.: Source location privacy considerations in wireless sensor networks. In: 4th International Symposium of Ubiquitous Computing and Ambient Intelligence, UCAmI 2010. CEDI, pp. 29–38 (2010)
6. Pfitzmann, A., Köhntopp, M.: Anonymity, unobservability, and pseudonymity - A proposal for terminology. In: Federrath, H. (ed.) Designing Privacy Enhancing Technologies. LNCS, vol. 2009, pp. 1–9. Springer, Heidelberg (2001)
7. Misra, S., Xue, G.: Efficient anonymity schemes for clustered wireless sensor networks. International J. Sensor Networks 1(1), 50–63 (2006)
8. Luo, X., Ji, X., Park, M.: Location privacy against traffic analysis attacks in wireless sensor networks. In: ICISA 2010. IEEE, Piscataway (2010)

9. Park, J.-H., Jung, Y.-H., Ko, H., Kim, J.-J., Jun, M.-S.: A privacy technique for providing anonymity to sensor nodes in a sensor network. In: Kim, T.-H., Adeli, H., Robles, R.J., Balitanas, M. (eds.) UCMA 2011, Part I. CCIS, vol. 150, pp. 327–335. Springer, Heidelberg (2011)
10. Chen, J., Du, X., Fang, B.: An efficient anonymous communication protocol for wireless sensor networks. Wireless Communications and Mobile Computing 12(14), 1302–1312 (2011)

Encryption/Decryption of X-Ray Images Using Elliptical Curve Cryptography with Issues and Applications

Vikas Pardesi and Aditya Khamparia

CSE Department, Lovely Professional University, Punjab, India
vikaspardesi@hotmail.com, aditya.khamparia88@gmail.com

Abstract. Elliptic Curve Cryptography is a public key cryptography scheme which is leading now days because of its great advantages (small key size, less time for encryption, no solution to Discrete Logarithmic problem, impossible time taken for brute force attack) for handheld, low memory, portable and small devices. Applying Elliptic Curve Cryptography on text and image gives almost equal performance. In this paper we are summarizing that how Elliptic Curve Cryptography encrypts and decrypt text data and image. Operations responsible for encryption are point multiplication, point addition, point doubling, and point subtraction are explained in detail. One more thing which is very necessary in decryption of the Elliptic Curve Cryptography that is Discrete Logarithmic Problem which is also explained briefly and many more comparative study about Elliptic Curve Cryptography advantages, Disadvantages, Elliptic Curve Cryptography attacks and its applications comparing with other Encryption Algorithms.

Keywords: Discrete Logarithmic Problem, Elliptic Curve Cryptography, Public Key Cryptography.

1 Introduction

In area of information security this is very important that the information we are sharing and transferring over network with each other that must be secured in the network. So in this era we have a challenge to secure our information whether it may be in any type of data whether it is raw data, text data, image data, and audio/video data whatever it is. So from earlier days there are many security algorithms and many encryption methods came but they may have some drawbacks in implementation or in overhead or in complexity. In this paper we are summarizing whole idea about Elliptic curve cryptography how it works on text data as well as on image.

Elliptic curve cryptography [1] is a public key cryptography proposed by koblitz and miller in year 1985. In the public key cryptography scheme generally those people who are taking part in the communication each one have their own secret key and each one's public key [2] i.e. transferred over network. In Elliptic Curve Cryptography, computational operations are defined over the elliptic curve i.e., $y^2=(x^3+ax+b) \bmod p$, where $4a^3+27b^2 \neq 0$. Elliptic curve vary with the value of 'a' and 'b'.

S.C. Satapathy et al. (eds.), *Emerging ICT for Bridging the Future – Volume 2*,
Advances in Intelligent Systems and Computing 338, DOI: 10.1007/978-3-319-13731-5_39

Now we are giving a brief introductory encryption and decryption of text and image. Procedure is equal likely here we are giving example of a single character (text Data) and one 2D image. In encryption process, firstly we generate point that lies on elliptic curve. From these points we take base point that is represented by B_P (x, y). Base point has the initial value (x, y) that satisfies the Elliptic curve. We take another value i.e., affine point A_M(x, y) it may be any point that lies on elliptic curve. Then generate public and private key of each user like if there is two user A and B then their public keys are PU_A and PU_B and private keys are PR_A and PR_B respectively. Private Key of users is a random number and public key of users is generated by performing multiplication operation with private key and base point. After generating key pairs, we take each pixel value one by one in an image or a single character ASCII value from a message. We perform multiplication operation with pixel value 'A' and affine point A_M i.e., S_M=A*A_M. In Elliptic Curve Cryptography point multiplication is the basic operation which consists of two operations point addition and point doubling. By performing multiplication operation encrypt the whole message and generate Cipher Text i.e. C_T = {(R*B_P), (S_M+R*PU_B). here (R*B_P) is a coordinate value it is (x1, y1) and (S_M+R*PU_B) is also a coordinate value it is (x2, y2) where S_M=A*A_M and 'A' is scalar value i.e., pixel value or ASCII value of data and A_M is affine point. And R is a random number and PU_B is public key of user B so finally cipher text is C_T {(x1, y1), (x2, y2)}.

In Decryption of the encrypted data we apply the private key of the user B i.e., PR_B on the first element i.e., R*B_P. This is subtracted from the other term (S_M+R*PU_B) to recover the S_M. In last we apply Discrete Logarithmic problem to recover our pixel value or ASCII value.

2 Previous Analysis and Work on Elliptic Curve Cryptography

Elliptic curve cryptography is a very wide area for finding many precious things which are remained unexplored. Since 1965 after proposing the Elliptic Curve Cryptography by the koblitz there are many researchers in this field contributed their work in those Alessandro Cilardo said about Elliptic Curve Cryptography engineering [8]. In this he told that in implementation of the ECC (Elliptic Curve Cryptography) there are many ECC attacks they are possible.

In recent years there are many research is going on ECC they said that side channel attack and its countermeasures can be prevented with some vast and huge set of instructions or algorithm, and now days we are optimizing that algorithms.

3 Operations in Elliptical Curve Cryptography

3.1 Point Multiplication

Point multiplication is the operation of elliptical curve cryptography which contains itself two operations i.e. point addition and point doubling. In point multiplication [7] operation we multiply one scalar value that may be your character ASCII value or image pixel value. E.g. point is p(x, y) that is multiplied by scalar a then we get another point k=a*p(x, y).

- **Point addition** – in this we add two points to get another point e.g. K(x, y) = L(x, y) + M(x, y).
- **Point doubling** – in this we add a point to itself to get another point e.g. K(x, y) = L(x, y) + L(x, y).

For example to understand the multiplication we have a scalar value i.e., 7 and point is P(x, y).

Multiply 7*P(x, y).

Q(x, y) = P(x, y) + 2(3(P(x, y))

Q(x, y) = P(x, y) + 2(P(x, y) + 2P(x, y))………………………………………….. (1)

so we got last 2P(x, y) in this we perform point doubling operation and we get another point say P1(x, y) put it in the equation (1) i.e., Q(x, y)=P(x, y)+2(P(x, y)+P1(x, y)) now in bracket we perform point addition operation between P(x, y) and P1(x, y) and get new point p2(x, y) now equation becomes Q(x, y)=P(x, y)+2(P2(x, y)). Then again we apply point doubling and addition then we get final point Q(x, y).

3.2 Point Addition

In point multiplication we again and again using point addition so we have a formula for computing the value i.e., say two points are $K(X_k, Y_k)$ and $L(X_l, Y_l)$ and $J(X_j, Y_j)=K(X_k, Y_k)+L(X_l, Y_l)$ now

$X_j= S^2-X_k-X_l \bmod p$

$Y_j= -Y_k+S(X_k-X_j) \bmod p$

$S=(Y_k-Y_l)/(X_k-X_l) \bmod p$

Where S is the slope of the line which is passing through point $K(X_k, Y_k)$ and $L(X_l, Y_l)$ and p is a prime number.

3.3 Point Doubling

In point doubling say one point $J(X_j, Y_j)$ that gives one point $L(X_l, Y_l)$ after doubling operation so $L(X_l, Y_l)=2J(X_j, Y_j)$

$X_l=S^2-2X_j \bmod p$

$Y_l=-Y_j+S(X_j-X_l) \bmod p$

$S=(3X_j^2+a)/(2Y_j) \bmod p$

Where S is the tagent at point J and a is one of the parameter chosen with the elliptic curve and p is a prime number.

3.4 Point Subtraction

Say two points $K(X_k, Y_k)$ and $L(X_l, Y_l)$ and $M(X_m, Y_m)=K(X_k, Y_k)-L(X_l, Y_l)$ then $M(X_m, Y_m)=K(X_k, Y_k)+(-L(X_l, Y_l))$

where $-L(X_l, Y_l)=(X_l, -Y_l \bmod p)$ where p is a prime number.

4 Procedure and Algorithm for ECC

There are two steps we have to follow for fulfillment of our implementation. First one is Encryption and Decryption and in decryption we have two process first one is decryption and another is Discrete Logarithmic Approach [5] for finding the solution.

Encryption- we have to encrypt so first of all we have to generate all points that satisfying the elliptical curve follow this function **GetPoints (a, b, p)**

Step 1 take x=0 or any other positive integer
Step 2 loop until $x<p$
I. $Y^2=(x^3+ax+b) \bmod p$
II. If y^2 is perfect square
display(x, square root (y))
Else
X++;
Step 3 End. Where p is selected prime number and x, y are co-ordinates.

Here we are not focusing on what should be the procedure for distribution of the key over public network but for better understanding of the approach here we are taking notation say User A and User B are two authorized users respectively U_A,U_B [3].

U_A have PU_A and PR_A key pair and same as user B contains pair of PU_B and PR_B.

Follow this function for ECC Encryption
EccEncrypt (A, A_M, PR_B, B_P)
Step 1-For all A (i.e. pixel value or ASCII character)
Find S_M=A *A_M // A is constant, A_M is random Affine point in elliptic curve.
Step 2- Find PU_B=PR_B*B_P //B_P is the base point Of Elliptic curve, PR_B is the private key
Step 3- End;
Encrypted data= (R*B_P, S_M+R*PB_P)

Decryption-
Follow this procedure for ECC decryption
Let R*B_P be the first point and (S_M+R*PU_B) be the second point PR_B *R*B_P=PR_B *first point
Calculate S_M= S_M+R*PB_P-PR_B*R*B_P
Calculate the scalar value of 'A' from S_M using discrete logarithmic problem [4].

Discrete Logarithmic Problem [4]- when we do decryption of the ECC then we are needed of Discrete Logarithmic Problem for finding our real scalar value that can be pixel value or character ASCII value. if we look carefully then we got at decryption that S_M=A*A_M for finding the value of A. we shift the A_M at left side i.e. a=S_M*(1/A_M). (1/A_M) we get from Discrete Logarithmic problem and then we again call point multiplication algorithm of ECC for S_M with (1/A_M). Then we get scalar value of a. In the discrete Logarithmic problem (1/A_M) we get from Extended Euclidian Distance problem solving Algorithm.

Example- We have two variable P, Q, R where we have to find Q and given equation is this P=Q*R

P*(1/R) = Q and (1/R) is solved through Extended Euclidian Algorithm Technique for finding the value of (1/R). After getting the value we just apply multiplication operation with P and we got our final decrypted Result i.e., scalar value or pixel value.

5 Implementation Results of Elliptic curve Cryptography on Text and Image

Before understanding the analysis first we have to understand the how encryption is flowing step by step in whole procedure.so here we are giving a flow chart for better understanding of the approach.

Now here is some implementation detail of ASCII character and image pixel value. Some screen shots are here that can better help you for understanding how procedure is going on and how this is following the elliptic curve algorithm for point generation [7] and encryption.

5.1 ECC Applied on a Single Character

According to elliptic curve $y^2=(x^3+ax+b) \bmod p$ where $4a^3+27b^2\neq 0$ we generate points that lies on elliptic curve. Let p=37, a=-1, b=1 on which 'a' and 'b' value satisfying above equation.

X	Y
1	1
1	36
3	13
3	24
6	1
6	36
14	3

Fig. 1. Look Up table for Base point and Affine point [12]

We have to generate the private key of user PR_A=13,
Private Key of user PR_B = 17,
Base point B_P= (1, 1).
P_M is affine point (6, 1).

Now encrypted text is= ($R*B_P$, S_M+R*PU_B)
R is a random number=13.
Now PU_B is PR_B*B_P=17(1, 1).
PU_B= (5.482019e+000, -1.300886e+001)
Approximate value is PU_B= (5.5,-13)(round-off)
Secret key $R*B_P$= 13(1, 1) = (9.867327e-001, -9.733738e-001).

Multiplication-

S_M=A*A_M ,39(1, 6)= (1.817513e+005, -7.748475e+007)

ASCII value 39 is corresponding to character ‘’’.,R*PU_B= (1.427888e+001, 5.407907e+001),S_M+R* PU_B= (1.427888e+001, 5.554202e+001)

Encrypted Data-

C_T= ((9.867327e-001, -9.733738e-001), (1.427888e+001, 5.554202e+001))

At decryption side- R*$B_{P*}PR_B$=1.427888e+001, 5.554202e+001

Subtraction s1, s2=1.817513e+005, -7.748475e+007

Now by the **Discrete Logarithmic Problem** f1, f2=1.453599+001, 5.554202e+001

A=1.871153e+001 and 1.963456e+001

So final approximate ASCII value is 19+20=39 i.e., character ‘’’.

5.2 ECC Applied on an Image

According to this equation $4a^3+27b^2 \neq 0$ we generate the elliptic curve p=37(assumed), a=-1, b=1 on which ‘a’ and ‘b’ value satisfying above equation. We have to generate the private key of user PR_A=13, Private Key of user B is PR_B = 17, Base point B_P is = (1, 1).A_M affine point is (6, 1).

Now encrypted text is= (R*B_P, S_M+R*PU_B).R is a random number=13,Now PU_B is PR_B*B_P=17(1, 1).PU_B= (5.482019e+000, -1.300886e+001),PU_B= (5.5,-13)
Secret key R*B_P=13(1, 1)= (9.867327e-001, -9.733738e-001).
S_M=A*A_M,8(6, 1) = (1.848905e+002, -2.514031e+002),Another pixel i.e., 9 and ,S_M= 9(6, 1) = 8.215585e+001, -7.448383e+002),R*PU_B= (1.427888e+001, 5.407907e+001),(S_M+R* PU_B) for pixel 8 =(2.740463e+001, 1.434947e+002),And (S_M +R* PU_B) for pixel 9 is= (4.199843e+001, 2.720861e+002).,And (R*B_P*S_M +R*PU_B) for 8= ((9.867327e-001, -9.733738e-001), (2.740463e+001, 1.434947e+002)),For pixel 9 is (R*B_P, S_M+R* PU_B) = ((9.867327e-001, -9.733738e-001), (4.199843e+001, 2.720861e+002)).

Original image

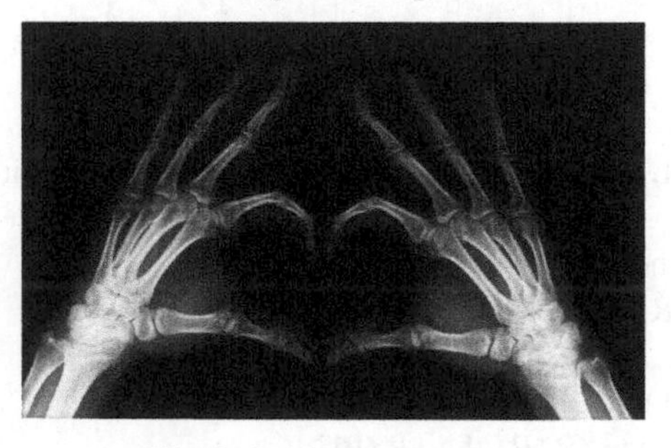

Fig. 2. Original Image [11]

Encrypted Image

Fig. 3. Image after encryption

Decrypted Image

Fig. 4. Image after Decryption

From results we are looking that we are not getting exactly image as we have supplied in original because if we look on encrypted points then we can analyze that they are not integer they are real numbers and they cannot give exactly same result so in the text ASCII character we can round off the results and get an exact ASCII value but not same in image.

6 Applications and Pros and Cons

In future ECC is going to be in demand because it has great advantage over other public cryptography scheme such as RSA and DES. In RSA and DES the main problem is key size and more over head for computation of Permutation and combination [6].

6.1 Advantages

a) Key size and digital signature that are generated through ECC are very shorter in size compare to other Cryptography scheme.
b) This is based on discrete logarithmic form so easily can be converted into elliptic curve form.
c) No time consumes for permutation and combination and less time taking for encryption.
d) Till date no solution found for breaking the Discrete Logarithmic approach so brute force attack on ECC takes too many years (uncountable).

6.2 Disadvantages

a) ECC uses curves generators fields' etc. This is more complex to calculate so this is not good for processor health.
b) ECC systems are much slower than RSA in large no. of public key generation and also works in Constrained Environment [8].

6.3 Applications

a) Simple Key generation by ECC is a great application in cryptography.
b) Shorter Certificate
c) Shorter Signature can also be generated with the help of ECC.

Now we are moving to the IPV6 because there are some drawbacks in IPV4 such as when we transfer our data then IPSEC protocol and IPV4 can't work simultaneously and it has to be improved for better working and secure transmission of the data and elliptic curve cryptography is the solution for this because when we transfer the data by the RSA (Rivest-Shamir-Adleman) algorithm then its key size is very long and data can be corrupted in middle of the way but elliptic curve cryptography contains very small key compare to RSA and when we transfer the data through ECC then corruption of data has less chances so we have to do more work in this field.

7 Conclusion

In this paper we summarizes that what is elliptic curve cryptography and what are the operations involved in the Encryption and decryption and we analyze the results of any text character and on an image. We also analyzed the look up table which contains the base point and all other affine points and more about elliptic curve, further we discussed about what are type of attacks are possible in very much brief on ECC cryptography. Their countermeasures are not discussed so much because that is not our concern. Elliptic curve cryptography applications are in the real world and also in constrained environment, some area are also present where we cannot use this cryptography techniques.

References

1. Kolbitz, N.: Elliptic Curve Cryptosystems. Mathematics of Computation 48, 203–209 (1987)
2. Stallings, W.: Cryptography and network Security, 5th edn. Prentice-Hall
3. Standard specifications for public key cryptography, IEEE standard, pl363 (2000)
4. Albirt, A.J.: Understanding & Applied Cryptography and Data Security. CRC Press, Pearson
5. Certicom, standards for Efficient Elliptic curve cryptography, SEC 1: Elliptic Curve Cryp-tography, Version 1.0 (September 2000), http://www.secg.org/download/aid-385/sec1_final.pdf
6. Certicom website, http://www.certicom.com/index.php?action=ecc_tutorial,home
7. Anoop, M.S.: Elliptic Curve Cryptography-An implementation Guide
8. Cilardo, A., Coppolino, L., Mazocca, N., Roman, L.: Elliptical Curve Cryptography Engineering. Proceedings of the IEEE 94(9), 395–406 (2006)

9. Kocher, P.C.: Timing attacks on implementations of Diffie–Hellman, RSA, DSS, and other systems. In: Koblitz, N. (ed.) CRYPTO 1996. LNCS, vol. 1109, pp. 104–113. Springer, Heidelberg (1996)
10. Kocher, P.C., Jaffe, J., Jun, B.: Differential power analysis. In: Wiener, M. (ed.) CRYPTO 1999. LNCS, vol. 1666, pp. 388–397. Springer, Heidelberg (1999)
11. http://mmattes.deviantart.com/art/X-Ray-Heart-121441422
12. Self Created table randomly for affine points

Network Quality Estimation – Error Protection and Fault Localization in Router Based Network

M. HemaLatha[1], P. Padmanabham[2], and A. Govarhan[3]

[1] Lakireddy Balireddy College of Engineering, Mylavaram, Krishna Dt, A.P, India
[2] Bharat Institute of Engineering and Technology, Hyderabad, India
[3] JNTUH, Hyderabad, India
lathamunnangi@gmail.com

Abstract. Devices of Network are used for Different purposes like Huge data transmission, easy to access and time saving are the applications of digitized communication system. Wireless communication systems consist of a number of routers and links. Processing speed, link failure, control of one router over another router as well extended delay causing huge problems in transmission. The proposed method is based on a three phased technique. The network parameter detection phase includes a protocol oriented technique for network parameter considerations and mutual node communication. To provide communication the second phase includes security inclusion in network correlated nodes. Here this paper proposes principal and credential sharing among a number of nodes present in the network who are neighbor to each other. The last phase of the paper includes the detection of the error present in the network. For a flow contained development, this paper first gives an introduction related to security in UMA network in the section one. The second section is a literature study of some previous related works. A well-structured and modularized proposal is given in section three followed by a simulation result in section four. At last an overall conclusion is given.

Keywords: reliability, latency, availability, failure rate, data rate, authentication, authorization.

1 Introduction

The world is connected through multiple data sharing centers. Long range and variation in operating regions imposes requirement of routers in networks. Routers are coming pressures like the failure of hardware [1], unwanted control of another node over it [2], losing of data or changing of data present in the data packet [3] [4] [5] or headers, used for data carrying. As the distance between the source and destination and the region of local area network (LAN) are increasing respective quantity, the risk of network failure is going on increasing [4]. Networks are growing exponentially in day to day life. This leads to the discovery of a suitable protocol, which is able to detect and replace data packet transmission paths. Although repeater and router in the network solving the coverage problem, it is enhancing the causes of errors. Some of the most common errors include flow of the data in a way [6] [7], which are not

S.C. Satapathy et al. (eds.), *Emerging ICT for Bridging the Future – Volume 2,*
Advances in Intelligent Systems and Computing 338, DOI: 10.1007/978-3-319-13731-5_40

coming under the routing protocols. Few unwanted nodes having some adverse effect on the networking system [8]. They try to alternate the flow of data. Cryptographic data for data transmission is one of the solutions present in network [9] [10]. Some solution present in the network focuses acknowledgement based transmission technique [3]. This technique is able to solve basic problems of the network. Still generation and transmission of acknowledgement put pressure on security threats detection method and on the transmission system. The reliability is decreasing as many as transmission occurs. Protocols are unable to detect errors as soon as possible. .

1.1 Architecture

The Internet routing infrastructure is also vulnerable to attacks. Because of the very nature of this infrastructure, the router can act a large number of hosts, entire networks, or even the global Internet [11]. The objectives of routing attacks can include black holing and loss of connectivity, track redirection to networks controlled by adversaries, track subversion and data interception, or persistent routing instability [12]. There are various approaches that have been used in IP trace back, and many of these can be broadly categorized under packet logging or packet marking schemes. Depending on such criteria as storage overhead either within the packet itself or at the nodes traversed, link speeds, or computational demands, among other mitigating factors; each category has its advantages and disadvantages. For example, the hash-based approach [13] is a logging method that can trace a single packet, unlike most packets marking schemes, which assume a reasonably large number of packets for a successful trace back. Another routing technology Stealth probing is a secure data plane monitoring tool that relies on the efficient symmetric cryptographic protection of the IPsec protocol suite that is applied in end-router-to-end-router fashion. One of the other protocol present is BGPmon, it uses XML to represent BGP messages, handling all attribute and element types, and various classes of data [14].

1.2 Applications

Cryptographic protocols security is of crucial importance due to their widespread use in critical systems and in day-to-day life [15]. Large open networks, where trusted and un trusted parties coexist and here messages transit through potentially “curious” if not hostile providers pose new advantages to the designers of communication protocols [16]. Network routers occupy a key role in modern data transport [17]. Modern ISP, enterprise, and data center networks demand reliable data delivery to support performance-critical services, thus requiring the data plane to correctly forward packets along the routing paths. Real-world incidents reveal the existence of compromised routers in the ISP and enterprise networks that sabotage network data delivery [18]. Network Scanner or Network Enumeration is a computer program used to retrieve user names, and info on groups, shares and services of networked computers [19].

1.3 Issues

It is important to initially emphasize that erasure security be relative [20]. Attacks that are hinged upon the guess-ability of initial TCP sequence numbers (ISN): so that an arbitrary host can exploit an address-based trust relationship to establish a client writes-only TCP session [21]. Securing IP routing is a task that is central in diminishing the Internet's liability to mascon gyrations and malicious attacks [22]. As there are numerous attacks on published protocols, designing AKE protocols is error prone [23]. In large and constantly evolving networks, it is difficult to determine how the network is actually laid out. This information is invaluable for network management, simulation, and server sitting [24]. Traditional topology discovery algorithms are based on SNMP, which is not universally deployed [25]. Compromised routers can drop, modify, mis-forward or reorder valid packets [26]. The BGP, routing protocol includes no mechanism for verifying either the authenticity (correct origin) or the accuracy of the routing information it distributes. A particularly problematic case is that of sophisticated malicious routers (e.g., routers that have been compared) [27].

2 Literature Review

In [2], the authors have argued that robust routing requires not only a secure routing protocol but also well-behaved packet forwarding. They have proposed an approach to robust routing in which routers, assisted by end hosts, adaptively detect poorly performs routes that appear suspicious, and use a secure trace *route* protocol to attempt to detect an offending router.

The authors of [28] say FD protocols require only pair wise participation of nodes, deployment of FD can proceed in an incremental fashion that is compatible with incentives for informing routing decisions at the network edge. However, when the authors consider the placement and selection of FD protocols, natural questions arise about the division of labor between the end host and the edge router. They argue that the placement of FD protocols depends on the parties responsible for providing confidentiality and driving routing decisions.

Pepper and Salt Probing may even be efficient enough to be deployed in the core of the Internet, as part of an architecture where core routers inform their routing decisions by running FD to destination networks.

The authors of [29] have designed and analyzed efficient path-quality monitoring protocols that give accurate estimates of path quality in a challenging environment where adversaries may drop, delay, modify, or inject packets. Their protocols have reasonable overhead, even when compared to previous solutions designed for the non-adversarial settings.

We are exploring how to compose multiple instances of our PQM protocols running over multiple paths simultaneously to determine whether the adversary resides on either the forward or reverse path, or to localize the adversary to particular nodes or links.

3 Problem Definition and Proposed Methodology

Our previous paper proposes information sharing and acknowledgement on error detection method to trace out the attack and error as soon as possible.

3.1 Problem Definition

The previous paper idea is quite clear about the network parameters. Location of the error is not tracked in a convenient way. Although the authentication and authorization is verified; it is still unable to explain the methodology of handling the principals and credentials. For handling the security and error detection at exact location, this paper proposes two way conformations for RTT and relative detection method in a single way transmission long range network. Mutual principal and credential verification will increase the security and error checking in a rapid manner.

The main issues are-

1. In the previous papers nearby node is not chosen through the quality constraints.
2. Security is not enforced through the digital parameterized technology.
3. Exact detection of error is not supported in the previous paper.

The proposed method is a three phased technique. The first phase includes a protocol oriented technique for network parameter considerations to have mutual node communication. The second phase includes security inclusion in network correlated nodes. Here this paper proposes principal and credential sharing among a number of nodes present in the network who are neighbors' to each other. The third phase of the paper includes the detection of the error present in the network.

3.2 Proposed Methodology

In this section this work is considering some phases for the node identification providing security and error detection.

From the diagram it is clear that there are a number of end users who are using a number of intermediate nodes for the long range data communication. The one way and two way flow is shown in the diagram.

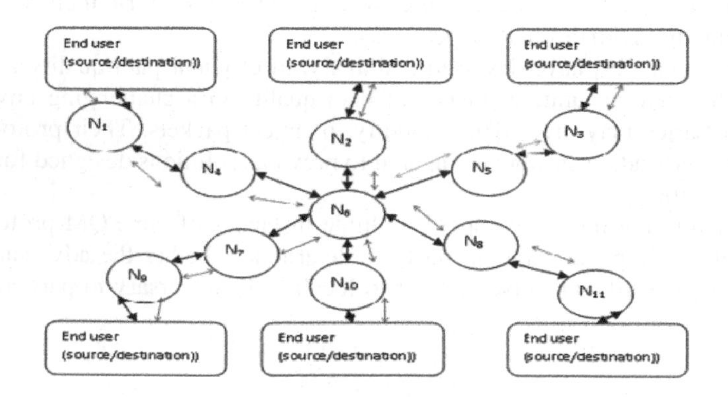

Fig. 1. Shows the architecture diagram of the proposed network

The below figure is about the proposed methodology presented over the current paper. This shows the three phased solution of the proposed methodology

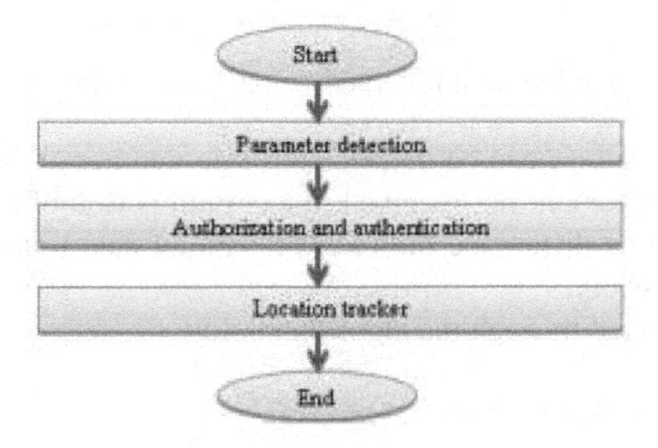

Fig. 2. Shows the architecture diagram for proposed methodology

3.2.1 Network Parameter Calculation Phase

The aim of this phase is to get the data in a dynamic and secured manner. A number of parameters that performs in service of the network are availability, latency, delivery, mean time between failure and mean time between restoring

Availability. For the availability, we are going for the formula required channels divided by available channels. Then we have to get the maximum value which is less than or equal to 1 for this choosing. The above parameters are detected though the current sender and intermediate nodes or intermediate nodes to intermediate nodes.

The availability is given as the below mathematical formula.

$$Av = \frac{rq_{ch}}{ch_{av}} \quad \ldots \ldots \ldots \ldots \ldots \ldots \ldots \ldots \ldots \ldots \quad (1)$$

Where rq_{ch}= required channel , ch_{av}= channel available

Latency. This Thus the minimum latency is detected for the next communication. For the latency detection we have to use a time stamp at each node's register.

The latency is given by the formula-

$$La = A_R - DP_s \; (per\ one\ communication) \ldots \ldots \ldots \ldots \ldots \ldots \ldots (2)$$

Where A_R=acknowledgement received time, DP_s =data packet sent time

Delivery Rate. Delivery rate is decided through data sent and acknowledgement received in a certain period of time. The best delivery rate is chosen for the next communication.

The delivery rate is given in the below formula.

$$DR = \frac{AK_R}{DP_s} \dots\dots\dots\dots\dots\dots\dots\dots\dots\dots\dots\dots\dots \quad (3)$$

Where AK_R= acknowledgement received, DP_s= total data packet sent

Failure Data Rate. Failure can be detected through number of data packets sent and number of acknowledgement not received in a certain period of time. Mean time can be detected as the time difference between first failed data transmission to the nest success data transmission.

The failure is detected through the algorithm below-

```
Failure data rate ()
{
    If (acknowledgement received)
          Continue;
    Else
    {
          Time t1=current time ();
          Continue the acknowledgement;
          Track the successful time;
    }
}
```

Failure data calculation the below formula is given-

$$FD = \frac{T_2 - T_1}{TP} \dots\dots\dots\dots\dots\dots\dots\dots\dots\dots\dots\dots\dots \quad (4)$$

Where T_1 and T_2 is detected as given in the above, and TP is the average time duration. TP is decided by the user.

Average Function. Then we will calculate a mathematically average to detect the most secure way for data transmission. We are applying a number of parameters. Some path or node may be the lack of something, but having advantages in other thing. So it depends on us to decide the best. Getting the average, we need not be worried about each individual factor.

For the average function, we will go for the below formula.

$$AF = (P_1 * Av + P_2 * FD + P_3 * DR)and\ (\min(L)) \dots\dots\dots\dots \quad (5)$$

Where P_1, P_2 and P_3 are the priorities decided by the user in choosing the best service scenario for individual services. As previously said; minimum latency has been the best case always in network [3] [5]. Then this paper is choosing the best average function for the network transmission. The data is transferred in the chosen pathway.

Most of the performance metrics are derived from the above mentioned quality parameters. Availability is done through the channel available to the required channel available. The functionality is decided through credentials and principles described in phase-2. Loss is a simple way of calculating acknowledgement received and data packet sent. One way the loss is consists of one way data packet transmission and one way acknowledgement transmission. Round trip loss is consists of two way calculation of data packet transmissions and acknowledgement transmissions. Delay is a measurement of time stamp. For this parameter this paper proposes a time attribute should be made at each node participating in the data transmission. The accepted transmission should be the first acknowledgement received time. Other transmissions are compared to the above accepted time. Utilization parameters are derived in the above paragraph as bandwidth, capacity. A fractional summation of bandwidth required to bandwidth available with channel available to channel required gives the utilization factor. Utilizing a node for few numbers of services is always reliable. All path qualities are useful in different conditions.

Algorithm for Network Parameter Calculation Phase

Step-1- start the procedure with some parameters.
Step-2- get the channel availability as given in equation-1.
Step-3- get the latency as given in equation-2.
Step-4- get the delivery data rate as given in equation-3.
Step-5- get the failure data rate as given in equation-4.
Step-6- get the average function as given in equation-5.
Step-7- choose the best way for the transmission of data and transmit the data.

3.2.2 Security Enhancement Phase

Credential and principals are specific for a node and for a communication. One to one relation is there among the nodes of a network. When a node sends its first request to the intermediate node the intermediate nodes acknowledge with a principal and credential. Principal is general identity and credentials are the hidden identity. This will be limited for a certain period of time. After the period the process will be renewed in the same way. This is the main procedure for security enhancement.

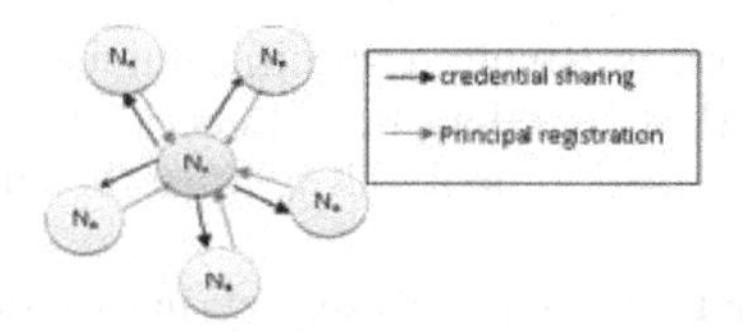

Fig. 3. Shows the principal and credential sharing among consecutive nodes

To make one principal for whole the process and credential has to be changed at a number of requests or a period of time. Here a node's identity is taken as the principal. The node's identity may be a MAC address or IP address or any given name. For the simplicity here it is proposed to have a given name. The name is alpha numeric. The credential is generated through a function F (k). It is like the password. It is generated by the receiver node for the sender node. The principal is giving authentication and credential is giving authorization. These methods enhance the security to another standard. Having a single principal and multiple credential authentications is achieved through a number of steps.

Generation of Credentials. For the generation of the credentials this paper is giving the below formula.

$$credential = F(k) * Rand() * specific\ number \ldots\ldots\ldots\ldots\ldots\ldots\ldots\ldots\ldots\ldots (6)$$

Where F (k) = a randomly generated function , $Rand()$ = function for randomly generated number, Specific number details are given in the next phase.

Algorithm for Security Enhancement Phase

Step-1- start the procedure with a group of nodes.
Step-2- every node is creating its own id as principal.
Step-3- this principal is shared with the next receiver node.
Step-4- the nest receiver node provides the credentials.
Step-5- at every unit of transmission the principal and credential are same.
Step-6- end the procedure.

3.2.3 Location of Error Detection Phase

For the exact location of error present in the network, it is proposed a specific data field should be carried in the data packet. The data packet carries all the intermediate nodes. An acknowledgement will be sent in the same way that the way data is flowed. Each intermediate node keeps a track of each node's data processed. Each data packet is having a summation field.

Fig. 4. Shows flow of data through nodes in linear media

Each node in the network is having a specific number. When the data packet is moving through the network it is adding the numbers. at the time of the acknowledgement transmission the summation field will subtract the number.

Table 1. Shows data packet style in data packet

Source	Destination	Current node's value	Summation value	Intermediate nodes	

As these two data packet and acknowledgement packet is flowing the same path, at a specific node both values will be same. If the value is different the path has to verify. That's the way we can get the data of failure correctly. We can get the exact location of path failure. At this condition the quality parameters are verified.

Generating the Specific Number (SN). The specific number is generated as

$$\mathrm{SN} = \mathrm{rand}(0.1) * \mathrm{n} \ldots\ldots\ldots\ldots\ldots\ldots\ldots\ldots\ldots\ldots\ldots\ldots\ldots\ldots (7)$$

Where n is the any positive multiplicand of 10.

Summation Field (SF). At the time of transmission - The summation field is generated with the below formula

$$SF = \sum_{i=1}^{x} SN_i \ldots\ldots\ldots\ldots\ldots\ldots\ldots\ldots\ldots\ldots\ldots\ldots\ldots\ldots\ldots\ldots\ldots\ldots (8)$$

Where 'i' represents all the intermediate node's number.
Summation field at the time acknowledgement-

$$SF = SF_{n+1} - SN_n \ldots\ldots\ldots\ldots\ldots\ldots\ldots\ldots\ldots\ldots\ldots\ldots\ldots\ldots\ldots (9)$$

Where SF_{n+1}= summation field till the next node. , SN_n= specific number of the current node.

Suppose a network is having five nodes a, b, c, d, e. The values of the nodes should be 1, 2, 4, 8, 16. When a data is moving from a sender to receiver the value after the first node is 1 and after b is 3, after c is 7, after d is 15, and after e is 31. When it will return it will return like 31, 15, 7, 3 and 0. If it is at sender it is no acknowledged as 0, the sender sends a verification message to the network. Suppose the value is coming negative. Then the data packet is passed through more nodes. If it is positive, but not zero, it is missed some node in the path. After identifying the node we can just add or subtract the effect of that particular node. So we can get the error free data.

Algorithm for Location of Error Detection Phase

Step-1- start the procedure with the sender as first or 0th position.
Step-2- every node having their unique ID.
Step-3- every nodes are generating their unique function as given equation.
Step-4- the summation field is generated as given in equation- .
Step-5- the data packet is generated as given in figure- .
Step-6- the summation field is generated reversely as given in equation-
Step-7- the error is detected as given in the above condition.

4 Simulation Result

4.1 Simulation Model and Parameters

The Network Simulator (NS2) [30], is used to simulate the proposed architecture. The simulation settings and parameters are summarized in table.

Table 2. Shows Simulation settings and parameters

No. of Nodes	60
Simulation Time	50 sec
Traffic Source	Exponential and TCP
Packet Size	512
Sources	5
Rate	1,2,3,4 and 5Mb
Lambda	30.0
Mu value	33
Flows	1,2,3,4 and 5

4.2 Performance Metrics

The proposed Network Quality Estimation and Error Protection (NQEEP) is compared with the Trace Route technique. The performance is evaluated mainly, according to the following metrics.

- **Packet Delivery Ratio:** It is the ratio between the number of packets received and the number of packets sent.
- **Packet Drop:** It refers the average number of packets dropped during the transmission
- **Throughput:** It is the number of packets received by the receiver.
- **Delay:** It is the amount of time taken by the nodes to transmit the data packets.

4.3 Results

Case-1 (Exponential scenario): Based on Rate. In our first experiment we vary the data transmission rate as 1,2,3,4 and 5Mb.

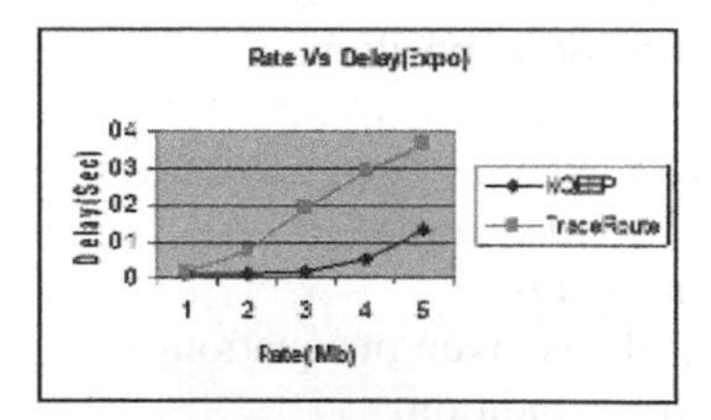

Fig. 5. Rate Vs Delay

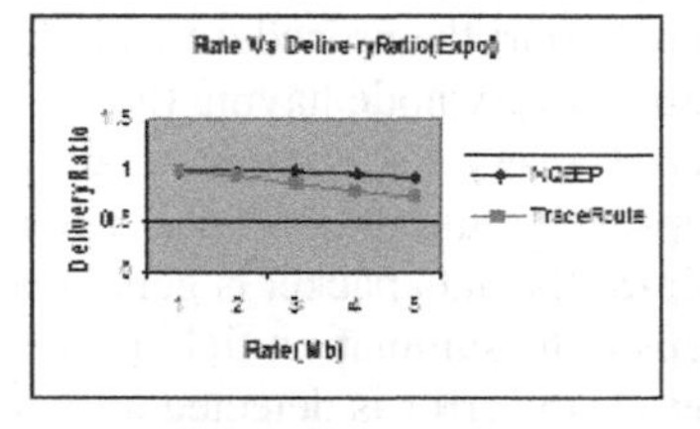

Fig. 6. Rate Vs Delivery Ratio

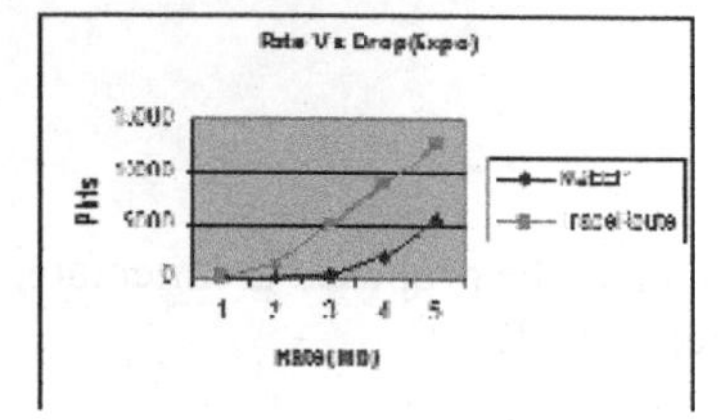

Fig. 7. Rate Vs Drop

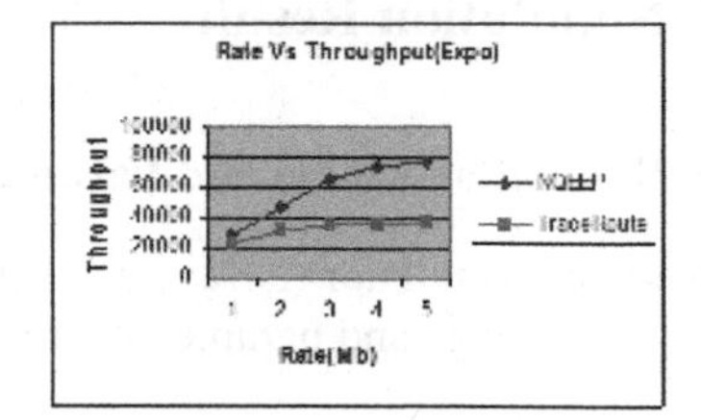

Fig. 8. Rate Vs Throughput

Figure 5 shows the delay of NQEEP and TraceRoute techniques for different rate scenario. We can conclude that the delay of our proposed NQEEP approach has 70% of less than TraceRoute approach.

Figure 6 shows the delivery ratio of NQEEP and TraceRoute techniques for different rate scenario. We can conclude that the delivery ratio of our proposed NQEEP approach has 11% of higher than TraceRoute approach.

Figure 7 shows the drop of NQEEP and TraceRoute techniques for different rate scenario. We can conclude that the drop of our proposed NQEEP approach has 84% of less than TraceRoute approach.

Figure 8 shows the throughput of NQEEP and TraceRoute techniques for different rate scenario. We can conclude that the throughput of our proposed NQEEP approach has 40% of higher than TraceRoute approach.

Case-2 (TCP scenario):Based on Flows. In our second experiment we vary the tcp flows as 1,2,3,4 and 5.

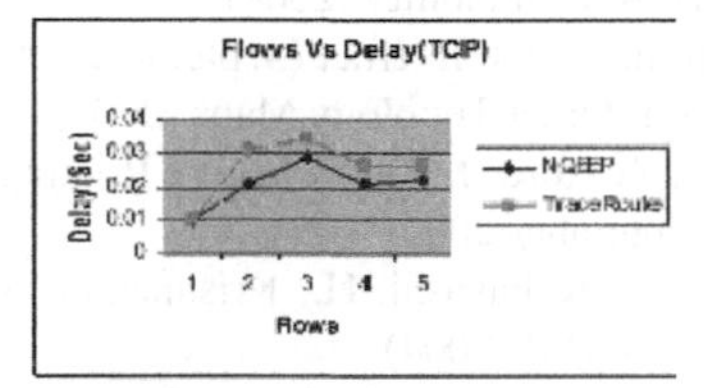

Fig. 9. Flows Vs Delay

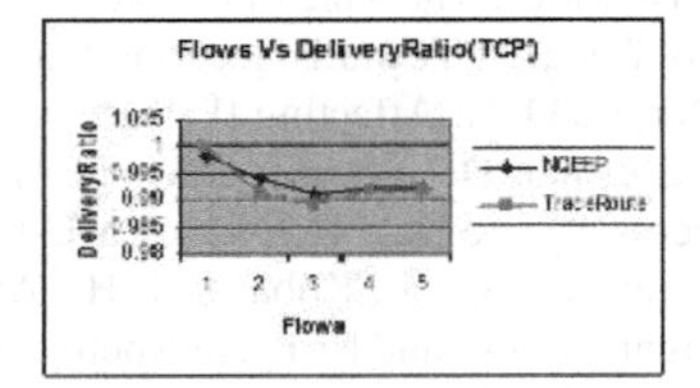

Fig. 10. Flows Vs Delivery Ratio

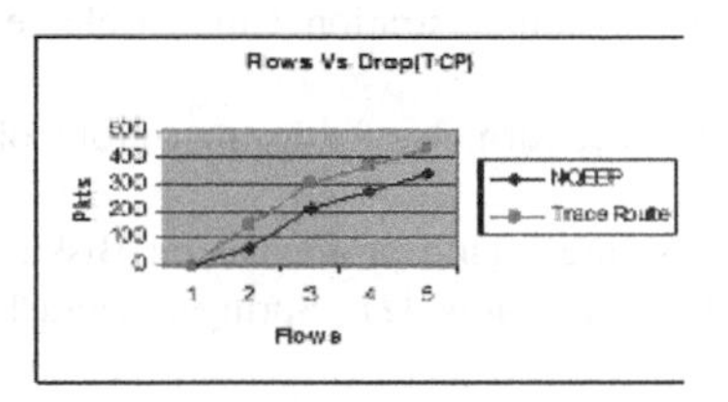

Fig. 11. Flows Vs Drop

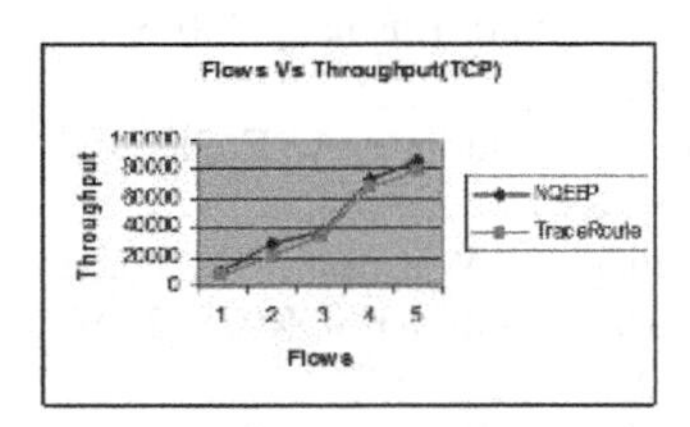

Fig. 12. Flows Vs Throughput

Figure 9 shows the delay of NQEEP and TraceRoute techniques for different flows scenario. We can conclude that the delay of proposed NQEEP approach has 18% of less than TraceRoute approach.

Figure 10 shows the delivery ratio of NQEEP and TraceRoute techniques for different flows scenario. We can conclude that the delivery ratio of our proposed NQEEP approach has 1% of higher than TraceRoute approach.

Figure 11 shows the drop of NQEEP and TraceRoute techniques for different flows scenario. We can conclude that the drop of our proposed NQEEP approach has 27% of less than TraceRoute approach.

Figure 12 shows the throughput of NQEEP and TraceRoute techniques for different flows scenario. We can conclude that the throughput of our proposed NQEEP approach has 14% of higher than TraceRoute approach.

5 Conclusion

Here in the paper all the network quality will take in to consideration. So we are able to provide a best networking data transfer method. Security is maintained through a principal and credential, which is having life of short time stamp providing high security in data transmission. The last phase is able to solve if there is any issue present in the network.

References

1. Goldberg, S., Xiao, D., Barak, B., Rexford, J.: Measuring Path Quality in the Presence of Adversaries: The Role of Cryptography in Network Accountability (2008)
2. Goldberg, S.: Towards Securing Inter domain Routing on the Internet (September 2009)
3. Martins, O.A.: Affecting IP Traceback with Recent Internet Topology Maps (2005)
4. Wendlandt, D., Avramopoulos, I., Andersen, D.G., Rexford, J.: Don't Secure Routing Protocols, Secure Data Delivery, CMU-CS-06-154 (September 2006)
5. Avramopoulos, I., Kobayashi, H., Avramopoulos, I., Kobayashi, H., Krishnamurthy, A.: Highly Secure and Efficient Routing. In: IEEE INFOCOM (2004)
6. Janic, M., Kuipers, F., Zhou, X., Van Mieghem, P.: Implications for QoS provisioning based on traceroute Measurements. In: Stiller, B., Smirnow, M., Karsten, M., Reichl, P. (eds.) QofIS/ICQT 2002. LNCS, vol. 2511, pp. 3–14. Springer, Heidelberg (2002)
7. Cisco, Small Business 300 Series Managed Switch Administration Guide Release 1.3 (2013)
8. Corin, R., Durente, A., Etalle, S., Hartel, P.: Using trace formulae for security protocol design (2001)
9. Cortier, V., Warinschi, B., Zălinescu, E.: Synthesizing secure protocols. In: Biskup, J., López, J. (eds.) ESORICS 2007. LNCS, vol. 4734, pp. 406–421. Springer, Heidelberg (2007)
10. Cortier, V., Warinschi, B.: Computationally Sound, Automated Proofs for Security Protocols. In: Sagiv, M. (ed.) ESOP 2005. LNCS, vol. 3444, pp. 157–171. Springer, Heidelberg (2005)
11. Padmanabhan, V.N., Simon, D.R.: Secure Traceroute to Detect Faulty or Malicious Routing, http://research.microsoft.com/crypto/dansimon/me.htm
12. Nordström, O., Dovrolis, C.: Beware of BGP Attacks (2005)
13. Goldberg, S., Xiao, D., Barak, B., Rexford, J.: Measuring Path Quality in the Presence of Adversaries: The Role of Cryptography in Network Accountability (2008)
14. Claffy, K.: Border Gateway Protocol (BGP) and Traceroute Data Workshop Report. ACM SIGCOMM Computer Communication Review 42(3) (July 2012)
15. Cortier, V., Warinschi, B., Zălinescu, E.: Synthesizing secure protocols, IST-2002-507932, JC9005 (2008)
16. Aiello, L.C., Aiello, L.C.: Verifying Security Protocols as Planning in Logic Programming. ACM Transactions on Computational Logic 2(4), 542–580 (2001)

17. Mızrak, A.T., Cheng, Y.-C., Marzullo, K., Savage, S.: Fatih: Detecting and Isolating Malicious Routers (2005)
18. Zhang, X., Lan, C., Perrig, A.: Secure and Scalable Fault Localization under Dynamic Traffic Patterns (2011)
19. Murali, G., Pranavi, M., Navateja, Y., Bhargavi, K.: Network Security Scanner. In: Pranavi, M., et al. (eds.) Int. J. Comp. Tech. Appl., IJCTA 2(6), 1800–1805 (November-December 2011), http://www.ijcta.com
20. Garfinkel, S., Shelat, A.: A Study of Disk Sanitization Practices. IEEE Security and Privacy (January-February 2003)
21. Daniels, T.E., Spafford, E.H.: Subliminal Trace route in TCP/IP. CERIAS Technical Report 2000/10
22. Avramopoulos, I., Rexford, J.: Stealth Probing: Securing IP Routing through Data-Plane Security (June 27, 2005)
23. Schmidt, B., Meier, S., Cremers, C., Basin, D.: Automated Analysis of Diffie-Hellman Protocols and Advanced Security Properties (2009)
24. Martins, O.A.: Affecting IP Traceback with Recent Internet Topology Maps (2005)
25. Siamwalla, R., Sharma, R., Keshav, S.: Discovering Internet Topology. In: IEEE INFOCOM 1999 (1999)
26. Lee, S., Wong, T., Kim, H.S.: Secure Split Assignment Trajectory Sampling: A Malicious Router Detection System (2006)
27. Padmanabhan, V.N., Simon, D.R.: Secure Traceroute to Detect Faulty or Malicious Routing, Microsoft Research, http://www.research.microsoft.com/epadmanab/, http://research.microsoft.com/crypto/dansimon/me.htm/
28. Goldberg, S., Xiao, D., Barak, B., Rexford, J.: A Cryptographic Study of Secure Internet Measurement. Technical Report (March 5, 2007)
29. Goldberg, S., Xiao, D., Tromer, E., Barak, B., Rexford, J.: Path-Quality Monitoring in the Presence of Adversaries (March 27, 2008)
30. Network simulator, http://www.isi.edu/nsnam/ns

Cryptanalysis of Image Encryption Algorithm Based on Fractional-Order Lorenz-Like Chaotic System

Musheer Ahmad[1], Imran Raza Khan[2], and Shahzad Alam[1]

[1] Department of Computer Engineering, Faculty of Engineering and Technology, Jamia Millia Islamia, New Delhi-110025, India
[2] Department of Computer Science and Engineering, Institute of Technology and Management, Aligarh-202001, India

Abstract. This paper provides break of an image encryption algorithm suggested by Xu *et al.* recently in [Commun Nonlinear Sci Numer Simulat 19 (10) 3735–3744 2014]. The authors realized a Laplace transformation based synchronization between two fractional-order chaotic systems to execute error-free encryption and decryption of digital images. The statistical analyses show the consistent encryption strength of Xu *et al.* algorithm. However, a careful probe of their algorithm uncovers underlying security shortcomings which make it vulnerable to cryptanalysis. In this paper, we analyze its security and proposed chosen plaintext-attack/known plaintext-attack to break the algorithm completely. It is shown that the plain-image can be successfully recovered without knowing secret key. The simulation of proposed cryptanalysis evidences that Xu *et al.* algorithm is not secure enough for practical utilization.

Keywords: Image encryption, security, fractional chaotic system, synchronization, chosen-plaintext attack.

1 Introduction

Due to recent advancements in information and communication technologies, the digital images have become indispensable mean of communication in the application areas of defense and military, multimedia-broadcasting, satellite-communication, tele-medicine, tele-education, weather forecasting, disaster management, etc,. The demand of secure and fast image-based communication has attracted growing attention of researchers worldwide. Consequently, enormous numbers of image-encryption proposals have been suggested using different techniques to serve the purpose. However, the underlying architecture of many proposals suffers from serious security flaws which make them susceptible to even classical cryptographic attacks. Many image encryption proposals have been successfully broken by cryptanalysts under various attacks and found insecure [1-10]. Cryptanalysis is the science of breaching cryptographic systems to recover plaintext without an access to secret key. The objective of an attacker is to find a way to recover secret key or plaintext in lesser time or storage than the brute-force attack [1]. It is practiced to find weaknesses, if any, in the security system that eventually may leads to the previous results [11, 12].

S.C. Satapathy et al. (eds.), *Emerging ICT for Bridging the Future – Volume 2*,
Advances in Intelligent Systems and Computing 338, DOI: 10.1007/978-3-319-13731-5_41

The new security systems are being designed to replace broken ones and new cryptanalytic techniques are invented to crack the improved security systems. In practice, both the cryptanalysis and cryptography are two equally significant aspects of a security system. It is recommended to design against possible cryptanalysis [13]. The four classical attacks in cryptanalysis [11], in context to image encryption, are: (1) *Ciphertext-only attack*: the attacker only has access to some ciphertext images that can be utilized to recover the plaintext image, (2) *Known-plaintext attack*: the attacker can obtain some plaintext images and corresponding ciphertext images to reveal the plaintext image, (3) *Chosen-plaintext attack*: the attacker can have temporary access to encryption machine and choose some specially designed plaintext images to generate corresponding ciphertext images, and (4) *Chosen-ciphertext attack*: the attacker can have temporary access to the decryption machine and choose some specially designed ciphertext images to obtain the corresponding plaintext images.

Very recently, Xu *et al.* [14] proposed an image encryption algorithm based on synchronization of two fractional-order chaotic systems. The dynamics of fractional-order chaotic systems has more complex behaviour than integer order systems. A Laplace transformation based synchronization of (Drive and Response) systems is realized. A 3D Lorenz-like fractional-order chaotic system is employed at the sender side to encrypt digital plain-images. The algorithm has statistical features of almost flat histograms, higher information entropy, low cross-correlation among adjacent pixels, high key-sensitivity and large key space. Our contribution includes careful security probe of algorithm to find underlying flaw of plain-image independency on the generation of decimal codes used during plain-image pixels encryption. Different plain-images yield same decimal codes if the secret key is kept unchanged. As a result, the algorithm fails to resist the proposed chosen-plaintext/known plaintext attacks which completely breaks the algorithm and recovers the plain-image.

The structure of the remaining paper is arranged as follows: Section 2 provides review of image encryption algorithm under probe. Section 3 analyzes and discusses the cryptanalysis of encryption algorithm under proposed two different attacks. The simulation of cryptanalysis is also demonstrated in the same Section. The conclusions of the work are provided in Section 4.

2 Xu *et al.* Algorithm

The Xu *et al.* image encryption algorithm is based on synchronization of fractional-order Lorenz-like chaotic systems in drive-response configuration via Pecora and Carrol (PC) control method. The fractional order chaotic system employed by Xu *et al.* in encryption algorithm is described as [15]

$$\begin{aligned} D^{\alpha_1} x &= a(x-y) \\ D^{\alpha_2} y &= bx - lxz \\ D^{\alpha_3} z &= -cz + hx^2 + ky^2 \end{aligned} \tag{1}$$

Where, α_1, α_2, α_3 are fractional derivative orders, x, y, z are state variables, and a, b, c, l, h, k are system parameters. The system (1) shows chaotic behaviour for α_1=0.97, α_2=0.98, α_3=0.99, a=10, b=40, c=2.5, l=1, h=2, k=2. The fractional-order derivative is solved using the Caputo fractional derivative method defined below [16]:

$$D^{\alpha} f(t) = \frac{1}{\Gamma(n-\alpha)} \int_0^t \frac{f^{(n)}(\tau)}{(t-\tau)^{\alpha-n+1}} d\tau \quad for \quad n-1 < \alpha < n \tag{2}$$

Where $\Gamma(z) = \int_0^{\infty} e^{-t} t^{z-1} dt$ is the Euler's Gamma function.

Xu *et al.* build a PC drive-response configuration: a drive system constitutes fractional order Lorenz-like system with (x, y, z) variables denoted by subscript m (for master), and a response system given by the subspace containing (y, z) variables. The chaotic signal x_m is adopted to drive the response system. The drive (master) system is defined by

$$\begin{aligned} D^{\alpha_1} x_m &= a(x_m - y_m) \\ D^{\alpha_2} y_m &= bx_m - lx_m z_m \\ D^{\alpha_3} z_m &= -cz_m + hx_m^2 + ky_m^2 \end{aligned} \tag{3}$$

The response (slave) system is defined by

$$\begin{aligned} D^{\alpha_2} y_s &= bx_m - lx_m z_s \\ D^{\alpha_3} z_s &= -cz_s + hx_m^2 + ky_s^2 \end{aligned} \tag{4}$$

The synchronization between the systems (3) and (4) using Laplace transformation theory is realized first before an error-free encryption and decryption of images at sender and receiver sides is performed. It has been shown that the error vectors ($e_1 = y_s - y_m$, $e_2 = z_s - z_m$) converge to zero and a complete synchronization is achieved after initial synchronization time of about five seconds. The readers are advised to go through the Ref. [14] for detailed description of synchronization.

The encryption algorithm suggested by Xu *et al.* to encrypt digital images is follows as:

Algorithm #1: *Encrypt-Xu( )*

Input : Plain-image P
Output : Encrypted image C

1. Read the plain-image matrix $P_{M\times N}$ (of size M×N) and convert the 2D matrix to 1D array $P = \{p_1, p_2, \ldots\ldots, p_{MN}\}$ (of size MN)
2. Initialize all fractional derivative orders, system variables and parameters.
3. Simulate the fractional-order system (3) and (4) to achieve complete synchronization.

4. Further iterate system (3) for MN times and capture the chaotic sequence $Z = \{z_1, z_2, \ldots\ldots, z_{MN}\}$

5. Preprocess the sequence Z to obtain decimal codes $D = \{d_1, d_2, \ldots\ldots, d_{MN}\}$, $d_i \in [0, 255]$ as

$$d_i = round(mod((abs(z_i - floor(abs(z_i))) \times 10^5, 256)) \quad i = 1, 2, \ldots\ldots, MN$$

6. Encrypt the pixels of plain-image p_i using d_i as

$$c_i = Bin2Dec(Dec2Bin(p_i) \oplus Dec2Bin(d_i)) \qquad i = 1, 2, \ldots\ldots, MN$$

7. Convert 1D array $C = \{c_1, c_2, \ldots\ldots, c_{MN}\}$ to 2D matrix of encrypted image C.

The decryption is performed in the same way as the encryption, except that the chaotic sequence Z is obtained from fractional-order system defined in eqn (4) with same set of secret keys. The schematic diagram of Xu *et al.* image encryption-decryption algorithm based on drive-response configuration is depicted in Figure 1.

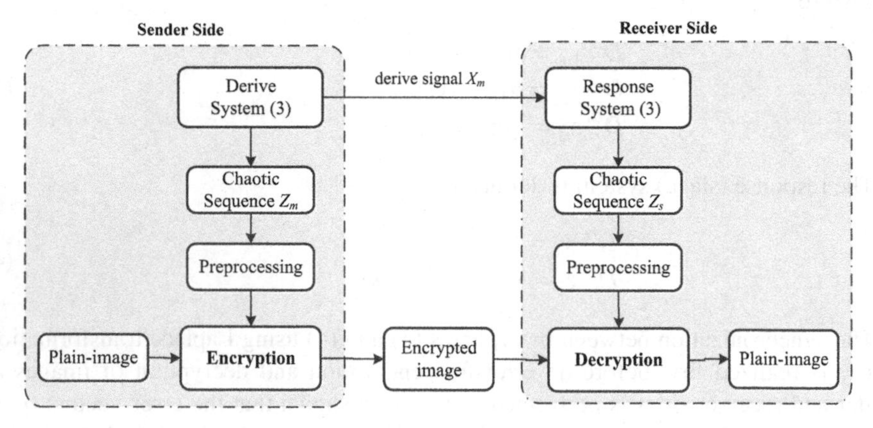

Fig. 1. Schematic diagram of Xu *et al.* image encryption-decryption algorithm

3 Cryptanalysis of Xu *et al.* Algorithm

In cryptography, there is an axiom stated by Auguste Kerckhoffs in 19-th century: "*A cryptographic system should be secure even if everything about the system, except the key, is public knowledge*" [17]. Kerckhoffs's axiom was reformulated by Claude Shannon as "*The enemy knows the system*" and acknowledged as Shannon's maxim [18]. Hence, everything about encryption algorithm including its implementation is public except the secret key which is private. In other words, the attacker has temporary access to encryption or decryption machines. Recovering the plain-image is as good as knowing the secret key.

The overall security of Xu *et al.* encryption algorithm relies on the secrecy of the initial conditions assigned to secret key components α_1, α_2, α_3, x, y, z, a, b, c, l, h, k. The design of their algorithm depicts that if, anyhow, attacker knows the generated decimal codes d_i, he/she can recover the plain-image from the received encrypted image. Thus, the generated decimal codes d_i are the equivalent keys of algorithm. So, instead of trying to know the actual initial conditions of key components, an attacker may design method to deduce the decimal codes. This can be achieved by exploiting the following inherent flaws of algorithm under probe.

- The decimal codes used to encrypt the plain-image always remain unchanged when different plain-images are encrypted. The generation of decimal codes is independent to the pending plain-image information.
- The algorithm has high key sensitivity which makes the brute-force attack infeasible. However, it has lack of, or practically no, plain-image sensitivity. Means, a minute change in the plain-image doesn't cause a drastic change in the encrypted content from security point of view.

The attacker utilizes above analytical information to execute attacks to reveal the plain-image. Assume that the attacker has gained temporary access to the encryption machine and encrypted image C which is to be decoded. Let P be the plain-image which is to be recovered from its received encrypted image C.

The chosen-plaintext (CPA) attack needs specially designed image to reveal decimal codes. A zero image Q (of size C) containing all pixels with zero gray values is designed for the purpose. The revelation of P from C under CPA attack is provided in algorithm **#2**. The method y=*Encrypt-Xu*(x) encrypts the input plain-image x according to Xu *et al.* algorithm and return corresponding encrypted image y.

$$Q = \{q_{1,1}, q_{1,2}, \ldots\ldots, q_{1,N}, q_{2,1}, q_{2,2}, \ldots\ldots, q_{2,N}, \ldots\ldots\ldots, q_{M,N-1}, q_{M,N}\}$$

Where gray-value of pixel at (i, j) is $q_{i,j} = 0$ for all $i = 1 \sim$ M, $j = 1 \sim$ N.

Algorithm #2: *CPA-attack( )*

Input	:	Zero image Q and received encrypted image C
Output	:	Recovered image P

begin

D = *Encrypt-Xu*(Q) $\quad // Q \oplus D = 0 \oplus D = D$

P = *Bitwise-ExOR*(C, D) $\quad // C \oplus D = (P \oplus D) \oplus D = P \oplus (D \oplus D) = P$

end

In order to illustrate the cryptanalysis under CPA attack, we give the simulation results in Figure 2.

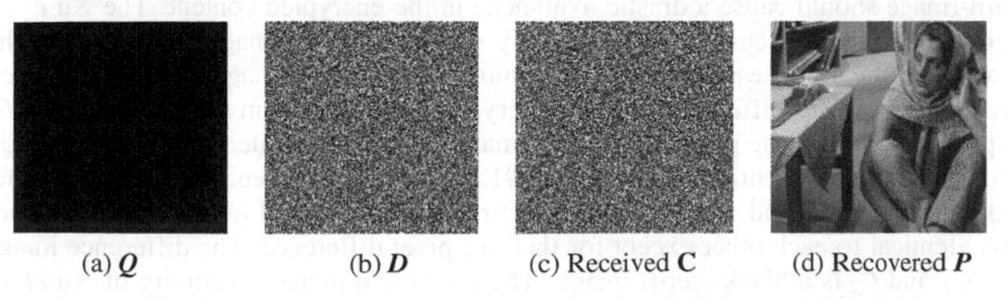

(a) $\boldsymbol{Q}$ (b) $\boldsymbol{D}$ (c) Received **C** (d) Recovered $\boldsymbol{P}$

Fig. 2. Simulation of cryptanalysis under chosen-plaintext attack

The known-plaintext (KPA) attack doesn't needs any specially designed image. Instead, it takes the access of pair of plaintext and ciphertext images and analyzes them to reveal the necessary information. Let the attacker has the access of plain-image P_1 and its encrypted image C_1. The successful revelation of P from C under KPA attack is provided in algorithm **#3**.

Algorithm #3: *KPA-attack( )*

Input	:	Pair of plain-image P_1 and its encrypted image C_1
Output	:	Recovered image P

begin

$D = Bitwise\text{-}ExOR(P_1, C_1)$ // $P_1 \oplus C_1 = P_1 \oplus (P_1 \oplus D) = (P_1 \oplus P_1) \oplus D = D$

$P = Bitwise\text{-}ExOR(D, C)$ // $D \oplus C = D \oplus (P \oplus D) = P \oplus (D \oplus D) = P$

end

The simulation results of cryptanalysis under KPA attack are provided in Figure 3.

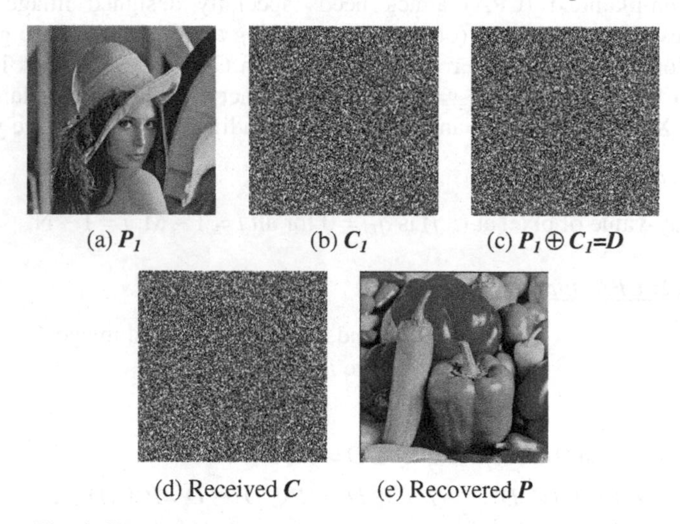

(a) P_1 (b) C_1 (c) $P_1 \oplus C_1 = D$

(d) Received C (e) Recovered P

Fig. 3. Simulation of cryptanalysis under known-plaintext attack

The Shannon's properties of confusion and diffusion for a strong cryptosystem necessitate a high sensitiveness to a minute change in plain-image. A small change in plain-image should cause a drastic avalanche in the encrypted content. The Xu *et al.* algorithm has poor sensitiveness to a tiny change in plain-image. To illustrate the severity of the weakness, we take two almost similar plain-images P_1 and P_2 which have just one pixel difference and are encrypted with algorithm #1 to get C_1 and C_2 respectively. Since, the generation of decimal codes is independent to the plain-image information, the execution of algorithm #1, for P_1 and P_2, generates same decimal codes to output C_1 and C_2. As a result, the resultant encrypted images C_1 and C_2 are also identical to each other except for that one pixel difference. The difference image J of C_1 and C_2 is a black (zero) image. The poor plain-image sensitivity of Xu *et al.* algorithm is simulated in Figure 4.

Algorithm #4: *Plain-image_Sensitivity( )*

Input : Plain-images P_1 and P_2 having only one pixel difference
Output : Difference image J

begin

$C_1 = Encrypt\text{-}Xu(P_1)$

$C_2 = Encrypt\text{-}Xu(P_2)$

$J = Bitwise\text{-}ExOR(C_1, C_2)$ // *since* $x \oplus x = 0$

end

(a) $\boldsymbol{P_1}$ (b) $\boldsymbol{P_2}$ (c) $\boldsymbol{J}$

Fig. 4. Simulation of poor plain-image sensitivity of Xu *et al.* algorithm

4 Conclusion

This paper presented a break of an image encryption algorithm suggested by Xu *et al.* very recently. Their algorithm exploited the features of fractional-order chaotic systems and based on the drive-response configuration. The drive system is synchronized with response system via Laplace transformation theory before the encryption/decryption process begins. The security probe of the algorithm unveils that the generation of decimal codes solely depends on secret key and independent to pending plain-image information. It yields same decimal codes sequence when different plain-images are encrypted. This flaw makes it susceptible to proposed attacks. It has been shown that plain-image can be recovered under chosen-plaintext/known-plaintext attacks without having any knowledge of secret key. It also highlighted the poor plain-image sensitivity of algorithm. Hence, the presented work demonstrated successful cryptanalysis and found that the Xu *et al.* encryption algorithm is not at all secure for practical utilization.

References

1. Hermassi, H., Rhouma, R., Belghith, S.: Security analysis of image cryptosystems only or partially based on a chaotic permutation. Journal of Systems and Software 85(9), 2133–2144 (2012)

2. Çokal, C., Solak, E.: Cryptanalysis of a chaos-based image encryption algorithm. Physics Letters A 373(15), 1357–1360 (2009)
3. Rhouma, R., Solak, E., Belghith, S.: Cryptanalysis of a new substitution-diffusion based image cipher. Communication in Nonlinear Science and Numerical Simulation 15(7), 1887–1892 (2010)
4. Li, C., Lo, K.T.: Optimal quantitative cryptanalysis of permutation-only multimedia ciphers against plaintext attacks. Signal Processing 91(4), 949–954 (2011)
5. Rhouma, R., Belghith, S.: Cryptanalysis of a spatiotemporal chaotic cryptosystem. Chaos, Solitons & Fractals 41(4), 1718–1722 (2009)
6. Ahmad, M.: Cryptanalysis of chaos based secure satellite imagery cryptosystem. In: Aluru, S., Bandyopadhyay, S., Catalyurek, U.V., Dubhashi, D.P., Jones, P.H., Parashar, M., Schmidt, B. (eds.) IC3 2011. CCIS, vol. 168, pp. 81–91. Springer, Heidelberg (2011)
7. Rhouma, R., Belghith, S.: Cryptanalysis of a new image encryption algorithm based on hyper-chaos. Physics Letters A 372(38), 5973–5978 (2008)
8. Özkaynak, F., Özer, A.B., Yavuz, S.: Cryptanalysis of a novel image encryption scheme based on improved hyperchaotic sequences. Optics Communications 285(2), 4946–4948 (2012)
9. Solak, E., Rhouma, R., Belghith, S.: Cryptanalysis of a multi-chaotic systems based image cryptosystem. Optics Communications 283(2), 232–236 (2010)
10. Sharma, P.K., Ahmad, M., Khan, P.M.: Cryptanalysis of image encryption algorithm based on pixel shuffling and chaotic S-box transformation. In: Mauri, J.L., Thampi, S.M., Rawat, D.B., Jin, D. (eds.) SSCC 2014. CCIS, vol. 467, pp. 173–181. Springer, Heidelberg (2014)
11. Schneier, B.: Applied Cryptography: Protocols Algorithms and Source Code in C. Wiley, New York (1996)
12. Bard, G.V.: Algebraic Cryptanalysis. Springer, Berlin (2009)
13. Military Cryptanalysis Part I- National Security Agency, `http://www.nsa.gov/public_info/_files/military_cryptanalysis/mil_crypt_I.pdf` (last access on August 30, 2014)
14. Xu, Y., Wang, H., Li, Y., Pei, B.: Image encryption based on synchronization of fractional chaotic systems. Communications in Nonlinear Science and Numerical Simulation 19(10), 3735–3744 (2014)
15. Li, R.H., Chen, W.S.: Complex dynamical behavior and chaos control in fractional-order Lorenz-like systems. Chinese Physics B 22(4), 040503 (2012)
16. Petráš, I.: Fractional-order nonlinear systems modeling, analysis and simulation. Springer, Heidelberg (2011)
17. Kerckhoffs's principle, `http://crypto-it.net/eng/theory/kerckhoffs.html` (last access on August 28, 2014)
18. Shannon, C.E.: Communication Theory of Secrecy Systems. Bell System Technical Journal 28, 662 (1949)

The Probabilistic Encryption Algorithm Using Linear Transformation

K. Adi Narayana Reddy[1] and B. Vishnuvardhan[2]

[1] Faculty of CSE, ACE Engineering College, Hyderabad, India
[2] Faculty of IT, JNTUH College of Engineering, Karimnagar, India

Abstract. The probabilistic encryption produces more than one ciphertext for the same plaintext. In this paper an attempt has been made to propose a probabilistic encryption algorithm based on simple linear transformation. The variable length sub key groups are generated using a random sequence. A randomly selected element is replaced by each element of the plaintext from the corresponding indexed sub key group. With this a cryptanalyst cannot encrypt a random plaintext looking for correct ciphertext. The security analysis and performance of the method are studied and presented.

Keywords: Hill cipher, Sub key groups, Pseudo Random number, Probabilistic encryption.

1 Introduction

The information to be transmitted must be secure against several attacks. This is achieved by using encryption and decryption techniques, which converts readable of information into unreadable form and vice versa. Many numbers of cryptographic algorithms are available to provide secured transformation of information, but the efficiency and strength of the algorithm is one of the most important aspects to be studied in the field of information security. With the development of probabilistic encryption algorithms a cryptanalyst cannot encrypt random plaintext looking for correct ciphertext because the encryption process produces more than one ciphertext for one plaintext. We consider linear transformation based cryptosystem which is a simple classical substitution cipher.

In 1929 Hill developed a simple cryptosystem based on linear transformation. It is implemented using simple matrix multiplication and it hides single character frequency and also hides more frequency information by the using large key matrix. But it is vulnerable to known plaintext attack and the inverse of every shared key matrix may not exist all the time. It is a simple traditional symmetric key cipher and the message is transmitted through the communication channel is divided into 'm' blocks, each of size 'n'. Assume that both 'n' and 'm' are positive integers and M_i is the ith block of plaintext. This procedure encrypt each of the block M_i, one at a time using secret key matrix. It maps each character with unique numeric value like A=0, B=1 ... to produce the 'n' characters in each of the block. The ith ciphertext block C_i can be obtained by encrypting the ith plaintext block M_i using the following equation (1)

S.C. Satapathy et al. (eds.), *Emerging ICT for Bridging the Future – Volume 2*,
Advances in Intelligent Systems and Computing 338, DOI: 10.1007/978-3-319-13731-5_42

$$C_i = M_i K \bmod m \tag{1}$$

In which K is an n x n key matrix. The plaintext can be obtained from the decrypted cipher text using following equation (2)

$$M_i = C_i K^{-1} \bmod m \tag{2}$$

In which K^{-1} is the key inverse and it exist only if the GCD (det K (mod m), m) = 1.

Many researchers improved the security of linear transformation based cryptosystem. Yeh, Wu et al. [16] presented an algorithm which thwarts the known-plaintext attack, but it is not efficient for dealing bulk data, because too many mathematical calculations. Saeednia [13] presented an improvement to the original Hill cipher, which prevents the known-plaintext attack on encrypted data but it is vulnerable to known-plaintext attack on permutated vector because the permutated vector is encrypted with the original key matrix. Ismail [5] tried a new scheme HillMRIV (Hill Multiplying Rows by Initial Vector) using IV (Initial Vector) but Rangel-Romeror et al. [10] proved that If IV is not chosen carefully, some of the new keys to be generated by the algorithm, may not be invertible over Z_m, this make encryption/decryption process useless and also vulnerable to known-plaintext attack and also proved that it is vulnerable to known-plaintext attack. Lin C.H. et al. [8] improved the security of Hill cipher by using several random numbers. It thwarts the known-plaintext attack but Toorani et al.[14, 15] proved that it is vulnerable to chosen ciphertext attack and he improved the security, which encrypts each block of plaintext using random number and are generated recursively using one-way hash function but Liam Keliher et al [7] proved that it is still vulnerable to chosen plaintext attack . Ahmed Y Mahmoud et al [1, 2, 3] improved the algorithm by using eigen values but it is not efficient because the time complexity is more and too many seeds are exchanged. Reddy, K.A. et al [11, 12] improved the security of the cryptosystem by using circulant matrices but the time complexity is more. Again Kaipa, A.N.R et al [6] improved the security of the algorithm by adding nonlinearity using byte substitution over GF (2^8) and simple substitution using variable length sub key groups. It is efficient but the cryptanalyst can find the length of sub key groups by collecting pair of same ciphertext and plaintext blocks. Later Adinarayana reddy [17] improved the security of [6] by dynamic byte substitution where the dynamic byte substitution shifts the static location to new secret location. Goldwasser and Micali introduced the probabilistic encryption algorithm [18] but is impractical to implement. In this paper randomness will be included to the linear transformation based cryptosystem to overcome chosen-plaintext and chosen-ciphertext attacks and to reduce the time complexity.

In this paper an attempt has been made to introduce the concept of probabilistic encryption. The detailed algorithm is presented in section-2 and its performance and security analysis are studied and presented in section-3.

2 Method

In this paper an attempt is made to propose a randomized encryption algorithm which produces more than one ciphertext for the same plaintext. The following sub sections explain the proposed method.

2.1 Algorithm

Let M be the message to be transmitted. The message is divided into 'm' blocks each of size 'n' where 'm' and 'n' are positive integers and pad the last block if necessary. Let M_i be the ith partitioned block (i = 1, 2, ... m) and size of each M_i is 'n'. Let C_i be ciphertext of the ith block corresponding to the ith of block plaintext. In this paper the randomness is added to the linear transformation based cryptosystem. Each element of the plaintext block is replaced by a randomly selected element from the corresponding indexed sub key group. The randomly selected element will not be exchanged with the receiver. In this method key generation and sub key group generation is similar to hybrid cryptosystem [3]. Choose a prime number 'p'. The following steps illustrate the algorithm.

1. **Step 1: Key Generation.** Select randomly 'n' numbers $(k_1, 1k_2 \cdots, k_n)$ such that $\text{GCD}(k_1, 1k_2 \cdots, k_n) = 1$. Assume $k_i \in Z_p$. Rotate each row vector relatively right to the preceding row vector to generate a shared key matrix K_{nxn}. The generated key matrix is called prime circulant matrix.
2. **Step 2: Sub Key Group Generation.** Let r = $\sum_{i=1}^{n} k_i$ mod p and generate a sequence of 'p' pseudo-random numbers S_i (i = 0, 1,... p-1) with initial seed value as r. Generate the sub-key group S_G as i from pseudo-random numbers as, j = (i + S_i) mod b, for all $i \in S_{G_j}$ and $b < \lfloor p/2 \rfloor$. (i.e. the sub-key groups are formed with the pseudorandom number sequence.)
3. **Step 3: Encryption.** The encryption process encrypts each block of plaintext using the following steps.
   3.1 Initially the transformation is applied as $Y = KM \bmod p$
   3.2 The index of very element of the vector Y is calculated as, Index = Y mod b (i.e. Y mod b = (y_1 mod b, … y_n mod b)) and corresponding to the vector Index an element is randomly selected from the corresponding sub key group S_G and that becomes C_2 and C_1 = (y_1 / b, … y_n / b) + (y_1 mod b, … y_n mod b) mod p i.e. C1 = Y/b + Index
   3.3 The pair of ciphertext (C_1, C_2) is transferred to other end.
4. **Step 4: Decryption.** The decryption process decrypts each of the received ciphertext pair (C1, C2) using the following steps

4.1 Receiver receives ciphertext pair (C_1, C_2) and searches for the index vector Index from C_2 then calculate Y as Y = (b*(C_1-Index) + Index) mod p

4.2 In order to obtain the plaintext the inverse key matrix is multiplied with the resultant vector Y as

$$M = K^{-1}Y \bmod p$$

The algorithm is explained through the following example

2.2 Example

Consider a prime number p as 53 and the set of relatively prime numbers as [5, 27, 13]. Generate shared key matrix K_{3x3}. Assume the plaintext block M= [12, 14, 3]. Generate a sequence of 'p' pseudo-random number with seed value as r = 45. Assume b = 5 and generate five sub-key groups (S_G) from the random number sequence. The sub key groups are random and of variable length.

S_G [0] = {0, 6, 17, 21, 24, 25, 31, 38, 50}
S_G [1] = {1, 4, 9, 12, 16, 29, 30, 34, 39, 40, 43, 44, 46, 48, 49}
S_G [2] = {2, 3, 13, 22, 23, 26, 37, 45, 51, 52}
S_G [3] = {7, 10, 15, 19, 20, 27, 33, 42}
S_G [4] = {5, 8, 11, 14, 18, 28, 32, 35, 36, 41, 47}
Y = KM mod p = KM mod 53 = [0, 42, 44]
Y/ b = (0, 8, 8)
Y% b = (0, 2, 4)

Now select elements randomly from the corresponding sub key groups and add the position of those elements to the corresponding quotient Y / b.

The possible ciphertext pairs are presented in table 1.

The same plaintext is mapped to many ciphertext pairs

After communicating the ciphertext pair (C_1, C_2) to the receiver, the decryption process outputs the plaintext as [12, 14, 3].

3 Performance Analysis

The performance analysis is carried out by considering the computational cost and security analysis which are to show the efficiency of the algorithm.

3.1 Computational Cost

The time complexity measures the running time of the algorithm. The time complexity of the proposed algorithm to encrypt and to decrypt the text is O (mn^2) which is shown in the equation (4), where 'm' is number of blocks and 'n' is size of each block, which is same as that of original Hill cipher. In this process T_{Enc} and T_{Dec} denote the running time for encryption and decryption of 'm' block of plaintext respectively.

$$T_{Enc}(m) \cong m(n^2)T_{Mul} + m(n^2)T_{Add}$$
$$T_{Dec}(m) \cong m(n^2)T_{Mul} + m(n^2)T_{Add} + mnT_s \quad (3)$$

In which T_{Add}, T_{Mul}, and T_s are the time complexities for scalar modular addition, multiplication, and search for the index respectively.

$$T_{Enc}(m) \cong m(n^2)c_1 + m(n^2)c_2 \cong O(mn^2)$$
$$T_{Dec}(m) \cong m(n^2)c_1 + m(n^2)c_2 + mnc_3 \cong O(mn^2) \quad (4)$$

Where c_1,c_2 and c_3 are the time constants for addition, multiplication and index search respectively. The running time of encryption our method and other methods are analysed and presented in the Fig 1. The running time our method is equal to the linear transformation based cipher. As the block size increases the time to encrypt is linearly increasing whereas other method is increasing exponentially. Similarly the time to decrypt also increases linearly as block size increases. This method outperforms comparing with other methods. Our method even reduced the space complexity because it needs only n memory locations to store the key as the shared key is circulant matrix.

3.2 Security Analysis

The key matrix is shared secretly by the participants. The attacker tries to obtain the key by various attacks but it is difficult because the random selection of elements from sub key groups. It is difficult to know the elements of the sub key groups because each sub key group is of variable length and generated by modulo which is an one-way function.

The proposed cryptosystem overcomes all the drawbacks of linear transformation based cipher and symmetric key algorithms. This is secure against known-plaintext, chosen-plaintext and chosen-ciphertext attacks because one plaintext block is encrypted many number of ciphertext blocks. This is due to the random selection of element from the corresponding sub key group. Therefore, the cryptanalyst can no longer encrypt a random plaintext looking for correct ciphertext. To illustrate this assume that the cryptanalyst has collected a ciphertext C_i and guessed the corresponding plaintext M_i correctly but when he/she encrypt the plaintext block M_i the corresponding ciphertext block C_j will be completely different. Now he/she cannot confirm M_i is correct plaintext for the ciphertext C_i. This makes the more secure and efficient. It also requires huge memory. It is also secure against brute force attack if the key size is at least 8 and prime modulo p is at least 53. It is free from all the security attacks.

Table 1. Ciphertext corresponding to Plaintext

	Plaintext M	C1	C2
Case 1	[12, 14, 3]	[0, 10, 12]	[6, 26, 41]
Case 2	[12, 14, 3]	[0, 10, 12]	[31, 51, 18]
Case 3	[12, 14, 3]	[0, 10, 12]	[50, 2, 18]

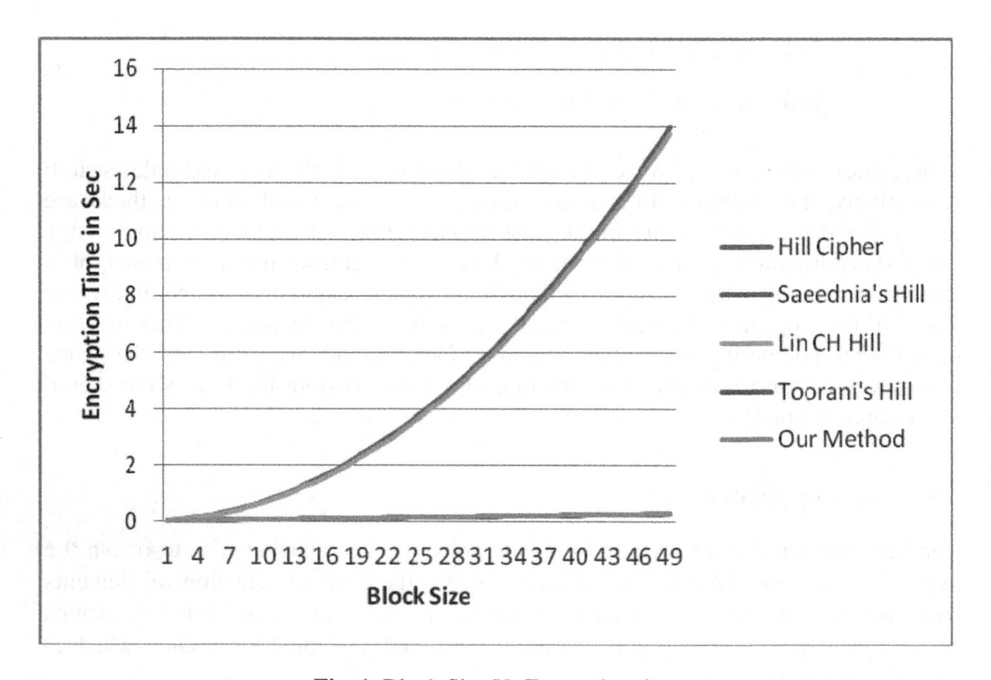

Fig. 1. Block Size Vs Encryption time

4 Conclusion

The structure of the proposed cryptosystem is similar to substitution ciphers i.e. initially the linear transformation is applied on the original plaintext block then the result is replaced by a randomly selected element from the corresponding sub key group. The sub key groups are of variable length and each sub key group is generated randomly using one-way modulo function. The proposed probabilistic encryption algorithm produces more than one ciphertext for one plaintext because each element of the block is replaced by a randomly selected element from the corresponding sub key group. The proposed cryptosystem is free from all the security attacks and it has reduced the memory size from n^2 to n, because key matrix is generated from the first row of the matrix and is simple to implement and produces high throughput.

References

1. Ahmed, Y.M., Chefranov, A.G.: Hill cipher modification based on eigenvalues hcm-EE. In: Proceedings of the 2nd International Conference on Security of Information and Networks, October 6-10, pp. 164–167. ACM Press, New York (2009), doi:10.1145/1626195.1626237
2. Ahmed, Y.M., Chefranov, A.: Hill cipher modification based on pseudo-random eigen values HCM-PRE. Applied Mathematics and Information Sciences (SCI-E) 8(2), 505–516 (2011)
3. Ahmed, Y.M., Chefranov, A.: Hill cipher modification based generalized permutation matrix SHC-GPM. Information Science Letter 1, 91–102
4. Hill, L.S.: Cryptography in an Algebraic Alphabet. Am. Math. Monthly 36, 306–312 (1929), `http://www.jstor.org/discover/10.2307/2298294?uid=3738832&uid=2129&uid=2&uid=70&uid=4&sid=21102878411191`
5. Ismail, I.A., Amin, M., Diab, H.: How to repair the hill cipher. J. Zhej. Univ. Sci. A. 7, 2022–2030 (2006), doi:10.1631/jzus.2006.A2022
6. Kaipa, A.N.R., Bulusu, V.V., Koduru, R.R., Kavati, D.P.: A Hybrid Cryptosystem using Variable Length Sub Key Groups and Byte Substitution. J. Comput. Sci. 10, 251–254 (2014)
7. Keliher, L., Delaney, A.Z.: Cryptanalysis of the toorani-falahati hill ciphers. Mount Allison University (2013), `http://eprint.iacr.org/2013/592.pdf`
8. Lin, C.H., Lee, C.Y., Lee, C.Y.: Comments on Saeednia's improved scheme for the hill cipher. J. Chin. Instit. Eng. 27, 743–746 (2004), doi:10.1080/02533839.2004.9670922
9. Overbey, J., Traves, W., Wojdylo, J.: On the keyspace of the hill cipher. Cryptologia 29, 59–72 (2005), doi:10.1080/0161-110591893771
10. Rangel-Romeror, Y., Vega-Garcia, R., Menchaca-Mendez, A., Acoltzi-Cervantes, D., Martinez-Ramos, L., et al.: Comments on "How to repair the Hill cipher". J. Zhej. Univ. Sci. A 9, 211–214 (2008), doi:10.1631/jzus.A072143
11. Reddy, K.A., Vishnuvardhan, B., Madhuviswanath, Krishna, A.V.N.: A modified hill cipher based on circulant matrices. In: Proceedings of the 2nd International Conference on Computer, Communication, Control and Information Technology, February 25-26, pp. 114–118. Elsevier Ltd. (2012), doi:10.1016/j.protcy.2012.05.016
12. Reddy, K.A., Vishnuvardhan, B., Durgaprasad: Generalized Affine Transformation Based on Circulant Matrices. International Journal of Distributed and Parallel Systems 3(5), 159–166 (2012)
13. Saeednia, S.: How to make the hill cipher secure. Cryptologia 24, 353–360 (2000), doi:10.1080/01611190008984253
14. Toorani, M., Falahati, A.: A secure variant of the hill cipher. In: Proceedings of the IEEE Symposium on Computers and Communications, July 5-8, pp. 313–316. IEEE Xplore Press, Sousse (2009), doi:10.1109/ISCC.2009.5202241
15. Toorani, M., Falahati, A.: A secure cryptosystem based on affine transformation. Sec. Commun. Netw. 4, 207–215 (2011), doi:10.1002/sec.137
16. Yeh, Y.S., Wu, T.C., Chang, C.C.: A new cryptosystem using matrix transformation. In: Proceedings of the 25th IEEE International Carnahan Conference on Security Technology, October 1-3, pp. 131–138. IEEE Xplore Press, Taipei (1991), doi:10.1109/CCST.1991.202204
17. Reddy, K.A., Vishnuvardhan, B.: Secure Linear Transformation Based Cryptosystem using Dynamic Byte Substitution. International Journal of Security 8(3), 24–32
18. Gold Wasser, S., Micali, S.: Probabilistic Encryption. Journal of Computer and System Sciences 28(2), 270–299, doi:10.1016/0022-0000(84)90070-9

Network Management Framework for Network Forensic Analysis

Ankita Bhondele[1], Shatrunjay Rawat[2], and Shesha Shila Bharadwaj Renukuntla[2]

[1] Bansal Institute of Science and Technology, Bhopal, India
ankitabhondele@gmail.com
[2] International Institute of Information Technology, Hyderabad, India
shatrunjay.rawat@iiit.ac.in,
sheshashila.bharadwaj@reserach.iiit.ac.in

Abstract. Tracing malicious packets back to their respective sources is important to defend the internet against attacks. Content based trace-back techniques have been proposed to solve the problem of source identification. It is not feasible to effectively store and query all the data stored in the devices for extended periods of time due to resource limitations in the network devices.

In this paper, we propose a management framework for network packet trace-back with optimum utilization of device storage capacity. We aim to remotely manage the devices and also to store large forensic data so that we can identify the source of even older attacks.

Keywords: Network Management Framework, Bloom filters, Network forensics.

1 Introduction

Computer network has experienced a rapid growth over the past years and with this growth have also come several security issues. To secure network from different types of attacks, attempts have been made at hardware and software levels. New security applications, Firewalls, Antivirus Softwares, Intrusion Detection Systems and many more are used to protect network and host from damage. To stop attacks and their side-effects, it is important to know the actual source of attack so that we may block the malicious system creating problem.

Previously, source machine was identified based on IP address in the packet header which can be altered or spoofed. This is not an effective way to identify the actual source of packet. Second way of source identification is based on packet payload. To perform trace-back in a network based on payload, payload attribution has to be performed by each network device through which packet passes. Payload attribution [1], [6] is an important element in network forensics, which makes the identification of source of packet possible, based on part of packet payload called *"excerpt"*. Practically, it is not possible to store the entire payload at every routing device due to storage and privacy concerns. To deal with the above problem, it is required that data

S.C. Satapathy et al. (eds.), *Emerging ICT for Bridging the Future – Volume 2*,
Advances in Intelligent Systems and Computing 338, DOI: 10.1007/978-3-319-13731-5_43

be stored in compressed form. Burton Howard Bloom, in 1970, proposed a bloom filter [4] which is a space efficient data structure that works on probabilistic algorithm. Bloom Filter has its application in various fields including network forensics [5].Various hash based techniques and their variations have been proposed [2] for storing packet content which are Source Path Isolation Engine (SPIE) [7], Block Bloom Filters (BBF) and Hierarchical Bloom Filters (HBF) to allow analysis of network events for investigation purposes.

Implementation of same payload attribution techniques on all routers is practically not possible as devices have different storage and processing capacity. New technique was proposed [3] which provides flexible environment for the implementation of these attribution techniques based on parameters like block size and false positive probability (FPP). This heterogeneous implementation saves storage space as well as reduces the processing burden on devices. With increase in network traffic, bloom filters required for storing forensic data at each device will also increase. In today's networks, it is difficult to store and query all the packet data for extended periods of time. To offer more protection to the network from damage, we will require forensic data to be stored for longer period which results into larger storage space requirement for devices.

In this paper, we propose a framework that solves the problem of storage requirement at devices during forensic analysis and allows remote management of devices. The proposed management framework uses Simple Network Management Protocol (SNMP) to communicate between Network Management System (Manager) and network device (Agent). During trace-back, each device in this framework will implement payload attribution method with varying block size and False Positive Probability as per requirement. We have proposed methods under Network Management Framework that allow us to store large amount of data for longer period so that we can investigate even an older attack and find out the source for the same. Due to limited storage, it is difficult for devices to store large amount of information. The Network Management System (NMS) based framework overcomes the problem of storage space requirement at each network device by storing all the information at NMS end.

2 Proposed Network Management Framework

For the source identification, content based trace-back techniques were introduced over traditional trace-back techniques. Content based techniques overcome the problem of storage requirement for devices by storing data in compact form using *"bloom filters"*. In a Network, multiple machines may face attacks. In a large network, it is not possible to trace-back using single process as it will consume time. We need an approach that will identify source of attack using multi-process attributionsystem so that network can be protected from attacks.We propose a framework that provides a set of management services which help during forensic analysis.

The proposed Network Management framework consists of two main components: Network Management System (manager) and Device (Agent). SNMP is used to remotely manage a large network. This management protocol is used between the NMS and the device to make management related communication possible. The Request-Response nature of the protocol plays a big role in management tasks. The framework will monitor network activity and manage network resources as well.

The proposed work will provide network visibility to the administrator, so that he can perform trace-back for any victim based on malicious content identified. We have also introduced methods for efficient utilization of available storage space at devices to make payload attribution system more feasible. We have aimed towards minimization of storage space required during trace-back process. The methods under proposed framework are classified based on the way forensic data is stored.

2.1 Methods for Storing Forensic Data in a Network

To investigate instance of attack, information stored in devices plays a big role. Thus, the process of storing forensic data needs more attention. In order to make payload attribution system more scale-able, we have given methods under the framework for storing forensic data during analysis. In addition to the reduction of storage space, proposed methods also reduce processing burden on devices. There are two methods in the NMS based management framework for storing and processing forensic data which are named as *Distributed Management* and *Centralized Management methods.*

Distributed Management Method. In a distributed management, as the host receives any malicious packet from network, it informs the central authority about an occurred attack as shown in Fig.1. Upon receiving an "excerpt" (part) of malicious packet from victim host, NMS takes the responsibility of taking required action. In trace-back mechanism, each device stores forensic data in a special data structure called "Bloom Filter" using parameterized approach. Depending upon the device capability, we can implement BBF and HBF.

NMS sends query to devices to perform traceback in network for identifying the attacker, using the excerpt of packet. Device receives the excerpt query from NMS and checks whether excerpt is present in the available Bloom filters or not. To determine the path taken by the malicious packet, the query has to be matched with the bloom filter content at every device during traceback process. If the query matches stored bloom filter content, it means that the malicious packet has passed through that device. Each device in a network is configured to send query response back to NMS. NMS records all these query responses for future use. After performing membership test, query is sent to another device. Based on received response from each device, NMS constructs the path traversed by malicious packet. Once the path and source of attack is identified by NMS, it can take appropriate action against the attacker for future security. In such a way, each device is checked during traceback by NMS in a distributed management framework for identifying the source of attack.

In distributed management method, payload attribution works in way similar to other techniques as discussed in section II, but under the supervision of Management system. This method does not work for the minimization of storage space requirement. If we have all the bloom filters stored at a single device, failure of the device may result in loss of all forensic data. To avoid such risk, it is beneficial to have distributed forensic data.

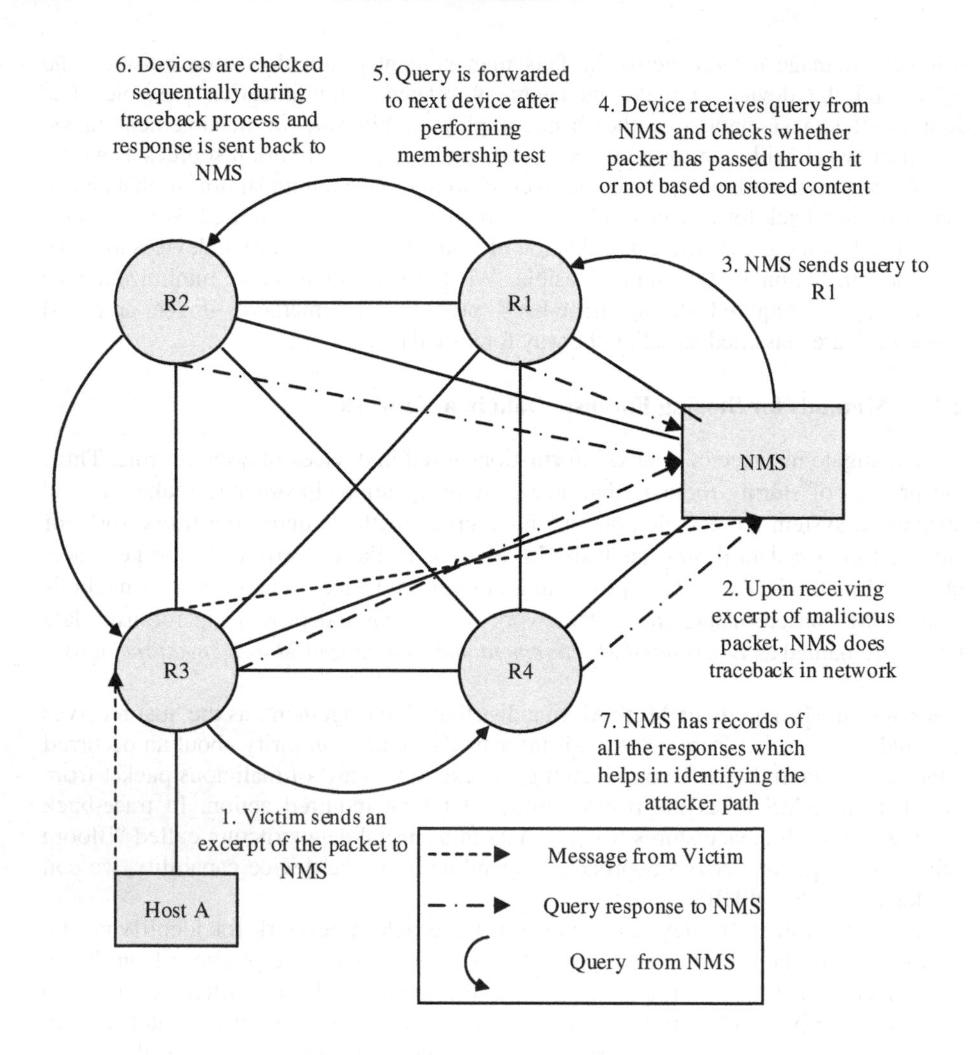

Fig. 1. Distributed Management Method

Centralized Management Method. Each device in second method, as shown in Fig. 2, is configured to store forensic data. For attack analysis, all the forensic data stored in devices is sent to a central storage to avoid distributed processing. Device sends stored forensic data periodically to NMS, thus we have all the bloom filters at one place. In a centralized management method, upon receiving an excerpt of malicious packet, NMS does not send query to network devices to check whether malicious packet has passed through it or not. Instead of querying network devices for forensic data, NMS itself processes all the queries and performs membership test on the data collected from network devices. All the distributed forensic data is recorded by NMS at one place and this recorded information helps NMS in reducing the packet and processing load of the network due to querying process.

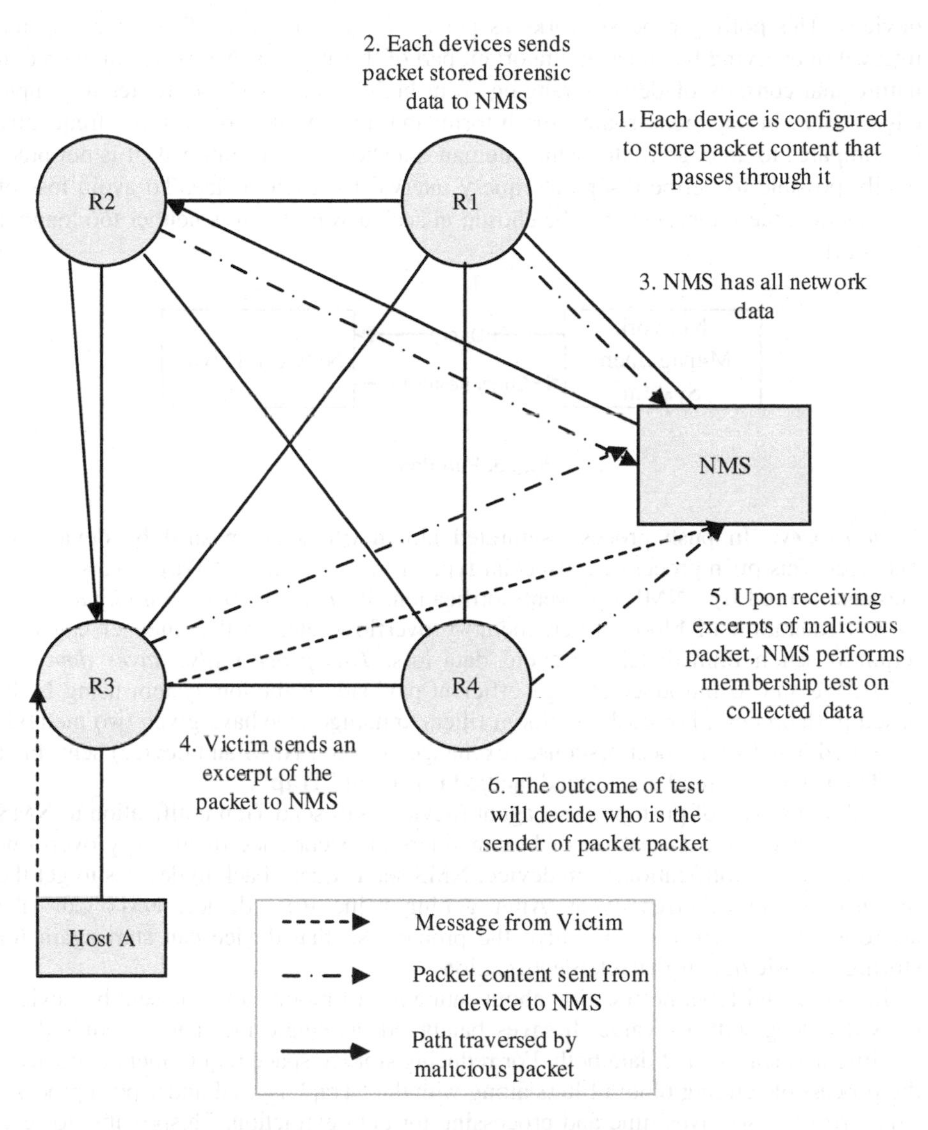

Fig. 2. Centralized Management Method

NMS performs the membership test for identifying the source of packet and test results will decide the path taken by the attack packet. For storing forensic data centrally, it is required to know how to extract stored forensic data from devices. For this purpose we have introduced methods so that the manager can easily extract information from agent (device). These methods are named as Pull Process and Push Process.

Pull Process. Once the bloom filter saturates, it needs to be moved to NMS storage. For the purpose of storing forensic data centrally, manager periodically polls the devices. This polling process works as per predefined intervals. Thus, deciding the interval of querying becomes an important part of this process. Network is dynamic in nature and consists of devices ranging from high speed backbone router to simple edge router. In backbone router, the information update takes place more frequently as compared to an edge router which demands a shorter query interval. It is not practically possible to define a separate query interval for each device. To avoid loss of information, the interval should be chosen in such a way that it is neither too longnor too short.

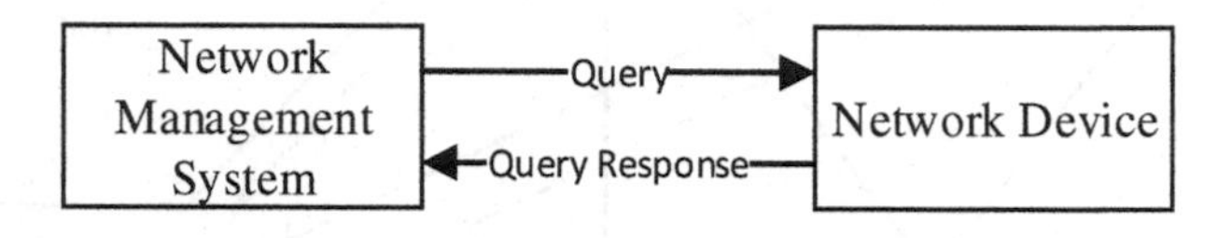

Fig. 3. Pull Process

Push Process. In push process, saturated bloom filters are pushed by devices to manager. This push process uses special type of message called "**Trap**". This type of message is used by SNMP for event notification. *We can use Trap to notify an event such*as saturation of bloom filter, memory overflow, etc. so that manger can take required action immediately to avoid data loss. *This process also saves time and effort.* We aim to introduce storage efficient payload attribution system using NMS based push process. For pushing bloom filters to manger, we have given two methods classified based on request-response (exchange between NMS and device) sent. First is "**Trap"** Only method and second method is Data in "**Trap**".

In the first type of push process, agent (device) will send Trap notification to NMS informing about the saturation of bloom filters or occurrence of memory overflow. Upon receiving notification from device, NMS sends query back to devices to get the instances of an occurred event. After getting value from device, NMS can take appropriate action in order to solve the problem so that device can start again for storing forensic data for attack identification.

In the second type, notification about saturation of bloom filters is sent by devices to NMS along with its value. It saves bandwidth as single message is required for sending notification and data both. For reducing storage space requirement at devices, the process of sending bloom filters along with the "Trap" is used under push process. This process also saves time and processing for data extraction. Thispossible to send bloom filter with in the "Trap" by placing its value in the variable-binding field.

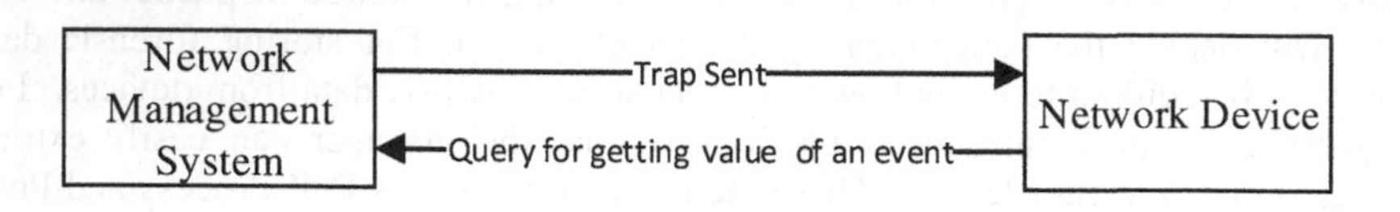

Fig. 4. TrapOnly

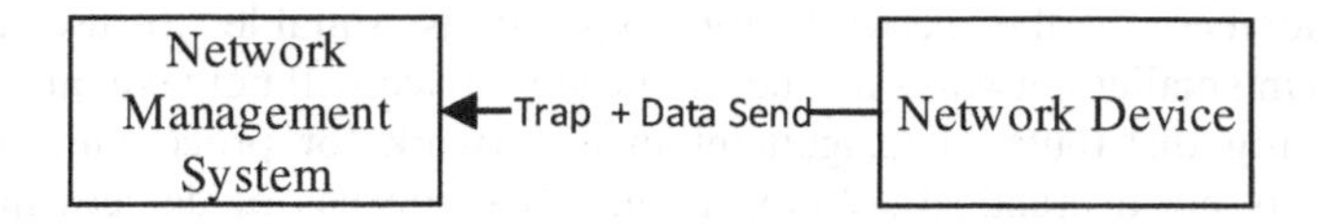

Fig. 5. Data in Trap

Benefits of Push Process over Pull Process. In a pull process under management framework, the manager polls devices for forensic data. Every time the manager needs an update, all the devices respond whether they have any updates or not. This results into burst of traffic, as manager sends a query for updated bloom filters to all network devices. By reducing the frequency of the polling process, we can reduce network traffic which in turn reduces network visibility. This reduced network visibility increases the risk of high forensic data loss, as most of the notifications go unanswered. Due to the loss of forensic data, we may fail to identify the source of attack. In a large network, it becomes impractical for manager to poll a large number of devices, which results in increased traffic and effort by the manager. The push process is a solution to the problem because it does not require a request from the manager. It sends message only when the device changes its state when bloom filters saturates. This process results in a substantial reduction in network traffic and allows data to be sent within a message. This also reduces storage space requirement at devices.

Comparison between Centralized and Distributed Management Methods

We have compared centralized and distributed methods, to analyze which method is more appropriate for effective payload attribution, based on few parameters listed below:

Storage and Processing burden: In distributed method, forensic data stored in the devices will remain in the device itself throughout the analysis process. For checking whether attack packet has passed through the device or not, device has to perform membership test which adds an extra processing burden on devices. Increase in the number of forensic queries will also result in increased processing load on devices during trace-back process. On other hand, in centralized management, all the forensic data is brought to NMS and processed there. This saves excess processing.

In distributed method, each device has to store large amount of forensic data in bloom filters for investigation purposes. Each device has limited memory and it cannot store forensic content for longer periods. The purpose of introducing centralized management method is to decrease the storage requirement by efficiently using available resources at device's end. NMS used in this method supports various functionalities and it can store huge amount of forensic data for desired period of time.

Network traffic: In distributed management,NMS has to send forensic queries to devices, as forensic data is in distributed form which results in increased network traffic.

*Time consumption:*It is difficult to implement distributed method in a larger network, as trace-back process may take lot of time as we have to check all the connected devices in the network topology. It is suitable to use distributed management in smaller networks as querying each device will not take much time.

Implementing distributed management in a network for processing and storing forensic data is not a practical solution as network consists of devices that vary in their processing and storage capacity. Thus, all the devices will not be able to store same amount of forensic data and will not support same processing speed. Processing speed and storage capacity are two important parameters and cannot be compromised. The centralized management method overcomes these problems of storing and processing data, and provides network visibility to the network administrator.

3 Conclusion

In this paper, we have shown how large storage space requirement during forensic analysis at devices can be reduced with the help of proposed framework. This framework consists of management system that manages all the network components and activities. SNMP is used for the information exchange and provides additional features that make the framework more effective for the identification of source of attack. We have given two methods under Framework namely distributed and centralized method. Using centralized method, we are able to store more data for forensic analysis. In this way, the proposed framework minimizes the processing and storage burden on network devices which was there in the previous methods. In future, this work can be extended by implementing the proposed methods to test its feasibility.

References

1. Shanmugasundaram, K., Bronnimann, H., Memon, N.: Payload attribution via hierarchical Bloom filters. In: CCS 2004: Proceedings of the 11th ACM Conference on Computer and Communications Security, pp. 31–41. ACM, New York (2004)
2. Ponec, M., Giura, P., Bronnimann, H., Wein, J.: Highly efficient technique for network forensics. In: The ACM Computer and Communication Security Conference (2007)
3. Shujath, M.S., Rawat, S.: Heterogeneous Configuration of Bloom Filter for Network Forensic Analysis. In: IEEE-CYBER (2012)
4. Bloom, B.: Space/Time Trade-Offs in Hash Coding with Allowable Errors. Comm. ACM 13(7), 422–426 (1970)
5. Broder, A.Z., Mitzenmacher, M.: Network applications of bloom filters: A survey. In: Fortieth Annual Allerton Conference on Communication, Control, and Computing, Coordinated Science Laboratory and the Department of Electrical and Computer Engineering of the University of Illinois at Urbana-Champaign (2002)
6. Shanmugasundaram, K., Memon, N., Savant, A., Bronnimann, H.: Fornet: A distributed forensics network. In: Workshop on Mathematical Methods, Models, and Architectures for Computer Networks Workshop, MMM-ACNS (2003)
7. Snoeren, A.C., Partridge, C., Sanchez, L.A., Jones, C.E., Tchakountio, F., Kent, S.T., Strayer, W.T.: Hash-based IP traceback. In: ACM SIGCOMM, SanDiego, California, USA (August 2001)

Fluctuating Pattern Size Handling for the Extracted Base Verb Forms from Participles and Tenses for Intelligent NLP

P.S. Banerjee[1], G. Sahoo[2], Baisakhi Chakraborty[3], and Jaya Banerjee[4]

[1] Department of Computer Science & Engineering
Jaypee University of Engineering & Technology, Guna, (MP) 473 226, India
[2] Department of Computer Science and Engineering,
B.I.T. Mesra, Ranchi, Jharkhand-835215, India
[3] Department of Information Technology, NIT Durgapur, West Bengal, India
[4] Aryabhatta Institute of Engineering and Management
{partha1010,jaya2008.banerjee}@gmail.com, gsahoo@bitmesra.ac.in,
baisakhi.chakraborty@it.nitdgp.ac.in

Abstract. Natural Language Processing (NLP) is a very important part of a conversation as well as that of a chatterbot. Complete vocabulary building for a chatterbot is a cumbersome and time intensive process and thus a self learning chatterbot is a much more efficient alternative. The learning process can be initiated by a multidimensional approach including individual words to phrases and the whole concepts. Verbs tend to be a constant feature since they have different forms, namely participles and tenses. Even though we will discuss the algorithm to derive the base verb from any participle or tense but for storing such data items it a special kind of data handling is required.

In this regard we will be discussing and proposing the fluctuating data handling strategy which will help us not just to understand the strategy of fluctuating data handling but also its correlation with data handling for smaller databases.

Keywords: smaller databases, fluctuating data, tense, participle, algorithm, pattern.

1 Introduction

Going through the work done on the development of the self-learning tele-text conversational entity, or chatterbot, called RONE, it was essential to develop suitable knowledge representation scheme and the appropriate sentence dissemination methods, [1]. Since not just the chatterbots but also any other NLP module are intended to not just converse with humans but also interpret and comprehend it hence a suitable form of Natural Language Processing (NLP) is needed to be implemented. Research's in NLP as well as machine learning indicates that it is possible to create lexical classes from data sets with some accuracy, [2], [3] and [4] but the very concept

S.C. Satapathy et al. (eds.), *Emerging ICT for Bridging the Future – Volume 2*,
Advances in Intelligent Systems and Computing 338, DOI: 10.1007/978-3-319-13731-5_44

of handling the fluctuating data set needs to be handled. Lexical classifications are a technique that accommodates certain critical NLP functions such as word sense disambiguation and parsing which are important in answering questions and information matching, [5], [6], [7], [8] and [9], all of which are critical to chatterbots and NLP modules. At this point we must think that even though the computing as well as the storage resources in light weights computing devices is increasing with every new device, optimum utilization of these limited resources is a must for such a database system. Now the lexical classes can be used to form generalizations and take into account the syntax of a sentence as formed by its individual words, [10] and [11], as opposed to purely semantic classes. In the initial phases most lexical classes have to be painstakingly defined manually during system development and as such are rarely very comprehensive. As such, increasing emphasis has been placed on automatic classification and since verbs are generally the main predicates in sentences, verbs are the focus of such systems, [2]. An example of an automatically built lexicon is VALEX.

1.1 Part of Speech Derivation

Since the NLP system was developed based on the rules of English Language Grammar, the next step after looking at individual words is to match those individual words to their respective parts of speeches. The individual parts of speech are stored in the SQL database that is the knowledge base component of the system. Unless we input a ridiculous number of words and their parts of speech into the knowledge base, there is no doubt that self learning NLP module will encounter a word that will not return a match from the knowledge base.

1.2 The Issue of New Verbs

When a new word is found to be a verb, it can pose a problem. Since RONE is tense specific and also recognizes the differences between basic verbs and their other forms such as participles, any new verbs must be decomposed into its basic form if it is not already as such, and then expanded into all the different tense and participle forms respectively for storage.

We noted during our study, that verbs fall into certain patterns; on the basis of this we developed an architecture to handle this increasing decreasing data size. Unfortunately the patterns though distinct were not fully consistent. Therefore the rules for the algorithms of decomposing the verbs needed to utilize general rules with multiple exceptions for the verbs that did not fall within the patterns.

1.3 Verb and the Determining the Tense of a Sentence

The verbs are processed for a match in the relevant past, present and future denotations. We have found that the key verb is always the first verb in a phrase, meaning that the tense of a sentence is always given away by the initial verb. Take for example the following sentences:

“Harry **ran** away from school.” = Past tense; “**Did** Harry run away from school?” = Past tense; “Harry **is** the youngest boy in his class.” = Present tense “What **is** today’s date?” = Present tense; “Harry **will** take his exam tomorrow.” = Future tense “What **will** Harry do tomorrow?” = Future tense

The initial verb for each of the sentences is in bold. Notice that the verb mentioned for each sentence is in the tense that the sentence is in regardless of all the following verbs. The exception to this rule is that should a time related noun be included then the tense of the sentence can be affected, for example, “What is Harry doing tomorrow?", which requests a piece of information of a future tense nature, but it is impossible to request for such a piece of information without having the initial verb of the sentence be in the relevant tense if no time related noun such as “yesterday”, “tomorrow” and “later”, is used.

1.4 Recognizing Verb Patterns

Assembling the participles and tenses of verb is performed based on the end patterns of the verb. Any participle or tense can be derived from any other participle or tense of the same base verb.There are a few exceptions that are ignored which are the following:

“did", "should", "must", "could", "have", "do", "does", "would", "can", "has", "will", "are", "is", "need", "shall", "was", "were", "had", "am". These are ignored since their participles and tenses are built into the system first since they are critical in determination of sentence tense, [11].

2 Literature Review

A relatively good amount of work has been done in formally handling the pattern size but no specific methodology has been prescribed for the same in this regard.

2.1 Determining the Base Verb

When considering the participle or tense, apart from the verb is, whose future tense is ‘will’ and future participle if ‘will be’, we observed that no other verb in the English language has a future tense or future participle. All the verbs simply attach the word “will” to their respective base verbs to become future tenses and participles. Sometimes the past participles are formed by adding the verb “was” to the base verb of the intended participle. For example the base verb “run” has a past tense “ran”, a past participle of “was running”, a present participle of “running”, a future tense of “will run” and a future participle of “will be running”. Thereby only the base verb, past tense, and present participle need to be determined in the algorithm [12].

2.2 Non Base Verb Markers

Converting a participle or tense to a base verb involves first determining if the word in question is in fact a not a base verb. We found that verb participles and tenses are clearly “marked” when looking at the verb backwards. All verb participles in the English Language end with “ing”, or “s”, and all verb past tenses end with most commonly an “ed”. There are some unique cases of verb past tenses ending in “id” or “ade” and etcetera [13].

Table 1. Patterns of Participles and Tenses of Verbs

Key: => becomes
! not (EXCEPTION to the rule) == equals
!= not equals

Classifications (ends with)	Sub-Classes (ends with)	Variants (ends with)	Action	Examples
S	ies	== dies, ties	Remove last character	Ties => tie
PRESENT PARTICIPLE		All others	replace last 3 characters with ‘y’	Carries => carry
	us	none	No action	focus
	es	Vowel + consonant + es	Remove last character	Scores = score
		others	Remove last 2 characters	
It *PAST TENSE*	it	== bit	Add ‘e’	Bit => bite
Ought	thought		Replace last 5 characters with ‘ink’	thought => think
PAST TENSE/ PAST PARTICIPLE	fought		Replace last 5 characters with ‘ight’	Fought => fight
	sought		Replace last 5 characters with ‘eek’	Sought => seek
	bought		Replace last 5 characters with ‘uy’	Bought => buy
	brought		Replace last 5 characters with ‘ing’	Brought => bring

3 Algorithm Usage

The input in Table 1 was checked for the end patterns using “if else” conditions. A list of some 100 000 verbs, [18], was run through using the patterns shown in Table 1. Usage of the patterns is based on the following algorithm, represented in first-order logic, where an input verb (represented by *a*) is compared to the Classification (represented by *x*) and to the Sub-classes (represented by *y*) and to the variant (represented by *z*) [13].

3.1 Extendable ID Based Storage Basic Strategy

In Domain Storage, in general a pointer having size p, mostly of 4 bytes, is used to refer to a particular domain value. This leads to (p*N) bytes being used for the attribute for which we set up a domain structure. Instead of having a pointer to point to the domain value, there can be a unique identifier for each of the domain values and store the identifier corresponding to the value in a separate table. [14].

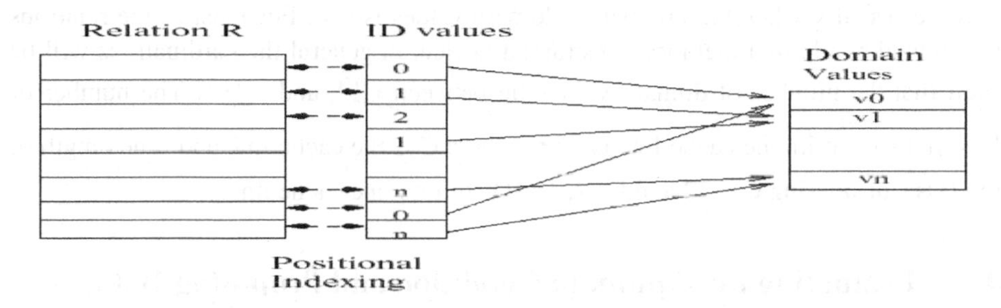

Fig. 1. ID Based Storage Model

This approach does not lead in saving of space since size of a pointer and a integer remains same for almost all models, both being typically 4 bytes.

The storage model (Figure 1) which is the base model on which our architecture stands uses flexible IDs, as in this case the length of the identifier grows and shrinks depending on the total count of the domain values. The main idea behind this model is that D distinct domain values can be distinguished by identifiers of length $\lceil \frac{\log_2 D}{8} \rceil$ bytes. Initially we take 1 byte to represent each identifier. With a single byte we can represent 2^8 domain values. Now with the increase in the number of domain values beyond 2^8, one byte will not be sufficient and we have to assign another byte for the identifiers. Now gradually when the domain value number is greater than 2^{16}, the ID length becomes 3. Hence we can observe that the length of the identifiers can be increased only under absolute need. When the number of domain values is above 2^{8l} where l is the length of the identifier, then we have to increase l. If there are deletions then the number of domain values will be less than 2^{8l}, then it becomes mandatorily essential to decrease l. Hence the length of the identifier is always $\lceil \frac{\log_2 D}{8} \rceil$ bytes which eases the data storage cost considerably.

The identifiers to the domain values are assigned in increasing order starting from 0 and adding one to the last value every time a new ID has to be assigned. The last identifier value that has been assigned is stored. If the value is one less than 2^{8l}, then the length of identifier has to be increased by one byte, or else length remains unchanged. Similarly when deletion occurs, we check whether the last ID value assigned was 2^{8l}. If yes, then we decrease the identifier length by one. In general the operations on the tuple like its insertion, creation and deletion for this model will be slightly more complex than Domain storage.

3.2 Advantages

ID based Storage has the following advantages. Instead of storing a pointer of size 4 bytes as a general course of action, the model uses extendable IDs. A lot of space is saved especially when the number of domain values is less. For most of the relations that would reside in a memory constrained device, in general the cardinalities will be such that the number of domain values lie between 2^8 and 2^{16}. The number of IDs that will fit in one cache line is $\frac{C}{l}$ where C is the cache line and l the length of the IDs.Considering C = 32 and l = 2, 16 IDs fit into one cache line.

4 Evaluating the Optimum Condition and Proposing Data Allocation Strategy for Increased Data size

Problem arises when we are forced to gradually increase the pointer size. Now its time to rethink that when the number of domain value is not greater than $2^{8(p-1)}$ (2^{24} for p=4) ID storage is always a better option. For the case when domain value is 2^{32} the very purpose of having this fraction gradually starts losing its efficacy. For this let the number of domain value not be greater than $2^{8(p-1)}$ (2^{24} for p=4) instead we can increase the number of Relation tables and by adopting two unique data allocation strategies can re-position the data.

Now considering increase the number of Relation tables and by adopting two unique data allocation strategies we hereby present our proposed model where a particular Classification is stored in the Domain table and the corresponding sub classes will be stored in the Relation table (Fig 1). To make this model more versatile what we have done is that instead of exceeding the optimum threshold of the pointer size we adopted a series of Relation table hovering around the centralized domain table (Figure 2). Our proposed model provides an advantage of adding flexibility to the storage model when we switch from a miniature database to a larger one. It provides extra compactness to the data storage model considering the cost and the benefit involved.

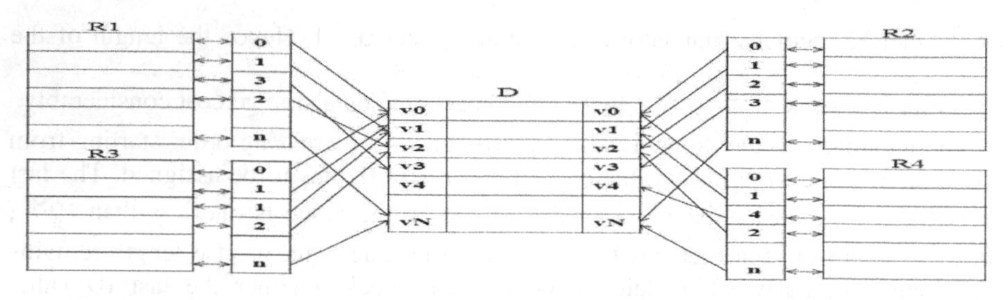

Fig. 2. Proposed Model for Distributed Data allocation Scheme for Intelligent NLP

The two strategies are: The Non Redundant best Fit Method & The Redundant All Beneficial Sites Method.

4.1 The Non Redundant best Fit Method [15]

The Non Redundant best Fit Method determines the single site most likely to allocate a fragment (which may be a file, table or subset of a table) based on maximum benefit. Where benefit may be interpreted as mean total query and update references. Ex, placing fragment Ri at the site S, where the number of local query and update references by all the user transactions is maximized. Hence forth a typical case study has been provided [15].

Table 2. System Parameter

Table	Size	Avg. local query (update) time (milliseconds)	Avg. remote query (update) time (milliseconds)
R1	300 KB	100 (150)	500 (600)
R2	500 KB	150 (200)	650 (700)
R3	1.0 MB	200 (250)	1000 (1100)

Table 3. User Transaction

Transaction	Site(s)	Frequency	Table accesses (reads, writes)
T1	S1, S4, S5	1	Four to R1 (3 reads, 1 write), two to R2 (2 reads)
T2	S2, S4	2	Two to R1 (2 reads), four to R3 (3 reads, 1 write)
T3	S3, S5	3	Four to R2 (3 reads, 1 write), two to R3 (2 reads)

Table 4. Local References for each table at each of the five possible sites

Table	Site	Transactions T1 (frequency)	T2 (frequency)	T3 (frequency)	Total local references
R1	S1	3 read , 1 write (1)	0	0	4
	S2	0	2 read (2)	0	4
	S3	0	0	0	0
	S4	3 read , 1 write (1)	2 read (2)	0	8 (max.)
	S5	3 read , 1 write (1)	0	0	4
R2	S1	2 read (1)	0	0	2
	S2	0	0	0	0
	S3	0	0	3 read, 1 write (3)	12
	S4	2 read (1)	0	0	2
	S5	2 read (1)	0	3 read, 1 write (3)	14 (max.)
R3	S1	0	0	0	0
	S2	0	3 read, 1 write (2)	0	8 (max.)
	S3	0	0	2 read (3)	6
	S4	0	3 read, 1 write (2)	0	8 (max.)
	S5	0	0	2 read (3)	6

4.2 The Redundant All Beneficial Sites Method [15]

The Redundant All Beneficial Sites Method selects all sites for a fragment allocation where the benefit is greater than the cost for one additional copy of that fragment. It can be assumed to start with no copy or one copy of each table or fragment of the table [15].

4.3 Benefit

The benefit is measured by the difference in elapsed time between a remote query and a local query, multiplied by the frequency of queries to fragment F originating from site S.

Benefit = No. of reads*frequency*(remote - local time) (ms)

4.4 Cost

The cost of additional copy is the total elapsed time for all the local updates for fragment F from transactions originating at site S, plus the total elapsed time for all the remote updates of fragment F at site S from transactions originating at other sites.

Table 5. Cost and benefit for each table located at five possible sites

Table	Site	Remote update (local update) transactions	No. of writes *freq*time(ms)	Cost (ms)
R1	S1	T1 from S4 and S5 (T1 from S1)	2*1*600 + 1*1*150	1350
	S2	T1 from S1, S4, S5	3*1*600	1800
	S3	T1 from S1, S4, S5	3*1*600	1800
	S4	T1 from S1 and S5 (T1 from S4)	2*1*600 + 1*1*150	1350
	S5	T1 from S1 and S4 (T1 from S5)	2*1*600 + 1*1*150	1350
R2	S1	T3 from S3 and S5	2*3*700	4200
	S2	T3 from S3 and S5	2*3*700	4200
	S3	T3 from S5 (T3 from S3)	1*3*700 + 1*3*200	2700
	S4	T3 from S3 and S5	2*3*700	4200
	S5	T3 from S3 (T3 from S5)	1*3*700 + 1*3*200	2700
R3	S1	T2 from S2 and S4	2*2*1100	4400
	S2	T2 from S4 (T2 from S2)	1*2*1100 + 1*2*250	2700
	S3	T2 from S2 and S4	2*2*1100	4400
	S4	T2 from S2 (T2 from S4)	1*2*250	2700
	S5	T2 from S2 and S4	2*2*1100	4400

Table 6. Benefit Table

Table	Site	Query (read) sources	No. of reads*frequency*(remote-local time)	Benefit (ms)
R1	S1	T1 at S1	3*1*(500 – 100)	1200
	S2	T2 at S2	2*2*(500 – 100)	1600
	S3	None	0	0
	S4	T1 and T2 at S4	(3*1+2*2)*(500 – 100)	2800
	S5	T1 at S5	3*1*(500 – 100)	1200
R2	S1	T1 at S1	2*1*(650 – 150)	1000
	S2	None	0	0
	S3	T3 at S3	3*3*(650 – 150)	4500
	S4	T1 at S4	2*1*(650 – 150)	1000
	S5	T1 and T3 at S5	(2*1+3*3)*(650 – 150)	5500
R3	S1	None	0	0
	S2	T2 at S2	3*2*(1000 – 200)	4800
	S3	T3 at S3	2*3*(1000 – 200)	4800
	S4	T2 at S4	3*2*(1000 – 200)	4800
	S5	T3 at S5	2*3*(1000 – 200)	4800

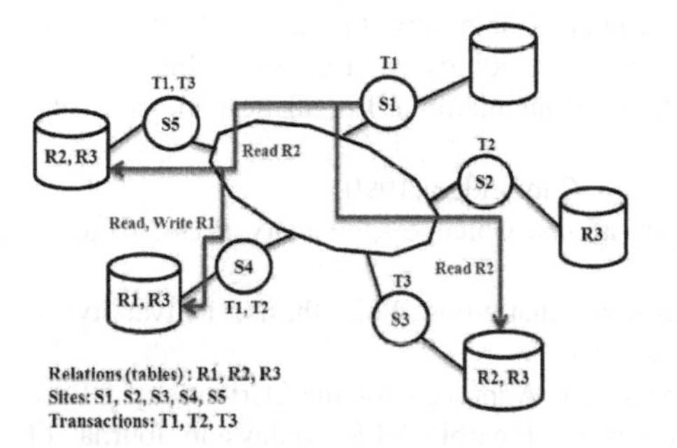

Fig. 3. Data Allocation Diagram for Redundant All Beneficial Sites Method

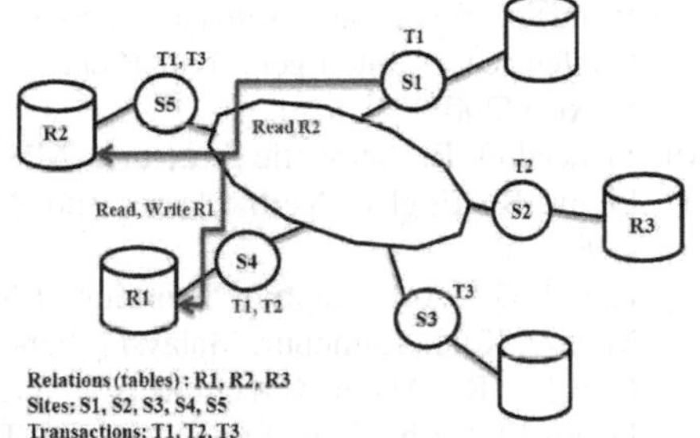

Fig. 4. Data Allocation Diagram for Non-Redundant Best Fit Method

5 Conclusion

Although the tests so far have indicated that any verb can be decomposed correctly into its base state, should any verb be decomposed incorrectly, the correction can be performed by adding another condition to Table 1. For memory constrained computing devices the model works well for a particular number of domain values but for larger data size at initial stages identifier based storage does have an advantage over others but after an optimum level it's not. After a particular pointer size it's better to fragment the data and have a distributed allocation.

References

[1] Raj, R.G.: Using SQL databases as a viable memory system for a learning AI entity. In: Proceedings of the 3rd International Conference on Postgraduate Education (ICPE-3 2008), Penang, Malaysia (2008) (Paper ID: BRC 160)

[2] Merlo, P., Stevenson, S.: Automatic verb classification based on statistical distributions of argument structure. Computational Linguistics 27(3), 373–408 (2001)
[3] Korhonen, A., Krymolowski, Y., Collier, N.: Automatic classification of verbs in biomedical texts. In: Proceedings of the 21st International Conference on Computational Linguistics and the 44th Annual Meeting of the ACL, pp. 345–352 (2006b)
[4] Schulte im Walde, S.: Experiments on the automatic induction of german semantic verb classes. Computational Linguistics 32(2), 159–194 (2006)
[5] Dorr, B.J.: Large-scale dictionary construction for foreign language tutoring and interlingual machine translation. Machine Translation 12(4), 271–322 (1997)
[6] Prescher, D., Riezler, S., Rooth, M.: Using a probabilistic class-based lexicon for lexical ambiguity resolution. In: Proceedings of the 18th International Conference on Computational Linguistics, Saarbrücken, Germany, pp. 649–655 (2000)
[7] Swier, R., Stevenson, S.: Unsupervised semantic role labeling. In: Proceedings of the 2004 Conference on Empirical Methods in Natural Language Processing, Barcelona, Spain, pp. 95–102 (August 2004)
[8] Dang, H.T.: Investigations into the Role of Lexical Semantics in Word Sense Disambiguation. Ph.D. thesis, CIS, University of Pennsylvania (2004)
[9] Shi, L., Mihalcea, R.: Putting pieces together: Combining FrameNet, VerbNet and WordNet for robust semantic parsing. In: Proceedings of the Sixth International Conference on Intelligent Text Processing and Computational Linguistics, Mexico City, Mexico (2005)
[10] Jackendoff, R.: Semantic Structures. MIT Press, Cambridge (1990)
[11] Levin, B.: English Verb Classes and Alternations. Chicago University Press, Chicago (1993)
[12] Raj, R.G.: An Adaptive Learning Tele-Text Chatterbot. PhD thesis, University of Malaya, Kuala Lumnpur, Malaysia (February 2011)
[13] Gopal, R.R., Abdul-Kareem, S.: A Pattern Based Approach for the Derivation of Base Forms of Verbs from Participles and Tenses for Flexible NLP. Malaysian Journal of Computer Science 24(2), 24–34 (2011)
[14] Sen, R., Ramamritham, K.: Efficient Data Management on Lightweight Computing Devices. M.Tech thesis, Indian Institute of Technology, Mumbai-400016, India (February 2004)
[15] Teorey, T.J.: Database modeling and Design (2000)

Enhanced Encryption and Decryption Gateway Model for Cloud Data Security in Cloud Storage

D. Boopathy and M. Sundaresan

Department of Information Technology, Bharathiar University,
Coimbatore, Tamilnadu, India
{ndboopathy,bu.sundaresan}@gmail.com

Abstract. The Cloud computing technology is a concept of providing online multiple resources in the scalable and reliable method. The user will access cloud service as per their requirement of computing level. There is no capital expenditure involved in computing resources for cloud user. The users have to pay as per their storage, network and service usage. The cloud computing have some highlighted issues like compliance, cross border data storage issues, multi-tenant and down time issues. The most important issues are related to storage of data in cloud. Data confidentiality, data integrity, data authentication and regulations on data protection are major problems that affect user's business. This paper discusses about encryption and decryption data security issues in cloud and its safety measures and comes out with a novel research work of cloud data security model.

Keywords: Cloud Security, Cloud Computing, Cloud Data Security, Data Security, Cloud Data Storage, Compliance Issues.

1 Introduction

The cloud service providers are widely spreading their service with their own business models. This own business model is basically designed and extracted from some open source model [1]. But the core things of cloud computing has never changed [3] [4]. The user's data will be stored outside the user's premises so the users easily lose their control over the data which are stored in cloud storage. When the user loses their control over their data [5], then the user is systematically locked in with their cloud service provider or third party service vendor [7].

There are no standardized international laws and regulations to protect the user's data stored in cloud. Some countries framed rules and regulations to protect their country data, but the existing rules and regulations are not sufficient to protect the data in cloud.

2 Problem Statement

The cloud security has many phases and they are identified as availability issues, data security issues, identify and access management issues, government issues and compliance issues [8].

S.C. Satapathy et al. (eds.), *Emerging ICT for Bridging the Future – Volume 2*,
Advances in Intelligent Systems and Computing 338, DOI: 10.1007/978-3-319-13731-5_45

There are number of cloud data security related issues associated with cloud storage.

- The data holding or storing service provider may illegally use the client data to develop their business.
- The cloud service providers give access to their local government and their authorities as per their jurisdiction regulations and standards.
- The service provider may bankrupt or vanish from the service providing market.
- The unauthorized access cannot be avoided in an effective manner.

While the user, must ensure that the provider has taken the proper security measures to protect their information or data in cloud.

3 Cloud Computing

National Institute of Standards and Technology defined cloud computing as follows:

"Cloud computing is a model for enabling convenient, on-demand network access to a shared pool of configurable computing resources (e.g., networks, servers, storage, applications, and services) that can be rapidly provisioned and released with minimal management effort or service provider interaction [2]."

3.1 Cloud Service Models

Software as a Service (SaaS) is a model provided to the consumer to use the cloud service provider's applications on a cloud infrastructure. The cloud running applications are accessible from different client devices through an interface such as a web browser [2].

Platform as a Service (PaaS) is a model provided to the consumer to deploy onto the cloud infrastructure consumer created or else consumer acquired applications, created using multiple programming languages and different tools supported by the cloud service providers [2].

Infrastructure as a Service (IaaS) is a model provided to the consumer to provisions like processing, storage, networks and other fundamental computing resources where the consumer is able to deploy and run based on random choice software, which can include different operating systems and multiple applications [2].

3.2 Cloud Deployment Models

Private Cloud is the cloud infrastructure model operated solely for an organization [6]. It may be managed or controlled by the deployed organization or a third party service providers and may exist on premise or off premise model [2].

Public Cloud is the cloud infrastructure made available to the general public or a large industry group purpose and it is owned and managed by an organization selling cloud services [2].

Community Cloud is the cloud infrastructure model shared by several organizations and supports a specific community that has shared concerns. It may be managed by the organizations or a third party and may exist on premise or off premise [2].

Hybrid Cloud is the cloud infrastructure model that is a composition of two or more clouds (i.e. private, community, or public) that remain unique entities but are bound together by standardized or technology proprietary [2].

4 Encryption and Decryption Gateway Server (E&DGS)

The Encryption & Decryption Gateway Server (E&DGS) is used to encrypt the data storing in cloud storage before crossing the data ownership country border. The schema used for encryption and encrypted keys are stored within the data ownership country border itself. If any user try to access the encrypted data stored in cloud storage, that users request is redirected to the Encryption & Decryption Gateway Server (E&DGS). If the users have credentials to access the encrypted data then the

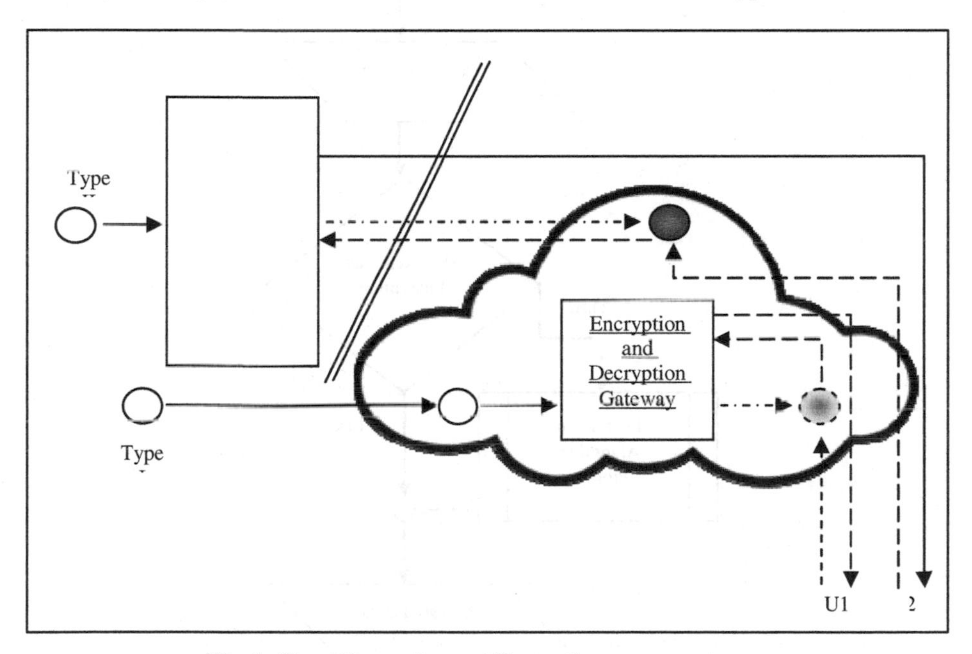

Fig. 1. Cloud Encryption and Decryption processes types

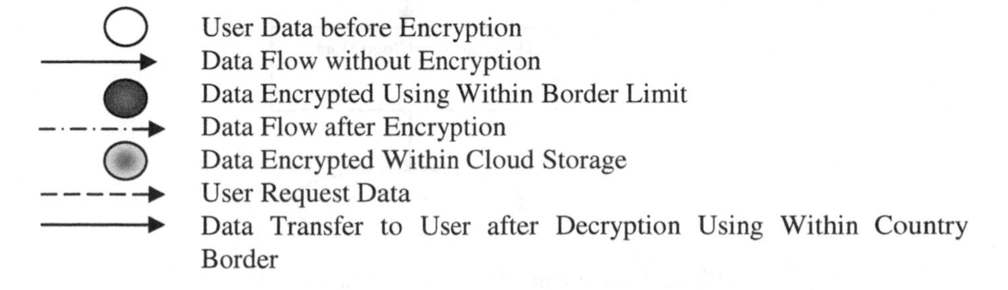

encrypted data will be decrypted and sent to the data request user. This method avoids the unauthorized data usage without data owner's knowledge. If any government legally has rights to view the data stored in their country territory, they also need permission from the Encryption & Decryption Gateway Server (E&DGS) to view the data which is stored in their country. The process of this Encryption & Decryption Gateway Server (E&DGS) model reduces the security threats on data stored in cloud storage and multi-tenant model.

Procedure for Data Storage Process and the Process Explained in Figure 2

1. Process started.
2. Data transferred from user to E&DGS.

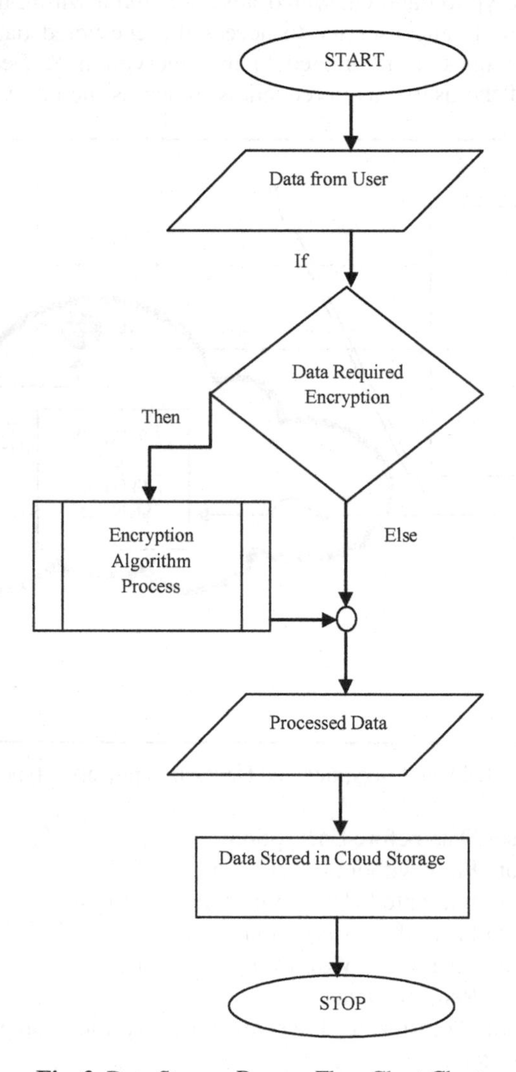

Fig. 2. Data Storage Process Flow Chart Chart

3. If data requires any encryption, then the data is transferred to the encryption algorithm process and then transferred to processed data state.
4. Else the data is transferred to processed state.
5. Data gets stored in cloud storage
6. Process stopped.

Procedure for Data Retrieval Process and the Process Explained in Figure 3

1. Process started.
2. Data access request from user.
3. Requested data retrieved from the cloud storage.
4. If the selected data requires any decryption then it is transferred to decryption algorithm process. Else the data sent to the request person as a processed data.
5. In decryption process, first it checks the user credentials. Whether the user has rights to decrypt it and access it or not.

   When the user is valid then the decryption takes place and then processed data transferred to the data request person.
6. Else the user is not valid then the request is cancelled and the process stops.
7. Process stopped.

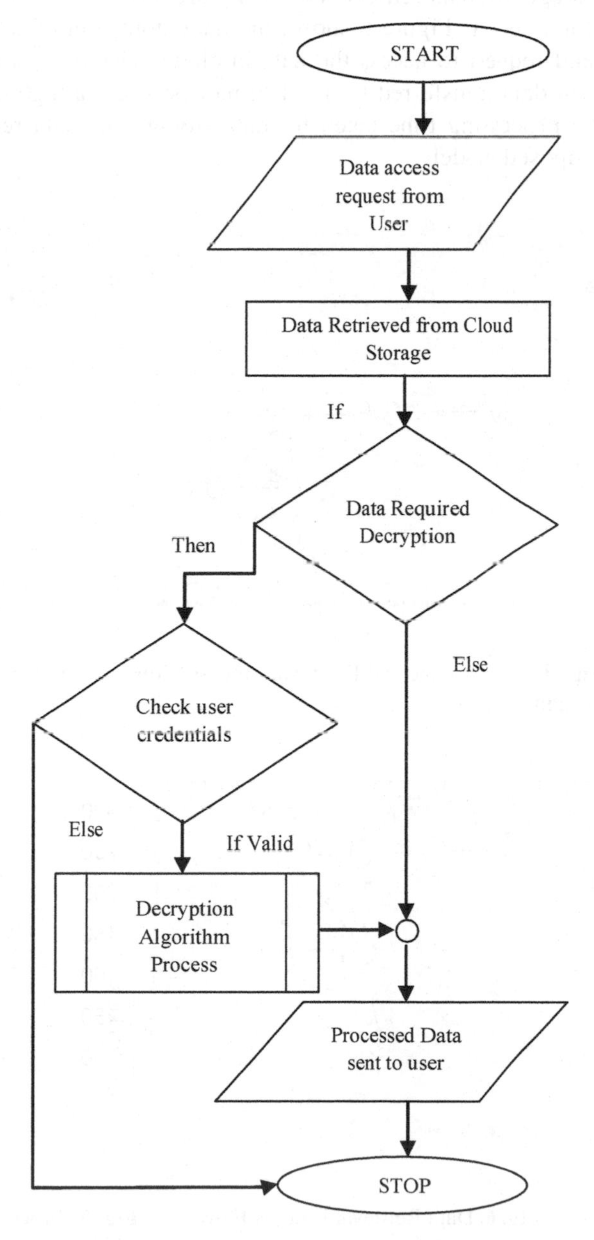

Fig. 3. Data Retrieval Process Flow Chart

5 Discussion

The Network Simulator 2 (NS2) is used to simulate the Encryption and Decryption Gateway Server Model (E&DGS). Figure 4 shows the data flow design of data storage and data retrieval model. Figure 5 shows the data transfer from the E&DGS to cloud server. Figure 6 shows the data storage in cloud multiple server and the user send request to access the data in cloud. That request is transferred to E&DGS and then data transferred to the data request user through cloud service. Figure 7 shows the processing time taken for data storage and data retrieval in cloud storage using proposed model.

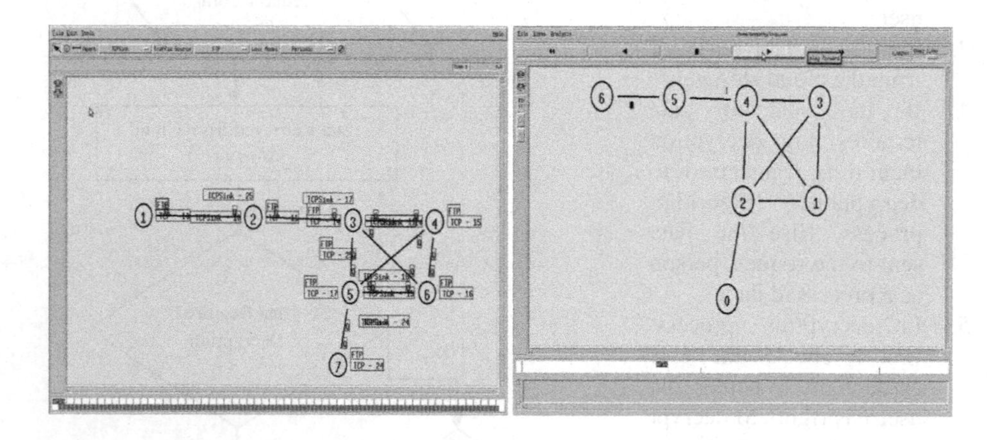

Fig. 4. Data storage and Retrieval Process Flow Design

Fig. 5. Data storage Process Flow

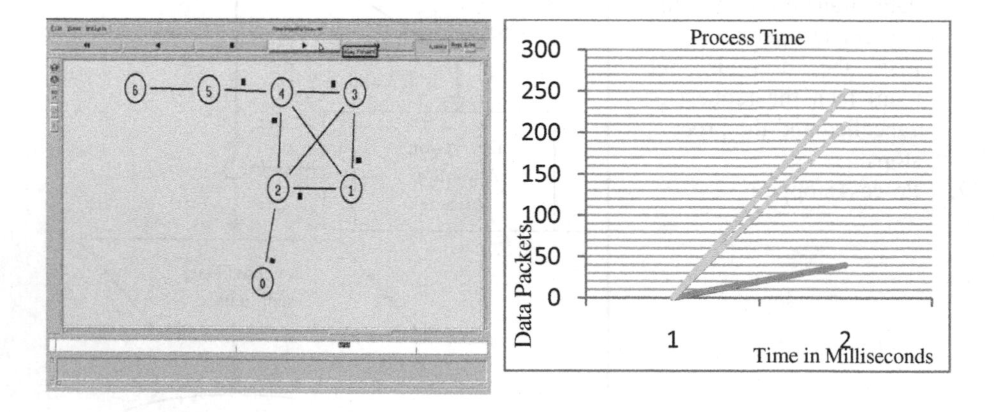

Fig. 6. Data Retrieval Process Flow

Fig. 7. Processing time taken for data storage and retrieval in proposed model

6 Conclusion and Future Scope

The proposed Encryption and Decryption Gateway Server (E&DGS) model works effectively on data encryption and decryption during data transfer across the country. If the user's request is redirected to the E&DGS model, it will avoid unauthorized and illegal data access and usage. This model works very effectively when additional security check points like digital watermark allocation and verification added to it. Despite these security check points, it is also necessary to make some revisions regarding the standards, rules and regulations on cloud service to provide maximum level of security for cloud service providers and cloud service users.

References

[1] Yu, X., Wen, Q.: A View About Cloud Data Security From Data Life Cycle. In: International Conference on Computational Intelligence and Software Engineering (CiSE), December 10-12, pp. 1–4 (2010)
[2] NIST Definition (February 02, 2014), http://csrc.nist.gov/publications/drafts/800-146/Draft-NIST-SP800-146.pdf
[3] Boopathy, D., Sundaresan, M.: Data Encryption Framework Model with Watermark Security for Data Storage in Public Cloud Model. In: International Conference on Computing for Sustainable Global Development, March 5-7, pp. 903–907 (2014)
[4] Boopathy, D., Sundaresan, M.: Location Based Data Encryption Using Policy and Trusted Environment Model for Mobile Cloud Computing. In: Second International Conference on Advances in Cloud Computing, September 19-20, pp. 82–85 (2013)
[5] Kaur, S.: Cryptography and Encryption In Cloud Computing. VSRD International Journal of Computer Science & Information Technology, VSRD-IJCSIT 2(3), 242–249 (2012); Kaur, A., Bhardwaj, M.: Hybrid Encryption For Cloud Database Security. International Journal of Engineering Science & Advanced Technology 2(3), 737–741 (2012)
[6] Cloud Deployment Models, http://bizcloudnetwork.com/defining-cloud-deployment-models (retrieved March 05, 2014)
[7] Tripathi, A., Yadav, P.: Enhancing Security of Cloud Computing using Elliptic Curve Cryptography. International Journal of Computer Applications (0975 - 8887) 57(1), 26–30 (2012)
[8] Vulnerability Rethink, http://www.cbronline.com/news/studycallsforvulnerability_030209 (retrieved January 02, 2014)

A New Approach for Data Hiding with LSB Steganography

G. Prashanti and K. Sandhyarani

Department of Computer Science and Engineering,
VLITS, Vadlamudi, Guntur Dist, Andhra Pradesh, India
{Prashanti.G,Sandhyarani.K,prashantiguttikonda77}@gmail.com

Abstract. Steganography is the art of hiding data into a media in such a way that the presence of data can't be detected when the communication is taking place. This paper provides a review of recent achievements of LSB based spatial domain steganography that have an improved steganography's ultimate objectives, which are undetectable, robustness and capacity of hidden data. These techniques can help researchers in understanding about image steganography and various techniques of hiding data in an image. Along with this, two new methods are proposed one for hiding secret message into cover image and the second is hiding a grey scale secret image into another grey scale image. These methods uses 4-states #table that generates pseudo random numbers which are used for hiding secret information. These methods provide higher security because secret information is hidden at different position of LSB of image depending on pseudo numbers generated by the #table.

1 Introduction

Problems in dealing with illicit interception, eavesdropping, confidentiality and unauthorized copying of digital media on the Internet have become the focus of serious attempts at prevention. Cryptography, steganography, watermarking, finger printing are some technologies used for securing the secrecy of communication. Steganography and Cryptography are cousins where the cryptography technique is scrambling of data with an encryption key to an obscured form to prevent others from understanding it. Steganography is a technique of embedding secret information into a cover object that prevents the intruder from suspecting the secret information in the cover object. It is favorable over encryption because encryption only obscures the meaning of the information, steganography hides the information, and thus provides security.

2 Recent Techniques in LSB Based Image Steganography

The idea of stegnography is to hide text in image with the conditions that the image quality is retained along with the size of the image. Hence, it is often necessary to employ sophisticated techniques to improve embedding reliability. In this paper we

S.C. Satapathy et al. (eds.), *Emerging ICT for Bridging the Future – Volume 2*,
Advances in Intelligent Systems and Computing 338, DOI: 10.1007/978-3-319-13731-5_46

have discussed some recent achievements for hiding data based on spatial domain that are proposed by some scholars which provides good security issue and PSNR value than general LSB based image Stegnography method.

N Sathisha [3] introduced a Spatial Domain technique in which difference between consecutive pixels and the mean of median values is determined to embed payload in 3bits of Least Significant Bit (LSB) and one bit of MSB in chaotic manner. S. M. Masud Karim [4] here, the secret key is used for distributing the secret information into different position of LSB image. R. L. Tataru [5], proposed that the cover image is divided in two pixel blocks and a chaotic algorithm is applied to find the new indexes for the working blocks where the information is hidden. D.Badrinath [6] proposed approach works by determining the smallest value of the Most Significant Bit (MSB) of the three planes and embed the data in Least Significant Bits (LSB) of the planes other than the plane that has minimum MSB value. Another approach is layout of pixels from an image in various ways, and according to the logical sequence of pixels, data can be hidden in the LSBs, up to four positions at max [8].

Dr. Diwedi Samadhi [7] had proposed a method called raster scan principle in which pixels from alternate horizontal lines are used for substituting the secret information. Ching-Sheng Hsu, [9] has applied the Ant Colony Optimization Algorithm to construct an optimal LSB substitution matrix. Manoj Kumar Ramaiya [10] proposed a steganography model that is based on Simplified DES encryption algorithm. Shabir A. Parah [11] presented a high capacity steganographic technique in which cover image is breaking into constitute bit planes and the data is embedded in the first three LSB planes under the control of a private key.

METHOD	IMAGE	PSNR
Chaos based Spatial Domain Steganography using MSB	Coverimage: Eight.tif payload image: Pears.png	42.532
MSB constrain based variable embedding		46.0118
Layout management scheme		3.0
Shape based		4.1
Color based		10.1
Finding optimal LSB substitution using Ant Colony Optimization Algorithm		34.58

3 Proposed System

We are proposing two different methods. One of them is hiding secret message into an image and the other is hiding a secret image into another cover image. One dimensional cellular automata is used for generating pseudo random numbers [1]. Quality of pseudo random numbers depends on CA rules applied. So rule 30 CA has been used for generating good quality pseudo random numbers In this algorithm different truth tables referred as #tables are used that work on 4-states (0, 1, 2, 3) as shown in figure 3 [2]. These # tables decide how many bits of the secret message are to be hidden in each pixel. Also the bits of secret message is hidden either in LSB of green or red components of each pixel so as to provide more security.

#0	0	1	2	3
0	3	2	1	0
1	2	3	0	1
2	1	0	3	2
3	0	1	2	3

#3	0	1	2	3
0	1	0	3	2
1	0	1	2	3
2	3	2	1	0
3	2	3	0	1

#2	0	1	2	3
0	2	3	0	1
1	3	2	1	0
2	0	0	2	3
3	1	1	3	2

#1	0	1	2	3
0	0	1	2	3
1	1	0	3	2
2	2	3	0	1
3	3	2	1	0

Truth table for # operation

Fig. 1. # tables

3.1 Hiding Secret Message into Color Image

Our proposed method can be given as

$$\text{Cover image} + \#\text{table} + \text{hidden information} = \text{stego image}$$

A simple secret message hiding stegnographic technique is shown in figure 2.

Fig. 2. Simple steganographic method for hiding secret message

Encoding Function

Input: secret message in bits, a 24 bit color image.

Output: stego image.

1. Get red component of each pixel of cover image and divide into 4 parts p1, p2, p3, p4 each containing 2 bits.
2. The value of p1 represents among the 4 #tables which #table to be used for mapping.
   If the bits of p1 are 00 then #0table is used for mapping.
   If the bits of p1 are 01 then #1table is used for mapping.
   If the bits of p1 are 10 then #2table is used for mapping.
   If the bits of p1 are 11 then #3table is used for mapping.
3. The other two parts p2 and p3 define the row and column number in the specified table where the cross point of them gives the new value.
4. The new value from #table represents the number of secret message bits to be hidden. If the new value is 3 then 3 secret message bits are hidden into blue component or green component. If the new value is 2 then 2 secret message bits are hidden into blue component or green component. If the new value is 1 then 1 secret message bit is hidden into blue component or green component. If the new value is 0 than no secret message bits are hidden.
5. The bits of p4 decide whether the secret message is to be hidden in green component or blue component of the pixel. If the bits of p4 are 00 or 01 data is hidden in LSB of green component. If the bits of p4 are 10 or 11 data is hidden in LSB of blue component. Continue from step1 to step 5 for each pixel in cover image until the entire message is hidden.
6. End.

Decoding Function

Input: stego image

Output: secret image

1. Get red component of each pixel of stego image and divide into 4 parts p1, p2, p3, p4 each containing 2 bits.
2. The value of p1 represents among the 4 #tables which #table to be used for mapping.
3. The other two parts p2 and p3 define the row and column number in the specified table where the cross point of them gives the new value.
4. The new value from #table represents the number of secret message bits to be derived from the pixel. The value of p4 decides whether the secret message is to be extracted from the green component or from the blue component of the pixel.
5. Extract the message from the pixel. Continue from step1 to step 6 for each pixel in cover image until the entire message is extracted.
6. End.

3.2 Hiding Secret Image into Cover Image

The simple steganographic system for image steganography is shown in fig. 3.

Fig. 3. Steganographic method for hiding secret image

Encoding Function

Input: Grey scale secret image, Grey scale cover Image

Output: Stego image

Steps:

1. Get the pixel from the secret image. Since it is a grey scale image each pixel consists of 8 bits which is to be hidden in cover image.
2. Get 2 consecutive pixels from the cover image the first pixel determines the number of bits of the pixel of the secret image is to be hidden in the second pixel.
3. Take the first pixel and divide into four parts p1, p2, p3, p4 each with 2 bits. The value of p1 decides whether the bits of the secret image are to be hidden in the second pixel or not. If the bits of p1 are 00 or 01 then no secret image bits are hidden in the second pixel. If the value of p1 is 10 or 11 then secret image bits are hidden in LSB of the second pixel.
4. The value of p2 represents among the 4 #tables which #table to be used for mapping.
5. The other two parts p3 and p4 defines the row and column number in the specified table where the cross point of them gives the new value.
6. The new value from #table represents the number of secret image bits to be hidden. If the new value is 3 than 3 secret image bits are hidden into the second pixel. If the new value is 2 than 2 secret image bits are hidden into LSB of the second pixel. If the new value is 1 or 0 than 1 secret image bit is hidden. Continue from step1 to step 6 for next two pixels (3, 4) in cover image and repeat it until the entire secret image is hidden.
7. End

Decoding Function

Input: stego image

Output: secret image

1. Get 2 consecutive pixels from the cover image.
2. Take the first pixel and divide into four parts p1, p2, p3, p4 each with 2 bits
3. The value of p1 decides whether the bits of the secret image are to be extracted from the second pixel or not.
4. The value of p2 represents among the 4 #tables which #table to be used for mapping. The other two parts p3 and p4 defines the row and column number in the specified table where the cross point of them gives the new value.
5. The new value from #table represents the number of secret image bits to be extracted .Extract the secret image bits from the LSB of the second pixel. Continue from step1 to step7 for next two pixels (3, 4) in cover image and repeat it until the entire secret image is extracted.
6. End.

4 Result Analysis

Encoding Secret Message into Cover Image

For Example the secret message is attack, convert into ASCII values and then into binary as shown below.

a - 97-1100001 t- 116-1110100 t- 116-1110100 a-97-1100001 c-99-1100100 k-107-1101100

Finally the secret message will be 1100001 1110100 1110100 1100001 1100100 1101100. Suppose the Cover image is shown in fig 4:

First Pixel 1: **11110001 11001000 00000111**

Take out the red component and divide into 4 parts.

11 11 00 01

p1 p2 p3 p4

Following the steps 2,3,4 of encoding function of a color image the altered pixel is

Original pixel 1: **11110001 11001000 00000111**

Altered pixel 1: **11110001 11001011 00000111**

Similarly for second pixel

Original pixel 2: **00111110 01011011 11001011**

Altered pixel 2: **00111110 01011011 11001000**

Continue this process until the entire secret message is hidden. Finally, we get a stego image consisting of secret message as shown in figure 5.

Image quality is one of the issues because degradation in the image quality may cause suspection to the intruder that something is embedded in the image. The more LSB bits are replaced by the message degradation increases. To improve this, a technique called randomization is used which selects some of the bytes in the image for embedding.

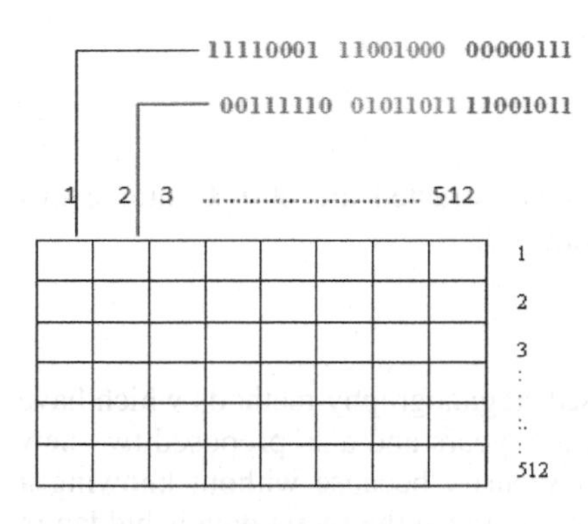

Fig. 4. Cover Image

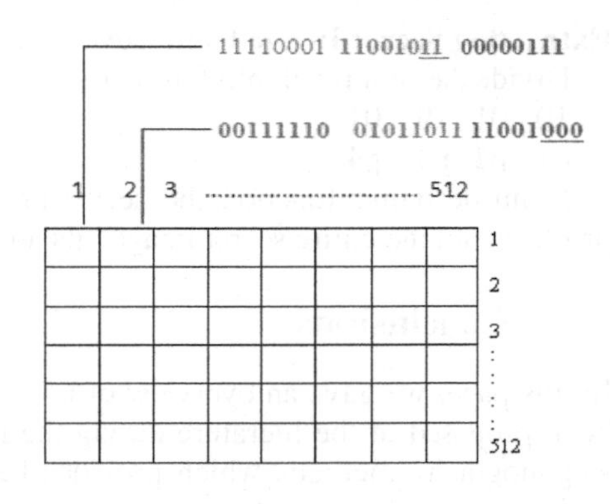

Fig. 5. Stego Image

Decoding Secret Image from the Stego Image

To get secret message from the stego image, take the first pixel.

Stego image pixel 1:11110001 **11001011_ 00000111**

11 11 00 01

p1 p2 p3 p4

extract the secret message by using the steps 2,3,4and finally we get 11.

Similarly for pixel2, **00111110 01011011 11001000**

By following decoding function we get the secret message as 000. Continue, this process until the entire message is extracted.

Hiding Grey Scale Secret Image into Grey Scale Cover Image

First we take the pixels one by one from the secret image and convert it into binary and these binary bits are hidden into cover image.

Divide the first pixel into four parts

10 01 10 01

P1 p2 p3 p4

By following the steps 2,3,4 in the algorithm 3 secret image bits are hidden into the LSB of the second pixel. The altered pixel will be 01110000

Continue this process and finally get the stego image as shown in figure 7.

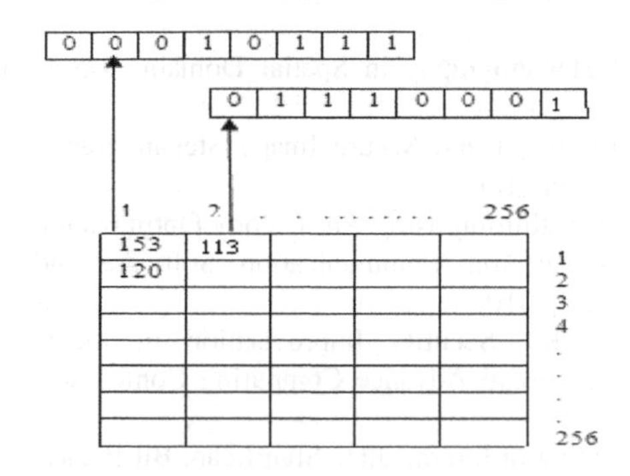

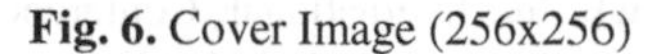

Fig. 6. Cover Image (256x256)

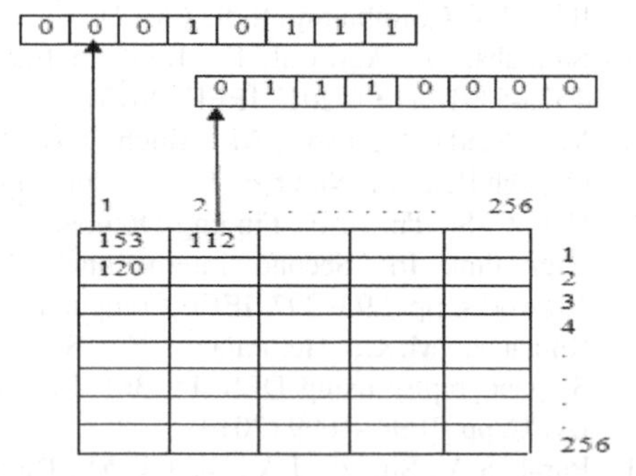

Fig. 7. Stego Image (256x256)

Extracting Secret Image from Cover Image

Divide the first pixel into four parts

10 01 10 01

p1 p2 p3 p4

From decoding function the secret image bits extracted are 000. Continue this process until the entire secret image bits are extracted.

5 Conclusions

In this paper we gave an overview of LSB based steganography methods which have been proposed in the literature during the last few years and also proposed two new steganography methods which provides better security because without knowing # tables the intruder cannot extract the information and also the secret data is hidden in the LSB of the pixels so image distortion is less and image quality is maintained.

References

1. Wahab, H.B.A., Rahma, A.M.S.: Proposed New Quantum Cryptography System using Quantum Description Techniques for Generated Curves. In: International Conference on Security and Management, Las Vegas, USA, pp. 658–664 (2009)
2. Al-Neaimi, A.M.A., Hassan, R.F.: New Approach for Modifying Blowfish Algorithm Using 4-States Keys. In: 5th International Conference on Information Technology, pp. 21–26 (2011)
3. Sathisha, N., Madhusudan, G.N., Bharathesh, S., Suresh Babu, K., Raja, K.B., Venugopal, K.R.: Chaos Based Spatial Domain Steganography using MSB. In: 5th International Conference on Industrial and Information Systems, ICIIS, India, pp. 177–182 (2010)
4. Masud Karim, S.M.: Md. Saifur Rahman., Md. Ismail Hossain.: A New Approach for LSB Based Image Steganography using Secret Key. In: 14th International Conference on Computer and Information Technology, ICCIT, American International University, Dhaka, Bangladesh, pp. 286–291 (2011)
5. Tataru, R.L., Battikh, D., Assad, S.E., Noura, H., Deforges, O.: Enhanced Adaptive Data Hiding in Spatial LSB Domain by using Chaotic Sequences. In: Eighth International Conference on Intelligent Information Hiding and Multimedia Signal Processing, pp. 85–88 (2012)
6. Badrinath, R., Anand, P.S.: MSB Constrain Based Variable Embedding. In: IEEE-20180, ICCCNT, Coimbatore, India (2012)
7. Samidha, D., Agrawal, D.: Random Image Steganography in Spatial Domain, 978-1-4673-5301-4/13 ©2013 IEEE (2013)
8. Manchanda, S., Dave, M., Singh, S.B.: Customized and Secure Image Steganography through Random Numbers Logic. Signal Processing 1(1)
9. Hsu, C.-S., Tu, S.-F.: Finding Optimal LSB Substitution using Ant Colony Optimization Algorithm. In: Second International Conference on Communication Software and Networks, pp. 293–297. IEEE Computer Society (2010)
10. Ramaiya, M.K., Hemrajani, N., Saxena, A.K.: Security Improvisation in Image Steganography using DES. In: 3rd IEEE International Advance Computing Conference, IACC, pp. 1094–1099 (2013)
11. Parah, S.A., Sheikh, J.A., Bhat, G.M.: Data Hiding in Intermediate Significant Bit Planes, A High Capacity Blind Steganographic Technique. In: International Conference on Emerging Trends in Science, Engineering and Technology, pp. 192–197 (2012)

Generation of 128-Bit Blended Key for AES Algorithm

S. Sridevi Sathya Priya[1], P. Karthigaikumar[2], and N.M. SivaMangai[1]

[1] Department of Electronics and Communication Engineering, Karunya University, Coimbatore, Tamilnadu, India

[2] Department of Electronics and Communication Engineering, Karpagam College of Engineering, Coimbatore,Tamilnadu, India

{s.d.s.priya,p.karthigaikumar}@gmail.com

Abstract. The AES algorithm is most widely used algorithm for various security based applications. Security of the AES algorithm can be increased by using biometric for generating a key. To further increase the security, in this paper a 128 bit blended key is generated from IRIS and arbitrary key. An IRIS based 128 bit key is generated from IRIS features. Generated key is concealed using arbitrary key to form a blended key using Fuzzy Commitment scheme. Brute force attack is widely used against the encrypted data. This attack searches the entire key space. If the key is more random , then chances for getting attack is less. In this paper Generated key randomness is verified and compared with biometric key randomness. Blended key is 10% more random than IRIS based biometric key.

Keywords: Blended key, AES algorithm, IRIS based key, Biometric key.

1 Introduction

Cryptography converts the information into unreadable form (cipher text) using a Cryptographic key. Decipher convert cipher text into plain text using same key. The security of the encryption depends mainly on the key. Depending on key the cryptographic encryption algorithms divided into two types symmetric key encryption and Asymmetric key encryption. Symmetric encryption algorithm uses single key for both encryption and decryption. Whereas asymmetric uses two keys one key for encryption and the other key is for decryption. One key is considered as a private key and the other key is considered as a public key. Symmetric algorithms faster than Asymmetric algorithms, because CPU cycles needed for symmetric is fewer than asymmetric algorithm. So symmetric algorithm is used widely. AES, DES, Triple DES, RC2 and RC6 are some symmetric algorithms. AES is a faster algorithm. This has low memory requirement [2], due this well suitable for embedded applications. Number of rounds involved in AES algorithm is higher than the other algorithms [3]. This ensures the security. The AES algorithm uses 128-block length and key lengths of 128, 192, and 256 bits [1]. The traditional cryptographic algorithm is depends mainly on length and size of the key. Maintaining a cryptographic key is very important for getting back the original Data. Cryptographic key are lengthy key. It is very difficult to remember keys. It should be kept at a secured place. If the keys are shared identification original user is difficult task. To solve this problem biometrics can be used for key generation. This is further secured by concealing the key using an arbitrary key.

S.C. Satapathy et al. (eds.), *Emerging ICT for Bridging the Future – Volume 2*,
Advances in Intelligent Systems and Computing 338, DOI: 10.1007/978-3-319-13731-5_47

2 AES Algorithm

The generated 128 bit key is well suitable for AES algorithm. Due to the combination of security, performance, efficiency, and flexibility.[3]. This has better resistance against the existing attacks. AES algorithm consist of 128 block length of bits and supports128, 192 and 256 key length of 11, 13, or 15 sub-keys, each 128-bits long. One sub-key corresponds to each AES processing round, thus each sub-key is referred to as a round key. The Add Round Key step is a transformation that combines the current state data block and the round key corresponding to the specific round using an XOR function. The Sub Bytes step replaces each state data byte with an entry in a fixed lookup table. The Shift Rows step rotates the four bytes of state data in each row in the state data matrix. The Mix Columns step performs a transformation on the four bytes of state data in each column in the state data matrix.AES algorithm steps are clearly given in Fig 1.

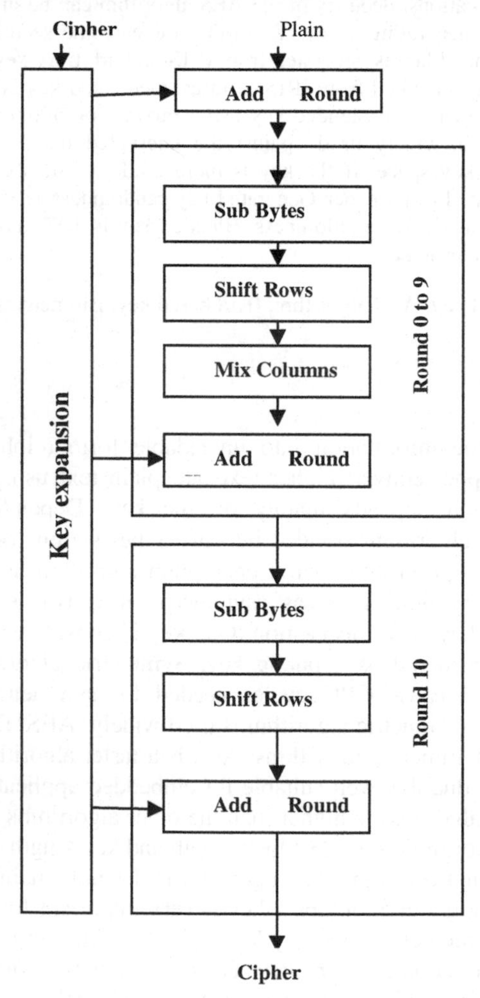

Fig. 1. AES Steps

3 Biometrics

Biometrics is unique to the individual [5]. This is a physiological and behavioral characteristic. These characteristics are compared using seven basic parameters [6] Universality ,Uniqueness ,Permanence ,Collectability ,Performance , Acceptability and Circumvention. This comparison is shown in Table 1.It shows the advantages of IRIS biometric. Iris is selected for generating biometrics because this has high universality, Distinctiveness, Permanence, Performance.

Table 1. Comparison of different biometrics

Biometric	Universality	Distinctiveness	Permanence	Collectability	Performance	Acceptability	Circumvention
Face	H	L	M	H	L	H	H
Finger Print	M	H	H	M	H	M	M
Hand Geometry	M	M	M	H	M	M	M
Iris	H	H	H	M	H	L	L
Keystroke	L	L	L	M	L	M	M
Signature	L	L	L	H	L	H	H
Voice	M	L	L	M	L	H	H

4 Related Work

Many research is done for the effective usage of biometrics in cryptography. In [7] a complete system is proposed for encrypt and secure the private key of any public key infrastructure is given. Flexible ICA algorithm is used for extracting features from IRIS and Hamming distance is used for matching. In [8], Arun et.al discussed about the multimodal biometric to overcome the problems in unimodal biometric system. Jagadeesan et.al[9] has explained procedure for generating Secured Cryptographic Key from Multimodal Biometrics for cryptographic algorithms. In multimodal biometric the fusion is possible in three stages, fusion at the feature level, fusion at the match level and fusion at the decision level[9]. Feature level fusion results in feature set where contain rich information about the raw biometric data than the match level. Feature level fusion is explained by Abhishek et.al[10] , in which the author used fuzzy vault and fuzzy commitment systems for key generation. Fuzzy Vaults system which depends on statistical analysis of IRIS is explained[11] . The characteristic of fuzzy commitment scheme is useful for applications such as biometric authentication systems, in which data is subject to random noise. Because the scheme is tolerant of error, it is capable of protecting biometric data just as conventional cryptographic techniques.[24] so Fuzzy commitment scheme is useful to protect biometric key which is generated from biometrics.

5 Methodology

IRIS based biometric key is generated from IRIS features. IRIS based key is concealed using arbitrary key. This process is done by using fuzzy commitment scheme[12]. This key increases the randomness, which will reduce the brute force attack. This is explained in Fig2. The process of generating blended key has 3 steps

Step 1. Extraction of features from iris
Step 2. IRIS Based key generation from IRIS extracted features
Step 3. Generation of Blended Key using Fuzzy commitment Scheme.

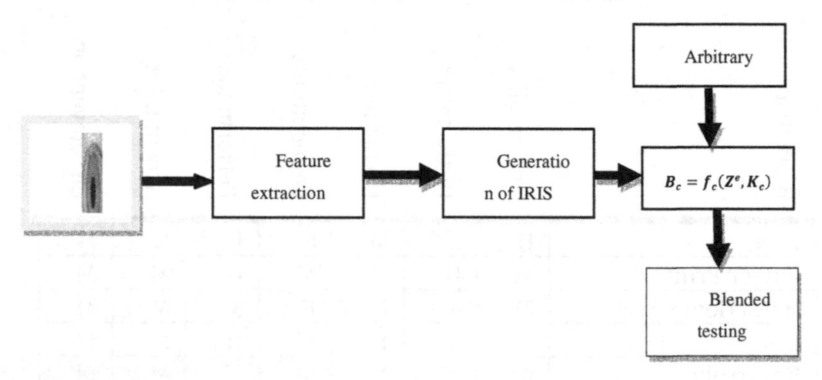

Fig. 2. Block diagram for generating blended Key

5.1 Generation of Iris Based Key

IRIS has reliable features like distributed and unshaped microstructures.Due to this, iris features are used for generating key. This extraction has segmentation, Normalization and Extraction of Iris Textures steps.

5.2 Segmentation

Localization and Isolation is done in segmentation. In localization, Inner and outer boundaries are identified and from identification Iris region is isolated in a digital eye image. To detect the Iris boundaries canny edge detection is used. This produces the edge map[13] Linear filtering is used to calculate the gradient of the image intensity .Thinning and thresholding is used to achieve a binary map of edges. Using traditional Hough transform, the identification of lines in the image is achieved, then Hough transform lines in the images are enhanced to find positions of arbitrary shapes. The Hough transform locates contours in an n-dimensional parameter space by examining whether they lie on curves of a specified shape. These parameters are the centre coordinates x and y, and the radius r that are capable to describe every circle in accordance with the following equation (1) $x^2 + y^2 = r^2$ (1) Normalization transforms the iris image into rectangular sized fixed image. Daugman's Rubber Sheet Model is employed for this transformation.

5.3 Daugman's Rubber Sheet Model

Normalization process includes unwrapping the iris and transforming it into its polar equivalent [9]. It is performed, by utilizing Daugman's Rubber sheet model and is depicted in the Fig 3.

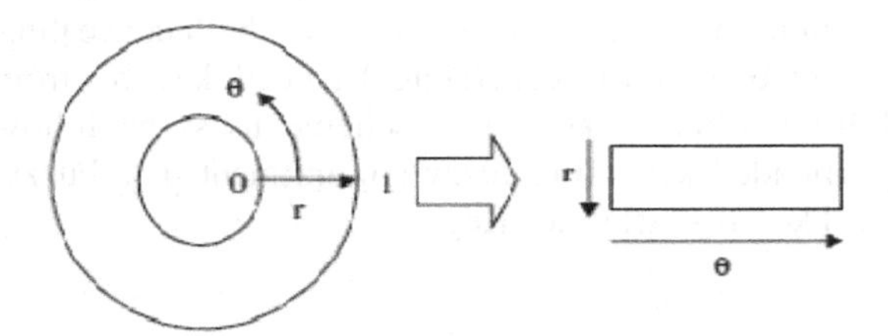

Fig. 3. Daugman's Rubber sheet model

Daugman's Rubber sheet model assigns a pair of real coordinates to each point on the iris, irrespective of its size and pupillary dilation. On polar axes, for each pixel in the iris, its equivalent position is found out. The process consists of two resolutions (i) Radial resolution and (ii) Angular resolution. The former is the number of data points in the radial direction whereas, the later part is the number of radial lines produced around iris region. Utilizing the following equation, the iris region is transformed to a 2D array by making use of horizontal dimensions of angular resolution and vertical dimension of radial resolution. $I[x(r,\theta), y(r,\theta)] \rightarrow I(r,\theta)$ (2) Where, in equation (2) I (x, y) is the iris region, (x, y) and (r, θ) are the Cartesian and normalized polar coordinates respectively. The range of θ is [0 2π] and r is [0 1] . x(r, θ) and y(r, θ) are described as linear combinations set of pupil boundary points. To perform the transformation, the formulae are given in the following equation (3)

$$
\begin{aligned}
x(r,\theta) &= (1-r)xp(\theta) + xi(\theta)\\
y(r,\theta) &= (1-r)yp(\theta) + yi(\theta)\\
xp(\theta) &= xp0(\theta) + rp\cos(\theta)\\
yp(\theta) &= yp0(\theta) + rp\sin(\theta)\\
xi(\theta) &= xi0(\theta) + ri\cos(\theta)\\
yi(\theta) &= yi0(\theta) + ri\sin(\theta)
\end{aligned} \quad (3)
$$

Where, in equation 3 (xp , yp) and (xi , yi) are the coordinates on the pupil and iris boundaries along the θ direction. (xp0 , yp0) , (xi0 , yi0) are the coordinates of pupil and iris centre .

5.4 Extraction of Iris Texture

The normalized 2D form image is disintegrated up into 1D signal, and these signals are made use to convolve with 1D Gabor wavelets. The frequency response of a Log-Gabor filter is given in equation (4),

$$G(f) = \exp\left(\frac{-(\log(\frac{f}{f0}))^2}{2(\log(\frac{\sigma}{f0}))^2}\right) \quad (4)$$

Where f0 indicates the centre frequency, and σ provides the bandwidth of the filter The Log-Gabor filter generates the biometric feature (texture properties) of the iris.The texture properties obtained from the log-gabor filter are complex numbers (a + ib) . Iris texture features are stored in the form of two various vectors Vector C_1 includes the real part of the complex numbers and vector C_2includes the imaginary part of the complex numbers. From C_1 and C_2 128 bit IRIS based AES encryption key is generated. Fuzzy commitment scheme can be used for both concealing and binding. In this paper this scheme used for generating Blended key by from IRIS based Cryptographic and from arbitrary key. This scheme uses the following function $B_c = f_c(Z^e, K_c)$ B_c –Blended key using fuzzy commitment, f $_c$- Fuzzy commitment function Z^c-IRIS based key K_c- Arbitrary Key.

6 Randomness Test

Various statistical tests a be applied to a sequence to attempt to compare and evaluate the sequence to a truly random sequence(16). Randomness is a probabilistic property.A statistical test is formulated to test a specific null hypothesis (H0). Null hypothesis under test is that the sequence being tested is random. Associated with this null hypothesis is the alternative hypothesis (Ha) which, for this document, is that the sequence is not random. For each applied test, a decision or conclusion is derived that accepts or rejects the null hypothesis, i.e., whether the generator is producing random values, based on the sequence that was produced. For each test, a relevant randomness statistic must be chosen and used to determine the acceptance or rejection of the null hypothesis. If the test statistic value exceeds the critical value, the null hypothesis for randomness is rejected. Otherwise, the null hypothesis is not rejected .The test is used to calculate a P-value that summarizes the strength of the evidence against the null hypothesis. For these tests, each P-value is the probability that a perfect random number generator would have produced a sequence less random than the sequence that was tested, given the kind of non-randomness assessed by the test. If a P-value for a test is determined to be equal to 1, then the sequence appears to have perfect randomness. A P-value of zero indicates that the sequence appears to be completely nonrandom. The NIST Test Suite is a statistical package consisting of 10 tests that were developed to test the randomness of binary sequences produced by either hardware or cryptographic random or pseudorandom number generators[16]. The 10 tests are The Frequency Test, Frequency Test within a Block , The Runs Test, Test for the Longest-Run-of-Ones in a Block, The Binary Matrix Rank Test, The Discrete Fourier Transform Test, The Non-overlapping Template Matching Test, The Overlapping Template Matching Test, The Serial Test,The Cumulative Sums Test.

7 Experimental Results

The Fig 3 a. show the original iris image and the Fig 3 b. points out the centre point of the image.An important feature of the Canny operator is its optimality in an aging noisy images as the method to link between strong and weak edges of the image by relating the weak edges in output. The inner boundary of the iris, forming the pupil, can be accurately determined by exploiting the fact that the boundary of the pupil essentially a circular edge. Then the image is localized and normalized. The former refers to limit the search regions i.e. pupil centre . IRIS segmentation is done in an iris image for the feature extraction process. The generated Keys are shown in Fig 3.

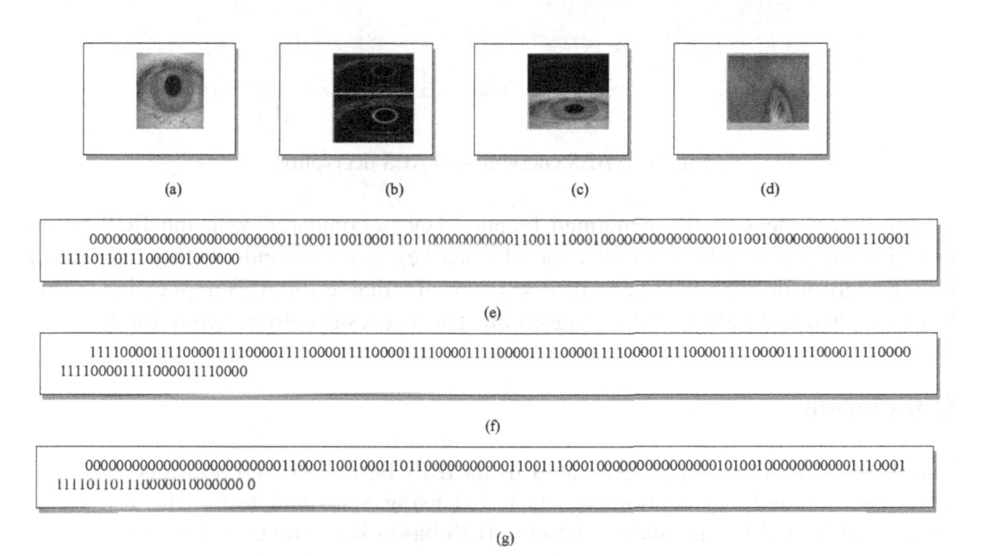

Fig. 4. (a) Input Image (b) Canny Edge Detection and Hough Transform (c) Radial Suppressed and localized Image (d) Normalized image (e) IRIS based Cryptographic Key (f).Arbitrary key (g) Blended key.

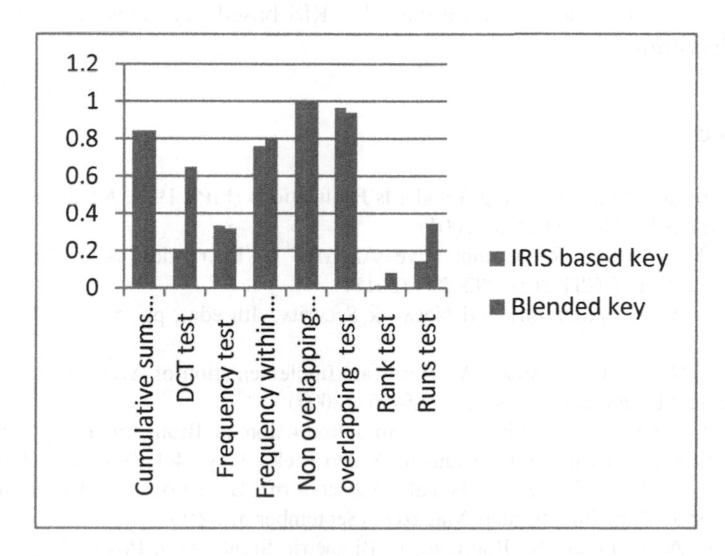

Fig. 5. IRIS based key and blended key Randomness comparison chart

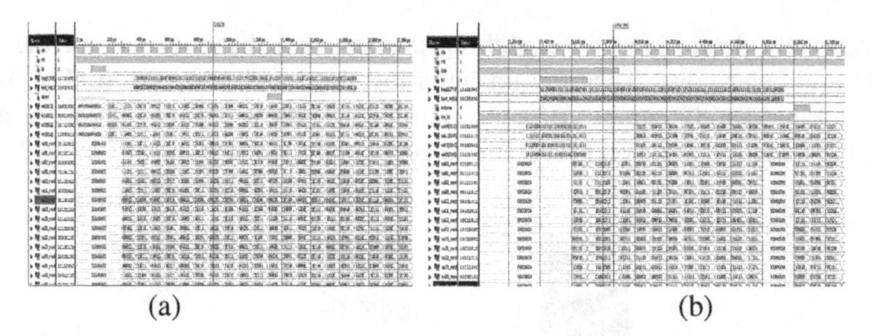

(a) (b)

Fig. 6. a) AES encryption b) AES decryption

The randomness of the generated blended key is compared with the IRIS based key. The results clearly show that the blended key is more random than IRIs based key it is given in Fig4. The generated key is well suitable for AES applications. This key is applied and tested for AES algorithm. The test results are shown in Fig 5.

Conclusion

The 128 bit Blended key is generated using IRIS based key and arbitrary key. The generated blended key randomness is tested using statistical tests. The test result shows that blened key is random than the IRIS based key. The guessing and cracking complexity of the cryptographic key can be increased using a biometric key generation method. So IRIS based key is generated. To increase the security, IRIS based key is concealed using arbitrary key. The key is cracked by many crackers using brute force attacks. This attack concentrates on randomness of the key. The generated key is 10% more random than the IRIS based key. This key is best suited for AES algorithm.

References

1. Federal Information Processing Standards Publications (FIPS 197), Advanced Encryption Standard (AES) (November 26, 2001)
2. Seth, S.M., Mishra, R.: Comparative Analysis of Encryption Algorithms for Data Communication. IJCST 2(2), 292–294 (2011)
3. Stallings, W.: Cryptography and Network Security, 4th edn., pp. 58–309. Prentice Hall (2005)
4. Mali, M., Novak, F., Biasizzo, A.: Hardware Implementation of AES Algorithm. Journal of Electrical Engineering 56(9-10), 265–269 (2005)
5. Jain, A.K., Ross, A., Prabhakar, S.: An Introduction to Biometric Recognition. IEEE Transactions on Circuits and Systems for Video Technology 14(1) (January 2004)
6. Schuckers, M.E.: Some Statistical Aspects of Biometric Identification Device Performance. Submitted to Stats Magazine (September 5, 2001)
7. Boukhari, A., Chitroub, S., Bouraoui, I.: Biometric Signature of Private Key by Reliable Iris Recognition Based on Flexible-ICA Algorithm. Int. J. Communications, Network and System Sciences 4, 778–789 (2011)

8. Ross, A., Jain, A.K.: Multimodal Biometrics An Overview. In: Proceedings of the 12th European Signal Processing Conference, pp. 1221–1224 (2004)
9. Jagadeesan, A., Duraiswamy, K.: Secured Cryptographic Key Generation from Multimodal Biometrics Feature Level Fusion of Fingerprint and Iris. International Journal of Computer Science and Information Security, IJCSIS 7(2), 28–37 (2010)
10. Nagar, A., Nandakumar, K., Jain, A.K.: Multibiometric Cryptosystems Based on Feature-Level Fusion. IEEE Transactions on Information Forensics and Security 7(1), 255–268 (2012)
11. Fatangare, M., Honwadkar, K.N.: A Biometric Solution to Cryptographic Key Management Problem using Iris based Fuzzy Vault. International Journal of Computer Applications (0975 – 8887) 15(5) (February 2011)
12. Juels, A., Wattenberg, M.: A fuzzy Commitment Scheme. In: Proceedings of the 6th ACM Conference on Computer and Communication Security, pp. 28–36 (November 1999)
13. Hao, F., Anderson, R., Daugman, J.: Combining Crypto with Biometrics Effectively. IEEE Transactions on Computers 55(9), 1081–1088 (2006)
14. Boukhari, A., Chitroub, S., Bouraoui, I.: Biometric Signature of Private Key by Reliable Iris Recognition Based on Flexible-ICA Algorithm. Int. J. Communications, Network and System Sciences 4, 778–789 (2011)
15. Daugman, J.: How Iris Recognition Works. IEEE Transactions on Circuits and Systems for Video Technology 14(1), 21–30 (2004)
16. Rukhin, A., Soto, J., Nechvatal, J., Smid, M., Barker, E., Leigh, S., Levenson, M., Vangel, M., Banks, D., Heckert, A., Dray, J., Vo, S.: A Statisticel Test Suite For Random And Pseudorandom Number Generators For Cryptographic Applications. NIST Special Publication 800-22 (May 2001)

Designing Energy-Aware Adaptive Routing for Wireless Sensor Networks

K. Seena Naik[1], G.A. Ramachandra[1], and M.V. Brahmananda Reddy[2]

[1] Department of CSE, S.K University, Anantapur, India
[2] Department of CSE, GITAM University, Bangalore, India
bramhareddy999@gmail.com

Abstract. Many energy-aware routing protocols take into account the residual battery power in sensor nodes or/and the energy required for transmission along the path. In the deployment of an environmental sensor network, we observed that applications may also impose requirements on routing, thus placing higher demands on protocol design. We demonstrated our approach to this issue through FloodNet, a flood warning system, which uses a predictor model developed by environmental experts to make flood predictions based on readings of water level collected by a set of sensor nodes. Because the model influences the node reporting frequency, we proposed the FloodNet adaptive routing (FAR) which transmits data across nodes with ample en- ergy and light reporting tasks whilst conserving energy for others low on battery power and heavily required by the monitoring task. As a reactive protocol, FAR is robust to topology changes due to moving obstacles and transient node failure. We evaluate the FAR performance through simulation, the result of which corresponds with its anticipated behavior and improvements.

Keywords: Adaptive sampling, energy awareness, environmental sensor networks.

1 Introduction

Wireless sensor networks (WSNs) consist of a number of sensor nodes equipped with sensing, data processing and communications capabilities for monitoring purposes. As microelectronic devices, these nodes have limited supply of energy. However, most WSN applications require very long unattended periods of operation. Also, where hazardous conditions limit human access, replenishment of power resources or replacement of nodes becomes impossible. In a multihop WSN, the malfunctioning of any node may lead to topology changes, or in worse cases, degraded network performance. Efficient energy conservation and management techniques are therefore of primary importance. In some applications, such as ours, the energy replenishment may be at its lowest precisely when the sensor readings are needed with highest priority.

To extend WSN lifetime, many routing protocols have been proposed with energy awareness being an essential design consideration. These schemes involve selecting the optimal path between a data originator and a data collecting unit. The selection

S.C. Satapathy et al. (eds.), *Emerging ICT for Bridging the Future – Volume 2*,
Advances in Intelligent Systems and Computing 338, DOI: 10.1007/978-3-319-13731-5_48

criteria take into account the available power in nodes or the energy required for transmission in the links along the path [1]. Some other optimality criteria include hop count, delay, forwarding cost, etc. WSNs may differ depending on application areas as the latter would impose domain-specific requirements on design and management. These requirements should not be underestimated especially if they have a direct impact on the behavior of sensor nodes. Consequently, the resulting routing protocol should have more complex optimality criteria for path selection, coupling application requirements with energy conditions in the network.

We illustrate our approach through FloodNet, a flood early warning system, which deploys a set of intelligent sensor nodes and uses a predictor model [2] to make flood predictions based on sensor readings. The research focus is to leverage the need for timely data and the need to conserve battery power. As a result, the system is made adaptive—the behavior of sensor nodes (i.e. the reporting frequency) varies according to application needs imposed upon by the predictor model. Reducing activity of individual sensor nodes and minimizing the data volume required help prolong network lifetime (definition in Sect. 3-C). Moreover, FloodNet achieves energy awareness through its routing algorithm: by carefully routing messages across nodes with ample energy and light reporting tasks, the FloodNet adaptive routing (FAR) can conserve energy for sensor nodes low on battery power or/and heavily required by the monitoring task. Delivery of data messages carrying sensor readings has no dependency on any specific sensor node and the optimal path is computed on demand. Hence, FAR is robust to topological changes due to transient node and link failure. In our study, nodes are not mobile and the network topology is static. We assume constant transmission power for sensor nodes. The energy consumption during idle time is not included in design and the simulation as we assume a situation under which the energy consumption due to interest diffusion, neighbor status maintenance and data report delivery (see Sect. 3) is dominant. Note that due to the inherent nonscalability of the flooding technique, the proposed solution in its current form may not be scalable enough for direct application to large scale sensor networks.

The remainder of the paper is structured as follows. The FloodNet scenario is described in Sect. 2 and the design of FAR is described in Sect. 3. In Sect. 4 we evaluate the FAR performance. Section 5 discusses the difference between FAR and other related efforts on energy aware routing. Finally, we outline some conclusions and future work in Sect. 6.

2 The FloodNet Scenario

The FloodNet project at University of Southampton is currently investigating the use of pervasive computing and Grid computing to provide early warning of possible flood. When a flood occurs, there is a clear correlation between the cost of flood damage and both the water depth and the amount of advance warning time given. By deploying wireless sensor nodes on the floodplain, FloodNet aims to offer better data to a flood predictor model which is used to make flood predictions, and to improve warning times. Sensor nodes collect information about water level which is feed through a message broker, hosted by a gateway, back to University of Southampton. The incoming data is stored on a grid and will be further utilized by the predictor model (see Fig. 1).

The project has deployed a set of intelligent sensor nodes at a stretch of river on the east coast of the UK, chosen as a test site on the basis of its tidal behavior (see Fig. 2). Sensor nodes are battery-powered, with solar cells attached. Hence, the sensor network will be in face of energy shortage if the solar cells have not recharged the battery for a long period of time in cloudy and winter conditions during which the battery voltage drops significantly. This creates a challenge for the design of the network communication protocol, i.e. to best accomplish the monitoring and information dissemination task[1] whilst delaying the occurrence of energy depletion.

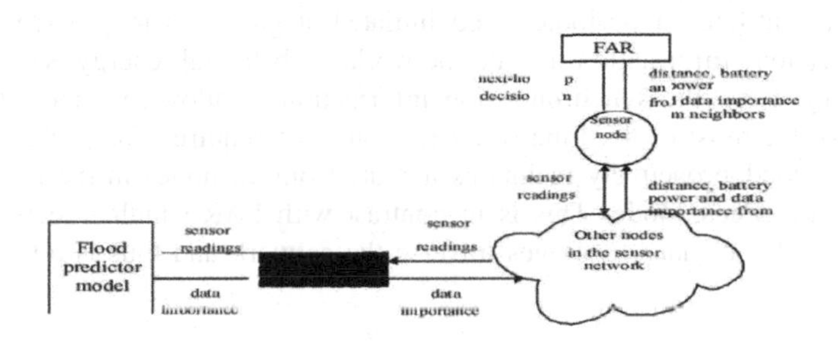

Fig. 1. The FloodNet information flow

The need to conserve battery power for sensor nodes, however, conflicts with the need for timely data, as the more data is sent the more energy is consumed. To prolong the lifetime of the sensor network, sensor nodes must enter into periods of reduced activity when running on low battery voltage [3]. The FloodNet system is therefore made adaptive in which the reporting rates of sensor nodes vary accord- ing to need. We have the flood predictor model carry out extensive processing in a short period of time (currently 1 hour). Upon each model iteration2 the network changes its behavior, altering the reporting rates for each node in the following model iteration according to the data importance placed upon it by the predictor model. The data importance reflects how important the data from a specific sensor node is.

The greater the importance value, the higher the reporting rate. The adaptability of reporting rates enables closer monitoring to be achieved in anticipation of possible flood events. The data importance may differ from one node to another because the location of nodes varies from one another—some are deployed in the channel while others are on the floodplain, and therefore the data collected from each node reflects the unique feature of water depth at the region where the node is situated. Sensor nodes will periodically switch on sensors and transmitters, take readings of water depth, and transmit the data at periodic time intervals. The diversity in behavior of nodes requires load balancing as sensor nodes with ample battery and a low reporting rate will encounter energy depletion far behind those with lower battery power and a high reporting rate.

FloodNet relies on the cooperative effort of all nodes to disseminate information across the network. The communication protocol therefore should be robust to the failure of sensor nodes and transmission medium, thus ensuring the functioning of the sensor network. We summarize the main objectives of FAR as follows and describe it in the following section:

- To best accomplish the monitoring and information dis- semination task of the sensor network.
- To delay the energy depletion which causes data loss.
- To conserve the battery of nodes with important data.
- To be robust to the failure of sensor nodes and transmis- sion medium.

Pulse [15] was designed for multi-hop wireless infrastruc- ture access to mobile users which utilizes a periodic flood initiated at gateways to provide routing and synchronization information to the network. Substantial energy savings can be achieved by using the synchronization information to allow idle nodes to power off their radios for most of the time when they are not required for packet forwarding. The Pulse flood proactively maintains a route from all nodes in the network to the infrastructure access node. This is in contrast with FAR which selects the routing path on-the-fly as data messages traverse the network and thus is robust to temporary failure.

3 Conclusions and Future Work

We believe ignoring requirements imposed upon by ap- plications may prevent energy-aware routing protocols from being most desirable, and have demonstrated our solution by developing an energy-aware adaptive routing protocol, FAR, for FloodNet and related projects. FAR takes into account the distinct behavior of individual nodes and uses the weight function as well as a set of rules in selecting a neighboring node to forward data messages. It allows messages to be carefully routed across sensor nodes with ample energy and light reporting tasks whilst conserving energy for those which have low energy level and heavy reporting tasks. FAR is robust to temporary failure as the routing algorithm will always make an effort to find alternative forwarding neighbors. Simulation results confirm the anticipated behavior of FAR and its improvements over EAR in the simulated environments. FAR has achieved all of its design objectives.

As introduced in Sect. 3-A, we currently employ a straight- forward function to convert a data importance to one of the three reporting rates allowed. Equation 2 shows that, given the same transmission capability c, weight w is more sensitive to variations of data importance t when the latter is small. This indicates that we could use alternative conversion functions if necessary. We will need to investigate the appropriate conversion functions by comparing predictions given by the flood predictor model with real water depth measurements for which the predictions is made to adjust the ranges for reporting rates, so as to provide more precise flood warnings. Further, we will explore how FAR works with sensor nodes that have the ability to control the power of their radio transmissions.

Acknowledgement. FloodNet is funded by the UK Department of Trade and Industry under the Next Wave Programme, as part of the En- viSense Centre for Pervasive Computing in the Environment. We also acknowledge the input of the Equator project (EPSRC Grant GR/N15986/01).

References

[1] Akyildiz, I.F., Su, W., Sankarasubramaniam, Y., Cayirci, E.: A Survey on Sensor Networks. IEEE Communications Magazine 40(8), 102–114 (2002)
[2] Neal, J., Atkinson, P., Hutton, C.: Real-time flood modelling using spatially distributed dynamic depth sensors. In: Brox, C., Kruger, A., Simonis, I. (eds.) Geosensornetzwerke - von der Forschung zur Praktischen Anwendung. Institut für Geoinformatik, Universität Münster, vol. 23, pp. 147–159. IfGI Prints, Germany (2005)
[3] Sinha, A., Chandrakasan, A.: Dynamic power management in wire- less sensor network. IEEE Design & Test of Computers 18(2), 62–74 (2001)
[4] Shah, R.C., Rabaey, J.: Energy Aware Routing for Low Energy Ad Hoc Sensor Networks. In: Proceedings of IEEE Wireless Communications and Networking Conference (WCNC), Orlando, FL, vol. 1, pp. 350–355 (March 2002)
[5] Chang, J.-H., Tassiulas, L.: Maximum lifetime routing in wireless sensor networks. IEEE/ACM Transactions on Networking 12(4), 609–619 (2004)
[6] Li, Q., Aslam, J., Rus, D.: Online power-aware routing in wireless ad-hoc networks. In: Proceedings of the 7th Annual International Conference on Mobile Computing and Networking, Rome, Italy, pp. 97–107 (2001)
[7] Singh, S., Woo, M., Raghavendra, C.S.: Power-aware routing in mobile ad hoc networks. In: Proceedings of the 4th Annual ACM/IEEE International Conference on Mobile Computing and Networking, Dallas, Texas, USA, pp. 181–190 (1998)
[8] Stojmenović, I., Lin, X.: Power-aware localized routing in wireless networks. IEEE Transactions on Parallel and Distributed Systems 12(10), 1–12 (2001)
[9] Yu, Y., Govindan, R., Estrin, D.: Geographical and Energy Aware Routing: a recursive data dissemination protocol for wireless sensor networks. UCLA Computer Science Department, Technical Report UCLA/CSD-TR-01-0023 (May 2001)
[10] He, T., Stankovic, J.A., Lu, C., Abdelzaher, T.F.: A spatiotemporal communication protocol for wireless sensor networks. IEEE Transactions on Parallel and Distributed Systems 16(10), 995–1006 (2005)
[11] Heinzelman, W., Chandrakasan, A., Balakrishnan, H.: Energy- efficient communication protocol for wireless microsensor networks. In: Proceedings of the 33rd Hawaii International Conference on System Sciences, vol. 8, pp. 3005–3014 (January 2000)
[12] Lindsey, S., Raghavendra, C.S.: PEGASIS: Power-Efficient Gathering in Sensor Information Systems. In: IEEE Aerospace Conference Proceedings, vol. 3, pp. 1125–1130 (March 2002)

A Dynamic and Optimal Approach of Load Balancing in Heterogeneous Grid Computing Environment

Jagdish Chandra Patni[1], M.S. Aswal[2], Aman Agarwal[3], and Parag Rastogi[3]

[1] Centre for Information Technoloy, COES, UPES Dehradun, India
patnijack@gmail.com
[2] Faculty of Engineering, Gurukula Kangri Vishwavidyalaya, Haridwar, India
mahendra8367@gmail.com
[3] Department of IT, Bharat Institute of Technology, Meerut, India

Abstract. Grid computing is some sort of distributed computing which shares the resources, processor and network in the organization or between the organizations for accomplish task. It involves huge amounts of computational task which require resource sharing across multiple computing domains. Resource sharing needs an optimal algorithm; to enhance the performance we should focus on how to increase the global throughput of computational grid. Load balancing in grid which distributes the workloads across various computing nodes to achieve optimal resource utilization, reduce latency, maximize throughput and to avoid any node by overload and under load. Several existing load balancing methods and techniques only interested in distributed systems those are having interconnection between homogeneous resources and speedy networks, but in Grid computing, these methods and techniques are not feasible due the nature of grid computing environment like *heterogeneity*, *scalability* and *resource selection* characteristics. To consider the above problem we need to develop such an algorithm which optimally balances the loads between heterogeneous nodes. It is based on tree structure where load is managed at different levels such as neighbor-based and cluster based load balancing algorithms which reduces complexity can and less number of nodes required for communication during load balancing.

Keywords: Load Balancing, Heterogeneous, Homogenous, Resource selection, static Load Balancing, Dynamic Load Balancing, Execution time, processor capability, communication delay, latency, Gridsim.

1 Introduction

At the present time the problems are growing very complex in the modern computing technology and hence require more computing power and more storage space. Based on these requirements, an organization requires higher computational resources when dealing with latest technology. The various technologies such as distributed computing, parallel computing and other are not much more comfortable for recent advancement in computing problems since the modern computing problems needs very large amounts of data which utilize more processing power and high storage of

S.C. Satapathy et al. (eds.), *Emerging ICT for Bridging the Future – Volume 2*,
Advances in Intelligent Systems and Computing 338, DOI: 10.1007/978-3-319-13731-5_49

data. To solve the problems with high computational requirement the grid computing is proposed as an effective resource sharing method. In grid computing, the network status and the resource status are to be managed effectively. In grid computing, the user can use thousands of computers without any geographical limitation to utilize the resources in effective and efficient manner. The Grid Architectures serving as a middleware technology for various purposes like resource allocation management, job scheduling, data management, security and authorization.

Grid computing is a form of distributed computing that involves coordinating and sharing computational power, network resources and storage available in dynamic form and anywhere without any geographical limitation [3].Due to imbalance loads and unequal computing capacities and capabilities, computers in one site may be heavily loaded while other site may be less loaded. Workload represents the amount of task to be performed and all resources have different processing speed and storage. It is required to implement an effective and efficient load balancing technique for proper allocation and sequencing of tasks on the computing nodes [2].

First, we propose a distributed load balancing model, transforming any Grid topology into a tree structure. Second, we develop a two level strategy to balance the load among resources of computational Grid. Our study will focus on primarily local rather than global load balancing in order to reduce the average response time of tasks and communication cost induced by the task transferring. Literature survey is present in section 2, Section 3 defines the proposed algorithm and method and section 4 contains experimental results.

2 Literature Survey

A number of research activities are identified in the grid environment, in which the task scheduling and load balancing are the major focused issues. Many scheduling algorithms for grid are having different computing requirements because of that some nodes may be heavily loaded or lightly loaded but we have to ensure that every node should be equally loaded, optimizing their utilization, throughput and response [12]. To achieve these goals, the load balancing algorithm should equally distribute the loads across all the nodes connected to the grid structure by which we can get the optimal output.

In decentralized approach, all resources in the distributed system are involved in making load balancing decisions [16]. Since load balancing decisions are distributed, so it is not easy to obtain the dynamic state information of the resources in the structure. The resources of higher levels are scheduled at higher priority and lower level smaller entities are scheduled at lower priority of the scheduler hierarchy. The above model is known by centralized approach and decentralized approach [20].

Load balancing algorithms can be further categorized as static, dynamic \or adaptive algorithms based on the load balancing decisions are made [11]. Static algorithms assume that all information like characteristics of jobs, the resources and communication network are known in advance. This static approach is simple and has minimal runtime overhead. However, it has two major disadvantages. Firstly, it is not easy to predict the workload distribution in advance and assume that all information remain constant, which may not apply to a grid environment whereas, dynamic load balancing algorithms take the decisions immediate for a newly arriving job without any previous knowledge hence it reduces processing time. However, this involves the additional cost of collecting and maintaining load information. Adaptive load balancing adapt their activities by dynamically changing their parameters, or even their policies, to suit the changing system state [19].

Xu [14] proposed a neighbor-based iterative dynamic load balancing algorithm for multicomputer in which processes migrate one step at a time each step according to a local decision made by the intermediate processor concerned [18]. Lin and Keller [10] presented a demand driven gradient model for load balancing [24]. In this approach a job continues to transfer until it reaches an under-loaded computing node or it reaches a computing node for which no neighboring computing nodes reports shortest distance to the nearest lightly loaded computing node. Shu [12] put forward an adaptive contracting within neighborhood (ACWN) method, which applied a saturation control technique on the number of hops that the job had to travel before reaching its destination. ACWN performed consistently better than the gradient model but did not take into account a computing node's own load before making a job migration decision.

Hui and Chanson [7] proposed an intuitive approach based on a hydrodynamic analogy, for a heterogeneous environment characterized by different computing powers and uniform communication. A diffusion schemes for a computational environment characterized by uniform computing powers and different communication parameters [14]. A highly decentralized, sender initiated [19] and scalable algorithm for scheduling jobs and load balancing resources in heterogeneous grid environments. This approach is not applicable for system comprised of heterogeneous computing nodes separated by a wide-area broadband network.

A neighbor-based load balancing algorithm in which each cluster is limited to load information from within its own domain, which consists of itself and its neighbors[6]. For low heterogeneous systems, the algorithm reduces load at weaker computing nodes and increases load at most powerful computing nodes. A decentralized load balancing algorithm for computational grid in which load updates between resources are done periodically, which leads to high messaging overhead[17].A sender initiated decentralized dynamic load balancing algorithm is proposed for computational grid environment [5]. Both authors juxtaposed the strong points of neighbor-based and cluster-based load balancing algorithms similar to us.

3 Proposed System

Growth of grid computing over the Internet, we require an optimal load balancing algorithm, that solve the issues of heterogeneity, communication, network, availability, resource selection and job characteristics.

1. In this paper we discuss the load balancing algorithm for optimal resource selection and processing power of multiple nodes those are connected to Grid environment.

2. in this approach we reduces the complexity of a site to collect current state information throughout the grid since real time monitoring of grid will cause system overhead.

3. An optimal information exchange scheduling scheme is adopted to enhance the efficiency of the load balancing model.

In a grid based computing environment where heterogeneity exists among nodes, it is necessary to prevent the processors from suffering of overloaded work and also to minimize time of processors. It is also important to equal distribution of load among the processor or as per their capability. Transfer the loads to lightly loaded nodes from heavily loaded nodes.

In proposed algorithm whenever user submit their tasks, they assigned resources according to their processing capability at first come first serve basis. So firstly it balance the load in the grid, based on the processing capacity or power where job comes first allocated to resources which consume less execution time among all available nodes and an queuing approach will be followed for each resource which keeps jobs waiting in queue for getting assigned and transfer the job to the resource taking less execution time.

Table 1. Shows how proposed algorithm works

Tasks (in MI)		Resources (in MIPS)		
Task\Resources		R1(250)	R2(450)	R3(350)
Name	(MI of tasks)	Execution Time taken by resource is calculated as=(MI of tasks)/(MIPS of resources)		
T1	35000	140	78	100
T2	45000	180	100	128.5
T3	65000	260	144.44	185.7
T4	75000	300	166.6	214.30
T5	60000	240	133.4	171.43
T6	55000	220	122	157

When a task arrives, processor computes the execution time using exact information about the current load in queue for resource and on all other processor, its arrival rate and service rate, waiting rate, total completion time. The processor selects a neighbor processor with the less execution time and immediately transfer task on that processor, if it can complete the job earlier than this processor. Thus it balance the load in the grid by carefully allocating optimal resource to task using queuing approach by checking and summing execution time of pending tasks in queue. The propose algorithm is implemented using GridSim Toolkit and simulated and tested in GridSim and compared with existing algorithm such as threshold based approach, adaptive load balancing algorithm.

Algorithm:-

1. Input the value for number of Resources & MIPS for each resource.
2. Initialize the GridSim toolkit.
3 Input the no of task
4 Create the length or size of task in MI (Millions of instructions).
5 Get the availability of all registered Resources.
6 N is total no of tasks arrived at System model.
7 Initialize the Next User = 0 and Queue Length of each resource = 0.

8 Find the execution time to be taken of each task on available resources by following.

(i) Execution time=MI [jobs]/MIPS [Resources]

(ii) Initially allocate the task to resources on FCFS basis.

(iii)For next incoming tasks, pending tasks in queue for each resource is checked and by summing up execution time of all tasks, their total execution time is calculated.

(iv)For each resource (calculated execution time is compared)

9 Find the resource 'R' having less execution time.

10 Thus Allocates the next incoming task to R and Submission Time_ task = GridSim.Clock ( ).

11 Next task = Next task + 1.

12 Queue Length R = Queue Length_ R + 1.

13 Check for the arrival of ay task 'J' from Resource R' after completing the execution.

14 If no resource arrival exist then goto step 17.

15 QueueLength_R' = QueueLength_R' – 1.16 ExecutionTime_J = GridSim.Clock ( ) - SubmissionTime_J.

17 If Next taskb <= N then goto step 6.

18 Print ExecutionTime_J of all tasks.

The main steps that proposed load balancing algorithm takes to make a transfer decision for tasks are

a. Monitoring: - It maintains the resource load state whenever a new set of tasks are joining.

b. Synchronization-Estimate load state information and compare the execution time taken by each resources.

c. Rebalancing Criteria- Calculating the new work distribution and making transfer decision on the basis of processing node which takes less execution time.

d. Job Migration- The job placed at the right position where it is transferred.

4 Experimental Results

In this section, we judge and check the performance of proposed algorithm on different system parameters as mentioned in Table 2 via simulations using GridSim for the problem described in Table1, proposed algorithm results shown in Table4. In proposed algorithm, it has been shown that when different tasks comes as task id, checked by its execution time taken on each resource, then got allocated to different resources indicated as resource id. Arrival time of each job is taken into consideration and their submission time waiting time, end time and total completion time job has taken to execute is calculated and printed.

Table 2.

Simulation Parameter	Value
Simulation runs	2
No. of Resources	2-20
No. of tasks	05-50
Grid Size (in MI)	10,000-75000
Processing Power of resources (in MIPS)	200-500

The parameters are used during simulation of optimal approach for load balancing algorithm.

The algorithm is simulated and its impact on the Grid environment is analyzed. It distributes the grid workload based on the resource's processing capacity which leads to minimize the overall task mean response time and maximize the system utilization.

The algorithm is tested under various load conditions in terms of task length varying from 10000-75000(MI) and obtained results as shown in Table 3 below which gives submission, execution, ending time, waiting time, total completion time for correspondent tasks.

Table 3.

Job	Res ID	Arrival-Time	Sub Time	Exec Time	End Time	Wait Time	Comp Time
T1	R2	1`	1	75	76	0	75.0
T2	R3	2	2	133	135	0	133.3
T3	R2	3	78	150	228	75	225.0
T4	R1	4	4	350	354	0	350.0
T5	R3	5	138	183	321	133	316.6
T6	R2	6	231	125	356	225	350.0

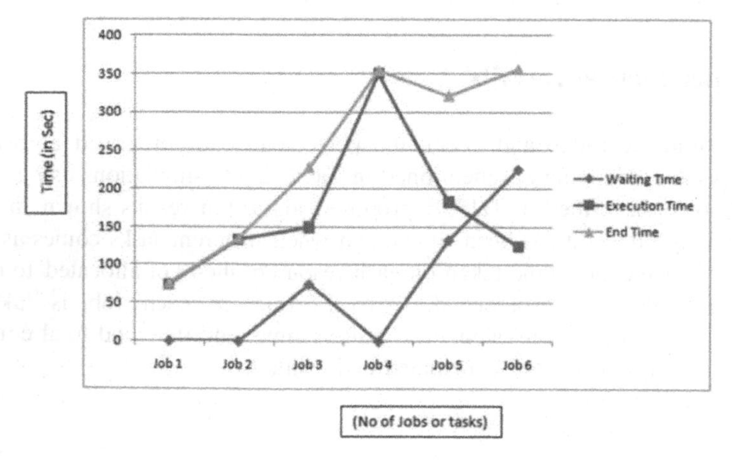

Fig. 1. Graph for waiting time, Execution time, and End time for no of jobs when no of resources is 3

The existing load balancing algorithms usually find the under-loaded node by their status information exchange between the nodes. In the proposed algorithm, load balancing is achieved by assigning workload to different processor according to their computing power which find under loaded node by looking into a queue where pending job's finish time is calculated for each resource by summing up execution time of all tasks, their total execution time is calculated and compared. Next task is allocated to one having less execution time.

Table 4.

Performance Metrics	Execution Time for Optimal Load Balancing Approach	Execution Time for Threshold Based Approach of Load Balancing
T1	75.0	145
T2	133.33	227
T3	225	261
T4	350	500
T5	316	533
T6	350	583

comparative table for same no of jobs and resources in form of execution time taken by proposed Optimal and queuing Based Approach and threshold queuing based approach is shown in Table4.

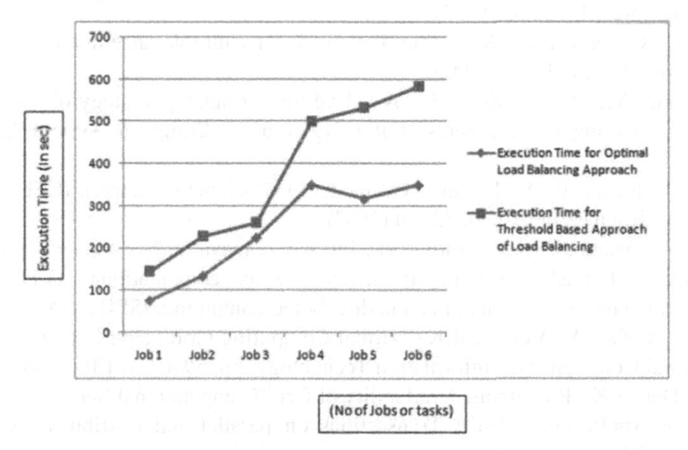

Fig. 2. The graph shows that the execution time of jobs under threshold based approach t is more than that of execution time of jobs with Optimal

5 Conclusion

This paper has addressed the problem of load balancing for tasks in grid system model. In this paper shown how we can achieve better performance when high amount of workload comes. It has been shown that proposed load balancing policy depicts a higher success, in comparison to the existing load balancing techniques. The effect of load balancing on job in terms of execution time is analyzed. Results show that the execution time of proposed algorithm is reduced.

References

1. Diekmann, R., Frommer, A., Monien, B.: Efficient schemes for nearest neighbor load balancing. Parallel Computing 25(7), 789–812 (1999)
2. Elsasser, R., Monien, B., Preis, R.: Diffusion schemes for load balancing on heterogeneous networks. Theory of Computing Systems 35(3), 305–320 (2002)
3. Feng, Y., Li, D., Wu, H., Zhang, Y.: A dynamic load balancing algorithm based on distributed database system. In: Proceedings of the Fourth International Conference on High- Performance Computing in the Asia-Pacific Region, pp. 949–952 (2000)
4. Ghosh, S., Melham, R., Mosse, D.: Fault-tolerance through scheduling of aperiodic tasks in hard real-time multiprocessor systems. IEEE Trans. Parallel and Distributed Systems 8(3), 272–284 (1997)
5. Grosan, C., Abraham, A., Helvik, B.: Multiobjective evolutionary algorithms for scheduling jobs on computational Grids. Journal of Parallel and Distributed Computing 65(9), 1022–1034 (2005)
6. Hu, Y.F., Blake, R.J.: An improved diffusion algorithm for dynamic load balancing. Parallel Computing 25(4), 417–444 (1999)
7. Hui, C.C., Chanson, S.T.: Theoretical analysis of the heterogeneous dynamic load balancing problem using a hydrodynamic approach. Journal of Parallel and Distributed Computing 43(2), 139–146 (1997)
8. Hwang, S., Kesselman, C.: A flexible framework for fault tolerance in the Grid. Journal of Grid Computing 1, 251–272 (2003)
9. Li, Y., Yang, Y., Ma, M., Zhou, L.: A hybrid load balancing strategy of sequential tasks for grid computing environments. Future Generation Computer Systems 25, 819–828 (2009)
10. Lin, F.C.H., Keller, R.M.: The gradient model load balancing method. IEEE Transactions on Software Engineering 13(1), 32–38 (1987)
11. Legrand, A., Marchal, L., Casanova, H., Lu, K., Zomaya, A.Y.: A hybrid policy for job scheduling and load balancing in heterogeneous computational Grids. In: Sixth international symposium on parallel and distributed computing, ISPDC 2007 (2007)
12. Rajan, A., Rawat, A., Verma, R.K.: Virtual Computing Grid using Resource Pooling. In: International Conference on Information Technology, pp. 59–64. IEEE (2008)
13. Roy, N., Das, S.K.: Enhancing Availability of Grid Computational Services to Ubiquitous Computing Applications. IEEE Transactions on parallel and distributed systems 20(7), 953–967 (2009)
14. Luo, S., Peng, X., Fan, S., Zhang, P.: Study on Computing Grid Distributed Middleware and Its Application. In: International Forum on Information Technology and Applications, pp. 441–445 (2009)
15. Huang, K.-C.: On Effects of Resource Fragmentation on Job Scheduling Performance in Computing Grids. In: 10th International Symposium on Pervasive Systems, Algorithms, and Networks, pp. 701–705. IEEE (2009)

16. Wang, L., Laszewski, G.V., Chen, D., Tao, J., Kunze, M.: Provide Virtual Machine Information for Grid Computing. IEEE Transactions on Systems, Man, and Cybernetics-Part A: System and Humans 40(6)
17. Tiburcio, P.G.S., Spohn, M.A.: Ad hoc Grid: An Adaptive and Self-Organizing Peer-to-Peer Computing Grid. In: 10th International Conference on Computer and Information Technology (CIT), pp. 225–232. IEEE (2010)
18. Bai, L., Hu, Y.-L., Lao, S.-Y., Zhang, W.-M.: Task Scheduling with Load Balancing using Multiple Ant Colonies Optimization in Grid Computing. In: Sixth International Conference on Natural Computation (ICNC), pp. 2715–2719. IEEE (2010)
19. Murata, Y., Egawa, R., Higashida, M., Kobayashi, H.: A History-Based Job Scheduling Mechanism for the Vector Computing Cloud. In: 10th Annual International Symposium on Applications and the Internet, pp. 125–128. IEEE (2010)
20. Iosup, A., Epema, D.: Grid Computing Workloads. IEEE Internet Workloads, 19–26 (March/April 2011)
21. Zhang, S., Zhang, S.: The Comparison between Cloud Computing and Grid Computing. In: International Conference on Computer Application and System Modeling (ICCASM), pp. 72–75. IEEE (2010)

Analysis on Range Enhancing Energy Harvester (Reach) Mote Passive Wake-Up Radios for Wireless Networks

N. Shyam Sunder Sagar[1], P. Chandrasekar Reddy[2], Chavali Sumana[3], Naga Snigdha[3], A. Charitha[3], and Abhuday Saxena[3]

[1] Dept of ECE,GITAM University Patancheru, Hyderabad, India
Shyam428@gmail.com
[2] Dept of ECE, JNTUH University, Hyderabad, India
[3] Dept of ECE, GITAM University, Hyderabad, India
{chavalisumana,naga.snigdha207,charitha.inha, abhyuday.saxena93}@gmail.com

Abstract. Today, the wireless sensor networks are used increasingly in many applications fields and in parallel, these are used by scientific researchers to improve and accelerated the features of such networks. These designs are very challenging and are sustainable. On the above, these energy-dependent sensors are relied to run for long periods. These Sensor nodes are usually battery-powered and thus have very shorter lifetime. In this paper, we introduce a novel passive wake-up radio device called REACH (Range Enhancing Energy Harvester) Mote for a traditional sensor node, which uses the energy harvester circuit combined with an ultra-low-power pulse generator to trigger the wake-up of the mote. And this approach aims to reduce the latency without increasing energy consumption.

Keywords: REACH Mote, Energy harvesting, Data latency, S-MAC.

1 Introduction

A sensor is an energy limited device that features capabilities of sensing and data processing, storing and transmitting. A wireless sensor network(WSN) of spatially distributed autonomous sensors to monitor physical or environmental conditions ,such as temperature, sound, pressure etc., and to cooperatively pass their data through the network to a main location in the form of packets. The major problems in this type of network are energy consumption and latency. In-order to decrease the energy consumption, the radio sensor is put into sleep/idle mode. This strategy causes a delay for the transmission of the sensor readings to the base station and, hence the processing tends to reduce in real time. Numerous methods have been proposed to optimize power consumption in battery limited sensor networks but none of the proposed solutions is universally acceptable as there is a trade-off between power consumption and data latency.

In this paper we would like to draw attention on various mechanisms involved with the passive radio receivers which include energy consumption and harvesting, range extension, hardware and software simulations, applications etc.,

S.C. Satapathy et al. (eds.), *Emerging ICT for Bridging the Future – Volume 2*,
Advances in Intelligent Systems and Computing 338, DOI: 10.1007/978-3-319-13731-5_50

2 Energy Harvesting and Range Extension

The Wake up ability of WISP (Wireless identification and sensing platform) has been increased by RF harvesting circuit .In this section, the basic description and design of the energy harvesting circuit and its interfacing principles has been introduced. The properties of the circuit components have also been mentioned.

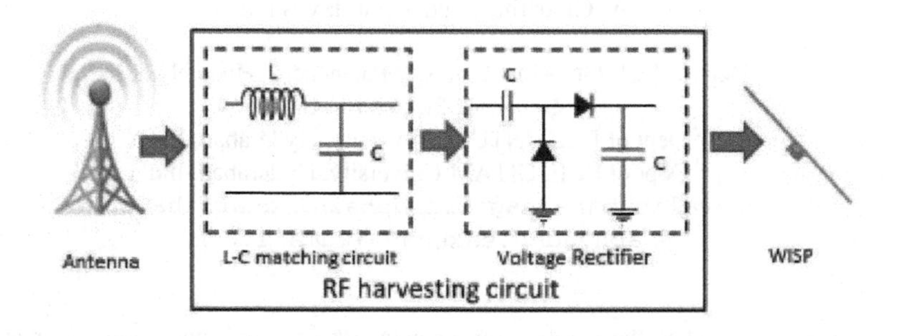

Fig. 1. Architectural view of the circuit and connections

2.1 Selection of Circuit Components

The overall conclusion of our design is to maximize the energy conversion from the front-end antenna to the WISP-Mote. For this, we use a matching circuit for balancing the as shown in Fig. 1, we carefully tune at the matching input impedance that results from the antenna side with the circuit load (i.e., the WISP and Tmote combination), as well as use a voltage rectifier that also functions as a multiplier. The multiplier circuit is based on the classical Dickson's voltage multiplier circuit as shown in Fig 2. It has a number of stages which are connected in parallel .Each stage is a combination of a capacitor and a diode that are connected in parallel. The major advantage of the circuit is the capacitor arrangement with respect to each other. Because of this capacitor arrangement the effective circuit impedance is reduced. This helps in matching the antenna side and the load side in a simpler manner.

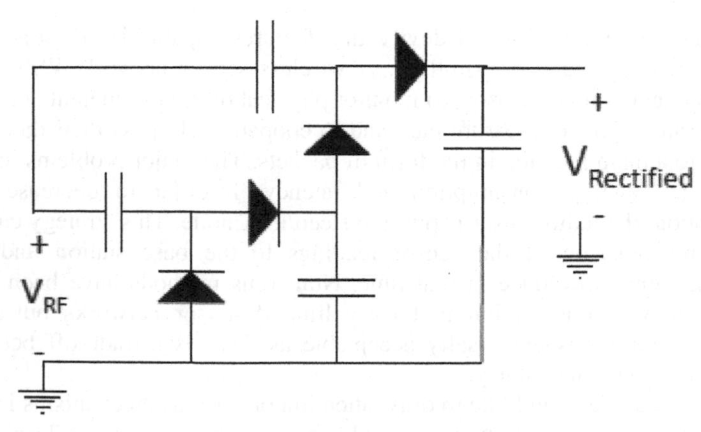

Fig. 2. Dickson diode based multiplier

The diodes having the lowest possible voltages are more preferable as the peak voltage of the AC signal obtained at the antenna is generally much smaller than the diode threshold. The diodes with a very fast switching time need to be used since the energy harvesting circuit operates at a very high frequency in MHz range .The metal-semiconductor junction is used instead of a semiconductor-semiconductor junction in the Schottky diodes. This allows the junction to operate much faster manner and gives a forward voltage drop as low as $0.15V$. We employ diodes from Avago Technologies like HSMS−2852 which have a turn-on voltage of $150mV$, measured at $0.1mA$. Because of this HSMS-2852 specific diode is suitable for operating in the low power region which is typically considered at a range of power between $-20dBm$ and $0dBm$.

The major influence on the output voltage of the energy harvesting circuit is based upon the number of multiplier stages. Here the output voltage is directly proportional to the number of stages used in the energy harvesting circuit. It progressively reduces the current drawn by the load that in turn impacts the overall charging time. The output voltage of the circuit to drive is set at $915MHz$. The number of stages are set to 10 in order to ensure the sufficient voltage to the output circuit.

2.2 Design of EH-WISP-Mote

A special kind of mote by the name EH-WISP-Mote (Enhanced-WISP-Mote),which is a combination of energy harvesting circuit along with WISP mote is built in order to extend the wake-up range of the WISP-Mote, where the output of harvester circuit is connected to the power supply of WISP mote(+Vcc) and the ground pins of both harvester circuit and WISP mote are connected together. While the ability to perform ID-based wakeup is retained, this parallel connection of WISP mote and harvester circuit is used to extend the range of the wake up receivers.

3 Design of REACH-Mote

3.1 Wake-Up Circuit

A sleeping mode which is in low power is used to awaken the Tmote sky node by either rising or falling edge and hence it is necessary to design a wake up circuit for triggering an interrupt onto the Tmote Sky node . As soon as the WISP receives wakeup signal WISP mote and EH mote make use of the MCU ,inorder to generate this pulse. Although an ultra low power microprocessor, MCU still consumes $250\mu A$ at $2.2V$ power supply.The wake-up range of the WISP-Mote is approximately constrained to $13ft$ [1] due to the limited energy harvested by the wake up radio receiver. To increase the range of the radio receiver it is necessary to construct a low energy consuming trigger generator when compared to the WISP.

The constraints included in building a trigger circuit are:

• Little energy must be consumed by the circuit in-order to increase the wake-up range.

• A rising/falling edge generated by the circuit must be capable of driving the Tmote Sky wakeup.

•The output voltage level of energy harvesting circuit is generally unstable and hence the trigger circuit must be able to work on variable voltages.

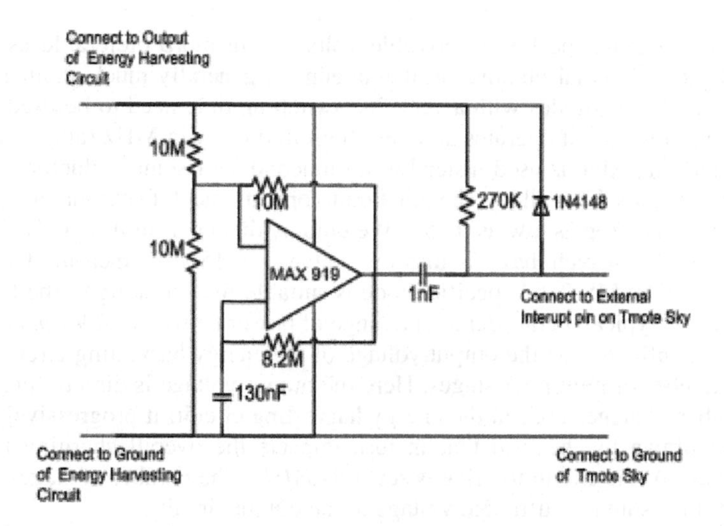

Fig. 3. Wake-up circuit of the REACH-Mote

Fig. 3 shows the wake-up circuit we built to trigger the Tmote Sky mote [2]. Experimental results show that the circuit can operate at a voltage range from $1.5V$ to $5V$, and it can generate a pulse of $100\mu s$ duration once per second. Moreover, the current consumed in this circuit is less than $1\mu A$, which is low enough to be driven by the energy harvester output even with very low power reception. Thus this circuit can achieve the design requirements of an ultra low power wake-up radio trigger.

We combine the RF energy harvesting circuit and the wakeup circuit as well as the Tmote Sky to build a new wake-up radio sensor node, which we called a REACH-Mote (Range Enhancing energy Harvester Mote). Fig. 4 shows a block diagram of the REACH-Mote and the actual components. In a wireless sensor network, the REACH-Mote works as described in Fig. 5. The REACH-Mote is put in sleep mode before the WuTx transmits the wake-up signal, i.e., the MCU on the Tmote Sky, which is an MSP430F1611 is put to LPM3 sleep mode [3] and the radio on the Tmote Sky is in the sleep mode. When a wake-up signal is sent by the WuTx, the energy harvesting circuit outputs a DC voltage. The wake-up circuit starts to generate the pulse once the DC voltage is higher than $1.5V$, and this will trigger the mote and put the mote's MCU into active mode in $5ms$ [4]. After waking up, the data transfer is started if the mote has data to send. The REACH Mote goes back to the sleep mode after the data transfer is finished.

In this work flow, the energy harvesting circuit is a passive component that does not consume energy from the battery. The wake-up circuit is powered by the energy harvesting circuit, so the wake-up circuit also does not drain energy from the battery. Thus, all of the energy provided by the REACH-Mote battery is used for sensing and data communication.

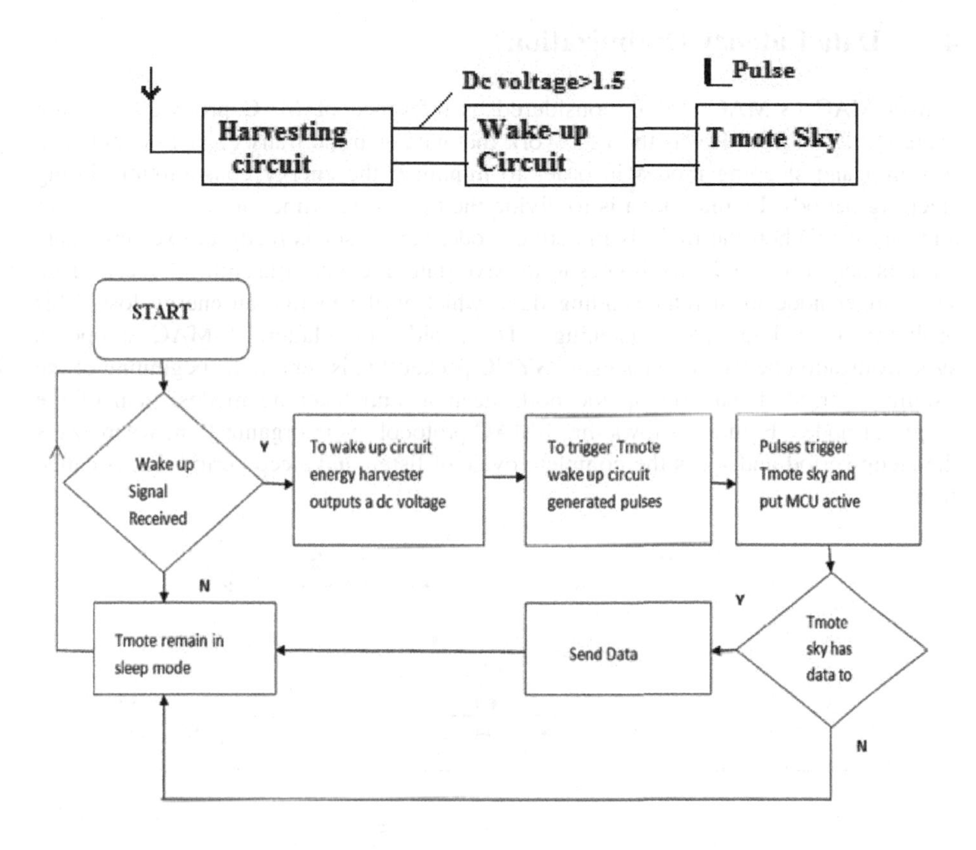

Fig. 4. REACH Mote Flow chart

3.2 Comparison of the REACH-Mote and the WISP-Mote

Every mote designed can be used to achieve passive wake ups with different approaches. But the REACH mote and the WISP mote are more superior with some advantages namely:

• The WISP-Mote and EH-WISP Mote, because of MCU and radio either receiving or decoding hardware can achieve ID-based wake-up along with RFID reader when compared to the REACH mote as the wake-ups are eliminated when the sink is not interested in data from all sensor nodes.

• The flexibility in MAC protocol design can be improved due to the preprocessing of WISP's MCU.

• Most of the advantages of the WISP mote and EH mote are due to WISP's MCU. But because of this the energy consumption increases and hence REACH mote is preferred.

• REACH mote is more economical when compared to WISP mote.

•Due to the low energy consumption of REACH-Mote wake-up circuit ,it has an edge over WISP mote in wake-up range and wake-up delay.

4 Data Latency Optimization

Sensor-MAC (S-MAC) [5] is considered as reference of MAC protocols that are designed for WSNs. It was the first work that managed the transceiver time between listening and sleeping modes in order to minimize the energy consumption during sleeping periods. Its main idea is to divide the transceiver time into two parts: active and passive. When the radio is in active mode, the sensor is ready to exchange data with its neighbors, and once it goes to passive state, it cannot transmit or receive data. The sender node must retry sending data, which will generate an energy loss. This problem is called deaf listening. To avoid this latter, S-MAC proposes synchronization between nodes using SYNC packet that is sent in the beginning of the listening period. It can set up the both sleeping and listening modes, inall of the neighbor nodes. Figure 5 shows the S-MAC protocol cycle organization, where *Ts* is the sleep period and *Fn* is the complete cycle of listen and sleep period that is called frame.

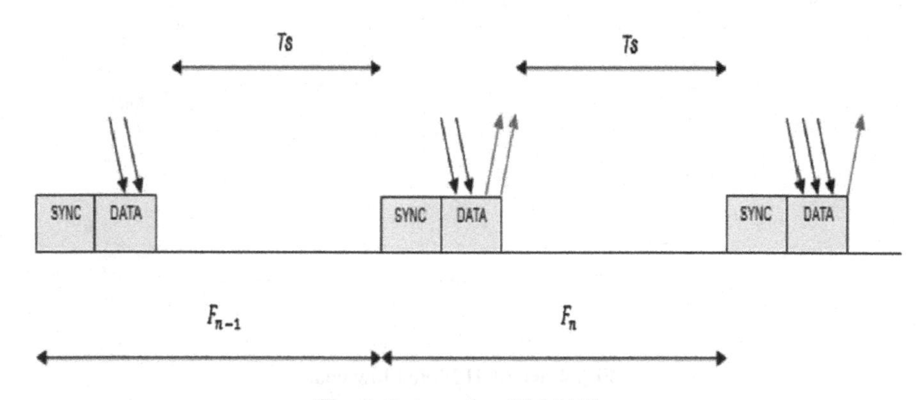

Fig. 5. Duty cycle of S-MAC

The OOM-MAC Algorithm

In S-MAC protocol, the nodes are set to the passive/active mode periodically, the both parts of the cycle, are fixed previously, that presents the major drawback of this protocol. In our algorithm, we enhance the management of the sleep schedule. Note that, we keep the same basic features of the SAC protocol [6]; nodes follow the same synchronization rules of S-MAC protocol, the collision avoidance by RTS/CTS control messages, message passing and overhearing avoidance.

So in our algorithm we are following the steps discussed below:

Step1: Do a sampling time in the interval •0.. *Ts*••Where the interval depends on the duty cycle.

Example: In our simulations As the clock frame is 1.603s and we considered 10% as our duty cycle the sleep period will be 1.442s. So, in this case we have to sample the interval •0.. 1.4*s*•.

Step2: we calculate the probability of transition from OFF to ON state for each sample.

Step 3: we take a number, *K*, randomly in the interval [0, 1].

Step4: Now, compare it with the table of probability, calculated previously, if the number K is smaller than Pi. ti), then the next activation moment of the transceiver is ti.

Thus, the duration of the OFF state is •0.. ti•, with a transition probability Pi. Once it is estimated the time activation of the next cycle, the node broadcasts its schedule to the neighbors by the SYNC packet.

5 Simulation Results

5.1 Simulation Parameters

The simulations are carried out using the simulator NS2(network simulator2), that is among of the most used simulators to study the performances of network protocols [7]. At the application layer we used the ON/OFF Exponential traffic source that will show the effectiveness of OOM-MAC protocol. To evaluate just the MAC layer, we use a static routing protocol that is the NOAH protocol [8]. The values used for the simulation parameters are summarized inTable 1.The topology of our simulation consists of 11 nodes with 10 hops. The first node, considered as the traffic source, and the last node is the sink. We consider that, just the neighbors of one hop are in the same area of scope. In other words, a node can exchange messages just with its immediate neighbors.

5.2 Measurement of Energy Consumption

The energetic criterion is the most important factor that we must take into consideration when designing a MAC protocol for WSN, due to its low energy resources. In OOM-MAC protocol, we aim to reduce the latency without increasing energy expenses, and the results of simulations show the effectiveness of OOM-MAC at these points. illustrates the difference between the energy consumption of OOM-MAC protocol and S-MAC one.

Simulations are performed according to the parameters mentioned in the previous subsection (Table 1). To evaluate the energy performance, we calculate the energy consumed by all nodes on the network, for different traffic levels. During ON periods (burst time), the source node sends packets with an inter-arrival time that varies from $0.5s$ up to $5s$. The horizontal axis indicates the inter arrival time of packet in the ON period, and the vertical axis shows the total of energy consumed at all nodes on the network for the transmission of 20 packet. The results of S-MAC protocol, twice higher, as compared to those given by OOM-MAC protocol. The reason of this difference, is that the cycle of the OOM-MAC protocol is based on dynamic sleep schedule. In the OOM-MAC protocol, we estimate the period of sleeping state according to transition probability that based on ON/OFF Markov model. So, this makes the MAC layer more suitable for bursty traffic load, thus, the transmission becomes twice as fast between the source node and destination, versus the S-MAC protocol.

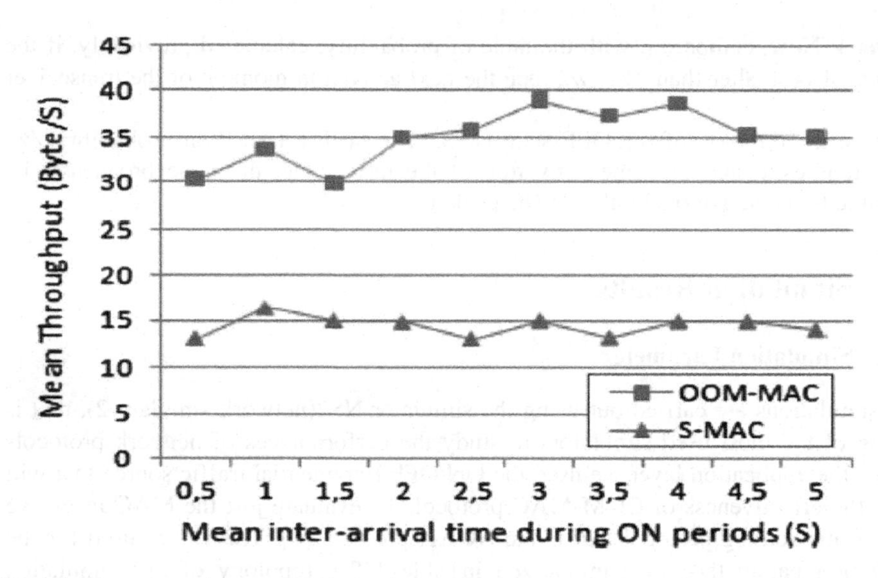

Fig. 6. Mean throughput in OOM-MAC and S-MAC protocol, with ON/OFF traffic source in high traffic load

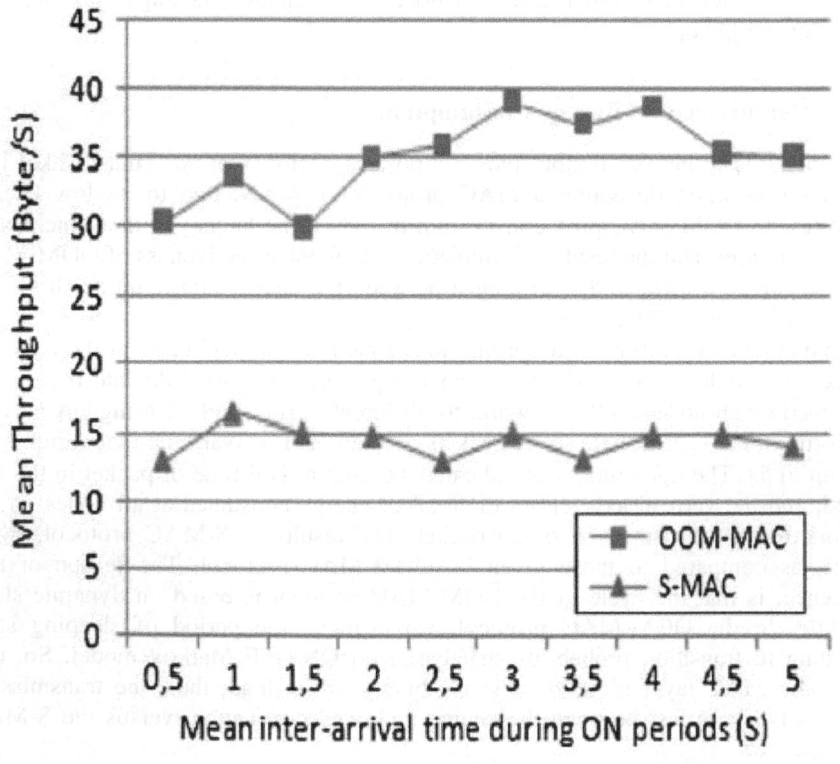

Fig. 7. Mean throughput in OOM-MAC and S-MAC protocol, with ON/OFF traffic source in different traffic load

6 Conclusions and Future Work

In this paper, the concept of REACH-Mote has been discussed in order to increase the wake up range of passive wake up radio sensor nodes .When compared to other traditional WISP motes it clearly shows that REACH mote has 20% better coverage range. We also discussed about the data latency of OOM-MAC protocol using Markrov model and its simulation results generated using NS-2. We plan to evaluate the performance of REACH-Motes in different communication scenarios and compare the benefits and drawbacks of long range broadcast wake-ups achieved with REACH-Motes and shorter range ID-based wake-ups achieved with EH-WISP-Motes

References

1. Ba, H., Demirkol, I., Heinzelman, W.: Feasibility and Benefits of Passive RFID Wake-up Radios for Wireless Sensor Networks. In: IEEE Globecom (2010)
2. http://www.maximintegrated.com/app-notes/index.mvp/id/1186
3. http://www.ti.com/product/msp430f1611
4. http://www.ti.com/tool/msp430-3p-motei-tmotesky-dsgkt
5. Ye, W., Heidemann, J., Estrin, D.: An energy-efficientmac protocol forwireless sensor networks. In: Proceedings of the IEEE Twenty-First Annual Joint Conference of the IEEE Computer and Communications Societies, INFOCOM 2002, vol. 3, pp. 1567–1576 (2002), doi:10.1109/INFCOM.2002.1019408
6. Ye, W., Heidemann, J., Estrin, D.: Medium access control with coordinated adaptive sleeping for wireless sensor networks. IEEE/ACM Transactions on Networking 12(3), 493–506 (2004), doi:10.1109/TNET.2004.828953
7. The network simulator - ns-2, http://www.isi.edu/nsnam/ns/
8. NO ad-hoc routing agent (NOAH), http://icapeople.epfl.ch/widmer/uwb/ns-2/noah/

UOSHR: UnObservable Secure Hybrid Routing Protocol for Fast Transmission in MANET

Gaini Sujatha and Md. Abdul Azeem

MVSR Engineering College, Nadergul, Hyderabad, India
{gaini.sujatha,abdulazeem77}@gmail.com

Abstract. A Mobile ad hoc network (MANET) is a collection of independent mobile nodes that are self organized and self configured. Routing in MANET is a critical issue due to mobility of node and openness of network, so an efficient routing protocol makes the MANET secure and reliable. UOSPR deals with the concept of anonymity, unlinkability and unobservability as well as uses BATMAN proactive routing in which every node has knowledge of best hop details instead of maintaining entire network topology. In UOSPR there is latency in data transmission if route is not found in the routing table at a particular time interval. We know that ad hoc network has limited battery power, but already half of the power used for security in USOR and UOSPR, remaining power has to be used efficiently for calculating routes and data transmission. For this purpose UOSHR is introduced by combining reactive (USOR) and Proactive (UOSPR) routing protocols. This technique eliminates latency in finding route, achieves fast and secure transmission of data. NS2 is used to implement and validate the effectiveness of the design.

Keywords: Anonymity, unlikability, unobservability, privacy, routing, BATMAN, USOR.

1 Introduction

Mobile ad hoc network (MANET) is a collection of wireless mobile hosts, form a temporary network without aid of stand-alone or central administration or access point. Nodes in the network not only act as host but also as a router to find route to other nodes which are not in direct range. Mostly, MANETs are used in military battlefield, commercial sectors like emergency rescue operations and personnel area network etc. Due to limited battery power and insecure mobile nodes, an efficient routing protocol is required. Since MANETs can be set up anywhere and can allow any node into network, a privacy preserving routing protocol is must.

MANET routing protocols are classified into three types: Reactive routing protocols (Non-table driven, on demand), Proactive routing protocols (Table driven) and Hybrid routing protocols.

In Reactive routing protocol, routes are discovered through route request and route reply packets, only when source want to send data to the destination like AODV and DSR. In Proactive routing protocol, every node maintains the network topology

S.C. Satapathy et al. (eds.), *Emerging ICT for Bridging the Future – Volume 2*,
Advances in Intelligent Systems and Computing 338, DOI: 10.1007/978-3-319-13731-5_51

information by periodically exchanging HELLO packet with its neighbors like DSDV and OLSR. The combination of reactive and proactive routing protocols is a hybrid routing protocol. It takes the advantages of two protocols, as a result route is found quickly in the routing zone. UOSHR can find route without any delay. Hybrid routing protocol requires the properties called adaptive, flexible, efficient and practical for successful deployment Ex: ZRP.

With regard to privacy preserving routing protocols, the terminology called anonymity, unlinkability, and unobservability should be followed which is discussed in [3]. These can be achieved by combing group signature [4] and ID-based encryption [5]. Privacy is also maintained by protecting data and control packets including their headers.

A group signature allows a member of group to anonymously sign a message on behalf of the group. Sign is verified by all group members without knowing the person's identity. The group manager has ability to reveal the original signer in the event of disputes. ID-Based encryption is a type of public key encryption, the PKG (Private key Generator) first publishes a master public key and retains the corresponding master private key. It allows any party to generate a public key from known identity value and master key. PKG generates the corresponding private keys.

The rest of the paper is organized as follows: Section 2 describes existing secure routing protocols for MANET. Section 3 describes method of UOSHR and its algorithm. Section 4 furnishes performance analysis of UOSHR by comparing with related work. Finally, we summarize and conclude the paper in section 5.

2 Related Work

Normal AODV uses on-demand nature to find route to the destination. When malicious node is entered into the network, packet delivery ratio becomes less and packet delay becomes more, so to avoid attacking malicious node to the network, privacy has to be maintained.

Allowing adversaries to trace routes make serious threat to network. This can be avoided by route anonymity, location privacy and identity anonymity. The anonymous routing protocol, ANODR [6] gives location privacy for each forwarding node to prevent enemy from tracing route from start to end node of network. This design is completely based on broadcast with trapdoor assignment by embedding trapdoor information to the packet, known only to the receiver. Identity anonymity and strong location privacy is not provided in ANODR, they are important properties of anonymity.

Allowing adversaries to trace routes make serious threat to network. This can be avoided by route anonymity, location privacy and identity anonymity. The anonymous routing protocol, ANODR [6] gives location privacy for each forwarding node to prevent enemy from tracing route from start to end node of network. This design is completely based on broadcast with trapdoor assignment by embedding trapdoor information to the packet, known only to the receiver. Identity anonymity and strong location privacy is not provided in ANODR, they are important properties of anonymity.

Weak location privacy and route anonymity achieved in SDDR [13]. ANODR is efficient than SDDR at data transmission range. ASR [10] uses one time public/private key pair concept to achieve stronger location privacy, identity privacy and route anonymity by ensuring nodes on the route having no information on their distance to the source/destination node. ASR abandons the onion routing to reduce computation overhead, but generation of public/private key for each RREQ is expensive.

A new cryptography technique called as pairing enables neighboring nodes to authenticate each other anonymously by dynamically changing pseudonym of nodes instead of real identities. MASK [7] allows neighboring nodes to establish pair wise secrete key for anonymous route discovery and data forwarding. When comparing with AODV, MASK outperforms under heavy traffic load. SDAR [14] and ODAR [15] uses long-term public key at each node for anonymous key establishment.

The drawback of ANODR, ASR, SDAR and MASK is that a global adversary can trace RREP packet by following the flow from destination back to source. ARM [11] addresses this problem by enabling private communication between nodes which makes tougher for adversaries to focus on attacks by making use of secrete key and pseudonym. ALARM [12] makes use of public key cryptography and group signature to preserve privacy. It uses nodes current location to construct a secure MANET map. Based on the current map, each node can decide to which node it wants to make the communication. Node privacy under this framework is preserved even if portion of nodes are stationary, or if speed of the motion is not very high. ALARM leaks a lot of sensitive information such as network topology and location of a node.

The above all anonymous routing protocols achieve complete anonymity and partial unlinkability, but unobsevability is not achieved. Due to this, attacker can easily trace packet flow and control traffic. For the first time USOR [1] has introduced which is privacy preserving on demand routing protocol which achieved anonymity, unlinkability and content unobservability in MANET. Group signature is used for anonymity and unlikability. It uses ID based encryption and pseudonyms for unobservability. In USOR both control packets like route request, route replay packets and data packets including their headers were encrypted, which were indistinguishable from dummy packets of outside adversaries.

In a Simple pragmatic approach to routing using BATMAN [8], includes comparison of BATMAN and OLSR. Traffic overhead of OLSR is almost 90% higher than BATMAN. Since BATMAN only maintains control packet of OGM, but OLSR has to maintain HELLO message and topology control messages. Android Application uses BATMAN routing [9], implemented to provide free telephony services. This method has four scenarios: single hop scenario, Join/Leave scenario, multihop scenario, and network reconfiguration scenario.

USOR concentrates on security, but there is latency in finding route to the destination due to on demand nature. To reduce latency and to improve efficiency of data delivery, proactive routing protocol is introduced, i.e. UOSPR [2]. BATMAN routing is a proactive routing protocol maintains best hop details among all neighbors instead of maintaining complete network topology which avoids computation overhead and provides fast and secure transmission. This routing protocol also establishes anonymous keys to bring security to the data and achieves anonymity, unlikability and content unobsevability.

3 Proposed Work

UOSHR technique incorporates unobservable security with hybrid routing scheme. Hybrid routing consist of BATMAN (Better Approach to Mobile Ad hoc Network) proactive routing and unobservable secure on demand routing (USOR). UOSHR executes in two phases: Anonymous key establishment and Hybrid routing.

3.1 Anonymous Key Establishment

In this phase, every node in the ad hoc network communicates with its direct neighbors within its radio range for anonymous key establishment. Suppose there is a node A with a private signing key gsk_A and a private ID-based key K_A in the ad hoc network. It is surrounded by a number of neighbors within its power range. Node A does the following the anonymous key establishment procedure:

(1) Node A generates a random number $r_A \in Z^* q$ and computes r_AP, where P is the generator of G1. It then computes a signature of r_AP using its private signing key gsk_A to obtain $SIGgsk_A(r_AP)$. Anyone can verify this signature using the group public key gpk. It broadcast (r_AP, $SIGgsk_A$ (r_AP)) within its neighborhood.

(2) A neighbor X receives message from A, verifies the signature with gpk. If the verification is successful, X chooses a random number $r_X \in Z^*q$ and computes r_XP. X computes a signature $SIGgsk_X(r_AP|r_XP)$ using its own signing key gsk_X. X computes the session key $k_{AX} = H_2(r_Ar_XP)$, and replies to A with message (r_XP, $SIGgsk_X$ ($r_AP|r_XP$),E_{kAX} (k_X*$|r_AP|r_XP$)), where k_X* is X's local broadcast key.

(3) Upon receiving the reply from X, A verifies the signature inside the message. If the signature is valid, A proceeds to compute the session key between X and itself as $k_{AX} = H_2(r_Ar_XP)$. A also generates a local broadcast key k_A*, and sends E_{kAX} (k_A*$|k_X$*$|r_AP|r_XP$) to its neighbor X to inform about the established local broadcast key.

(4) X receives the message from A and computes the same session key as $k_{AX} = H_2(r_Ar_XP)$. It then decrypts the message to get the local broadcast key k_A*.

3.2 Hybrid Routing Scheme

Route discovery is done by combination of BATMAN and USOR. Initially, routing starts with BATMAN, every node updates routing information for every periodic interval. If a node wants to send data it checks in the routing table for a concern route node then sends data to that node. If route is not found, instead of waiting for next interval, node discovers route with USOR (on demand) through RREQ and RREP later forwards the data. This way UOSHR eliminates latency in finding route and improves the data delivery ratio.

In BATMAN, all the participating nodes periodically broadcast originator messages (OGMs) to its neighbors. An OGM consists of originator address, sending node address and unique sequence number. Upon receiving the OGMs, the neighbor nodes change the sending address to their own address and re-broadcast the message, and so on and so forth. Thus, the network is flooded with originator messages. The sequence numbers are used to determine the freshness of an OGM, in order to allow nodes to re-broadcast OGMs at once. See [8] and [1] for more information about the workings of the algorithm of BATMAN and USOR respectively.

Suppose source A wants to send data to the destination D with B is an intermediate node. The following are the packet formats of RREQ, RREP, HELLO and DATA respectively.

$Nonce_A$, Nym_A, E_{KA*}(RREQ N_A E_D(S D r_AP), seqno)
$Nonce_B$, Nym_{AB}, E_{KAB*}(RREP N_A E_A(S D r_AP r_DP), seqno)
$Nonce_A$, Nym_A, E_{KA*}(HELLO N_A E_A(S D r_AP), Org, SA, seqno, H)
$Nonce_A$, $Nygm_A$, E_{KAB*}(DATA, N_A, seqno, E_{AD}(Payload))

Here $Nonce_A$ is the alternative name of node A, $Nym_A = H_3(K_{A*} | Nonce_A)$, RREQ is packet type, N_A is another random number as route pseudonym, which is used as index to specific route entry, Org is originator address of HELLO packet, SA is sending address and H is hop count.

This routing scheme requires two tables, one is called neighbor table which keeps track of OGM of BATMAN, another is called routing table which stores route found through RREQ and RREP packets of USOR for simplicity. The following algorithm explains the implementation of UOSHR.

USOHR Algorithm

1. Initialize the nodes as follows
   a. Key server: it can share the key at initial time
   b. Normal node: normal mobile node
2. Key server initially sends the Group ID key to all the mobile nodes
3. If normal node receives the GID it stores it into memory
4. If node having GID
   a. Request key server for private key
   b. key server sends private key, public key, pseudonym
5. If node not having GID
   a. It Cannot access the request
6. Checks for the session availability
   a. If session available
      i. Start the communication with LB Key
   b. If no secure session
      i. Send the session request to neighbor with Digital sign
      ii. Neighbor checks the sign and establish the session key to requester
      iii. Receives the session key from neighbor and wait for small interval
      iv. Establish local broad-cast key and send to secured neighbors
7. After authentication is over, sender digitally signs on HELLO packet and periodically broadcast to find best hop and stores in neighbor table.

8. If node (i) wants to communicate with another node
   a. It checks for route in the neighbor table
   b. If found
      i. Sends data to destination
   c. If not found
      i. Checks in routing table
      ii. If found
         A. Sends data to destination
      iii. If not found
         A. Discovers route with USOR

4 Analysis and Comparison to Related Work

We have tested MANET with AODV and malicious AODV. We implemented USOR, UOSPR and UOSHR then compared the performance of them.

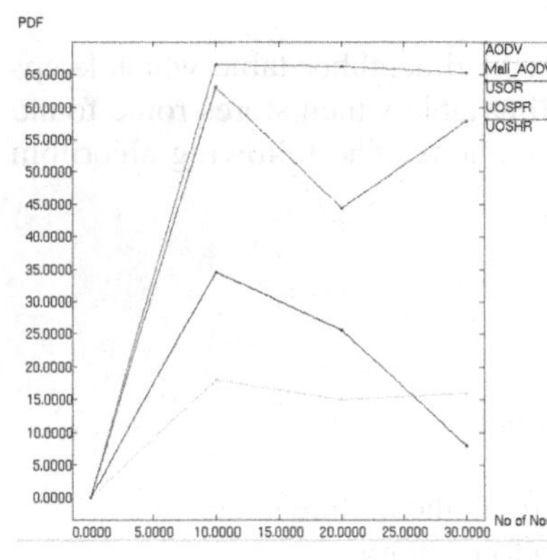

Fig. 1. Packet Delivery Ratio Comparison of AODV, malicious AODV. USOR, UOSPR and UOSHR

Table 1. Packet Delivery Ratio Comparison of AODV, malicious AODV. USOR, UOSPR and UOSHR

nodes	aodv %	Maodv %	usor %	uospr %	uoshr %
10	66.33	22.5	34.58	63.79	64.55
20	66.35	16.50	26.66	46.35	47.35
30	65.35	15.06	7.98	57.77	58.12

PDF comparison of four protocols has shown in fig and table 1. The packet delivery fraction that is the ratio of received packets to the send packets, of normal AODV is high among all, but due to lack of security if malicious node enters into network, PDF becomes very low. To avoid this privacy preserving routing is implemented, i.e. USOR, which is implemented against all attacks, but due to on demand nature of routing PDF is less compared to UOSPR. PDF is improved in UOSPR than USOR. UOSHR has high PDF than UOSPR.

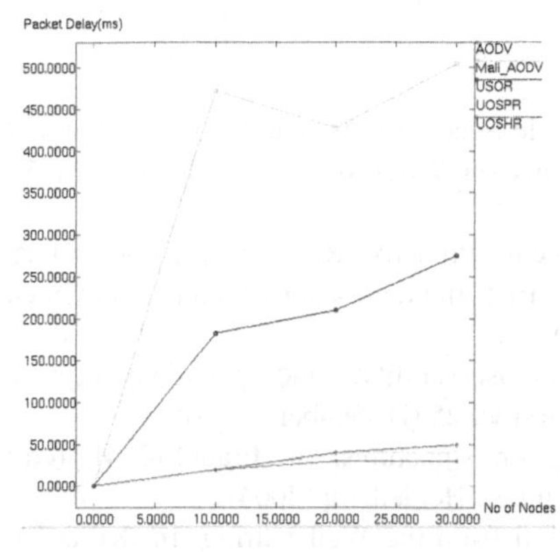

Fig. 2. Packet Delay Comparison of AODV, malicious AODV, USOR, UOSPR and UOSHR

Table 2. Packet Delay Comparison of AODV, malicious AODV, USOR, UOSPR and UOSHR

nodes	aodv	Maodv	usor	uospr	uoshr
10	0.02s	4.73s	0.18s	0.02s	0.02s
20	0.03s	4.28s	0.21s	0.04s	0.04s
30	0.03s	50.51s	0.27s	0.04s	0.05s

In AODV, packet delay is very less since security is not there, but if malicious node enters, delay become high as shown in the Fig and table 2 which decreases the performance of the network. To avoid that secured privacy preserving routing protocol USOR is implemented. Due to the packet encryption and decryption, delay becomes high compared to normal AODV. There is latency in route discovery of USOR, UOSPR made pre route discovery to reduced latency. As PDF is more in UOSHR, PD is also more compared to UOSPR.

5 Conclusion and Future Work

In this paper, we proposed an unobservable secure hybrid routing by combining USOR and UOSPR. When route is not found within periodic interval by using UOSPR, instead of waiting for next interval it finds route using USOR. This way UOSHR achieves efficient data transmission without delay in finding the route to the destination. PDF of UOSHR is more compared to UOSPR and USOR. Through anonymous key establishment, session key and local broadcast keys were generated, with those keys, packets were encrypted before transmitting to another node in a network. Hence, unobservability, anonymity and unlinkability are maintained in the network. We can enhance UOSHR by testing with wormhole attack as well as block hole attack.

Acknowledgments. Authors would like to thank all anonymous reviewers for their valuable comments. We also thank all technical staff of CSED, MVSREC for extending their support and giving valuable suggestions. And also thanks to parents for their support. Finally thanks to almighty god for being with me in each and every moment.

References

1. Wan, Z., Ren, K., Gu, M.: USOR: An Unobservable Secure On-Demand Routing Protocol for Mobile Ad Hoc Networks. IEEE Transactions on Wireless Communications 11(5), 1922–1932 (2012)
2. Balaji, S., Prabha, M.: UOSPR: UnObservable Secure Proactive Routing Protocol for Fast and Secure transmission using B.A.T.M.A.N. In: IEEE International Conference on Green High Performance Computing, May 14-15 (2013)
3. Pfitzmann, A., Hansen, M.: Anonymity, unobservability, and pseudonymity: a consolidated proposal for terminology. draft, version v0.25 (December 6, 2005)
4. Boneh, D., Boyen, X., Shacham, H.: Short group signatures. In: Franklin, M. (ed.) CRYPTO 2004. LNCS, vol. 3152, pp. 41–55. Springer, Heidelberg (2004)
5. Boneh, D., Franklin, M.: Identity-based encryption from the Weil pairing. In: Kilian, J. (ed.) CRYPTO 2001. LNCS, vol. 2139, pp. 213–229. Springer, Heidelberg (2001)
6. Kong, J., Hong, X.: ANODR: anonymous on demand routing with untraceable routes for mobile ad-hoc networks. In: Proc. ACM MOBIHOC 2003, pp. 291–302 (2003)
7. Zhang, Y., Liu, W., Lou, W.: Anonymous communications in mobile ad hoc networks. In: 2005 IEEE INFOCOM (2005)
8. Johnson, D., Ntlatlapa, N., Aichel, C.: Simple pragmatic approach to mesh routing using BATMAN. In: 2nd IFIP International Symposium on Wireless Communications and Information Technology in Developing Countries (2008)
9. Sicard, L., Markovics, M., Manthios, G.: An Ad hoc Network of Android Phones Using B.A.T.M.A.N. In: The Pervasive Computing Course, Fall 2010. The IT University of Copenhagen (2010)
10. Zhu, B., Wan, Z., Bao, F., Deng, R.H., KankanHalli, M.: Anonymous secure routing in mobile ad-hoc networks. In: Proc. 2004 IEEE Conference on Local Computer Networks, pp. 102–108 (2004)
11. Seys, S., Preneel, B.: ARM: anonymous routing protocol for mobile ad hoc networks. In: Proc. 2006 IEEE International Conference on Advanced Information Networking and Applications, pp. 133–137 (2006)
12. Defrawy, K.E., Tsudik, G.: ALARM: anonymous location-aided routing in suspicious MANETs. IEEE Trans. Mobile Comput. 10(9), 1345–1358 (2011)
13. El-Khatib, K., Korba, L., Song, R., Yee, G.: Secure dynamic distributed routing algorithm for ad hoc wireless networks. In: International Conference on Parallel Processing Workshops, ICPPW 2003 (2003)
14. Boukerche, A., El-Khatib, K., Xu, L., Korba, L.: SDAR: a secure distributed anonymous routing protocol for wireless and mobile ad hoc networks. In: Proc. 2004 IEEE LCN, pp. 618–624 (2004)
15. Sy, D., Chen, R., Bao, L.: ODAR: on-demand anonymous routing in ad hoc networks. In: 2006 IEEE Conference on Mobile Ad-hoc and Sensor Systems (2006)

An ANN Approach for Fault Tolerant Wireless Sensor Networks

Sasmita Acharya[1] and C.R. Tripathy[2]

[1] Department of Computer Applications, VSS University of Technology, Burla, India
talktosas@gmail.com
[2] Department of Computer Science and Engineering, VSS University of Technology, Burla, India
crt.vssut@yahoo.com

Abstract. Wireless Sensor Network (WSN) is an emerging technology that has revolutionized the whole world. This paper proposes an artificial neural network model for a reliable and fault-tolerant WSN based on an exponential Bi-directional Associative Memory (eBAM). An eBAM has higher capacity for pattern pair storage than the conventional BAMs. The proposed model strives to improve the fault tolerance and reliability of packet delivery in WSN by transmitting small-sized packets called vectors. The vectors are associated with the original large-sized packets after encoding the associations between the hetero-associative vectors for the given problem space. This encoding is done by the application of the evolution equations of an eBAM. The performance characteristics of the proposed model are compared with other BAM models through simulation.

Keywords: Wireless Sensor Network, Reliability, Fault Tolerance, Bi-directional Associative Memory, Spurious memory.

1 Introduction

The Wireless Sensor Networks (WSNs) consist of a large number of low-cost, limited energy, small, inexpensive and autonomous sensor nodes that are deployed in vast geographical areas for remote operations. The WSNs are used in many applications like battlefield surveillance, forest fire monitoring, search and rescue operations, environmental monitoring, prediction and detection of natural calamities and structural health monitoring etc [1]. The WSNs may be classified based on communication functions, data delivery models and network dynamics [2-6]. The resource constraints of sensors and their dynamic topology pose many technical challenges in network discovery and routing, network control, collaborative information processing and querying etc.

The Computational Intelligence (CI) combines elements of learning, evolution, adaptation and fuzzy logic to create intelligent machines [9]. There are various CI paradigms like neuro-computing, reinforcement learning, and evolutionary computing, fuzzy computing and neuro-fuzzy. The CI techniques have been successfully used by researchers to address many challenges in WSNs [7-10].

S.C. Satapathy et al. (eds.), *Emerging ICT for Bridging the Future – Volume 2*,
Advances in Intelligent Systems and Computing 338, DOI: 10.1007/978-3-319-13731-5_52

This paper presents an artificial neural network (ANN) model for the design of a reliable and fault-tolerant WSN using an eBAM. The basics of artificial neural networks are discussed in Section 2. The eBAM and its evolution equations are discussed in Section 3. The Section 4 outlines the system model for a fault-tolerant and reliable WSN using an eBAM. The Section 5 of the paper discusses the simulation results. Finally, Section 6 concludes the paper.

2 ANN and WSN

The analogy between a biological neuron, an artificial neuron and a sensor node is shown in Fig. 1. A sensor node is analogous to a biological or artificial neuron. A sensor gathers information from the environment and converts it to an electrical signal. A sensor node has to go through a pre-processing step called filtering (using a function f(x)). This filtering in a sensor node corresponds to weights in artificial neurons or synapses in biological neurons. The next processing within the processor corresponds to the chemical processing done by the soma in case of a biological neuron or applying the activation function h(x) as in the case of an artificial neuron. After processing, the modified reading of the sensor is sent out via the radio link which is similar to an axon in case of a biological neuron.

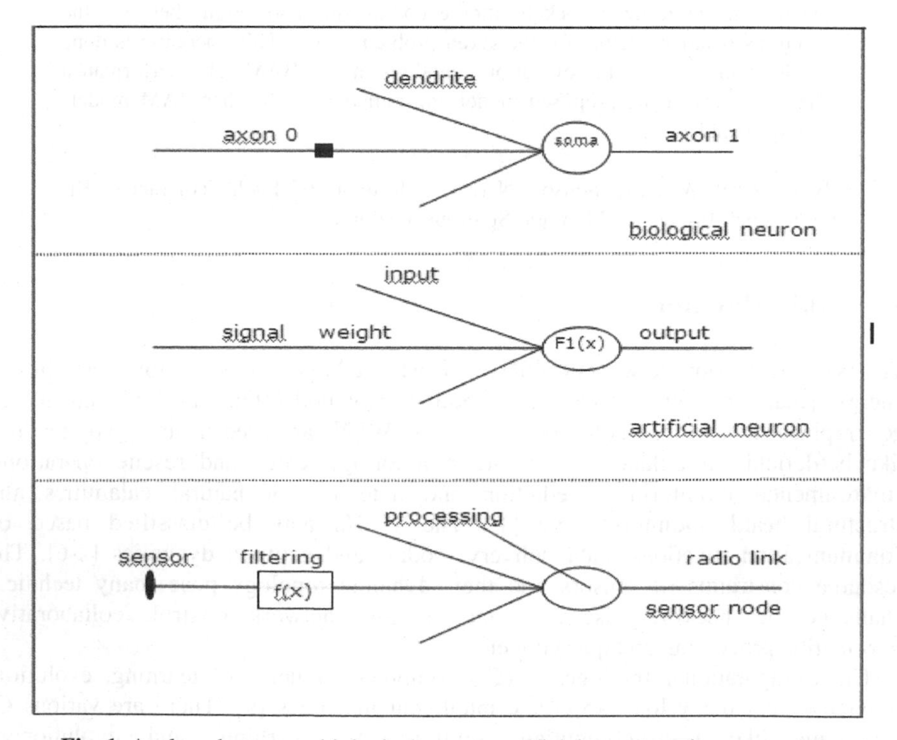

Fig. 1. Analogy between a biological neuron, artificial neuron and a sensor node

An artificial neural network (ANN) consists of a network of neurons that are organized as input, hidden and output layers. The ANNs exchange information between neurons frequently. Based on their connectivity, they are classified into – feed forward and recurrent networks. A two-level ANN architecture for WSNs can be devised by viewing the whole sensor network as a neural network. There could be a neural network within each sensor node inside the WSN to decide on the output action.

3 Overview of an Exponential BAM (eBAM)

An eBAM has higher capacity for pattern pair storage than the conventional BAMs. The model takes advantage of the exponential nonlinearity in the evolution equations of an eBAM causing a significant increase in the performance of the system. As the recall process is in progress, the energy decreases which ensures stability to the system.

3.1 Evolution Equations for eBAM

Let there are N training pairs given by $\{(A_1,B_1), (A_2,B_2), \ldots, (A_n, B_n)\}$ where $A_i = (a_{i1}, a_{i2}, \ldots, a_{in})$ and $B_i = (b_{i1}, b_{i2}, \ldots, b_{in})$.

Let X_i and Y_i be the bipolar modes of the training pattern pairs A_i and B_i respectively given by $X_i \in \{-1, 1\}^n$ and
$Y_i \in \{-1, 1\}^p$.

The following equations are used in the recall process of an eBAM.

$$y_k = 1 \text{ if } \sum_{i=1}^{N} y_{ik}\, b^{Xi.X} \geq 0 \quad (1)$$

$$y_k = -1 \text{ if } \sum_{i=1}^{N} y_{ik}\, b^{Xi.X} < 0 \quad (2)$$

$$x_k = 1 \text{ if } \sum_{i=1}^{N} x_{ik}\, b^{Yi.Y} \geq 0 \quad (3)$$

$$x_k = -1 \text{ if } \sum_{i=1}^{N} x_{ik}\, b^{Yi.Y} < 0 \quad (4)$$

Here, ‘b’ is a positive number, b > 1 and “.” represents the inner product operator of X and X_i, Y and Y_i.

4 System Model for Fault-Tolerant and Reliable WSN Using eBAM

The proposed model strives to improve the fault tolerance and reliability of packet delivery in WSN. It assumes that the sensor node has to sense a packet from a set P =

$\{P_1, P_2, P_3, \ldots, P_{10}\}$ where the size of each packet is of k-bits. There are more chances of faulty delivery of a larger size packet because of noise. So, a possible solution is to transmit a small-sized packet (vector) which is associated to the original large-sized packet after encoding the associations between these hetero-associative vectors. This ensures better reliability and fault tolerance.

An eBAM is used for encoding the associations. It consists of two layers – layer-1 with ‘n’ units and layer-2 with ‘m’ units. The architecture of the proposed eBAM is shown in Fig. 2. It is a fully connected recurrent network where the inputs are different from the outputs. After training the network, if vector A is presented over the network, then it recalls the associated vector B and vice-versa.

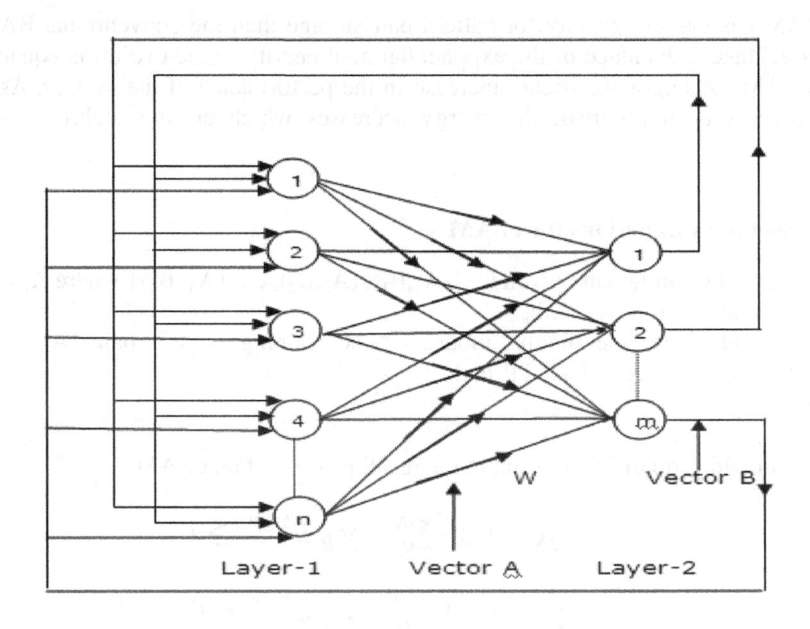

Fig. 2. Architecture of the proposed eBAM

4.1 Training Procedure (Encoding the Associations)

This is done to encode the association between the hetero-associative patterns for the given problem space. An associated packet (vector) is generated for each original data packet. These small-sized vectors are then sent for data transmission.

Original Packet	Associated Packet for transmission (Vector)
Binary Versions :	
$A_1 = [\,1\,0\,1\,0\,1\,0\,0\,1\,]$	$B_1 = [\,1\,1\,1\,]$
$A_2 = [\,1\,1\,0\,0\,1\,1\,0\,1\,]$	$B_2 = [\,1\,0\,0\,]$
$A_3 = [\,1\,0\,0\,1\,0\,1\,0\,1\,]$	$B_3 = [\,1\,0\,1\,]$

Bi-polar versions :

$X_1 = [\ 1\ -1\ 1\ -1\ 1\ -1\ -1\ 1\ ]$ $Y_1 = [\ 1\ 1\ 1\ ]$

$X_2 = [\ 1\ 1\ -1\ -1\ 1\ 1\ -1\ 1\ ]$ $Y_2 = [\ 1\ -1\ -1\ ]$

$X_3 = [\ 1\ -1\ -1\ 1\ -1\ 1\ -1\ 1\ ]$ $Y_3 = [\ 1\ -1\ 1\ ]$

4.2 Retrieval of a Vector Corresponding to a Pattern

To retrieve vector Y_2 corresponding to pattern X_2, first, all the inner products are computed as follows:

$$< X_2 . X_1 > = 2\ ,\ < X_2 . X_2 > = 8 \text{ and } < X_2 . X_3 > = 2$$

Then, choosing b = 2,

$$b^{< X_2 . X_i >} \text{ for } i = 1, 2, 3 = (\ 2^2, 2^8, 2^2\ ) = (\ 4, 256, 4\ )$$

$$\text{So, } Y = [\ (1 \quad 1 \quad 1) \begin{pmatrix} 4 \\ 256 \\ 4 \end{pmatrix}, (1 \quad -1 \quad -1) \begin{pmatrix} 4 \\ 256 \\ 4 \end{pmatrix}, (1 \quad -1 \quad 1) \begin{pmatrix} 4 \\ 256 \\ 4 \end{pmatrix}]$$

$$= (\ 264, -256, -248\ )$$

$$= (\ 1, -1, -1\ )$$

$$= Y_2$$

So, the vector corresponding to pattern X_2 is Y_2.

4.3 Recall of Noisy Pattern by eBAM

Suppose the 1st, 6th and 7th bits of bi-polar version of X_2 get distorted due to noise and the following noisy pattern X is received at a sensor node.

$$X = [\ -1\ 1\ -1\ -1\ 1\ -1\ 1\ 1\ ]$$

First, all the inner products are computed as follows:

$$< X . X_1 > = 0\ ,\ < X . X_2 > = 2 \text{ and } < X . X_3 > = -4$$

Then, choosing b = 2,

$$b^{< X_2 . X_i >} \text{ for } i = 1, 2, 3$$

$$= (\ 2^0, 2^2, 2^{-4}\ ) = (\ 1, 4, 1/16\ )$$

$$\text{So, } Y = [\,(1 \;\; 1 \;\; 1)\begin{pmatrix}1\\4\\1/16\end{pmatrix},\ (1 \;\; -1 \;\; -1)\begin{pmatrix}1\\4\\1/16\end{pmatrix},\ (1 \;\; -1 \;\; 1)\begin{pmatrix}1\\4\\1/16\end{pmatrix}]$$

$$= (\,81/16, -49/16, -47/16\,)$$

$$= (\,1, -1, -1\,)$$

$$= Y_2$$

So, the correct data packet is X_2 and is given by, $X_2 = [\,1 \;\; 1 \;\; -1 \; -1 \;\; 1 \;\; 1 \; -1 \;\; 1\,]$.

So, the eBAM could easily recall the correct stored pattern for a noisy pattern whose data bits were distorted due to noise by simple calculations using the evolution equations.

5 Simulation Results

The purpose of this simulation was to compare the performance of the proposed eBAM with that of a symmetrical BAM (sBAM) and a generalized asymmetrical BAM (GABAM). The simulation was carried out in MATLAB 7.5.0. The output function parameter was set to 0.1 and the learning parameter was set to 0.01. To limit the simulation time, the number of output iterations before each weight matrix update was set to 1.

5.1 Learning and Recall of Bipolar Patterns

The learning was carried out according to the following procedure:

a) random selection of a pattern pair.
b) computation of transpose matrices X^T and Y^T according to output function.
c) computation of the weight matrix update.
d) repetition of steps a) to c) until the weight matrix converges (about 2000 learning trials).

Then the performance of the proposed eBAM was compared with that of a GABAM and a sBAM with respect to their recall capability for noisy inputs, storage capacity and percentage of spurious memory recall.

5.2 Recall Performance

After completing the learning phase, the performance of the network was tested on a noisy recall task. The task was to recall the correct associated pattern from a noisy input obtained by randomly flipping bits in the input pattern. The no. of bits flipped varied from 0 to 10, corresponding to a noise proportion of 0 to 20%. The network recall performance of an eBAM was compared with that of a sBAM and a GABAM by testing with different noisy inputs. The proposed eBAM recall performance clearly outperformed that of a sBAM and was slightly less than that of a GABAM as shown in Fig. 3.

5.3 Storage Capacity Performance

The network storage capacity was evaluated by using random 10-dimensional bipolar vectors and varying the number of learning pairs from 1 to 15, generating different test sets and comparing with that of a sBAM and a GABAM. The simulation results are shown in Fig. 4. It clearly indicates that an eBAM has slightly better storage capacity than a sBAM or a GABAM.

5.4 Spurious Memory Recall Performance

The proportion of spurious attractors was determined by calculating the number of vectors that stabilized in a spurious state in relation to the total number of vectors. 1000 number of random vectors was generated to conduct the experiment. The simulation results for the recall of spurious memory are shown in Fig. 5. The figure shows that with the proposed eBAM, the network developed only 35 % spurious attractors while with GABAM, it was 70 % and with sBAM, the number of spurious recalls was 95 %. So, the proposed eBAM model performs significantly better than a GABAM or a sBAM with respect to the percentage of spurious memories during recall. This greatly enhances the reliability of packet delivery in WSNs.

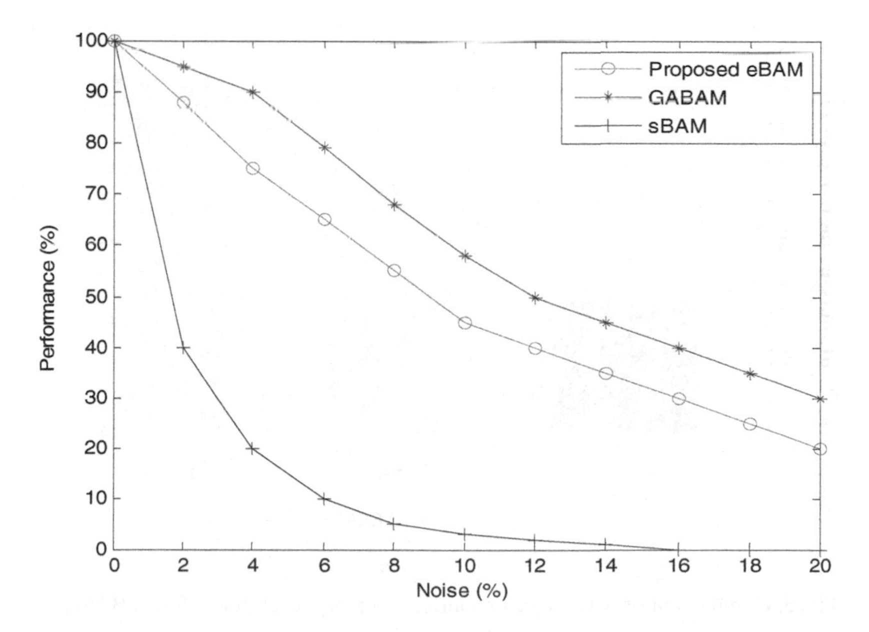

Fig. 3. Comparison of Recall Capability for different BAMs

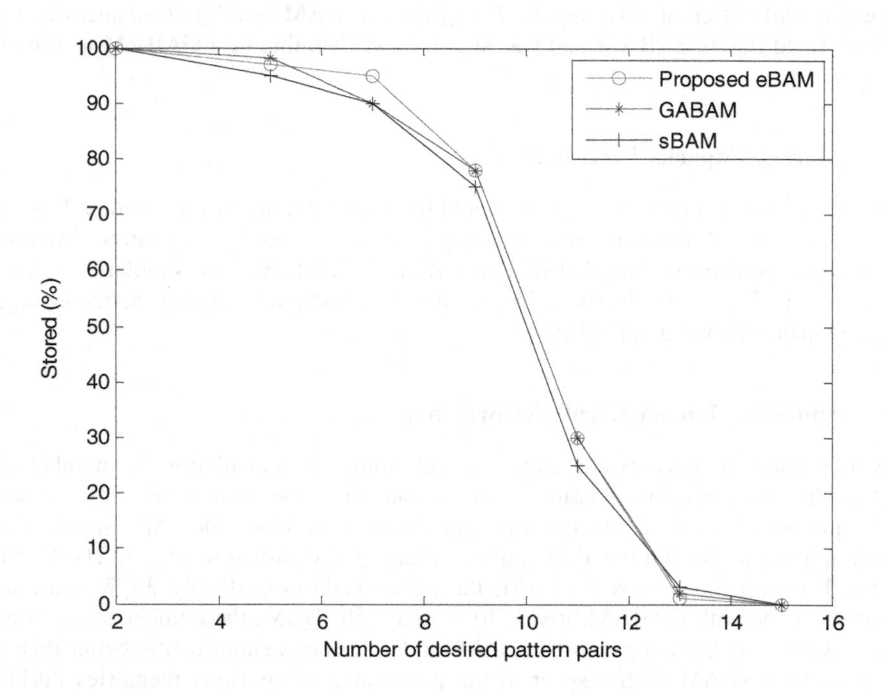

Fig. 4. Comparison of Storage Capacity for different BAMs

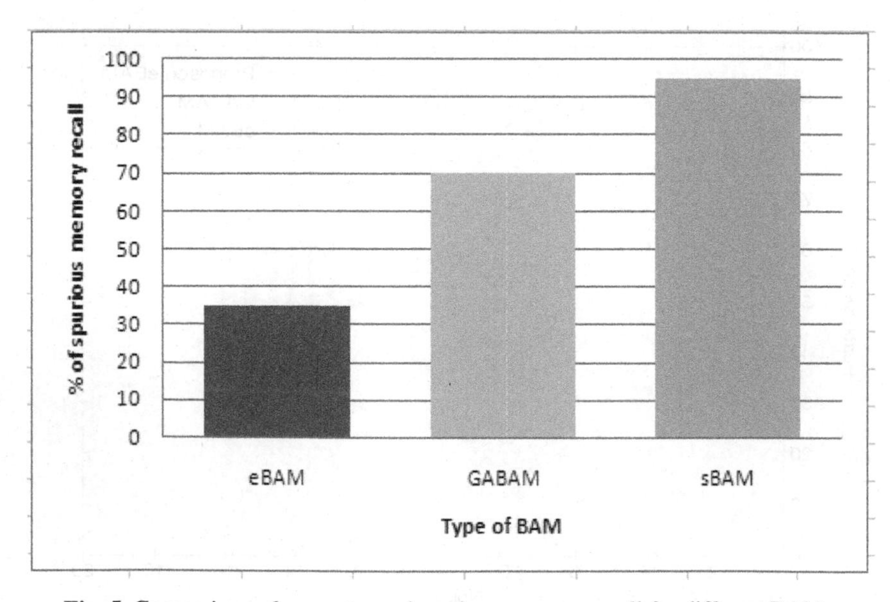

Fig. 5. Comparison of percentage of spurious memory recall for different BAMs

6 Conclusion

This paper presented an ANN model for the design of a reliable and fault-tolerant WSN based on eBAM's evolution equations. An eBAM was embedded at each sensor node of the system model to ensure reliable and timely delivery of packets. This was done considering the eBAM's powerful features of retrieving the original data packet despite noisy associated patterns within its fault tolerance limit. The performance of the proposed eBAM was compared with that of GABAM and sBAM and it was found that the proposed eBAM is more efficient than other BAMs in pattern recall, storage capacity and spurious memory recall thus giving stability to the system. Thus, efficient ANN implementations using simple computations can replace the traditional signal processing algorithms in WSNs as well as provide WSNs more powerful features like massive parallelism, adaptivity, learning and generalization ability, fault tolerance, reliability and low computation.

References

1. Akyildiz, I.F., Su, W., Sankarasubramaniam, Y., Cayirci, E.: A survey on sensor networks. IEEE Comm. Mag. 40(8), 102–114 (2002)
2. Culler, D., Estrin, D., Srivastava, M.: Overview of sensor networks. IEEE Computer 37(8), 41–49 (2004)
3. Romer, K., Mattern, F.: The design space of wireless sensor networks. IEEE Wireless Comm. 11(6), 54–61 (2004)
4. Heinzelman, W.B., Murphy, A.L., Carvalho, H.S., Perillo, M.A.: Middleware to support sensor network applications. IEEE Network 18(1), 6–14 (2004)
5. Intanagonwiwat, C., Govindan, R., Estrin, D.: Directed diffusion: a scalable and robust communication paradigm for sensor networks. In: Proc. of ACM/IEEE MobiCom (2000)
6. Intanagonwiwat, C., Govindan, R., Estrin, D., Heidemann, J., Silva, F.: Directed diffusion for wireless sensor networking. IEEE/ACM Trans. on Networking 11(1), 2–16 (2003)
7. Kumar, V., Patel, R.B., Singh, M., Vaid, R.: A Neural Approach for Reliable and Fault Tolerant Wireless Sensor Networks. International Journal of Advanced Computer Science and Applications 2(5), 113–118 (2011)
8. Kosko, B.: Bidirectional Associative Memories. IEEE Transactions on Systems, Man and Cybernetics 18(1), 49–60 (1988)
9. Kulkarni, R.V., Forster, A., Venayagamoorthy, G.K.: Computational Intelligence in Wireless Sensor Networks: A Survey. IEEE Communications Surveys and Tutorials 13(1), 68–95 (2011)
10. Chartier, S., Boukadoum, M.: A Bidirectional Heteroassociative Memory for Binary and Grey-Level Patterns. IEEE Transactions on Neural Networks 17(2), 385–396 (2006)

An Intelligent Stock Forecasting System Using a Unify Model of CEFLANN, HMM and GA for Stock Time Series Phenomena

Dwiti Krishna Bebarta[1], T. Eswari Sudha[1], and Ranjeeta Bisoyi[2]

[1] Dept of CSE, GMR Institute of Technology, Rajam, AP, India
bebarta.dk@gmrit.org, sudha.eswari@gmail.com
[2] Multidisciplinary Research Cell, S.O.A. University, Bhubaneswar, Odisha, India
ranjeeta.bisoi@gmail.com

Abstract. The aim of this work to suggest and apply a unify model by combining the Computationally Efficient Functional Link Artificial Neural Networks (CEFLANN), Hidden Markov Model (HMM), and Genetic Algorithms (GA) to predict future trends from a highly uncertainty stock time series phenomena. We present a framework of an intelligent stock forecasting system using complete features to predict stock trading estimations that may consequence in better profits. Using CEFLANN architecture, the stock prices are altered to independent sets of values that become input to HMM. The trained and tested HMM output is used to identify the trends in the stock time series data. We apply different methods to generate complete features that raise trading decisions from stock price indices. We have used population based optimization tool genetic algorithms (GAs) to optimize the initial parameters of CEFLANN and HMM. Finally, the results achieved from the unified model are compared with CEFLANN and other conventional forecasting methods using performance assessment techniques.

Keywords: HMM, Genetic Algorithm, FLANN, CEFLANN, MAPE, AMAPE.

1 Introduction

The most important factor for being successful in trading in stock market is the ability to predict future market fluctuations appropriately. The goal of every investor is to buy stocks when market is low and sell when market trend is high to gain profit. Analyst's and researchers speaks related to the efficient market supposition, it is practically impossible to predict accurately based on historical stock market data. Professional traders use two major types of analysis to make accurate decisions in financial markets: fundamental and technical. Fundamental analysis is to consider company's overall condition of the economy, type of industry and overall financial strengths. Whereas technical analysis depends on charts, comparison tables, technical indicators and historical prices of stocks.

S.C. Satapathy et al. (eds.), *Emerging ICT for Bridging the Future – Volume 2*,
Advances in Intelligent Systems and Computing 338, DOI: 10.1007/978-3-319-13731-5_53

Presently forecasting technique has improved the accuracy by introducing vital forecasting methodologies. Though, the technical analysis traditionally started with various statistical models as a tool for forecasting future data with the help of historical data. Different models introduced like Auto Regressive Integrate Moving Average (ARIMA) [1], a flexible autoregressive conditional heteroskedasticity (ARCH) model [2] and Multiple Linear Regression (MLR) [3] are used for forecasting the time series phenomena. However such models are not good enough to predict efficiently when the stock time series data is considered as highly non-linear, high degree of vagueness, and volatile. Presently technical analysis is carry out by using advanced machine intelligence methods. Artificial Neural Network (ANN) [4], [5], [6] is suggesting one of the efficient soft computing methods for stock market prediction. ANN's are data driven model and are not efficient because of the multiple layers between the input and output layers. To improve further in accuracy and efficiency in forecasting researcher have also used Fuzzy logic, which is based on expert knowledge. Fuzzy logic can be used either autonomously [7], [8] or hybridized [9], [10] with other methods for forecasting stock data. These ANN's and hybrid network systems involves more computational complexity during training and testing for forecasting stock time series data. HMM [11], [12], [13] is a form of probabilistic finite state system used extensively for pattern recognition and classification problems. Using HMM for predicting future stock trend based on past data is not simple. HMM that is trained on the past dataset of the chosen stocks is used to search for the variable of interest data pattern from the historical dataset. In order to accomplish better prediction than the other models, the intellectual Functional Link Artificial Neural Network (FLANN) [14], [15], [16] is used. These networks are computationally efficient, involve less computational complexity and provide superior prediction performance compared to other standard models. In this paper we have proposed a unify model combining computationally efficient functional link artificial neural network (CEFLANN), Hidden Markov Model (HMM) and genetic algorithms (GAs) to achieve the better forecasting results. We have explored three different stock data for forecasting to exhibit our model. The prediction results obtained from our model is further compared with the fusion model of ANN, HMM, and GA discussed in paper [17].

In this paper, we imply and develop a unify model consisting of computationally efficient Functional Link Artificial Neural Network (CEFLANN) structural design [18], [19] with HMM and a population based optimization tool genetic algorithms (GAs) [20] for better prediction of prices of leading stock market phenomena. The structure of CEFLANN model is simple and with proper jumbles of technical and fundamental parameters prediction accuracy is improved. A Hidden Markov Model (HMM), which is built on the probabilistic structure for modeling time series annotations to predict stock prices. Further by using population based optimization tool GAs for initializing the parameters of the CEFLANN and HMM.

The rest of the paper is planned as follows: section 2, the details of the unify model provided; section 3, study of datasets; section 4, experimental setup, performance estimation and results; and lastly section 5 tenders a brief conclusion and advance researches are given.

2 Overview of the Unify Model

2.1 Computationally Efficient Functional Link Artificial Neural Network (CEFLANN) Model

The architecture of CEFLANN described in the paper [18], [19] is a nonlinear adaptive model and can approximate stock time series phenomena through a supervised learning. The algorithm used in the CEFLANN is an optimum method for computing the gradients of the error by the help of back-propagation learning method. This algorithm usually treated as robust with respect to annoyance or unstable data. Fig. 1 shows the block diagram representation of CEFLANN architecture.

This CEFLANN architecture uses a functional expansion block where all the inputs are used to enhance the dimension of input space as described below.

$$\underbrace{X = [x_1, x_2, \ldots\ldots, x_m, x_0, x_{p+1}, \ldots. x_{p+m}]}_{Input Vector} \qquad \underbrace{[x_{m+1}]}_{Desired Vector} \tag{1}$$

Where $[x_1, x_2, \ldots\ldots, x_m]$ are from real stock data, $[x_{p+1}, \ldots. x_{p+m}]$ enhanced input using Functional Expansion Block (FEB) and x_{m+1} is target data taken from real stock data.

The trigonometric basic function used to enhance the input pattern is based on tan hyperbolic as specified below.

$$\begin{aligned} x_{p+1} &= \tanh(a_{01} + a_{11}x_1 + a_{21}x_2 + a_{31}x_3 + \ldots. + a_{m1}x_m) \\ x_{p+2} &= \tanh(a_{02} + a_{12}x_1 + a_{22}x_2 + \ldots\ldots\ldots\ldots\ldots + a_{m2}x_m) \\ &\ldots\ldots\ldots\ldots\ldots\ldots\ldots\ldots\ldots\ldots \\ x_{p+m} &= \tanh(a_{0m} + a_{1m}x_1 + a_{2m}x_2 \ldots\ldots\ldots\ldots\ldots + a_{mm}x_m) \end{aligned} \tag{2}$$

Where, the associate parameter given in matrix [A] eq. (3) and input matrix X in eq. (1) is used in above eq. (2)

$$A = \begin{bmatrix} a_{01} & a_{11} & a_{21} \ldots\ldots\ldots\ldots & a_{m1} \\ a_{02} & a_{12} & a_{22} \ldots\ldots\ldots\ldots\ldots & a_{m2} \\ \ldots\ldots\ldots\ldots\ldots\ldots \\ a_{0m} & a_{1m} & a_{2m} \ldots\ldots\ldots\ldots & a_{mm} \end{bmatrix} \tag{3}$$

$$\begin{aligned} &O = W * X^T \\ &where \\ &W = [w_1, w_2, w_3, \ldots\ldots, w_m, w_0, w_{p+1}, \ldots\ldots, w_{p+m}] \\ &X = [x_1, x_2, \ldots\ldots, x_m, x_0, x_{p+1}, \ldots. x_{p+m}] \end{aligned} \tag{4}$$

The weight vector [W] and associated parameters [A] are updated in the direction of the negative gradient of the performance function using back-propagation learning algorithm by computing the error between the target output and the expected output is written in eq. (5)

$$e = x_{m+1} - O \tag{5}$$

By taking gradient of the error with respect to the weights, the weight adjustment is obtained in eq. (6)

$$W(t+1) = W(t) + \frac{\mu \tanh(\beta e) * X}{\lambda + X^T X} \tag{6}$$

$$a_{j+m-1} = a_{j+m-1} + P$$

where

$$P = C * P1 * x_{j+m}$$

where

$$P1 = (\mathrm{sech}(a_{j-1} + a_j * x_j + a_{j+1} * x_{j+2} + \ldots + a_{j+m-1} * x_{j+m-1})^2$$

2.2 Hidden Markov Model

A Hidden Markov Model (HMM) is like a finite state machine that represents the statistical regularities of sequences. In HMMs, the states are hidden and observations are probabilistic function of state. Steps to define HMM more formally:

Step-1: N, the number of states in the model.

$$S = \{s_1, s_2 \ldots s_N\}$$

Step-2: M, the number of distinct observation sequences.

$$Q = \{q_1, q_2 \ldots q_m\}$$

Step-3: The state transition probability distribution matrix, $T = \{t_{ij}\}$

Where,

$$t_{ij} = \Pr ob(c' = s_j \; or \; c = s_i)$$

$$where \quad 1 \leq i \; and \; j \leq N$$

Where, c' is the next state, c is the current state, and S_j is the jth state.

Step-4: Observation Probability distribution matrix, $B = [b_{s_j}(q_t)]$

Where,

$$b_{s_j}(q_t) = \Pr ob(q_t \, \mathrm{I} \, c = s_j)$$

$$where \quad 1 \leq j \leq N \; and \; 1 \leq t \leq M$$

Step-5: The initial state distribution vector [π] and

$$\pi_i = \Pr ob(q_0 = s_i)$$

Where, $i \leq N$ and q_0 is initial state

Step-6: The final HMM model is λ= (T, B, π)

In this study, the Baum–Welch algorithm [13] is attempted to re-estimate the HMM parameters A, B and π so that the HMM model best fits in the training data set. An algorithm Forward- Backward method is used to compute the probability $P(Q/\lambda)$ of observation sequence $Q= q_1, q_{2,.....}q_m$ for the given model λ, where $\lambda = (A, B, \pi)$.

2.3 Integration of CEFLANN with HMM

The proposed unify model consisting of CEFLANN and HMM cascaded as shown in given Fig. 1 to forecast stock prices. We have employed CEFLANN as a tool to bring out an observed sequence. These observed sequences are further introduced to HMM to improve a better accuracy in the prediction result. The GA is used to find out optimal initial parameters for both CEFLANN and HMM.

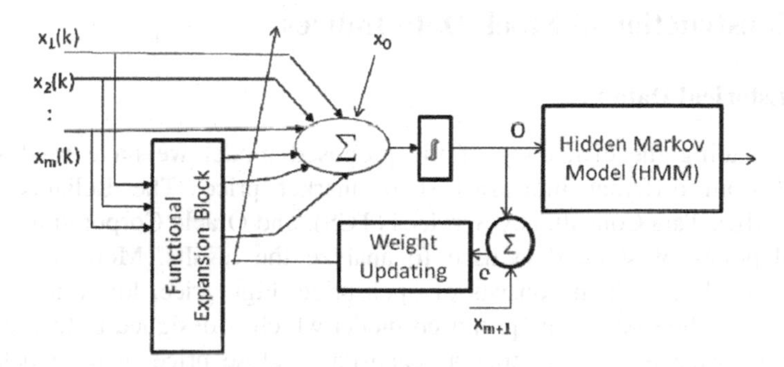

Fig. 1. The UNIFY MODEL

To combine CEFLANN and HMM to make our unify model the integration method states below:

a. We have created the structure of the CEFLANN having 'm' inputs of real historical data and 'm' number of enhanced input using FEB. One bias term is also introduced for reducing error during training the network.

b. We have initialized the respective parameters of the network.

c. The output of the network 'O' resulted from CEFLANN is fed into the HMM as input to get the final output 'Y'.

d. GA method is used to optimize the initial parameters of CEFLANN and HMM since performance of these methods depends on the initial values.

2.4 Using GA to Optimize the Parameters of HMM

There are three parameters in HMM (T, B, π) which are needed to optimize by using GA instead of considering random values. The initial parameters of GA are chosen suitably. The initial parameters of GA are as follows:

For Observation probability distribution and Transition probability matrix

Chromosome size	16
Population type	Double
Population size	20
Elite parent selection	2
Crossover fraction	0.8
Migration fraction	0.2
Generation	100
Fitness limit	-infinity
Initial Population	Random
Fitness scaling	Rank Scaling
Selection	Stochastic Uniform

For Initial probability distribution

Chromosome size	4
Population type	Double
Population size	20
Elite parent selection	2
Crossover fraction	0.8
Migration fraction	0.2
Generations	100
Fitness limit	-infinity
Initial Population	Random
Fitness scaling	Rank Scaling
Selection	Stochastic Uniform

3 Construction of Stock Data Indices

3.1 Historical Data Set

For investigating the efficiency of the proposed model we have used stock data collected from different industries share market price. The Reliance Industries Limited (RIL), Tata Consultancy Services (TCS), and Oracle Corporation (ORACLE) historical prices of stock data used to analyze the results. More than 2000 data patterns considered which consists of open price, high price, low price, close price and volume of the stocks. The proposed model which is designed to forecast the next day closing price by taking input as each day's close price in the stock index of datasets. Following 80:20 ratio, we have used closing price of 200 days for training and next 30 days of closing price is used for testing the model. The MATLAB implementation has done for forecasting using the proposed unify model.

3.2 Data Normalization

To form an input vector for the unify model to learn, the range of values is 0 to 1. The historical past data collected from web is used for training and testing the model. The normalization formula in the following equation (7) is shaping by stating the data in terms of maximum and minimum values of the stock data.

$$NV = \frac{x_i - x_{\min}}{x_{\max} - x_{\min}} \tag{7}$$

Where, NV is normalized value and x_i is actual value, x_{min} and x_{max} represents minimum and maximum values of original stock data set.

4 Experiments and Results

4.1 Simulation Results

The comparison of target vs. predicted stock closing price of RIL stock during testing of CEFLANN only and with HMM is given in Figs. 2 & 3. The stock market indices for one day in advance for a period of 30 days are showcased.

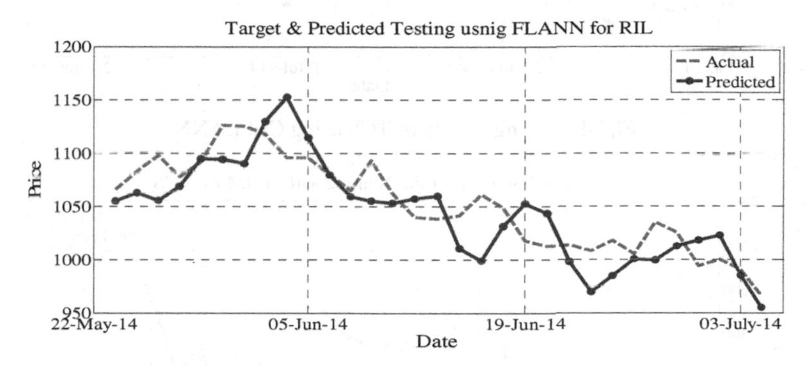

Fig. 2. Testing results of RIL using CEFLANN

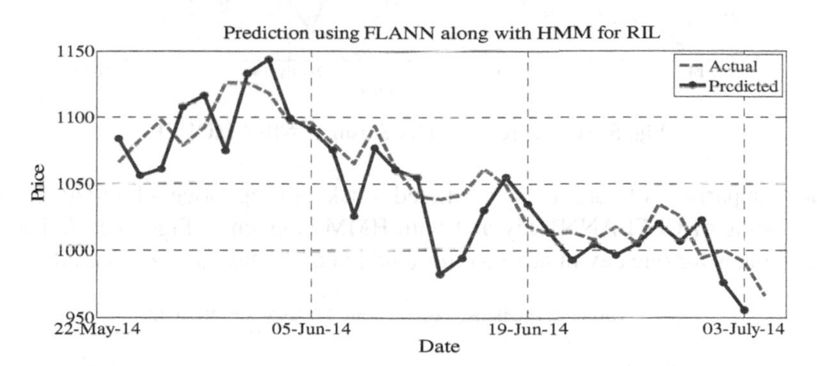

Fig. 3. Testing results of RIL using UNIFY MODEL

The comparison of target vs. predicted stock closing price of TCS stock during testing of CEFLANN only and with HMM is given in Figs. 4 & 5. The stock market indices for one day in advance for a period of 30 days are showcased.

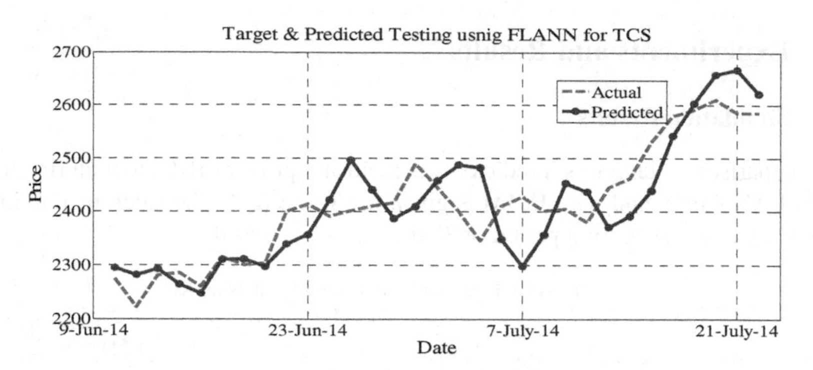

Fig. 4. Testing results of TCS using CEFLANN

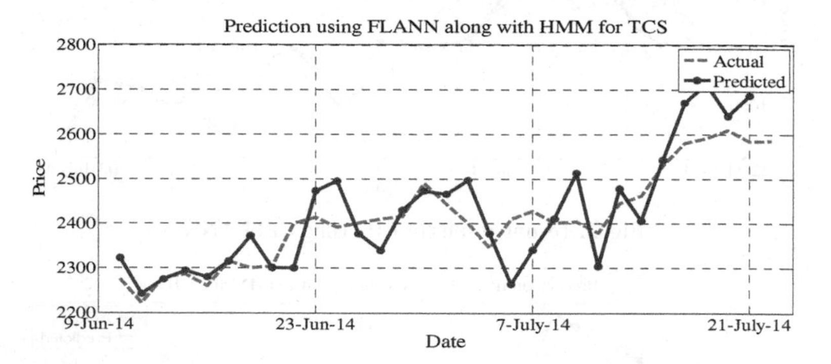

Fig. 5. Testing results of TCS using UNIFY MODEL

The comparison of target vs. predicted stock closing price of ORACLE stock during testing of CEFLANN only and with HMM is given in Figs. 6 & 7. The stock market indices for one day in advance for a period of 30 days are showcased.

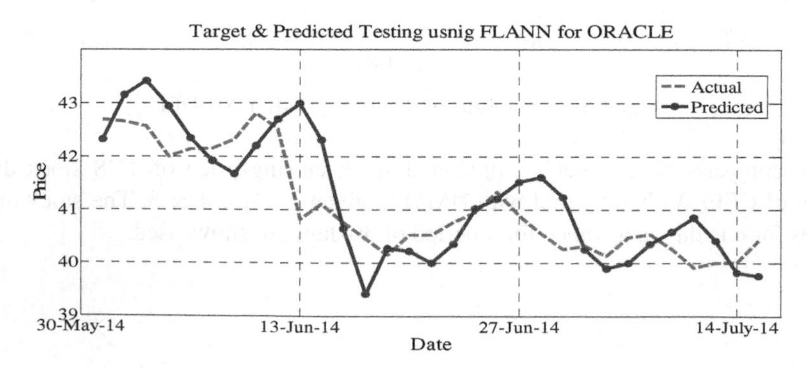

Fig. 6. Testing results of ORACLE using CEFLANN

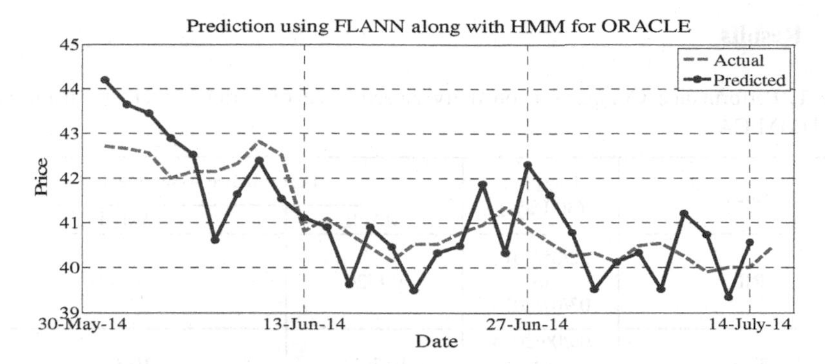

Fig. 7. Testing results of ORACLE using UNIFY MODEL

4.2 Performance Metric

To evaluate the performance of the proposed model there are various criterions can be used. Popularly the accuracy measures like Mean Absolute Percentage Error (MAPE) and Average Mean Absolute Percentage Error (AMAPE) is used to study the accuracy of the model. The above named error calculation methods are mentioned below in eq. (8) and eq. (9).

$$MAPE = \left(\frac{1}{N} \sum_{j=1}^{N} \frac{abs(e_j)}{y_t} \right) \times 100\% \tag{8}$$

$$AMAPE = \left(\frac{1}{N} \sum_{j=1}^{N} \left[\frac{abs(e_j)}{\overline{y}} \right] \right) \times 100\%$$
$$where \quad \overline{y} = \frac{1}{N} \sum_{j=1}^{N} [y_t] \tag{9}$$

Where $e = y_t - y$, y_t & y represents the actual and forecast values; $\overline{y}$ is the average value of actual price.

4.3 Results

Table 1. Performance Comparison on daily closed price of various stocks by using simple ANN-HMM-GA

Stock	Duration (30 Days)	Performance Assessments	
		MAPE	AMAPE
RIL	22/05/2014 to 03/07/2014	9.1720	9.3324
TCS	09/06/2014 to 21/07/2014	10.3611	10.3810
ORACLE	30/05/2014 to 14/07/2014	6.9921	7.3630

Table 2. performancecomparison on daily closed price of various stocks by using only CEFLANN

Stock	Duration (30 Days)	Performance Assessments	
		MAPE	AMAPE
RIL	22/05/2014 to 03/07/2014	8.7212	8.7324
TCS	09/06/2014 to 21/07/2014	7.2600	7.2814
ORACLE	30/05/2014 to 14/07/2014	5.8512	5.8669

Table 3. Performance Comparison on daily closed price of various stocks by using unify model

Stock	Duration (30 Days)	Performance Assessments	
		MAPE	AMAPE
RIL	22/05/2014 to 03/07/2014	6.1854	6.1795
TCS	09/06/2014 to 21/07/2014	5.6401	5.662
ORACLE	30/05/2014 to 14/07/2014	5.6569	5.6713

5 Conclusion

The main idea of this paper was to build a unify model that can be used in predicting the stock time series phenomena in the decision making process and that would execute similar to an actual investor. We have proposed a high-tech method that can forecast one-day ahead market values by using our unify model. Our experimental analysis concluded that, by implementing and testing the proposed method on real stock data, we could estimate the market trend and make acceptable decisions. Our unify model consists an adaptive CEFLANN, a HMM and GAs to forecast stock data indices. We find the performance of the unify model is better than that of the basic models (Hassan, Baikunth and Michael, 2007) and our adaptive model (D K Bebarta, Birendra Biswal and P K Dash, 2012). The adaptive CEFLANN based on back propagation learning method used combining with HMM by the help of optimized parameter using GAs to forecast three different stocks. The results obtained from all models suggested our unify model gives best accuracy over other methods. Further this work can be suitably extended to get prediction result with improved MAPE by using high end learning methods like case based dynamic window and adaptive evolutionary optimized method differential evolution (DE) algorithm.

References

1. Zhang, G.P.: Time series forecasting using a hybrid ARIMA and neural network model. Neurocomputing 50, 159–175 (2003)
2. Cheong, C.W.: Modeling and forecasting crude oil markets using ARCH-type models. Energy Policy 37, 2346–2355 (2009)
3. Jahandideh, S., Jahandideh, S., Asadabadi, E.B., Askarian, M., Movahedi, M.M., Hosseini, S., Jahandideh, M.: The use of artificial neural networks and multiple linear regression to predict rate of medical waste generation. Waste Management 29, 2874–2879 (2009)
4. Chen, A.S., Leung, M.T., Daouk, H.: Application of Neural Networks to an emerging financial market: Forecasting and trading the Taiwan Stock Index. Comput. Operations Res. 30, 901–923 (2003)
5. Chang, P.-C., Liu, C.-H., Lin, J.-L., Fan, C.-Y., Ng, C.S.P.: A neural network with a case based dynamic window for stock trading prediction. Expert Systems with Applications 36(3, Pt 2), 6889–6898 (2009)
6. Giles, C.L., Lawrence, S., Tsoi, A.C.: Noisy Time Series Prediction using Recurrent Neural Networks and Grammatical Inference. Machine Learning 44, 161–183 (2001)
7. Dostál, P.: Forecasting of Time Series with Fuzzy Logic. In: Zelinka, I., Chen, G., Rössler, O.E., Snasel, V., Abraham, A. (eds.) Nostradamus 2013: Prediction, Model. & Analysis. AISC, vol. 210, pp. 155–162. Springer, Heidelberg (2013)
8. Karnik, N.N., Mendel, J.M.: Applications of type-2 fuzzy logic systems to forecasting of time-series. Information Sciences 120, 89–111 (1999)
9. Teoh, H.J., Chen, T.-L., Cheng, C.-H., Chu, H.-H.: A hybrid multi-order fuzzy time series for forecasting stock. Expert Systems with Applications 36, 7888–7897 (2009)
10. Castillo, O.: Hybrid Intelligent Systems for Time Series Prediction Using Neural Networks, Fuzzy Logic, and Fractal Theory. IEEE Transactions on Neural Networks 13, 1395–1408 (2002)

11. Rafiul, M., Nath, B.: StockMarket Forecasting Using Hidden Markov Model: A New Approach. In: IEEE Conference on Intelligent Systems Design and Applications (ISDA 2005), pp. 192–196 (2005), doi:10.1109/ISDA.2005.85
12. Park, S.-H., Lee, J.-H., Song, J.-W., Park, T.-S.: Forecasting Change Directions for Financial Time Series Using Hidden Markov Model. In: Wen, P., Li, Y., Polkowski, L., Yao, Y., Tsumoto, S., Wang, G. (eds.) RSKT 2009. LNCS, vol. 5589, pp. 184–191. Springer, Heidelberg (2009)
13. Rabinar, L.R.: A tutorial on hidden Markov models and selected applications in speech recognition. Proceedings of IEEE 77, 257–285 (1989)
14. Pao, Y.H., Takefji, Y.: Functional-Link Net Computing. IEEE Computer Journal, 76–79 (1992)
15. Majhi, R., Panda, G., Sahoo, G.: Development and performance evaluation of FLANN based model for forecasting of stock markets. Experts Systems with Applications An International Journal (2008)
16. Bebarta, D.K., Biswal, B., Rout, A.K., Dash, P.K.: Forecasting and classification of Indian stocks using different polynomial functional link artificial neural networks. In: IEEE Conference, INDCON, pp. 178–182 (2012), doi:10.1109/INDCON.2012.6420611
17. Hassan, M.R., Nath, B., Kirley, M.: A fusion model of HMM, ANN and GA for stock market forecasting. Expert Systems with Applications 33, 171–180 (2009)
18. Bebarta, D.K., Biswal, B., Dash, P.K.: Comparative study of stock market forecasting using different functional link artificial neural networks. Int. J. Data Analysis Techniques and Strategies 4(4), 398–427 (2012)
19. Rout, A.K., Biswal, B., Dash, P.K.: A hybrid FLANN and adaptive differential evolution model for forecasting of stock market indices. International Journal of Knowledge-Based and Intelligent Engineering Systems 18, 23–41 (2014)
20. Chau, C.W., Kwong, S., Diu, C.K., Fahrner, W.R.: Optimization of HMM by a genetic algorithm. In: IEEE International Conference on Acoustics, Speech, and Signal Processing, pp. 1727–1730 (1997)

Subset K-Means Approach for Handling Imbalanced-Distributed Data

Ch.N. Santhosh Kumar[1], K. Nageswara Rao[2], A. Govardhan[3], and N. Sandhya[4]

[1] Dept. of CSE, JNTU- Hyderabad, Telangana., India
santhosh_ph@yahoo.co.in
[2] PSCMR college of Engineering and Technology, Kothapet, Vijayawada, A.P., India
pricipal@pscmr.ac.in
[3] CSE, SIT, JNTU Hyderabad, Telangana, India
govardhan_cse@jntuh.ac.in
[4] CSE Department,VNR Vignana Jyothi Institite of Engineering & Technology, Hyderabad, India
sandhyanadela@gmail.com

Abstract. The effectiveness of clustering analysis relies not only on the assumption of cluster number but also on the class distribution of the data employed. This paper represents another step in overcoming a drawback of K-means, its lack of defense against imbalance data distribution. *K*-means is a partitional clustering technique that is well-known and widely used for its low computational cost. However, the performance of *k*-means algorithm tends to be affected by skewed data distributions, i.e., imbalanced data. They often produce clusters of relatively uniform sizes, even if input data have varied cluster size, which is called the "uniform effect." In this paper, we analyze the causes of this effect and illustrate that it probably occurs more in the *k*-means clustering process. As the minority class decreases in size, the "uniform effect" becomes evident. To prevent the effect of the "uniform effect", we revisit the well-known K-means algorithm and provide a general method to properly cluster imbalance distributed data.

The proposed algorithm consists of a novel under random subset generation technique implemented by defining number of subsets depending upon the unique properties of the dataset. We conduct experiments using ten UCI datasets from various application domains using five algorithms for comparison on eight evaluation metrics. Experiment results show that our proposed approach has several distinctive advantages over the original k-means and other clustering methods.

Keywords: data, *k*-means clustering algorithms, oversampling, K-Subset.

1 Introduction

Cluster analysis is a well-studied domain in data mining. In cluster analysis data is analyzed to find hidden relationships between each other to group a set of objects into clusters. One of the most popular methods in cluster analysis is k-means algorithm. The popularity and applicability of k-means algorithm in real time

S.C. Satapathy et al. (eds.), *Emerging ICT for Bridging the Future – Volume 2*,
Advances in Intelligent Systems and Computing 338, DOI: 10.1007/978-3-319-13731-5_54

applications is due to its simplicity and high computational capability. However, further investigation is the need of the hour to better understand the efficiency of k-means algorithm with respect to the data distribution used for analysis.

A good amount of research had done on the class balance data distribution for the performance analysis of k-means algorithm. For skewed-distributed data, the k-means algorithm tend to generate poor results as some instances of majority class are portioned into minority class, which makes clusters to have relatively uniform size instead of input data have varied cluster of non-uniform size. In [1] authors have defined this abnormal behavior of k-means clustering as the "uniform effect". It is noteworthy that class imbalance is emerging as an important issue in cluster analysis especially for k-means type algorithms because many real-world problems, such as remote-sensing, pollution detection, risk management, fraud detection, and especially medical diagnosis are of class imbalance. Furthermore, the rare class with the lowest number of instances is usually the class of interest from the point of view of the cluster analysis.

Liu et al. [2], proposed a multiprototype clustering algorithm, which applies the k-means algorithm to discover clusters of arbitrary shapes and sizes. However, there are following problems in the real applications of these algorithms to cluster imbalanced data. 1) These algorithms depend on a set of parameters whose tuning is problematic in practical cases. 2) These algorithms make use of the randomly sampling technique to find cluster centers. However, when data are imbalanced, the selected samples more probably come from the majority classes than the minority classes. 3) The number of clusters k needs to be determined in advance as an input to these algorithms. In a real dataset, k is usually unknown. 4) The separation measures between sub-clusters that are defined by these algorithms cannot effectively identify the complex boundary between two sub-clusters. 5) The definition of clusters in these algorithms is different from that of k-means. Xiong et al. [1] provided a formal and organized study of the effect of skewed data distributions on the hard k-means clustering. However, the theoretic analysis is only based on the hard k-means algorithm. Their shortcomings are analyzed and a novel algorithm is proposed.

2 Class Imbalance Learning

One of the most popular techniques for alleviating the problems associated with class imbalance is data sampling. Data sampling alters the distribution of the training data to achieve a more balanced training data set. This can be accomplished in one of two ways: under sampling or oversampling. Under sampling removes majority class examples from the training data, while oversampling adds examples to the minority class. Both techniques can be performed either randomly or intelligently.

The random sampling techniques either duplicate (oversampling) or remove (under sampling) random examples from the training data. Synthetic minority oversampling technique (SMOTE) [3] is a more intelligent oversampling technique that creates new minority class examples, rather than duplicating existing ones. Wilson's editing (WE) intelligently under-samples data by only removing examples that are thought to be noisy.

Finding minority class examples effectively and accurately without losing overall performance is the objective of class imbalance learning. The fundamental issue to be resolved is that the clustering ability of most standard learning algorithms is significantly compromised by imbalanced class distributions. They often give high overall accuracy, but form very specific rules and exhibit poor generalization for the small class. In other words, over fitting happens to the minority class. Correspondingly, the majority class is often over generalized. Particular attention is necessary for each class. It is important to know if a performance improvement happens to both classes and just one class alone.

3 Related Work

In this section, we first review the major research about clustering in class imbalance learning and explain why we choose oversampling as our technique in this paper.

Tapas Kanungo et al., [4] have presented a simple and efficient implementation of Lloyd's k-means clustering algorithm, which stores the multidimensional data points in a kd-tree. A kd-tree is a binary tree, which represents a hierarchical subdivision of the point set's bounding box using axis aligned splitting hyperplanes. Renato Cordeiro de Amorim et al., [5] have proposed a variation of k-means for tackling against noisy features using feature weights in the criterion. Serkan Kiranyaz et al., [6] have proposed a framework using exhaustive k-means clustering technique for addressing the problem in a long term ECG signal, known as Holter register. The exhaustive K-means clustering is used in order to find out (near-) optimal number of key-beats as well as the master key-beats. The expert labels over the master key-beats are then back-propagated over the entire ECG record to obtain a patient-specific, long-term ECG classification.

Haitao xiang et al., [7] have proposed a local clustering ensemble learning method based on improved AdaBoost (LCEM) for rare class analysis. LCEM uses an improved weight updating mechanism where the weights of samples which are invariably correctly classified will be reduced while that of samples which are partially correctly classified will be increased. Amuthan Prabakar et al., [8] have proposed a supervised network anomaly detection algorithm by the combination of k-means and C4.5 decision tree exclusively used for portioning and model building of the intrusion data. The proposed method is used mitigating the Forced Assignment and Class Dominance problems of the k-Means method. Li Xuan et al., [9] have proposed two methods, in first method they applied random sampling of majority subset to form multiple balanced datasets for clustering and in second method they observed the clustering partitions of all the objects in the dataset under the condition of balance and imbalance at a different angle. Christos Bouras et al., [10] have proposed W-k means clustering algorithm for applicability on a corpus of news articles derived from major news portals. The proposed algorithm is an enhancement of standard k-means algorithm using the external knowledge for enriching the ''bag of words'' used prior to the clustering process and assisting the label generation procedure following it.

P.Y. Mok et al., [11] have proposed a new clustering analysis method that identifies the desired cluster number and produces, at the same time, reliable clustering solutions. It first obtains many clustering results from a specific algorithm, such as Fuzzy C-Means (FCM), and then integrates these different results as a judgment matrix. An iterative graph-partitioning process is implemented to identify the desired cluster number and the final result.Luis A. Leiva et al., [12] have proposed Warped K-Means, a multi-purpose partition clustering procedure that minimizes the sum of squared error criterion, while imposing a hard sequentiality constraint in the classification step on datasets embedded implicitly with sequential information. The proposed algorithm is also suitable for online learning data, since the change of number of centroids and easy updating of new instances for the final cluster is possible. M.F.Jiang et al., [13] have proposed variations of k-means algorithm to identify outliers by clustering the data the initial phase then using minimum spanning tree to identify outliers for their removal.

Jie Cao et al., [14] have proposed a Summation-bAsed Incremental Learning (SAIL) algorithm for Information-theoretic K-means (Info-Kmeans) aims to cluster high-dimensional data, such as images featured by the bag-of-features (BOF) model, using K-means algorithm with KL-divergence as the distance. Since SAIL is a greedy scheme it first selects an instance from data and assigns it to the most suitable cluster. Then the objective-function value and other related variables are updated immediately after the assignment. The process will be repeated until some stopping criterion is met. One of the shortcomings is to select the appropriate cluster for an instance. Max Mignotte [15] has proposed a new and simple segmentation method based on the K-means clustering procedure for applicability on image segmentation. The proposed approach overcome the problem of local minima, feature space without considering spatial constraints and uniform effect.

4 Framework of k-Subset Algorithm

This section presents the proposed algorithm, whose main characteristics are depicted in the following sections. Initially, the main concepts and principles of k-means are presented. Then, the definition of our proposed K-subset is introduced in detail.

K-means is one of the simplest unsupervised learning algorithms, first proposed by Macqueen in 1967, which has been used by many researchers to solve some well-known clustering problems [16]. The technique classifies a given data set into a certain number of clusters (assume k clusters). The algorithm first randomly initializes the clusters center. The next step is to calculate the distance between an object and the centroid of each cluster. Next each point belonging to a given data set is associated with the nearest center. The cluster centers are then re-calculated. The process is repeated with the aim of minimizing an objective function knows as squared error function given by:

$$Jv = {}_{i=1}^{C}\,{}_{j=1}^{Ci}\left(\left\|x_i - v_j\right\|\right)^2 \tag{1}$$

Where, $(\|x_i - v_j\|)$ is the Euclidean distance between the data point x_i and cluster center v_j , ci is the number of data points in cluster and c is the number of ith cluster centers.

The entire process is given in the following algorithm,

4.1 Dividing Majority and Minority Subset

An easy way to sample a dataset is by selecting instances randomly from all classes. However, sampling in this way can break the dataset in an unequal priority way and more number of instances of the same class may be chosen in sampling. To resolve this problem and maintain uniformity in sample, we propose a sampling strategy called weighted component sampling. Before creating multiple subsets, we will create the number of majority subsets depending upon the number of minority instances.

4.2 Identifying Number of Subsets of Majority Class

In the next phase of the approach, the ratio of majority and minority instances in the unbalanced dataset is used to decide the number of subset of majority instances (T) to be created.

T= no. of majority inst (N)./no. of minority inst (P).

4.3 Combing the Majority Subsets and Minority Subset

The so formed majority subsets are individual combined with the only minority subset to form multiple balanced sub datasets of every dataset. The number of balanced sub datasets formed depends upon the imbalance ratio and the unique properties of the dataset

4.4 Averaging the measures

The subsets of balanced datasets created are used to run multiple times and the resulted values are averaged to find the overall result. This newly formed multiple subsets are applied to a base algorithm; in this case k-means is used to obtain different measures such as AUC, Precision, F-measure, TP Rate and TN Rate.

Algorithm : K-Subset

1: {Input: A set of minor class examples *P*, a set Of major class examples *N*, *jPj* <*jN j*, and *T*, the number of subsets to be sampled from *N*.}
2: $i \leftarrow 0$, T=N/P. repeat
3: $i = i + 1$
4: Randomly sample a subset *Ni* from *N*, *jNij* = *jPj*.
5: Combine P and Ni to form NPi
6: Apply filter on a NPi

7: Train and Learn on a base Algorithm (k-means) using NPi. Obtain the values of AUC,TP,FP,F-Measure
8: until *i* = *T*
9: Output: Average Measure;

5 Datasets

In the study, we have considered 10 binary data-sets which have been collected from the KEEL [18] and UCI [17] machine learning repository Web sites, and they are very varied in their degree of complexity, number of classes, number of attributes, number of instances, and imbalance ratio (the ratio of the size of the majority class to the size of the minority class). The number of classes' ranges up to 2, the number of attributes ranges from 8 to 60, the number of instances ranges from 57 to 3772, and the imbalance ratio is up to 15.32. This way, we have different Imbalance Ratios (IRs): from low imbalance to highly imbalanced data-sets. Table 1 summarizes the properties of the selected data-sets: for each data-set, S.no, Dataset name, the number of examples (#Ex.), number of attributes (#Atts.), class name of each class (minority and majority) and the IR. This table is ordered according to the name of the datasets in alphabetical order. We have obtained the AUC metric estimates by means of a 10-fold cross-validation. That is, the data-set was split into ten folds, each one containing 10% of the patterns of the dataset. For each fold, the algorithm is trained with the examples contained in the remaining folds and then tested with the current fold. The data partitions used in this paper can be found in UCI-dataset repository [17] so that any interested researcher can reproduce the experimental study. The algorithms used in the experimental study and their parameter settings, which are obtained from the KEEL [18] and WEKA [19] software tools.

Table 1. Summary of benchmark imbalanced datasets

S.no	Datasets	# Ex.	# Atts.	Class (_,+)	IR
1.	Breast_w	699	9	(benign; malignant)	1.90
2.	Colic	368	22	(yes; no)	1.71
3.	Diabetes	768	8	(tested-potv; tested-negtv)	1.87
4.	Ecolic	336	7	(cp; oml)	2.33
5.	Hepatitis	155	19	(die; live)	3.85
6.	Ionosphere	351	34	(b;g)	1.79
7.	Labor	57	17	(bad; good)	1.85
8.	Sick	3772	30	(negative; sick)	15.32
9.	Sonar	208	60	(rock ; mine)	1.15
10.	Vote	435	17	(democrat ; republican)	1.58

Several clustering methods have been selected and compared to determine whether the proposal is competitive in different domains with the other approaches. Algorithms are compared on equal terms and without specific settings for each data problem.

6 Experimental Results

Table 2-9 presents the performance of each clustering technique averaged across all learners and data sets. These tables give a general view of the performance of each technique using each of the eight performance metrics. Tables 2-9 provide both the numerical average performance (Mean) and the standard deviation (SD) results. If the proposed technique is better than the compared technique then '•' symbol appears in the column. If the proposed technique is not better than the compared technique then '○' symbol appears in the column. The mean performances were significantly different according to the T-test at the 95% confidence level.

We carry out the empirical comparison of our proposed algorithm with the benchmarks. Our aim is to answer several questions about the proposed learning algorithms in the scenario of two-class imbalanced problems.

1) In first place, we want to analyze which one of the approaches is able to better handle a large amount of imbalanced data-sets with different IR, i.e., to show which one is the most robust method.

2) We also want to investigate their improvement with respect to classic clustering methods and to look into the appropriateness of their use instead of applying a unique preprocessing step and training a single method. That is, whether the trade-off between complexity increment and performance enhancement is justified or not. Given the amount of methods in the comparison, we cannot afford it directly. On this account, we compared the proposed algorithm with each and every algorithm independently. This methodology allows us to obtain a better insight on the results by identifying the strengths and limitations of our proposed method on every compared algorithm. The clustering evaluations were conducted on ten widely used datasets. These real world multi-dimensional datasets are used to verify the proposed clustering method. Table 2, 3, 4, 5, 6, 7, 8 and 9 reports the results of Accuracy, AUC, Precision, Recall, F-measure, Specificity, FP Rate and FN Rate respectively for all the ten datasets from UCI.

A two-tailed corrected resampled paired t-test is used in this paper to determine whether the results of the cross-validation show that there is a difference between the two algorithms is significant or not. Difference in accuracy is considered significant when the p-value is less than 0.05 (confidence level is greater than 95%). The results in the tables show that K-Subset has given a good improvement on all the clustering measures. Two main reasons support the conclusion achieved above. The first one is the decrease of instances in majority subset, has also given its contribution for the better performance of our proposed K-Subset algorithms. The second reason, it is well-known that the resampling of synthetic instances in the minority subset is the only way in oversampling but conduction proper exploration – exploitation of prominent instances in minority subset is the key for the success of our algorithm. Another reason is the deletion of noisy instances by the interpolation mechanism of K-Subset.

Table 2. Summary of tenfold cross validation performance for Accuracy on all the datasets

Datasets	K-Means	Density	FF	EM	Hier	K-Subset
Breast_w	95.82±2.26○	96.22±2.19○	84.94±6.96●	93.75±2.79●	65.52±0.44●	94.64±3.15
Colic	60.57±11.89○	65.30±10.85○	58.67±9.91●	66.13±7.11○	63.05±1.13○	76.96±10.2
Diabetes	65.42±5.87○	65.60±5.68○	65.16±3.42○	64.67±5.74○	65.11±0.34○	63.07±6.22
Ecolic	55.86±6.77●	56.37±6.72●	62.41±6.56●	60.60±5.33●	70.00±0.00○	81.74±5.47
Hepatitis	71.09±12.58○	73.15±12.16○	72.14±12.77○	73.83±10.53○	79.38±2.26○	76.65±16.03
Ionosphere	70.80±6.71●	73.06±6.35○	62.75±6.65●	73.08±6.47○	64.10±1.35●	69.70±10.3
Labor	65.45±22.84●	69.12±21.69○	74.73±13.17○	55.67±21.78●	64.67±3.07●	64.37±25.06
Sick	73.75±7.86●	71.28±8.74●	87.29±6.06○	50.01±26.34●	93.88±0.08○	79.19±5.16
Sonar	52.43±10.28○	50.12±10.40	50.94±8.28	49.59±9.55●	51.78±3.41	70.10±14.44
Vote	85.73±5.30	87.22±4.64○	84.54±8.10●	60.59±7.74●	61.38±0.81●	88.19±8.84

Table 3. Summary of tenfold cross validation performance for AUC on all the datasets

Datasets	K-Means	Density	FF	EM	Hier	K-Subset
Breast_w	.950±0.027○	.966±0.021○	.785±0.098●	.951±0.022○	.500±0.000●	.947±0.030
Colic	.628±0.108○	.678±0.092○	.570±0.114●	.691±0.068○	.500±0.000●	.766±0.105
Diabetes	.608±0.067●	.617±0.068○	.520±0.044●	.670±0.070○	.502±0.006●	.634±0.062
Ecolic	.534±0.067●	.535±0.066●	.521±0.057●	.567±0.051●	.500±0.000●	.921±0.054
Hepatitis	.753±0.136○	.781±0.122○	.670±0.163●	.800±0.101○	.500±0.000●	.768±0.160
Ionosphere	.706±0.080●	.743±0.079○	.530±0.067●	.771±0.058○	.500±0.000●	.703±0.102
Labor	.631±0.237●	.668±0.233○	.657±0.169○	.586±0.196●	.500±0.000●	.640±0.254
Sick	.574±0.157○	.567±0.154○	.481±0.038●	.516±0.086●	.500±0.000●	.768±0.054
Sonar	.521±0.103○	.498±0.104	.513±0.082○	.497±0.096●	.500±0.000○	.719±0.148
Vote	.871±0.053○	.885±0.047○	.855±0.083●	.759±0.053●	.500±0.000●	.877±0.089

Table 4. Summary of tenfold cross validation performance for Precision on all the datasets

Datasets	K-Means	Density	FF	EM	Hier	K-Subset
Breast_w	.961±0.024●	.989±0.015○	.823±0.071●	.998±0.007○	.655±0.004●	.921±0.048
Colic	.784±0.107○	.821±0.083○	.719±0.120●	.838±0.069○	.630±0.011●	.765±0.130
Diabetes	.725±0.047○	.734±0.051○	.665±0.044●	.821±0.072○	.652±0.004●	.592±0.054
Ecolic	.727±0.052○	.727±0.053○	.712±0.035○	.746±0.036○	.700±0.000○	.810±0.093
Hepatitis	.426±0.150●	.453±0.156●	.405±0.233●	.457±0.136●	.000±0.000●	.787±0.220
Ionosphere	.557±0.147●	.573±0.145●	.309±0.369●	.585±0.068	.000±0.000●	.821±0.131
Labor	.474±0.389●	.523±0.388○	.532±0.484○	.364±0.437●	.000±0.000●	.631±0.327
Sick	.952±0.026●	.952±0.028●	.936±0.005●	.958±0.037○	.939±0.001●	.845±0.068
Sonar	.493±0.133●	.460±0.140●	.422±0.238●	.459±0.108●	.110±0.198●	.866±0.142
Vote	.952±0.048○	.960±0.042○	.929±0.077●	.996±0.017○	.614±0.008●	.906±0.131

Table 5. Summary of tenfold cross validation performance for Recall on all the datasets

Datasets	K-Means	Density	FF	EM	Hier	K-Subset
Breast_w	.961±0.024●	.989±0.015○	.823±0.071●	.998±0.007○	.655±0.004●	.921±0.048
Breast_w	.976±0.022○	.953±0.029●	.992±0.033○	.907±0.042●	1.000±0.000○	.977±0.030
Colic	.541±0.231○	.582±0.195○	.635±0.229○	.576±0.100○	1.000±0.000○	.735±0.183
Diabetes	.760±0.091○	.747±0.089○	.957±0.106○	.594±0.075●	1.000±0.000○	.772±0.083
Ecolic	.595±0.108●	.606±0.112●	.778±0.121●	.664±0.081●	1.000±0.000○	.982±0.080
Hepatitis	.824±0.225○	.865±0.190○	.583±0.319●	.906±0.153○	.000±0.000●	.781±0.251
Ionosphere	.702±0.187	.787±0.195○	.185±0.284●	.912±0.074○	.000±0.000●	.594±0.178
Labor	.555±0.413●	.588±0.412○	.345±0.338●	.330±0.403●	.000±0.000●	.710±0.363
Sick	.779±0.122●	.755±0.139●	.957±0.078○	.719±0.169●	1.000±0.000○	.772±0.087
Sonar	.471±0.201●	.471±0.215●	.524±0.392●	.525±0.194●	.235±0.425●	.654±0.190
Vote	.810±0.069○	.829±0.064○	.814±0.098●	.931±0.062○	1.000±0.000○	.837±0.133

Finally, we can make a global analysis of results combining the results offered by Tables from 2–9:

Our proposals, K-Subset is the best performing one when the data sets are of imbalance category. We have considered a complete competitive set of methods and an improvement of results is expected in the benchmark algorithms i;e K-means, Density, FF, EM and Hier. However, they are not able to outperform K-Subset. In this sense, the competitive edge of K-Subset can be seen.

Table 6. Summary of tenfold cross validation performance for F-measure on all the datasets

Datasets	K-Means	Density	FF	EM	Hier	K-Subset
Breast_w	.968±0.017●	.971±0.017●	.898±0.045●	.950±0.024●	.792±0.003●	.947±0.030
Colic	.608±0.162●	.662±0.137○	.638±0.141●	.678±0.082○	.773±0.008○	.729±0.143
Diabetes	.739±0.053●	.737±0.050●	.779±0.049●	.685±0.055●	.789±0.003○	.668±0.056
Ecolic	.649±0.074○	.655±0.074○	.739±0.066○	.700±0.052○	.824±0.000○	.876±0.079
Hepatitis	.549±0.157●	.582±0.153●	.451±0.226●	.598±0.131	.000±0.000●	.755±0.197
Ionosphere	.617±0.155●	.660±0.157○	.173±0.222●	.711±0.059○	.000±0.000●	.661±0.145
Labor	.481±0.358○	.522±0.357○	.404±0.370●	.328±0.387○	.000±0.000●	.640±0.303
Sick	.851±0.066○	.834±0.077○	.945±0.044○	.807±0.109○	.968±0.000○	.804±0.066
Sonar	.462±0.149●	.447±0.159●	.414±0.257●	.480±0.133●	.149±0.270●	.724±0.151
Vote	.873±0.048○	.888±0.042○	.864±0.075●	.961±0.037○	.761±0.006●	.862±0.102

Table 7. Summary of tenfold cross validation performance for Specificity on all the datasets

Datasets	K-Means	Density	FF	EM	Hier	K-Subset
Breast_w	.968±0.017○	.979±0.029○	.578±0.196●	.996±0.012○	.000±0.000●	.911±0.055
Colic	.608±0.162●	.773±0.157●	.506±0.336●	.807±0.093○	.000±0.000●	.799±0.148
Diabetes	.739±0.053○	.487±0.155●	.082±0.163●	.746±0.145○	.000±0.000●	.500±0.094
Ecolic	.649±0.074○	.464±0.149○	.264±0.145●	.470±0.100○	.000±0.000●	.861±0.074
Hepatitis	.549±0.157○	.696±0.143○	.757±0.154○	.695±0.125○	1.000±0.000○	.755±0.263
Ionosphere	.617±0.155●	.699±0.110●	.875±0.215○	.629±0.098●	1.000±0.000○	.812±0.162
Labor	.481±0.358●	.749±0.281○	.968±0.117○	.854±0.257○	1.000±0.000○	.570±0.380
Sick	.851±0.066○	.401±0.379○	.033±0.095●	.500±0.434○	.000±0.000●	.852±0.063
Sonar	.462±0.149○	.528±0.198○	.502±0.357○	.470±0.171○	.768±0.420○	.784±0.251
Vote	.873±0.048●	.941±0.075	.895±0.122●	.996±0.017○	.000±0.000●	.918±0.116

Table 8. Summary of tenfold cross validation performance for FP Rate on all the datasets

Datasets	K-Means	Density	FF	EM	Hier	K-Subset
Breast_w	.076±0.049	.021±0.029•	.422±0.196○	.004±0.012•	1.000±0.000○	.083±0.054
Colic	.285±0.240○	.227±0.157•	.494±0.336○	.193±0.093•	1.000±0.000○	.201±0.148
Diabetes	.544±0.145○	.513±0.155•	.918±0.163○	.254±0.145•	1.000±0.000○	.500±0.094
Ecolic	.526±0.143○	.536±0.149○	.736±0.145○	.530±0.100○	1.000±0.000○	.138±0.074
Hepatitis	.318±0.146•	.304±0.143•	.243±0.154•	.305±0.125•	.000±0.000•	.245±0.263
Ionosphere	.289±0.110○	.301±0.110○	.125±0.215•	.371±0.098○	.000±0.000•	.188±0.162
Labor	.292±0.313○	.251±0.281•	.032±0.117•	.146±0.257•	.000±0.000•	.430±0.380
Sick	.614±0.366•	.599±0.379•	.967±0.095○	.500±0.434•	.000±0.000•	.148±0.062
Sonar	.429±0.195•	.472±0.198•	.498±0.357•	.530±0.171•	.232±0.420•	.216±0.251
Vote	.068±0.078○	.059±0.075	.105±0.122○	.004±0.017•	1.000±0.000○	.083±0.116

Table 9. Summary of tenfold cross validation performance for FN Rate on all the datasets

Datasets	K-Means	Density	FF	EM	Hier	K-Subset
Breast_w	.024±0.022•	.047±0.029○	.008±0.033•	.093±0.042○	.000±0.000•	.023±0.030
Colic	.459±0.231•	.418±0.195•	.365±0.229•	.424±0.100•	.000±0.000•	.265±0.183
Diabetes	.240±0.091•	.254±0.089•	.043±0.106•	.406±0.075○	.000±0.000•	.229±0.083
Ecolc	.405±0.108○	.394±0.112○	.222±0.121○	.336±0.081○	.000±0.000•	.019±0.080
Hepatitis	.176±0.225•	.135±0.190 •	.417±0.319○	.094±0.153•	1.000±0.000○	.219±0.251
Ionosphere	.298±0.187•	.213±0.195•	.815±0.284○	.088±0.074•	1.000±0.000○	.407±0.178
Labor	.445±0.413•	.413±0.412•	.650±0.340○	.540±0.436○	1.000±0.000○	.290±0.363
Sick	.221±0.122○	.245±0.139○	.043±0.078•	.281±0.169○	.000±0.000•	.278±0.132
Sonar	.529±0.201○	.529±0.215○	.476±0.392○	.475±0.194○	.765±0.425○	.346±0.190
Vote	.190±0.069•	.171±0.064•	.186±0.098•	.069±0.062•	.000±0.000•	.163±0.133

Considering that K-Subset behaves similarly or not effective than K-means shows the unique properties of the datasets where there is scope of improvement in minority subset and not in majority subset. Our K-Subset can mainly focus on improvements in majority subset which is not effective for some unique property datasets. The Accuracy, AUC, Recall, F-measure, TN Rate, FP Rate and FN Rate measure have shown to perform well with respect to K-Subset. The strengths of our model are that K-Subset only increases the number of majority subsets thereby strengthens the minority class. One more point to consider is our method tries to remove the noisy instances from both majority and minority set if any applicable. Finally, we can say that K-Subset is one of the best alternatives to handle class imbalance problems effectively. This experimental study supports the conclusion that the a prominent recursive oversampling approach of majority subsets can improve the class imbalance behavior when dealing with imbalanced data-sets, as it has helped the K-Subset methods to be the best performing algorithm when compared with four classical and well-known algorithms: K-means, Density, FF, EM and a well-established Hierarchical algorithm.

7 Conclusion

In this paper, a novel clustering algorithm for imbalanced distributed data has been proposed. This method uses unique oversampling technique to almost balance dataset such that to minimize the "uniform effect" in the clustering process. Empirical results have shown that K-Subset considerably reduces the uniform effect while retaining or improving the clustering measure when compared with benchmark methods. In fact, the proposed method may also be useful as a frame work for data sources for better clustering measures.

References

1. Xiong, H., Wu, J.J., Chen, J.: K-means clustering versus validation measures: A data-distribution perspective. IEEE Trans. Syst., Man, Cybern. B, Cybern. 39(2), 318–331 (2009)
2. Liu, M.H., Jiang, X.D., Kot, A.C.: A multi-prototype clustering algorithm. Pattern Recognit. 42, 689–698 (2009)
3. Chawla, N., Bowyer, K., Kegelmeyer, P.: SMOTE: Synthetic minority over-sampling technique. J. Artif. Intell. Res. 16, 321–357 (2002)
4. Kanungo, T., Mount, D.M., Netanyahu, N.S., Piatko, C.D., Silverman, R., Wu, A.Y.: An Efficient k-Means Clustering Algorithm: Analysis and Implementation. IEEE Transactions on Pattern Analysis and Machine Intelligence 24(7) (July 2002)
5. de Amorim, R.C., Mirkin, B.: Minkowski metric, feature weighting and anomalous cluster initializing in K-Means clustering. Pattern Recognition 45, 1061–1075 (2012)
6. Kiranyaz, S., Ince, T., Pulkkinen, J., Gabbouj, M.: Personalized long-term ECG classification: A systematic approach. Expert Systems with Applications 38, 3220–3226 (2011)
7. Xiang, H., Yang, Y., Zhao, S.: Local Clustering Ensemble Learning Method Based on Improved AdaBoost for Rare Class Analysis. Journal of Computational Information Systems 8(4), 1783–1790 (2012)
8. Muniyandi, A.P., Rajeswari, R., Rajaram, R.: Network Anomaly Detection by Cascading K-Means Clustering and C4.5 Decision Tree algorithm. In: International Conference on Communication Technology and System Design 2011. Procedia Engineering, vol. 30, pp. 174–182 (2012)
9. Li, X., Chen, Z., Yang, F.: Exploring of clustering algorithm on class-imbalanced data
10. Bouras, C., Tsogkas, V.: A clustering technique for news articles using WordNet. Knowl. Based Syst. (2012), http://dx.doi.org/10.1016/j.knosys.2012.06.015
11. Mok, P.Y., Huang, H.Q., Kwok, Y.L., Au, J.S.: A robust adaptive clustering analysis method for automatic identification of clusters. Pattern Recognition 45, 3017–3033 (2012)
12. Leiva, L.A., Vidal, E.: Warped K-Means: An algorithm to cluster sequentially-distributed data. Information Sciences 237, 196–210 (2013)
13. Jaing, M.F., Tseng, S.S., Su, C.M.: Two Phase Clustering Process for Outlier Detection. Pattern Recognition Letters 22, 691–700 (2001)
14. Cao, J., Wu, Z., Wu, J., Liu, W.: Towards information-theoretic K-means clustering for image indexing. Signal Processing 93, 2026–2037 (2013)
15. Mignotte, M.: A de-texturing and spatially constrained K-means approach for image segmentation. Pattern Recognition Lett. (2010), doi:10.1016/j.patrec, 09.016

16. Maimon, O., Rokach, L.: Data mining and knowledge discovery handbook. Springer, Berlin (2010)
17. Blake, C., Merz, C.J.: UCI repository of machine learning databases. Machine-readable data repository. Department of Information and Computer Science, University of California at Irvine, Irvine (2000), http://www.ics.uci.edu/mlearn/MLRepository.html
18. http://www.keel.com
19. Witten, I.H., Frank, E.: Data Mining: Practical machine learning tools and techniques, 2nd edn. Morgan Kaufmann, San Francisco (2005)

Dynamic Recurrent FLANN Based Adaptive Model for Forecasting of Stock Indices

Dwiti Krishna Bebarta[1], Ajit Kumar Rout[1], Birendra Biswal[2], and P.K. Dash[3]

[1] Dept of CSE, GMR Institute of Technology, Rajam, AP, India
[2] Dept of ECE, GMR Institute of Technology, Rajam, AP, India
[3] Multidisciplinary Research Cell, S.O.A. University, Bhuabneswar, Odisha, India
{bebarta.dk,ajitkumar.rout,birendra.biswal}@gmrit.org, pkdash.india@gmail.com

Abstract. Prediction of future trends in financial time-series data are very important for decision making in the share market. Usually financial time-series data are non-linear, volatile and subject to many other factors like local or global issues, causes a difficult task to predict them consistently and efficiently. This paper present an improved Dynamic Recurrent FLANN (DRFLANN) based adaptive model for forecasting the stock Indices of Indian stock market. The proposed DRFLANN based model employs the least mean square (LMS) algorithm to train the weights of the networks. The Mean Absolute Percentage Error (MAPE), the Average Mean Absolute Percentage Error (AMAPE), the variance of forecast errors (VFE) is used for determining the accuracy of the model. To improve further the forecasting results, we have introduced three technical indicators named Relative Strength Indicator (RSI), Price Volume Change Indicator (PVCI), and Moving Average Volume Indicator (MAVI). The reason of choosing these three indicators is that they are focused on important attributes of price, volume, and combination of both price and volume of stock data. The results show the potential of the model as a tool for making stock price prediction.

1 Introduction

The price variation of stock market is a dynamic system and the disordered behavior of the stock price movement duplicates complication of the price prediction; however, the highly non-linear, dynamic complicated domain knowledge inherent in the stock market makes it very difficult for investors to make the right investment decisions promptly. It is necessary to develop an intelligent system to get real-time pricing information, reduce one obsession of investors and help them to maximize their profits. Financial Time series is considered to be a difficult problem in forecasting, due to the many complex features frequently present in such series are irregularities, volatility, trends and noise. Many researchers in the past have applied various soft computing techniques to predict the movements of the stock markets. Even though a lot of research going on the field of stock market prediction [1], [2], [3]; still it remains to be a big question whether stock market can be predicted accurately.

Tools based on artificial neural network (ANN) have gained popularity due to their inherent capability to approximate any non-linear function, less sensitive to error,

S.C. Satapathy et al. (eds.), *Emerging ICT for Bridging the Future – Volume 2*,
Advances in Intelligent Systems and Computing 338, DOI: 10.1007/978-3-319-13731-5_55

tolerate noise, and chaotic components. Familiar ANN models like Multilayer Perceptron (MLP) [4], Radial Basis Function (RBF) [5], [6], Recurrent Neural Network (RNN) [7], Multi Branch Neural Networks (MBNN), Local Linear Wavelet Network (LLWN) [8], and hybrid models like Genetic-Neural Network [9], [10] are also demonstrated for stock price forecasting.

Further it is well known that the artificial neural network (ANN) suffers from slow convergence, local minimum, over fitting and generalization and number of neurons in the hidden layer are chosen by trial and error. Thus to overcome from this FLANN based simple network may be used for the prediction of time series data with a less computational overhead than the ANN network. Unlike the earlier FLANNs where each input is expanded to have several nonlinear functions of the input itself, only a few functional blocks comprising nonlinear functions of all the inputs is used in this paper thereby resulting in a high dimensional input space for the neural network.

The FLANN, originally proposed by Y.H.Pao and Y.Takefji [11] comprises a single layer neural network in which nonlinearity is introduced as a functional block, thus giving rise to a higher dimension input space. The normal practice of choosing the functional block with trigonometric functions such as cos(x), sin(x), cos(πx), sin(πx), [12] etc. or polynomials [13], [14], [15].

In this paper, we have suggested the Dynamic Recurrent Functional Link Artificial Neural Network (DRFLANN) architecture for prediction of prices of leading Indian stock indices. Our proposed DRFLANN structure is very simple with no hidden layer. The main advantage of the DRFLANN is the reduced computational cost in the training stage, while maintaining a good performance of approximation. The use of dynamic elements known to be adaptive parameters used in our proposed model along the connection strengths helps to improve the convergence speed of the network.

2 Dynamic Recurrent FLANN (DRFLANN) Model

This new proposed architecture of DRFLANN has much less computational complexity and causes high convergence speed. Fig.1 shows the single layer computationally efficient DRFLANN architecture.

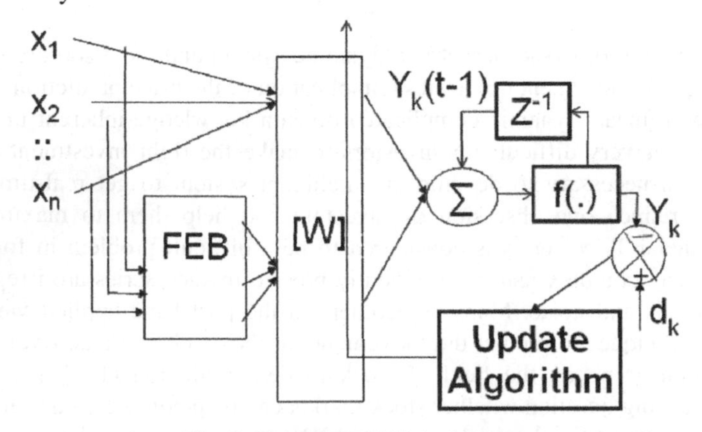

Fig. 1. The DRFLANN Architecture

Let the input vector pattern and target is given in eq. (1) from a selected data set of n number of elements.

$$X = \begin{bmatrix} x_1, x_2, \ldots\ldots, x_m \\ \vdots \\ x_{m+i}, x_{m+i+1}, \ldots\ldots, x_{n-1} \end{bmatrix} \begin{bmatrix} x_{m+1} \\ \vdots \\ x_n \end{bmatrix} \tag{1}$$

Input Vector *Desired Vector*

There are $[x_1, x_2, \ldots\ldots, x_n]$ data set out of which the above matrix suggested the split of input pattern and desired vector for training and testing the model. We have used 70:30 ratios of the selected data for training and testing. Each input pattern is then applied to the Functional Expansion Blocks (FEBs) where the elements are functionally expanded through a set of tan hyperbolic trigonometric function as mentioned in eq. (2).

$$\begin{aligned} x_{m+1} &= \tanh(a_{01} + a_{11}x_1 + a_{21}x_2 + a_{31}x_3 + \ldots + a_{m1}x_m) \\ x_{m+2} &= \tanh(a_{02} + a_{12}x_1 + a_{22}x_2 + \ldots\ldots + a_{m2}x_m) \\ &\ldots\ldots \\ x_M &= \tanh(a_{0M} + a_{1M}x_1 + a_{2M}x_2 \ldots\ldots + a_{mM}x_m) \end{aligned} \tag{2}$$

Where the adaptive parameters used to enhance the input pattern at the functional expansion block is represented in a matrix form in eq. (3)

$$A = \begin{bmatrix} a_{01} & a_{11} & a_{21} \ldots\ldots & a_{m1} \\ a_{02} & a_{12} & a_{22} \ldots\ldots & a_{m2} \\ \ldots & & & \\ a_{0M} & a_{1M} & a_{2M} \ldots\ldots & a_{mM} \end{bmatrix} \tag{3}$$

The final input at the input layer is now presented in eq. (4)

$$I = [x_1, x_2, \ldots\ldots, x_{m+1}, x_{m+2}, \ldots\ldots, x_M] \tag{4}$$

The estimated output of the model is calculated using eq. (5 & 6)

$$Y(t) = f(s) = \frac{1}{(1 + e^{(-0.5*S)})} \tag{5}$$

$$where \quad s = I(t)]^T * W \tag{6}$$

W is the vector used as connection strength of the network given in eq. (7)

Where $$W = [w_1, w_2, w_3, \ldots\ldots, w_m, w_{m+1}, \ldots\ldots, w_M] \tag{7}$$

The weight vector [W] and the adaptive parameters vector [A] are updated in the direction of the negative gradient of the performance function. Randomly the values are chosen to assign weight vector [W] and the adaptive parameter vector [A] to train the network. These random values are between 0 and 1.

Weight vector is updated using back propagation (BP) learning algorithm. The following eq. (8) is given as

$$w_{t+1} = w_t + e_t \frac{\partial o_t}{dw} \tag{8}$$

The weight adjustment is computed by eq. (9)

$$w_{t+1} = w_t + \alpha * e_t * (1 - o_t)^2 * I(t) \tag{9}$$

3 Analysis of Datasets and Selection of Inputs and Assessment Methods

The stock market or equity market time series is a financial measure in the world economy. For the purpose of analysis, we have used the data collected from Reliance Industries Limited (RIL), Bombay Stock Exchange (BSE), and International Business Machines Corp. (IBM) stocks for our studies. In this paper we have presented our IBM results collected from 20/05/2003 to 14/06/2012, a total number of 2300 trading days of data. Following 70:30 ratios, we have used 2000 days of trading data for training and remaining 300 days of trading day data for validating the model. The data used for simulating the model to forecast is on the closing price of index on each day The MATLAB implementation has done to simulate the forecasting model.

The entire input data patterns including technical indicators are normalized to values between 0.0 through 1.0. The normalization formula in eq. (10) is used to express the data in terms of the minimum and maximum value of the dataset.

$$u = (x_i - x_{\min}) / (x_{\max} - x_{\min}) \qquad (10)$$

Where, 'u 'and 'x' represents normalized and actual value respectively.

3.1 Forecasting Analysis with Technical Indicators

We carry out forecasts by using each of the technical indicators listed below combining with inputs to the network. Technical indicators are any class of metrics whose value is derived from generic activity in a stock or asset. The Technical indicators look to predict the future price value by looking at past patterns. A brief explanation of each indicator is mentioned here.

To reduce the over learning of the proposed model and to reduce the complexity of the architecture, we have tried with three important technical indicators namely RSI (5), PVC (5), and MAVI (5, 20). The reason of choosing these three variables is that they are focused on two prime aspects stock data i.e. price, and volume. The selected three indicators mention below in eq. (11).

$$RSI(5) = \sum_{i=1}^{n} Max(\Delta x_i, 0) \Big/ \sum_{i=1}^{n} |\Delta x_i| \qquad PVC(5) = \sum_{i=1}^{n} (\Delta x_i \times \Delta v_i) \Big/ \sum_{i=1}^{n} (|\Delta x_i \times \Delta v_i|) \qquad MAVI(5,20) = MV_5 / MV_{20} \qquad (11)$$

Where x_i is the closing price on the ith day, $\Delta x_i = x_i - x_{i-1}$, $\Delta v_i = v_i - v_{i-1}$, MV5 is 5 day moving average of trading volume, and MV20 is 20 day moving average of trading volume.

3.2 Assessment of Forecasting Results

The proposed DRFLANN architecture is authenticated with testing results based on the various issues discussed before to study the accuracy of forecasting results. The Mean Absolute Percentage Error (MAPE) in eq. (12) represents the absolute average prediction error between actual and forecast value. This gives the repulsive effect of very small prices; the Average Mean Absolute Percentage Error (AMAPE) in eq. (13) is adopted and compared. The impact of the model uncertainty needs to be measured on forecast results, the variance of forecast errors (σ^2_{Err}) in eq. (14) is used. The smaller the variance, the more accurate is the forecast results and more confidence is the model.

$$MAPE = \left(\frac{1}{N} \sum_{j=1}^{N} \frac{abs(e_j)}{y_t} \right) \times 100\% \tag{12}$$

$$AMAPE = \left(\frac{1}{N} \sum_{j=1}^{N} \left[\frac{abs(e_j)}{\bar{y}} \right] \right) \times 100\% \quad where \quad \bar{y} = \frac{1}{N} \sum_{j=1}^{N} [\, y_t \,] \tag{13}$$

$$\sigma^2_{Err} = \left(\frac{1}{N} \sum_{j=1}^{N} \left[\frac{abs(e_j)}{\bar{y}} \right] - AMAPE \right)^2 \tag{14}$$

Where $e = y_t - y$, y_t & y represents the actual and forecast values; $\bar{y}$ is the average value of actual price; σ^2_{Err} variance of forecast error and N is the forecasting period.

4 Simulation Results

Simulation study is carried out using the data set IBM stock. The below listed case studies are reported in the paper to show the limited impacts on price forecast errors. We have shown a set of figures based on testing results and comparison tables to show the impact of case studies on to our model.

Computer simulations for training and testing results were plotted using different case studies to validate the model. Finally the performance evaluation in stock price prediction is given in table1. For a fair comparison, same data set is used for the forecasting the prices.

Case Study 1
In this section, the DFLARNN model is trained with the data set mentioned above and the simulation testing result portrayed in Fig. 2 without using technical indicators.
Case Study 2 - A
In this section, the DFLARNN model is trained with the data set mentioned above using technical indicator Relative Strength Indicator (RSI)
Case Study 2 - B
In this section, the DFLARNN model is trained with the data set mentioned above using technical indicator Relative Strength Indicator (RSI) and Price Volume Change Indicator (PVC).
Case Study 2 - C
In this section, the DFLARNN model is trained with the data set mentioned above and the simulation testing result portrayed in Fig. 3 using technical indicator Relative Strength Indicator (RSI), Price Volume Change Indicator (PVC).and Moving Average Volume Indicator (MAVI)

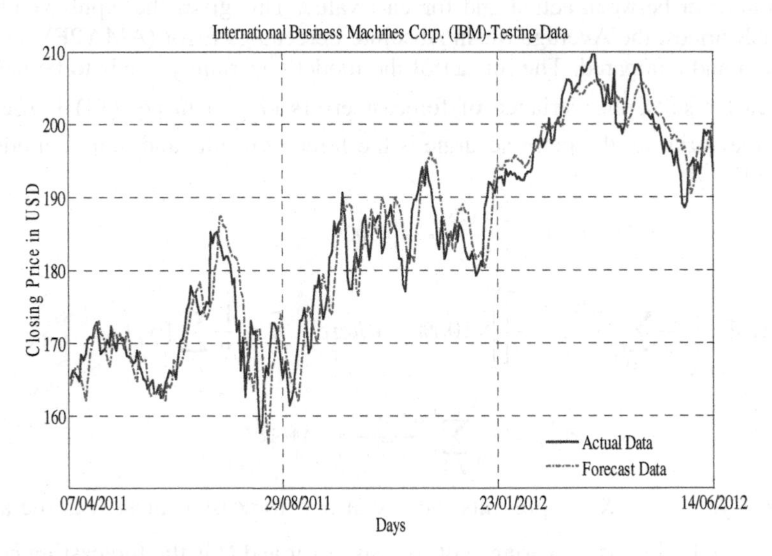

Fig. 2. International Business Machines Corp. (IBM) Stock Testing results

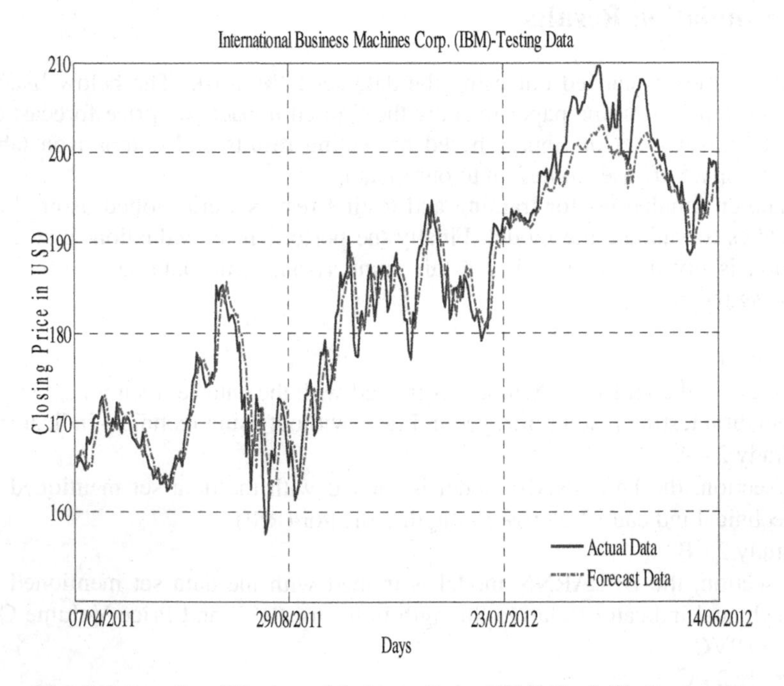

Fig. 3. International Business Machines Corp. (IBM) Stock Testing results

Table 1. Performance Comparison on daily closed price of various case studies

Method	Duration (300 Days)	Performance Assessments		
		MAPE	AMAPE	Variance (σ^2_{Err})
Case-1	07/04/2011 to 14/06/2012	1.810284	1.814413	0.032266
Case-2 (A)	07/04/2011 to 14/06/2012	1.579375	1.581133	0.024502
Case-2 (B)	07/04/2011 to 14/06/2012	1.523785	1.528979	0.022913
Case-2 (C)	07/04/2011 to 14/06/2012	1.434505	1.463944	0.021005

5 Conclusions

Predicting the stock market index is of great interest because successful prediction of stock prices may be guaranteed benefits. The task is very complicated and very difficult for the trader's to make decisions on buying or selling an instrument. In this study, Dynamic Recurrent FLANN Based Adaptive Model is adopted to predict the stock market return on the various stock indices like BSE, IBM etc. The study on IBM stock data shows that the performance of stock price prediction can be significantly enhanced. This model based on stock market prediction model employing the RMS based weight update mechanism is introduced to make our model computationally much more efficient. Further this proposed DRFLANN model, composed of data pattern and technical indicators, improves the performance of forecast results. Further to achieve better robust result using this model we will try different learning methods along with optimization procedure.

References

1. Brownstone, D.: Using the percentage accuracy to measure neural network predictions in stock market movements. Neurocomputing 10, 237–250 (1996)
2. Chen, A.S., Leung, M.T., Daouk, H.: Application of Neural Networks to an emerging financial market: Forecasting and trading the Taiwan Stock Index. Comput. Operations Res. 30, 901–923 (2003)
3. Chang, P.-C., Liu, C.-H., Lin, J.-L., Fan, C.-Y., Ng, C.S.P.: A neural network with a case based dynamic window for stock trading prediction. Expert Systems with Applications 36(3, Pt 2), 6889–6898 (2009)
4. Zhang, Y., Wu, L.: Stock market prediction of S&P 500 via combination of improved BCO approach and BP neural network. Expert Systems with Applications 36(5), 8849–8854 (2009)

5. Gorriz, J.M., Puntonet, C.G., Salmeron, M., de la Rosa, J.J.G.: A new model for time-series forecasting using radial basis functions and exogenous data. Neural Comput. & Applic., pp. 101–111 (2004), doi:10.1007/s00521-004-0412-5
6. Feng, H.-M., Chou, H.-C.: Evolutional RBFANs prediction systems generation in the applications of financial time series data. Expert Systems with Applications 38, 8285–8292 (2011)
7. Giles, C.L., Lawrence, S., Tsoi, A.C.: Noisy Time Series Prediction using Recurrent Neural Networks and Grammatical Inference. Machine Learning 44, 161–183 (2001)
8. Chen, Y., Dong, X., Zhao, Y.: Stock Index Modeling using EDA based Local Linear Wavelet Neural Network. In: International Conference on Neural Networks and Brain, pp. 1646–1650 (2005), doi:10.1109/ICNNB.2005.1614946
9. Hao, H.-N.: Short-term Forecasting of Stock Price Based on Genetic-Neural Network. In: Sixth International Conference on Natural Computation (ICNC 2010). IEEE Conference Publications (2010)
10. Wu, L.: A Hybrid Model for Day-Ahead Price forecasting. IEEE Transactions on Power Systems 25(3), 1519–1530 (2010)
11. Pao, Y.H., Takefji, Y.: Functional-Link Net Computing. IEEE Computer Journal, 76–79 (1992)
12. Majhi, R., Panda, G., Sahoo, G.: Development and performance evaluation of FLANN based model for forecasting of stock markets. Experts Systems with Applications An International Journal 36(3, Pt 2), 6800–6808 (2008)
13. Bebarta, D.K., Biswal, B., Dash, P.K.: Comparative study of stock market forecasting using different functional link artificial neural networks. Int. J. Data Analysis Techniques and Strategies 4(4), 398–427 (2012)
14. Bebarta, D.K., Biswal, B., Rout, A.K., Dash, P.K.: Forecasting and classification of Indian stocks using different polynomial functional link artificial neural networks. In: IEEE Conference, INDCON, pp. 178–182 (2012), doi:10.1109/INDCON.2012.6420611
15. Patra, J.C.: Chebysheb Neural Network-Based Model for Dual-Junction Sollar Cells. IEEE Transactions on Energy Conversion 26, 132–139 (2011)

MPI Implementation of Expectation Maximization Algorithm for Gaussian Mixture Models

Ayush Kapoor[1], Harsh Hemani[2], N. Sakthivel[3], and S. Chaturvedi[4]

Bhabha Atomic Research Centre, Autonagar, Visakhapatnam, India
{ayush21011991,shakthiveln,shashankvizag}@gmail.com,
harshhemani@live.com

Abstract. Gaussian Mixture Model (GMM) is a mathematical model represented by a mixture of Gaussian distributions, each with their own mean and variance parameters. In applications like speaker-recognition, where the GMM is to be trained on hundred thousand of data points for several different speakers, training a GMM becomes computationaly intensive and requires long time to train. One way to solve this problem is by implementing a parallel algorithm for training the GMM. In this paper, we present a parallel algorithm for Expectation Maximization using the message passing paradigm, laying the emphasis on data parallelism. Our results show that our parallel algorithm when applied to an input of 200,000 points for a 8 dimensional Gaussian mixture model having 128 components takes 81.86 seconds on 128 cores which is about 92 times faster than the serial implementation.

1 Introduction

A Gaussian Mixture Model (GMM) is a probability density function consisting of multiple underlying gaussian components [1,2,6]. It is represented as a convex combination of Gaussian components. The probability density function of a Gaussian mixture model, consisting of k components is given by the equation:

$$\Pr(x|\lambda) = \sum_{i=1}^{i=k} w_i \mathcal{N}(x|\mu_i, \Sigma_i), \tag{1}$$

where x is a D-dimensional data vector (feature vector), $\mathcal{N}(x|\mu_i, \sigma_i)$ are the densities of the individual component gaussians,w_i,i=1, ..., k are the weights of individual components .These weights signify prior probability of the components. Each component density is a D variate Gaussian probability density function of the form,

$$\mathcal{N}(x|\mu_i, \Sigma_i) = \frac{1}{(2\pi)^{D/2}|\Sigma_i|^{1/2}} \exp\{-\frac{1}{2}(x-\mu_i)' \, \Sigma_i^{-1}(x-\mu_i)\}, \tag{2}$$

S.C. Satapathy et al. (eds.), *Emerging ICT for Bridging the Future – Volume 2*,
Advances in Intelligent Systems and Computing 338, DOI: 10.1007/978-3-319-13731-5_56

with mean vector μ_i of dimension D and covariance matrix Σ_i of dimension D*D. The weights of the individual gaussian components satisfy the following constraints

$$0 \leq w_i \leq 1$$

and,

$$\sum_{i=1}^{i=m} w_i = 1.$$

The Gaussian mixture model is characterized by the mean vectors, covariance matrices and mixture weights from all component densities.These three parameters will be collectively represented as:

$$\lambda = \{w_i, \mu_i, \Sigma_i\} \qquad i = 1, \ldots, k.$$

Apart from speaker recognition ,Gaussian Mixture Models are widely used in speech segmentation[3],speech recognition[4], and image segmentation [5].

2 Maximum Likelihood Parameter Estimation

Given a set of input data vectors $X = \{x_1, x_2, \ldots, x_n\}$ drawn from an unknown distribution, problem is to estimate the parameters λ of the Gaussian Mixture Model that best fit the data.There are many methods for estimating the GMM parameters,out of which the most well established method is the Maximum likelihood estimation (MLE). MLE finds the model parameters that maximizes the likelihood of the Gaussian mixture model given the training data.

For a set of data X, likelihood $L(\lambda|X)$ can be written as,

$$L(\lambda|X) = \Pr(X|\lambda) = \prod_{k=1}^{k=n} \Pr(x_k|\lambda)$$

$$= \Pr(x_1|\lambda)\Pr(x_2|\lambda)\ldots\Pr(x_n|\lambda)$$

Taking log of likelihood

$$\log \Pr(X|\lambda) = \Sigma_{k=1}^{k=n} \log \Pr(x_k|\lambda)$$
$$= \sum_{k=1}^{k=n} \log\{\sum_{i=1}^{i=m} w_i \mathcal{N}(x_k|\mu_i, \Sigma_i)\} \tag{3}$$

Goal of MLE is to find the parameter λ which maximizes equation (3). Maximization of this term is obtained by Expectation Maximization Algorithm.

3 Expectation Maximization Algorithm

Expectation maximization algorithm is a statistical learning technique used for maximum likelihood parameter estimation of models that rely upon latent variables[7,8,10].
EM algorithm works iteratively, performing two steps namely: Expectation step (E) and Maximization step (M). The expectation (E) step calculates the expectation of log-likelihood of the complete data based on the current estimate of the model parameters and the observed data,that is,it computes

$$E\left(\log\Pr\left(h|\lambda\right)|x,\lambda_i\right)$$

where h is the complete data consisting of observered data x and unobserved data z ,that is, h=(x,z) and the maximization (M) step evaluates new model parameters by maximizing the expected log-likelihood found on the E step ,that is ,it computes,

$$\lambda^{i+1} = \max\left(E\left(\log\Pr\left(\left(h|\lambda\right)|x,\lambda^i\right)\right)\right)$$

To apply the EM algorithm for ML estimation in a GMM, we need to introduce a latent variable. In this case, hidden variable Z is a matrix whose entry z_{ij} is equal to 1 if component i produced measurement j, otherwise it is zero.

4 Serial Expectation Maximization Algorithm

1:Initialize the means μ_i,covariance matrix Σ_i and the weights w_i $i = 1, \ldots, k$ [9].
2:**E step**:Calculate the responsibilities(posterior probabilities) and log probabilities of data points using the current parameter values ,that is,means,covariance, weights.

$$\gamma\left(z_{ni}\right) = \frac{w_i\mathcal{N}\left(x_n|\mu_i,\Sigma_i\right)}{\sum_{j=1}^{j=k} w_j\mathcal{N}\left(x_n|\mu_j,\Sigma_j\right)} \quad i = 1, \ldots, k \tag{4}$$

$$N_i = \sum_{n=1}^{n=N} \gamma\left(z_{ni}\right) \quad i = 1, \ldots, k \tag{5}$$

3:**M step**:Evaluate the parameters using the current responsibilities computed in E-step

$$\mu_i^{new} = \frac{1}{N_i}\sum_{n=1}^{n=N} \gamma\left(z_{ni}\right)x_n \quad i = 1, \ldots, k \tag{6}$$

$$\Sigma_i^{new} = \frac{1}{N_i}\sum_{n=1}^{n=N} \gamma\left(z_{ni}\right)\left(x_n - \mu_i^{new}\right)\left(x_n - \mu_i^{new}\right)^T \quad i = 1, \ldots, k \tag{7}$$

$$w_i^{new} = \frac{N_i}{N} \quad i = 1, \ldots, k \tag{8}$$

4:**Check for convergence**:

$$\log \Pr(X|\mu, \Sigma, w) = \sum_{n=1}^{n=N} \log\{\sum_{i=1}^{i=k} w_i \mathcal{N}(x_n|\mu_i, \Sigma_i)\}$$

and check for convergence of likelihood.If convergence criteria not satisfied then return to step 2.Otherwise,stop.

5 Parallel Expectation Maximization Algorithm

Let us consider a Gaussian Mixture model with k components .Let there be n data points with each data point being a D dimensional vector.In this implementation we have assumed the covariance matrix to be diagonal.

1: Initialize the means μ_k which is obtained by running Kmeans clustering algorithm on n data points ,covariance Σ_k which is obtained by taking the variance of all data points ,weights π_k which is equal to 1/number of components.

2: Distribute the points among the processors such that all processors are load-balanced

3: **E step**: Let each processor compute the log probability of their set of points and responsibilities of components for their set of points ,that is,probability of each data point (x) being generated by component(k) according to equation(4).

4: **M step**: Each processor computes the sum of responsibilities of each components for all of data point belonging to that processor. Each Processor send the sum of responsibilities it had computed to processor 0 which compute the sum and returns it back to all the processors.In this way N_k is computed in parallel for all the k components.After having computed N_k weights are computed for each processor according to equation(8)
Each processor sends the weighted sum which is the product of responsibilities(it contains the responsibilities of each components for data points belonging to that processor) and data points belonging to that processor to processor 0.Processor 0 computes the sum and returns it back to all the processor .After computing the weighted sum then each processor computes means,variance according to equation (6) and (7).

5:**Checking for Convergence**: Each processor sends the log probability of each data points belonging to that processor(computed in step 3) to processor 0.Processor 0 computes log likelihood and checks for the convergence .If convergence criterion not satisfied ,then return to step 2.
Convergence criteria: Program stops when change in log likelihood is less than threshold .

6 Experimental Results

Parallel Expectation Maximization algorithm was implemented in MPI(Message passing interface)[11]. Program was run on cluster consisting of 60 nodes to determine the maximum possible speedup that is achievable. Problem was run

Table 1. Execution Time and speedup for problem having 50000 data points

Data set:50,000		
Number of Processes	Execution time	Speedup
1	2066.13438296	1.000
8	480.659826994	4.304
16	151.223110914	13.662
24	89.7638151646	23.017
32	62.1220431328	33.259
64	29.9413180351	69.006
128	35.2424361706	58.626

Table 2. Execution Time and speedup for problem having 100000 data points

Data set:100,000		
Number of Processes	Execution time	Speedup
1	3450.16113782	1.000
8	962.537559986	3.584
16	471.808099985	7.313
24	269.467671871	12.803
32	157.656415939	21.884
64	71.6118109226	48.178
128	45.9984891415	75.006

on at most 128 cores. Three data sets were used each one containing a different number of points. Each feature vector in the data sets was 8-dimensional. Tests were run on a Gaussian mixture model having 128 components. A typical problem size for Gaussian Mixture Model ranges from 35000-200,000 points in a speaker recognition application. Following results were achieved:

Table 3. Execution Time and speedup for problem having 200,000 data points

Data set:200,000		
Number of Processes	Execution time	Speedup
1	7475.72441196	1.000
8	2226.11574793	3.358
16	1100.89717317	6.791
24	721.7227211	10.358
32	492.45282197	14.655
64	162.59663415	45.977
128	81.8623049259	91.621

Table 1 shows that execution time decreases with the increase in the number of processes. But as we go beyond 64 cores execution time starts increasing because communication time is more than computation time.Table 2 and Table 3 shows that execution time decreases with increase in number of processes.

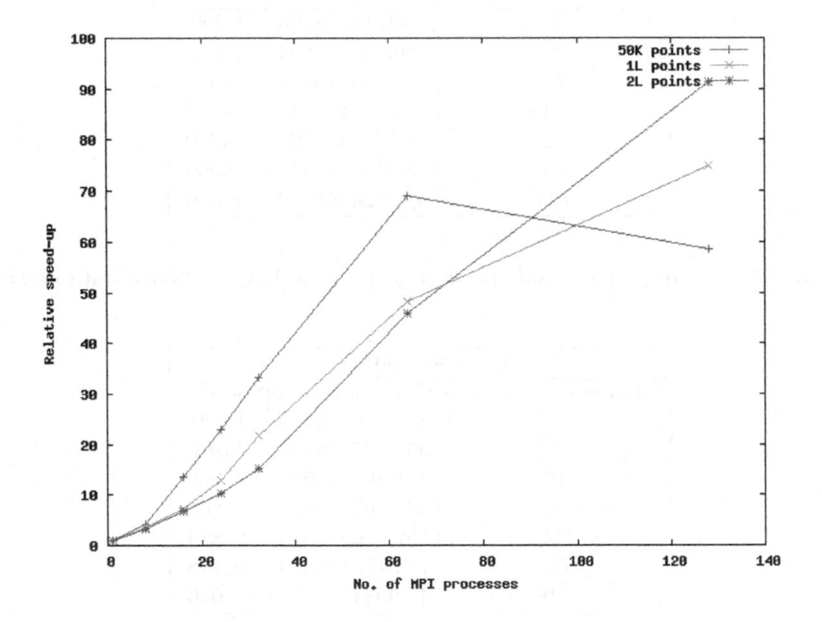

Fig. 1. Speed-up vs No of MPI processes

For problem having 50,000 points Figure 1 shows that as we increase the number of cores,speedup increases. But after 64 processes, communication time becomes more dominant than computation time ,so speedup decreases.For problems having 100,000 and 200,000 points ,speedup increases with increase in number of processes.

7 Conclusion

Parallel Implementation of Expectation Maximization Algorithm was done using Message Passing Interface.The effect of increasing the number of MPI processes on Expectation Maximization algorithm was studied. It was found out that speed-up increases with the increase in number of cores. For a small data set it was observed that as we increase the number of processes speed-up was increasing but afterwards it started decreasing as communication among processors became more dominant over computation. The results show that MPI implementation of Expectation maximization algorithm when applied to an input of 200,000 points for 8 dimensional Gaussian Mixture Model having 128 components takes 81.862 sec on 128 cores which is about 92 times faster when compared to serial implementation.The future work involves developing parallel Expectation Maximization algorithm for hybrid cluster.

References

1. Reynolds, D.A.: Gaussian Mixture Models. Encyclopedia of Biometric Recognition. Springer (2008)
2. Allili, M.S.: A Short Tutorial on Gaussian Mixture Model. University of Québec in Outaouais, Canada (2010)
3. Park, K., Park, J.C., Oh, Y.H.: GMM Adaptation based Online Speaker Segmentation for Spoken Document Retrieval. IEEE Transactions on Consumer Electronics 56, 1123–1129 (2010)
4. Stuttle, M.N.: A Gaussian Mixture Model Spectral Representation for Speech Recognition. Ph.D Dissertation Cambridge University (2003)
5. Farnoosh, R., Zarpak, B.: Image segmentation using Gaussian Mixture Model. IUST International Journal of Engineering Science 19, 29–32 (2008)
6. Marsland, S.: Machine Learning An Algorithmic Perspective. CRC Press (2009)
7. Wikipedia, `http://en.wikipedia.org/wiki/Expectation-Maximization_algorithm`
8. Do, C.B., Batzoglou, S.: What is the expectation maximization algorithm? Nature Biotechnology 26, 897–899 (2008)
9. Bishop, C.M.: Pattern Recognition and Machine Learning. Springer (2006)
10. Blume, M.: Expectation-Maximization:A Gentle Introduction. Technical University of Munich Institute for Computer Science (2002)
11. Dalcin, L.: MPI for Python, `http://mpi4py.scipy.org`

Effect of Mahalanobis Distance on Time Series Classification Using Shapelets

M. Arathi and A. Govardhan

School of Information Technology, JNT University Hyderabad, Hyderabad, India
arathi.jntu@gmail.com, govardhan_cse@yahoo.co.in

Abstract. The sequence of values that are measured at time intervals equally spaced is time series data. Finding shapelets within a data set as well as classifying that data based on shapelets is one of the most recent approaches to classification of this data. In the classification using shapelets, Euclidean distance measure is adopted to find dissimilarity between two time series sequences. Though the Euclidean distance measure is known for its simplicity in computation, it has some disadvantages: it requires data to be standardized and it also requires that the two data objects being compared be of the same length. It is sensitive to noise as well. To overcome the problem, Mahalanobis distance measure can be used. In the proposed work, classification of time series data is performed using time series shapelets and used Mahalanobis distance measure which is the measure of distribution between a point and distribution. Correlations between data set is considered. It does not depend on scale. The cost complexity pruning is performed on decision tree classifier. The Mahalanobis distance improves the accuracy of algorithm and cost complexity pruning method reduces the time complexity of testing and classification of unseen data. The experimental results show that the Mahalanobis distance measure leads to more accuracy and due to decision tree pruning the algorithm is faster than existing method.

Keywords: Time series classification, Shapelets, Mahalanobis distance measure, Decision trees, Information gain, Cost complexity pruning.

1 Introduction

For a decade, there have been a number of papers on time series classification. Spaced at equal time intervals, it is an ordered sequence of values. The analysis of time series data includes various methods that try to perceive such data. That is, time series analysis includes either understanding the underlying data context or forecasting. Applications of its classification is not limited to: scientific investigations, economic and sales forecasting, study of natural phenomena, engineering experiments, analysis of customer behavior, stock market analysis, medical treatments.

The major interest of research in the mining of time series data covers classification, indexing, summarization, clustering and anomaly detection. In classification, an unlabeled time series should be assigned to a class predefined. In indexing, for a given

S.C. Satapathy et al. (eds.), *Emerging ICT for Bridging the Future – Volume 2*,
Advances in Intelligent Systems and Computing 338, DOI: 10.1007/978-3-319-13731-5_57

sequence and measure similarity/dissimilarity, we need to retrieve sequences that are akin to those of time series. In summarization, given a time series object Q containing n data points where n is a large number, an estimation of Q retaining its primary properties is created. In clustering, the time series objects need to be grouped under some similarity/dissimilarity measure. In detecting anomalies, given a time series Q along with some model of normal behavior, all sections of Q containing abnormal behavior are found.

Time series classification algorithms are broadly categorized as distance-based, feature-based and, model-based. In the algorithms that are based on distance, the classification of the data is based on the distance between the objects. The distance measures used to compare the time series data are : Euclidean distance, Dynamic Time Warping (DTW), Longest Common Subsequence (LCSS), Edit Distance on Real sequence (EDR), Edit Distance with Real Penalty (ERP), Sequence Weighted Alignment model (Swale), search of similarity based on Threshold Queries (TQuEST), Spatial Assembling Distance (SpADe). In feature-based classification, the time series sequence is converted to feature vector on which the usual classification methods are applied. Here, feature selection plays an important role. Some of the most popular feature selection/data reduction techniques are: Discrete Fourier Transform (DFT) [1], Discrete Wavelet Transform (DWT) [2], Discrete Cosine Transformation (DCT) [3], Singular Value Decomposition (SVD) [1], Piecewise Aggregate Approximation (PAA) [4], Adaptive PAA [5], ChebyShev Polynomial [6], Symbolic Aggregate Approximation (SAX) [7], and Indexable Piecewise Linear Approximation [8]. The model-based methods construct a model for the data within a class and classify new data according to the model that best fits it. In the classification step, a new sequence is assigned to the class with the highest likelihood.

A most promising recent approach of classifying time series data is finding shapelets within a data set [9]. A shapelet is a subsequence of time series data which represents a particular class. The algorithms that are based on shapelets are interpretable, accurate and faster than existing classifiers [10, 11].

There are two types of classification algorithms: algorithms that consider the whole (single) time series sequence (global features) for classification and algorithms that consider a portion of a single time series sequence (local features) for classification. The shapelets are local features. In classification using shapelets, a shapelet that represents a particular class is identified and then, the classification is done based on the shapelet information. Because shapelets are small in size compared to the original data, algorithms that use shapelets for classification, result in less time and space complexity. Shapelets have successfully been used in many other applications like early classification [12], gesture recognition [13], and as a filter transformation for TSC [14].

For classification with shapelets, decision trees (binary) are used, where each nonleaf node represents a shapelet and leaf nodes represent class labels. To know how well the shapelet classifies the data, information gain [15] is used. Apart from this, the measures, such as, the Wilcoxon signed-rank test [16], Kruskal-Wallis [17], and Mood's Median [18] can also be used. The information gain/entropy measure is a better choice among other measures, because, early entropy pruning can be done to avoid unnecessary computations performed when finding the shapelet.

To compare two time series data, a metric distance measure should be used. A distance measure is metric, if it satisfies the: 1) Positive definiteness 2) Symmetry 3) Triangle Inequality.

The rest of the paper is organized as follows. In Section 2, related work is reviewed. The definition and comparison of the distance measures is discussed in Section 3. The decision tree pruning is discussed in Section 4. The experimental results are discussed in Section 5. The paper is concluded in Section 6.

2 Related Work

The closest work is that of time series classification using shapelets [9]. Here, the authors classify the time series data using shapelets. It generates all possible subsequences of all possible lengths where a subsequence is subset of consecutive values of the time series sequence. Each subsequence is tested to see how well it can classify the data. For this, it generates an object histogram which contains all the time series objects distances to the given subsequence. To find the distance between two time series sequence or between a time series sequence and subsequence, Euclidean distance measure is used. The time series objects in the histogram are in increasing order of distance. An optimization in computing the distance between the time series and subsequence is performed. That is, instead of computing the final distance value between the subsequences of a given time series data and the given subsequence, the distance calculations can be stopped when the partial computation is more than the least distance. This is early abandon [19]. To find the best shapelet, information gain is used.

Another optimization is performed to reduce the time complexity called entropy pruning. This is done during object histogram computation. Once a time series sequence is added to object histogram, it is checked to see if the remaining calculations of other time series objects with the given subsequence can be pruned. For this, the partially computed object histogram is taken. The remaining objects (for which the distance has not been computed to the given candidate) of one class are added to one end of the histogram and the objects of other class are added to the other end of the histogram and vice versa. Now, the information gain is computed. If it is greater than the best known so far, then the histogram computation is continued, otherwise the remaining calculations with the candidate are pruned.

The classification of time series data with shapelets along with their corresponding split point produces a binary decision. Hence, binary decision trees are used. Because one shapelet is not sufficient to classify the entire time series data, a number of shapelets are used which clearly distinguishes one class from other. The shapelets are used along with the distance threshold (split point), which divides the data into two sets. The non leaf nodes of the decision tree specify shapelet and the distance threshold; and leaf nodes specify the class label. To predict class label, the time series sequence is fed into classifier, which moves it from the root to the leaf node. It gives the predicted label. While moving from root to leaf node, the sequence is compared with every shapelet on the path using Euclidean distance measure.

It is possible to have same the best information gain for different subsequences especially for small datasets. Such ties can be broken in favour of the longest subsequence, the shortest subsequence or the subsequence that clearly distinguishes one class from another.

Usually, the time series data are very large with many innumerable values in a sequence. Hence, it is very expensive and difficult to compare two time series. Several methods have been identified for comparing time series sequences. Some of the distance measures as mentioned in Section I are: Euclidean distance, DTW [20, 21], LCSS [22], EDR [23], ERP [24], Swale [25], TQuEST [26], SpADe [27]. The Euclidean distance is summed up by the Euclidean measure between corresponding points in each time series. The metric is most intuitive for comparing time series data. It is very simple to compute with time complexity as O(n). But, the problem with it is that it is sensitive to noise and needs the two sequences of the equal length. It also needs standardization of the time series data, if scales differ. DTW is an elastic measure. The two sequences need not be of equal length. Time shifting is done between the two series by repeating the elements. It is based on dynamic programming, hence has quadratic time complexity. It is sensitive to noise. A threshold value, ϙ, is introduced by the LCSS technique. The scoring technique handles the noise. If the distance between two sequences is less than ϙ in each dimension, then they are supposed to match and are given a match reward of 1. If the distance is not less that ϙ in some dimension, they do not match, and therefore there is no reward. Hence, it is sturdy. It rewards matches, but does not penalize mismatched parts. It is also based on dynamic programming. EDR only scores gaps and mismatches, but do not reward matches. It is robust in presence of noise and time shifting. It is based on dynamic programming. ERP is similar to L_1-norm, but also supports time shifting locally. It is a metric distance measure. It is sensitive to noise. DTW, LCSS, and EDR can handle time shifting locally. But the problem with them is that they are not metric. ERP is based on dynamic programming. Swale is similar to LCSS, but it also penalizes dissimilar parts. TQuEST specifies a threshold query which comprises a query time series TQ and a threshold *th*. The database time series are decomposed into time intervals of subsequent elements where the values are above *th*. The query sequence TQ is also decomposed in such a way. Each interval is considered as a point in two-dimensional space with x as starting time and y as the end of interval. How similar two sequences are, is computed by using Minkowski distance measure. Now, the threshold query returns all the sequences of the database that have a similar interval sequence. All the above distance measures show poor performance if there is shifting and scaling in amplitude dimensions which can be handled by SpADe distance measure.

Our focus is on seeing the performance of Mahalanobis distance measure in time series classification using shapelets and to study the effect of cost complexity pruning on the proposed method. To the best of our knowledge, the proposed method gives more accurate and faster classification of time series data than the existing method.

3 Proposed Method

The techniques of evaluating the similarity of time series sequences have attracted the attention of the database researchers. The selection of a distance function to find the similarity between two sequences is a challenging issue for researcher. There are two numeric measures to compare data objects: similarity & dissimilarity. The similarity measure tells about the extent of similarity of two objects. The value is more when objects have greater similarity. The value is often in the range [0, 1]. The dissimilarity measure specifies the difference between the two data objects. The difference is less when the objects have greater similarity. Zero is often the minimum dissimilarity. In this paper, the dissimilarity/distance measure is used to compare two data objects. Using the Mahalanobis Distance measure instead of the Euclidean distance measure improves the accuracy of the algorithm.

3.1 Euclidean Distance Measure

The Euclidean measure is geometric based on the Pythagorean formula, given as,

$$dist = \sqrt{\sum_{k=1}^{n} (p_k - q_k)^2} \tag{1}$$

where n stands for the number of dimensions and p_k, and q_k are the kth components of objects, p and q, respectively.

The advantage of the Euclidean distance measure is its simplicity in computation. But it has some disadvantages. Firstly, it requires that when variables are of different scales, the data need to be standardized. Since, it is incorrect to compare time series data with different offsets and amplitudes, they must be normalized/standardized so that the mean is 0 and standard deviation is 1. The normalization of time series data can be performed by subtracting mean from each value of time series data and dividing the result by standard deviation. Consider a regression problem which makes use of class information regarding, age, test scores as well as time. If all are on different scale, then they cannot be compared. This issue can be surmounted by a normalized Euclidean measure. But it incorporates only variances and not covariances unlike Mahalanobis measure which covers both of them. The next disadvantage is that it requires that both the sequences under comparison must be of the same length, and thirdly, it is sensitive to noise.

3.2 Mahalanobis Distance Measure

The Mahalanobis measure is the one between a point and distribution as presented by P. C. Mahalanobis in the year 1936 [28]. It is a unitless measure used to identify and gauge the similarity of an unknown sample set to a known sample. It differs from the Euclidean by considering the equivalence of the data set and is invariant of scale.

Given a time series $x^{(k)}$, let the ith data point be $x_i^{(k)}$. First, compute the (sample) covariance matrix $C = (c_{ij})$ of a family of time series $x^{(1)}, x^{(2)}, \ldots, x^{(N)}$ of lengths n by $c_{ij} = \frac{1}{N-1} \sum_{k=1}^{N} (x_i^{(k)} - \bar{x}_i)(x_j^{(k)} - \bar{x}_j)$. N is the number of instances and $\bar{x}_i$ is the average of the ith data point of the time series ($\bar{x}_i = \frac{1}{N} \sum_{k=1}^{N} x_i^{(k)}$).

The Mahalanobis measure is case of the generalized ellipsoid measure $D_M(x, y) = (x - y)^T M (x - y)$. M is proportional to inverse of covariance matrix i.e., $M \alpha C^{-1}$. Though the Mahalanobis distance measure is often defined by setting M to the inverse of the covariance matrix ($M = C^{-1}$), it is convenient to normalize it when possible so that the determinant of the matrix M is one: $M = (\det(c))^{\frac{1}{n}} c^{-1}$ where n is the time series sequence length. The Mahalanobis distance measure minimizes the sum of distances between time series $\sum_{x,y} D_M(x, y)$ subject to a regularization constraint on the determinant ($\det(M) = 1$). In this sense, it is optimal.

When the covariance is non-singular ($\det(C) \neq 0$), then the covariance is positive definite, and so is the matrix M: it follows that the square root of the generalized ellipsoid distance measure is a metric. That is, $D_M(x, y) = 0 \Leftrightarrow x = y$, it is symmetric, non-negative, and also satisfies the inequality of triangle.

3.3 Euclidean vs. Mahalanobis Distance Measure

Mahalanobis measure takes the co-variances of data objects into consideration leading to elliptic decision boundaries in the 2D case, whereas the Euclidean distance measure leads to circular boundaries. In statistics, the distance is measured by the scale of the data. Standard deviation is scale. For univariate data, an object, one standard deviation away from the mean, is closer to the mean than the one which is five standard deviations away. By computing *z-score*, the distance from the mean can be specified, for normally distributed data. The z-score of x is computed as $z = (x-\mu)/\sigma$, where μ is mean of the time series data and σ is the standard deviation. This is dimensionless quantity.

The graph in Fig. 1 shows simulated bivariate normal data overlaid with prediction ellipses. The ellipses are 10%, 20%, and so on till 90%. The prediction ellipses are the outlines of the bivariate normal density function. For ellipses near the origin, the probability density is high, and for farther ellipses, it is low.

In Fig. 1, there are 2 red marks. One is at (4, 0), and the other at (0, 2). To see which mark is closer to the origin, let us consider the two distance measures. The Euclidean values are 4 and 2. Hence, according to the Euclidean distance measure, the point at (0, 2) is more nearer to the origin. For this distribution, variance in X axis is more than that in Y axis. Therefore, the point (4, 0) is fewer standard deviations away from the origin than the point (0, 2). Hence, it is less likely to see an observation near (0, 2) than at (4, 0). Thus, according to Mahalanobis distance, the point at (4, 0) is more nearer to the origin than the point at (0, 2).

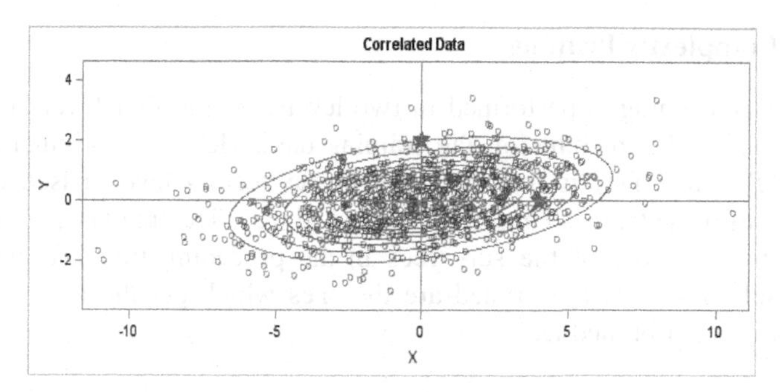

Fig. 1. Bivariate normal data with predicted ellipses

The prediction ellipses is a multivariate generalization of units of standard deviation. The bivariate probability outlines can be used to compare distances to the bivariate mean. Point *a* is nearer to the origin than point *b* if the ellipse containing *a* is enclosed within the ellipse containing *b*.

Mahalanobis measure has the qualities as given below: 1) The variances are different for each dimension. 2) The covariance between variables is considered. 3) For uncorrelated variables with unit variance, its performance is similar to that of Euclidean measure.

4 Decision Tree Pruning

The decision tree classifier is built for time series dataset using shapelets as explained in Section II. It has been observed that while a decision tree is being built, the branches reflect anomalies in the training data owing to noise or outliers. The methods of pruning the tree tackle the data overfitting problem. These methods normally use numerical measures to remove the branches that are least reliable. Pruned tree is small and simple and easy to understand. Pruned trees are fast and good at classifying than unpruned trees.

In decision tree induction process, if a tightly stopping criteria is used, it will lead to small and underfitted trees. But, if loosely halting criteria is used, it will lead to generation of giant decision trees overfitted to the training data set. Many methods to prune decision trees have been introduced to solve the later problem [29]. The two methods of tree pruning are prepruning and post pruning. In the first method, a tree is pruned by halting its construction early. Then the node becomes a leaf holding the most recurrent class among the subset tuples or the probability distribution of those tuples. In post pruning approach, some of the subtrees are withdrawn from a fully generated one. A subtree at a given node is pruned by removing its branches and substituting it with a leaf. That is, given a decision tree classifier C and an inner (non-root, non-leaf) node t. Then pruning of C with respect to t is the deletion of all successor nodes of t in C which makes t a leaf node. The leaf is labeled with the most recurrent class among the subtrees that are substituted. This process is repeated on all nonleaf nodes. The removal of the subtree should not result in reduction of the accuracy of the decision tree. Hence it leads to a smaller and accurate decision tree.

4.1 Cost Complexity Pruning

Cost-complexity pruning is performed in two levels. At the first level, a set of trees DT_0, DT_1, . . . , DT_k are built on the training data. Here, the original tree before pruning is DT_0 while DT_k is the root tree. In the second level, based on its error appraisal, one of these trees is chosen as the pruned tree. The tree DT_{i+1} is procured by substituting one or more of the sub–trees in the preceding tree DT_i with suitable leaves. The sub–trees that are pruned are the ones which get the lowest increase in obvious error rate per pruned leaf:

$$err = \frac{er(prd(DT, dt), Smp) - er(DT, Smp)}{|leaf(DT)| - |leaf(prd(DT, dt))|} \tag{2}$$

where *er(DT, Smp)* indicates the error rate of the tree *DT* over the sample *Smp* and |leaf(DT)| indicates the number of leaves in *DT*. *prd(DT,dt)* indicates the tree secured by substituting the node *dt* in *DT* with a suitable leaf.

In the second level, the error of each tree that is pruned DT_0, DT_1, . . . , DT_k is appraised as leading to the selection of the pruned tree that is best ever.

5 Experimental Results

The experiments are conducted on standard datasets such as wheat, mallet, coffee, gun, projectile points, historical documents, beef, car etc. [30]. On all the datasets, our proposed method has shown around 10% – 15% increase in accuracy and 15% – 22% decrease in time complexity.

The wheat dataset contains 775 spectrographs of samples of wheat, which were grown in Canada between 1998 and 2005. There are various kinds of wheat, such as Soft White Spring, Canada Western Red Spring, and Canada Western Red Winter. The wheat grown in a particular year is the class label. For this dataset, the proposed method has shown 12% increase in accuracy as shown in Fig. 2. And there was 20% increase in speed (due to decision tree pruning) during testing phase and classification of unseen data.

There has been extensive study on Gun/NoGun motion capture time series dataset [10], [31]. This data has two classes: Gun and No Gun. The classification algorithm should be able to identify whether the actor is holding gun or not. The difference between the two classes can be identified if the time series data of the actor is observed: how he puts his hand down by his side. The proposed method has shown 8% increase in accuracy for Gun/NoGun problem as shown in Fig. 3. Hence, the proposed method has more accuracy than the existing method. And there was 16% improvement in speed (due to decision tree pruning) during testing phase and during classification of unseen data. Hence, the proposed method is more accurate and fast than existing method.

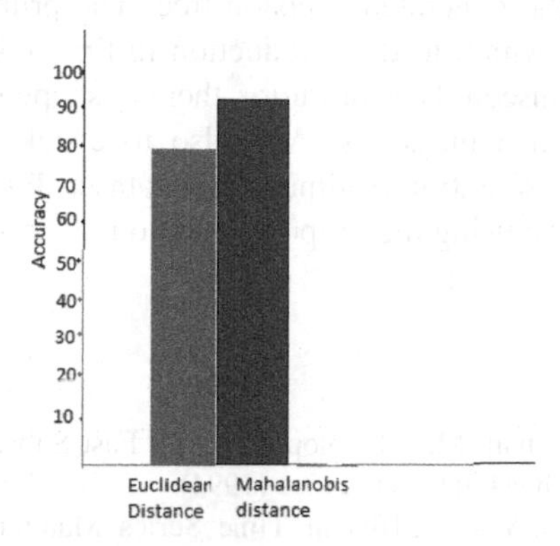

Fig. 2. Accuracy for wheat dataset using Euclidean vs Mahalanobis distance

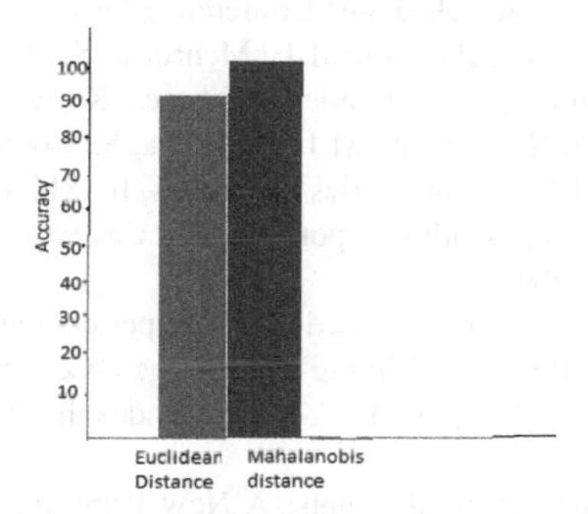

Fig. 3. Accuracy for Gun/NoGun dataset using Euclidean vs Mahalanobis distance

6 Conclusion and Future Scope

The time series dataset is classified using shapelets. The shapelets are time series subsequences and are highly representative of a class. Because one shapelet is not sufficient to classify the data, a number of shapelets are used which clearly distinguishes one class from other. The shapelets are used along with distance threshold, which divides the data into two sets. The decision tree is used as classifier. The non leaf nodes of the decision tree specify shapelet and distance threshold; and leaf nodes specify the class label. To classify a time series data, it is fed into decision tree classifier, which moves it from the root node to leaf node, which in turn gives the predicted class label. While moving from root to leaf node, the time series data is compared with every shapelet on the path using Mahalanobis distance measure. Mahalanobis distance measure is a good choice for classification as it takes the correlation of data items into consideration and is scale in-variant. Hence, it is obvious that Mahalanobis distance measure will give more accurate results. The experimental results have also shown that the distance measure results in more accuracy than the Euclidean distance measure. And we have performed cost complexity pruning on the generated decision tree. The pruning method reduces the size of the decision tree which leads to reduction in time taken in testing phase and also in classification of unseen data. In future, there is scope to compare the proposed method with other distance measures. And also to check how the algorithm will perform on reduced representation of time series dataset. Further, there is also scope to do signature verification using the proposed method.

References

1. Faloutsos, C., Ranganathan, M., Manolopoulos, Y.: Fast Subsequence Matching in Time Series Databases. In: SIGMOD Conference (1994)
2. Pong Chan, K., Fu, A.W.-C.: Efficient Time Series Matching by wavelets. In: ICDE (1999)
3. Korn, F., Jagadish, H.V., Faloutsos, C.: Efficiently supporting ad hoc queries in large Datasets of time sequences. In: SIGMOD Conference (1997)
4. Keogh, E.J., Chakrabarti, K., Pazzani, M.J., Mehrotra, S.: Dimensionality Reduction for Fast Similarity Search in Large Time Series Databases. Knowl. Inf. Syst. 3(3) (2001)
5. Keogh, E.J., Chakrabarti, K., Pazzani, M.J., Mehrotra, S.: Locally Adaptive dimensionality Reduction for Indexing Large Time Series Databases. In: SIGMOD Conference (2001)
6. Cai, Y., Ng, R.T.: Indexing spatio-temporal trajectories with chebyshev polynomials. In: SIGMOD Conference (2004)
7. Lin, J., Keogh, E.J., Wei, L., Lonardi, S.: Experiencing SAX: a novel symbolic representation of time series. Data Mining Knowledge Discovery 15(2) (2007)
8. Chen, Q., Chen, L., Lian, X., Liw, Y., Yu, J.X.: Indexable PLA for Efficient Similarity Search. In: VLDB (2007)
9. Ye, L., Keogh, E.: Time Series Shapelets: A New Primitive for Data Mining. In: KDD 2009, June 29-July 1 (2009)
10. Ding, H., Trajcevski, G., Scheuermann, P., Wang, X., Keogh, E.: Querying and Mining of Time Series Data: Experimental Comparison of Representations and Distance Measures. In: Proc of the 34th VLDB, pp. 1542–1552 (2008)
11. Keogh, E., Kasetty, S.: On the need for Time Series Data Mining Benchmarks: A Survey and Empirical Demonstration. In: Proc. of the 8th ACM SIGKDD, pp. 102–111 (2002)
12. Yu, P., Wang, K., Xing, Z., Pei, J.: Extracting interpretable features for early classification on time Series. In: Proc. 11th SDM (2011)
13. Hartmann, B., Link, N.: Gesture recognition with inertial sensors and optimized DTW prototypes. In: Proc. IEEE SMC (2010)
14. Lines, J., Davis, L., Hills, J., Bagnall, A.: A shapelet transform for time series classification. Tech. report, University of East anglia, UK (2012)
15. Han, J., Kamber, M.: Data Mining: Concepts and Techniques, 2nd edn. Elsevier Publisher,
16. Wilcoxon, F.: Individual Comparisons by Ranking Methods. Biometrics 1, 80–83 (1945)
17. Kruskal, W.H.: A Nonparametric test for the several sample problem. The Annals of Mathematical Statistics 23(4), 525–540 (1952)
18. Mood, A.M.F.: Introduction to the theory of statistics (1950)
19. Keogh, E., Wei, L., Xi, X., Lee, S., Vlachos, M.: LB_Keogh Supports Exact Indexing of Shapes under Rotation Invariance with Arbitrary Representations and Distance Measures. In: The Proc. of 32nd VLDB, pp. 882–893 (2006)
20. Geurts, P.: Pattern Extraction for Time Series Classification. In: Siebes, A., De Raedt, L. (eds.) PKDD 2001. LNCS (LNAI), vol. 2168, pp. 115–127. Springer, Heidelberg (2001)

21. Keogh, E.J., Ratanamahatana, C.A.: Exact indexing of dynamic time wraping. Knowl. Inf. Syst. 7(3) (2005)
22. Gunopulos, D., Kollios, G.: Discovering similar multidimensional trajectories. In: ICDE (2002)
23. Chen, L., Özsu, M.T., Oria, V.: Robust and fast similarity search for moving object trajectories. In: Sigmod Conference (2005)
24. Chen, L., Ng, R.T.: On the marriage of Lp-norms and edit distance. In: VLDB (2004)
25. Morse, M.D., Patel, J.M.: An efficient and accurate method for evaluating time series similarity. In: SIGMOD Conference (2007)
26. Aßfalg, J., Kriegel, H.-P., Kröger, P., Kunath, P., Pryakhin, A., Renz, M.: Similarity search on time series based on threshold queries. In: Ioannidis, Y., et al. (eds.) EDBT 2006. LNCS, vol. 3896, pp. 276–294. Springer, Heidelberg (2006)
27. Chen, Y., Nascimento, M.A., Oosi, B.C., Tung, A.K.H.: SpADe: On Shape-based Pattern Detection in Streaming Time Series. In: ICDE (2007)
28. Mahalanobis, P.C.: On the generalized distance in statistics. Proceedings of the National Institute of Sciences of India 2(1), 49–55 (1936)
29. Breiman, L., Friedman, J., Olshen, R.A., Stone, C.J.: Classification and regression trees. Wadsworth (1984)
30. Datasets, http://www.cs.ucr.edu/~eamonn/time_series_data/
31. Xi, X., Keogh, E., Shelton, C., Wei, L., Ratanamahatana, C.A.: Fast Time Series Classification using Numerosity Reduction. In: The Proc. of the 23rd ICML, pp. 1033–1040 (2006)
32. Quinlan, J.R.: Simplifying Decision Trees. International Journal of Man-Machine Studies (1987)

Text Summarization Basing on Font and Cue-Phrase Feature for a Single Document

S.V.S.S. Lakshmi[1], K.S. Deepthi[1], and Ch. Suresh[2]

[1] Department of Computer Science and Engineering, ANITS, Sangivalasa, Visakhapatnam-531162, AP, India
lakshmi.it@anits.edu.in
[2] Department of Information Technology, ANITS, Sangivalasa, Visakhapatnam-531162, AP, India

Abstract. In recent times owing to the magnitude of data present digitally across networks over a wide range of databases the need for text summarization has never been higher. The following paper deals with summarization of text derived from the syntactic and semantic features of the words in the document. We apply the technique of calculating a threshold value from both the attribute and semantic structure of the individual words. The algorithm helps in calculating the threshold value in order to give weightage to a particular word in a document. Initially the document undergoes the preprocessing techniques; the obtained data will be kept in a data set, then on that data we will apply the proposed algorithm in order to get a summarized data.

Keywords: summarization, threshold, semantic, extraction, cue word, statistical, preprocessing.

1 Introduction

In recent times the documents we see and come across are exceeding our capacity and time frame to read them. Hence the need for automatic text summarization has never been greater. This enables us to understand and comprehend the idea and thought behind a document in a compressed form. For this purpose we extract sentences and form an effective summary using our proposed algorithm. The summary that is obtained remains unchanged in its meaning. The summary should meet the major concepts of the original document set, should be redundant-less and ordered. The summarization of text is lies on sentences either being extracted or generated. Extracting a sentence basically derives upon accessing words that have peculiar attributes i.e. italic, bold, and underlined, title word, start word etc.

Generating sentences has possessed far too many problems owing to its complex analytical nature and profound knowledge required for calculating and generating words basing on the documents provided. The paper focuses on extracting sentences based upon the syntactic and semantic features of the content in the text. It is argued that summaries generated automatically are far more inferior to the ones generated by humans.

S.C. Satapathy et al. (eds.), *Emerging ICT for Bridging the Future – Volume 2*,
Advances in Intelligent Systems and Computing 338, DOI: 10.1007/978-3-319-13731-5_58

1.1 Summarization

Text mining is the analysis of data contained in natural language text. Text mining works by transposing words and phrases in unstructured data into numerical values which can then be linked with structured data in a database and analyzed with traditional data mining techniques.Text summarization (TS) is the process of identifying the most salient information in a document or set of related documents and conveying it in less space (typically by a factor of five to ten) than the original text. Identifying the redundancy is a challenge that hasn't been fully resolved yet. Method consists of selecting important sentences, paragraphs etc. from the original document and concatenating them into shorter form without changing the original concept or meaning of the document.It is very difficult to achieve consistent judgments about summary quality from human judges. This fact has made it difficult to evaluate (and hence, improve) automatic summarization.

Summarization task is done in two different methods, i.e. extractive and abstractive. An extractive summarization of sentences is decided based on statistical and linguistic features of sentences. An Abstractive summarization [9] [10] attempts to develop an understanding of the main concepts in a document and then express those concepts in clear natural language. It uses linguistic methods to examine and interpret the text and then to find the new concepts and expressions to best describe it by generating a new shorter text that conveys the most important information from the original text document.

1.2 Summarization Features

The **font based feature** i.e. bold, italic, underlined and all the combination of these are considered to be more important when calculating the weight for ranking the sentences of the document. For this reason the accuracy rate of our system is more than that of MS-Word automatic text summarization in most cases. **cue-phrase feature i**s based on the hypothesis that the relevance of a sentence is computed by the presence or absence of certain cue words in the cue dictionary. Sentences containing any cue phrase (e.g. "in conclusion", "this letter", "this report", "summary", "argue", "purpose", "develop", "attempt" etc.) are m**ost** likely to be in summaries.**Sentence location feature** is usually first and last sentence of first and last paragraph of a text document are more important and are having greater chances to be included in summary.**Sentence length feature** is very large and very short sentences are usually not included in summary.**Proper noun feature** is name of a person, place and concept etc. Sentences containing proper nouns are having greater chan**ces** for including in summary.**Upper-case word feature** finds sentences containing acronyms **o**r proper names are included.Title method finds the sentence weight is computed as a sum of all the content words appearing in the title and (sub-) headings of a text. you have more than one surname, please make sure that the Volume Editor knows how you are to be listed in the author index.

Single document summarization is the process of creating a summary from a single text document. Multi- document summarization shortens a collection of related documents; into single summary.The proposed algorithm mainly focuses on single document summarization. If we apply each algorithm separately i.e. font based feature and cue-phrase feature we will get summarized data with less accuracy, so we propose an algorithm which is combination of both the above mentioned algorithms in order to get summarized document with maximum accuracy.

2 Proposed System

The above mentioned techniques can individually produce summarized data but with less accuracy. Different combinations of algorithms were developed to produce summarized data but they could not able to produce with maximum accuracy.so we have developed a methodology which is combination of font feature based and cue-phrase based feature, So our algorithm succeed to produce summarized data with maximum accuracy.The steps are following :

a) Select the Features (FONT & CUE WORD)

b) Identify the Tokens based on Threshold value & assign the ranks

c) Select the Sentences Based on ranking

d) Generate a Summary

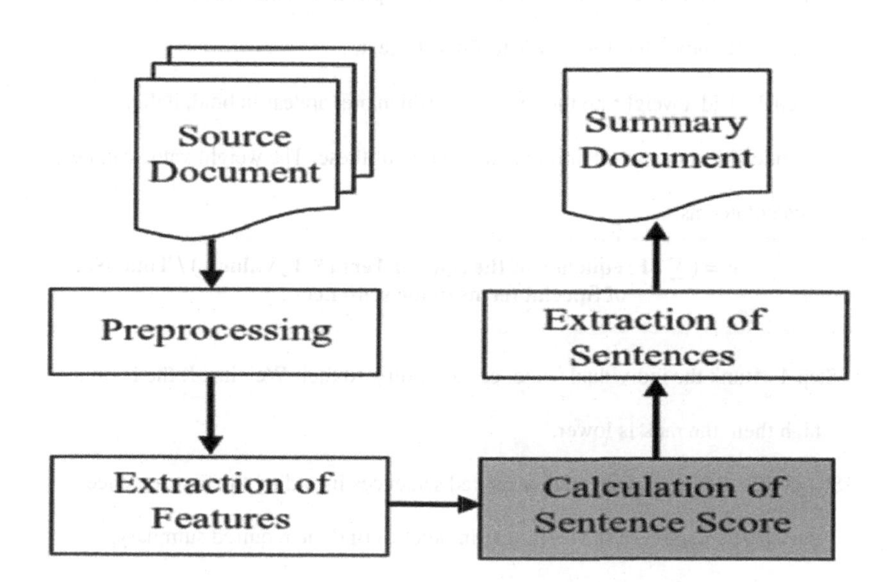

Fig. 1. Text summarization based on Features

The summarized data will be obtained by using an algorithm which is combination of both font based and cue-phrase technique. Initially the document undergoes preprocessing techniques then we apply proposed technique on the obtained data after preprocessing.

2.1 Threshold Value Calculation

A value beyond which there is a change in the manner is called Threshold value. Where T_Value is taken as follows:

- For F b, F I, F u, F c, **T_Value=1**
- For F b – F I , F I – F u , F b – F u , F b –F c , F I -F c ,F u - Fc **T_Value=2**
- For F b –F u –F c, F b – F I – F c , F I - F u – F c **T_Value=3**
- For F b – F I - Fu- F c **T_Value =4**

2.2 Algorithm

Input: A text in .txt or .rtf format.

Output: A relevant summarized text which is shorter than the original text

Step 1: Read a text in .txt or .rtf format and split it into individual tokens.

Step 2: Remove the stop words to filter the text.

Step 3: Add a weight to the sentences which are appear in bold, italic, underlined, cue word or any combination of these. The weight value can be calculated as:

F = (∑ (Frequency of the Special Term * T_Value)) / Total No. of Special terms in the sentence

Step 4 : Rank the individual sentences according to their Weights. If the F value is high then the rank is lower.

Step 5: Finally, extract the higher ranked sentences including the first sentence of the first paragraph of the input text in order to find the required summary.

3 Results and Discussion

We had tested our algorithm by passing 20 documents; we got summarized data with different accuracy. The average accuracy we got is 75%. The graph is shown below:

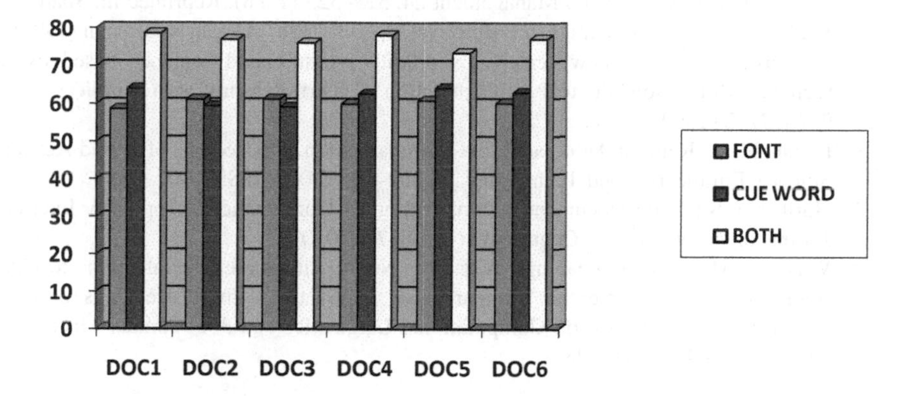

Fig. 2. Summarization Accuracy Graph

4 Conclusion and Future Work

We had presented an algorithm which produces a summarized data using combination of font based and cue- phrase algorithm. The algorithm produces an efficient summarized data when compared to word pruning and using individual algorithms of above mentioned. Finally we could produce accurate summarized data using our proposed methodology.

As our proposed algorithm is combination of above mentioned two techniques, so we could produce able to produce summarized data with 75% accuracy.so future work is to generate summary using abstractive method and we can also use all the features which are mentioned above to generate a summary using extractive approach to get 100% accurate summarized data .

References

1. Agarwal, R.: Semantic features extraction from technical texts with limited human intervention. Ph.D Thesis, Mississippi State Univ, U.S. (1995)
2. D' Avanzo, E., Magnini, B., Vallin, A.: ITC-irst. Keyphrase Extraction for Summarization purposes: The LAKE system at DUC-2004. In: Document Understanding Conference (2004)
3. Edmundson, H.P.: New methods in automatic extracting. Journal of the Association for Computing Machinery 16(2), 264–285 (1969)

4. Kruengkrai, C., Jaruskulchai, C.: Generic Text Summarization Using Local and Global Properties. In: Proceedings of the IEEE/WIC International Conference on Web Intelligence (2003)
5. Jezek, K., Steinberger, J.: Automatic Text Summarization (the state of the art 2007 and new challenges). In: Znalosti 2008, pp. 1–12 (2008)
6. Salton, G., Buckley, C.: Term-weighting approaches in automatic text retrieval. Information Processing and Management 24, 513–523 (1988); Reprinted in: Sparck-Jones, K., Willet, P. (eds.) Readings in I. Retrieval, pp. 323–328. Morgan Kaufmann (1997)
7. Aliguliyev, R.M.: A new sentence similarity measure and sentence based extractive technique for automatic text summarization. Expert Systems with Applications 36(4), 7764–7772 (2009)
8. Fattah, M.A., Ren, F.: Automatic Text Summarization. Proceedings of World Academy of Science, Engineering and Technology 27, 192–195 (2008) ISSN 1307- 6884
9. Hariharan, S.: Multi Document Summarization by Combinational Approach. International Journal of Computational Cognition 8(4), 68–74 (2010)
10. Wasson, M.: Using Leading Text for News Summaries: Evaluation results and implications for commercial summarization applications. In: Proceedings of the 17th International Conference on Computational Linguistics and 36th Annual Meeting of the ACL, pp. 1364–1368 (1998)

A Novel Feature Selection and Attribute Reduction Based on Hybrid IG-RS Approach

Leena H. Patil and Mohammed Atique

Department of Computer Science and Engineering,
Sant Gadge Baba Amravati University, Amravati, India
harshleena23@rediffmail.com,
mohd.atique@gmail.com

Abstract. Document preprocessing and Feature selection are the major problem in the field of data mining, machine learning and pattern recognition. Feature Subset Selection becomes an important preprocessing part in the area of data mining. Hence, to reduce the dimensionality of the feature space, and to improve the performance, document preprocessing, feature selection and attribute reduction becomes an important parameter. To overcome the problem of document preprocessing, feature selection and attribute reduction, a theoretic framework based on hybrid Information gain-rough set (IG-RS) model is proposed. In this paper, firstly the document preprocessing is prepared; secondly an information gain is used to rank the importance of the feature. In the third stage a neighborhood rough set model is used to evaluate the lower and upper approximation value. In the fourth stage an attribute reduction algorithm based on rough set model is proposed. Experimental results show that the hybrid IG-RS model based method is more flexible to deal with documents.

Keywords: Document Preprocessing, Feature Selection, Information Gain, Roughset.

1 Introduction

Now a day the number of text document on the internet is increasing tremendously. To deal the large amount of data, data mining becomes an important technology. Text documents are growing rapidly due to the increasing amount of information available in electronic and digitized form [1][2], such as electronic publications, various kinds of electronic documents, e-mail, and the World Wide Web. Document Preprocessing and feature selection is very useful tool in today's world where large amount of documents and information are stored and retrieved electronically. To solve this problem of dimensionality different technique such as document preprocessing, feature selection and attribute reduction approaches are used. Document preprocessing is a process that extracts a set of new terms [3][4] from the original document/ terms into some distinct key term set. Feature selection process a subset which selects from the original set based some criteria of feature importance. In a

S.C. Satapathy et al. (eds.), *Emerging ICT for Bridging the Future – Volume 2,*
Advances in Intelligent Systems and Computing 338, DOI: 10.1007/978-3-319-13731-5_59

wide range of text categorization many feature selection methods are used. [14] Proposed information gain is the most effective method compared to other five feature selection methods [16] such as IG, term strength, mutual information, X^2 statistic, document frequency.

Feature Selection, called as attribute reduction is a major problem in the field of data mining, pattern recognition and machine learning. Features may be relevant or irrelevant; it may have different discriminatory or predictive power. To apply measure to calculate uncertainty from fuzzy approximation spaces and [14][15] used it to reduce heterogeneous data. However, it becomes time consuming to generate a fuzzy equivalent relation. Therefore to be more effective a hybrid method information gain and Neighborhood Rough set model has been proposed.

2 Document Preprocessing

To Organize and browse thousands of documents smoothly, document preprocessing becomes a most important step, which affects on the result [5] where thousands of words are present in a document set; the aim of this is to reduce dimensionality for having the better accuracy for classification. Document preprocessing is divided into following stages:

1. Each sentences gets divided into terms
2. Stop words removal
3. Word Stemming
4. WordNet Senses
5. Global Unique words and frequent word set gets generated.

3 Feature Selection

Feature selection becomes an important task in machine learning, pattern recognition and data mining. It focuses on most important relevant features instead of irrelevant features [7] which makes more difficult in knowledge discovery process. Feature subset selection finds an optimal subset feature of a database based on some criteria, so that an efficient classifier with a highest accuracy can be generated. For text categorization [13][14] compared other five feature selection method which includes Information Gain (IG), X2 statistic document frequency, term strength, and mutual information. [15[16] Proposed that IG is the most effective method as compared to other feature selection method.

3.1 Feature Ranking with Information Gain

Information Gain is used as a significance measures based on entropy. For feature selection information gain is used which constitutes a filter approach. Information gain is an attribute selection measure and is based on entropy. Feature selection depends on the IG value of the feature; it also determines which feature to be selected. In machine learning field information gain is the most popular feature selection method. The information gain of a given feature $t_k$with respect to the class c_i is the reduction in uncertainty about the value of c_i when we know the value of t_k Information gain of a feature tk toward a category c_i is labeled as follows

$$IG\ (t_k, c_i) = \sum_{c\in\{c_i, c_i\}} \sum_{t\in\{t_k, t_k\}} P\ (t, c) \log \frac{p(t,c)}{p(t)p(c)}$$

Where $p(c)$ is the probability that category c occurs $p(t,c)$ is the probability that documents in the category c contains the word t. $p(t)$ is the probability that the term t occurs. The larger information gain of a feature owns the more important the feature is for categorization using WordNet.

3.2 Rough Set Theory

In this section we review several basic concepts in rough set theory and attribute reduction. Throughout this paper we suppose that the universe data used is denoted by information system $(IS) = < U, A >$, where U is a non empty and finite set of samples $\{x_1, x_2, x_3, .., x_n\}$ called an universe. A is a set of attributes to characterize the samples. $< U, A >$ is called as decision table. If $A = C \cup D$, where C is an condition attribute and D is an decision attribute. For Example given an arbitrary variable xi ∈ U and B $\subseteq$ C, the neighborhood $\delta_B\ (x_i)$ of x_i in feature space B is defined as

$\delta_B\ (x_i) = \{x_j | x_j \in U,\ \Delta^B\ (x_i, x_j) \leq \delta\}$, where Δ is a distance function. For $\forall_{x1,x2,x3} \in U$, it usually satisfies:

1. $\Delta\ (x_1, x_2) \geq 0, \Delta\ (x_1, x_2) = 0$ if and only if x1=x2;
2. $\Delta(x_1, x_2) = \Delta\ (x_2, x_1)$;
3. $\Delta(x_1, x_3) \leq \Delta\ (x_1, x_2) + \Delta\ (x_2, x_3)$; The three different metric distance functions are most widely used in machine learning and pattern recognition. A general metric names minkowsky distance can also be defined.

If the set of objects and the neighborhood relation N over U is called as neighborhood approximation space. For any $\boldsymbol{X} \subseteq \boldsymbol{U}$, two objects called lower and upper approximation of x in $< \boldsymbol{U}, \boldsymbol{N} >$ are defined as

$\underline{NX} = \{x_i \mid \delta\ (x_i) \subseteq X, x_i \in U\}$,

$\overline{NX} = \{x_i \mid \delta\ (x_i) \cap X \neq \emptyset, x_i \in U\}$

Obviously $\underline{NX} \subseteq X \subseteq \overline{NX}$. The boundary region of X in the approximation space is defined as $BNX = \overline{NX} - \underline{NX}$. The size of the boundary effects on the degree of roughness of X in the approximation space$< \boldsymbol{U}, \boldsymbol{N} >$. The size of the boundary region depends on attribute X to hold U and threshold $\boldsymbol{\delta}$.

3.3 Decision System Based on Rough Set

An neighborhood information system also called a neighborhood decision system denoted by $NDT = < U, C\ U\ D, N >$, if there are two kinds of attribute: condition and decision, and there at least exists a condition attribute for the neighborhood relation.

Definition 1: Consider the neighborhood decision system $NDT = < U, C\ U\ D, N >$, $x_1, x_2, .., x_n$ are the objects with decisions 1 to N, $\delta_B(x_i)$ is the neighborhood information granule generated by attribute $B \subseteq C$, the lower and upper approximations of decision D with respect to attribute B defined as

$$\underline{N_B}\,D = \cup_{i=1}^{N} \underline{N_B} X_i\ ,\ \overline{N_B} D = \cup_{i=1}^{N} \overline{N_B} X_i \text{ Where } \underline{N_B}\,X = \{x_i | \delta_B\,(xi) \subseteq X\,, xi\ \in U\}$$

$$\overline{N_B}\,X = \{x_i | \delta_B\,(x_i) \cap X \neq \emptyset\,, x_i\ \in U\}$$

The decision boundary region of D with respect to attribute B is defined as

$$BN\,(D) = \overline{N_B}\,D - \underline{N_B}\,D$$

Definition 2: Given any subset $X \subseteq U$ in neighborhood approximation space <U, A, N, we define variable precision Lower and upper approximation of X as

$\underline{N^k}\,X = \{x_i | I\,(\delta\,(x_i), X) \geq k, x_i\ \in U\}, \underline{N^k}\,X = \{x_i | I(\,\delta\,(x_i), X) \geq\ 1 - k, x_i\ \in U\}$, Where $1 \geq K \geq 0.5$.

Definition 3: Given the $NDT =< U, C\ \cup\ D, N >$, the distance function Δ and neighborhood size δ, the dependency degree of D to B is defined as $\gamma_B D = \frac{|POS_B\,(D)|}{|U|}$ Where $|\cdot|$ is the cardinality of a set. $\gamma_B(D)$ is the ability of B to approximate D. As $POS_B\,(D) \subseteq\ U$, we have $0 \leq\ \gamma_B(D) \leq 1$. we say D completely depends on B and the decision system is consistent in terms of Δ andδ. If $\gamma_B(D) =\ 1$; otherwise, it can be D depends on B in the degree of γ.

Definition 4: Given a neighborhood decision system $NDS =< U, C\ \cup\ D, N >, B \subseteq C, \forall\,a\ \in\ B$, it defines the significance of a in B as $Sig_1\,(a, B, D) = \gamma_B(D) - \gamma_{B-a}(D)$ The significance of an attribute is based on three variables: a, B, and D. The above definition is for backward feature selection. In this redundancy features are eliminated from original feature one by one. The significance measure for forward selection is $Sig_2\,(a, B, D) = \gamma_{BUa}(D) - \gamma_B(D)\ \ \forall\,a\ \in A - B$ As $0 \leq \gamma_B(D) \leq 1$ and $\forall\,a\ \in B:\ \gamma_B(D) \geq\ \gamma_{B-a}(D)$, we have $0 \leq\ Sig_1(a, B, D) \leq 1, 0\ \leq\ Sig_2(\,a, B, D) \leq 1$. As attribute a is superfluous in B with respect to D if $Sig_1\,(a, B, D) =\ 0$, otherwise, a is indispensible in B. With these proposed measure a forward greedy search algorithm for attribute reduction based on rough set theory has been proposed as follow

From the Algorithm 1 it shows that the positive region of decision becomes monotonous with the attribute. So to increase the speed a fast forward algorithm has been proposed. An algorithm named fast forward algorithm has been proposed and explained as follow.

Algorithm 1. Forward greedy search algorithm for attribute reduction based on rough set theory.

Input: 1. <U, C U D, F> 2. Delta 3. Size of the neighborhood

Output: reduct R

1. $\emptyset \rightarrow R$; //R contains all selected attributes
2. For each $a_i \in C - R$
3. Compute $\gamma_{R \cup a_i}(D) = \frac{|POS_{B \cup a_i}(D)|}{|U|}$
4. Compute $SIG(a_i, R, D) = \gamma_{R \cup a}(D) - \gamma_R(D)$
5. End. Consider the attribute a_k satisfying $SIG(a_k, R, D) = \max_i (SIG(a_i, R, D))$
6. If $SIG(a_k, R, D) > €$, where € is the positive number used for convergence.
7. $R \cup a_k \rightarrow R$ Goto step 2 Else Return R
8. End if

Algorithm 2. Fast forward attribute reduction based on rough set.

Input: <U,C U D> Delta, the size of the neighborhood

Output: reduct R

1. $\emptyset \rightarrow R, U \rightarrow S$; R is the selected attributes and S is the set of samples out of positive region.
2. While $S \neq \emptyset$
3. for each $a_i \in A - R$ Generate an decision table $DT_i = < U, R \cup a_i, D >$
4 $\emptyset \rightarrow POS_i$ for each $O_j \in S$ compute delta in NDT
5. if delta belongs to Xk $POS_i \cup O_j \rightarrow POS_i$
6. End if
7. End for
8. Find a_k If $POS_k \neq \emptyset$
9. $R \cup a_k \rightarrow R$ $S \rightarrow POS_k \rightarrow S$
10. Else
16. Exit While
17. End if
18. End While
19. return R

20 End

To consider discrete subsets an algorithm is modified as:

Algorithm 3. Fast forward discrete attribute reduction based on rough set.

Input: <U,C U D>, Delta, the size of the neighborhood
Output: reduct R
1. $\emptyset \rightarrow R, U \rightarrow S$; R is the selected attributes and S is the set of samples out of positive region.
2. While $S \neq \emptyset$
3. for each $a_i \in A - R$
4. Generate an decision table $DT_i = < S, R \cup a_i, D > \quad \emptyset \rightarrow POS_i$
5. for each $O_j \in S$ compute delta in NDT
6. if delta belongs to Xk $POS_i \cup O_j \rightarrow POS_i$
7. End if
8. End for
9. Find a_k If $POS_k \neq \emptyset$
10. $R \cup a_k \rightarrow R \quad S \rightarrow POS_k \rightarrow S$
11. Else
12. Exit While
13. End if
14. End While
15. Return R
16. End

4 Experimental Analysis

The data sets used in the experiments is outlined in Table 1. All the experiments have been carried out on a personal computer with Windows 7, Inter(R) Core (TM) i7 CPU (2.66 GHz) and 4.00 GB memory. The software used is MATLAB R2010b. .

Table 1. The general description of Reuters 21578 dataset

Data set	Feature samples	Numerical attributes	Class
Reuters 21578	1328	24818	04

Table 2 shows the number of selected features where the information gain feature selection technique has applied.

Table 2. Feature reduced based on Information Gain

Data set	Feature samples	Numerical attributes	Information Gain	% of features reduced
Reuters 21578	1328	24818	9167	36.93

Table 3. shows the number of feature selected for Rough set by considering the delta interval from [0.001, 0.015,0.01]

Table 3. Number of Features Selected for Rough set

Data set	Algorithm	No. of Feature samples	Numerical attributes	RS		
				δ=0.001	δ =0.015	δ=0.01
Reuters 21578	Algorithm 1	1328	24818	497	483	477
	Algorithm 2	1328	24818	523	517	509
	Algorithm 3	1328	24818	538	524	518

Figure.1 presents the number of features selected for rough set approach.

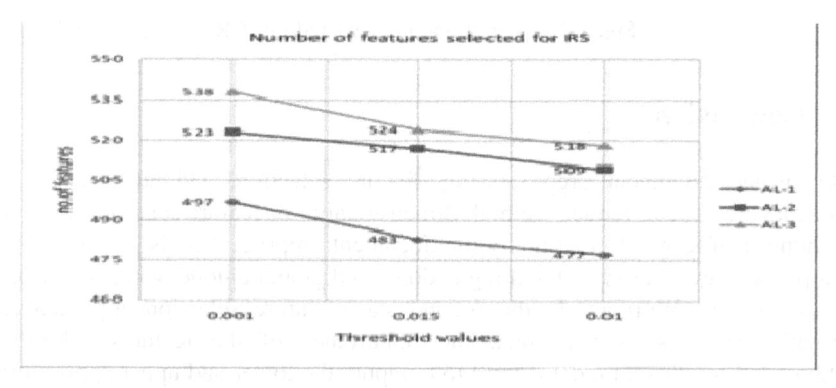

Fig. 1. Number of Features selected for Rough set

Table 4 shows the number of features selected using IG-RS technique for delta interval of [0.001, 0.5].

Table 4. No. of Features selected for IGRS

Data set	Algorithm	No. of Feature samples	Numerical attributes	IGRS		
				δ =0.001	δ =0.015	δ =0.01
Reuters 21578	Algorithm 1	1328	24818	115	110	98
	Algorithm 2	1328	24818	117	114	109
	Algorithm 3	1328	24818	121	117	111

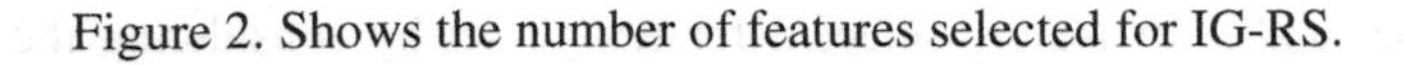

Figure 2. Shows the number of features selected for IG-RS.

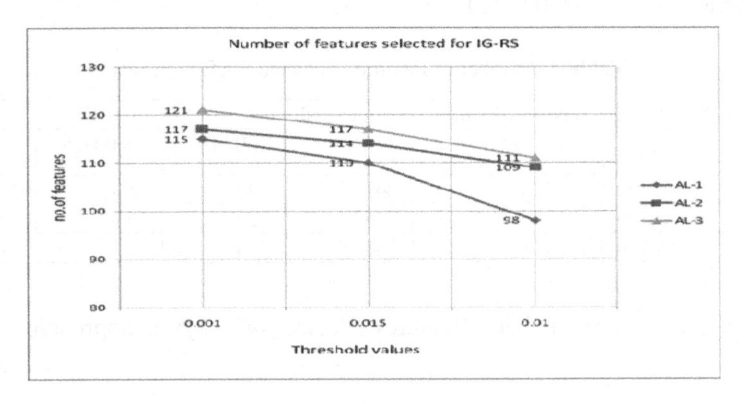

Fig. 2. Number of features selected for IGRS

5 Conclusion

In this paper, document preprocessing, feature selection, and attribute reduction approaches are used to reduce the high dimensionality of feature space composing the large number of terms. In this firstly the document preprocessing is performed where the stop words are removed, stemming is done, and global unique words are generated with the help of Wordnet. In the second stage feature selection approach called information gain is used to rank the importance of the features. Thirdly, an neighborhood rough set model is used to compute the lower and upper approximation value. Lastly attribute reduction algorithms are applied on rough set approach. In this a hybrid approach IG-RS shows the superior performance for attribute reduction feature selection.

References

[1] Steinbach, M., Karypis, G., Kumar, V.: A comparison of document clustering techniques. In: Proc. of the 6th ACM SIGKDD Int'l Conf. on Knowledge Discovery and Data Mining, KDD (2000)

[2] Fung, B., Wang, K., Ester, M.: Hierarchical document clustering using frequent item sets. In: Proc. of SIAM Int'l Conf. on Data Mining, SDM, pp. 59–70 (May 2003)

[3] Beil, F., Ester, M., Xu, X.: Frequent term-based text clustering. In: Proc. of Int'l Conf. on Knowledge Discovery and Data Mining, KDD 2002, pp. 436–442 (2002)

[4] Chen, C.L., Tseng, F.S.C., Liang, T.: An integration of fuzzy association rules and WordNet for document clustering. In: Proc. of the 13th Pacific-Asia Conference on Knowledge Discovery and Data Mining, pp. 147–159 (2009)

[5] Chen, C.-L., Tseng, F.S.C., Liang, T.: An integration of WordNet and fuzzy association rule mining for multi-label document clustering. Data & Knowledge Engineering 69, 1208–1226 (2010)

[6] Chen, C.-L., Tseng, F.S.C., Liang, T.: Mining fuzzy frequent itemsets for hierarchical document clustering. Information Processing and Management 46, 193–211 (2010)
[7] Yang, J., Liu, Y., Zhu, X., Liu, Z., Zhang, X.: A new feature selection based on comprehensive measurement both in inter-category and intra-category for text categorization. Information Processing and Management (2012)
[8] Xu, Y., Wang, B., Li, J.-T., Jing, H.: An Extended Document Frequency Metric for Feature Selection in Text Categorization. In: Li, H., Liu, T., Ma, W.-Y., Sakai, T., Wong, K.-F., Zhou, G. (eds.) AIRS 2008. LNCS, vol. 4993, pp. 71–82. Springer, Heidelberg (2008)
[9] Yang, Y., Pedersen, J.O.: A comparative study on feature selection in text categorization. In: Proceedings of the 14th International Conference on Machine Learning, pp. 412–420 (1997)
[10] Pawlak, Z., Skowron, A.: Rough Sets: Some Extensions. Information Sciences 177, 28–40 (2007)
[11] Jensen, R., Shen, Q.: Semantics-preserving dimensionality reduction: rough and fuzzy-rough-based approaches. IEEE Transactions of Knowledge and Data Engineering 16, 1457–1471 (2004)
[12] Jensen, R., Shen, Q.: Fuzzy-rough sets assisted attribute selection. IEEE Transactions on Fuzzy Systems 15(1), 73–89 (2007)
[13] Uğuz, H.: A two-stage feature selection method for text categorization by using information gain, principal component analysis and genetic algorithm. Knowledge-Based Systems 24, 1024–1032 (2011)
[14] Hu, Q., Yu, D., Liu, J., Wu, C.: Neighborhood rough set based heterogeneous feature subset selection. Information Sciences 178, 3577–3594 (2008)
[15] Wang, H.: Nearest neighbors by neighborhood counting. IEEE Transactions on PAMI 28, 942–953 (2006)

A Novel D&C Approach for Efficient Fuzzy Unsupervised Classification for Mixed Variety of Data

Rohit Rastogi, Saumya Agarwal, Palak Sharma, Uarvarshi Kaul, and Shilpi Jain

CSE-Dept-ABES Engg. College, Ghaziabad (U.P.), India
rohit.rastogi@abes.ac.in,
{som.roxs,saniakaul27,shilpijain474}@gmail.com,
sharma.palak595@ymail.com

Abstract. The Clustering or unsupervised classification has variety of requirements in which the major one is the capability of the chosen clustering approach to deal with scalability and to handle with the mixed variety of data set. Data sets are of many types like categorical/nominal, ordinal, binary (symmetric or asymmetric), ratio and interval scaled variables.

The present scenario of Variety of latest approaches of unsupervised classification are Swarm Optimization based, Customer Segmentation based, Soft Computing methods like GA based, Entropy based and Fuzzy based methods and hierarchical approaches have two serious bottlenecks…Either they are hybrid mathematical techniques or large computation demanding which increases their complexity and hence compromises with accuracy.

The proposed methodology deals with this problem with a newly and efficiently generated algorithm and shapes a better, lesser complex and computationally demanding pseudo codes which may lead in future a revolutionary approach. We work upon multivariate data set. In case of nominal variables, we quantify the dataset by different methods to construct combined category quantifications and plot the object scores.

Here we have proposed an iterative procedure to calculate the cluster centers and the object memberships. To support our approach, a numerical experiment has been demonstrated. For all mixed variety of attributes Binary, Interval and Ratio scaled along with ordinal type attributes, an efficient methodology by Divide and Conquer (D&C) approach has been designed and has been given in this write up.

We will separately calculate the grouping criteria and objective function and will sum up them to get combined function. The better groups may be obtained my maximizing this function.

Keywords: Data mining, clustering, mixed type attributes, k-means, Fuzzy, Correspondence.

1 Introduction

[4] A cluster is a collection of objects which are "similar" among them and are "dissimilar" to the objects belonging to other clusters. In segmentation we attempt to

S.C. Satapathy et al. (eds.), *Emerging ICT for Bridging the Future – Volume 2*,
Advances in Intelligent Systems and Computing 338, DOI: 10.1007/978-3-319-13731-5_60

determine which components of a data set naturally belong together. This is the key idea of clustering. A loose definition of clustering could be "the process of organizing objects into groups whose members are similar in some way". A good clustering method produces high quality clusters with high intra class similarity and low interclass similarity.

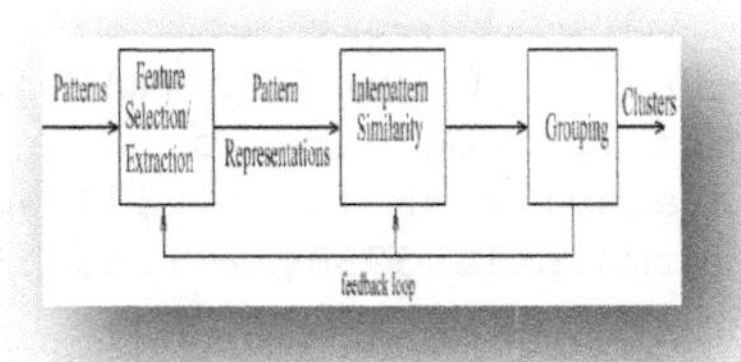

Fig. 1. Stages in Clustering

Fuzzy clustering methods are among today evolving methods but it has also an additional drawback that it is less computationally sensitive for mixed variety of dataset. It works well with high dimensionality and scalability for large categorical dataset but when all the Data sets of many types come in picture for a real life application like volcanic eruption or earthquake or demographic mapping and allocation in city planning with spatial dataset, it comes under trouble.

To measure the Quality of Clustering, we generally calculate the dissimilarity/similarity metrics where similarity is expressed in terms of a distance function, which is typically metric: $d(i, j)$.

In this paper, we have tried to find out a concept to introduce a fuzzy criterion to measure the heterogeneity of clusters. The paper organization is as follows. Section 2 contains the brief description of the related works. Section 3 describes basic framework of the proposed idea and illustrates the basic differences with respect to already available other different categorical clustering algorithm and inspiration of work. Experiment and the results with discussion have been put in section 4 and the possible future scopes and concluding remarks are shown in section 5.

2 Already Existing Fuzzy Methods for Grouping and Related Works

Fuzzy Clustering can be used in Knowledge Discovery in databases (KDD) or DM tasks and it deals with categorical multivariate data. It is defined by Cross Classification table, contingency table or concurrence matrix.

2.1 Most Well Known Algorithms

Fuzzy C- means: based on minimization of objective function for fuzzy partitioning**[28]**.-**Fuzzy C-lines** (FCL)**[9]**, **Fuzzy C- varieties** (FCV)**[10]**, **Fuzzy-c regression[11]** are approaches which are based on modification of prototypes of clusters. Their features are that these have different shapes of cluster centers, prototypes or clusters. Similarity/dissimilarity criteria are conducted with distances from cluster centers to data points. We may use – Euclidian, Mahalanobis, Manhattan or Minkowski Distances.-**FCV:** clusters are shown by prototypical linear variety and clustering criteria and we have to compute sum of distance b/w data samples and prototypes, uses eigen value solution of generalized scatter matrix.-**Local Linear principal component Analysis (PCA):** It estimates the local linear models by partitioning the data set into several groups.-**Linear Fuzzy Clustering:** based on local subspace estimation**[17]** and **[18]**.-**Fuzzy C- Ellyptotypes (FCE):** Extracts ellepsoidal shape clusters**[19]** by tuning the priority of two clustering criteria.FCE is Hybrid of –FCM and FCV where FCM-Used to partition samples into spherical shape clusters by minimizing distance b/w data points and cluster centers and FCV-searches linear shape clusters**[5]**. Limitation- Not works well in mixed data sct (nominal and categorical and other types).

2.2 Data Used and Its Characteristics

Categorical variable data set can be defined by Binary Indicator, Frequency or scaled variable and correspondence analysis is used to quantify the multi categorical data to maximize the correlation among data**[8]**. It is used for dimension reduction. This technique**[6]** consists of clusters that are generated as the distance of data points and proto types having limitations that it is difficult to get accurate answers when data set consists of nominal variables also along with numerical variables. The clustering results by the proposed method show similarity to those of correspondence Analysis or Hayashi's Quantification method **[7]**.

2.3 Some Recent Inspirational Works

[13] Yamakawa et al.- proposed Hybrid of Fuzzy Clustering and Correspondence Analysis which finds the relationship among data and scatter diagram and Inoue et al. propose the Fuzzy Clustering algorithm for categorical multivariate data. Need to calculate eigen vectors and so computation is high **[14]**. Ryoke and Nakamori **[15]** proposed an agent-based clustering approach that performs data partitioning followed by the extraction of rules. In the agent-based approach, data partitioning is performed by the agents that act following their own criteria or models inside. Therefore, there is no global objective function.

[28]Fuzzy c-Means Clustering of Mixed Databases Including Numerical and Nominal Variables**[2]**.Linear Fuzzy Clustering for Mixed Databases based on Optimal Scaling**[3]**.Optimal Scaling Fuzzy clustering **[25]**,**[26]**,**[27]** has features that uses FCE and we have to find the correlation b/w the variables and their mutual dependencies, uses least square method, 2 phase technique, Model Estimation Phase and Optimal Scaling phase. Here we calculate the numerical scores for nominal data set so that they suit the FCE clustering**[23]**.

Correspondence Analysis: To discover Relations, quantifies the individuals and the categories by solving an eigen value problem**[24]**. After the quantification, we can plot the individuals and the categories on one or two dimensional space. Then, they are divided in some groups in accordance with the coordinates and characteristics of the data set are detected. between individuals and categories**[21]**.

This paper has been inspired From Fuzzy Clustering for Categorical Multivariate Data **[1]** by Chi-Hyon Oh, Katsuhiro Honda and Hidetomo Ichihashi**[22]**. Its main Features are that it is only fuzzy Clustering and calculates only simple objective function which represents degree of aggregation of each cluster.

3 The Methodology and Benefits of Our Proposed Approach

So, by above discussion, we may deduce that grouping target has certain problems and challenges which is considerable progress has been made in scalable clustering methods but current clustering techniques do not address all the requirements adequately, still an active area of research.

3.1 Our Proposed Algorithm

3.1.1 Calculation of Objective Function

It is an iterative procedure through necessary condition of local minima and we can obtain clusters for overall dataset**[20]**. Objective function can be obtained by using the Entropy maximization method defined by Miyamoto et al. After the fuzzy clustering, we are supposed to obtain memberships for individuals and categories each and all.

3.1.2 Benefits of This Approach

Looking into obtained clusters mixed up with individuals and categories, the similar result of data analysis to the correspondence analysis can be derived since our proposed method can easily provide fuzzy clusters. Solving simple algebraic equations that are far easier. The Eigen value problems and doesn't require the calculation of cluster centers. It provides us with useful way to analyze categorical multivariate data Numeric examples show the usefulness of our method.

3.2 The Methodology of Proposed Approach

The real life dataset can be represented by cross classification table, contingency table or co occurrence matrix where the rows represent the different objects or individuals and columns are categories/attributes/properties or features. In above tabular structure, M individuals described by set of qualitative variables d_{ij} with N categories. D_{ij} are collected by some questionnaire or concurrence relations or from UCI website of machine repository.

Table 1. An example of Categorical Multi-variant Data set [3]

	1	2	...	J	...	N
1	d_{11}	d_{12}	...	d_{1j}	...	d_{1N}
2	d_{21}	d_{22}	...	d_{2j}	...	d_{2n}
. . .	. . .	. . .	...	. . .	...	. . .
I	d_{i1}	d_{i2}	...	d_{ij}	...	d_{1N}
. . .	. . .	. . .	...	. . .	...	. . .
M	d_{M1}	d_{M2}	...	d_{Mj}	...	d_{MN}

3.3 EMVDFG: Efficient Multi Variate Dataset Fuzzy Grouping Algorithm

$$\sum_{c=1}^{C} u_{ci} = 1,\ u_{ci} \in [0,1], i = 1, \ldots, M \tag{1}$$

$$\sum_{j=1}^{N} (w_{cj} + w_{cr} + w_{cb} + w_{co} + w_{cn}) = 1 \tag{2}$$

$$\text{where}(w_{cj} + w_{cr} + w_{cb} + w_{co} + w_{cn}) \in [0,1], c = 1, \ldots C$$

where u_{ci} is the membership of the i-th individual for the c-th cluster and w_{cj} is that of the j-th category for the c-th cluster. C denotes the number of clusters. Though it seems that u_{ci} and w_{cj} have the same constraints since the memberships sum to one, they are different. Let w_{cb}, w_{co}, w_{cr}, w_{cn} define the respective membership of the c-th cluster to the binary, ordinal, ratio-scaled and numeric data sets in the database.

(I)For u_{ci}, the memberships of the i-th individual to the clusters in totality has to be one.

On the other hand

(II)$w_{cj}+w_{cb}+w_{co}+w_{cr}+w_{cn}$ shows the total membership of the c-th cluster to all the data types should be one.

(III) Divide and Combine then Conquer approach is applied on each of the data set and their degree of aggregation and simple objective function is separately calculated and then combined function is maximized.

Degree of Aggregation in clustering criterion of the EMVDFG.

$$\sum_{i=1}^{M} \sum_{j=1}^{N} u_{ci} (w_{cj} + w_{cr} + w_{cb} + w_{co} + w_{cn})\, d_{ij}, c = 1, \ldots, C. \tag{3}$$

If we define the total amount of memberships w_{cj}, w_{cb}, w_{co}, w_{cr}, w_{cn} of the j-th category to the clusters as one, in such a way as in above equation, we are unable to obtain proper clusters. Then the maximized degree of aggregation is obtained by putting

$$\sum_{j=1}^{C} (w_{cj} + w_{cr} + w_{cb} + w_{co} + w_{cn}) = 1 \tag{4}$$

$$.(w_{cj} + w_{cr} + w_{cb} + w_{co} + w_{cn}) \in [0,1], j = 1, \ldots N$$

3.3.1 Objective Function Using D & C Approach

The EMVDFG can be driven by optimization of an objective function to maximize the degree of aggregation. We use Lagrange's method of indeterminate multiplier to derive the objective function for the EMVDFG. The objective function can be written as follows:

$$\max L = \sum_{c=1}^{C} \sum_{i=1}^{M} \sum_{j=1}^{N} u_{ci}(w_{cj} + w_{cr} + w_{cb} + w_{co} + w_{cn})\, d_{ij} - T_u \sum_{c=1}^{C} \sum_{i=1}^{M} u_{ci} \log u_{ci} - T_w \sum_{c=1}^{C} \sum_{j=1}^{N} (w_{cj} + w_{cr} + w_{cb} + w_{co} + w_{cn}) \log(w_{cj} + w_{cr} + w_{cb} + w_{co} + w_{cn}) + \sum_{i=1}^{M} \lambda_i \left(\sum_{c=1}^{C} u_{ci} - 1\right) + \sum_{c=1}^{C} \gamma_c \left(\sum_{j=1}^{N} (w_{cj} + w_{cr} + w_{cb} + w_{co} + w_{cn}) - 1\right) \tag{5}$$

where Lamda and Gamma are Lagrangian multipliers respectively. The second and third terms in (5) represent entropy maximization as a regularization which was introduced in Fuzzy c-Means by Miyamoto et al.**[28]** for the first time. It enables us to obtain fuzzy clusters. Tu and Tw are the weighting parameters which specify the degree of fuzziness**[12]**. The remaining terms describe the constraints of memberships. From the necessary conditions for the optimality of the objective function L, i.e., Del L/Del $u_{ci} = 0$ and Del L/ Del $(w_{cj} + w_{cb} + w_{co} + w_{cr} + w_{cn}) = 0$,

So, we have the following equations.

$$(u_{ci}) = \frac{\exp\left(\sum_{j=1}^{N}\left((w_{cj} + w_{cr} + w_{cb} + w_{co} + w_{cn}) d_{ij}/T_u\right)\right.}{\sum_{c=1}^{C} \exp\left(\sum_{j=1}^{N}\left((w_{cj} + w_{cr} + w_{cb} + w_{co} + w_{cn}) d_{ij}/T_u\right)\right.} \tag{6}$$

$$(w_{cj} + w_{cr} + w_{cb} + w_{co} + w_{cn}) = \frac{\exp\left(\sum_{i=1}^{M} (u_{ci} d_{ij}/T_w)\right.}{\sum_{j=1}^{N} \exp\left(\sum_{i=1}^{M} (u_{ci} d_{ij}/T_w)\right.} \tag{7}$$

The optimization algorithm is based on Picard iteration through necessary conditions for local minima of the objective function.

3.3.2 The EMVDFG Algorithm

- Set values of parameters C, Tu, Tw and Epsilon.
- Randomly initialize the memberships u_{ci}.
- Calculate the Objective Function using D&C Approach means separately get the objective function for different data sets and combine them to get the overall function.
- Maximize the Objective function by First Differential Calculus.
- Update membership w_{cj} .
- Update memberships u_{ci} using equation prev. given.
- If max $| u_{ci}^{New} - u_{ci}^{Old} | <$ epsilon, then stop.
- Otherwise, return to Step 2.

4 Experiments and Results and Discussion

Numerical Example-Lets have a Shelf-Book Data Set

- We have used the following values of parameters for the EMVDFG.
- The number of clusters C: 2
- The degree of fuzziness Tu: 0.1
- The degree of fuzziness Tw : 1.5
- Stopping condition of the EMVDFG E (epsilon)= 0.0001

Table 2. Membership of Shelves

Shelf	Cluster1	Cluster2
1	0.338	<u>0.662</u>
2	0.011	<u>0.989</u>
3	0.011	<u>0.989</u>
4	0.002	<u>0.998</u>
5	0.141	<u>0.859</u>
6	<u>0.894</u>	0.106
7	<u>0.988</u>	0.012
8	<u>0.996</u>	0.004
9	<u>0.973</u>	0.027

Table 3. Membership of Books

Books	Cluster1	Cluster2
1	0.044	<u>0.066</u>
2	0.044	<u>0.066</u>
3	0.044	<u>0.066</u>
4	0.039	<u>0.146</u>
5	0.035	<u>0.311</u>
6	0.038	<u>0.075</u>
7	0.038	<u>0.075</u>
8	0.035	<u>0.083</u>
9	<u>0.067</u>	0.043
10	<u>0.237</u>	0.024
11	<u>0.250</u>	0.023
12	<u>0.129</u>	0.022

4.1 Result Explanation

In Table 3 and Table 4(shown above), we underlined larger memberships of Shelves and Books. We assume that shelves and books are more likely to belong to the cluster to which they have larger memberships. From Table 3, we can see that shelves are divided into { 1,2,3,4,5} an d {6,7,8,9}. On the one hand, books are partitioned into {1,2,3,4,5,6,7,8), and (9, 10,11,12}. These results are reasonable in accordance with Table 2.Figure 1(coming next) shows the result of the Correspondence Analysis applied to Table 2 and represents the scatter diagram of shelves and books after quantification.

Table 4. Shelf-Book Retrieval dataset

B Vs S	B 1	B 2	B 3	B 4	B 5	B 6	B 7	B 8	B 9	B 10	B 11	B 12
S1	1	1	1	0	0	0	0	0	0	0	0	0
S2	0	0	1	1	1	1	1	0	1	0	0	0
S3	0	1	0	1	1	0	0	1	0	0	0	0
S4	1	0	0	0	2	0	0	1	0	0	0	0
S5	0	0	0	1	0	1	1	0	0	0	0	0
S6	0	0	0	0	0	0	0	0	0	1	0	0
S7	0	0	0	0	0	0	0	0	0	1	1	0
S8	0	0	0	0	0	0	0	0	0	1	1	1
S9	0	0	0	0	0	0	0	0	1	0	1	1

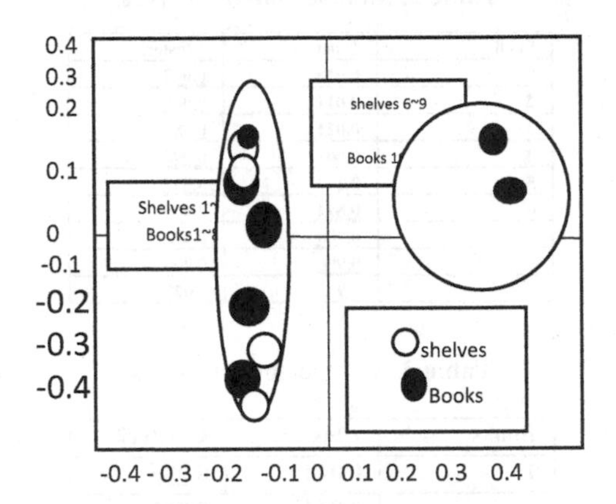

Fig. 2. The Result of the Correspondence Analysis

The values corresponding to the first and second eigen values were plotted on the diagram. The horizontal axis corresponds to the first eigen value and the vertical axis does the second one. In Figure 1, ■ and 0 and indicate shelf and books respectively. We can divide shelves and books into two groups according to the observation of Figure 1.One is shelves (1,2,3,4,5} and books {12,3,4,5,6,7,8 } and the other is shelves {6,7,8,9} and books {10,11,12}. Only the book 9 belongs to neither group. The two groups are circled in Figure 1. Comparing the result of the Correspondence. Analysis with that of the proposed method,we can observe that the similar results are obtained except for the book 9.

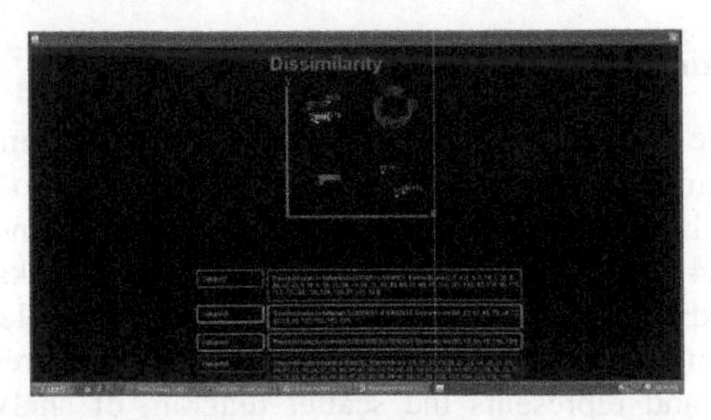

Fig. 3. The Resultant colored graphical representation of the input elements as per the given thresholds values and their pictorial representation

5 Conclusions

This paper proposed a new approach to the FCE clustering that can handle nominal variables based on the optimal scaling approach. Nominal variables are transformed into numerical scores in each iteration step so that they suit the current fuzzy partition. Then, cluster prototypes and fuzzy memberships are estimated using only numerical data. So, it can be said that the proposed method performs the FCE clustering in a single numerical data space using numerical scores. By the way, the alternate least squares methods with optimal scaling have been proven to be monotone-convergent. The study on the convergence property of the proposed method is remained in future work.

- The EMVDFG was applied to the Shelf-Book retrieval dataset which was a kind of category.
- Correspondence Analysis was also applied to the data set and compared with the EMVDFG.
- The EMVDFG showed the similar result to that of the Correspondence Analysis.
- While the Correspondence Analysis requires solving eigen value problem which is computationally demanding, the EMVDFG needs simple algebraic calculations.
- Therefore, we can conclude that the EMVDFG is not only a fuzzy clustering algorithm handling categorical multivariate data but also a simple alternative of the Correspondent Analysis.

6 Future Scope and Possible Applications of The Research Work

In Divide and Conquer approach, the method can be extended accurately for other type of mixed variety of data like continuous, ordinal, discrete categorical or exponentially distributed ratio-scaled data set in real life applications. The future of data mining lies in predictive analytics and probability can be an important tool for it. From the time of Darwin, predictive analysis is used for consolidation and stability.

In Backtracking based approach, the algorithm can be practically implemented for large variety of data. In future, the outliers can be treated as seed value point of new clusters so that dynamism can be more improved upon and accuracy is ensured.

In Backtracking based approach with k-medoids/k-modes concept, the method can be extended accurately for other type of mixed variety of data like continuous, ordinal, discrete categorical or exponentially distributed ratio-scaled data set in real life applications. Therefore, data partitioning is performed in a numerical data space minimizing a single objective function. Considering the similarity between the proposed objective function of Eq.(5) and the least squares criterion for linear fuzzy clustering, the proposed method can be extended to the linear fuzzy clustering of mixed data set.

Acknowledgment. This work was supported in part by the National Institute of Technology and Technical Teacher's Training and Research, Chandigarh, affiliated by MHRD, Govt. of India.

The authors would like to thank the reviewers for their valuable suggestions. They would also like to thank HOD-CSE, Prof. Dr. R. Radhakrishnan and HOD-IT, Prof. A.K. Sinha along with Prof. Jagdish Singh for their motivation and support in early stage of the paper. Last but definitely not the least we would thank the Almighty God without whose grace this paper would not have achieved success.

References

1. Oh, C.-H., Honda, K., Ichihashi, H.: Fuzzy Clustering for Categorical Multivariate Data, pp. 2154–2159, 0-7803-7078-3/0u$l0.0(0C)u)O lEEE
2. Honda, K.: Fuzzy c-Means Clustering of Mixed Databases Including Numerical and Nominal Variables. In: Proceedings of the 2004 IEEE Conference on Cybernetics and Intelligent Systems, Singapore, December 1-3, pp. 558–562. IEEE (2004) 0-7803-86434/04/$20.00 Q 2004 IEEE
3. Uesugi, R., Honda, K., Ichihashi, H.: Linear Fuzzy Clustering for Mixed Databases Based on Optimal Scaling. In: 2006 IEEE International Conference on Fuzzy Systems Sheraton Vancouver Wall Centre Hotel, Vancouver, BC, Canada, Vancouver, BC, Canada, July 16-21, pp. 778–782 (2006) 0-7803-9489-5/06/$20.00/©2006 IEEE
4. Han, J., Kamber, M.: Data Mining Concepts and Technique
5. Adriaans, P., Zantinge, D.: Data Mining. Addison Wesley Longman (1996)
6. Berry, M.J.A., Linoff, G.S.: Dare Mining Techniques. John Wiley & Sons (1997)
7. Tenenhaus, M., Young, E.W.: An analysis and synthesis of multiple correspondence analysis, optimal scaling, dual scaling, homogeneity analysis and other methods for quantifying categorical multivariate data. Psychornetrika 50(1), 91–119 (1985)
8. Hayashi, C.: On the prediction of phenomena from qualitative data and the quantification of qualitative data from the mathematical statistical point of view. Annals of the Insrimre of Statistical Mathematics 3, 69–98 (1952)
9. Bezdek, J.C.: Puffem recognition with fuzzy objective function algorithms. Plenum Press, New York (1981)
10. Bezdek, J.C., Coray, C., Gundenon, R., Watson, J.: Detection and characterization of cluster substructure. I. linear structure, fuzzy c-lines. SIAM J. Appl. Math. 40(2), 339–357 (1981)
11. Bezdek, J.C., Coray, C., Gundenon, R., Watson, J.: Detection and characterization of cluster substructure. II. fuzzy c-varieties and convex combinations thereof. SIAM J. Appl. Math. 40(2), 358–372 (1981)
12. Hathaway, R.J., Bezdek, J.C.: Switching regression models and fuzzy clustering. IEEE Trans. on Fuzzy System 1(3), 195–204 (1993)
13. Yamakwa, A., Kanaumi, Y., Ichihashi, H., Miyoshi, T.: Simultaneous Application of Clustering and Correspondence Analysis. In: Proc. of IJCNN 1999, Paper #625, pp. 1–6 (1999)
14. Inoue, K., Urahama, K.: Fuzzy Clustering Based on Co occurrence Mamx and Its Application to Data Retrieval. Transactions of IEICE D-If J-81-DII(12), 957–966 (2000) (in Japanese)
15. Miyamoto, S., Mukitidono, M.: Fuzzy c-means as a regularization and maximum entropy approach. In: Proc. of lFSA 1997, vol. lI, pp. 86–92 (1997)

16. Landauer, T.K., Dumais, S.T.: The latent semantic analysis theory of acquisition, induction and representation of knowledge. Psychot! Rev. 104(2), 211–240 (1997)
17. Hayashi, C.: On the prediction of phenomena from qualitative data and the quantification of qualitative dam from the mathenlatical statistical point of view. Institute of Statistical Mathematics 3, 69–98 (1952)
18. Gifi, A.: Nonlinier Multivariate analysis. Wiley (1990)
19. Bond, J., Michailidis, G.: Homogeneity analysis in Lisp-Stat. Journal of Statistical Software 1(2) (1996)
20. Whittle, P.: On principal components and least square methods of factor analysis. Skmd. Akt. 36, 223–239 (1952)
21. Honda, K., Nakamun, Y., Ichihashi, H.: Simultaneous application of fuzzy clustering and quantification with incomplete categorical data. Journal of Advanced Computational Intelligence and Intelligence Informaticst 8(4), 397–402 (2004)
22. Honda, K., Ichihashi, H.: Linear fuzzy clustering techniques with missing values and their application to local principal component analysis. IEEE Transactions on Fuzzy Systems 12(2), 183–193 (2004)
23. Oh, C.-H., Honda, K., Ichihashi, H.: Fuzzy clustering for categorical multivariate data. In: Proc. of Joint 9th IFSA world Congress and 20th Nafips International Conference, pp. 2154–2159 (2001)
24. Umayahara, K., Miyamoto, S., Nakamori, Y.: Formulations of Fuzzy Clustering of categorical data. In: Proc. of International Workshop of Fuzzy Systems and Innovational Computing, pp. 344–349 (2004)
25. Ryoke, M., Nakamori, Y.: Agent-based clustering and rule extraction. In: Proc. of SCIS & ISIS 2002, #22Q3-5 (2002)
26. Fisher, R.A.: The use of multiple measurements in taxonomic problems. Annual Eugenics 7(Pt 11), 179–188 (1936)
27. Honda, K., Sugiura, N., Ichihashi, H., Araki, S.: Collaborative filtering using principal component analysis and fuzzy clustering. In: Zhong, N., Yao, Y., Ohsuga, S., Liu, J. (eds.) WI 2001. LNCS (LNAI), vol. 2198, pp. 394–402. Springer, Heidelberg (2001)
28. Honda, K., Ichihashi, H.: Component-wise robust linear fuzzy clustering for collaborative filtering. International Journal of Approximate Reasoning 37(2), 127–144 (2004)

Automatic Tag Recommendation for Journal Abstracts Using Statistical Topic Modeling

P. Anupriya[1] and S. Karpagavalli[2]

[1] PSGR Krishnammal College for Women, Coimbatore, India
pri_una@yahoo.com
[2] GR Govindarajulu School of Applied Computer Technology, Coimbatore, India
karpagam@grgsact.com

Abstract. Topic modeling is a powerful technique for unsupervised analysis of large document collections. Topic models conceive latent topics in text using hidden random variables, and discover that structure with posterior inference. Topic models have a wide range of applications like tag recommendation, text categorization, keyword extraction and similarity search in the broad fields of text mining, information retrieval, statistical language modeling.

In this work, a dataset with 200 abstracts fall under four topics are collected from two different domain journals for tagging journal abstracts. The document model is built using LDA (Latent Dirichlet Allocation) with Collapsed Variational Bayes (CVB0) and Gibbs sampling. Then the built model is used to find appropriate tag for a given abstract. An interface is designed to extract and recommend the tag for a given abstract.

Keywords: Topic modeling, Latent Dirichlet Allocation, Gibbs sampling, Tag Recommendation.

1 Introduction

The massive amount of information stored in unstructured texts cannot simply be used for further processing by computers, which typically handle text as simple sequences of character strings. For that reason, specific pre-processing methods and algorithms are needed in order to mine useful patterns. Text mining refers to the process of extracting interesting information and knowledge from unstructured text. It aims at disclosing the concealed information with the large number of words and structures in natural.

In the most text mining tasks, the texts are represented as a set of independent units such as unigrams, bigrams or multi-grams which construct the feature space, and the text is normally represented by the assigned values such as binary, Term Frequency (TF) or Term Frequency Inverse Document Frequency (TF-IDF) [1]. In such case, most of the features occur only a few times in each context, the representation of vectors tend to be very sparse and very similar contexts may be represented by different features in the vector space. These challenges demand effective representative techniques and algorithms. Blei, Ng and Jordan et.al, [2] proposed LDA model which is a

S.C. Satapathy et al. (eds.), *Emerging ICT for Bridging the Future – Volume 2*,
Advances in Intelligent Systems and Computing 338, DOI: 10.1007/978-3-319-13731-5_61

generative probabilistic model of a corpus and the documents are represented as weighted relevancy vectors over latent topics, where a topic is characterized by a distribution over words. These topic models are a kind of hierarchical Bayesian models of a corpus [3].

2 Related Work

Tagging, in recent times emerged as a popular way to sort out large and vibrant web content. It usually refers to the action of associating with or assigning some keyword or entity to a piece of information. Tagging helps to portray an item and allows it to be found again by browsing or searching. Researchers have developed different techniques and algorithms for tagging of documents for various applications.

Griffiths et al [4] analyzed abstracts from PNAS by using Bayesian model selection to establish the number of topics and showed that the extracted topics capture meaningful structure in the data. They outlined the applications of the analysis as tagging abstracts and identifying hot and cold topics. Xiance et al. [5] proposed a scalable and real-time method for tag recommendation. They modeled document, words and tags using tag-LDA model. An evaluation on the dataset showed significant improvement over a baseline method.

Krestel et al. [6] introduced an approach to personalize tag recommendation that combines a probabilistic model of tags from the resource with tags from the user. They investigated simple language models as well as LDA and experiments on a real world dataset crawled from a big tagging system showed that personalization improves tag recommendation. Dredze, Mark, et al. [7] developed an unsupervised learning framework for selecting summary keywords from emails using latent representations of the underlying topics in a user's mailbox. They demonstrated that summary keywords generated using latent concept models using LDA served as a good approximation of message content.

In the proposed work, document modeling is done on journal abstracts data using LDA with CVB0 sampling and Gibbs sampling. It involves pre-processing of documents, document modeling using LDA, an interface to tag the abstract to a specific topic.

3 Latent Dirichlet Allocation

LDA is a generative probabilistic model for collections of discrete data such as text corpora. It is a three-level hierarchical Bayesian model that can infer probabilistic word clusters, called topics, from the document-word matrix [2]. LDA models each document as a mixture of topics and the model generates automatic summaries of topics in terms of a discrete probability distribution over words for each topic, and further infers per-document discrete distributions over topics [8]. LDA has no exact inference methods because of loops in its graphical representation. Variational Bayes (VB) and Gibbs sampling (GS) have been two commonly-used approximate inference methods for learning LDA.

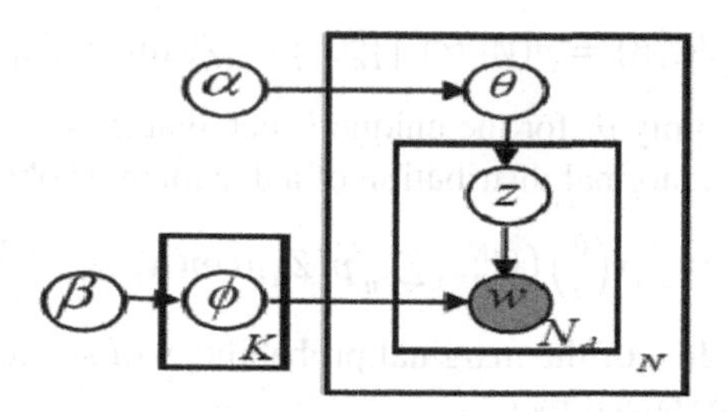

Fig. 1. Graphical model representation of LDA

Formally, A ***word*** is the basic unit of discrete data, defined to be an item from a vocabulary indexed by *{1, . . . , V}*. The words are represented using unit-basis vectors that have a single component equal to one and all other components equal to zero. Thus, using superscripts to denote components, the *v*th word in the vocabulary is represented by a *V*-vector ***w*** such that ***wv*** **= 1** and ***wu*** **= 0** for *u _= v*. A ***document*** is a sequence of *N* words denoted by ***w*** = (*w*1; *w*2, . . . , *wN*), where *wn* is the *n*th word in the sequence. A ***corpus*** is a collection of *M* documents denoted by ***D*** = *{***w**1; **w**2; . . . ; ***wM****}*. LDA assumes the following generative process for each document **w** in a corpus *D*:

1. Choose *N* ~Poisson (ξ).
2. Choose *N* ~ Dir (α).
3. For each of the *N* words
   (a) Choose a topic z_n ~ Multinomial (θ).
   (b) Choose a word w_n from $p\ (w_n / z_n, \beta\ )$ a multinomial probability conditioned on the topic z_n

The dimensionality *k* of the Dirichlet is assumed known and fixed. The word probabilities are parameterized by a *k* *V* matrix β where $\beta_{ij} = p\ (w^j = 1 |\ z^i = 1)$, which for now treated as a fixed quantity that is to be estimated. Finally, the Poisson assumption is not critical to anything that follows and more realistic document length distributions can be used as needed. Furthermore, note that *N* is independent of all the other data generating variables ($\theta$and z). It is thus an ancillary variable and will be generally ignored its randomness in the subsequent development. A *k*-dimensional Dirichlet random variable θ can take values in the (*k*−1)-simplex (a *k*-vector θ lies in the (*k*−1)-simplex if $\theta_i \geq 0$, $\sum_{i=1}^{k} \theta_i$=1), and has the following probability density on this simplex (1):

$$p(\emptyset\ /\propto) = \frac{\Gamma(\sum_{i=1}^{K} \alpha_i)}{\prod_{i=1}^{k} \Gamma(\alpha_i)}\ \theta_1^{\alpha_1 - 1}\ .....\theta_k^{\alpha_k - 1}, \quad (1)$$

where the parameter $\alpha$is a *k*-vector with components α_i >0, and where (*x*) is the Gamma function. The Dirichlet is a convenient distribution on the simplex —it is in the exponential family, has finite dimensional sufficient statistics, and is conjugate to the multinomial distribution. Given the parameters α and β, the joint distribution of a topic mixtureθ, a set of *N* topics **z**, and a set of *N* words **w** is given by (2):

$$p(\emptyset, z, w/\alpha, \beta) = p(\emptyset \mid \propto) \prod_{n=1}^{N} p(z_n/\theta) p(w_n/z_n, \beta), \quad (2)$$

where $p(z_n \mid \theta)$ is simply θ_i for the unique i such that $z_n^i = 1$. Integrating over θ and summing over z, the marginal distribution of a document is obtained by (3):

$$p(w|\alpha, \beta) = \int p\left(\frac{\theta}{\alpha}\right) \left(\prod_{n=1}^{N} \sum_{z_n} p(Z_n|\theta) p(w_n|z_n, \beta)\right) d\theta \quad (3)$$

By taking the product of the marginal probabilities of single documents, the probability of a corpus is obtained by (4):

$$p(D|\alpha, \beta) = \prod_{d=1}^{M} \int p(\theta_d/\alpha) \left(\prod_{n=1}^{N_d} \sum_{Z_{dn}} p(Z_{dn}|\theta_d) p(W_{dn}/Z_{dn}, \beta)\right) d\theta_d \quad (4)$$

The parameters $\propto$ and β are corpus level parameters, assumed to be sampled once in the process of generating a corpus. The variables θ_d are document-level variables, sampled once per document. Finally, the variables z_{dn} and $w_{dn}$are word-level variables and are sampled once for each word in each document.

3.1 LDA Estimation with CVB0 Sampling

Collapsed Variational Bayes is computationally efficient, easy to implement and significantly more accurate than standard Variational Bayesian inference for LDA. It is possible to marginalize out the random variables θ_{kj} and $\emptyset_{wk}$from the joint probability distribution. Following a variational treatment, the variational posteriors over z variables which is once again assumed to be factorized: Q(z) = $\prod_i q(z_i)$. This collapsed variational free energy represents a strictly better bound on the (negative) evidence than the original VB [9]. The derivation of the update equation for the $q(z_i)$ is slightly more complicated and involves approximations to compute intractable summations. It is given below by (5).

$$\gamma_{ijk} \propto \frac{N_{wk}^{\Gamma ij} + \eta}{N_k^{\Gamma ij} + W\eta} \left(N_{kj}^{ij} + \alpha\right) \exp\left(-\frac{V_{kj}^{ij}}{2(N_{kj}^{ij} + \propto)^2}\right) - \frac{V_{wk}^{ij}}{2(N_{wk}^{ij} + \eta)^2} + \frac{V_k^{ij}}{2(N_{kj}^{ij} + W\eta)^2} \quad (5)$$

$N_{wk}^{\Gamma ij}$ denotes the expected number of tokens in document j assigned to topic k (excluding the current token), and can be calculated as follows: $N_{wk}^{\Gamma ij} = \sum_{i' \neq i} \gamma i' jk$. A further approximation can be made by using only the zeroth-order information and the approximate algorithm is referred as CVB0.

3.2 LDA Estimation with Gibbs Sampling

Gibbs Sampling is a special case of Markov-chain Monte Carlo (MCMC) [10] and often yields relatively simple algorithms for approximate inference in high-dimensional models such as LDA [11]. Let $\underset{w}{\rightarrow}$ and $\underset{z}{\rightarrow}$ be the vectors of all words and their topic assignment of the whole data collection W. The topic assignment for a particular word depends on the current topic assignment of all the other word positions. More specifically, the topic assignment of a particular word t is sampled from the following multinomial distribution (6).

$$p(z_i = k \mid \vec{z}_{\neg i}, \vec{w}) = \frac{n_{k,\neg}^{(t)} + \beta_t}{\left[\sum_{v=1}^{V} n_k^{(v)} + \beta_v\right] - 1} \frac{n_{k,\neg}^{(t)} + \alpha_k}{\left[\sum_{j=1}^{K} n_m^{(j)}\right] - 1}, \tag{6}$$

Where $n_{k,\neg}^{(t)} + \beta_t$ is the number of times the word t is assigned to topic k except the current assignment. $\left[\sum_{v=1}^{V} n_k^{(v)} + \beta_v\right] - 1$ is the total number of words assigned to topic k except the current assignment. $n_{k,\neg}^{(t)} + \alpha_k$ is the number of words in document m assigned to topic k except the current assignment. $\left[\sum_{j=1}^{K} n_m^{(j)}\right] - 1$ is the total number of words in document m except the current word t. In normal cases, Dirichlet parameters $\vec{\alpha}$ and $\vec{\beta}$ are symmetric, that is, all αk (k = 1....K) are the same, and similarly for βv (v = 1....V).

4 Experiment and Results

4.1 Data Set

The dataset contains 200 abstracts of different research papers published in Computer Science and Medical domains. The 4 topics covered in the data set are machine learning, computer networks in computer science field and cancer and cardiovascular disease in the medical field. For computer science field, the first 50 abstracts are collected from Journal of Machine Learning Research and remaining 50 abstracts are taken from Journal of Networks. For medical field, the first 50 abstracts are collected from American Journal of Cancer Research and the remaining 50 abstracts are taken from American Journal of Cardiovascular Disease of NCBI PubMed Central Library.

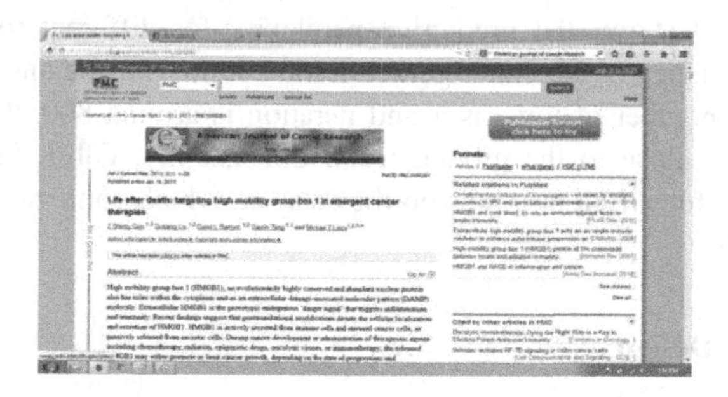

Fig. 2. Screen Shot of Sample Journal Abstract

4.2 Pre Processing

Scala NLP API is used to pre-process the documents. The documents are tokenized and lowercased. Non-words and non-numbers are ignored and terms less than three characters are removed. Standard English stop words from the document are removed. Some common stop words which is usually used in journal abstracts like abstract, paper, method, study, result, etc. are also removed from the documents.

4.3 Model Building

Stanford Topic Modeling implementation is used to build the model. LDA learning algorithms such as CVB0 and Gibbs Sampling with same fixed hyper-parameters $\propto = \beta = 0.01$, are employed and found that Gibbs sampling outperforms CVB0 sampling for building the model. The entire data set is used as the training set to build the model with number of topic is 4 and iteration as 1000. Perplexity is the most common measure to assess the strength of built model. A lower perplexity score indicates better generalization performance.

Table 1. Perplexity of samplings as a function of number of topics

Number of Topics	Perplexity	
	CVB0 Sampling	Gibbs Sampling
2	1001.91	1000.91
3	1015.31	953.91
4	834.25	814.75
5	1031.62	884.19
6	921.62	1027.59

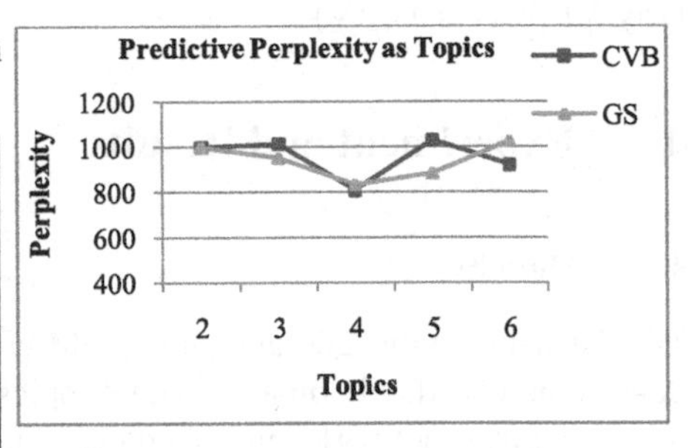

Fig. 3. Predictive Perplexity as Topics

Tab.1.and Fig.3 shows the predictive perplexity for different topics, where the lower perplexity indicates the better generalization ability for the unseen test set. As a result, when the number of topic is 4 and iteration is around 800, the model is converging, which is close to the ground truth. Consistently, Gibbs Sampling has the lowest perplexity for different topics on the data set, which confirms its effectiveness for learning LDA.

4.4 Interface Design

The interface is designed using Microsoft .NET Framework to automatically tag the abstracts. The built model is used in the interface to tag the abstracts. The screen shots of the interface are shown in the fig 4, fig 5 and fig 6.

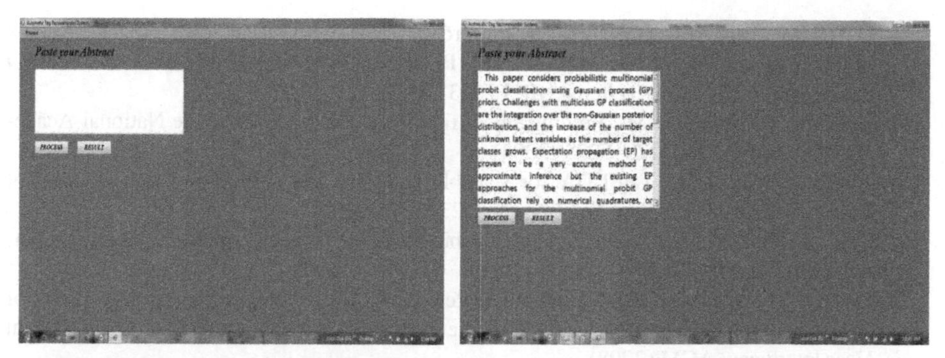

Fig. 4. Interface for providing Abstract **Fig. 5.** Pasted Abstract

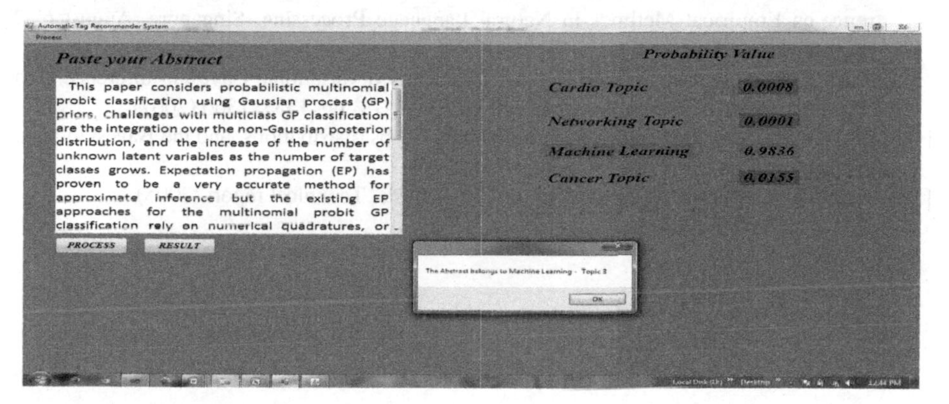

Fig. 6. Pasted Abstract belongs to Machine Learning Topic

5 Conclusion

Topic models have seen many successes in recent years, and are used in a variety of applications, including analysis of news articles, tag assignment for documents, topic-based search interfaces and summarization of documents. In this proposed work, the use of Latent Dirichlet Allocation is examined to recommend appropriate tags for journal abstracts. Abstracts are analyzed by using LDA with CVB0 and Gibbs sampling to establish the number of topics. The results showed that Gibbs sampling outperforms CVB0 sampling and the interface is designed to tag the journal abstracts into appropriate category.

References

[1] Jones, K.S.: A statistical interpretation of term specificity and its application in retrieval. Journal of Documentation 28(1), 11–21 (1972)

[2] Blei, D.M., Ng, A.Y., Jordan, M.I.: Latent Dirichlet allocation. J. Mach. Learn. Res. 3, 993–1022 (2003)

[3] David, M., Blei, T.L., Griffiths, M.I., Jordan, J.B.: Hierarchical topic models and the nested Chinese restaurant process. In: Proceedings of the Conference on Neutral Processing Information Systems NIPS 2003 (2003)
[4] Griffiths, T., Steyvers, M.: Finding scientific topics. Proceedings of the National Academy of Sciences 101, 5228–5235 (2004)
[5] Xiance, S., Sun, M.: Tag-LDA for scalable real-time tag recommendation. Journal of Computational Information Systems 6(1), 23–31 (2009)
[6] Ralf, K., Fankhauser, P.: Personalized topic-based tag recommendation. Neurocomputing 76(1), 61–70 (2012)
[7] Dredze, M., Wallach, H.M., Puller, D., Pereira, F.: Generating summary keywords for emails using topics. In: Proceedings of the 13th International Conference on Intelligent User Interfaces. ACM (2008)
[8] Ramage, D., Hall, D., Nallapati, R., Manning, C.D.: Labeled LDA: A supervised topic model for credit attribution in multi-labeled corpora. In: Proceedings of the 2009 Conference on Empirical Methods in Natural Language Processing, Singapore, August 6 -7, pp. 248–256 (2009)
[9] Teh, Y.W., Newman, D., Welling, M.: A collapsed variational Bayesian inference algorithm for latent Dirichlet allocation. In: NIPS, vol. 19, pp. 1353–1360 (2007)
[10] Geman, S., Geman, D.: Stochastic relaxation, Gibbs distributions, and the Bayesian restoration of images. IEEE PAMI 6, 721–741 (1984)
[11] Heinrich, G.: Parameter estimation for text analysis. Technical report (2005)

Diabetic Retinal Exudates Detection Using Extreme Learning Machine

P.R. Asha[1] and S. Karpagavalli[2]

[1] Department of Computer science,
PSGR Krishnammal College for Women, Coimbatore, India
[2] GR Govindarajulu, School of Applied Computer Technology Coimbatore, India
ashamscsoft@gmail.com, karpagam@grgsact.com

Abstract. Diabetic Retinopathy is a disorder of the retina as a result of the impact of diabetes on the retinal blood vessels. It is the major cause of blindness in people like age groups between 20 & 60. Since polygenic disorder proceed, the eyesight of a patient may commence to deteriorate and causes blindness. In this proposed work, the existence or lack of retinal exudates are identified using Extreme Learning Machine(ELM). To discover the occurrence of exudates features like Mean, Standard deviation, Centroid and Edge Strength are taken out from Luv color space after segmenting the Retinal image. A total of 100 images were used, out of which 80 images were used for training and 20 images were used for testing. The classification task carried out with classifier extreme learning machine (ELM). An experimental result shows that the model built using Extreme Learning Machine outperforms other two models and effectively detects the presence of exudates in retina.

Keywords: Color Space, Extreme Learning Machine; Fuzzy C-means; Histogram Specification.

1 Introduction

Diabetic retinopathy (DR) may be considered as complications on microvascular system in the retina due to prolonged hyperglycaemia. Diabetic Retinopthy is the leading cause of visual impairment in the western world, particularly among persons of working age. It is estimated that DR develops in more than 75% of diabetic patients within 15-20 yrs of diagnosis of diabetes. According to the latest World Health Organization (WHO) report, India has 31.7 million diabetic subjects, and the number is expected to increase to a staggering 79.4 million by 2030. [1]. The presence of DR in the Chennai Urban Rural Epidemiology (CURES) Eye Study in southern India was 17.6 percent, significantly less than age-matched western counterparts. On the other hand, resulting from the large number of diabetic subjects , DR is going to produce a public health problem in India. CURES Eye study clearly explained that the primary systemic threat causes for beginning and also improvement of DR are severity of diabetes, amount of glycaemic control and hyperlipidaemia [2].

S.C. Satapathy et al. (eds.), *Emerging ICT for Bridging the Future – Volume 2*,
Advances in Intelligent Systems and Computing 338, DOI: 10.1007/978-3-319-13731-5_62

Asian Young Diabetes Research (ASDIAB) research, disclosed the occurrence of DR in 724 young diabetic subjects of age group 12-40 yr with existence of diabetes < 12 months in 7 centers of four Asian countries. It will be attractive to notice that DR occurrence was minimum among Indians (5.3%) rather than other ethnic groups including Malays (10%) and also Chinese (15.1%) [2].

2 Feature Extraction

2.1 Preprocessing

Preprocessing of retinal images need to be performed for effective feature extraction. In preprocessing optic disc removal, HSV color space conversion, local contrast enhancement and histogram specification are performed.

2.2 Segmentation

Image segmentation can be a technique of partitioning image pixels depending on one or a lot of handpicked image possibilities and also during this situation the preferred segmentation attribute is color. Fuzzy C-means (FCM) bunch allows pixels to remain in various categorizations with multiple degrees of membership. In this paper FCM segmentation technique will depend on a rough and a fine stage. The coarse phase is responsible for analyzing Gaussian smoothed histograms of each and every color band of the image, in order to produce associate degree primary classification into various classes along with the centre for each and every cluster. Within the fine phase, FCM clustering assigns any specific other unclassified pixels to the nearby class supported the minimization of associate degree objective function.

2.3 Feature Extraction

In this proposed work features of Luv color space has been extracted from the segmented image. Luv color space which split chrominance and luminance is considered to be better color space for extracting features. After analyzing various attributes 18 features were chosen. They are Mean Luv value inside the region (1-3), Mean Luv value outside the region (4-6), Standard deviation of Luv value inside the region (7-9), Standard deviation of Luv value outside the region (10-12), Luv values of region Centriod (13-15), Region size (16), Region compactness (17), and Region edge strength (18).

4 Extreme Learning Machine

Extreme Learning Machine (ELM) is a new learning algorithm for Single-hidden Layer Feed forward neural Networks (SLFNs) with supervised batch learning [7] which provides good generalization performance for both classification and regression problems at highly fast learning speed. The output function of the generalized SLFN is given by, $F(x) = \sum_{i=1}^{L} \beta_i h_i(x)$

Where $h_i(x)$ is the output of the ith hidden node. The steps in ELM algorithm are, Given a training set $N = \{(X_k t_k)|X_k \in R^n, t_k \in R^m, k = 1, \ldots, N\}$,an activation function g(x) and the number of hidden neurons $\widetilde{N}$,

- Randomly assign input weights $w_i$and biases b_i according to some continuous probability density function.
- Calculate hidden layer of output matrix H.
- Calculate the output weights

In kernel based ELM, If the hidden layer feature mapping h(x) is unknown to users, users can be described a kernel function for ELM. ELM Kernel function is given by, $KELM\ (xi, xj) = 1/H\ f\ (xi).f\ (xj)$. That is, the data has feed through the ELM hidden layer to obtain the feature space vectors, and their co-variance is then calculated and scaled by the number of hidden units.

5 Experiment and Results

The experiment has been carried out using the Retinal images from the DIARETDB0 and DIARETDB1 database [12]. Each database contains 130 and 89 images respectively. In this experiment, a total of 100 images with both the exudates and non-exudates were taken for processing.

In the dataset, for training 80% of the data used and 20% used for testing. Extreme Learning Machine learning algorithm has been implemented in MATLAB. The result shows that Extreme Learning Machine has the accuracy of 90% with the sensitivity and specificity as 100 and 87 respectively. To detect the presence or absence of exudates GUI interface has been developed in MATLAB. To detect exudates two buttons are used. When the exudates detection button is clicked, if the image has exudates, the red button will be displayed as affected image which is shown in figure 3 and if the image is normal without exudates, the green button will be displayed which is shown in figure 4.

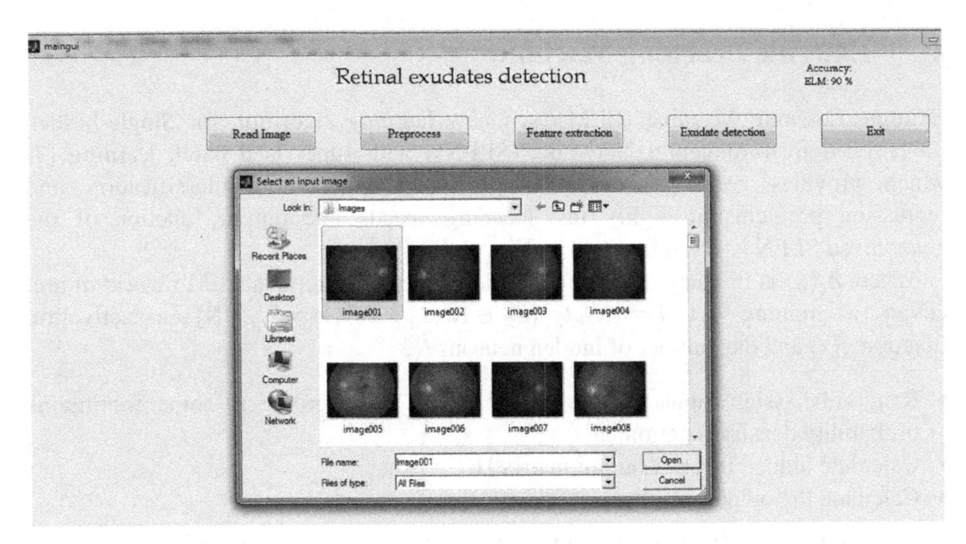

Fig. 1. Snapshot of reading image

Figure 1 allows the user to enable images from the different folders.

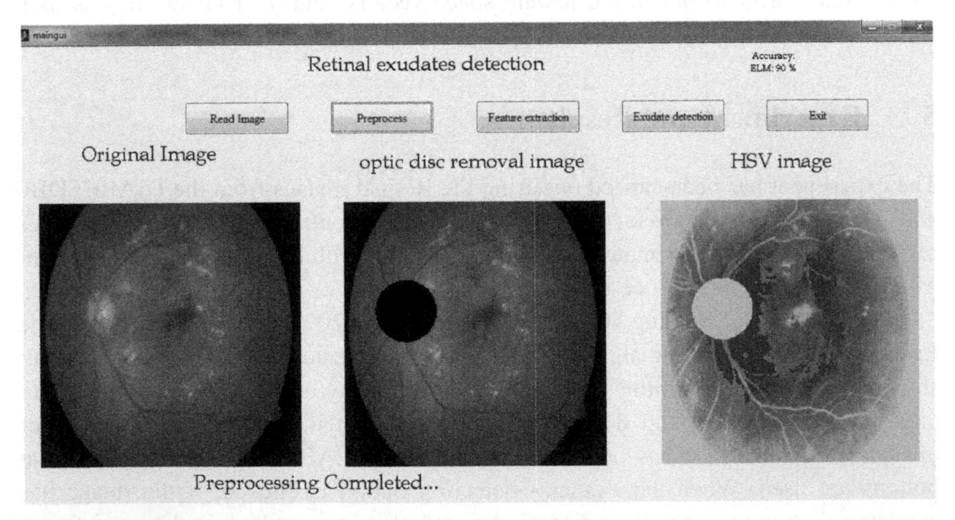

Fig. 2. Snapshot of Pre-processing the image

Figure 2 shows the pre-processing operations. The images shown in this for pre-processing are optic disc removal and color conversion from RGB to HSV. Pre-process button enables the user to perform two operations.

Once pre-process is done 18 features like mean, standard deviation, centroid, size and edge strength are extracted from Luv color space. Feature extraction button enables the user to view these features.

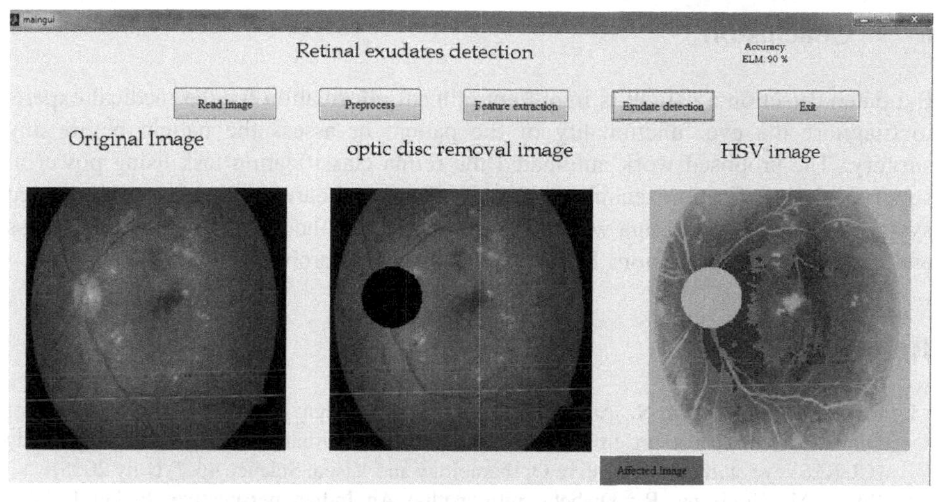

Fig. 3. Snapshot of Exudates Detection

After extracting features, the presence or absence of exudates is detected. Exudate detection button enables the user to identify exudates in an easy manner. If the exudates are present it will show as affected image which is shown in figure 3.

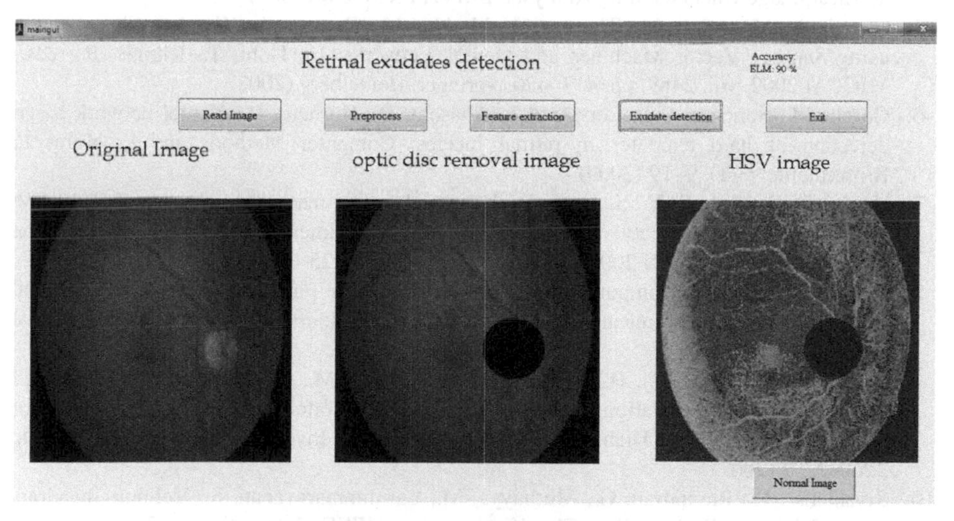

Fig. 4. Snapshot of Exudates Detection

After extracting features, the presence or absence of exudates is detected. Exudate detection button enables the user to identify exudates in an easy manner. If the exudates are not present it will show as normal image as shown in figure 4.

6 Conclusion

Exudates detection a patient is important clinical information for the medical experts to diagnose the eye functionality of the patient or assess the patient before any surgery. The proposed work automated the retina classification task using powerful supervised classification technique namely Extreme Learning Machine. Automated systems to classify the retina will enable the doctors in their decision-making process and to take effective decisions for the patients with eye problems.

References

1. Rema, M., Premkumar, S., Anitha, B., Deepa, R., Pradeepa, R., Mohan, V.: Prevalence of Diabetic Retinopathy in urban india: The chennai urban rural epidemiology study (CURES) eye study. Investigative Opthamology and Visual Science 46(7) (July 2005)
2. Rema, M., Pradeepa, R.: Diabetic retinopathy: An Indian perspective. Indian J. Med. Res. 125 (March 2007)
3. Priya, R., Aruna, P.: Diagnosis of Diabetic Retinopathy using Machine Learning Techniques. Journal on Soft computing 3(04) (July 2013)
4. Osareh, A., Mirmehdi, M., Thomas, B.: Automatic recognition of exudative maculopathy using fuzzy c-means clustering and neural networks. In: Claridge, E., Bamber, J. (eds.) Medical Image Understanding Analysis, BMVA Press, UK (2001)
5. Osareh, A., Mirmehdi, M., Thomas, B., Markha, M.: Comparative Exudate Classification using Support Vector Machines and Neural Networks. In: Dohi, T., Kikinis, R. (eds.) MICCAI 2002, vol. 2489, pp. 413–420. Springer, Heidelberg (2002)
6. Garcia, M., Sanchez, C.I., Lopez, M.I., Abasolo, D., Hornero, R.: Neural network based detection of hard exudates in retinal images. Computer Methods and Programs in Biomedicine 93(1), 9–19 (2009)
7. Huang, G.-B., Zhu, Q.-Y., Siew, C.-K.: Extreme learning machine: a new learning scheme of feed forward neural networks. In: Proceedings of the International Joint Conference on Neural Networks (IJCNN 2004), Budapest, Hungary, pp. 25–29 (July 2004)
8. Rashid, S.: Shagufta: Computerized Exudate Detection in Fundus Images using Statistical Feature Based Fuzzy C-means Clustering. International Journal of Computing and Digital Systems 3(2) (2013)
9. Niemeijer, M., Ginneken, B.V., Russell, S.R., Suttorp, M., Abramoff, M.D.: Automated Detection and Differentiation of Drusen, Exudates and Cotton-wool spots in Digital Color Fundus Photographs for Diabetic Retinopathy Diagnosis. Invest. Ophthalmol, Vis. Sci. 48, 2260–2267 (2007)
10. Sivakumar, R., Ravindran, G., Muthayya, M., Lakshminarayanan, S., Velmurughendran, C.U.: Diabetic Retinopathy Classification. In: IEEE International Conference on Convergent Technologies for the Asia-Pacific Region, vol. 1, pp. 205–208 (2003)
11. Walter, T., Klein, J.C., Massin, P., Erginay, A.: A contribution of image processing to the diagnosis of diabetic retinopathy-detection of exudates in color fundus images of the human retina. IEEE Transactions on Medical. Imaging 21(10), 1236–1243 (2002)
12. `http://www2.it.lut.fi/project/imageret`

Spatial Data Mining Approaches for GIS – A Brief Review

Mousi Perumal, Bhuvaneswari Velumani, Ananthi Sadhasivam, and Kalpana Ramaswamy

Department of Computer Applications, Bharathiar University, Coimbatore, Tamilnadu, India
mousiperumal@gmail.com, bhuvanes_v@yahoo.com, ananthikalps@gmail.com, kalpanacbe2009@gmail.com

Abstract. Spatial Data Mining (SDM) technology has emerged as a new area for spatial data analysis. Geographical Information System (GIS) stores data collected from heterogeneous sources in varied formats in the form of geodatabases representing spatial features, with respect to latitude and longitudinal positions. Geodatabases are increasing day by day generating huge volume of data from satellite images providing details related to orbit and from other sources for representing natural resources like water bodies, forest covers, soil quality monitoring etc. Recently GIS is used in analysis of traffic monitoring, tourist monitoring, health management, and bio-diversity conservation. Inferring information from geodatabases has gained importance using computational algorithms. The objective of this survey is to provide with a brief overview of GIS data formats data representation models, data sources, data mining algorithmic approaches, SDM tools, issues and challenges. Based on analysis of various literatures this paper outlines the issues and challenges of GIS data and architecture is proposed to meet the challenges of GIS data and viewed GIS as a Bigdata problem.

Keywords: GIS, SDM, Geodatabases, Bigdata, Spatial and non-spatial data, Topology.

1 Introduction

Geographic Information System (GIS) has emerged as a new discipline due to the development of communication technologies. GIS is applied in various domains to infer information with respect to location. Enormous amount of data is generated in the form of image, fat files from sources like satellite imaginary sensors and other devices. Understanding of information stored in these large databases requires computational analysis and modeling techniques. Spatial data mining has emerged as a new area of research for analysis of data with respect to spatial relations. SDM techniques are widely used in GIS for inferring association among spatial attributes, clustering, and classifying information with respect to spatial attributes.

The objective of this paper is to provide with a brief summary of GIS data models data sets, data sources to provide better understanding of GIS for analyzing data

S.C. Satapathy et al. (eds.), *Emerging ICT for Bridging the Future – Volume 2*,
Advances in Intelligent Systems and Computing 338, DOI: 10.1007/978-3-319-13731-5_63

analysis using data mining techniques. This paper is organized as follows the section 2 provides with a detailed over view of GIS data sources, data representations. Section 3 provides with description of SDM tasks applied in various domains of GIS data. This paper also presents with the overview of SDM tools for GIS. Section 4 describes the issues and challenges with respect to GIS data set are discussed and architecture is proposed for the same finally drawn by conclusion in section 5.

2 Overview of GIS

The development of information and communication technologies in GIS Domain has generated huge volume of data representing spatial information of water bodies, forest reserves, urbanization, etc., GIS databases stores spatial and non spatial data received from heterogeneous components connected, with each other such as sensors, laptop, mobile etc. Analysis of data deposited in GIS has gained importance in domains related to knowledge management and data mining. Recent widespread use of spatial databases has lead to the studies of Spatial Data Mining (SDM), Spatial Knowledge Discovery (SKD), and the development of SDM techniques. GIS can be viewed as collection of components such as Data, Software, Hardware, Procedures and methods used by people for analysis and decision making with respect to location. Fig 1, represents the components of GIS. The focus of this section is to provide with an overview of GIS data source, data formats, trends and Data Mining applications in GIS.

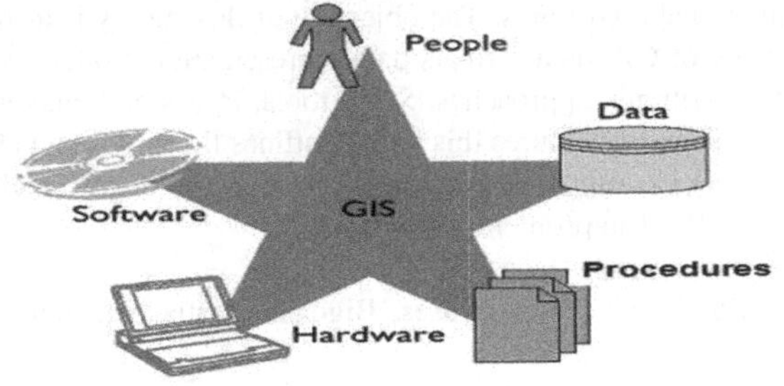

Fig. 1. Components of GIS

Spatial Data mining techniques combined with widely used in various studies to mine interesting facts associated in domains Transport, Tourism, Soil quality monitoring, water resource monitoring, and deforestation [7], [9], [13], [17], [20-22], [24-25], [27-29], [31], [40-42], [44-45], [48-49], [54], [60], [62-63], [73], [75], [78], [82], [87].

2.1 Data Sources

The Geodatabase is used as a "container" used to hold a collection of datasets for representing GIS features. The features of GIS systems are stored in form of tables and raster images. The various dataset of GIS available are listed in Table 1 with its description.

Table 1. GIS Data Sources

Data source	Site	Description	O/P
Yahoo! BOSS	http://www.yahooapis.com	BOSS (**Build your Own Search Service**). Provides a facility of Place finder & Place Spotter to make location aware.	*P
CityGrid	http://developer.citygridmedia.com http://docs.citygridmedia.com/display/citygridv2/Getting+Started	Incorporates local content into web and mobile applications.	*O
Geocoder.us	http://geocoder.us	Provides latitude & longitude of any **US** address , Geocoding for incomplete address ,Bulk Geocoding and Calculates distances	O
GeoNames	http://www.geonames.org	It contains geographical names, populated places and alternate names. All categorized into one out of nine feature classes.	O
US Census	http://www.census.gov http://www.census.gov/geo/www/tiger	Offers several file types for mapping geographic data based on data found in our **MAF/TIGER** database. **MAF**-**M**aster **A**ddress **F**ile. **TIGER** -- **T**opologically **I**ntegrated **G**eographic **E**ncoding.	O
Zillow	http://www.zillow.com	Zillow is a home and real estate marketplace.	O
Natural Earth	http://www.naturalearthdata.com	Natural Earth is a public domain map dataset available at 1:10m, 1:50m, and 1:110 million scales.	O
OpenStreet Map	http://www.openstreetmap.org	OpenStreetMap is a free worldwide map, created by many people.	O
MaxMind	http://www.maxmind.com	GeoIP - IP Intelligence databases and web services minFraud-transaction fraud detection database	O
ArcGis, MapGis.	http://www.esri.com/software/arcgis	Helps to organize and analyze geographic data	O
OpenEarthquake Data	http://earthquake.usgs.gov/earthquakes/map	Gives the databases related to earthquake	O

*P - Licensed Data Source, *O – Open Source Data

2.2 Data Representation in GIS

GIS Systems collects data from various heterogeneous data sources from wide range of communicating devices. The data from the communicating devices is available in different representation and file formats. IN GIS the data representation from these devices is classified into two main categories as Raster and Vector Data types. Fig 2 represents the visual representation of GIS data type. Raster is a two dimensional data type, which stores the value of pixel colors of raster images in a cell and the attribute values are continuous in nature. Raster data type is used to represents information

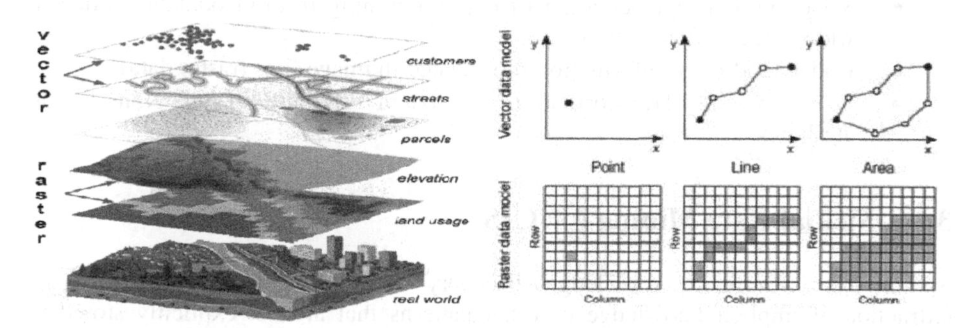

Fig. 2. Visual representation of GIS data type (ref. 91)

Fig. 3. Data models of GIS

from sources such as air photos, scanned maps, elevation layers; remote sensing data. Vector data types are used to represent discrete features in GIS and have a layered architecture representing point, line, and polygon. Vector data types are used to represents information from sources such as roads, rivers, cities, lakes, park boundaries with a layered hierarchy. The data model of GIS is given in Fig 3.

Table 2. File Formats in GIS

Raster Data		Vector Data	
Description	**File format**	**Description**	**File format**
Arc/Info ASCII Grid, Binary Grid,	prj, adf	ESRI Generate Line, shapefile	arc, shp
ADRG/ARC Digitilized Raster Graphics	gen,thf, gen, jpeg, tif	MicroStation Design Files	Dgn
Magellan BLX Topo	blx, xlb	Digital Line Graphs	Dlg
Bathymetry Attributed Grid	bag	Autodesk Drawing, eXchange Files	dwg,dxf
Microsoft Windows Device Independent Bitmap	bmp	ARC/INFO interchange file	e00
BSB Nautical Chart Format	kap	Geography Markup Language	Gml
VTP Binary Terrain Format	bt	ISOK	kf85
Spot DIMAP (metadata.dim)	dim	MapInfo Interchange Format	mif,mid
First Generation , New Labelled USGS DOQ	doq	Spatial Data Transfer System	Sdts
Military Elevation Data (.dt0, .dt1, .dt2)	dt0, dt1, dt2	Scalable Vector Graphics	Svg
Arc/Info Export E00 GRID	adf ,shx	Topologically Integrated Geographic Encoding and Referencing Files	Tiger
ECRG Table Of Contents (TOC.xml)	xml	Vector Product Format	Vpf
ERDAS Compressed Wavelets (.ecw)	ecw	Idrisi32 ASCII vector export format	Vxp
Eir – erdas imagine raw	bl,raw	Microsoft Windows Metafile	Wmf

2.3 Challenges

Analysis of GIS databases representing spatial information is complex due to the varied formats, representation and data sources. The various challenges involved mining spatial data from Geodatabases are listed below:

- Need help of domain experts to relate and understand Spatial and Non-Spatial Data.
- Selection and representation of data for mining from Geodatabases due to wide range of file formats.
- Understanding of information represented in image files (raster data).
- Selection and Transformation of spatial attributes from Non spatial attributes.

3 Spatial Data Mining for GIS

Spatial Data Mining or knowledge discovery in spatial databases refers to the extraction of implicit knowledge or other patterns that are not explicitly stored in spatial database [47][34][72][56]. The word spatial refers to the data associated with the geographic location of the earth. A large amount of spatial data has been collected in various applications, ranging from remote sensing to GIS, computer cartography, environmental assessment and planning. The collected data is huge in such a way that it's far of human knowledge to analyze it, new and efficient methods are needed to discover knowledge from large spatial databases [38]. Spatial data mining is the analysis of geometric or statistical characteristics and relationships of spatial data. The advance in spatial data has enabled efficient querying of large spatial databases.

Over the last few years, spatial data mining has been often used in many applications, like Marine Ecology [89].remote sensing[29], space exploration[79], traffic analysis, climatic change [17]NASA Earth Observing System (EOS), Census Bureau, National Inst. of Justice, National Inst. of Health etc. there is a need of evaluating the structural and topological consistency among multiple representations of complex regions with broad boundaries, mining the frequent trajectory patterns in a spatial–temporal database [4], extracting the spatial association rules from a remotely sensed database [3], generating polygon data from heterogeneous spatial information [71], and analyzing the change of land use [2]. Extraction of spatial rules is one of the main targets of spatial data mining [72], [83] and has been used in many real time applications. [14] Proposed a model to that selects the locations of land-use by using the decision rules generated by nearest neighbours.

Mining of Spatial data set is found to be complex as the spatial data is not represented explicitly in geodatabases [52]. Analysis of Spatial data has become important for analysing information with respect to location. So, analysis of spatial data requires mapping of spatial attributes with non spatial attributes for effective decision making. Spatial data is also known as geospatial data contains information about a physical object that can be represented by numerical values in a geographic coordinate system. Spatial data are multidimensional and auto correlated. Spatial data includes location, shape, size and orientation. [18]. Non spatial data is also called as attribute or characteristic data which is independent of all the geometric considerations. Non spatial data includes height, mass and age, etc. The spatial attributes are classified in three major relations as Distance relation, Direction relation and Topological relation. Topological relation is always non spatial data, so it requires spatial mapping to convert non spatial to spatial data.[57],[47]. Table 3 provides with an overview of spatial relations for modelling attributes.

Table 3. Classification of Spatial Relations

Topological relation	Over laps, Contains, Touches, Disjoint, Covers, Equals, Coveredby 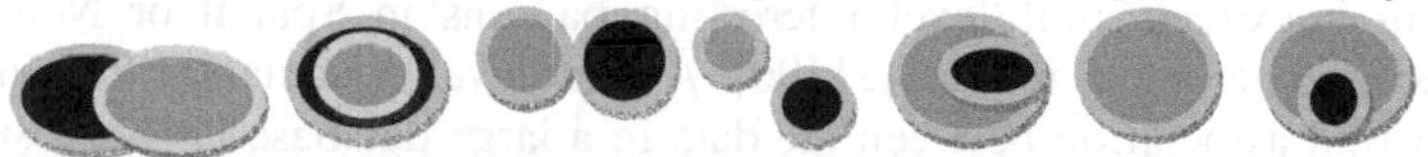
Distance Relation	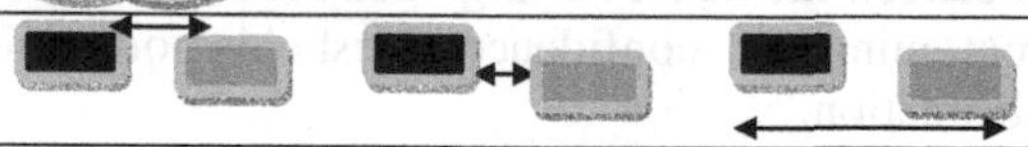
Direction Relation	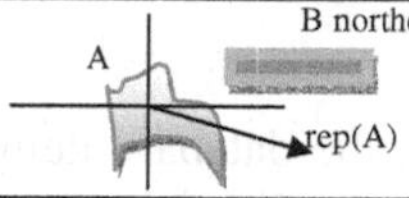 A; B northeast of A; rep(A)

3.1 Spatial Data Mining Tasks

Brief overview of Spatial Data Mining tasks are discussed in the section below. Mining of Spatial Data using data mining techniques such as association, classification, clustering, and trend detection generates interesting facts associated in various domains. Spatial Data Mining tasks are generally an extension of data mining tasks in which spatial data and criteria are combined [50],[51],[71] to form various tasks to find class identification, to find association and co-location of Spatial and Non-Spatial data, make the clustering rules to detect the outliers and to detect the deviations of trends.

3.1.1 Spatial Classification

The spatial object has been classified by using its attributes. Each classified object is assigned a class. Spatial classification is the process of finding a set of rules to determine the class of spatial object[11]Spatial classification methods extend the general-purpose classification methods to consider not only attributes of the object to be classified but also the attributes of neighboring objects and their spatial relations. The spatial classification techniques such as Decision trees (C4.5), Artificial Neural Networks (ANN), remote sensing, Spatial Autoregressive Regression to find the group the spatial objects together. The classification problem is applied in the area of transporting for dividing spatial locations based on the area.

3.1.2 Spatial Association Rule

Association Rule Mining is the process of finding frequent patterns, associations, correlations among sets of items or objects in transaction databases, relational databases, and other information repositories Frequent pattern is described as set of items, sequence etc., that occurs frequently in a database the main motivation is to finding regularities in data[23] ,[61][67],[68].An association among different sets of spatial entities that associate one or more spatial objects with other spatial objects. The association rule is used to find the frequency of items occurring together in transactional databases. Spatial Association relation is based on the topological relation. Association Rule is an expression of the form X ==> Y. A spatial Association Rule describes a set of features describing another set of features in spatial databases. Spatial association is a rule A->B where A, B are set of predicates, the predicates can be Spatial or Non Spatial but needs at least one Spatial predicate [3],[18],[47].

Spatial Association is used to find positive and negative Association Rule Mining which extracts multilevel interesting patterns in Spatial or Non-Spatial predicates using topological relation[2],[90]. A multilevel Association Rule has been generated to find association between the data in a large database [14] and suggested a method by applying different minimum confidence thresholds for mining associations at different levels of abstraction.

3.1.3 Spatial Clustering

Clustering is a process of grouping the database items into clusters. All the members of the cluster have similar features. Spatial Clustering task is an automatic or unsupervised classification that yields a partition of a given dataset depending on a similarity function. Spatial clustering is based on the distance and direction relation. Spatial clustering techniques used ranges from partitioning method, hierarchical method, and density-based method to grid-based method. Similarity can be expressed in terms of a distance function.

3.1.4 Trend Detection

A spatial trend is a regular change of one or more non-spatial attributes when spatially moving away from a start object. Spatial trend detection is a technique for finding patterns of the attribute changes with respect to the neighbourhood of some spatial object. One of the trend detection techniques is kriging to predict the location from outside the sample. Table 4 describes spatial data mining tasks and its techniques used in various domains such as Transportation, Tourism management, Environmental and agriculture from various literatures

Table 4. Spatial Data Mining Technique

SDM Domains	Techniques/Methods	Usage	References
Transport	Add-on Environmental Modelling System (TRAEMS)	Add-on module to existing transport plan, provides rapid information based on traffic related outcomes	[20],[24],[40]
	GIS-based DSS (Classification Technique)	evaluates urban transportation policies, provides estimates of road traffic to traffic policies	[12],[48],[77]
	ArcView GIS 3.3 and ArcInfo (Clustering technique)	Represents geometry of streets and establish the connection between the GIS street data and the roadway links.	[41],[84]
	Transportation object-oriented modelling (TOOM)	Gathers data from the GPS trace, matches GIS and mathematical algorithms to propose a new schedule.	[24],[74],[78],[62],[85]
Tourism Management	Association technique	Used to integrate the ICT technologies with tourism	[21]
	Web GIS	Provides a new generation interface and expands the ways in which travel information can be accessed.	[31],[74]
	Tourism GIS	Used to provide users with a quick and convenient travel information query method	[7],[25],[26]
	Unified GIS database on the cycle infrastructure (UDCI)	Provide information about the overall cycle trail network on tourism	[53],[73],[78]
	spatial tourism interaction model or gravity model	Used to provide about green tourism potential, human resources, and the shortest distances among villages.	[74],[9],[28],[6],[21],[45]
Ecological model	Indicates the water catchments by integrating geographic area.	Maintains the Bio-Diversity	[18],[22],[75]
Argiculture	GIS and Saptial Data mining	Used to access the soil quality, water resource management	[69],[82],[86],[87]
land allocation (MOLA)	Location prediction	Provides the selection of land use models, illegal land fills	[8],[15],[16]
Environmental Decision Support System	Site selection, Location prediction, clustering	Provides environmental support of assessment of various relates feature with environmental degradation etc.,	[18],[22],[25],[44],[50],[59],[63],[65],[76],[89]
Health Care	Accessing of resource	Provides with an decision making support for various health related domains	[8],[14],[17]

3.2 Spatial Data Mining Tools

Spatial Data Mining tools are used by researchers to mine spatial relations among spatial datasets in various application domains. There exists many numbers of tools as opens source and propriety software. It is found that the tools are input specific in terms of file formats. So common file formats is not available for any specific tools which is a challenge for choosing the correct tool. Table 5 describes about various Spatial Data Mining Tools.

Table 5. Spatial Data Mining tools

Spatial Data Mining Tools	O/ P	Developer	Language used	Main Features and Data Source uses
DBMiner (DBlearn)	O	Data mining research group, Simon Fraser University, Canada	Data Mining Query Language	Supports data mining functions including (https://www.dbminer.com)
GeoMiner	O	Data mining research group, Simon Fraser University, Canada	Geo-Mining Query Language	Supports the data mining tasks (http://www.downloadcollection.com)
GeoDA(ESDA, STARS)	O	Dr. Luc Anselin	Python	Supports spatial autocorrelation statistics, spatial regression (http://geodacenter.asu.edu)
Weka-GDPM	O	University of Waikato, N Z	Java	Supports several standard data mining tasks (http://weka.software.informer.com)
R language (sp, rgdal, rgeos)	O	Ross Ihaka and Robert Gentleman at the University of Auckland,NZ	C, FORTRAN, Python	Used to analyze statistical and graphical techniques, (http://www.r-project.org/)
Descrates	O	Jankowski and Andrienko	Python	Used to visualize and analyse source data and results of the classification on the map.(https://www.descartes.com)
ArcGIS (ArcView, ArcInfo, ArcEditor)	p	Environmental Systems Research Institute (ESRI)	Python, Web API, .NET	Supports spatial analysis and modeling features including overlay, surface, proximity, suitability, and network analysis, as well as interpolation analysis and other geo statistical modeling techniques. (http://www.esri.com/software/arcgis)

4 Issues and Challenges

The issues and challenges in applying Spatial Data Mining in GIS can be viewed in terms of Integration of data and Mining huge volume of data. Architecture is proposed to address the issues of data integration and volume of data based on the analysis of data of GIS from literature is given in Fig 4.Data warehousing technology is used as a tool for data integration and stores summarized data. Currently a Bigdata approach has gained attention for mining data parallel using architectures like Hadoop and Mapreduce. In this paper we propose an architecture which can have a Bigdata platform modeled for representing a data warehouse. The other challenge in integration of data is semantic representation of information organized from various data sources. An ontology layer is proposed for semantic representation of data [13]. The data mining techniques can be applied above these layers by modifying the algorithmic approaches suitable for BigData architecture.

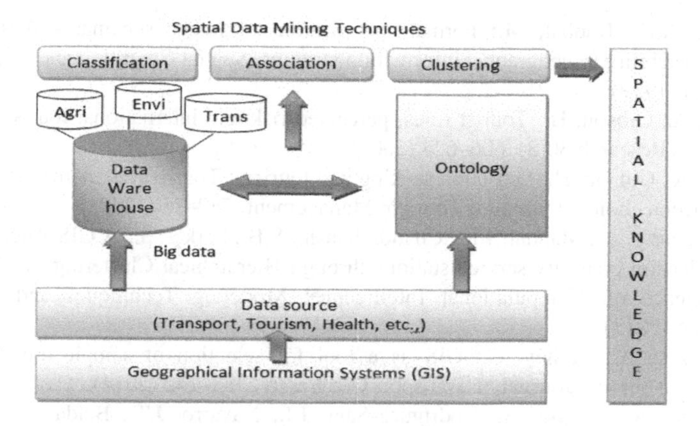

Fig. 4. A Big Data Approach – Integration of GIS Data

5 Conclusion

This paper provides with a detailed survey on spatial data bases and its characteristics. A detailed analysis and description of GIS Data sources, data formats and data representation is presented from various literatures. The classical data mining algorithm applied for various applications in GIS are also discussed. On analysis it is identified that semantic integration of GIS datasets is necessary for analyzing spatial attributes with respect to non spatial attributes in all domains. The other major challenge of GIS databases can be viewed as volume and data formats in this work we have proposed an architecture for data integration using data ware house approach and ontology in GIS. The other major issue in GIS is huge volume of data generation for which we have proposed to view the problem as Bigdata and represented the same in our proposed architecture design. In future the proposed architecture would be implemented and tested for any specific domain. SDM tools would also be presented with a brief overview discussing the merits and limitations

References

1. Gatrell, A.C., Bailey, T.C., Diggle, P.J., Rowlingson, B.S.: Spatial Point Pattern Analysis and Its Application in GeographicalEpidemiology. Transactions of the Institute of British Geographers, New Series 21(1), 227–256 (1996)
2. Du, S., Qin, Q., Wang, Q., Ma, H.: Evaluating structural and topological consistency of complex regions with broad boundaries in multi-resolution spatial databases. Inform. Sci. 178, 52–68 (2008)
3. Lee, A.J.T., Hong, R.W., Ko, W.M., Tsao, W.K., Lin, H.H.: Mining spatial associationrules in image databases. Inform.Sci. 177, 1593–1608 (2007)
4. Lee, A.J.T., Chen, Y.A., Ip, W.C.: Mining frequent trajectory patterns in spatial–temporal databases. Inform. Sci. 179, 2218–2231 (2009)

5. Frank, A.U., Raubal, M.: Formal specification of image schemata—A step towards interoperability in geographic information systems. Spatial Cognition and Computation 1, 67–101 (1999)
6. Lepp, A., Gibson, H.: Tourist roles, perceived risk and international tourism. Journal of Tourism Research 30(3), 606–624 (2003)
7. Lepp, A., Gibson, H.: Sensation seeking and tourism: Tourist role, perception of risk and destination choice. Journal of Tourism Management, 740–750 (2008)
8. Chakraborty, A., Mandal, J.K., Chandrabanshi, S.B., Sarkar, S.: A GIS Anchored system for selection of utility service stations through Hierarchical Clustering. In: International Conference on Computational Intelligence: Modeling Techniques and Application, CIMTA (2013)
9. Fraszczyk, A., Mulley, C.: GIS as a tool for selection of sample areas in a travel behaviour survey. Journal of Transport Geography, 233–242 (2014)
10. Zaragozí, A., Rabasa, A., Rodríguez-Sala, J.J., Navarro, J.T., Belda, A., Ramón, A.: Modelling farmland abandonment: A study combining GIS and data mining techniques. Journal of Agriculture, Ecosystems and Environment 155, 124–132 (2012)
11. Brinkoff, T., Kriegel, H.-P.: The Impact of Global Clustering on Spatial Database Systems. In: Proceedings of the 2Uth VLDB Conference, Santiago, Chile, pp. 168–179 (1994)
12. Brown, A.L., Affum, J.K.: A GIS-based environmental modelling system for transportation planners. Journal of Computers, Environment and Urban Systems, 577–590 (2002)
13. Boomashanthini, S.: Gene ontology similarity metric based on DAG using diabetic gene. Compusoft on International Journal of Advances Computer Technology, ISSN: 2320 0790
14. Pope III, A., Burnett, R.T., Thurston, G.D., Thun, M.J., Calle, E.E., Krewski, D., Godleski, J.J.: Cardiovascular Mortality and Long-Term Exposure to Particulate Air Pollution. Circulation 109, 71–77 (2004)
15. Silva, J.G., Ferreira, S.B., Bricker, T.A., DelValls, M.L., Martín-Díaz, E.: Site selection for shellfish aquaculture by means of GIS and farm-scale models, with an emphasis on data-poor environments. Journal of Aquaculture 318, 444–457 (2011)
16. Jones, C.B., Alani, H., Tudhope, D.: Geographical information retrieval with ontologies of place. In: Montello, D.R. (ed.) COSIT 2011. LNCS, vol. 2205, pp. 322–335. Springer, Heidelberg (2001)
17. Wettescherek, D., Aha, D.W., Mohri, T.: A Review and empirical evaluation of feature Weighting Methods for a class of lazy Learning Algorithms. Artificial Intelligence Review 10, 1–37 (1997)
18. McKinney, D.C., Cai, X.: Linking GIS and water resources management models: an object-oriented method. Journal of Environmental Modelling & Software 17, 413–425 (2002)
19. Gavalas, D.: Mobile recommender systems in tourism. Journal of Network and Computer Applications, 319–333 (2014)
20. Badoe, D.A., Miller, E.J.: Transportation land-use interaction: empirical findings in North America, and their implications for modelling. Journal of Transportation Research, 235–263 (2000)
21. Buhalis, D., Law, R.: Progress in information technology and tourism management: 20 years on and 10 years after the Internet—The state of eTourism research. Journal of Tourism Management (2008)

22. Elewa, H.H., Ramadan, E.M., El-Feel, A.A., Abu El Ella, E.A., Nosair, A.M.: Runoff Water Harvesting Optimization by Using RS, GIS and Watershed Modellin. International Journal of Engineering
23. Fonseca, M.E.: Using ontologies for integrated geographic information systems. Transactions in GIS 6, 231–257 (2002)
24. Marzolf, F., Trépanier, M., Langevin, A.: Road network monitoring: algorithms and a case study. Journal of Computers & Operations Research, 3494–3507 (2006)
25. Ferri, F., Rafanelli, M.: GeoPQL: A Geographical Pictorial Query Language That Resolves Ambiguities in Query Interpretation of forest cover in the Annapurna Conservation Area, Nepal. Journal of Applied Geography, 159–168 (2013)
26. Fonseca, F., Rodríguez, M.A.: GeoSpatial Semantics. In: 2007 Proceedings Second International Conference, GeoS 2007, Mexico City, Mexico, November 29-30 (2007)
27. Chunchang, F., Nan, Z.: The Design and Implementation of Tourism Information System based on GIS. Journal of Physics Procedia, 528–533 (2012)
28. Arampatzis, C.T., Kiranoudis, P., Scaloubacas, D.: A GIS-based decision support system for planning urban transportation policies. European Journal of Operational Research, 465–475 (2004)
29. Tradigoa, G., Veltria, P., Grecob, S.: Geomedica: managing and querying clinical data distributions on geographical database systems. Procedia Computer Science 1, 979–986 (2012)
30. Ravikumar, G., Sivareddy, M.: An Effective Analysis of Spatial Data Mining Methods Using Range Queries. Journal of Global Research in Computer Science 3(1) (2012)
31. Chang, G., Caneday, L.: Web-based GIS in tourism information search: Perceptions, tasks and trip attributes. Journal of Tourism Management, 1435–1437 (2011)
32. Cai, G.: Contextualization of Geospatial Database Semantics for Human–GIS Interaction. Geoinformatica. Geoinformatica 11, 217–237 (2007), doi:10.1007/s10707-006-0001-0
33. Stuckenschmidt, H., van F, H.: Information Sharing on the Semantic Web. ACM Subject Classification (1998) ISBN 3-540-20594-2
34. Bai, H., Ge, Y., Wang, J., Li, D., Liao, Y., Zheng, X.: A method for extracting rules from spatial data based on rough fuzzy Sets. Journal of Knowledge Based Systems 57, 28–40 (2014)
35. Kang, I.-S., Kim, T.-W., Li, K.-J.: A Spatial Data Mining Method by Delaunay Triangulation
36. Han, J., Fu, Y.: Discovery of multiple-level association rules from large databases. In: Proceedings of the 21st VLDB Conference, p. 420
37. Komorowski, L., Polkowski, Z., Skowron, A.: Rough sets: a tutorial. Springer, Heidelberg (1999)
38. Niebles, J.C., Wang, H., Fei-Fei, L.: Unsupervised learning of human action categories using spatial–temporal words. Int. J. Comput. Vis 79, 299–318 (2008)
39. Dreze, J., Khera, R.: Crime, Gender, and Society in India: Insights from Homicide Data. Population and Development Review, Population Council, Stable 26(2), 335–352 (2000), http://www.jstor.org/stable/172520
40. Thill, J.-C.: Geographic information systems for transportation in Perspective. Journal of Transportation Research, 3–12 (2000)
41. Armstrong, J.M., Khan, A.M.: Modelling urban transportation emissions: role of GIS. Journal of Computers, Environment and Urban Systems, 421–433 (2004)
42. Rogalsky, J.: The working poor and what GIS reveals about the possibilities of public transit. Journal of Transport Geography (2010)

43. Yao, J., Murray, A.T., Agadjanian, V.: A geographical perspective on access to sexual and reproductive health care for women in rural Africa. Journal of Social Science & Medicine 96, 60–68 (2013)
44. Lee, J.-S., Ko, K.-S., Kim, T.-K., Kim, J.G., Cho, S.-H., Oh, I.-S.: Analysis of the effect of geology, soil properties, and land use on groundwater quality using multivariate statistical and GIS methods. Chinese Journal of Geochemistry 25(Suppl.) (2006)
45. Noguera, J.M., Barranco, M.J., Segura, R.J., Martínez, L.: A mobile 3D-GIS hybrid recommender system for tourism. Journal of Information Sciences, 37–52 (2012)
46. Smith, R., Mehta, S.: The burden of disease from indoor air pollution in developing countries: comparison of estimates. International Journal of Hygiene and Environmental Health 206(4-5), 279–289 (2003)
47. Koperski, K., Han, J.: Discovery of spatial Association Rules in Geographic Information Databases. In: Egenhofer, M.J., Herring, J.R. (eds.) Advances in Spatial Databases. LNCS, vol. 951, pp. 47–66. Springer, Heidelberg (1995)
48. Tang, K.X., Waters, N.M.: The internet, GIS and public participation in transportation planning. Journal of Planning in Progress, 7–62 (2005)
49. Kenneth, J., Dueker, J., Butler, A.: A geographic information system framework for transportation data sharing. Journal of Transportation Research, 13–36 (2000)
50. Guarino, L., Jarvis, A., Hijmans, R.J., Maxted, N.: Geographic Information Systems (GIS) and the Conservation and Use of Plant Genetic Resources. In: IPGRI 2002 (2002)
51. Anselin, L.: Exploring Spatial Data with GeoDaTM: A Workbook, Copyright 2004-2005 Luc Anselin, All Rights Reserved (March 6, 2005)
52. Ester, M., Kriegel, H.-P., Sander, J.: Spatial data mining: a database approach. In: Scholl, M., Voisard, A. (eds.) Advances in Spatial Databases. LNCS, vol. 1262, pp. 47–66. Springer, Heidelberg (1997)
53. Goodchild, M., Egenhofer, R., Fegeas, C.: Kottman. Interoperating Geographic Information Systems. Kluwer, Boston (1998)
54. Kwan, M.: Interactive geovisualization of activity-travel patterns using three-dimensional geographical information systems: a methodological exploration with a large data set. Transportation Research Part C: Emerging Technologies 8(1-6), 185–203 (2000)
55. Liebhold, M.: The geospatial web: A call to action: What we still need to build for an insanely cool open geospatial web. O'Reilly Network, Sebastopol (2005)
56. Yuan, M., Buttenfield, B., Gahegan, M., Miller, H.: Geospatial data mining and knowledge discovery. In: McMaster, R.B., Usery, E.L. (eds.) A Research Agenda for Geographic Information Science, pp. 365–388. CRC, Boca (2004)
57. Egenhofer, M.: Spatial SQL A Query and Presentation Language. IEEE Transactions and Data Engineering 6, 86–95 (1994)
58. Egenhofer, M.J., Rashid, A., Shariff, B.M.: Metric details for natural-language spatial relations. ACM Transactions on Information Systems 16, 295–321 (1998)
59. Barroeta-Hlusicka, M.E., Buitrago, J., Rada, M., Pérez, R.: Contrasting approved uses against actual uses at La Restinga Lagoon National Park, Margarita Island, Venezuela. A GPS and GIS method to improve management plans and rangers coverage. J. Coast. Conserv. 16, 65–76 (2012), doi:10.1007/s11852-011-0170-3
60. Scotch, M., Parmanto, B., Gadd, C.S., Sharma, R.K.: Exploring the role of GIS during community health assessment problem solving: experiences of public health professionals. International Journal of Health Geographics
61. Egenhofer, M.J.: Categorizing Binary Topological Relations Between Regions. Lines, and Points in Geographic Databases (2000)

62. Bil, M., Bilova, M., Kube, J.: Unified GIS database on cycle tourism infrastructure. Journal of Tourism Management (2012)
63. Bhargavi, P., Jyothi, S.: Soil Classification Using Data Mining Techniques: A Comparative Study. International Journal of Engineering Trends and Technology (2011)
64. Compieta, P., Di Martino, S., Bertolotto, M., Ferrucci, F., Kechadi, T.: Exploratory spatio-temporal data mining and visualization. Journal of Visual Languages and Computing 18, 255–279 (2007)
65. Antunes, P., Santos, R.: The application of Geographical Information Systems to determine environmental significance. Journal of Environmental Impact Assessment Review 21, 511–535 (2001)
66. Kuba, P.: Data structures for spatial data mining. Publications in the FI MU Report Series
67. Agrawal, R., Srikant, R.: Fast Algorithms for Mining Association Rules in Large Databases. In: Proceedings of the 20th International Conference on VLDB, pp. 478–499 (1994)
68. Agrawal, R., Imielinski, T., Swami, A.: Mining association rules between sets of items in large databases. Proceedings of the ACM
69. Guting, R.H.: An Introduction to spatial Database System. VLDB Journal 3, 357–400 (1994)
70. Chintapalli, S.M., Raju, P.V., Abdul Hakeem, K., Jonna, S.: Satellite Remote Sensing and GIS Technologies to aid Sustainable Management of Indian Irrigation Systems. In: International Archives of Photogrammetry and Remote Sensing, vol. XXXIII, Part B7. Amsterdam (2000)
71. Schockaert, S., Smart, P.D., Twaroch, F.A.: Generating approximate region boundaries from heterogeneous spatial information: an evolutionary approach. Inform. Sci 181, 257–283 (2011)
72. Shekhar, S., Zhang, P., Huang, Y., Vatsavai, R.R.: Trends in spatial data mining. In: Kargupta, H., Joshi, A. (eds.) Data Mining: Next Generation Challenges and Future Directions. AAAI/MIT (2003)
73. Lee, S.-H., Choi, J.-Y., Yoo, S.-H., Oh, Y.-G.: Evaluating spatial centrality for integrated tourism management in rural areas using GIS and network analysis. Journal of Tourism Management, 14–24 (2013)
74. Shaw, S.-L., Xin, X.: Integrated land use and transportation interaction: a temporal GIS exploratory data analysis approach. Journal of Transport Geography, 103–115 (2003)
75. Yammani, S.: Groundwater quality suitable zones identification: application of GIS. Journal of Environ Geol 53, 201–210 (2007)
76. Brinkhoff, T., Kriegel, H.P., Schneider, R., Seeger, B.: Multistep processing of Spatial Joins. In: Proc. 1994 ACM-SIGMOD Conf. Management of Data, Minneapolis, Minnesota, pp. 197–208 (1994)
77. Nyerges, T.I., Montejano, R., Oshiro, C., Dadswell, M.: Group-based geographic information systems for Transportation improvement site selection. Journal of Transport Geography 5(6), 349–369 (1997)
78. Hsu, T.-K., Tsai, Y.-F., Wu, H.-H.: The preference analysis for tourist choice of destination: A case study of Taiwan. Journal of Tourism Management, 288–297 (2009)
79. Fayyad, U.M., Smyth, P.: Image database Exploration: Progress and Challenges. In: Proc. 1993 Knowledge Discovery in Data Bases, wsashington, pp. 14–27 (1993)
80. Vieira, V.M., Webster, T.F., Weinberg, J.M., Aschengrau, A.: Spatial-temporal analysis of breast cancer in upper Cape Cod. International Journal of Health Geographics (2008)
81. Bhuvaneswari, V., Rajesh, R.: A Genetic Algorithmic Survey of Multiple Sequence Alignment. In: International Conference on Systemics, Cybernetics and Informatics

82. Gu, W., Wang, X., Ziébelin, D.: An Ontology-Based Spatial Clustering Selection System. Journal of Procedia Engineering 32 (1999)
83. Liu, X.M., Xu, J.M., Zhang, M.K., Huang, J.H., Shi, J.C., Yu, X.F.: Application of geostatistics and GIS technique to characterize spatial variabilities of bioavailable micronutrients in paddy soils. Journal of Environmental Geology 46, 189–194 (2004)
84. Wang, X.: Integrating GIS, simulation models, and visualization in traffic impact analysis. Journal of Computers, Environment and Urban Systems, 471–496 (2005)
85. Wang, X.: Integrating GIS, simulation models, and visualization in traffic impact analysis. Journal of Computers, Environment and Urban Systems 29, 471–496 (2005)
86. Du, Y., Liang, F., Sun, Y.: Integrating spatial relations into case-based reasoning to solve geographic problems. Known.-Based Syst (2012)
87. Vagh, Y.: The application of a visual data mining framework to determine soil, climate and land use relationships. Journal of Procedia Engineering 32, 299–306 (2012)
88. Qin, Y., Jixian, Z.: Integrated application of RS and GIS to agriculture Land use planning. Journal of Geo-spatial Information Science 5(2), 51–55 (2002)
89. Kemp, Z., Lee, H.T.K.: A Marine Environmental System for Spatiotemporal Analysis. In: Proc. of 8th Symp. on Spatial Data Handling SDH 1998, Vancouver, Canada, pp. 474–483 (1998)
90. Zhang, S., Zhang, J., Zhang, B.: Deduction and application of generalized Euler formula in topological relation of geographic information system (GIS). Science in China Ser. D Earth Sciences 47(8), 749–759 (2004)
91. Image Source, http://serc.carleton.edu/eyesinthesky2/week5/intro_gis.html

Privacy during Data Mining

Aruna Kumari[1,3], K. Rajasekhara Rao[2,3], and M. Suman[1,3]

[1] ECM, Electronics and Computer Engineering, KL University, Guntur, Andhra Pradesh, India
[2] Prakash Engineering College, Tuni, India
[3] CSI life members

KL University Vaddeswaram, Guntur(dist), 522502, AP, India
aruna_D@kluniversity.in, krr_it@yahoo.co.in, suman.maloji@kluniversity.in

Abstract. The large amounts of data stored in computer files are increasing at a very remarkable rate. It is analysed and evaluated that the amount of data in the world is increasing like water coming into the ocean. At the same time, the users or a person who operates these data are expecting more sophisticated Knowledge. The Languages like Structure Query Languages are not adequate to support this increasing demand for information. Data mining makes an effort to solve the problem. We proposed a new approach for maintaining privacy during mining the knowledge from the data stores. Our approach is based on vector quantization, it quantizes the data to its nearest neighbour values that uses two algorithms one LBG and Modified LBG in codebook generation process.

Keywords: Distortion, accuracy, codebook, privacy.

1 Introduction

Data Mining is a process of extracting and inheriting hidden and trivial knowledge from huge volumes of data stores. As data contains personal information, there is a threat to privacy if the individual data is mined. This paper introduces two algorithms for securing mining of useful data information from large data sets.

Data mining is widely used in many applications such as follows:

1. Medical Data Analysis
2. Stock Market Analysis
3. Trend Analysis
4. Direct Marketing Analysis
5. Fraud Detection Analysis
6. Supply Chain Optimization
7. National Security
8. Education System
9. Web Education
10. Credit Scoring
11. Product Marketing and future directions
12. Analyzing of any Company Data for further Development

S.C. Satapathy et al. (eds.), *Emerging ICT for Bridging the Future – Volume 2*,
Advances in Intelligent Systems and Computing 338, DOI: 10.1007/978-3-319-13731-5_64

2 Related Work

Data modification deals with modifying the original data by adding some random noise or by swapping the attribute values or by suppressing the attribute values,...etc.

Secure multiparty computations: - The data is distributed over one or more parties and if they want analyse the data by sharing the together, there is threat to privacy while sharing the information. Cryptography is mainly used approach for secure multi-party computation[12].

Data Transformations deals with rotation, scaling and translation of attribute values. Our new approach deals with transformation of original to new with the help of codebook generation algorithms.

Table 2 shows a 4-anonymous table derived from the table in Table 1 (here "*" denotes a suppressed value so, for example, "pin code = 1485*" means that the pin code is in the range [14850–14859] and "age=3*" means the age is in the range [30-39]. Note that in the 4-anonymous table, each tuple has the same values for the quasi-identifier as at least three other tuples in the table. Because of its conceptual simplicity, k-anonymity has been widely seen as a viable definition of privacy in data mining[11][13].

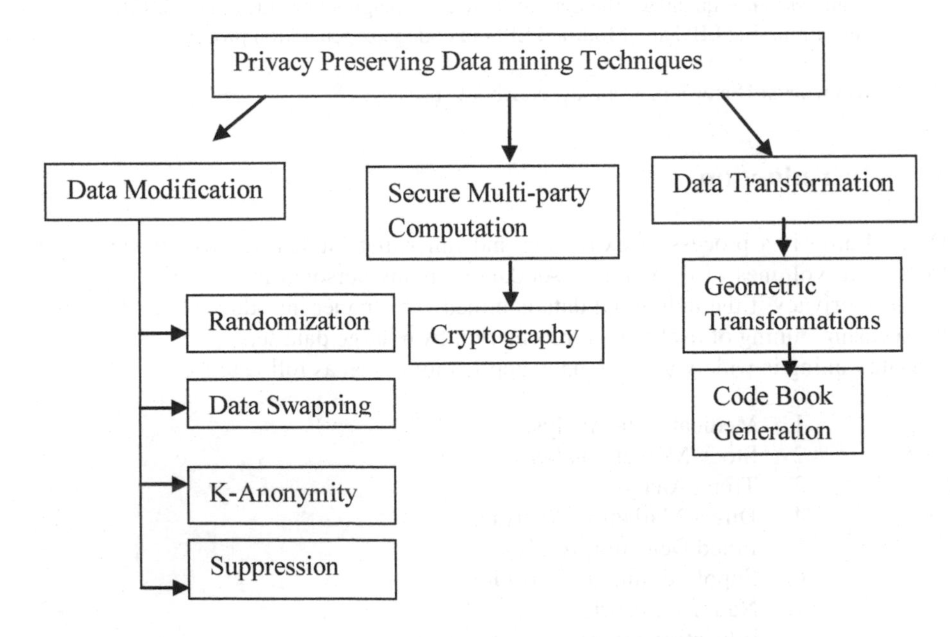

Fig. 1. Taxonomy of Privacy Preserving Data Mining

Table 1. Inpatient Micro data

	Non Sensitive			Sensitive
	Pin Code	Age	Nationality	Disease
1	13053	28	Russian	Heart Disease
2	13068	29	Pakistani	Heart Disease
3	13068	21	Chinese	Viral Infection
4	13053	23	Pakistani	Viral Infection
5	14853	50	Indian	Cancer
6	14853	55	Russian	Heart Disease
	14850	47	Pakistani	Viral Infection
8	14850	49	Pakistani	Viral Infection
9	13053	31	Pakistani	Cancer
10	13053	37	Indian	Cancer
11	13068	36	Chinese	Cancer
12	13068	35	Pakistani	Cancer

Table 2. 4-anonymous Inpatient Micro data

	Non Sensitive			Sensitive
	Pin Code	Age	Nationality	Disease
1	130**	<30	*	Heart Disease
2	130**	<30	*	Heart Disease
3	130**	<30	*	Viral Infection
4	130**	<30	*	Viral Infection
5	1485*	≥40	*	Cancer
6	1485*	≥40	*	Heart Disease
7	1485*	≥40	*	Viral Infection
8	1485*	≥40	*	Viral Infection
9	130**	3*	*	Cancer
10	130**	3*	*	Cancer
11	130**	3*	*	Cancer
12	1308*	3*	*	Cancer

3 Privacy Preserving Data Mining

Privacy preserving data mining aims to provide valid data mining results by not revealing the underlined sensitive information. Figure 2 and 3 shows the Proposed Architectures of privacy preserving data mining.

When the techniques are applied, it not only extracts useful data, may also reveal sensitive information. So as to provide protection for sensitive information some privacy preserving techniques can be applied on original data then mining can be performed.

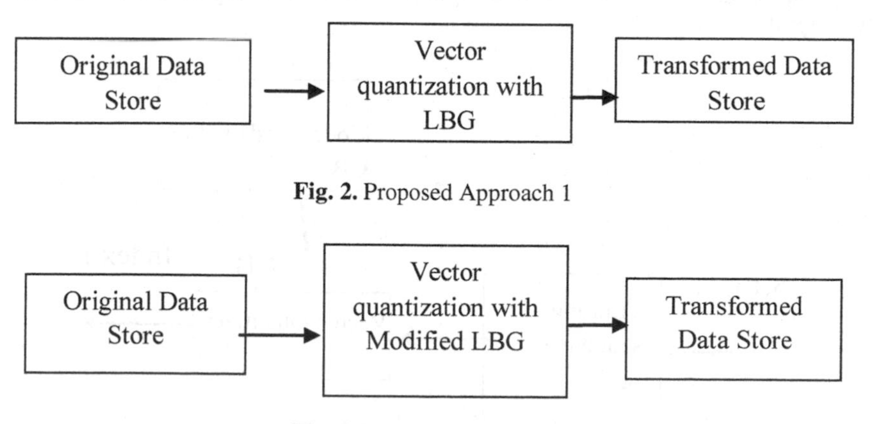

Fig. 2. Proposed Approach 1

Fig. 3. Proposed Approach 2

Figute: 2 and 3 explores the sequence of steps to be followed for achieving secure data mining results. This work is proposed to perform clustering task on both original data called as R1 and on transformed data called as R2. Finally R1 and R2 will be observed and analysed for evaluating the performance of proposed approach. This work provides one of the solutions for privacy preserving data mining (central data warehouse not on distributed databases) and the performance is measured in terms of accuracy of data mining result and privacy of sensitive data.

This work proposes Data Transformation using Vector quantization with two codebook generation algorithms LBG and Modified LBG; these algorithms will be explained in detail in next sections.

3.1 Steps Involved in ppdm Using Vector Quantization

In this work, a codebook[10] will be constructed from training data from which reconstructed data will be generated by approximating each point of data to its nearest value. Because original data should not be revealed once the transformation is performed, transformed data set represents approximate data not exact and original data thus privacy is preserved.

Vector quantization consists of following steps:

1. Generating Codebook from Training Data.
2. Encoding the original data with the help of Codebook by nearest neighbor search;
3. Using the index table, data will be reconstructed by looking up in the codebook.

Vector quantization [1,2] is a process whereby the elements of a vector of k data samples are jointly quantized. Vector quantization (VQ) is generally used for data compression. In previous days, the design methodology of a vector quantizer (VQ) is treated as a big problem in terms of the need for multi-dimensional integration. Linde, Buzo, and Gray (LBG) introduced an algorithm for Vector quantization design based on training sequence. A VQ that is designed based on this algorithm are referred as LBG-VQ[6].

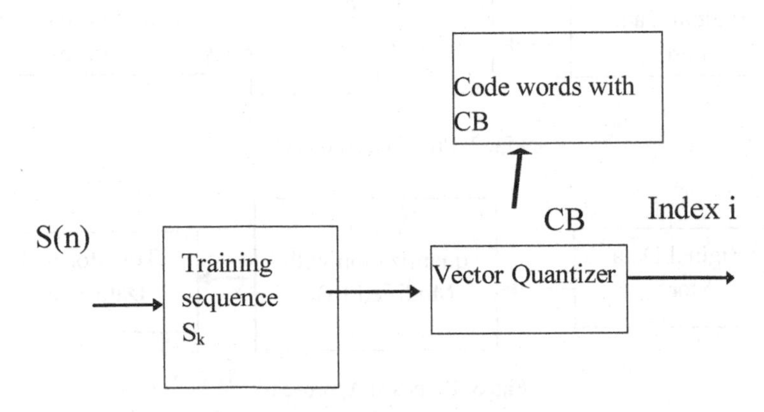

Fig. 4. Components of Vector Quantizer

3.1.1 Proposed Approach 1

Step 1: Take training sequence from real world data and UCI repository water treatment data set have been taken, experiments conducted on two types of data sets.

Step 2: Construct code book using LBG by taking input data from training sequence call it as D

Step 3: Now Code book contains set of all centroids (or code words)
Code book data can be compressed version of original training sequence; we can call it as D'

Step4: clustering is performed on both data sets D and D`. Results were compared.

3.1.2 Proposed Approach 2

Step 1: Take training sequence from Real world data and UCI repository water treatment data set.

Step 2: Construct code book using Modified LBG by taking input data from training sequence call it as D.

Step 3: Now Code book contains set of all centroids (or code words).
Code book data can be compressed version of original training sequence; we can call it as D'.

Step4: clustering is performed on both data sets D and D`. Results are compared.

3.2 Optimality Criteria

Let N_N and C_n are the two parameters that can be used for denoting minimization problem, and it should satisfy the two conditions such as nearest neighbour condition and centroid condition.

3.2.1 Nearest Neighbor Condition

$$N_N = \{V : \| V - C_n \|^2 \leq \| V - C_n' \|^2 \; \forall n' = 1, 2, ...N\} \tag{3.1}$$

Where ER_n, encoding region and C_n is the code vector (centroid). The Nearest neighbor condition says that each encoding region ER_n should consist of set of vectors that are near to C_n than any of the other code vectors.

3.2.2 Centroid Condition (or Code Vector)

$$C_n = \frac{\sum_{X_m \in S_n} V_m}{\sum_{X_m \in S_n} 1} \tag{3.2}$$

This condition says that the code vector or centroid C_n is average of all the training vectors that are in encoding region ER_N. In general, minimum one training vector should belongs to each encoding region ER_N (so that the denominator in the above equation will be never 0). Once transformed data set is generated, a data owner releases the transformed data to the data miners for extracting hidden knowledge from the data store.

4 Experimental Results

In this research, an implementation of employing LBG (Approach 1) Modified LBG (Approach 2) has been developed for code book generations, from which Quantization have been Performed. As mentioned above, the purpose of Vector Quantization is to Transform Original data to other form before sending it to the Data Miners. Furthermore, the results have also proved that Vector Quantization Techniques can be tuned to provide optimal performance with respect to Accuracy of Data mining result and Privacy of Sensitive Data.

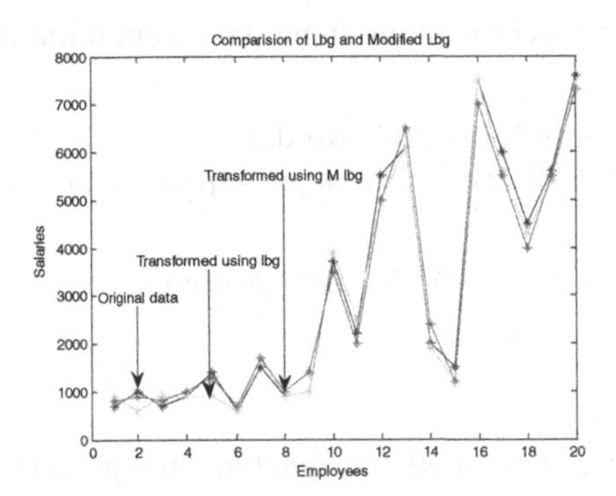

Fig. 5. Comparison of LBG and M-LBG

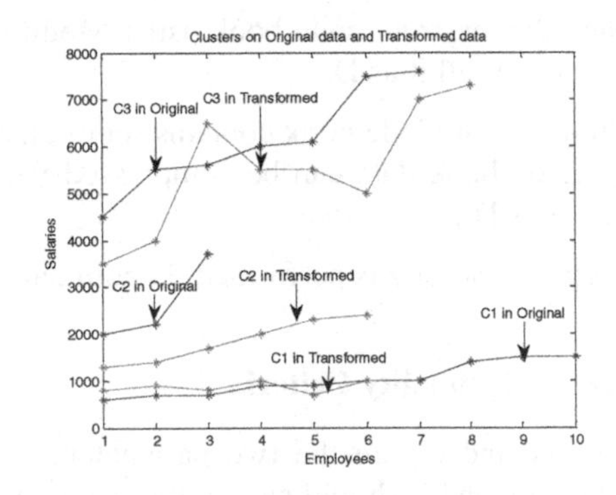

Fig. 6. Clusters between original data and transformed data

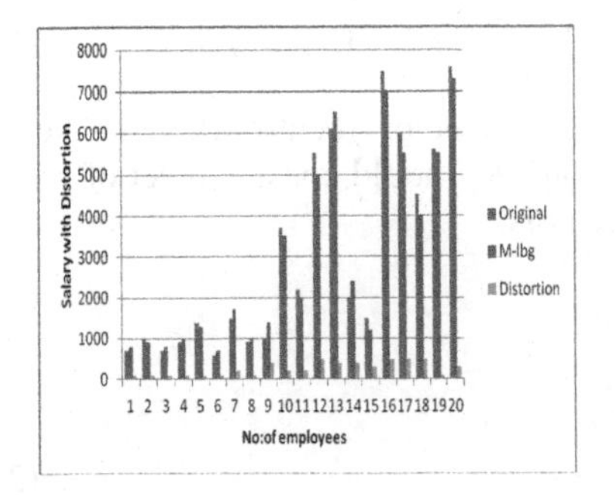

Fig. 7. Distortion between Original and M-LBG

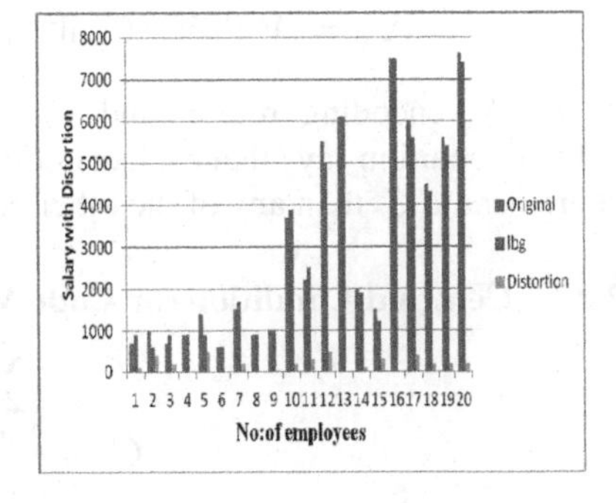

Fig. 8. Distortion Between original and LBG

Figure 5 and 6 represents the original data and transformed data. Figure 7 and Figure 8 shows the differences between original data and transformed data. The main aim of this paper to transform original data to sanitized data. so that, sanitized data does reveal individual information. LBG algorithm and M-lbg algorithms are used in codebook generations phase of vector quantization. When M- lbg is used for codebook generation, we get nearest values in the transformed data compared to Lbg.

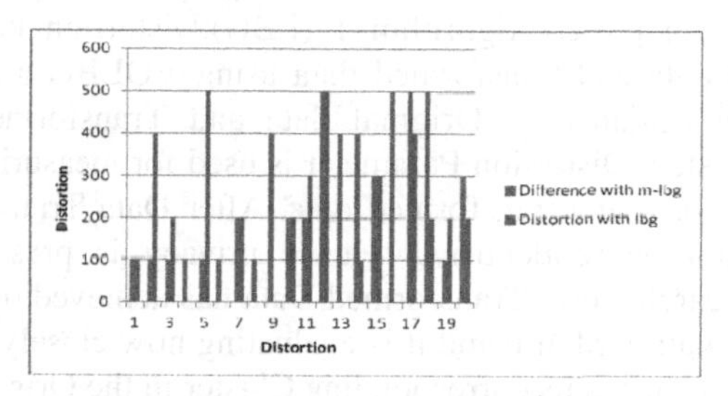

Fig. 9. Distortion between the VQ-LBG and VQ-M LBG

Table 3. Distortion comparision with M-LBG

Original	M-LBG	Distortion
700	800	100
1000	900	100
700	800	100
900	1000	100
1400	1300	100
600	700	100
1500	1700	200
900	1000	100
1000	1400	400
3700	3500	200
2200	2000	200
5500	5000	500
6100	6500	400
2000	2400	400
1500	1200	300
7500	7000	500
6000	5500	500
4500	4000	500
5600	5500	100
7600	7300	300

Table 4. Distortion comparision with M-LBG

Original	LBG	Distortion
700	900	100
1000	600	400
700	900	200
900	900	0
1400	900	500
600	600	0
1500	1700	200
900	900	0
1000	1000	0
3700	3900	200
2200	2500	300
5500	5000	500
6100	6100	0
2000	1900	100
1500	1200	300
7500	7500	0
6000	5600	400
4500	4300	200
5600	5400	200
7600	7400	200

Accuracy looks for similarity between the clusters and it compares how closely the objects of one cluster in the original data are matching with objects of other cluster in the Transformed dataset. It can be calculated by measuring information loss with the following formula. Table 3 and Table 4 finds the dissimilarities between the original data and transformed using lbg and m-lbg and achieved using bellow formula.

$$Infor_L = \frac{1}{N}\sum_{i=1}^{K} Cluster_i(D) - Cluster_i(D')$$

5 Conclusions

In case of an objective where the requirement is to preserve privacy with the data mining results, we can prefer Algorithm 1 (LBG).Distortion between the Cluster variance of original data and transformed data using VQLBG are tested. Distortion Between the cluster variance of Original data and Transformed data using VQ Modified LBG are tested. Distortion Parameter is used for measuring the dissimilarity between the original data and transformed data. After Data Transformation, Salaries of Individuals cannot be re-identified. Hence privacy is preserved for sensitive attribute. In this research work, Transformed Data has achieved desired Data Quality using Accuracy Evaluation Metric and it is evaluating how closely each Cluster in the Transformed dataset matches its corresponding Cluster in the Original data set.

References

1. Latha, M.M., Ram, M.S.S., Siddaiah, P.: Multi Switched Split Vector Quantizer. International Journal of Computer Information and Systems Science and Engineering, IJCSE 2(1) (2010)
2. Latha, M.M., Ram, M.S.S., Siddaiah, P.: Multi Switched Split Vector Quantization. Proceedings of World Academy of Science Engineering and Technology 27 (February 2008) ISSN 1307-6884
3. Linde, Y., Buzo, A., Gray, R.M.: An algorithm for vector quantizer design. IEEE Trans. COM-28(1), 84–95 (1980)
4. Agrawal, R., Srikant, R.: Privacy Preserving Data Mining. In: Proc. of ACM SIGMOD Conference on Management of Data (SIGMOD 2000), Dallas, TX (2000)
5. Privacy and Security Aspects of Data Mining in conjunction with ICDM 2004, Brighton, UK (November 2004)
6. Sinha, B.K.: Privacy preserving clustering in data mining
7. Verykios, V.S., Bertino, E., Fovino, I.N.: State-of-the-art in Privacy Preserving Data Mining. SIGMOD Record 33(1) (March 2004)
8. Tsai, C.W., Lee, C.Y., Chiang, M.C., Yang, C.S.: A Fast VQ Codebook Generation Algorithm via Pattern Reduction. Pattern Recognition Letters 30, 653–660 (2009)
9. Vaidya, J., Clifton, C.: Privacy-Preserving Decision Trees over vertically partitioned data. In: Jajodia, S., Wijesekera, D. (eds.) Data and Applications Security 2005. LNCS, vol. 3654, pp. 139–152. Springer, Heidelberg (2005)
10. Somasundaram, K., Vimala, S.: A Novel Codebook Initialization Technique for Generalized Lloyd Algorithm using Cluster Density. International Journal on Computer Science and Engineering 2(5), 1807–1809 (2010)
11. Clifton, C., Kantarcioglou, M., Lin, X., Zhu, M.: Tools for privacy-preserving distributed data mining. ACM SIGKDD Explorations 4(2) (2002)
12. Kantarcioglu, M., Clifton, C.: Privacy-Preserving Distributed Mining of Association Rules on Horizontally Partitioned Data. IEEE TKDE Journal 16(9) (2004)
13. Kantarcioglu, M., Vaidya, J.: Privacy-Preserving Naive Bayes Classifier for Horizontally Partitioned Data. In: IEEE Workshop on Privacy-Preserving Data Mining (2003)

A Memory Efficient Algorithm with Enhance Preprocessing Technique for Web Usage Mining

Nisarg Pathak[1], Viral Shah[2], and Chandramohan Ajmeera[3]

[1] Narsinhbhai Institute of Computer Studies and Management, Sarva Vidyalaya Campus, Kadi - 382 715, Gujarat, India
nisarg.pathak@gmail.com

[2] SAP BI-BO, Accenture India, Mumbai, India
viral54@gmail.com

[3] Department of Computer Science and Engineering, Osmania University, Hyderabad, Telangana, India
cm.ajmeera@gmail.com

Abstract. Huge amount of data is generated daily by billions of web users. The usage pattern of the web data could be very prized to the company in the field of understanding consumer behavior. Web usage mining includes three phases namely preprocessing, pattern discovery and pattern analysis. The focus of this paper is to establish an algorithm for pattern discovery based on the association between the users accessed web pages. We have proposed a complete preprocessing methodology to identify the distinct users. The foundation of the algorithm is to find the frequently accessed web pages. The biggest constrain for mining web usage patterns are computation overhead and memory overhead. The performance evaluation of algorithm shows that our algorithm is efficient and scalable.

Keywords: Web Usage Mining, Clustering, Association rules mining, Server Log, User Identification.

1 Introduction

Web Mining is applications of data mining techniques on web server log data to automatically retrieves, extract and analyze information. To discover the knowledge regarding website usage it is essential to mine the web usage files generated on web servers [1], [2]. The web mining is roughly categories in three different horizons as follows:

Web Content Mining. Mining the information contain in the web page i.e. text, image, audio or video data in the web.

Web Structure Mining. It is mining the website's hyperlink structure. It uses the *graph theory* to analyze the graph structure of web.

Web Usage Mining. It is used to discover the usage pattern of web surfing, to understand the surfing patterns. It uses the secondary data retrieve from the interaction of users to the web server while surfing the website. The

S.C. Satapathy et al. (eds.), *Emerging ICT for Bridging the Future – Volume 2*,
Advances in Intelligent Systems and Computing 338, DOI: 10.1007/978-3-319-13731-5_65

user interaction generates *registration data, user session, cookies, user URL request, & mouse clicks* etc... The web usage mining is three step process:

- Preprocessing: It cleans web log data by removing log entries that are not needed for the mining process, data integration, identify users, & sessions.
- Pattern discovery: Different data mining methods alike path analysis, association rule, sequential patterns, cluster & classification are applied to discover patterns.
- Pattern analysis: The patterns are analyzed using various tools to filter out the potentially useful rules/patterns [9].

The discovered rules and patterns can then be used for improving the system performance for making modifications to the web site. The web usage mining is an aid to improve the website, to attract visitors, or to give regular users a personalized and adaptive services.

1.1 Pattern Discovery Phase

Association rule mining is an important and fundamental task of data mining to find the correlation among data items [10]. We have proposed the algorithm here where we have tried to apply the *Clustered Association Rule Mining* to the web usage data. We have obtain the association among web pages that are frequently appear together in user's session. At the end of the algorithm execution we have the result like *A.html, B.html* $\rightarrow$ *C.html.* The interpretation of the result is *The user who have visited page A and page B, it is quite possible that during the same session user will visit the page C.*

The information contained in association rules can be used to learn about website visitor behavior patterns, enhance website structure making it more effective for the visitors, or improve web marketing campaigns [3] [4]. The web page association is differs from the general association mining on transactional databases. The web usage data contains a large number of tightly correlated items due to the link structure of a website [5]. The web pages in the same user sessions gives *hard association rules* which may not be the potentially useful [6]. Some of the proposed association rule interestingness measures are all-confidence [7], collective strength [8], conviction and lift.

1.2 Web Log Data

It's the data which points the information about website visitor activity. Any user query is register as a log information in server whenever a user request for any page, text, image etc...

Location of Web Log File: Web log file can be located either from *Web server logs* or *Web proxy server* or can be from *Client's browser.*

Type of Web Log File: Server logs are also found in different types. They are *Access log file*, *Error log file*, *Agent log file* and *Referrer log file.*

Web Log File Format: The file is simple plain text file. The display of log files data are either in *W3C Extended format* or *NCSA common and IIS format.*

A web server log file records are requests in chronological order. The most popular log file formats are the *Common Log Format* and the *Extended Common Log Format.* A common log format file is created by the web server to keep track of the requests that occur on a web site. A standard log file has the following format:

$$< \text{IP addr} > < \text{base URL} > < \text{date} > < \text{method} > < \text{file} >$$

$$< \text{Protocol} > < \text{code} > < \text{bytes} > < \text{referrer} > < \text{user agent} >$$

2 Web Usage Mining Architecture

The basics of web usage mining is described in section 1. The steps of web usage mining is described as follows:

2.1 Data Collection

The for web usage mining is collected from web server log [11].

2.2 Pre-processing

The phase cleans the data and present it in structured form like relational database of XML to provide better support to log query for pattern discovery. Web usage mining requires the use of special heuristics functions. The normal procedure of data pre-processing includes following 5 steps:

1. **Data Cleansing:** This step removes irrelevant items from web log data [12]. The following two types of information are irrelevant: The *first is graphics, video and formate information* and *second is HTTP status code.* The status codes over 299 or under 200 of every records in web logs are removed.
2. **User Identification:** We need to identify each distinct user. This can be easily done if the user gives identity like username and password each time user accesses the website. Another way to identify the user is the user of cookies, but abuse of cookies information can lead to violations of privacy. Only IP Addresses are not sufficient for user identification, the introduction of proxy server has greater the difficulties. The researchers have proposed heuristic method and user identification through navigational patterns. To design the efficient website we need to identify the unique webuser. We have adopted the rule for user identification: *Different IP addresses distinguish the users, if the IP address is same then different browsers and OS distinguish the users.*
3. **User Session Identification:** A session is a sequence of webpage navigation through website. If all of the IP address, browsers and operating systems are same, the referrer information should be taken into account. The simplest methods are time oriented in which one method based on total session

time and the other based on single page stay time. The second method depends on page stay time which is calculated with the difference between two timestamps. Third method based on navigation uses web topology in graph format. We have used 30 minute timeout for sessions timeout property value.
4. **Path Completion:** Another pre-processing task which is usually performed after sessionization is path completion. The task uses referring URLs and website topology. Client or proxy-side caching can often result in missing access references to those pages or objects that have been cached. However, by examining the site topology and the referrer field it is possible to partially reconstruct the path followed by the user. At the end of this phase, the user session file is ready.

2.3 Association Transactions and Their Identification

The user session file is now prepared for data mining. The file is a collection of page reference grouped by user sessions. The task can be accomplished by any of the following methods:

- time window
- reference length
- maximal forward reference

The task varies with the techniques used for web usage mining. We have, in this research paper, applied Clustered Association Rule Mining.

2.4 Pattern Analysis

The final step is to analyze the patterns to eliminate the irrelevant rules or patterns.

3 Proposed Algorithm of Clustered Association Rule

In the association rule mining sometimes all the generated rules are not actionable. Sometimes it may be misleading too. To eliminate such misleading rule, at the time of association rule generation set the minimum support and minimum confidence at such a level. It is obvious that if the minimum support is too small all the rules will be taken in consideration and if the minimum support is very large the rules may not get generated. For any organization just to get the web page association may not be as meaningful. Different group of people access the website, the rule extracted from the weblog may not give desired knowledge for the particular group.

3.1 Solution to Limitation of Association Rule Mining

So the obvious solution is to group the webuser based on their access patterns. Clustering is one such process which groups the webuser in same cluster with similar access patter. This reduces the size of data set to specific user group for association rule mining. As the rule generation is done targeted user group the generated rules are highly relevant and meaningful.

3.2 Proposed Algorithm

The proposed approach for web usage mining is a three step process.

1. The very first step is to collect weblog file, as discussed in section 2.1, followed by data pre-processing, as discussed in section 2.2. At the completiion of the pre-processing the refined weblog file is stored.
2. The second step is to find the cluster within the weblog data. Here we have used two different kinds of clusting algorithm one is EM-algorithm and other is DBSCAN algorithm.
3. The association rule generation process is carried-out by our algorithm.
   - Our algorithm is mainly for the association rule by generating frequently accessed unique pages occurred in the transaction database and the maximum length that can be possible in web usage mining.

As per the proposed approach clustering will be performed and every user will be assigned to specific cluster according to behaviour and access patterns. If we apply the association rule mining technique on this clustered data, it will require less computing time, less memory usage and better accuracy.

4 Experimental Results

The proposed algorithm is tested on the university information system accessed by students, academicians and other curious users. To get the proper information the web pages of the information system is divided in two categories, either it could be content page or supporting.

Step 1. The collected log file has 5,82,815 entries. After the preprocessing phase, as discussed in section 2.2, the log file entries are reduced to 32,378. As discussed, we kept the User Session Identification time threshold to 30 minutes and 10,289 sessions were identified.

Step 2. To segment the webusers in clusters EM clustering is used.

The Table 1 shows two segments of visitors. The result is very clear the *Cluster-1* corresponds to the users who have accessed the content pages more, obviously they are either students or faculties, and the *Cluster-2* corresponds to the users who have accessed the supporting pages more, they are the website visitors who are keen to get information about the university.

Table 1. User Access Pattern

	Content	Supporting
Cluster 1	1	0.0323
Probability = 0.7012	D = 0.4463	D = 0.1768
Cluster 2	0	1
Probability = 0.2988	D = 0.4463	D = 0.4574

Table 2. Navigation Pattern for 3 Sessions

	Access-1	Access-2	Access-3
Cluster 1	Supporting	Supporting	Supporting
Probability = 0.6825	350 (376.03)	374.97 (376.03)	375.03 (376.03)
Cluster 2	Content	Content	Content
Probability = 0.3175	174.97 (175.97)	174.97 (175.97)	174.97 (175.97)

Table 3. Navigation Pattern for 4 Sessions

	Access-1	Access-2	Access-3	Access-4
Cluster 1	Content	Content	Content	Content
Probability = 0.676	92.95 (93.95)	92.9 (93.95)	87.82 (93.95)	86.71 (93.95)
Cluster 2	Supporting	Supporting	Supporting	Supporting
Probability = 0.2601	34.47(36.51)	35.49 (36.51)	35.51 (36.51)	35.49 (36.51)
Cluster 3	Supporting	Supporting	Supporting	Supporting
Probability = 0.0639	10.01(11.54)	8.46 (11.54)	10.35 (11.54)	9.27 (11.54)

For the sake of experiment we have taken the users session with sequential page access is 3-Pages, 4-Pages or 5-Pages.

The Table 2 displays the user segmentation where users have access at least 3 web pages. As per th result in Table 2 probability of Cluster-1 is higher. This depicts any user access of less than 3 pages are the outside person who has not reached to any particular content page.

The webuser with 4 sequential pages visits were segmented in Table 3. The largest cluster, cluster-1, corresponds to the access of content pages. Other clusters, Cluster-2 and Cluster-3, with lesser probability corresponds to supporting pages. This may lead us to a vital information that the visitor of the website may not reaches to a particular information that was looking.

Step 3. We now apply our algorithm on the particular user cluster to determine the Association Rules. Here in our example we may find-out the association rules of Cluster-1 in 3-webpage access users. After getting the access patterns we may act upon the pages from where user abandons the website. Similarly for the users with 4-webpage access we may come-up with the navigational rules to construct a knowledge regarding users information requirement and based on that we may refine contents of the webpages.

5 Performance Evaluation

We have executed our algorithm on clustered data and performed the comparative analysis with the very basic Association Rule Mining algorithm, *Apriori*. We have segmented the clustered weblog data into smaller parts to compare the accuracy and performance of our proposed algorithm with apriori algorithm. The result is presented in Figure 1. By looking at the figure 1 we can clearly say that the accuracy is improved at the great extent. We were able to achieve

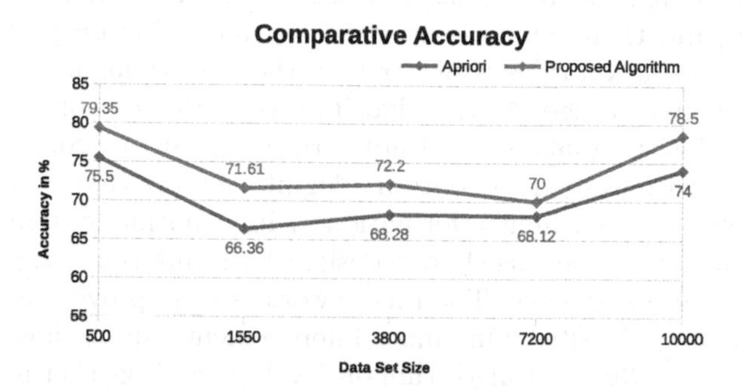

Fig. 1. Comparative Accuracy of Proposed Algorithm with ARM

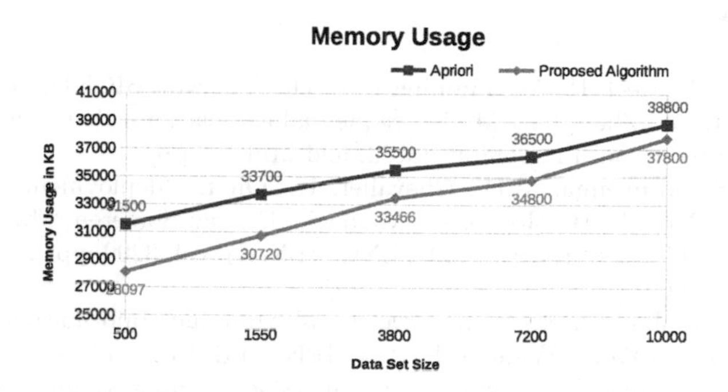

Fig. 2. Comparative Memory Usage

almost 80% of accuracy in rule generation. As per the figure 1 we can say that increase in the data set also increases the accuracy of the result. The memory usage of algorithm is total memory required to execute any algorithm. The maximum amount of memory required to execute the algorithm is considered as *Peak Memory Required.* The Figure 2 shows the Peak Memory Required for Apriori and our algorithm. It can be observed from figure 2 that the Peak Memory Requirement of algorithm is dependent on the size of the input data. From the figure 2 comparative analysis shows that though the Peak Memory Requirement is increasing with the size of data, still the memory required to execute the algorithm is lesser then apriori algorithm.

6 Conclusion

We can describe the web usage mining as the sighting and scrutiny of user based on their access patterns. For web usage mining one of the easiest and simple algorithm is the Apriori algorithm. But to overcome the drawback of association rule mining we have proposed a new algorithm of clustered association rule mining. Which creates cluster of the web user at the first stage based on their access patterns and then generates association rules. Our proposed algorithm for generating association rules is better then the Apriori algorithm in terms of accuracy and memory usage. As the algorithm is carried out on clustered data, the probability of getting false or misleading rules is near to zero. To validate the algorithm an exercise was carried out on a log file of more than 6,00,000 entries. The exercise revels astonishing information of information systems users. The derived information may be used to redesigned the information system pages to improve the user experience. The future work is to improve the algorithm to tackle the dynamic web sites. One more improvement can be done at the clustering stage by using different algorithm or developing altogether new algorithm which is domain specific.

References

1. Kosala, R., Blockeel, H.: Web mining research: A survey. SIGKDD Explorations
2. Wang, Y., Li, Z., Zhang, Y.: Mining sequential association-rule for improving Web document prediction. In: Compu. Intell. and Multi. App.
3. Anand, S.S., Mulvenna, M.D., Chevalier, K.: On the deployment of web usage mining. In: Berendt, B., Hotho, A., Mladenič, D., van Someren, M., Spiliopoulou, M., Stumme, G. (eds.) EWMF 2003. LNCS (LNAI), vol. 3209, pp. 23–42. Springer, Heidelberg (2004)
4. Cooley, R., Mobasher, B., Srivastava, J.: Web mining: Information and pattern discovery on the World Wide Web. Proc. IEEE Intl. Conf. Tools with AI
5. Huang, X., Cercone, N., Aijun, A.: Comparison of interestingness functions for learning web usage patterns. In: Proc. Intl. Conf. Info. Knowl. Managt. (2002)
6. Huang, X.: Comparison of interestingness measures for web usage mining: An empirical study. IJITDM
7. Omiecinski, E.R.: Alternative interest measures for mining associations in databases. IEEE Transac. on Knowl. and Data Engi.
8. Aggarwal, C.C., Yu, P.S.: A new framework for itemset generation. In: Symposium on Principles of Database Systems
9. Dong, D.: Exploring on Web Usage Mining and its Application. In: 5th World Congress on Intelligent Control and Automation, June 15-19 (2004)
10. Dunham, M.H.: Data Mining Introductory and Advanced Topics. Person Educations, India (2006 edition)
11. Pitkow, J., Bharat, K.K.: WebViz, A tool for world wide web access log analysis. In: 1st International WWW Conference (1994)
12. Aye, T.: Web log Cleaning for mining of web usage patterns, 978-1-61284-840-2/11/2011 IEEE

A Systematic Literature Review on Ontology Based Context Management System

Rajarajeswari Subbaraj and Neelanarayanan Venkatraman

VIT University, Chennai, Tamil Nadu, India
{rajarajeswari.s,neelanarayanan.v}@vit.ac.in

Abstract. In ubiquitous environments, context modeling components and personalization engines make systems adaptable to the user behavior. Current research in context management focused on specific domain or environment. To the best of our knowledge, there are no open standardized interfaces or standard representation of context models exist. Recently, ontology based Context Management System (CMS) has been proposed in research for the design of context aware system in ubiquitous computing. The objective of this research work is to systematically identify relevant research work from various electronic data sources using systematic literature review (SLR) method. Our searches identified 188 papers, out of which we found 31 for inclusion in this study. The identified CMS have been analyzed in order to identify research gaps in ontology based Context Management (CM) research domain. The findings show that most ontology based CMS do exist which are predominantly focused on specialized or specific application domain. Considering the findings in the context of previously proposed research agendas, some of the key challenges identified in ontology based CMS are: lack of consistency checker component for verifying or ensuring consistency of sensed or gathered context information against the context model and unsecured means of transferring context information between various components in a distributed environment.

Keywords: Ontology, Context Management System, Context Management Framework, Context-Aware Computing.

1 Introduction

Ubiquitous computing is an emerging trend where computers are made available for human need. To emphasize more on this, ubiquitous computing plays a major role in many areas like human computer interaction, artificial intelligence, context aware computing etc. Out of the three, we shall focus more on context aware computing. Context is nothing but a way of describing or characterizing the current status of an entity or object. An entity may be anything like person, location, activity, time, relation among them. This is termed as context information. Context information can be an emotional status of the user or event or position or orientation or date or time or objects which are used by the people. Context information can be obtained from either physical or logical sensors. Context aware computing begins with acquisition of

S.C. Satapathy et al. (eds.), *Emerging ICT for Bridging the Future – Volume 2*,
Advances in Intelligent Systems and Computing 338, DOI: 10.1007/978-3-319-13731-5_66

context data, then understanding the context and triggering the events based on the context. Ubiquitous computing, or pervasive computing, supports the vision in which computing is transparently integrated into our living environment and daily lives. Everyday objects are empowered with computational capabilities in order to enable users to interact with computing devices more naturally and casually than we currently do with desktop computers. Context aware applications are capable of autonomously adapting their behavior in response to context changes. Based on a systematic literature review approach, we identify the requirements for modeling context and for building a generic and flexible platform to support context-aware application development. Kitchenham et al. suggest Systematic Literature Review (SLR) [2] as a main method to undergo complete study on any research problem. The aim of an SLR is not just to aggregate all existing evidence on a research question but to provide appropriate software solution in a specific context. Searching can be done in two ways either by automatic search, i.e. search for a text or string in Electronic Data Sources (EDS) or by manual search, where searching is done manually by browsing the journals or conference proceedings. Considering the growing amount of literature on ontology based Context Management System (CMS), timely summary of all relevant knowledge about modeling context information using ontologies, and collection and provision of context information becomes the existing knowledge on ontology based CMS. This paper is organized as follows: Section 2 contains the research methodology applied; it describes all the details explaining how the systematic literature review has been applied and the information necessary to replicate the study; Section 3 discusses about the existing context management system based on ontology and comparison of context management system; and Section 4 presents the summary of results and discussion.

2 Methodology

2.1 Research Method

SLR is a review of research solutions. SLR is not just a method to gather all existing system or solution for a research question; it provides support for practitioners in the development by providing the existing solution. A protocol has been defined in which the research questions were detailed and the process followed to report the results. The screening phase has been broken down into two separate phases. In the initial stage relevant papers were identified by analyzing publications title and abstract and in the second stage, a full-text analysis was undertaken to discover and record the concrete technologies reported in each of the relevant papers. After the second screening stage, the primary studies were subsequently filtered by keywords. The first author was responsible for designing and conducting the study; while, the second author, was in charge of supervising every stage of the study. Tasks undertaken by the second author therefore ranged from the validation of the protocol, the assessment of decisions taken at each stage, the verification of a random subset of papers at each inclusion/exclusion step, and the complete assessment of the final set of primary studies.

2.2 Information Resources

The search has used various electronic sources as mentioned below to gather information:

- ACM Digital Library
- IEEE Explore
- Science Direct
- Springer

These resources contain both the conference proceedings and journals. The searches resulted in duplication of papers, and duplication can be removed manually.

2.3 Search Criteria

The search process was an automated search of specific online electronic sources as given in Section 2.2. A critical phase in performing an automatic search is the identification of the keywords to be used when building a search query string. A pilot search was performed on IEEE Xplore to assess the quality of the search string against the identified set of known relevant studies. The original query was modified accordingly resulting in the following search string: (A1 [in abstract] AND A2 [in document title]).

A1: Ontology
A2: Context Management System

Each document in the search result was reviewed by first author. The first author applied the detailed inclusion and exclusion criteria to make the decision on including or excluding the document. The second author checked all the documents that are included and excluded at this stage to verify the correctness. When difference of opinion existed, both authors discussed their point of view and a final decision on inclusion of the document was made. The discussions and final outcome were recorded for future reference.

2.4 Study Selection

To assess the relevance of each document, the documents were analyzed based on their title and abstract and, when necessary to clear any doubts, their full text by first author. Then the same process was repeated by the second author. Then the final decision was made by both authors after a discussion. Based on titles, duplicate and irrelevant papers are removed from the study in the first stage. At the end of the first stage, 41 papers are included for the next stage. In the next stage, 41 papers were analyzed based on the content in abstract. In the next stage, the full paper was analyzed. After second and third stage, 31 papers were included for the final study. To be included in this study a research paper needed to satisfy two requirements: (i) the document should discuss the design and representation of a context management system and (ii) the context management system must use context model based on ontology.

3 Ontology Based Context Management Systems

Context management system in [13] supports context-aware applications in a smart home. The system collects, observes, interprets, and provides the context data to the applications. Further, the system provides interoperability between system components using semantic level. This system uses meta data about the context information to remove the uncertainty in context data, to ensure the freshness and to understand the context clearly. It uses rule based and ontology based reasoning system.

CANDEL [30], a generic context representation framework that considers the context as product line which consists of set of context primitives [30]. In this approach context management composed of entities, called Context Proxy Components [30] which are distributed in the environment, and it deliver the context information as a feature model [30]. Adaptations of applications according to context is described using petrinet [30] based approach.

Zhou Zhong et al. [32] address the interoperability between CMSs at application level. They propose a context management system which comprises of cross task management and context search engine to support scheduling when cross context happens. Context search engine which derived from semantic searching accepts context, filters the results using queries. Cross task management performs scheduling, monitoring and cross space management of cross context.

Stefan Pietschmann et al. present CROCO [27] a context management service based on ontology, which allows context gathering across different application and context modeling. Cross application Context management Service (CROCO) uses the blackboard approach, which is a data-centric approach where external events can post information on a blackboard, or subscribe for notifications for changes. CROCO performs functions like (context) data management, consistency checking, reasoning, and updating context data.

Dejene Ejigu et al. propose a hybrid approach for collaborative context management [9]. The system uses conceptual context representation model based on ontology and relational database. This system uses ontology for representing the context information and relational schema for the context information management. They propose EHRAM conceptual model for context representation.

Carlos Baladrón et al. present a global context management framework [5] for converged context management and its functioning in a future Internet for the integration of context data from all sources and publish to client applications. Further, the solution allow inference and context prediction with the help of context intelligent approach. The open telco operator infrastructure can be used with this framework for user provisioning and security features.

The architecture for a context aware system for mobility management [19] includes a Global Context Server, some Local Proxy and Adaptation Servers, and the Mobile Nodes [19]. Mobile nodes have a middleware for context management for collection and processing context data and some context information are transferred to the local servers. Initial level decisions on mobility are done in mobile node and high level decisions on mobility are assessed in local servers [19].

HyCoRE [4], a generalized hierarchical Hybrid Context Reasoning Engine design consists logical layers for the following processes: gathering context information which includes sensing the data and preprocessing it, context consumption, processing the context information and context reasoning; knowledge storage and knowledge retrieval. The reasoning engine proposed here supports both probabilistic reasoning and also non-probabilistic reasoning to improve the reasoning mechanism in new or existing CMS.

Herma van Kranenburg et al. designed a generic framework for context processing in tailored applications and provide services in a multi domain mobile environment [16]. This framework supports context which are application specific, cross-layer context processing, exchange of context information in heterogeneous environment, and multi domain context sources are accessed through single interface.

Holger Schmidt et al. [17] propose a context service which provides a generic Web service interface performing functionality like gathering, discovering and managing of context at runtime. The Web Ontology Language (OWL) is used for describing the context and the SPARQL and RDF Query Language are used for discovery.

Beatriz Fuentes et al. [3] present a novel framework for the context management of future networks, that is able to collect, transform and reason on context information. This framework comprised of Context Source, Context Provider, Context Information Base, Context Broker, Context Ontology, Context Client, and Context Quality Control Function. The framework focuses on autonomic context discovery, monitoring, diagnosis and prediction. The proposed framework aims at improving the context awareness level in the network and, thus facilitating the realization of autonomic management mechanisms.

Jinhwan Lee and Kwei-Jay Lin [20] present the context management framework for real-time SOA by using the inteLLigent Accountability Middleware Architecture (LLAMA) . LLAMA is an open-source middleware framework to manage service composition, run-time monitoring, problem analysis, and reconfiguration and optimization [20]. The context agent, which is available in the user's device, collects context information and transfer to the context manager. The context manager incorporates with Qbroker [20], trust broker and context agent for service composition and reconfiguration of services

ConServ is a context management web service which enables sharing of context information and management of context information between heterogeneous smart spaces [15]. ConServ is a REST based web service which uses model-view-controller architectural pattern [15]. RDF data store is used to store user details. The RDF data and normal data is segregated for the effective performance of ConServ and data which is not to be shared or which is not to be used for reasoning is stored in the relational data store.

Amel Bouzeghoub et al. propose a hybrid architecture combining a process oriented context management system COSMOS [1], COntext entitieS coMpositiOn and Sharing) and an ontology-based context management system (MUSE, Multi ontology based User Situation awarenEss). The situation identification is performed at two levels. The first level identifies the situation from the data collected from devices and sensors. The next level applies knowledge base. The hybrid approach leads to scalability and efficiency.

Essa Basaeed et al. proposes a knowledge-centered framework [12] based on ontology which provides the context aware model which identifies learner's goal and activities and mapping the activities to available learning resources and services.

Dexter H. Hu et al. present a context management framework based on semantic technology named Context Torrent [11], which provides semantically searchable and sharable context information among different context-aware applications. In this framework context is modeled using OWL and RDF.

Daniela Nicklas et al. proposes a Nexus semantic services [8] which carry out rules represented using first order logic for reasoning. This approach is explained with the help of a smart environment scenario in the Smart Room, the Conference Guard [8].

Dejene Ejigu et al. propose a collaborative context aware service platform (CoCA) [10] which is independent and used for context-aware application development. CoCA proposes an architecture using ontology and collaborative approaches for context-aware services with reasoning mechanism.

Maria A. Strimpakou et al. describes the COMANTO [25] ontology which is a generic upper level ontology to be used in the context management system and enable stakeholders to share, collaborate and augment their services and context knowledge.

Run Yang et al. provide Context-Aware General Service Model (CAGSM) [26] for use in next generation network NGN. The CAOSM is structured in multiple layers. Through the general description language in CAGSM we can model the context information. The developers can easily use context for applications creation by selecting appropriate description language, as well as discovering and binding with various context service.

Federica Paganelli et al. [14] propose a context management system using ontology which includes a service oriented framework for handling and monitoring the patient health condition. This framework can be integrated with the home care network for handling critical situations of patient and providing alarm.

Josue Iglesias et al. present an inference system PIRAmIDE [21] for context management, for reasoning tasks in resource-constrained devices. The rule-based reasoning is used for retrieving the information from the inference system.

M. Khedr and A. Karmouch present ACAI [24], an innovative Agent-based Context Aware Infrastructure based on ontology, which maintains spontaneous applications both locally and across different domains. A multiagent framework was proposed for assisting development and runtime provisioning of spontaneous applications.

Cinzia Cappiello et al. provide a framework for context representation and context management in a common environment which is adaptable in nature and supports multichannel access. This framework was explained with MAIS [7] architecture and services related to the tourism domain. The MAIS architecture provides required services with appropriate features automatically and efficiently from service providers.

Chiung-Ying Wang et al. proposes UbiPaPaGo [6], a mechanism for finding path in spatial conceptual map (SCM) [6] using genetic algorithm (GA) [6]. The SCM model is used to represent the map of the environment. UbiPaPaGo [6] automatically finds the optimal path that satisfies the requirements of user.

Tarak Chaari et al. [28] propose a context model based on ontology which can be reused. Context model is of two levels: generic level and domain specific level. They propose a generic framework which is adaptable in nature where applications are adaptable to the context in surrounding environment.

Tobias Rho et al. present a context-management system based on RDF which provides reasoning mechanism with tight integration with programming languages. It supports flexible context queries using Context Query Language and it implemented the query library for JCop [29] language.

Zakwan Jaroucheh et al. propose infinitum [31], a middleware architecture that incorporates the Google Wave Federation Protocol [31] which uses ontology-based context models. Infinitum context management system contains a set of context server, which stores context information available in a predefined domain. This architecture builds a cross-domain scalable context management and collaboration framework, which is implemented and evaluated in a real time application of SMART University [31] to support virtual team collaboration.

Korbinian Frank et al. [22] propose the inference system based on probabilistic logic for reasoning and it also includes special inference rules extending Bayesian networks.

Hong-Linh Truong et al. [18] propose a context representation for disaster management domain. This representation is extensible and supports interoperability to include entities existing in disasters and relationships among them.

Lobna Nassar et al. proposed VANET IR-CAS which is a context aware system that utilizes information retrieval (IR) techniques, such as indexing, document scoring and document similarity [23] to enhance context aware information dissemination in VANET. It uses a hybrid context model; spatial model for service filtering, ontology model for context reasoning [23] and knowledge sharing, markup model for file exchange, and situational model for safety and convenience services.

3.1 Criteria for Evaluating Context Management System

Earlier research work in Context management system defines how sensor data can be gathered, processed and evaluated for static environment. To support dynamic adaptation, generic and reusable context management systems are required. The context management is continuously evolving towards a higher intelligence with the better security and privacy management. Various criteria can be identified to evaluate the context management system as discussed below:

- Consistency of context information
- Security of the context information
- Reasoning mechanism
- Privacy of context data

4 Results and Discussions

We found that earlier work on context management system focused on domain specific and location based applications. Further work on context management system supports various domain and different types of context information. Recently ontology based context management system has been widely adapted as it provides extensibility and reasoning mechanism. The goal of this study was to analyze various context management systems which use the ontology for modeling context. The papers were obtained in a search of electronic resources which includes journals, conference and workshop proceedings. Then a systematic selection process was undergone to identify papers related to context management system based on

Table 1. Comparison of Context Management Systems

Paper ID	Consistency	Privacy	Security	Reasoning
1				
3				✓
4				✓
5				✓
6				
7				
8				✓
9				✓
10				
11				
12				✓
13				✓
14				✓
15			✓	
16				✓
17				
18				
19				
20				✓
21				✓
22				
23	✓			✓
24				
25				
26				
27	✓			✓
28				
29				
30	✓			✓
31		✓		
32				

ontology. The results tabulated in Table 1 give a good understanding of context management system. Of 188 papers are analyzed in the systematic review, 31 papers were identified as an ontology based context management system. Considering the findings in the context of previously proposed research agendas, some of the key challenges in ontology based CMS are (i) Lack of consistency checker component for verifying or ensuring consistency of sensed or gathered context information against the context model (ii) Unsecure means of transferring context information between various components in a distributed environment.

References

1. Bouzeghoub, A., Taconet, C., Jarraya, A., Do, N.K., Conan, D.: Complementarity of Process-oriented and Ontology-based Context Managers to Identify Situations, pp. 222–229. IEEE (2010)
2. Kitchenham, B., Pearl Brereton, O., Budgen, D., Turner, M., Bailey, J., Linkman, S.: Systematic literature reviews in software engineering – A systematic literature review. Information and Software Technology 51, 7–15 (2009)
3. Fuentes, B., Bantouna, A., Bennacer, L., Calochira, G., Ghader, M., Katsikas, G., Yousaf, F.Z.: On Accomplishing Context Awareness for Autonomic Network Management. IIMC International Information Management Corporation (2012)
4. Beamon, B., Kumar, M.: HyCoRE: Towards a Generalized Hierarchical Hybrid Context Reasoning Engine, pp. 30–36. IEEE (2010)
5. Baladrón, C., Aguiar, J.M., Carro, B., Calavia, L., Sánchez-Esguevillas, A.: Framework for Intelligent Service Adaptation to User's Context in Next Generation Networks. IEEE Communications Magazine, 18–22 (2012)
6. Wang, C.-Y., Hwang, R.-H., Ting, C.-K.: UbiPaPaGo: Context aware path planning. Expert Systems with Applications 38, 4150–4161 (2011)
7. Cappiello, C., Comuzzi, M., Mussi, E., Pernici, B.: Context Management for Adaptive Information Systems. Electronic Notes in Theoretical Computer Science 146, 69–84 (2006)
8. Nicklas, D., Grossmann, M., Mínguez, J., Wielandy, M.: Adding High-level Reasoning to Efficient Low-level Context Management: a Hybrid Approach. In: Sixth Annual IEEE International Conference on Pervasive Computing and Communications, pp. 447–452 (2008)
9. Ejigu, D., Scuturici, M., Brunie, L.: Semantic Approach to Context Management and Reasoning in Ubiquitous Context-Aware Systems. IEEE (2007)
10. Ejigu, D., Scuturici, M., Brunie, L.: CoCA: A Collaborative Context-Aware Service Platform for Pervasive Computing. In: International Conference on Information Technology (2007)
11. Hu, D.H., Dong, F., Wang, C.-L.: A Semantic Context Management Framework on Mobile Device. In: International Conferences on Embedded Software and Systems, pp. 331–339 (2009)
12. Basaeed, E., Berri, J., Benlamri, R., Zemerly, J.: M-Learning Activity-based Context Management. IEEE (2007)
13. Kim, E., Choi, J.: A Context Management System for Supporting Context-Aware Applications. In: IEEE/IFIP International Conference on Embedded and Ubiquitous Computing, pp. 577–582 (2008)

14. Paganelli., F., Giuli, D.: An Ontology-Based System for Context-Aware and Configurable Services to Support Home-Based Continuous Care. IEEE Transaction on information Technology in Biomedicine 15(2), 324–333 (2011)
15. Hynes, G., Reynolds, V., Hauswirth, M.: Enabling Mobility between Context-Aware Smart Spaces. In: International Conference on Advanced Information Networking and Applications Workshops, pp. 255–260 (2009)
16. van Kranenburg, H., Bargh, M.S., Iacob, S., Peddemors, A.: A Context Management Framework for Supporting Context-Aware Distributed Applications. IEEE Communications Magazine, 67–74 (2006)
17. Schmidt, H., Flerlage, F., Hauck, F.J.: A Generic Context Service for Ubiquitous Environments. IEEE (2009)
18. Truong, H.-L., Manzoor, A., Dustdar, S.: On Modeling, Collecting and Utilizing Context Information for Disaster Responses in Pervasive Environments. In: CASTA 2009, Amsterdam (2009)
19. Pedrasa, J.R., Seneviratne, A.: A Proposed Architecture for Context-Aware Mobility Management. In: International Symposium on Communications and Information Technologies, pp. 721–734 (2007)
20. Lee, J., Lin, K.-J.: A Context Management Framework for Real-Time SOA, pp. 559–564. IEEE (2010)
21. Iglesias, J., Bernardos, A.M., Álvarez, A., Sacristán, M.: A light reasoning infrastructure to enable context aware mobile applications. In: IEEE/IFIP International Conference on Embedded and Ubiquitous Computing (2010)
22. Frank, K., Kalatzis, N., Roussaki, I., Liampotis, N.: Challenges for Context Management Systems Imposed by Context Inference. In: MUCS 2009, Spain (2009)
23. Nassar, L., Karray, F., Kamel, M., Sattar, F.: VANET IR-CAS: Utilizing IR Techniques in Developing Context Aware System for VANET. In: DIVANet 2012, Paphos (2012)
24. Khedr, M., Karmouch, A.: ACAI: agent-based context-aware infrastructure for spontaneous applications. Journal of Network and Computer Applications 28, 19–44 (2005)
25. Strimpakou, M.A., Roussaki, I.G., Anagnostou, M.E.: Context Ontology for Pervasive Service Provision. In: Proceedings of the 20th International Conference on Advanced Information Networking and Applications, AINA 2006 (2006)
26. Yang, R., Yang, Y., Meng, X., Mi, Z.: The study on the context-aware application architecture in NGN. In: ICWMMN 2006 Proceedings (2006)
27. Pietschmann, S., Mitschick, A., Winkler, R., Meißner, K.: CROCO: Ontology-Based, Cross Application Context Management. In: Third International Workshop on Semantic Media Adaptation and Personalization, pp. 88–93 (2008)
28. Chaari, T., Ejigu, D., Laforest, F., Scuturici, V.-M.: A comprehensive approach to model and use context for adapting applications in pervasive environments. The Journal of Systems and Software 80, 1973–1992 (2007)
29. Rho, T., Appeltauer, M., Lerche, S., Cremers, A.B., Hirschfeld, R.: A Context Management Infrastructure with Language Integration Support. ACM (2011)
30. Jaroucheh, Z., Liu, X., Smith, S.: CANDEL: Product Line Based Dynamic Context Management for Pervasive Applications. In: International Conference on Complex, Intelligent and Software Intensive Systems, pp. 209–216 (2010)

31. Jaroucheh, Z., Liu, X., Smith, S.: An approach to domain-based scalable context management architecture in pervasive environments. Pers. Ubiquit. Comput. 16, 741–755 (2012)
32. Zhong, Z., Gu, J., Zhang, Y., Lin, X.: Enhancing applications interoperability in context management for practice tasks. In: IEEE International Multi-Disciplinary Conference on Cognitive Methods in Situation Awareness and Decision Support (CogSIMA), San Diego, pp. 122–129 (2013)

Improving Classification by Outlier Detection and Removal

Pankaj Kumar Sharma, Hammad Haleem, and Tanvir Ahmad

Department of Computer Engineering, Faculty of Engineering and Technology, Jamia Millia Islamia, New Delhi 110025, India
sharmapankaj1992@gmail.com, hhaleem@connect.ust.hk, tahmad2@jmi.ac.in

Abstract. Most of the existing state-of-art techniques for outlier detection and removal are based upon density based clustering of given dataset. In this paper we have suggested a novel approach for iteratively pruning of outliers based upon the non-alignment with model created in a n-dimensional hyperspace. The technique could be used for any classification problem as a pre-processing step, regardless of the classifier used. We have tested our hypothesis with Support Vector and RandomForest classifiers and have obtained significant improvement in results for both these classifiers. The effectiveness of this novel method has also been verified by improvements in results while performing classification using these standard classifiers. When pruned with our method standard classifiers like SVC and RandomForest classifier showed an improvement up to 4 percent.

Keywords: Outlier detection, Outlier removal, Supervised learning, Classification, preprocessing.

1 Introduction

In the field of information retrieval, especially in the domain of classification problems there has been a lot of ongoing research. A large number of researchers have done research work in various applications of classification algorithms. Many algorithms have been developed using very abstract ideas adopted from other fields like Statistics, Probability, Neural networks and Vectors. These methods have fared quite well and have been able to deliver satisfactory results. These classification algorithms have been used widely in to (but not limited to) text classification, web page classification, image classification etc. Most classification algorithms can be easily classified into two broad subsets namely supervised learning and unsupervised or semi-supervised learning methods.

Supervised learning based classification algorithms use class labels, present along the dataset which embarks their belonging to one or more classes. The necessary conditions involved with supervised learning based models is that the class labels must be present for each and every point present in the dataset. This could be done either by manually assigning class label to each point or by doing with a help of some non supervised learning mechanism. An unsupervised

S.C. Satapathy et al. (eds.), *Emerging ICT for Bridging the Future – Volume 2*, Advances in Intelligent Systems and Computing 338, DOI: 10.1007/978-3-319-13731-5_67

classification algorithm, does not require class labels. These algorithms can form clusters of similar points and could be used for embarking clusters to those class labels rather than marking each point individually. Lastly the semi supervised algorithm makes the best use of supervised and unsupervised learning algorithms. These methods might require a very small amount of labelled dataset.

In recent past all this ongoing research had been successfully able to produce a variety of state of art classification methods. These methods generally deliver quite satisfactory results on a variety of classification problems. Still almost all of these state-of-art classification techniques are vulnerable to and suffer loss in efficiency in the presence of outliers or noisy data points in the dataset. In a variety of previous researches it has been proved that a pruned or less noisy dataset is able to deliver better classification results compared to a dataset with noises and outliers.

An outlier in a dataset is an observation or a point that is considerably dissimilar to or inconsistent with the remainder of the data. Outlier is a data object that deviates significantly from from the rest of the objects.Detection of such outliers is important for many applications.

Outlier detection and removal algorithms find multiple applications within data mining. Data cleansing requires that aberrant data items be identified and dealt with appropriately. For example, outliers are removed or considered separately in regression modelling for achieving improved accuracy. In many applications outliers are more interesting than non aberrant data, fraud detection in online transactions is a classic example where attention gets focused on the outliers because these are more likely to represent cases of fraud.

With a relatively large amount of features present in a dataset, at times it becomes a challenging task to identify the outliers present. In this paper we have proposed an efficient and effective outlier detection and removal algorithm, specifically useful in supervised classification problems. In the proposed method we create a n-dimensional vector hyperspace based model. The model considers a specific feature as a dimension in the hyperspace. When these individual feature vectors are summed we generate a n-dimensional vector for the dataset. With the iterative pruning method proposed in the subsequent section, it become really easy computationally to remove those outliers and subsequently their effect from the training dataset.

In this paper the technique presented for outlier pruning from the training dataset could improve the results obtained by state of the art methods. The technique provides a promising advantage, regardless of the classification technique used underneath. The proposed hypothesis was implemented and tested with the Support Vector Classifier (SVC) and RandomForest classifier and significant improvement in results for both classifiers was registered.

This paper has been divided into 6 sections. Section 2 attempts to identify the related work. Section 3 consists of our proposed method. Section 4 consists of details of experiments that we carried while testing our proposed method, whose results are explained in detail in section 5. Section 6 finally concludes our work, followed by references.

2 Related Work

In one of the most recognized work related to outlier detection and removal [1], the authors suggest that outliers may be considered as the data points which does not belong any to any particular cluster but cannot be considered as noise. A very significant effort in this field has been made by Hodge and Austin in [2]. Their work have documented the development in this field well. Further another review of outlier detection algorithms using statistical techniques is given by presented by Petrovskiy [3]. Another robust technique of identifying outliers efficiently on very large data sets based on distance measured are given in [4-13]. Another class of outlier detection methods where a cluster of small sizes can be considered as clustered outliers [14-19]. Hu and Sung [20], who proposed a method to identify both high and low density pattern clustering, further partition this class to hard classifiers and soft classifiers. A number of other methods which used mathematical models for outlier detection and removal have been research. Another related class of methods consists of detection techniques for spatial outliers [15, 18, 21].

3 Proposed Method

The proposed technique can be divided into following steps.

The algorithm uses the training datasets for two purposes. Firstly for creating and iterating over the n-dimensional categories vectors, thereby pruning the dataset by identifying and removing outliers. Here n in n-dimensional category vector is equal to the number of features involved in the classification and total number of categories vectors generated is equal to the number of categories involved (two in case of a binary classification problem). Secondly the dataset is also used for determining the accuracy of the model generated from the pruned dataset.

To obtain more precise results training dataset is divided into two parts:

1. A generator set for pruning the dataset
2. An analysis set for measuring the correctness of the model obtained from pruned dataset.

The ratio of size of generator set and analysis set is kept in the ratio of 9:1 and both these sets are obtained by random sampling of training set at each iteration.

After obtaining the generator and analysis sets, we first use our generator set for creating categories vectors (CVs) and then iteratively pruning these sets by removing the outliers. This process is discontinued when we either notice a no change or negative change in the accuracy of model obtained from pruned generator sets. The accuracy of the model generated is measured using analysis set.

The algorithm uses method get_accuracy which provides the accuracy of given classifier C which uses our pruned Generator set GS for training and Analysis set AS for testing of the classifier. Another algorithm used is Categories Vectors generator (Algorithm 2). This algorithm is used for generating a vector related to each label by combining the data samples belonging to those labels.

3.1 Analysis of the Technique

Assuming that running the classifier on a set of category vector is M. Then the total time complexity of the algorithm for selection of best iteration value can be said to O (n * M * V). Where n, is number of iterations and V is the time complexity for generating a category vector.

For generation of a single category vector we need to calculate the cosine distance of individual feature from the vector. Assuming number of features be V , then time complexity of each vector would be O(V) , Now suppose we have n such vectors then the overall complexity of generation of the category vector for whole dataset would be O(I * n) = O (N). Assuming number of Iterations is quite small compared to number of features in the dataset.

Algorithm 1. Main Algorithm: Pruned_generator_set Algorithm

Data: Generator set GS, Analysis set AS, number of iterations before stopping k nz

Result: Pruned generator set with lesser or no outliers GS

1 Pruned generator set GS = GS
2 Saved generator set SGS = {}
3 Accuracy at each iteration = [ ]
4 n = 0
5 $accuracy_i$ = get_accuracy(GS, AS, C) `// Initial accuracy`
6 Categories Vectors CV_s = generate_categories_vectors(GS)
7 $accuracy_o$ = get accuracy(GS, AS, C)
8 outliers = get_outliers(CV_s, GS) `// Find outliers`
9 GS = GS - outliers `// Remove identified outlier`
10 $accuracy_n$ = get_accuracy(GS, AS, C)
11 △ accuracy = $accuracy_n$ - $accuracy_o$
12 **if** $\triangle accuracy \leq 0$ **then**
13 | n = n+1
14 **else**
15 | n = 0
16 **end**
17 **if** *if n ¡ k* **then**
18 | goto step 6
19 return GS

Space Complexity for the Algorithm. The overall space complexity of the algorithm is not so large. For generation of a single class vector, We compare all existing feature vectors with current category vector. The overall, space requirement is of 1 for storing the distance generated from the cosine distance function. Overall space complexity of the algorithm is O(N)

Algorithm 2. Categories Vectors generator

```
   Data: Pruned generator set GS
   Result: Categories Vectors CVs
 1 CVs = {} foreach data sample ds ∈ GS do
 2 |  label = get_label(ds) // returns class label of the ds
 3 |  if has_key (CVs, label) then
 4 |  |  CV = {} // checks if label is in CVs
 5 |  else
   |  |  // Vector corresponding to label
 6 |  |  CV = CVs [label]
 7 |  end
 8 end
 9 foreach feature and weight pair (f, w) in ds do
10 |  if has_key(CV, f) then
11 |  |  CV[f] = CV[f] + w
   |  |  // Update weight
12 |  else
13 |  |  CV[f] = w
14 |  end
15 |  CVs[label] = CV
16 end
17 return CVs
```

4 Experimental Setup

4.1 Description of Dataset

In our experiments we have used Titanic data set [21] given by Kaggle. The table gives the detailed view of various features which have been provided in the dataset for making a correct predictions.

The dataset corresponds to a binary classification problem in which only two class labels were involved (1 for survived and 0 for not-survived). During the preprocessing step, we converted non integer values like Gender into integer values. Discrete values like fare were converted into discrete values based upon the distribution of values of these features. Finally some features (name, passenger id, etc.) were removed and missing values of remaining features were filled by using suitable regression techniques.

4.2 Benchmarking Dataset with Existing Techniques

After cleaning the dataset, we divided the dataset into two subsets, namely Generator set and Analysis set in the ratio of 9:1. These subsets were chosen without replacement randomly from the training dataset.

VARIABLE DESCRIPTIONS:	
survival	Survival (0 = No; 1 = Yes)
pclass	Passenger Class (1 = 1st; 2 = 2nd; 3 = 3rd)
name	Name
sex	Sex
age	Age
sibsp	Number of Siblings/Spouses Aboard
parch	Number of Parents/Children Aboard
ticket	Ticket Number
fare	Passenger Fare
cabin	Cabin
embarked	Port of Embarkation (C = Cherbourg; Q = Queenstown; S = Southampton)

To test our hypothesis we have used two very different classification techniques, namely Support Vector Classifier (SVC) [23] and Random Forests Classifier [24]. Former is based on the support vector machine, while latter is based on the decision tree. We have use sklearn pythons library [22] implementation of these classifiers. We were able to improve our results on both these algorithm, therefore we can safely assume that the technique could be used for improving classification results for any classifier.

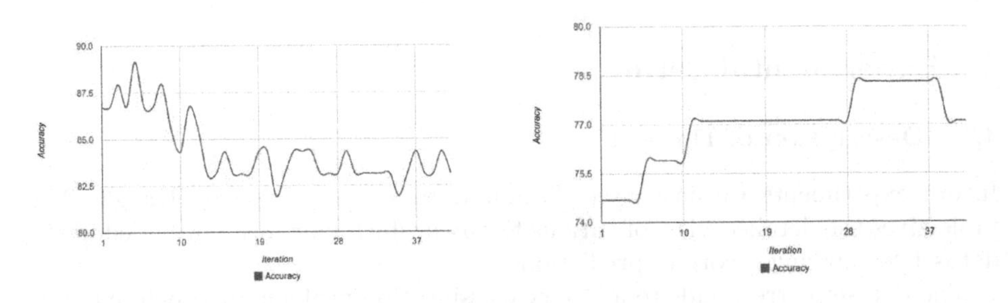

Fig. 1. Accuracy v/s iteration for (a) RandomForest Classifier and (b) SVM Classifier

For RandomForest classifier, initial accuracy observed for first iteration, i.e., without any outlier removal was near 86.00% which with subsequent iterations improved. The accuracy reached a maxima of 89 % showing an increment of 3 percent accuracy with state-of-art RandomForest classifier. With further iterations, the associated accuracy decreased rapidly as multiple iterations led to removal of those data points which were not outliers. This trend was observed for both the classifiers. In case of SVC initial accuracy without outlier removal was observed to be near 74.5% which increased to a good 79.00% after few iterations.

5 Results

In this paper, we presented and analysed a novel method for detection and further removal of outliers. The method was well tested with a standard dataset and two state-of-art classifiers. A significant improvement in the classifier accuracy was observed when outlier removal method was adopted.

For the dataset taken into consideration, the standard SVC and The Random forest classification methods produced results with accuracy of 74.5% and 86% respectively. But when the proposed approach for data cleansing was applied a quite appraisal in the accuracy for the standard methods was recorded. Both of the methods showed an increment of 3-6% in the accuracy. Resulting to an observed overall-all accuracy of 79% and 89% respectively.It may appear that improvement in the accuracy are quite small and easily neglected, but on a contrary these have been recorded over two state of art methods and hencc arc quite encouraging.

6 Conclusion and Future Work

This paper is an attempt of using supervised learning based methodology for outlier detection and removal. The work done in this paper demonstrate that supervised learning could be used effectively in outlier detection and improving the accuracy of classification techniques. The improvement observed in accuracy of both the standard classification techniques taken into consideration i.e. the support vector and RandomForest classifiers suggests that the proposed methodology could be used for improving the performance of any classifier. As the classification techniques selected for comparing results utilizes quite disparate approaches for classification, we can consider the proposed method to be applicable with any of existing classification methods.

In future this technique could be further extended incorporate negative training, i.e. we can have a method which can identify wrongly detected outliers and negate the effect of such removals from the dataset. This can be done by re-adding those feature which were removed due to incorrect judgement by outlier detection technique. Also, work can be done to identify context specific outliers and identify semantic meaning associated with the outliers for effective judgement on removal from dataset.

References

1. Aggarwal, C.C., Philip, S.Y.: Outlier detection for high dimensional data. ACM Sigmod Record 30(2) (2001)
2. Hodge, V.J., Austin, J.: A survey of outlier detection methodologies. Artificial Intelligence Review 22(2), 85–126 (2004)

3. Petrovskiy, M.I.: Outlier detection algorithms in data mining systems. Programming and Computer Software 29(4), 228–237 (2003)
4. Knorr, E.M., Raymond, T.N.: A Unified Notion of Outliers: Properties and Computation. In: KDD (1997)
5. Knox, E.M., Raymond, T.N.: Algorithms for mining distancebased outliers in large datasets. In: Proceedings of the International Conference on Very Large Data Bases (1998)
6. Fawcett, T., Foster, P.: Adaptive fraud detection. Data Mining and Knowledge Discovery 1(3), 291–316 (1997)
7. Williams, G.J., Zhexue, H.: Mining the knowledge mine. In: Sattar, A. (ed.) Canadian AI 1997. LNCS, vol. 1342, pp. 340–348. Springer, Heidelberg (1997)
8. Knorr, E.M., Raymond, T.N., Tucakov, V.: Distance-based outliers: algorithms and applications. The International Journal on Very Large Data Bases 8(3-4), 237–253 (2000)
9. Jin, W., Anthony, K.H.T., Han, J.: Mining top-n local outliers in large databases. In: Proceedings of the Seventh ACM SIGKDD International Conference on Knowledge Discovery and Data Mining. ACM (2001)
10. Breunig, M.M., et al.: LOF: identifying density-based local outliers. ACM Sigmod Record 29(2) (2000)
11. Williams, G., et al.: A comparative study of RNN for outlier detection in data mining. In: 2013 IEEE 13th International Conference on Data Mining. IEEE Computer Society (2002)
12. Hawkins, S., He, H., Williams, G.J., Baxter, R.A.: Outlier detection using replicator neural networks. In: Kambayashi, Y., Winiwarter, W., Arikawa, M. (eds.) DaWaK 2002. LNCS, vol. 2454, pp. 170–180. Springer, Heidelberg (2002)
13. Bay, S.D., Schwabacher, M.: Mining distance-based outliers in near linear time with randomization and a simple pruning rule. In: Proceedings of the Ninth ACM SIGKDD International Conference on Knowledge Discovery and Data Mining. ACM (2003)
14. Kaufman, R.L., Rousseeuw, P.: Finding groups in data: An introduction to cluster analysis. Hoboken NJ John Wiley & Sons Inc., PJ (1990)
15. Raymond, T.N., Han, J.: Efficient and Effective Clustering Methods for Spatial Data Mining. In: Proc. of. (1994)
16. Ramaswamy, S., Rastogi, R., Shim, R.: Efficient algorithms for mining outliers from large data sets. ACM SIGMOD Record 29(2) (2000)
17. Barbara, D., Chen, P.: Tracking Clusters in Evolving Data Sets. In: FLAIRS Conference (2001)
18. Shekhar, S., Chawla, S.: A tour of spatial databases (2002)
19. Acuna, E., Rodriguez, C.: The treatment of missing values and its effect on classifier accuracy. In: Classification, Clustering, and Data Mining Applications, pp. 639–647. Springer, Heidelberg (2004)
20. Hu, T., Sung, S.Y.: Detecting pattern-based outliers. Pattern Recognition Letters 24(16), 3059–3068 (2003)
21. Data set for training and testing, `http://www.kaggle.com/c/titanic-gettingStarted/data`
22. Python Sklearn Library with implementations of popular algorithms, `http://scikit-learn.org/`
23. Suykens, J.A.K., Vandewalle, J.: Least Squares Support Vector Machine Classifiers. Neural Processing Letters 9(3), 293–300 (1999)
24. Breiman, L.: Random Forests, Statistics Department University of California, Berkeley, CA 94720

An Efficient Approach to Book Review Mining Using Data Classification

Harvinder, Devpriya Soni, and Shipra Madan

Department of CSE and IT, Jaypee Institute of Information Technology, Noida, India
{harvinder,devpriya.soni,shipra.madan}@jiit.ac.in

Abstract. With the growing usage of internet and popularity of opinion-rich resources like online reviews, people are actively using information technology to know and form their opinion. These reviews give vital information about a product and can impinge on its demand in cyber space. In this paper we have presented a hybrid technique combining TF-IDF method with opinion analysis using multinomial Naïve Bayes Classification algorithm to mine data of online reviews of books thereby improving the review results. TF-IDF method uses weighted technique to compute the weight of a word in a document and opinion analysis gives the polarity about a particular product. We amalgamate both techniques to improvise the efficiency of the results which may consequently be used by recommender systems for better recommendations.

1 Introduction

Advancement of web based technologies has led to storage of abundant information on internet. Looking at the e-commerce prospective, all major products are now-a-days available on internet on various reputed websites such as Amazon, Flipkart, Google etc. These websites contain detailed description of the products including their price, availability, functionality, customer ratings, feedbacks, blogs etc. A new buyer can utilize these digital words of mouth to draft a viewpoint about a specific product. For example, a book is given ratings by various consumers who have previously bought and read it. Collectively, these reviews may be used to form a firm opinion about how is the book, what genre of people like or do not like it, if it is better than its prequel etc. For instance, a given book XYZ may have around 500 reviews where review stars range from one to five and the textual content may vary from "Very nice" to "Not Worth reading". One option is that a reader goes through all the provided reviews and analyze whether she should read the book or not based upon analysis done through those reviews. This process of opinion formation is clearly time consuming and tiresome especially if the number of reviews is large and diverse in nature. Alternatively, automated machine learning computational techniques can be used for the purpose of opinion building and determining the sentiment direction of online review text provided on the website.

The idea of harnessing opinions of millions of people in online community is not new and was conceived in early 1990s. New dimensions to retrieval of data are being

S.C. Satapathy et al. (eds.), *Emerging ICT for Bridging the Future – Volume 2,*
Advances in Intelligent Systems and Computing 338, DOI: 10.1007/978-3-319-13731-5_68

added up to the existing techniques very frequently. With the rising popularity and availability of opinion-rich resources such as blogs, feedbacks, online review sites new challenges are coming up as people can now actively use information technologies to look for and comprehend the opinions of others. This sudden upsurge of opinion mining and sentiment analysis has given birth to evolvement of new computational techniques dealing with the textual and sentimental context simultaneously.

Assembling massive amount of data to extract what people think has become an interesting and challenging aspect of information retrieval. Further, recommending the products based on the analysis adds up a new aspect to the opinion mining and sentimental clustering. Recommendation systems are used to recommend a particular product to a specific user based on his previous likings and disliking. Such systems use user preferences and past search history for the purpose of suggesting items to purchase or scrutinize a product. Now a days they are becoming extremely popular for providing suggestions that effectively snip large information spaces so that users are directed towards those items that best meet their needs, preferences and location. Individualized or customized nature of such systems makes them a desired candidate for various assorted applications. Major challenge is to minimize the gap between automatically computed text and user's semantic control to evaluate the features [1].

In this paper, we have used word frequency alongside star based popularity of a book to efficiently mine the data and improvise the review results. Our work tends to focus on proficiently using the data from a review dataset of books. Various sites s.a Goodreads, Amazon, Flipkart, Google books etc that provide the rating of a product along with customer comments for a given book. A customer provides reviews to a book by giving it stars. This star count determines the popularity of a book. Along with the stars, a user also provides some comments on how did he/she like the book. These comments may consist of varied opinions like Excellent, Loved it, Worth reading once, Waste of Time etc. and thus provide a vital amount of information about a particular book. Comments are visible to us on any website, but to make their use more fruitful, they can be used to depict the polarity of a given book. Blend of comments, star based reviews and user profile can be used to infer various non-trivial facts about any given book.

This paper proposes a mining technique which combines the traditional approach of star based rating on content based text mining with opinion based approach based on sentimental classification. The approach is used to give a clearer vision of a book review using computational techniques which may further be used by recommendation systems. The aim of our work is to make predictive analysis of online book reviews collected from internet. The massive data on internet includes reviews, feedback, ratings, and blogs etc which potentially facilitate us in determining quality of a product. However, our dataset only includes limited set of reviews of books for experimental purposes. Our goal is to predict desired results more accurately and efficiently based on content written in conjunction with the star-based ratings facilitating us to put forward large amount of purposeful and serviceable data.

The rest of the paper is organized as follows: First section comprises the literature survey on text mining, classification and opinion mining techniques. Following sections explain the process of assembling, extracting and classifying the dataset to perform scale and sentiment based computational analysis approach using machine learning algorithms. Subsequently, details about experimental setup and results are illustrated with examples used. Finally, conclusions on the proposed research methodology are drawn and further research is suggested.

2 Related Work

Text mining or knowledge discovery from textual databases refers generally to the process of extracting interesting and non-trivial patterns or knowledge from unstructured text documents. It can be viewed as an extension of data mining or knowledge discovery from (structured) database. [8] Most of the techniques in text mining are based on the weight of individual words calculated using TF-IDF (Term Frequency -Inverse Document Frequency) method. This weight is a statistical measure used to evaluate how important a word is to a document in a collection or corpus. [5][6]We have applied TF-IDF to raw data to accurately extract frequently occurring words in a document. Classification techniques may further be used to divide this data into groups and get a clearer picture for information retrieval from corpus. Han and Camber [7] put forward remarkable sets of classification techniques to cluster data using a variety of mining techniques. These techniques vary according to user as well as usage field of the data.

Content based text mining is one of the methods used to learn individual profiles from description of examples. Mining text gives subjective data about an individual, based on which various conclusions can be drawn for a particular profile. Machine learning algorithm for text categorization [9] has earlier been discussed for content based recommendation where content of data is taken into account for evaluating further parameters for mining data. However, Studies reveal that reviews with extreme opinions prove to be more helpful than those with mixed or neutral opinions. [4] Further, with rise in popularity and availability of ample opinion resources like blogs, feedback, reviews etc, new pathways are emerging to extract relevant information from the clouds of data. Growing emergence of opinion based resources, including personal profile based feedback, new challenging areas have aroused to extract useful pieces of information out of opinion based data.
This eruption is giving opinions and sentiments a computational angle. Pang and Lee have applied machine learning techniques to classify movie reviews according to sentiments. They employed Naive Bayes, Maximum Entropy, and SVM classifiers for classification and observed remarkable results where machine learning techniques outperformed human-produced baselines [10].

Mining data to find out interesting patterns can be done by applying classification techniques such that polarity of data may decide the inclination of a product to either a positive or a negative side. This classification helps in creating a clear implication of a given product making it either suitable or unsuitable for an individual's interest. [11] We've applied a document level classification technique including score analysis, voting and multinomial Naïve Bayes classification for enhancing the prediction accuracy of reviews.

3 Data Extraction

3.1 Assemble the Data

A book review dataset of 2000 reviews was collected from an existing dataset [14] for our experimental purpose. This collection of data contained an anonymous user id, isbn number of a book, comments in English language, book names, star based reviews represented as numeric value ranging from 5 to 1. Score 0 is not considered as it implies that user may not have rated the product. We've only used user id, ratings and comments for our purpose of analysis. English language contains words that can be used as sentiment indicators to know the sentimental direction of a sentence. Major adjectives, adverbs and other strong sentiment indicator words are filtered out to further put in the training data set [12]. This training data set is then used to extract attributes and the extracted attributes direct us towards results. The process is explained below in detail.

3.2 Preprocessing and Filtering of Data

The data assembled in the previous step contains polarity data based on English words and scale data based on star ratings data of book reviews. This corpus is then converted into vector form to perform further operations as listed below.

i. **Tokenization and stemming:** Process of tokenization involves dividing a given document into a set of words. The sentences are broken and further converted into a set of individual words. Next, Stemming of the above set is done in which all the overlapping or duplicate words are omitted. Adjectives, nouns, spelling errors etc of same word are treated as one entity. For example, wonderful, wonderrr, wondering etc are treated as one single word "wonder". Next the string data is converted into vectors for algorithmic dispensation.

ii. **Word to vector conversion:** Till now, we've performed the stemming of words to filter out contender words for further classification. All the significant words contributing to sentiment analysis are filtered out separately and vector transformation of obtained set is then done. In further steps, word frequency is counted and classification is done on above vectors.

iii. **TF-IDF method:** Next, we use TF-IDF technique to retrieve frequently occurring words in a document. For instance, given a document collection D, a word w, and an individual document d ϵ D, we can calculate weight of word as

$$w_d = f_{(w,\,d)} * \log\left(|D|/f_{(w,\,D)}\right)$$

where $f_{(w,\,d)}$ equals the number of times w appears in d, |D| is the size of the corpus, and $f_{(w,\,D)}$ equals the number of documents in which w appears in D (Salton & Buckley, 1988, Berger, et al, 2000). For example a book contains the word "excellent" 10 times in 20 reviews out of total 50 reviews given.

Then,

$$\mathbf{f}_{(\mathbf{w},\, \mathbf{d})} = (10/50) \text{ and}$$
$$\mathbf{w}_{\mathbf{excellent}} = 0.2*\log(50/20) \approx 0.39$$

Eventually, weights of all the words are calculated using the above formula. This step provides all the most frequently occurring words in the document. Higher the weight of a word, more valuable it is for consequent processing of corpus.

iv. Extracting relevant attributes: Above data filtering gives a list of various words out of which only some may be helpful for classification. For example consider the sentence "This book is simply wonderful and I am waiting for its next series". This sentence contains words like this, book, is, and, for, its etc which are quite irrelevant for the work , whereas words like use, excellent, waiting etc are relevant with respect to further classification. This extracted data obtained now requires human - intervened pruning to be done to sort out all unusable words. As explained above, words like the, is, an, for etc. carry highest weightage but are not at all helpful in drawing any results. All the unproductive words categorized as stopwords are removed from dataset before sentiment analysis and classification.

4 Classification

We have used a supervised training technique known as Naïve Bayes Classification algorithm for Text Classification to classify the text in review database. A Naive Bayes classifier is a simple probabilistic classifier based on applying Bayes' theorem with strong naïve independence assumptions. It uses the method of maximum likelihood to classify the polarity of any given word [14]. This classification technique is used to train the classifiers about the polarity of data based on input vectors of strings. The algorithm is applied on two types of datasets. First dataset includes only the review comments, user id and book id while the second set contains first set along with star based ratings.

After suitable features are shortlisted, the classifier is then trained on the training dataset. The process of training is an iterative process and we have used 70-30 ratio for training-test dataset. To further improvise results, we repeatedly applied stop word removal, word stemming, and input features filtering in an effort to produce a better output. The efficiency of the classifier trained on the training data is then assessed on the test dataset. The results obtained after classification signify that using star based rating with polarity gives a more accurate result than using only polarity measure individually.

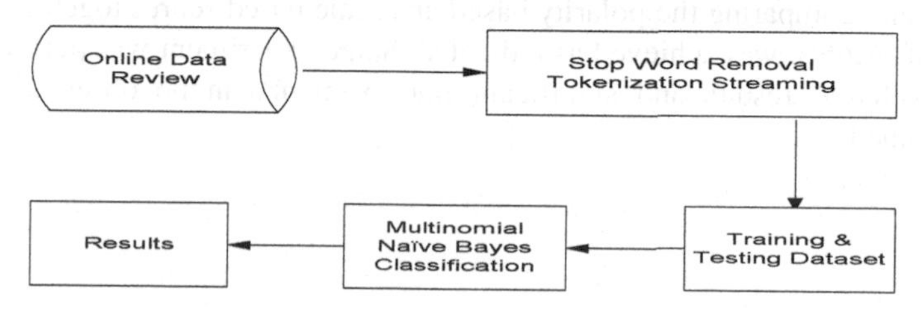

Fig. 1. General Process Overview to show the steps being followed in procedure

Steps Followed

i. Data Extraction is done from pre-parsed online reviews datasets available [13].
ii. Preprocessing of data including important steps of tokenization, stemming, and stopwords removal is done to make data equipped for input to next step.
iii. Data is extracted in two separate files. One file including star based rating and polarity based words and other file contains only polarity based words.
iv. Word frequency in each word is calculated using TF-IDF method. This gives the total count of a word in a document.
v. To classify polarity of words, categorization of words,both positive and negative words is gathered from SentiWordnet [15].
vi. Multinomial Naïve Bayes classifier is used to classify data into different categories based on their scores.

5 Experimental Results

We have used the dataset of 2000 reviews for performing predictive analysis of book reviews with an additional precision. The snap shot taken for the result is shown in Fig. 2. The results shown in Table 1 depict that the star based ratings together with polarity measure gives better results as compared to the individual polarity estimating technique.

```
=== Stratified cross-validation ===
=== Summary ===

Correctly Classified Instances        1433               71.65   %
Incorrectly Classified Instances       567               28.35   %
Kappa statistic                          0.433
Mean absolute error                      0.2843
Root mean squared error                  0.5277
Relative absolute error                 56.858  %
Root relative squared error            105.5363 %
Total Number of Instances             2000
```

Fig. 2. Results of experiments depicting accuracies of hybrid approach

To test our approach, we applied this technique to a dataset of N=2000. This unigram classification gives us results in four categories as shown in the Table I below. The depiction evidently illustrates the accuracy levels achieved while evaluating and comparing the polarity based and scale based scores together. The best accuracy of 71.65% was achieved as indicated. Same experiment was performed only on polarity based results and significant improvement can be observed from the results obtained.

Table 1. Illustration of comparison of result using both approaches

Task Method	**Feature**	**User Training Set for Polarity based classification**	**Total Data Dimension (2000)**	**Attained Accuracy**
Naïve Bayes	Rating+ Polarity Scores	Correctly Classified Instances	1546	71.6
Naïve Bayes	Rating+ Polarity Scores	Incorrectly Classified Instances	454	22.7 %
Naïve Bayes	Positivity/ Negativity Scores	Correctly Classified Instances	1377	68.85 %
Naïve Bayes	Positivity/ Negativity Scores	Incorrectly Classified Instances	623	31.15 %

This evidently suggests that words may give an accurate output but we can always improve the accuracy by combining them with defined syntactic sets.

6 Conclusion and Future Work

Our paper focuses on predicting mining results of book reviews with additional precision. The work emphasizes on the importance of using scale based rating adjoined with sentiment based results. The algorithm shows that a bit of improvement can occur if we affix the star based ratings with the polarity measure. It has been observed that increasing the number of review parameters lead to more accurate results. Presently, our work is focused only on unigrams and we intend to extend this work to n-grams with the expectation of more accurate results. Also datasets containing same word with different meaning and linguistically different interpretations play important role in every sphere therefore; semantic direction is another way to pursue the work.

References

1. Burke, R.: Hybrid Recommender Systems: Survey and Experiments 12(4), 331–370 (2002)
2. Adomavicius, G., Tuzhilin, A.: Toward the Next Generation of Recommender Systems: A Survey of the State-of-the-Art and Possible Extensions. IEEE Transactions on Knowledge and Data Engineering 17(6), 734–749 (2005)
3. Jannach, D., Zanker, M., Felfernig, A., Friedrich, G.: Recommender Systems. Cambridge University Press (2010)

4. Caoa, Q., Duanb, W., Gana, Q.: Exploring determinants of voting for the "helpfulness" of online user reviews: A text mining approach. Decision Support Systems 50(2), 511–521 (2011)
5. Salton, G., Buckley, C.: Term-weighting approaches in automatic text retrieval in Information Processing and Management 24(5), 513–523 (1988)
6. Salton, G.: Automatic Text Processing: The Transformation, Analysis, and Retrieval of Information by Computer Reading. Addison-Wesley (1989)
7. Han, J., Kamber, M.: Data Mining, Concepts and Techniques 3rd edn. (2011)
8. Feldman, R., Dagan: Knowledge discovery in textual databases (KDT). In: Proceedings of the First International Conference on Knowledge Discovery and Data Mining (KDD-95), Montreal, Canada, August 20-21, pp. 112–117. AAAI Press (1995)
9. Mooney, R.J., Roy, L.: Content-based Book Recommending Using Learning for Text. In: Proceedings of the Fifth ACM Conference on Digital Libraries, pp. 194–204 (2000)
10. Pang, B., Lee, L.: Opinion Mining and Sentiment Analysis. Foundations and Trends in Information Retrieval 2(1-2), 1–135 (2008)
11. Li, G., Liu, F.: A Clustering based Approach on Sentimental analysis. In: IEEE International Conference on Intelligent System and Knowledge Engineering, Hangzhou, China, pp. 331–337 (2010)
12. Benamara, F., et al.: Sentiment Analysis: Adverbs and Adjectives Are Better than Adverbs Alone. In: Proceedings of International Conference Web-logs and Social Media, ICwsm 2007 (2007)
13. Blitzer, J., Dredze, M., Pereira, F.: Biographies, Bollywood, Boom-boxes and Blenders: Domain Adaptation for Sentiment Classification. Association of Computational Linguistics, ACL (2007)
14. Introduction to Naïve Bayes Classification, https://www.princeton.edu/~achaney/tmve/wiki100k/docs/Naive_Bayes_classifier.html
15. SentiWordNet: A Publicly Available Lexical Resource for Opinion Mining, http://sentiwordnet.isti.cnr.it/

Classification of Tweets Using Text Classifier to Detect Cyber Bullying

K. Nalini[1] and L. Jaba Sheela[2]

[1] Bharathiyar University, Coimbatore, Tamil Nadu, India
immanuelsamen@rediffmail.com
[2] Panimalar Engineering College, Chennai, Tamil Nadu, India
sujitha14@hotmail.com

Abstract. Cyber bullying and internet predation threaten minors, particular teens and teens who do not have adequate supervision when they use the computer. The enormous amount of information stored in unstructured texts cannot simply be used for further processing by computers, which typically handle text as simple sequences of character strings. Therefore, specific (pre)processing methods and algorithms are required in order to extract useful patterns. Text Mining is the discovery of valuable, yet hidden, information from the text document. Text classification one of the important research issues in the field of text mining. We propose an effective approach to detect cyber bullying messages from Twitter through a weighting scheme of feature selection.

Keywords: Cyber bullying, Text mining, Data sets, Text classification, Twitter.

1 Introduction

A number of life threatening cyber bullying experiences among young people have been reported internationally, thus drawing attention to its negative impact. Bullying as a form of social turmoil has occurred in various forms over the years with the WWW and communication technologies being used to support deliberate, repeated and hostile behavior by an individual or group, in order to harm others. Cyber bullying is defined as an aggressive, intentional act carried out by a group or individual, using electronic forms of contact repeatedly and over time, against a victim who cannot easily defend him or herself.

Detection of cyber bullying and the provision of subsequent preventive measures are the main courses of action in combating cyber bullying. The aim of this paper is to propose an effective method to detect cyber bullying activities on social media. Our detection method can identify cyber bullying messages, predators and victims. The proposed method is divided into two phases. The first phase aims to accurately detect harmful messages. We present a new way of feature selection, namely semantic and weighted features. Semantic features use the Latent Dirichlet Allocation (LDA) [1] algorithm to extract latent features, whereas weighted features are the bullying-like features which include static bad word vocabulary. LDA is a topic-modeling approach; it uses probabilistic sampling strategies to describe how words are

S.C. Satapathy et al. (eds.), *Emerging ICT for Bridging the Future – Volume 2*,
Advances in Intelligent Systems and Computing 338, DOI: 10.1007/978-3-319-13731-5_69

generated within a document based on the latent topics. It finds the most suitable set of latent variables. Given a set of messages, we use LDA to understand the semantic nature of word usage to detect underlying topics (bullying or non bullying) automatically.

The second phase aims to analyze tweets to identify predators and victims through their user interactions, and to present the results in a graph model. In Twitter, in the form of connected graphs (nodes and edges) allows users (nodes) to be connected with other users (nodes) directly or indirectly. In an online community, users may publicize anything. The publicized material against the target user can be accessed repeatedly, over time, by the target as well as any number of other users. Moreover, those who post hurtful messages can be anonymous. The anonymity of the bully might create a sense of fear within the target user as they are dealing with an unknown identity. This handicaps the target user while at the same time advantages the anonymous bully with a feeling of power and control over the target victim [2].

We make three contributions in this paper. Firstly, we propose a novel statistical detection approach, which is based on the weighted TFIDF scheme on bullying-like features. It also efficiently identifies latent bullying features to improve the performance of the classifier. Secondly, we present a graph model to detect the most active predators and victims in social networks. Besides identifying predators and victims, this graph model can be used to classify users in terms of levels of cyber bullying victimization, based on their involvement in cyber bullying activities. Thirdly, our experiments demonstrate that the proposed approach is effective and efficient. The rest of this paper is organized as follows. In Section 2, Introduction to Classification and Twitter are discussed. Section 3, Literature review on cyber bullying detection is presented. Section 4 explains the proposed methodology. Section 5 describes how the experiments are performed and results are discussed. The conclusion is presented in Section 6.

2 Text Classification

Text classification is an area where classification algorithms are applied on documents of text. The task is to assign a document into one (or more) classes, based on its content. Typically, these classes are handpicked by humans. For a classifier to learn how to classify the documents, it needs some kind of ground truth. For this purpose, the input objects are divided into training and testing data. Training data sets are those where the documents are already labeled. Testing data sets are those where the documents are unlabeled. The goal is to learn the knowledge from the already labeled training data and apply this on the testing data and predict the class label for the test data set accurately. Hence, the classifier is built of a Learner and an actual Classifier.

The classifier then uses this classification function to classify the unlabeled set of documents. This type of learning is called supervised learning because a supervisor serves as a teacher directing the learning process [3]. The choice of the size of the training and testing data set is very important.

2.1 Introduction to Twitter

Twitter is a social networking application which allows people to micro-blog about a broad range of topics. Micro-blogging is defined as a form of blogging that lets you write brief text updates (usually less than 200 characters) about your life on the go and send them to friends and interested observers via text messaging, instant messaging (IM), email or the web. Twitter helps users to connect with other Twitter users around the globe.

2.1.1 Architecture of Twitter

We will briefly discuss about how Twitter works before delving into the concepts and entities of Twitter. The Twitter API consists of three parts: two REST APIs and a Streaming API. The API currently provides a streaming API and two discrete REST APIs. Through the Streaming API users can obtain real-time access to tweets in sampled and filtered form. The API is HTTP based, and GETS, POST, and DELETE requests can be used to access the data. An interesting property of the streaming API is that it can filter status descriptions using quality metrics, which are influenced by frequent and repetitious status updates etc. The API used basic HTTP authentication and requires valid Twitter account. Data can be retrieved as XML or the more succinct JSON format. The format of the JSON data is very simple and it can be parsed very easily because every line, terminated by a carriage return contains one object.

3 Related Work

In [8], the authors present their observations of the micro-blogging phenomena by studying the topological and geographical properties of Twitter social network. People use micro-blogging to talk about their daily activities and to seek or share information. They also analyze the user intentions associated at a community level and show how users with similar intentions connect with each other. In paper [9], the authors consider the problem of detecting spammers on Twitter. They first collected a large dataset of Twitter that includes more than 54 million users, 1.9 billion links, and almost 1.8 billion tweets. Compared to this our dataset consisted of 1.8 million tweets. Using tweets related to three famous trending topics from 2009, they construct a large labeled collection of users, manually classified into spammers and non-spammers. The authors then identified a number of characteristics related to tweet content and user social behavior, which could potentially be used to detect spammers on twitter. In paper [5], the authors study how micro-blogging can be used for sentiment analysis purposes. They show how to use Twitter as a corpus for sentiment analysis and opinion mining. They use a dataset formed of collected messages from Twitter.

In [10], the authors illustrate a sentiment analysis approach to extract sentiments associated with negative or positive polarity of specific subjects in a document, instead of classifying the whole document as positive or negative. The essential issues in sentiment analysis are to identify how sentiments are expressed in texts and whether the expressions indicate positive (favorable) or negative (unfavorable) opinions toward the subject. Paper [6] delivers a new Twitter content classification framework based on sixteen existing Twitter studies and a grounded theory analysis of a personal Twitter history. It expands the existing understanding of Twitter as a multifunction tool for personal, profession and commercial communications with a split level classification scheme that offers broad categorization and specific sub categories for deeper insight into the real world application of the service.

4 Methodology

4.1 Classification Model for Harmful Posts Detection

A feature selection is an important phase in representing data in feature space. Twitter data are noisy, thus pre-processing has been applied to improve the quality of the research data and subsequent analytical steps; this includes converting uppercase letters to lower case, stemming, removing stop words, extra characters and hyperlinks. We propose the following types of features generated through the LDA [1] topic model and weighted TFIDF scheme. At the first step, we apply semantic features for the detection of harassing, abusive and insulting posts. In harassment detection [11] the appearances of pronouns in the harassing post were illustrated. Similarly in this work we used three types of feature sets: i) all second person pronouns 'you', 'yourself', etc. are counted as one term; ii) all other remaining pronouns 'he', 'she', etc., are considered together as another feature; iii) foul words such as 'f**k', 'bullshit', 'stupid', etc., which make the post cruel are grouped in another set of features. The list of bad words is available from noswearing.com.

4.2 Predator and Victim Identification

We considered a communication network of the users, which includes predators and victims. We used Gephi, a Graphical interface to visualize a user's connectivity based on the bullying posts in a network. The bullying network is a user group extracted based on the bullying posts by applying modularity theorem [4, 7], to measure the strength of partition of a network into sub graphs or communities. For a given graph, modularity is defined as the summation of the weight of all the edges that fall within the given subgroups minus the expected fraction if edges were distributed at random.

Now, to find predators and victims, we formally present a ranking module using the HITS method. A user can be a predator and a victim based on the messages he/she sends or receives. Therefore a user will be assigned a predator as well as a victim score. Following are two equations to compute predator and victim scores respectively:

$$p(u) \rightarrow \sum_{u \rightarrow y} v(y) \tag{1}$$

$$v(u) \rightarrow \sum_{y \rightarrow u} p(y) \tag{2}$$

Where p(u) and v(u) depict the predator and victim scores respectively. $u \rightarrow y$ indicates the existence of the bullying message from u to y, whereas, $y \rightarrow u$ indicates the existence of the bullying message from y to u. These two equations are an iteratively updating pair of equations for calculating predator and victim scores. They are based on our assumption that the most active predator links to the most active victims by sending bullying messages, and that the most active victim is linked by the most active predators by receiving bullying messages. In essence, if the user (u) is linked with another user with a high victim score, the user's predator score increases, and if the user (u) is linked through received messages to a user with a high predator score, the user's victim score increases. In each iteration, scores are calculated through in degrees and out degrees, and associated scores; this may result in large values. Thus scores are normalized to unit length, i.e., each predator and victim scores is divided by the sum of all predator and victim scores respectively.

Now, we define to rank predators and victims in a bullying network. For simplicity and to explain a real scenario, we select only five users as an example. The identification of the most active predators and victims in a bullying network. It is a weighted directed graph G= (U, A) with |U| nodes and |A| arcs where,

- each node u1 ∈ U is a user involved in the bullying conversation,
- each arc(u_i , u_j) ∈ A, is defined as a bullying message sent from u_i to u_j
- The weight of arc (u_i, u_j) denoted as w_{ij}, is defined as summation of in degrees.

Predators and victims can be identified from the weighted directed graph G: The victim will be the nodes with many incoming arcs and the predator will be the nodes with many outgoing arcs. This paper attempts to find if a user is the most active user (a predator or a victim).

4.2.1 Cyber Bullying Matrix

To identify a predator and victim based on their respective scores, we formulate a cyber bullying matrix (w).

$$W_{ij} = n \; if \; there \; exist \; n \;\; bullying \; posts \; from \; u_i \; to \; v_i \text{ , 0 otherwise} \qquad (3)$$

Since each user will have a victim as well as a predator score, scores are represented as the vectors of n*1 dimension , where i^{th} coordinate of the vector represent both the scores of the ith user, say p_i and v_i respectively. To calculate scores, equations p (u) and v (u) are simplified as the victim and predator updating matrix-vector multiplication equations. For the first iteration, p_i and v_i are initialized at 1. For each user (say i = 1 to N) predator and victim scores are as follows:

$$p(u_i) = \; w_{i1}v_1 + w_{i2}v_2 \ldots w_{in}v_n \qquad (4)$$

$$y(u_i) = \; w_{i1}p_1 + w_{i2}p_2 \ldots w_{in}p_n \qquad (5)$$

Then these equations converge at a stable value (say k), it provides the final predator and victim vector of each user. Finally, we calculate the eigenvector to get the predator and victim scores.

Algorithm 1 gives a general framework of identification of the top ranked most active predators and victims. In the Algorithm, N is a total number of users and Top is a threshold value, which is set manually.

Algorithm 1. Predators and victims identification.

Input: Set of users involved in the conversation with bullying post, N, Top.
Output: Set of Top Victim and Top Predator.
1. Extract senders and receivers from N.
2. Initialize predator and victim vector for each N.
3. Create adjacent matrix W using formula 3.
4. Calculate Predator and Victim vectors using iterative updating equations 4 and 5, and normalize until converge at stable value k.
5. Calculate Eigen vectors to find Predator and Victim scores.
6. Return Top ranked Predators and Victims.

5 Experiments and Results

5.1 Dataset

In order to gather a representative data set underlying our analysis, we facilitate the following data collection methodology. We crawled Twitter via the Twitter Filter API [12] which allows for filtering the public Twitter stream for given keywords. To particularly filter for tweets related to bullying, we searched the Twitter stream for tweets containing the strings that contains offensive words. In total, we were able to gather more than 1laks tweets between Jan 1st 2013 and Jan 30th 2014. The Filter API delivers all tweets matching the given query up to a rate limiting equal to the rate limiting of the public streaming API (approximately 1% of all tweets). As the number of tweets matching our query constantly was below this limit (maximum number of tweets crawled per day: more than 30,000), we were able to crawl all tweets matching the given filter keywords during the given time period.

SVM was applied to the two class classification problem using a linear kernel. Each post is an instance; positive classes contain bullying messages and negative classes contain non-bullying messages. Tenfold cross validation was performed in which the complete dataset was partitioned ten times into ten samples; in every round, nine sections were used for training and the remaining section was used for testing. The F-1 measure was not considered because of the large number of negative cases. Identifying bullying is a very critical issue because of false positive and false negative cases. Identifying non-bullying instance as bullying itself is a sensitive issue (false positive); on the other hand, system should not bypass bullying post as normal post (false negative). Therefore, false positive and false negative both are critical. Thus, precision, recall and F-1 measures are considered for the performance evaluation metric. Also, we report identified false positive and false negative cases by the classifier. In literature various strategies are proposed under imbalance text classification—we used oversampling of minority cases to improve the training of classifiers.

The performance of the classifier was evaluated on precision, recall and F-1 measure based on the top ranked features generated through LDA method against the truth set, as tested on the datasets. Precision: the total number of correctly identified true bullying posts out of retrieved bullying posts. Recall: number of correctly identified bullying cases from total number of true bullying cases. F-1 measure: the equally weighted harmonic mean of precision and recall. The table 1 shows the classifier performance.

Table 1. Classifier performances based on different feature reduction methods

Method	Precision	Recall	F-Measure
DF+SVM	0.8471	0.7770	0.8105
PCA+SVM	0.8397	0.7870	0.8125
LDA+SVM	0.8896	0.8554	0.8724

5.2 Comparison of Weighted TFIDF with Baseline Method

We compare the proposed weighted TFIDF method with the work done in for harassment detection using TFIDF, sentiment and contextual features on three different datasets. The proposed feature selection method using weighted TFIDF provided better performance as shown in Table 2.

Table 2. Comparison of Proposed Approach with Baseline Method on Other Datasets

		Kongregate	Slashdot	MySpace	Twitter
Baseline	Precision	0.35	0.32	0.42	0.62
Baseline	Recall	0.60	0.28	0.25	0.52
Baseline	F-1 measure	0.44	0.30	0.31	0.58
Weighted TFIDF	Precision	0.87	0.78	0.86	0.87
Weighted TFIDF	Recall	0.97	0.99	0.98	0.96
Weighted TFIDF	F-1 measure	0.92	0.87	0.92	0.97

5.3 Victim and Predator Identification

In identifying cyber bullying predators and victims we determine the most active predators and the most attacked users, victims through the sent and received bullying messages, and the density of the badness of the message. In this paper, we identified the most active predators and victims, and compare the involvement of users in a bullying relationship as shown in the Table 3. Table 3 shows that in some cases there are more than one user at the same rank. Therefore, users with the same rank are grouped together. We also noted that predators flagged at Rank I are also identified as a victim at Rank II. Similarly Rank II predators are Rank VII victims too.

Table 3. Performance of Graph Model: Predators and Victims Identification

Rank	I	II	III	IV	V	VI	VII	VIII
Number of users (Predators)	4	2	1	1	2	7	3	2
Number of users (Victims)	8	4	7	2	2	1	9	8

6 Conclusion

In this paper we propose an approach for cyber bullying detection and the identification of the most active predators and victims. To improve the classification performance we employ a weighted TFIDF function, in which bullying-like features are scaled by a factor of two. The overall results using weighted TFIDF outperformed other methods. This captures our idea to scale-up inductive words within the harmful posts. Also, throughout our experiments, we note that comparatively better performance was observed for false negative compared to false positive cases in individual and combined datasets. This is because of the fewer positive cases available for classifier's training. Therefore advance methods, which are capable of dealing with a few training sets in automatic cyber bullying detection, and to reduce false positive and false negative cases need to be developed, In addition, we proposed a cyber bullying graph model to rank the most active users (predators or victims) in a network. It can also be used to detect the level of cyber bullying victimization for decision making in further investigations.

References

1. Blei, D.M., Ng, A.Y., Jordan, M.I.: Latent Dirichlet Allocation. Journal of Machine Learning Research 3, 993–1022 (2003)
2. Butler, D., Kift, S., Campbell, M.: Cyber Bullying In Schools and the Law: Is There an effective Means of Addressing the Power Imbalance. Murdoch University Electronic Journal of Law 16, 84–114 (2009)
3. Christopher Manning, D., Raghavan, P., Schütz, H.: Introduction to Information Retrieval. ACM Digital Library 13, 192–195 (2008)
4. Newman, M.E.J.: Fast algorithm for detecting community structure in networks. Physical Review E 69 (2004)
5. Pak, P.: Twitter as a Corpus for Sentiment Analysis and Opinion Mining. In: Proceedings of the International Conference of Language Resource and Evaluation, LREC, pp. 1320–1326 (2009)
6. Dann, S.: Twitter content classification. Journal of First Monday 15 (2010)
7. Newman, M.E.J.: Modularity and community structure in networks. Proceedings of National Academy of Sciences 103, 8577–8696 (2006)

8. Java, Song, Finin, Tseng: Why-We-Twitter-Understanding-Micro blogging-Usage-and-Communities. In: Proceedings of the Joint 9th WEBKDD, pp. 56–65 (2007)
9. Magno, G., Rodrigues, T.: Detecting Spammers on Twitter. In: Proceedings of the 18th Conference on USENIX Security Symposium (2009)
10. Nasukawa, Yi: Sentiment Analysis: Capturing Favorability Using Natural Language Processing Definition of Sentiment Expressions. In: Proceedings of the 2nd International Conference on Knowledge Capture, K-CAP 2003, pp. 70–77 (2003)
11. Yin, D., Davison, B.D., Xue, Z., Hong, L., Kontostathis, A., Edwards, L.: Detection of Harassment on Web 2.0. In: Proceedings of the Content Analysis in The Web 2.0, Spain (2009)
12. Barbosa, Feng, J.: Robust sentiment detection on Twitter from biased and noisy data. In: Proceedings of the 23rd International Conference on Computational Linguistics, pp. 36–44 (2010)

An Overview on Web Usage Mining

G. Neelima[1,*] and Sireesha Rodda[2]

[1] GMRIT, Rajam, Srikakulam, A.P, India
[2] GITAM University, Visakhapatnam, A.P, India
neelima.g@gmrit.org

Abstract. The prolific growth of web-based applications and the enormous amount of data involved therein led to the development of techniques for identifying patterns in the web data. Web mining refers to the application of data mining techniques to the World Wide Web. Web usage mining is the process of extracting useful information from web server logs based on the browsing and access patterns of the users. The information is especially valuable for business sites in order to achieve improved customer satisfaction. Based on the user's needs, Web Usage Mining discovers interesting usage patterns from web data in order to understand and better serve the needs of the web based application. Web Usage Mining is used to discover hidden patterns from weblogs. It consists of three phases like Preprocessing, pattern discovery and Pattern analysis. In this paper, we present each phase in detail, the process of extracting useful information from server log files and some of application areas of Web Usage Mining such as Education, Health, Human-computer interaction, and Social media.

Keywords: Web Usage Mining, Web server logs, Data Preprocessing, Pattern discovery.

1 Introduction

World Wide Web is a growing collection of large amount of information and usually a great portion of time is needed to identify the appropriate information, so various techniques are needed to analyze the data. One of the techniques used is Web mining. Using Web mining, we can analyze and discover the useful information from the web. Web Usage Mining (WUM) extracts useful information based on users' needs from web log information. Based on the user needs and likes, WUM gives the appropriate information using the web server logs. To extract and process the information, web usage mining follows two main steps by [1] [2]: Data preprocessing and Pattern discovery. The huge data present in the web is a collection of raw data, so to get the user needed information the web data preprocessing should be done.

The different phases in web usage mining include data cleaning, data preparation, user identification, session identification, data integration, data transformation, pattern

* Corresponding author.

S.C. Satapathy et al. (eds.), *Emerging ICT for Bridging the Future – Volume 2,*
Advances in Intelligent Systems and Computing 338, DOI: 10.1007/978-3-319-13731-5_70

discovery and pattern analysis. The data preprocessing is most critical phase in the WUM. The preprocessing of data can be done on the original data or on the data integrated from multiple sources. The purpose of web usage mining is to discover hidden information from weblog data, so we have to mine the data from log files. Log files provide information about the activity of user, viz., which web site he/she using, whom you send/receive e-mail etc. These files are maintained by the system administrator. This paper provides a comprehensive survey of web usage mining. Section 2 describes the various kinds of Web server log files available for application of web usage mining. Section 3 details the different phases present in web usage mining. Section 4 summarizes various applications existing in the domain of web usage mining. Section 5 discusses the challenges and concerns arising due to the application of web usage mining and Section 6 concludes the paper.

2 Web Server Logs

A web server log is a log file or simple text file which stores activities performed by the user and maintained by the server ie., it maintains a history of web page requests. Generally these log files cannot be accessible by the user, only the web administrator can handle them. The Log files in different web servers maintain different types of information. Consider the example log file, which contains

- The IP address of the computer making the request (i.e. the visitor)
- The identity of the computer making the request
- The login ID of the visitor
- The date and time of the hit
- The request method
- The location and name of the requested file
- The HTTP status code (e.g. file sent successfully, file not found, etc)
- The size of the requested file
- The web page which referred the hit (e.g. a web page containing a hyperlink which the visitor clicked to get here)

```
151.44.15.252 - - [25/May/2004:00:17:39 +1200] "GET /data/zookeeper/status.html
HTTP/1.1"    200    4195    "http://www.mediacollege.com/cgi-
bin/forum/commentary.pl/noframes/read/209" "Mozilla/4.0 (compatible; MSIE 6.0;
Windows NT 5.1; Hotbar 4.4.7.0)"
```

Fig. 1. Example Log file

These details of the log file are then used for web usage mining process. Web usage Mining is applied to identify the highly utilized web site. The utilization of a website would be the frequently visited web site or the web site being utilized for longer time duration. Therefore the quantitative usage of the web site can be found if the log file is analyzed.

A **Web log** is a file to which the Web server writes information each time a user requests a website from that particular server. A log file can be located in three different places:

1) Web Server Log files

The log file that resides in the web server notes the activity of the client who accesses the web server for a web site through the browser. The contents of the file will be the same as it is discussed in the previous topic. In the server which collects the personal information of the user must have a secured transfer?

2) Web Proxy Server Log files

A Proxy server is said to be an intermediate server that exist between the client and the Web server. Therefore if the Web server gets a request of the client via the proxy server then the entries to the log file will be the information of the proxy server and not of the original user.

These web proxy servers maintain a separate log file for gathering the information of the user.

3) Client Browsers Log files

This kind of log files can be made to reside in the client's browser window itself. Special types of software exist which can be downloaded by the user to their browser window. Even though the log file is present in the client's browser window the entries to the log file is done only by the Web server.

3 Web Usage Mining Process

Web usage mining, from the data mining aspect, is the task of applying data mining techniques to discover usage patterns from Web data in order to understand and better serve the needs of users navigating on the Web. As every data mining task, the process of Web usage mining also consists of two main steps[3] [4]:

(1) Data preprocessing phase,
(2) Pattern discovery phase

3.1 Phase- 1: Data Preprocessing

The first issue in the preprocessing phase is data preparation [1]. The data preparation process is often the most time consuming and computationally intensive step in the Web usage mining process. The process may involve preprocessing the original data, integrating data from multiple sources, and transforming the integrated data into a form suitable for input into specific data mining operations. This process by Kamika Chaudhary and Santosh Kumar Guptaet.al [5] is known as data preparation.

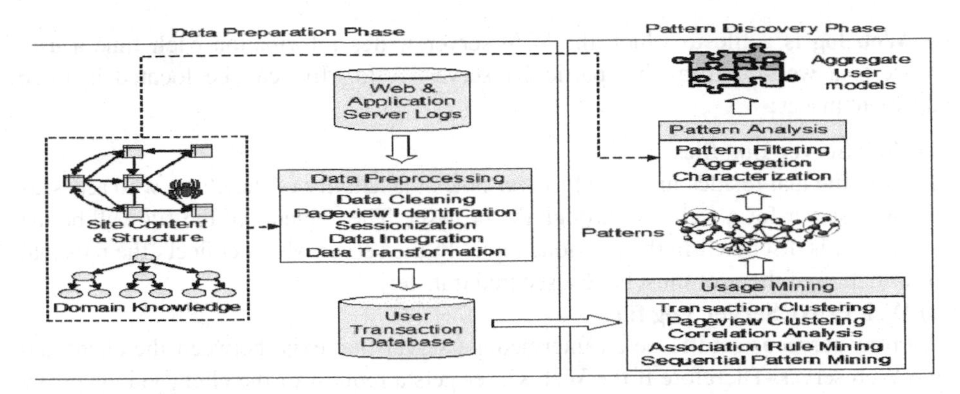

Fig. 2. Web Usage Mining Process

1. **Data preparation:**

Web data can be collected and used in the context of Web personalization [6][7]. These data are classified in four categories according to[8]:

1.1 Content data are presented to the end-user appropriately structured. They can be simple text, images, or structured data, such as information retrieved from databases.

1.2 Structure data represent the way content is organized. They can be either data entities used within a Web page, such as HTML or XML tags, or data entities used to put a Web site together, such as hyperlinks connecting one page to another.

1.3 Usage data represent a Web site's usage, such as a visitor's IP address, time and date of access, complete path (files or directories) accessed, referrers' address, and other attributes that can be included in a Web access log.

1.4 User profile data provide information about the users of a Web site. A user profile contains demographic information for each user of a Web site, as well as information about users' interests and preferences. Such information is acquired through registration forms or questionnaires, or can be inferred by analyzing Web usage logs.

2. **Preprocessing:** The information available in the web is heterogeneous and unstructured. Therefore, the preprocessing phase is a prerequisite for discovering patterns. The goal of preprocessing is to transform the raw click stream data into a set of user profiles. Data preprocessing presents a number of unique challenges which led to a variety of algorithms and heuristic techniques for preprocessing tasks such as merging and cleaning, user and session identification etc. Various research works are carried in this preprocessing area for grouping sessions and transactions, which is used to discover user behavior patterns[1][7].

2.1 Data Cleaning: Data Cleaning is a process of removing irrelevant items such as jpeg, gif files or sound files and references due to spider navigations. Improved data quality improves the analysis on it. The Http protocol requires a separate connection for every request from the web server. If a user request to view a particular page along with server log entries graphics and scripts are download in addition to the HTML file. An exception case is Art gallery site where images are more important. Check the Status codes in log entries for successful codes. The status code less than 200 and greater than 299 were removed.

2.2 User Identification: Identification of individual users who access a web site is an important step in web usage mining. Various methods are to be followed for identification of users. The simplest method is to assign different user id to different IP address. But in Proxy servers many users are sharing the same address and same user uses many browsers. An Extended Log Format overcomes this problem by referrer information, and a user agent. If the IP address of a user is same as previous entry and user agent is different than the user is assumed as a new user. If both IP address and user agent are same then referrer URL and site topology is checked. If the requested page is not directly reachable from any of the pages visited by the user, then the user is identified as a new user in the same address. Caching problem can be rectified by assigning a short expiration time to HTML pages enforcing the browser to retrieve every page from the server [9].

2.3 Session Identification: A user session can be defined as a set of pages visited by the same user within the duration of one particular visit to a web-site. A user may have a single or multiple sessions during a period. Once a user was identified, the click stream of each user is portioned into logical clusters. The method of portioning into sessions is called as Sessionization or Session Reconstruction. A transaction is defined as a subset of user session having

3.2 Phase-2: Pattern Discovery Phase

1. Pattern Discovery: Once user transactions have been identified, a variety of data mining techniques are performed for pattern discovery in web usage mining. These methods represent the approaches that often appear in the data mining literature such as discovery of association rules and sequential patterns and clustering and classification etc. Classification is a supervised learning process, because learning is driven by the assignment of instances to the classes in the training data. Mapping a data item into one of several predefined classes is done. It can be done by using inductive learning algorithms such as decision tree classifiers, naive Bayesian classifiers, Support Vector Machines etc., Association Rule Discovery techniques are applied to databases of transactions where each transaction consists of a set of items. By using Apriori algorithm by [4] is the biggest frequent access item sets from transaction databases that is the user access pattern are discovered. Clustering is a technique to group users exhibiting similar browsing patterns. Such knowledge is especially useful for inferring user demographics in order to perform market segmentation in Ecommerce applications or provide personalized web content to pages. Sequential Patterns are used to find inter-session patterns such that the presence of a set of items followed by another item in a time-ordered set of sessions. By using this approach, web marketers can predict future visit patterns which will be helpful in placing advertisements aimed at certain user groups.

2. Pattern Analysis: Pattern analysis is the final stage in web usage mining. Mined patterns are not suitable for interpretations and judgments. So it is important to filter out uninteresting rules or patterns from the set found in the pattern discovery phase. In this stage tools are provided to facilitate the transformation of information into knowledge. The exact analysis methodology is usually governed by the application for which Web mining is done. Knowledge query mechanism such as SQL is the most common method of pattern analysis. Another method is to load usage data into a data cube in order to perform OLAP operations[4].

4 Applications

Web usage mining is the application of data mining techniques to discover usage patterns from Web data, in order to understand and better serve the needs of Web-based applications. Web usage mining has seen a rapid increase in interest, from both the research and practice communities. Following are some of the applications.

1. Education:
[10] says that Nowadays, the application of web usage mining in educational systems is increasing exponentially. An interesting application of Web usage mining is link recommender systems (LRS). Their purpose is to facilitate the navigation of the users on a Web site and to help them not to get lost when they are browsing through hypertext documents. On the other hand, there is an increasing interest in applying data mining to educational systems, making educational data mining a new and growing research. EDM is an emerging discipline, concerned with developing methods for exploring the unique types of data that are obtained from different types of educational contexts. On the one hand, there are traditional face-to-face classroom environments such as special education and higher education. On the other hand, there are computer-based education and Web-based education such as well-known learning management systems the examples of which are WebCT, BlackBoard and Moodle, intelligent tutoring systems. Our web usage mining approach uses all the available usage information about students (profile and log information) in order to learn user routes or browsing pathways for personalized link recommendation. Web usage mining generally consists of three phases: data preparation, pattern discovery and recommendation. The first two phases are performed off-line and the last phase is performed online. In education, data preparation will transform Web log files and profiles into data with the appropriate format. Pattern discovery will use a data mining technique, such as clustering, sequential pattern and association rule mining. Finally, recommendation will use the discovered patterns to provide personalized links or contents.

2. Health informatics:
Internet can serve as the backbone for implementing supply chain solutions to add value to health care providers, their suppliers, and their patients. According to [11] Even though health care information systems in some hospitals and clinics have been linked together with a local area network or a wide area network, network based health care systems have not been popular until the advent of the Internet. The three primary Internet applications that the healthcare industry uses, to varying degrees, are the Internet, intranets, and extranets. Doctors can use the Internet to do more than

download information and communicate with other providers; it can also be used to send complex medical files across the Web.

3. Human-computer interaction:
Web usage mining is a kind of web mining, which exploits data mining techniques to discover valuable information from navigation behavior of World Wide Web users. Web usage mining (WUM) is a new research area which can be defined as a process of applying data mining techniques to discover interesting patterns from web usage data. Web usage mining provides information for better understanding of server needs and web domain design requirements of web-based applications. Web usage data contains information about the identity or origin of web users with their browsing behaviors in a web domain. Web pre-fetching, link prediction, site reorganization and web personalization are common applications of WUM. Web usage mining and its relationship with computer. [12] tells that All the three phases of Web Usage Mining provide good log file which is free from inconsistent, un-useful data. It helps in filtering unwanted access patterns/ web pages. The Web Structure Mining plays an important role with various benefits including , quick response to the web users, reducing lot of HTTP transactions between users and servers thus saving memory space of server, better utilization of bandwidth along with server processor time.

4. Social media:
Web usage mining also plays an important role in social networks analysis. [13] It is useful for the analysis of social networks extraction discussed in section 2 of this paper. The usage data and user communications on an on-line social networking website can be transformed into relational data for social-networks construction. In addition, web usage mining is also a tool for measuring centrality degree. social network analysis is finding the communities embedded in the social network datasets, and moreover, analyzing the evolutions of the communities in dynamic networks. The evolution pattern as one kind of temporal analysis aspect sometimes could provide us an interesting insight from the perspective of social behavior. Recently, a considerable amount researches have been done on this topic. In the field of social network, Web community is also used to mean a set of users having similar interests. In [14] Social networking products are flourishing. Sites such as MySpace, Facebook, and Orkut attract millions of visitors a day, approaching the traffic of Web search sites 2. These social networking sites provide tools for individuals to establish communities, to upload and share user generated content, and to interact with other users. In recent articles, users complained that they would soon require a full-time employee to manage their sizable social networks. Indeed, [15] take Orkut as an example. Orkut enjoys 100+ million communities and users, with hundreds of communities created each day. A user cannot possibly view all communities to select relevant ones.

5 Challenges

Web Usage Mining is the automatic discovery of user interactions with a web server, including web log. The web log files are collected from web server. WUM focuses on privacy concerns and is currently the topic of extensive debate. The knowledge gathered from WUM can be very useful in many Web applications such as Web caching, Web perfecting, intelligent online advertisements, and in addition to construct Web personalization. Most of the research challenges efforts for modeling personalization systems are clustering pages or user session, association rule generation and sequential pattern generation. These challenges are the most popular ones encountered in almost all the web usage mining research. And these problems have a huge impact on the success or failure of web usage mining research.

6 Conclusion

This paper has attempted to provide a review of rapidly growing area of web usage mining. Web usage mining makes use of this information in order to mine the desired information and make it available to user efficiently and efficaciously. Content and structure preprocessing allows raw data to be preprocessed along these dimensions also. The involvement of intelligent agents and knowledge query mechanisms improves the efficiency of pattern analysis. Process, Applications related with web usage mining are discussed in this paper. This paper has aimed at describing challenges, and the hope is that the research community will take up the challenge of addressing them.

References

[1] Patel, K.B., Patel, A.R.: Process of Web Usage Mining to find Interesting Patterns from Web Usage Data. In: International Journal of Computers & Technology Volume 3(1) (August 2012)

[2] Langhnoja, S., Barot, M.: Pre-Processing: Procedure on Web Log FileforWeb Usage Mining. International Journal of Emerging Technology and Advanced Engineering 2(12) (December 2012) Website: http://www.ijetae.com, ISSN 2250-2459, ISO 9001:2008 Certified Journal

[3] Mitharam, M.D.: Preprocessing in Web Usage mining. International Journal of Scientific & Engineering Research 3(2), 1 (2012) ISSN 2229-5518

[4] Sharma, A.: Web Usage Mining: Data Preprocessing, Pattern Discovery and Pattern Analysis on the RIT Web Data

[5] Chaudhary, K., Gupta, S.K.: Web Usage Mining Tools & Techniques: A Survey. International Journal of Scientific & Engineering Research 4(6), 1762 (2013) ISSN 2229-5518

[6] Srivastava, J., Cooley, R.: Web Usage Mining: Discovery and Applications of Usage Patterns from Web Data. SIGKDD Explorations (January 2000)

[7] Chitraa, V., Davamani, A.S.: A Survey on Preprocessing Methods for Web Usage Data. International Journal of Computer Science and Information Security (IJCSIS) 7(3) (2010)

[8] Langhnoja, S., Barot, M.: Pre-Processing: Procedure on Web Log FileforWeb Usage Mining. International Journal of Emerging Technology and Advanced Engineering 2(12) (December 2012) Website: http://www.ijetae.com, ISSN 2250-2459, ISO 9001:2008 Certified Journal

[9] Pani, S.K., Panigrahy, L.: Web Usage Mining: A Survey on Pattern Extraction from Web Logs. International Journal of Instrumentation, Control & Automation (IJICA) 1(1) (2011)
[10] Romero, C., Ventura, S., Zafra, A., de Bra, P.: Applying Web usage mining for personalizing hyperlinks in Web-based adaptive educational systems (received January 8, 2009) (received in revised form May 4, 2009) (accepted May 4, 2009)
[11] Siau, K.: Health Care Informatics. IEEE Transactions on Information Technology in Biomedicine 7(1) (March 2003)
[12] Geeta, R.B., Totad, S.G., Reddy, P.: Amalgamation of Web Usage Mining and Web Structure Mining. International Journal of Recent Trends in Engineering 1(2) (May 2009)
[13] Imran, M., Castillo, C., Diaz, F., Vieweg, S.: Processing Social Media Messages in Mass Emergency: A Survey (August 3, 2014), rXiv:1407.7071v2 [cs.SI]
[14] Raju, E., Sravanthi, K.: Analysis of Social Networks Using the Techniques of Web Mining 2(10) (October 2012)
[15] Zhang, Y.: Web Information Systems Engineering and Internet Technologies. Springer Science+Business Media, LLC (2011)

Author Index